THE SAFETY OF FOODS
SECOND EDITION

THE SAFETY OF FOODS
SECOND EDITION

Horace D. Graham, Ph.D.

Editor
Department of Chemistry
University of Puerto Rico
Mayaguez, Puerto Rico

AVI PUBLISHING COMPANY, INC.
Westport, Connecticut

© Copyright 1980 by
THE AVI PUBLISHING COMPANY, INC.
Westport, Connecticut

Library of Congress Cataloging in Publication Data

The Safety of foods.

  Includes index.
  1. Food contamination. I. Graham, Horace
Delbert, 1925–
TX531.S23 1980   614.3′1   79–18268
ISBN 0–87055–337–2

Printed in the United States of America

Preface to the Second Edition

The international symposium on "the Importance and Safety of Foods," held in Mayaquez, Puerto Rico in 1967, made possible the publication of the First Edition of this book. The great acceptance of those proceedings prompted the urge to prepare a second, revised edition.

Since the publication of the proceedings, there have been dramatic developments in, and awareness of, the safety of foods. Several vigilant consumer groups have been formed. Regulatory agencies have become more watchful and meticulous. Food producers and manufacturers have grown more aware of and concerned with meeting the demands of searching and critical consumers and are striving hard in the national efforts to maintain a well-fed and healthy population. Even school and church groups have taken action to make youths aware of good nutrition and the wholesomeness and safety of foods.

Revision of a conference proceedings can, realistically, be accomplished only by the convening of another conference. This was not possible. Transforming a conference proceedings into a text is an even more formidable task. After much hesitancy, the urgings of the publisher prevailed.

This second edition covers, as completely as is possible in a text of its size, the major areas which bear on the safety of foods. The multi-authored feature of the book assures that each chapter is written by persons who have been closely associated with the area covered. To the reader, this might reflect, at times, attention to too many details but such will be of value to advanced students, teachers and researchers. An additional jeopardy of any multiple-authored publication is the possibility of over-lapping and repetition. All effort has been made to keep this to a minimum. However, each author, being highly qualified in his specialty and frequently quite zealous over certain points or views, wishes to emphasize some area which is covered in some detail by another equally qualified writer. Considerable leeway was given to each author to express his views, even at the expense of some overlapping. Where overlapping appears obvious, therefore, the editor takes full responsibility realizing that in such cases, the material is generally presented in a different way, with different emphasis or in more detail. This approach, it is thought, though at times irritating to advanced students and scientists, may be helpful to students and layman groups who read the book.

This text is intended to serve students of Food Science and Technology and all others in fields related to the safety and wholesomeness of foods. That undergraduates, graduates, instructors in courses covering food toxicology and food safety, workers in the food industry, public health workers and administrators may find it helpful, is my utmost wish.

HORACE D. GRAHAM

July   1979

Preface to the First Edition

Technological advances in agriculture, food processing, and food distribution, as well as the rapid growth of population, particularly in certain areas of the Western Hemisphere, have resulted in greater awareness of the problems that may arise in the commercial inter-country movement of foods.

Adequacy of the food supply for an ever-increasing population has held the interest of scientists and laymen since the time of Malthus. Although various proposals have been forwarded to meet this challenge, approaches should be sought that are aimed not only at increasing the available food supply, but also at its better preservation and utilization.

The safety and quality of the food supply are of great concern to food scientists, nutritionists, and epidemiologists. Repeated, and in some cases, recent outbreaks of food-borne infections and/or intoxications traceable to foods involved in inter-country trading have focused renewed attention on food safety and food quality at the international level. Each country within the Western Hemisphere has promulgated its own food laws and regulations in an effort to achieve quality products. However, such regulations differ widely from one country to another and better coordination on an international level is not merely desirable, but urgently needed in order to facilitate the inter-country movement of foods.

This Conference was organized, not simply to review the problems inherent in the above-mentioned facets of food science and technology, but to evoke new, challenging, and concrete approaches to the main problems that must be solved in order that *our* Hemisphere will have a safe and nutritious food supply. It brought together, from all over the Hemisphere, researchers, administrators, and teachers in various aspects of food science, food technology and nutrition, food industry representatives, government officials, and others interested in food safety and food protection.

The Conference was organized by the Toxicology Study Section, U.S. Public Health Service, in cooperation with the Biology Department, University of Puerto Rico, Mayagüez, Puerto Rico. The committee members were: John C. Ayres, Iowa State University, Ames, Iowa; Frank Blood, Vanderbilt University, Nashville, Tennessee; C. O. Chichester, University of California, Davis, California; H. D. Graham, Chairman, University of Puerto Rico, Mayagüez, Puerto Rico; R. S. McCutcheon, National Institutes of Health; Bethesda, Maryland; J. J. Powers, University of Georgia, Athens, Georgia; B. S. Schweigert, Michigan State University, East Lansing, Michigan; and A. S. Stevens, U.S. Public Health Service, Bethesda, Maryland.

The Organizing Committee and the University of Puerto Rico, Mayagüez Campus, are deeply grateful to the United States Public Health Service, Division of Environmental Engineering and Food Protection for generous financial support. Special thanks are due to Lic. J. E. Arrarás, Chancellor of the University of Puerto Rico, Mayagüez Campus, for his personal interest in the Conference; to Dr. Cruz Perez and Dr. Bernandino Rodriguez Lopez for their able aid and advice as members of the local advisory committee; to Dr. J. Maldonado Capriles, who spearheaded the efforts for the role of the Biology Department in the Conference; to all members of the faculty and staff and others who assisted in translations and other organizational efforts or who, morally or otherwise, contributed toward the realization of the Conference.

H. D. GRAHAM

July 6, 1967

Dedication

Contributors

AVENS, J. S., Ph.D., Associate Professor of Food Technology, College of Agricultural Sciences, Colorado State University, Fort Collins, Colorado 80521

BERGDOLL, M. S., Ph.D., Food Research Institute, University of Wisconsin, Madison, Wisconsin 53706

BRADLEY, R. L., Ph.D., Professor of Food Science, Department of Food Science, University of Wisconsin, Madison, Wisconsin 53706

BRYAN, F. L., Ph.D., Chief, Foodborne Disease Training Instructional Services Division, Center for Disease Control, Public Health Service, Department of Health, Education and Welfare, Atlanta, Georgia 30333

COFFIN, W. E., Ph.D., Chemistry Division, Health Protection Branch, Health and Welfare Canada, Tunney's Pasture, Ottawa, Ontario, Canada K1A 0L2

COLLINS THOMPSON, DAVID L., Ph.D., Associate Professor. Department of Environmental Biology, University of Guelph. Guelph, Ontario, Canada NIG2WI

DO, J. Y., Ph.D., Department of Nutrition and Food Sciences, Utah State University, Logan, Utah 84322

DYMSZA, H. A., Ph.D., Professor, Department of Food Science and Technology, Nutrition and Dietetics, University of Rhode Island, Kingston, Rhode Island 02881

EWING, MARGARET, Ph.D., Professor, College of Veterinary Medicine, Oklahoma State University, Stillwater, Oklahoma, 74074

DIGIROLAMO, R., Ph.D., Director, Marine Resources Center, College of Notre Dame, Belmont, California 94002

GRAHAM, H. D., Ph.D., Professor, Department of Chemistry, University of Puerto Rico, Mayaguez, P.R. 00708

HAUSCHILD, A. H. W., Ph.D., Microbiology Research Division,

Bureau of Microbial Hazards, Health and Welfare, Canada, Ottawa, Ontario, Canada K1A 0L2

HUGUNEN, A. G., Ph.D., Professor of Food Science, Department of Food Science, University of Wisconsin, Madison, Wisconsin 53706

INGLE, G. W., Ph.D., Assistant Technical Director, Toxic Substances Control, Manufacturing Chemists Association, Washington, D.C. 20009

JORGENSEN, D. K., Ph.D., Commercial Development Manager, Monsanto Industrial Chemicals, Saint Louis, Missouri 63166

KEELER, R. F., Ph.D., Research Chemist, Poisonous Plant Research Lab, USDA Science and Education Administration, Federal Research, Western Region, Logan, Utah 84321

KILARA, ARUN, Ph.D., Assistant Professor, Department of Food Science, Pennsylvania State University, University Park, Pennsylvania 16802

LITCHFIELD, J., Ph.D., Manager, Bioengineering/Health Sciences Section, Battelle Memorial Institute, Columbus, Ohio 43201

MARTIN, S. E., Ph.D., Assistant Professor of Food Microbiology, Department of Food Science, University of Illinois, Urbana, Illinois

MASS, M. R., Ph.D., Food Research Institute, University of Wisconsin. Madison, Wisconsin 53706

MCKINLEY, W. P., Ph.D., Health Protection Branch, Health and Welfare Canada, Tunney's Pasture, Ottawa, Ontario, Canada KIA 0L2

MORGAN, P. M., Ph.D., Dr. P.H., Dean, College of Veterinary Medicine, Oklahoma State University, Stillwater, Oklahoma 74074

MORRISON, A. B., Ph.D., Assistant Deputy Minister, Health and Welfare Canada, Health Protection Branch, Tunney's Pasture, Ottawa, Ontario, Canada KIA 0L2

NAGEL, A. H., M. S., Manager, Safety and Compliance, Corporate Research Department, General Foods Corporation, Technical Center, White Plains, New York 10625

ORDAL, Z. J., Ph.D., Professor of Food Microbiology, Department of Food Science, University of Illinois, Urbana, Illinois 61801

OSER, B. L., Ph.D., Food and Drug Consultant, Bernard L. Oser Associates, 108-18 Queens Boulevard, Forest Hills, New York 11375

RUSSELL, F. E., Ph.D., University of So. California, Los Angeles, California

SALUNKHE, D. K., Ph.D., Professor, Department of Nutrition and Food Science, Utah State University, Logan, Utah 84322

SEN, N. P., Ph.D., Food Research Division, Health Protection Branch, Health and Welfare Canada, Tunney's Pasture, Ottawa, Ontario, Canada KIA 0L2

SHIMIZU, Y., Ph.D., Professor of Pharmacognosy, University of Rhode Island, Kingston, Rhode Island 02881

STOEWSAND, G. S., Ph.D., Associate Professor of Food Toxicology, Department of Food Science and Technology, New York Agricultural Experiment Station, Geneva, New York 14456

WOOD, R., D.V.M., Professor, College of Veterinary Medicine, Oklahoma State University, Stillwater, Oklahoma 74074

WU, M. T., Ph.D., Department of Nutrition and Food Sciences, Utah State University, Logan, Utah 84322

ZABIK, M., Ph.D., Professor, Department of Food Science and Human Nutrition, Michigan State University, East Lansing, Michigan 48824

Acknowledgements

In a cooperative task of this nature, it is impossible to enumerate all those who have contributed so much time, effort and thought to make it a worthwhile endeavor. To all the authors, I submit my most sincere thanks and appreciation for their splendid labors. Working with such a learned and dedicated group of specialists has been an invaluable education, particularly in human relations.

Special thanks are due to all the persons, agencies and publishers who so kindly and willingly granted permission for the use and/or reproduction of scientific material. The role of all of my co-workers in the Mayaquez conference has continued to be vital. Without that previous setting, this continuum would not have been possible. To these individuals, listed in the preface of the first edition, I owe a deep gratitude. Ltc. Duane Hilmas, contributed useful comments and Dr. Eugen Wierbicki of the U.S. Army Natick Laboratories provided literature and advice which were invaluable in the preparation of the chapter on wholesomeness and safety of irradiated foods.

Had it not been for the insistence of Dr. Tressler President of the Avi Publishing Company, this task would not have been attempted, much more completed. To him, I give my sincere thanks. Dr. Norman Desrosier, Editor-in-Chief, Avi Publishing Company, has always offered friendly advice and encouragement. His technical and scientific advice and editorial help in the preparation of the chapter on irradiated foods, warrant my utmost appreciation.

In the hours of need and frustrations, the members of the department of Chemistry at the University of Puerto Rico and of the department of Food Science and Technology, Nutrition and Dietetics of the University of Rhode Island, rendered help, encouragement and advice and provided stimulating discussions. My sincere thanks to all my colleagues, associates and friends in both institutions.

Last and most lasting is my wish to dedicate this book to the memory of Dr. Frank Blood, a friend, helper and colleague: His warmth and encouragement at the Mayaquez conference has left an indelible mark on me. I sincerely trust that this compilation will transmit at least a small portion of that unselfish and cheerful attitude displayed as he labored in his concern for the Safety of our Foods.

Horace D. Graham

May 1, 1979

Contents

PREFACE
Introduction and Overview *H. D. Graham*

1 Food Spoilage and Food-Borne Infection
   Hazards
   David L. Collins-Thompson
2 Sources of Food Spoilage Microorganisms
   Scott E. Martin and Z. John Ordal
3 Microbial Problems in Food Safety with
   Particular Reference to *Clostridium
   Botulinum*
   A. H. W. Hauschild
4 Staphylococcal Food Poisoning
   Merlin S. Bergdoll
5 Salmonella Food Poisoning
   John H. Litchfield
6 Viruses in Foods
   R. Di Girolamo
7 Mycotoxins in Foods and Feeds
   *D. K. Salunkhe, M. T. Wu, J. Y. Do and
   Melanie R. Maas*
8 Control of Foodborne Diseases
   Frank L. Bryan
9 Food-Borne Diseases of Animal Origin
   P. M. Morgan, W. W. Sadler and R. M. Wood
10 Nitrosamines *N. P. Sen*
11 Mercury in Foods *R. L. Bradley and
   A. G. Hugunen*
12 Trace Metal Problems with Industrial Waste
   Materials Applied to Vegetable Producing
   Soils *G. S. Stoewsand*

13   Polychlorinated Biphenyls and Polybrominated Biphenyls in Foods *Mary E. Zabik*

14   Sources of Pesticide Residues *D. E. Coffin and W. P. McKinley*

15   Antibiotics and Food Safety *Arun Kilara*

16   Safety and Wholesomeness of Irradiated Foods *Horace D. Graham*

17   Toxins in Plants *Richard F. Keeler*

18   Poisonous Marine Animals *H. A. Dymsza, Y. Shimizu, F. E. Russell and H. D. Graham*

19   The Need of Additives in Industry *D. K. Jorgensen*

20   The Proper Use of Food Additives *Horace D. Graham*

21   Regulating Additives From Food Contact Materials *G. W. Ingle*

22   Food Regulations in the Americas *Albert H. Nagel*

23   Food Control Under the Canadian Food and Drugs Act *A. B. Morrison*

24   Safety of Food Service Delivery Systems in Schools *John S. Avens*

     INDEX

Introduction and Overview

Horace D. Graham

Food is one of three absolute essentials for life, the other two being water and oxygen. Since time immemorial man, consciously or unconsciously, has been concerned about the safety of his food. Through knowledge or instinct, he selected wild plants, animals and fish for his table, skillfully avoiding, as far as possible, poisonous species. After fire was discovered, the process of cooking was developed and extended into commercial preservation by heat. Fermentation and freezing then joined thermal processing as means of conserving food.

Modifications and/or variations of these basic processes have been advanced to make food more available, safer and more economical. Within the last three decades, food irradiation has been proposed as a new method for food preservation.

To maintain a healthy life and healthy population, not only must there be an adequate supply of food but such food must be safe and wholesome—i.e., nutritious and free from components or agents which, when ingested, will cause illness of any type.

CHANGING TIMES

Over the past few years, especially since the end of World War II, many dynamic changes have occurred in the world which have greatly influenced production, processing, marketing and consumption of food. These include: Phenonomenal expansion of the habitable regions of the world with concomitant industrial development, increased population, growth of metropolitan areas, increased intra-

1

country trading of food materials, expansion and concentration of ethnic foods and food styles among migrants from other lands, mass feeding, almost incessant travelling, changing life styles and customs, expanded and modified agricultural practices, new and more rapid methods of food preparation, introduction of new formulated and fabricated foods, development of so-called convenience foods and heat-and-serve dishes. All of these factors have provided possibilities for changes in the composition of food products as well as in the nutritional and microbial quality of foods.

Aware of this, food scientists, government regulatory agencies, food manufacturers and consumer groups have become more concerned about the safety of foods. Although such concern is expressed more in countries which can afford the heavy expenditures involved in research, monitoring, education, legislation and enforcement, there is hardly a sector on the globe where attention to safety and wholesomeness of food is not given weighty consideration.

Within the last decade scientific bodies such as the Joint Task Force of the Southern Region Agricultural Station (1972) and the Food Safety Council (Anon, 1977) have been formed to specifically promote food safety. The objective of the former group is to pinpoint areas where research is needed to foster and assure the safety and wholesomeness of food under the changing trends of production, transportation, marketing, storage, preparation and consumption. The latter group aims, among other things, to (1) foster and develop criteria for the safety and wholesomeness of food and food ingredients and (2) to provide the public, consumers and food scientists with scientifically sound information. The group is comprised of representatives from government, industry, consumer organizations and academia.

DEFINITIONS

Consideration of the problem of food safety will be aided by definitions of food, hazard, toxicity and safety.

Simply defined, *food* is what we eat. It is any material, of plant or animal origin, which is ingested and can be digested by the organism (in this context, humans and animals) to provide nutrients which are absorbed and used for the growth, development and overall activities of the said organism.

Any consideration of food safety must include water, since this liquid is indispensable in the cultivation, processing, preservation and cooking of foods. Food quality is greatly influenced, or even determined by the quality of water with which it has come in contact. Polluted irrigation water can result in excessive uptake of hazardous minerals or pesticides by plants and animals or in their contamination by pathogenic organisms. Seafoods from polluted waters can cause severe health problems, also.

Hazard has been defined as the probability that injury (danger or damage) will result from the use of the substance (under consideration) in the proposed quantity and manner (Hall, 1978). Danger, injury or damage can be encountered only if there exists some hazard.

Toxicity is the capacity of a substance to produce injury. The term includes the capacity to induce teratogenic, mutagenic and carcinogenic effects.

Hall (1978) pointed out that hazard, as defined above, links toxicity (the inherent capacity to cause harm) and quantity. "The amount of substance present will determine if the potential for harm will be realized."

Safety is defined as freedom from danger, injury or damage. Hall emphasized that safety is the practical certainty that injury will not result from the substance when it is used in the manner and quantity proposed. Substance in this context, refers to food or food ingredients.

Wodica (1977) has listed hazards in the following order of priority. (1) microbiological (2) nutritional (3) environmental (4) natural toxicants (5) pesticide residues, and (6) food additives. Although the severity of the hazard posed by each of the above-listed factors may not follow that order in the opinion of some (Hall, 1978), it does serve as a useful base for discussion and elaboration.

Microbiological hazards in the form of food intoxications by *Clostridium botulinum* and *Staphylococcusaureus* and food infections caused by Salmonella species have been known for many years and have received much attention. These are discussed in Chapters 1-7 and recommendations for avoiding them are dealt with in Chapter 8. Research continues at the academic as well as at the industrial levels on the nature of the illnesses, the toxins and their elaborators. With the introduction of so many new foods, fabricated foods or formulated foods, fast foods, mass feeding in schools, colleges, industrial areas and metropolitan areas, processing technology and food preparation techniques have had to be modified in order to assure that such products meet specified microbial and nutrient standards, indicative of safe and wholesome food.

Hobbs (1977), in an illuminating article, discussed the importance of correct harvesting, storage and transportation in relation to the safety and keeping quality of grains. Microbial agents, including molds and mycotoxins, involved in food-borne diseases, epidemiology and microbiological specifications were critically analyzed, coordination between investigative and regulatory agents exhorted and areas needing future and accentuated attention pointed out.

A traditionally plentiful source of good food protein is seafoods. In addition to the fore-mentioned microbial hazards, marine foods can cause a variety of illnesses. Shellfish poisoning involves some of the most deadly toxins known. Toxicity of marine animals is discussed in Chapter 18.

NUTRITIONAL HAZARDS

Nutritional hazards differ drastically the world over. Obesity, a danger to health is, in many cases, reflective of the economic development of the area or of the group. It is discerned mainly in highly developed countries or population groups where economic means permit a large intake of food (calories) without an equivalent expenditure of energy. In the less developed areas of the world, malnutrition or even blatant hunger prevails. Attempts to alleviate this has resulted in a multiplicity of programs of aid, education, agricultural development and industrial and/or practical endeavors by organizations, civic groups, local and foreign governments and educational institutions.

The explosive increase in population while food production increases at a much less rapid rate has been often claimed as the reason for the scarcity of food in certain areas of the world. However, more intricate factors involving an interplay of economic, social, political, religious, anthropological and geographical considerations contribute significantly to the overall problem. Lack of protein is considered critical for the proper feeding of an ever-increasing world population.

Not only the quantity but also the quality of protein determines nutritional status and where either is lacking, severe disorders such as Kwashiorkor and lesser manifestations occur. Protein supplementation and amino acid fortification have been engaged in to avert these catastrophies. High protein sources such as soybean, soybean flour, peanut meal and flour, cottonseed meal and flour, sesame seed and flour and milk powder, and whey protein and their concentrates have been added to various formulas in order to improve the quality of foods of various groups and nationalities.

Often deficiencies such as of iodine, fluoride, minerals, (iron, Ca) and of vitamins occur in foods naturally or develop through processing or storage. These are usually corrected by supplementation or fortification. Darby and Hambraeus (1978) have pointed out that fortification, enrichment, restoration and imitation or fabricated products, if properly done, or formulated, can contribute much to global nutrition. However, they stressed that guidelines should be followed so that the maximum nutritional advantage of the practice or formulation is realized.

Few hazards due to excessive intake of vitamins exist, except in the case of vitamins A and D. Antivitamin factors such as thiaminase in fish, trypsin inhibitor in soybean and avidin in egg white can be readily eliminated by simple conventional processing or food preparation methods.

Food faddism has grown to a great extent over the last decade, and according to competent sources (Anon, 1974), can constitute a nutritional hazard. To quote this source "one of the most serious hazards of food faddism is that false promises of superior health and freedom from disease, which are believed to accrue from the use of health foods, delay individuals from obtaining necessary competent medical attention."

The regular intake of organic foods based on such unusual diets as the Zen macrobiotic diets and the liquid protein reducing diets, constitute major public health problems and are dangerous to those who adhere to them. The latter diets have been suspected of probably contributing to the death of several persons recently.

Organic foods, can be problematic too, microbiologically. At least one case of severe Salmonella contamination has been reported in alfalfa tablets prepared from vegetation grown on soil fertilized with composted sludge or organic fertilizer (Thomason et. al., 1977). Fortunately, the entire stock was destroyed. It should be pointed out, however, that organically grown foods, especially in home gardens, can and do contribute substantially to national health and welfare. False, extravagant claims are the principal criticisms.

ENVIRONMENTAL CONTAMINATION

Environmental contamination can represent serious, though completely controllable health problems in foods. If and when they occur, alarming consequences, even if not prolonged, can result. The incidences of mercury and of polybrominated biphenyls in fish and waters are examples. Swordfish containing excessive amounts of mercury, poultry containing PCB's and the catastrophic accident involving the latter in the dairy industry in Michigan created much fear among consumers. These topics are discussed in Chapters 11 and 13, respectively.

Considerable concern has been expressed over the disposal of domestic and industrial wastes and the possible entry of hazardous materials therein into food and feeds. A recent symposium and other articles (Smith, 1979; Sadoviski, *et al.*, 1978; Wiley and Westberger, 1969; ADSA-ASAS, 1978) have dealt with this topic and the public health implications.

PROBLEMS POSED BY SLUDGE

Jelnick and Braude (1978) have discussed the management of sludge use on land and the concern of the Food and Drug Administration over its public health responsibilities to ensure the safety and wholesomeness of the nation's food supply for humans and animals whose flesh and/or products (milk and eggs) are consumed by humans. It has been shown that sludges applied to land on which food and feed crops for animals are grown can contain pathogenic microorganisms, viruses, eggs of parasitic worms, toxic heavy metals, pesticides and industrial chemicals, such as PCB's and the chlorinated pesticides. The eggs (ova) of parasites are extremely hardy and will survive well in the most adverse environments.

Heavy metals of concern are cadmium and lead which present the greatest hazard to the safety of the food supply. High concentrations occur in sludge. Many plants, including grains, legumes and leafy vegetables take up cadmium quite readily from the soil. Lead is less readily translocated from soil to the edible portions of plants. Animals grazing on sludge-treated land may take in the toxicants through direct ingestion of the soil or through eating of vegetables to which sludge has adhered or which have taken up the metals from the soil. The toxic metals accumulate in the liver and kidney of animals. Soybean, corn, wheat, certain vegetables and other crops have been shown to accumulate high levels of metals from soil treated repeatedly with sludge. Grains have attracted particular attention because the large acreage cultivated apparently makes it attractive to treat such large tracts of land. However, it has been pointed out that grains and cereal products already supply about 23% of the total cadmium intake of the diet, hence any drastic increase in the cadmium content of such widely and heavily consumed commodity would pose serious health hazards.

The use of human waste, composted or otherwise treated, in Asiatic and other areas has been known for a long time. Data on the composition of foods grown on these lands are not as readily available as in the U.S. and other highly developed areas. However, similar hazards would most likely exist. The heavy metal content of such manures would most likely be less since, apparently, industrial wastes would be a minor component of the mixture or would be totally absent. The Food and Drug Administration (FDA) and the Environmental Protection Agency (EPA) have been working closely together to make sure that the application of sludge to land used for the production of food and feeds is properly controlled. The major concern is that with the growing interest of large municipalities, including vast industrial metropolitan areas interested in applying sludge to cropland, there is an increasing possibility of introducing hazardous amounts of residues into foods. The authors emphasized that maintaining such sludge contaminants at a low level is imperative so that the consumer of such foods are not exposed to unnecessary risks and secondly, that it will not be necessary to withdraw large acreages from production due to the presence of hazardous levels of contaminants in the soil as a result of unwise deposition of sludge on the land.

To protect the health of the public, the FDA has recommended certain limitations on the application of sludge to agricultural land. These are as listed:

(a) sludges should not contain more than 20 ppm cadmium, 100 ppm lead or 10 ppm PCB's, on the dry basis;

(b) crops which are customarily eaten raw should not be planted within 3 years after the last sludge application;

(c) crops such as green beans, beets, etc., which may contaminate other foods in the kitchen before cooking should not be grown on sludge-treated land unless the sludge gives a negative test for pathogens;

(d) because sewage can be regarded as filth, food physically contaminated with sludge can be considered adulterated even though there is no direct health hazard; hence, sludge should not be applied directly to growing or mature crops where sludge particles may remain in or on the food;

(e) commercial compost and bagged fertilizer products derived from sludges should be labeled properly to minimize any contamination of crops in human food chain which may result from their use.

HAZARDS FROM OTHER WASTES

The use of fly-ash on agricultural lands has also led to concern about the heavy metal content of food and food crops (Furr *et. al.* 1978). This topic is discussed at length in Chapter 12.

Environmental contingencies coupled with the urge to recycle residual nitrogen (protein) have spurred intense efforts to use chicken, cattle and pig waste as supplements in poultry and animal feed. As is the case with sludge, concern has been expressed here too about the public health hazards of such practices. Unless adequate precautions are taken, there exists the possibility for contamination by Salmonella, Clostridial species, viruses and parasites of various types, depending on the health status of the animals producing the refuse and the opportunity for rapid multiplication of pathogens therein. Public health aspects of recycling animal waste has been discussed by Bhattacharyca and Taylor (1975); Fontennot and Webb (1975); McCaskey and Anthony (1979); Moore et. al. (1976); Taylor and Geyer (1979) and Smith (1979).

NATURAL TOXICANTS

A variety of natural toxicants occur in products consumed by man and animals. They range from cyanogenetic glycosides (cassava, etc.) to hypoglycin (ackee) to vicine (favism from uncooked broad or faba bean, *Vicia faba*). Toxic elements like selenium are involved, also. These are discussed fully in Chapter 17.

The hazards involved here have been drastically reduced because man has learned to select, process and prepare his foods in ways so as to avoid or reduce toxicity due to the intake of such potentially poisonous substances. At times, the process is as simple as boiling and/or fermentation as in the case of the cassava (cyanide-containing) or collection at the proper stage of maturity, properly cleaning and removing areas of high concentration of toxin and proper cooking as in the case of the ackee (hypoglycin-containing). Plant breeding and/or selection can produce less toxic or toxin-free varieties as in the case of cottonseed meal and flour (gossypol content) or rapeseed meal (glucosinolates). Mushroom poisoning, an occasional danger, can be avoided if the harvester (gatherer) knows the edible species. Animals, likewise, have learned to select their feed and, given the proper option, will avoid certain toxic plants. However, under conditions of restricted pasturage, mixed pasturage or a general lack of feeding opportunity, they are liable to become intoxicated by poisonous plants.

Mycotoxins, produced by fungi are quite toxic (Wilson, 1978). Toxicity from such compounds can be avoided or minimized by the proper selection and storage of the susceptible products. This topic is discussed fully in Chapter 7.

Several other popular foods may contain so-called naturally occurring toxins. Honey, when made from nectar containing toxic material, may be cited: (White, 1973, 1978) In the USSR, Japan, Eastern and Pacific N.W. USA, it has been stated that toxicity from honey may be problematic. In New Zealand, honey may contain tutin and hyenanchin. Tutin from the tutu tree is picked up while the bee is feeding on the sweet excrement left on the leaves by an aphid, the vine hopper which feeds on the leaves of the tutu tree. The sweet honeydew is gathered by the bee and the tutin in it is partially hydroxylated to hyenanchin in the stomach of the bee (Hall, 1978).

Naturally, this type of hazard can be completely avoided by prior knowledge of the possible danger and not using the product from these areas or during the danger seasons or periods.

Avoidance of toxicity in foods can be effected by monitoring the source or agent and controlling it. The case of paralytic shellfish poisoning is pertinent here. Whenever an area is suspected of infection, fishing in this area is banned. Alternately, depuration is practiced to cleanse shellfish taken from suspicious areas prior to release for consumption. It has been recorded however, that certain populations knowing the seasonal or regional trends, skillfully avoided danger. Prakash *et al.* (1971) quote the following account on Indians in Nova Scotia.

"Loscarbot (1609) stated in his writings that the Indians at Port Royal on the Annapolis Basin, N.S., did not eat mussels. When they were starving, they would eat their dogs or bark of trees, but they would not eat these shellfish. To us, this food taboo suggests that Port Royal Indians must have had traditional knowledge of the hazard of eating shellfish from the open coast of the Bay or Fundy, where mussels are the dominant species of shellfish and are to this day responsible for illnesses and occasional deaths."

CHEMICAL RESIDUES

Pesticide residues in or on foods and feed may originate from treatment of agricultural crops and livestock, accidents, mislabelling, misuse of chemicals or often the application of municipal and animal waste to agricultural lands. Environmental contamination by the chemicals has resulted in heavy destruction of marine life (fish) and other wildlife. In some cases residues have been detected even in human milk (Hashemy, Tonkabony and Fateminassab 1977;

Anon, 1977). The topic of pesticide residues is discussed in Chapter 14.

Johnson (1977) pointed out how misuse in foods of useful chemicals can cause ill health and even death. In addition to the polybrominated biphenyl accident in Michigan, he mentioned the following: Alkyl mercury poisoning from consumption of treated grains, the well-known incidence of alkyl mercury poisoning from pork in New Mexico in 1969, as a result of feeding grain contaminated with methyl mercury fungicide, the Dutch margarine disease of 1960, due to the consumption of margarine containing an anti-spattering agent called ME 18, a triglyceride in which one fatty acid has a ring structure. Another example cited is the cardiac failure in beer drinkers during 1965 in Quebec City, Montreal, Canada, and in Omaha, Nebraska due to the use of cobalt sulfate to improve the "head" in the beer. The salient conclusion was that ignorance (e.g. mislabelling or labelling in a foreign language), human error or the mistaken use of apparently useful chemicals—for essentially trivial reasons without sufficient understanding of possible consequences—can result in serious illness, incapacitation and even death.

Animal products (meat, milk, eggs) are good vehicles for the transfer to man not only of noxious microorganisms, viruses and parasites (see Chapters 1–9), but also chemicals of various types—pesticides, PCBs and PBBs (see Chapter 13) antibiotics (Solomons, 1978; Van Houweling and Gainer, 1978; Chapter 15), hormones and radionucleotides (in milk). A recent symposium on the wise use of chemicals, drugs and additives by dairy producers delineates these possibilities (ADSA, 1978).

FOOD ADDITIVES

By far the most controversial aspect of food safety surrounds the use of food additives. This topic is discussed in Chapters 19–21. In the center of most of the controversial issues is the Food and Drug Administration, whose recent decisions on the use of cyclamates and saccharin in foods have provoked heated debates over the role and jurisdiction of the agency.

The Delaney clause which prohibits the use of any carcinogenic substance in foods has been the principal target of criticism by food manufacturers and others. The insistence on a "Zoro level" of such substances in food is thought by many to be unrealistic in this age of highly sophisticated and sensitive analytical equipment and techniques and the knowledge that traces of carcinogenics of one sort or another exist in almost all foods or foodstuffs.   Additionally, the exaggerated and sometimes unrealistic conditions under which cancer has been demonstrated in experimental animals has been questioned by many.

A more liberal approach to the regulation of food additives is thought to be the use of a risk vs. benefit approach when considering a particular chemical. This is discussed to some extent in Chapters 19 and 20. More extensive discussions have been presented by the IFT Expert Panel Scientific Status Summary (1978), by the FDA Commissioner Kennedy (1978), the Special Committee of the National Academy of Sciences and by Hanley and Wolfe (1977). Most of the advocates of less stricture on the part of the regulating agency point out that consumers make daily risk/benefit decisions in their life and should be given the same freedom with respect to their food.

The IFT Expert Panel's report (1978) has defined "benefit" as anything that contributes to an improvement in condition. Risk was divided into two categories, namely vital and non-vital. Vital risk is defined as being concerned with or manifesting life and necessary or essential to life.

Lack of foods would be considered a vital risk, since it is necessary to life, while loss of convenience, reduced satisfaction, increased cost or total removal from the market could be considered non-vital risks. However, the differentiation between vital and non-vital risks is not clearly drawn and is subject to interpretation and the particular situation.

The select panel of scientists on the National Academy of Science's committee recently suggested that the Food and Drug Administration should be allowed to set three levels of risk; high, moderate and low in order to provide a rating scale on which priorities could be determined. There were indications that governmental legislation dealing with food is confusing, overlapping and, at times, inadequate. Consumer groups, however, have objected to these and the other recommendations of the committee, claiming that their adoption would constitute giving control of the Federal Food regulations to the food industry.

In the meantime, the University of Wisconsin scientists, Drs. James and Elizabeth Miller, have reported that certain food additives, pesticides and vitamins can block the action of cancer-causing chemicals. Among the effective agents are BHA, BHT, certain chemicals occurring naturally in cabbage, Brussels sprouts and broccoli, vitamins C and E and certain vitamin A-related compounds. The influence of such findings on the "food additive debate" probably will be evident in the near future.

FOOD SAFETY EVALUATION

The safety evaluation of food ingredients has not slackened (Irving 1978, and Wodica 1977) and it is quite clear that continued research and national thinking and cooperation between government agencies, manufacturers and consumer groups will result in equitable legislation for all concerned, without forfeiting the "safety of foods."

In the meantime, it seems appropriate at this point to note the following modified words and quotations of Braude (1978):

"My approach to controversial, technical issues of public importance is to endeavor to produce and/or collect factual evidence and present it adequately, to refrain from speculations and philosophical indulgences and to assess, to the best of my ability, the hazards, disadvantages and benefits concerned with the relevant problem. My general attitude to the problems associated with the use of food additives*, can be best described by quoting Verdonk. "No scientist can protect the public from itself and from the strange complexion of this age in which obsession with absolute safety will cause people to sweep aside overriding benefits because they entail some risk," I would like to follow this quotation with another, which I believe is relevant to the subject: "When legal regulations are being designed, they must take into consideration the economics of both food production and research. If they do not, they may find themselves without products to sell, or alternatively, products which are of such low efficacy that they cannot play an important part in improving productivity and producing more food for man."

*Words added or substituted by author.

REFERENCES

AMERICAN DAIRY SCIENCE ASSOCIATION, AMERICAN SOCIETY OF ANIMAL SCIENCE, 1978. Symposium on Animal Waste Management. Abstracts, Joint Annual Meeting: Michigan State University, East Lansing, MI.

AMERICAN DAIRY SCIENCE ASSOCIATION (ADSA). 1978. Symposium: Wise use of Chemicals, Drugs and Additives by Dairy Producers: J. Dairy Science 61: pp. 660–682.

ANON, 1974. Food Faddism. Nutrition Reviews. Supplement. July, 1974. pp. 27 and 53.

ANON. 1977. Insecticides in breast milk: Nutrition Reviews. 34: No. 4 April, 1977. p. 72–73.

ANON, 1977. Representatives of industry, consumers, academia and government form Food Safety Council. Food Technology, 31:24.

BHATTACHARYA, A. N. and TAYLOR, J. C. 1975. Recycling animal waste as a feedstuff. A Review. J. Animal Sci. 41:1438–1457.

BRAUDE, R. 1978. Antibiotics in animal feeds in Great Britain: J. Animal Sci. 46:1425–1436.

DARBY, W. J. and HAMBRAEUS, L. 1978. Proposed nutritional guidelines for utilization of industrially produced nutrients. Nutrition Reviews. 36:65–71.

FONTENOT, J. P. and WEBB, K. E. 1975. Health aspects of recycling animal and waste by feeding. J. Animal Sci. 40:1267.

FURR, A. K., PARKINSON, T., GUTENMANN, W. H., PAKKALA, I. S. and LISK, D. J. 1978. Elemental content of vegetables, grains and forages, field grown on fly-ash amended soil. J. Agr. Food Chem. 26: 357–359.

HALL, R. 1978. Toxins, Aflatoxins, natural toxicants and antinutrients in foods. Safety data required for food additives. Perfumer and flavorist. 3:14–20.

HANLEY, J. W. and WOLFE, S. M. 1977. Is the law on food additives too strict? An interview of a company executive and the director of a public citizen health research group on pros and cons of the Delaney Clause. U.S. World and News Report, Inc. May 30, 1977. pp. 25–26.

HASHEMY-TONKABONY, S. E., and FATEMINASSAB, F. 1977. Chlorinated pesticide residues in milk of Iranian nursing mothers. J. Dairy Sci. 60:1858–1862.

HOBBS, B. C. 1977. Problems and solutions in food microbiology-14th Underwood-Prescott memorial award Lecture: Food Technology 31:90–96.

INSTITUTE OF FOOD TECHNOLOGISTS, 1976. The risk/benefit concept as applied to food. A scientific status summary by the Institute of Food Technologists' Expert Panel on Food Safety and Nutrition. Food Technology. March, 1976. 51–56.

INSTITUTE OF FOOD TECHNOLOGISTS, 1979. Developing public policy for food safety. 39th Annual meeting of the Institute of Food Technologists. St. Louis, MO. June 10–13, 1979.

IRVING, G. W. 1978. Safety evaluation of food substances called GRAS. Nutrition Reviews 36:321–356.

JELINEK, C. F. and BRAUDE, G. L. 1978. Management of sludge use on land. J. Food Protection: 41:476–480.

JOHNSON, PAUL E. 1977. Misuse in foods of useful chemicals. Nutrition Reviews. 35:225–229.

JOINT TASK FORCE OF SOUTHERN REGION AGRICULTURAL EXPERIMENT STATIONS AND USDA AGRICULTURE PERSONNEL. 1972. A Program of Research for the Southern Region in Food Safety: 71 pages.

KENNEDY, D. 1979. Should regulatory agencies take on the political dimension of making benefit/risk decisions? Food Product Development May, 1979. pp. 58 and 60.

McCASKEY, T. A., and ANTHONY, W. B. 1979. Human and animal health aspects of feeding livestock excreta. J. Animal Sci. 48:163–177.

MOORE, B. E., SAGIK, B. P. and SORBER, C. A. 1976. An Assessment of potential health risks associated with land disposal of residual sludges. p. 108–112. In proceedings of sludge management disposal and utilization. Informational Transfer, Inc., Rockville.

PRAKASH, A., MEDCOF, J. C. and TENNANT, A. D. 1971. Paralytic shellfish poisoning in Eastern Canada. Bulletin 177. Fisheries Research Board of Canada. Ottawa, Canada. 87 pages.

SADOVSKI, A. Y., FATTAL, B. and GOLDBERG, D. 1978. Microbial contamination of vegetables irrigated with sewage effluent by the drip method. J. Food Protection: 41:336–340.

SMITH, D. F. 1979. Update of national and state programs presently underway to curb point and nonpoint sources of water pollution from animal production facilities. J. Animal Sci. 48:223–227.

SOLOMONS, I. A. 1978. Antibiotics in Animal feeds. Human and safety issues. J. Animal Sci. 46:1360–1368.

TAYLOR, J. C. and GEYER, R. E. 1979. Regulatory considerations in the use of animal waste as feed ingredients. J. Animal Sci. 48:218–222.

THOMASON, B. M., CHERRY, W. B. and DODD, D. J. 1977. Salmonella in health foods. Applied and Environmental Microbiol. 34:602–603.

VANHOUWELING, C. D. and GAINER, J. H. 1978. Public health concern relative to use of subtherapeutic levels of antibiotics in animal feeds. J. Animal Sci. 46:1413–1424.

WHITE, J. W., JR. 1973. Toxic honeys. In Toxicants Occurring Naturally in Foods pp. 495–507. Comm. Food Prot. Natl. Acad. Sci. Washington, D.C.

WHITE, J. W., JR. 1978. Honey. Adv. Food Res. 24:287–374.

WILEY, B. B. and WESTERBERG, 1969. Survival of human pathogens in composted sewage. Applied Microbiology 18:994.

WILSON, B. J. 1978. Hazards of mycotoxin to public health. J. Food Protection. 41:375–384.

WODICA, V. O. 1977. Progress Report of the Food Safety Council's Scientific Advisory Committee on Food Ingredient Safety Criteria. Food Technology. 31:84–88.

Food Spoilage and Food-Borne Infection Hazards

David L. Collins-Thompson

Ideas of food spoilage and food-borne infection hazards have progressed a great deal since the timely appearance of the ghost of Jacob Marley. Scrooge's disbelief in seeing such an apparition was recorded as "a slight disorder of the stomach." Considering the nature of Scrooge and the sanitation problems of London at that time, it may well have been a reference to the spoilt meat syndrome or "ptomaine" poisoning. Such poisoning was considered to be due to the ingestion of bacterially spoilt meat. This belief persisted until the understanding of meat spoilage, the role of *Clostridium botulinum* and its toxins were discovered. A similar situation has developed in the field of food microbiology where "spoilt" was considered as unfit to eat. The term "food spoilage" now cannot be defined in simplistic terms, nor can one ignore the implications of such spoilage in relation to the world food supplies or food safety. In trying to define spoilage one has to take into consideration numerous factors both chemical and biochemical. Further, any definition would have to consider such terms as "spoilt" and include those changes in odor, texture or flavor which are undesirable. One even has to consider the sociological side of the question. Consider, for example the diet of the Inuit people. Some foods considered desirable or even delicacies by this group would be rejected by their southern counterparts. Even within the same ethnic group, problems arise with the meaning of spoilt food. Prime examples of this are sour wine (vinegar) or sour milk (buttermilk). Perhaps if one has to define food spoilage, the principle of exclusion applies. One must define what is not spoilage and anything that falls outside this criteria would be considered spoilt. Such a definition was used by Longrée (1972). In her definition or "fitness test" she includes such things as freedom from unavoidable extraneous matter including rodents, insects and parasites. The food should also be free from excess or unnecessary microbial activity or from such bacteria or parasites as are capable of causing food-borne illness.

Whichever way one cares to look at the areas covered by this definition, it is sufficient to note, despite our current knowledge or sophistication in defining terms, loss of food by the action of undesirable microorganisms is still significant. This fact in terms of economic loss and importance in world food shortages is now beginning to become a focal point for discussion. The estimated losses on a world wide basis run into millions of dollars each year. The estimated loss through microbial spoilage is difficult to estimate since this loss must be dependent on the application of technology available to the various countries. Such technology is not applied equally nor is it available since it may require a large capital input. Food spoilage is certainly a greater problem for emerging nations than those already developed. Available figures for food wastage, including plate waste, method of cooking, spoilage or preparing meals is given at 10% (FAO/WHO Nutritional Studies No. 28).

The relationship of spoilage to food safety is interesting. Conditions which permit spoilage organisms to develop are often the same as those which enable food poisoning organisms to develop. However, in some cases the presence of spoilage organisms is desirable and brings about a natural indicator as to the shelf life and safety of the product. A case in point is that of a meat filled product. The shelf life of such a product was considered terminated by the appearance of mold growth on the surface. An attempt to increase the shelf life of this product was undertaken by irradiating the surface with ultraviolet radiation. This treatment was found to be effective in preventing the surface mold growth, but the internal contents were more than overripe at the end of the extended shelf life. There are cases when the spoilage organisms actually assist in the growth of food poisoning bacteria. A reported study (Duitschaever et al., 1971) involving cheddar cheese demonstrates this point. Mold growth was observed on the surface of the cheese. Where this growth had taken place, coagulase positive *Staphyloccus aureus* was isolated (100–1,000/gram). These organisms were capable of producing enterotoxin D. The increase in pH at the surface of the cheese by the mold favored the localized growth of *S. aureus*.

FACTORS CONTROLLING FOOD SPOILAGE

The principle organisms involved in the spoilage of eight basic food groups are given in Table 1.1. This table shows several interesting features. One particular point is the central role which the *Pseudomonas* group plays in food spoilage. These organisms have been reported to be involved in food poisoning outbreaks. Likewise, food-borne outbreaks incriminating the streptococci and *Bacillus* have also been documented. Notwithstanding the comments on these organisms, the other group of importance is the *Lactobacillus* group. These organisms play an important role in food preservation, both by natural and unnatural means (starter cultures). An in-depth

TABLE 1.1.   PRINCIPAL ORGANISMS INVOLVED IN SPOILAGE OF SELECTED FOOD GROUPS

Food Group	Important Genera contributing to Spoilage
Bread and Cereals	Aspergillus, Bacillus Rhizopus, Fusarium, Penicillium.
Canned Foods	Bacillus, Clostridium, Streptococcus, Byssochlamys, Lactobacillus.
Cured Meats	Streptococcus, Lactobacillus Pseudomonas Achromobacter, Bacillus, Micrococcus Proteus.
Fish and Fish Products	Flavobacterium, Pseudomonas, Achromobacter, Proteus, Bacillus, Micrococcus Corynebacterium.
Milk Products	Streptococcus, Micrococcus, Alcaligenes, Bacillus, Clostridium, Pseudomonas, Lactobacillus, Penicillium, Torula, Geotricum.
Poultry and Poultry Products	Pseudomonas, Micrococcus, Achromobacter, Proteus, Penicillium Cladosporium, Aerobacter.
Red Meats	Pseudomonas Achromobacter, Lactobacillus, Leuconostoc Bacillus, Micrococcus Microbacterium, Thamnidium.
Vegetables and Fruit	Erwina, Botrytis, Pseudomonas Rhyzopus, Alternaria Trichoderma, Penicillium Fusarium.

discussion of individual foods and food spoilage can be found in Fraizer (1968), Nickerson *et al.*, (1974) and Weiser, *et al.*, (1971).

Understanding spoilage patterns in food is important, but in order to reason why such microbial successions take place, the underlying factors leading to spoilage must be determined. These factors also apply to the safety of foods. Mossel *et al.*, (1955) lists these factors as chemical, physical, and processing. Such factors can be termed intrinsic or extrinsic. Consider the simplified sequence of events in the spoilage of raw milk at 20°C (Fig. 1.1). There are a number of factors which play a role; i.e. water content, temperature and nutrient availability. These factors permit the outgrowth of certain organisms such as the coliforms, streptococci and lactobacilli. These organisms utilize the substrate lactose and in doing so, lower the pH from about 6.8 to about 5.0. When the pH reaches this lower value, curdling occurs. The lactic acid formed is then further metabolised by yeasts and molds. The breakdown of the proteins such as casein, and formation of amines causes a rise in pH and then the pseudomonads, *Bacillus* sp. and other proteolytic organisms can take over. These organisms continue the breakdown of the milk, reducing the oxidation/reduction (redox) potential, setting up anaerobic conditions. Such conditions permit the growth of *Clostridium* species and other anaerobic organisms to aid in the putrifactive process.

In this example we see the influence of these intrinsic or extrinsic factors. Those of major concern are listed in Table 1.2.

WATER ACTIVITY (a_w)

The concept of water activity is useful since it enables us to distinguish between free (available) and bound water. The available water or moisture in foods can be utilized by organisms for growth. In order to prevent growth one has to reduce this available moisture. This can be achieved either by drying or by adding solutes such as sucrose, salt or a gelling agent to bind the water. Any water bound by such solutes is considered unavailable unless further physical or biochemical breakdown occurs. This solute effect on water vapor pressure can be defined in mathematical concepts using Raoults' law.

$$\frac{P}{P_o} = \frac{n_2}{n_1 + n_2} = A_w$$

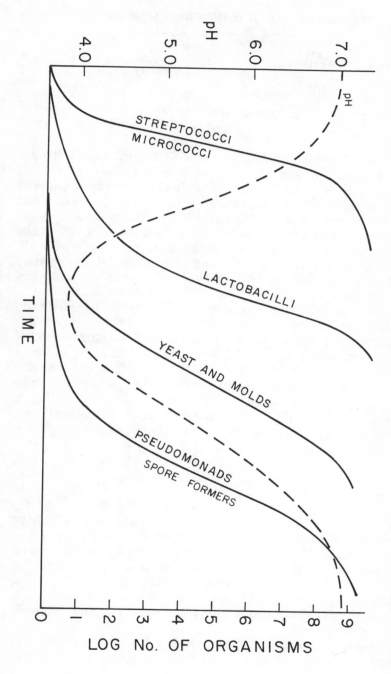

FIG. 1.1.   MICROBIAL SPOILAGE PATTERN OF RAW MILK AT 21°C.

TABLE 1.2. FACTORS INVOLVED IN DETERMINING SPOILAGE PATTERNS OF FOODS

Water activity (a_w)
pH
Temperature
Oxidation/Reduction Potential (Redox)
Substrates available
Microbial contaminants
Inhibitors

where P and P_o are the vapor pressure of the solution and pure solvent (water) and n_1 and n_2 are the number of moles of solute and solvent respectively in the solution. In a food system, pure water is assigned the value of 1.00 for water activity, i.e. the maximum. Another important concept is the relative humidity of the atmosphere in equilibrium with the substrate or food, i.e. a_w x 100. If the relative humidity surrounding a food is higher than the food itself, the a_w of the food would increase and, conversely, if the relative humidity is lower, the surface of the food would dry out thus lowering the a_w.

The importance of a_w to spoilage and food safety can be illustrated by studying the minimum levels in which organisms may grow. Most organisms of public health concern usually require an a_w of 0.90. The values shown in Fig. 1.2 demonstrate minimum values of some of these organisms. Foods in which the a_w is lower than the level for growth may still have organisms which are viable. The viability is also dependent on water activity. Studies by Higginbottom (1953), Scott (1958) and, more recently, Christian et al. (1973) have shówn the various effects of air and a_w on the survival of various types of bacteria in dried foods.

pH

The ability of pH or hydrogen ion concentration to affect or change microbial spoilage patterns has already been illustrated in the spoilage of milk (Fig. 1.1). Like other physical factors there is an optimum pH for the growth of microorganisms. The spectrum for growth is wide among microorganisms. Yeasts and molds have a preference for a pH in the acid region with some molds tolerating a pH of 2.0. Mold can grow over a wide pH range while most bacteria prefer a medium near 7.0.

Use is made of pH in the food industry for food preservation, preventing spoilage and food-borne hazards. Some acid-forming

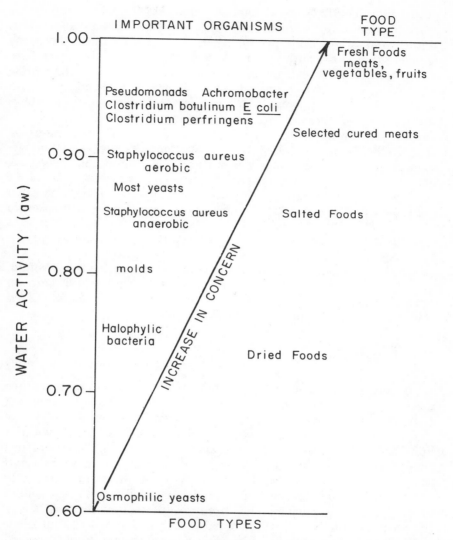

FIG. 1.2. MINIMUM a_w LEVELS OF IMPORTANT ORGANISMS INVOLVED IN SPOILAGE AND FOOD SAFETY

bacteria such as *Streptococcus lactis* and members of the *Lactobacillus* family produce lactic acid during their growth. During the fermentation the pH drops to about 5.5, and if the drop is rapid, it will prevent the outgrowth of *S. aureus*. This fact is used in cheese production. The lactic acid bacteria are important in a number of other fermentations such as preparation of cured meats, sauerkraut, yogurt and other milk products. In many cases these days, the fermentation is not natural but relies on the use of a starter culture added to a sterile substrate. Some foods are prepared by the simple addition of acid. An example of this is the use of citric or acetic acid in the pickling of fish, vegetables, and fruits. In terms of food safety, studies on non-sporing disease-producing organisms such as *S. aureus* and *Salmonella*, a pH value of about 4.5 prevents their outgrowth (Riemann, 1969). With spore formers such as *C. botulinum* a value of 4.5 and below is sufficient to inhibit the growth. A pH of 5.7 or below is reported to prevent the outgrowth of *C. perfringens*. In canned foods, a pH of 4.5 or above is often used as an indication of a low acid food and should therefore be processed to give a 12D botulinum cook. This entails cooking the product at a specific temperature above 100°C over a period of time to kill 1×10^{12} spore of *C. botulinum* (Nickerson *et al.*, 1974).

The mode of action of acid in the preservation of foods is twofold. The death or inhibition of bacteria is due in part to the hydrogen ion concentration. Also important is the presence of the undissociated acid or anion (Levine *et al.*, 1940).

TEMPERATURE

Like other factors, microorganisms have an optimum temperature for growth. In food spoilage, the psychrophils are bacteria which grow at low temperatures (5°C or lower). Groups such as the *Achromobacter* or *Pseudomonas* play a part in the breakdown of a number of foods at low temperature storage. The temperature ranges of psychrophils and other noted groups (i.e. mesophils and thermophils) are given in Table 1.3. The weakness in classifying organisms by their temperature requirements is that a great deal of overlapping occurs within the three groups. An example of this is *Streptococcus faecalis* which is capable of growing at 10°C but is not classified as a psychrophil. Likewise, some lactobacilli are involved in meat spoilage at 3.3°C but again are not normally classified as psychrophils (Pierson *et al.*, 1970). Many food poisoning organisms are capable of growth at temperatures as low as 6°C. Studies by Angelotti *et al.* (1961) have demonstrated that *Salmonella* can grow at pH 5.7 in chicken à la king at 6°C. Upper growth temperatures of such organisms as *Salmonella* and *S. aureus* have been recorded between 44 to 47°C.

TABLE 1.3.   CLASSIFICATION BY TEMPERATURE AND ROLE IN SPOILAGE OF IMPORTANT FOOD-BORNE ORGANISMS

Temperature Range of Growth °C	Examples of Classification	Organisms of Concern	Role in Spoilage
0–20		Pseudomonas Flavobacterium	Positive
	Psychrophilic	Achromobacter Micrococcus Cladosporium Penicillium	Positive
		Streptococcus (Group D)	Positive
20–45		Bacillus Clostridium	Varied
	Mesophilic	Staphylococcus Escherichia	
		Salmonella Vibrio	Minimum
45–60	Thermophilic	Clostridium Bacillus	Positive

The heat resistance of microorganisms varies from species to species (e.g. *Bacillus*). There is also variation in heat resistance during the growth stage of microorganisms. Young cells are less heat resistant than stationary phase cells, and the cocci tend to be more resistant than rod shaped cells. The spore occupies a very special place in food microbiology because of its high heat resistance. Many problems in canned food are a result of the survival of spores such as *Bacillus stearothermophilus*. This organism causes flat sour spoilage. Likewise, spores of *C. botulinum*, as a result of improper processing, can survive and lead to botulism.

High temperature above the maximum of growth eventually leads to the death of the organism. This fact is used in the process of pasteurization or sterilization. Such processes are aimed to kill the most heat resistant organisms either of health or spoilage concern, or both. During the heating process there is a period when some organisms exhibit an injured state before death is reached. This state of injury has been studied extensively (Hurst, 1977). Under such conditions, the cell has more exacting requirements and a longer lag phase for recovery and growth.

At the other extreme of temperature, i.e. freezing, the activity of microorganism is stopped. In contrast to heat, freezing controls the microorganisms rather than eliminates them. Temperatures of −20°C or lower over a period of time, has a lethal effect and can eliminate about 90% of its population. This effect can be compounded by a slow thaw cycle (Weiser *et al.*, 1945). Most sus-ceptible to freezing and thawing are the Gram negative micro-organisms. Another consideration in the destructive effect on bac-teria is the rate of freezing and the fluctuation of the temperature during cold storage. One cannot, however, rely on freezing as a method for elimination of spoilage or food poisoning organisms. Studies by Howard (1956) with *Salmonella* held at –10°C, showed they were still viable after 5 months.

Many of the temperature effects discussed above are influenced by the medium in which the microorganism is placed. Some sub-stances, such as acids, increase the lethal effect of heating or freezing while others such as starches or fats tend to have a pro-tective effect.

OXIDATION REDUCTION (REDOX) POTENTIAL

The growth of spoilage or pathogenic organisms can be strongly influenced by the availability of oxygen. Some organisms (aerobes) depend on oxygen for their metabolism, some can survive in low oxygen tension (facultative), while others such as the anaerobic group of organisms will grow only in the absence of oxygen. During spoilage of one food, all three conditions can exist in succession (Fig. 1.1). Aerobic organisms involved in food spoilage include the yeasts and molds, *Micrococcus* and *Pseudomonas*. Organisms such as the lactobacilli and some *Bacillus* species, tend to be facultative. Or-ganisms of the *Clostridium* group are anererobic, thus their im-portance in the spoilage of canned foods. Normally, fresh foods have a low redox potential within the tissues because of the presence of reducing substances (e.g. reducing sugars). This system favors growth of anaerobic organisms. Since most inner tissues are sterile, growth of bacteria usually starts at the surface where contamina-tion takes place easily and oxygen is available. A system which has been well studied is the spoilage of meats and one can see the influence of the redox potential on the type of spoilage (Ayres 1960, Ingram 1962). When meat is packaged in an oxygen permeable film, the *Pseudomonas/Achromobacter* compete very well and become the main spoilage group along with *Microbacterium theromosphactum* (Pierson *et al.*, 1971). If the packaging film is changed for an oxygen impermeable one, the lactobacilli become the main spoilage or-ganisms (Fig. 1.3). One should not ignore, however, the effects of respiration of the tissues in maintaining or trying to maintain a low redox potential, nor the lowering of the pH by the lactics in affect-

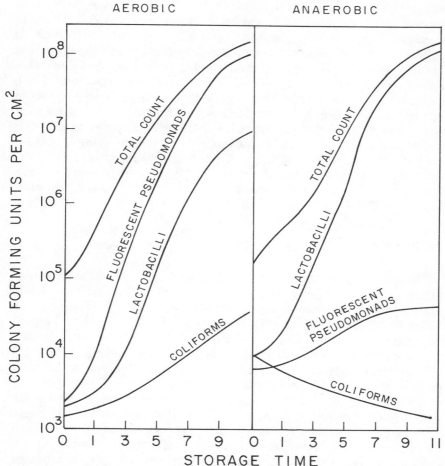

FIG. 1.3. THE EFFECT OF PACKAGING MATERIALS ON THE SPOILAGE OR-
GANISMS IN BEEF AT 4°C.

ing the spoilage of meat in the two systems. In aerobic spoilage of
meats, organisms such as *S. aureus* do not compete well, due in part
to the pH and the high oxygen demand by the *Pseudomonas/
Achromobacter* group. This demand can reduce the redox potential
at the growth surface and affect organisms less efficient in oxygen
utilization.

The redox potential also plays a part in the type of breakdown
products produced by spoilage organisms and in turn set up con-

ditions for pathogens (*C. botulinum*). If sufficient oxygen is present, theoretically, the food will eventually break down to end products such as carbon dioxide, water, ammonia and free fatty acids. If oxygen is limited, more intermediate products are produced such as lactic, acetic and formic acids and ethyl alcohol. Complete anaerobic conditions yield pungent smelling compounds such as putrescine and cadaverine. These amines were originally thought to be involved in ptomaine poisoning. Formation of such compounds arise as a result of amino acid decarboxylation (Thimann, 1963). Another interesting amine from the amino acid histidine, is histamine, which plays a part in allergic reactions and also gives rise to food poisoning. The microbial activity during the spoilage of fish can produce enough histamine to give rise to unpleasant reactions, if eaten by humans (Ienistae, 1973).

SUBSTRATE AVAILABILITY

The type of growth observed in the spoilage of foods is by saprophytic microorganisms, i.e. organisms capable of using organic matter and simple compounds. Again, an example of this is the breakdown of milk (Fig. 1.1). Compounds in this product help to determine the type of organism likely to succeed e.g. lactose fermentors. Some organisms are capable of breaking down proteins (proteolytic) using extracellular proteases; e.g. *Bacillus cereus*, and certain genera such as *Pseudomonas*, *Proteus* and *Clostridium*. A certain group of microorganisms produces enzymes to hydrolyse fats to fatty acids (lipases or glycerol hydrolase esterase). Spoilage organisms such as the *Microbacterium* possess these enzymes (Collins-Thompson, 1971). Likewise, genera like the *Pseudomonas* and *Micrococcus* are also capable of splitting fats and can be a cause of spoilage in such products as butter fat (Fraizer, 1967). There are several other groups, each with characteristic enzymes to utilize certain substrates, e.g. amylolyic and pectolytic organisms. Some foods may not provide the necessary requirements for growth in a very selective manner. Some lactic acid bacteria require vitamins of the B group for growth. Because of this need, bacteria such as *Lactobacillus leichmanni*, *Lactobacillus lactis* and *Streptococcus faecalis* have been used for vitamin assays. The levels of inorganic minerals such as Cu^{2+} or Fe^{3+} are usually sufficient in most foods to support growth. Energy requirements may also vary, since some organisms can use simple carbohydrates, while others can use proteins and amino acids as an energy source.

INHIBITORS

One can classify inhibitors in foods into 3 classes:

a. natural inhibitors
b. inhibitors as a result of microbial action
c. added inhibitors.

There are several reported natural inhibitors in foods, including benzoic acid in cranberries, oil in garlic and lysozyme in milk and eggs (Hurst, 1973). There are some inhibitors whose actions are not clearly defined and these include the antimicrobial properties of cocoa, citrus fruit oils and the lactenins in raw milk.

Inhibitory actions as a result of microbial action are well documented in the literature. Inhibition by lactic or acetic acid bacteria is considered to be due, in part, to a pH effect. However, some of the lactic acid bacteria are capable of producing antibiotic substances such as nisin and diplococcins (Hurst, 1972). Production of hydrogen peroxide by the lactic acid bacteria has also been demonstrated to be inhibitory to *C. botulinum* (Benjamin *et al.*, 1955). The antagonistic relationships between the *Pseudomonas* and gram positive bacteria such as *S. aureus* have been documented by Seminiano *et al.* (1966). Similar inhibition of *S. aureus* by *Proteus* and coliforms have also been reported by Graves *et al.* (1963).

There is also research which indicates inhibition between gram negative bacteria. One such case is with the *Pseudomonas* and *Achromobacter*, the psychrophilic organisms in refrigerated beef (Vanderzant *et al.*, 1968). Studies on specific foods for antagonism between microorganisms are numerous and one of the best reviews on this topic is by Hurst (1973).

The chemical preservatives that may be added to food include such acids as acetic, sorbic and benzoic. Compounds such as nitrites, proprionates and parabenzoates are also quite common. The permitted use of such additives will depend on the country where the food is prepared and eaten.

Sorbic acid is an additive whose inhibitory action ranges from yeast and molds to catalase positive bacteria (Emard *et al.*, 1952). The most suitable pH for such action is from 3 to 6. The type of foods which may contain sorbic acid include pickles and relishes, bread, cheese and wines. Benzoic acid is also a widely used additive. It is effective at a pH of 4.0 or lower, against yeast and molds, and as a result is used in fruit squashes and juices.

One of the most important and perhaps one of the most controversial additives at this time is sodium or potassium nitrite. This additive is important in the inhibition of outgrowth of spores of *C. botulinum* and toxin formation (Collins-Thompson *et al.*, 1974). Unfortunately, it has also been linked to the formation of nitrosamine, a potential carcinogen (see Chapter 10 on Nitrosamines).

Nitrite not only has a preserving action in meats but it also gives the meat it's characteristic color and flavor (Cho *et al.*, 1970).

The mechanism by which nitrite contributes to the inhibition of toxin formation by *C. botulinum* in meats is not clear. Various theories exist and have evoked several studies (Ingram, 1976).

Current research, in light of the nitrosamine problem, is directed at reducing or eliminating nitrite from such products. (See Chapter 10 on Nitrosamines).

FOOD-BORNE INFECTION HAZARDS

Earlier in this chapter it was stated that conditions which give rise to spoilage can easily give rise to food-borne infection hazards. This is because many of those factors permitting these conditions also govern the growth of food poisoning organisms (e.g. A_w, pH). There is also the fact that some of the pathogens are not spoilage organisms, in the sense that they do not betray their presence by overt spoilage e.g. *Salmonella*. Many of the food poisoning organisms do not compete well in a mixed flora situation and usually do their damage on processed foods when the threat of spoilage has been removed and cross contamination has taken place. In this situation, incorrect holding temperature sets the stage for an outbreak.

It should be clarified that the term infection applies to such organisms as *Salmonella* and not to such organisms as *S. aureus*. This latter organism causes food poisoning by the formation of a toxin in the food and is correctly termed a food-borne intoxication. The same is the case with the organism *C. botulinum*. Both of these organisms will be excluded from any further discussion along with *C. perfringens* (see Chapter 3 on Anaerobic Bacteria) although *C. perfringens* can be considered an infection. One should also consider the peculiar status of such genera as the *Pseudomonas*, *Streptococcus* and *Proteus* in food-borne infection hazards. Although there is some doubt as to their ability to cause food poisoning outbreaks, any discussion of food poisoning outbreaks would be incomplete without them.

Those organisms being considered in this section are listed in Table 1.4. Organisms such as *Listeria, Erysipelothrix* and *Mycobacteria*, play a limited role as agents of food-borne disease. In the past, cases have been reported with food as a vector.

SALMONELLA (See also Chapter 5 for more details)

There are, at this time, about 1500 serotypes of *Salmonella* pathogenic to man or animal and, judging from the sampling schemes for foods, it is the most sought after bacteria in foods. It is a Gram negative rod, usually motile and 1–3 microns in length. It has certain biochemical reactions which are used for classification or identification, of which H_2S production is one of the most important characteristics (Taylor *et al.*, 1969). The various species and strains of *Salmonella* can be identified from each other by serological techniques (Antigen-antibody reactions). In such a scheme one relies on the flagella antigens (H) and the somatic or body antigens(O). Further identification of strains within the same serotype may be obtained by phage typing. A less precise method for classification has been reported. In this scheme, one uses ecological adaption of the serotype to its host. Certain salmonellae are host specific, producing a disease in one particular animal species only, e.g., *Salmonella pullorum* in poultry. The type of infection noted here, however, follows a different etiology to the common food poisoning strains. The organisms associated with food infection are adaptable to both human and animals, of which the type species is *S. enteritidis*.

The main reservoir of salmonellae in the environment is the intestinal tract of animals. Animals used in the food chain are particularly important sources. These include chickens, turkeys, ducks and pigs. Other animals such as cattle and sheep can also

TABLE 1.4. GENERA OF BACTERIA INCRIMINATED IN FOOD-BORNE INFECTIONS

Gram +	Gram negative
Bacillus cereus	Salmonella/Arizona
Streptococcus	Shigella
Listeria	Escherichia coli
Erysipelothrix	Klebsiella
Mycobacteria	Vibrio parahaemolyticus
	Pseudomonas
	Proteus
	Yersina
	Brucella

contribute as carriers. Man, too, can act as a carrier, and as such, usually shows no signs of being ill. In this state he can excrete salmonellae for several weeks or months before his intestinal tract is finally free of them. It is this human source that plays an important role in cross contamination of cooked foods. Other sources of contamination of foods may arise from infected rodents and insects. Cockroaches and flies transmit salmonellae by contact rather than by infection and excretion. The use of raw sewage as fertilizer can also set up a cycle of infection. *Salmonella* can survive in contaminated soil for several months. Another important source injecting *Salmonella* into the food chain is animal foods. One such example is *S. agona* which was rarely encountered before 1969 in any country. At the end of 1972, this serotype was the 9th most common reported in the United States. The original vehicle of the organism was found to be imported fish meal. Similar reporting of increases in isolation from human and non-human sources were also reported in other countries.

With the multitude of sources, the ease with which salmonellae gain access to the food chain is not surprising. Nor is it difficult to see the problems in attempting to control such contamination in foods.

The consequences of eating contaminated foods may lead to illness characterized by nausea, headaches, vomiting, chills and diarrhea, usually 12–24 hrs after ingestion. The extent of the illness depends on several factors including dosage and sensitivity of the population infected. There is considerable discussion as to the number of cells required to cause an infection. In the case of *S. typhi*, only a few cells are required and the course of the disease is characterized by a septicemia. Such cases with food as a vector are rare, although there are cases reported. Fabian (1946) has noted outbreaks, as a result of contaminated cheese. In cases where the salmonellae are non host specific, recent reports by D'Aoust (1976) suggest that a dosage of 100 cells per gram may be sufficient to cause salmonellosis. Earlier reports (Committee on Salmonella, 1969) indicate that much higher levels (e.g. 10^6) are required. Such diversity in levels may well be accounted for by recognizing the difference in strains of salmonellae and the individual's sensitivity to the disease.

The fact that segments of the population such as the very young and the elderly are more sensitive to salmonellosis may explain the various reported dosages required to cause illness.

The types of food which may transmit *Salmonella* are numerous; among the chief sources with raw products are pork, poultry, egg and egg products and milk and milk products (Marth, 1967). There are a number of reports citing such products as raw pork sausages containing *Salmonella*. Levels over 50% of contaminated sausage have been reported (Committee on Salmonella, 1969). A great deal of interest has also been shown in processed poultry. In a survey of imported raw poultry into Japan, Suzuki reported levels of *Salmonella* contaminations from various exporting countries (Table 1.5). Studies in the United States have shown, however, levels of contamination may vary from 5.5 to 50%, depending on the processing plant (Wilder *et al.*, 1966). Similar studies in Canada have also revealed this variation (unpublished data).

One of the most interesting links with respect to poultry contamination and human salmonellosis, is the frequency in which those serotypes isolated from humans correspond to serotypes isolated from poultry. One has to be careful about these speculations, but such serotypes as *S. typhimurium, S. thompson, S. infantis, S. saint paul* and *S. heidelberg* are constantly isolated from poultry and also head up the top 10 serotypes isolated from humans in Canada and other parts of the world. The spread of human salmonellosis from powdered egg has also been well documented in England (Report, Medical Research Council 1947). Several of the serotypes found at this time were unknown in the U.K.

The types of products involved in salmonellosis in Canada are listed in Table 1.6. Such a list attests to the ability of salmonella to penetrate into our food supply. One other consideration that is frequently ignored in discussing salmonellosis is the economical cost. In Canada it has been estimated that the average stay for those requiring hospitalization is 14 days (E. Todd-Personal Communication). This does not account for the suffering and lost work days by those for whom the disease is self-limiting.

TABLE 1.5. INCIDENCE OF SALMONELLA CONTAMINATION OF POULTRY IMPORTED INTO JAPAN 1973

Country of Origin	% Contamination[+] with Salmonella
Denmark	3.8
Hungary	5.7
China	9.8
USA	10.7
Bulgaria	13.5
Canada	17.6
Netherlands	28.5

[+]6,523 Samples analyzed

TABLE 1.6. *FOOD-BORNE INCIDENCE IN CANADA INVOLVING SALMONELLA
1973/74

Food Type	Serotype	No. ill
Chocolate	S. eastebourne	21
Cracked eggs?	S. enteritidis	2 or more
Turkey?	S. saint paul	about 150
Coconut cream pudding	S. senftenberg	21
Smoked turkey	S. thompson	3
Chicken and turkey	S. thompson	75
Ham	S. typhimurium	6
Barbecued turkey	S. typhimurium	4
Turkey?	S. typhimurium	27
Pepper	S. weltevreden	?

(From *Food-borne disease in Canada, 1973)

ENTEROPATHOGENIC ESCHERICHIA COLI

E. coli is a normal inhabitant of the gut. The majority of these are
non-pathogenic, distributing themselves in the upper small intestine
and reaching their maximum numbers to 10^9 in the large intestine.
This organism has also been associated with a variety of infections.
Bray (1945) was one of the first to associate certain serotypes of E.
coli with infantile diarrhea. The classification of these serotypes is
based on certain biochemical and serological properties of its O
(somatic), K (envelope) and H (flagella) antigens. Of greatest concern are those serotypes termed enteropathogenic and these tend to
belong to the OK serogroups. The disease caused by such groups
manifests itself in the form of diarrhea, some times bloody with or
without fever (Sojka, 1973). Strains affecting humans can be divided
on the basis of whether they produce an enterotoxin or not. The two
categories are referred to as enterotoxigenic or enteroinvasive.
Those serotypes termed enterotoxigenic are sometimes referred to
as cholera type, or infantile diarrhea or Traveller's diarrhea. The
enteroinvasive E. coli is associated with "colitis" or a Shigella like E.
coli diarrhea. Those serotypes considered to be pathogenic to humans are listed by Bryan (1969). Many of these are host specific.
The main source of enteropathogenic E. coli is from faecal material
from animal and man. There is also evidence that strains capable of
causing human enteric disorders have been isolated from animals
(Shooter et al., 1970). Further epidemiological evidence is needed to
substantiate this point and to determine the extent to which such
strains play a role in food-borne diseases.

The problems of enteropathogenic E. coli in food have been outlined by Insalata (1973). One of the problems is the lack of informa-

tion of the true incidence of the toxic strains in foods. Some studies have been done with certain foods such as dairy products (Yang *et al.*, 1969), and meat products (Sakazaki 1971). These studies indicate levels of contamination were low in dairy products (3%) and about 10% for meat products. Such levels are not reflected in current statistics in food outbreaks (c.f. poultry and Salmonella). There are studies of dosage levels in adults required to give gastroenteritis by *E. coli* (Bryan, 1969). Such studies indicated that several million cells per gram are required. One should, however, bear in mind that, depending on the type or section of population (e.g. the very young), the required dose may vary. This is certainly reflected in the well-documented outbreak due to imported cheeses in the United States in 1971. Cell concentration of enteropathogenic *E. coli* O124:B17 found in the cheese varied from 10^3 to 10^7 per gram (Marier *et al.*, 1973). In this outbreak it was estimated that 387 persons in the 14 states became ill. The attack rate for exposed persons was estimated at 97%. The source of the infection was not determined, although contaminated river water entering the plant via a malfunctioning filtering system was suspected. *E. coli* O124 was isolated from plant equipment and from cheese undergoing the ripening process. Other countries have reported food poisoning cases due to *E. coli* and these include Japan, United Kingdom, and Australia.

One major problem in detecting enteropathogenic *E. coli* is the methodology, much of which is beyond the normal range of many small industrial laboratories. Many of these problems of isolation are outlined by Mehlman *et al.*, (1974). An essential part of such detection is the specialized serology training needed by laboratory personnel.

KLEBSIELLA

Klebsiella belongs to the family Enterobacteriaceae, (along with *Salmonella, Shigella, E. coli* and *Proteus*). They are gram negative, non-motile and capsulated. Among the various species listed is *K. pneumoniae*, an organism associated with bacterial pneumonias. Organisms of the *Klebsiella* group have been incriminated in foodborne outbreaks (Foster 1973; Bryan, 1969). Such outbreaks have been reported in Hungary but not in the United States or Canada.

The source for these organisms appears to be widespread. Reports indicate that they have been isolated from faecal material, forest environments, foodstuffs, natural waters and sugar cane.

Symptoms associated with a food-borne outbreak include nausea, headache, abdominal pain and diarrhea. An incubation time of 10 to 15 hours was recorded and all patients recovered within 24 hours.

The importance of *K. pneumoniae* in dairy products has been outlined by Schiemann (1976), suggesting that food may be an important source of this organism in hospitals.

SHIGELLA

Members of this genus belong to the family Enterobacteriacea. Many of their characteristics apart from their mobility, are similar to the genus *Escherichia*. Although *Shigella* is not considered an important source of food-borne outbreaks (2%), the disease itself can be much more severe than salmonellosis. The two principal serotypes of concern with regard to foods are *S. flexneri* and *S. sonnei*. *S. sonnei* appears to be more common in the United States and also has been reported extensively in the United Kingdom (Bryan, 1969).

From the food-borne outbreaks, the reported attack rate is high (50%) with a variable incubation period (7–40 hours). The disease is characterized by bloody stools with or without fever and abdominal cramps. Those persons infected by shigellosis can also became carriers and are frequently the source of food-borne outbreaks. Such carriers may continue to excrete *Shigella* for several weeks.

The number of cells required to give shigellosis is also difficult to determine. Some studies have been done which indicate that cell concentrates of 1–10 billion were required (Dack, 1956). This level is difficult to understand, since as pointed out, they would possibly have to be transmitted by human source. Evidence is also scant on the organism's ability to grow in foods. There are some survival data available on *Shigella flexneri* in foods. It has been shown to persist in eggs, clams, milk, flour, and oysters for various periods of time.

Various foods have been involved in outbreaks of shigellosis, including various salads (potato, shrimp and tuna), beans, and milk products (Bryan, 1969). Many of these outbreaks were transmitted by infected food handlers. Data from the United States have indicated that outbreaks involving salads are much higher in the last 10 years than any other food commodity where shigella has been definitely identified.

One interesting fact about *Shigella* is the high incidence involving the 1–4 year old range. This group appears to be the most susceptible one. There are several outbreaks also reported where the transmission of the disease has been via water.

VIBRIO PARAHAEMOLYTICUS

V. parahaemolyticus is a halophilic, gram negative, marine bacterium found in sea water, mud and plankton. It is also carried by fish and shellfish. Outbreaks usually occur in the summer months when temperatures of 32°C and above encourage the growth of this organism. This organism was first recognized as a cause of acute gastroenteritis in Japan. The Japanese custom of eating raw fish and shellfish in the form of "sushi" or "shirasuboshi" was one of the reasons for such outbreaks. *V. paraheomolyticus* was first isolated by Fujino *et al.* in 1953.

It was not until 10 years later, however, that the name *V. parahaemolyticus* was finally adopted (Sakazaki, 1969). Although this organism accounts for a large percentage of food-borne outbreaks in Japan, outbreaks in other countries have also been reported and the organism isolated from numerous marine sources. There are several important characteristics of this organism including its ability to grow well in 2–4% sodium chloride, with a maximum growth at 37.5°C. There are also about 54 different serological types based on O and K antigens. Many strains are able to produce lysis in human blood when isolated from persons who had suffered the disease. Strains isolated from marine sources were unable to cause such lysis. Such differences are utilized to distinguish potential food poisoning strains from less virulent strains. This test is referred to as the Kanayawa reaction (Barrow *et al.*, 1976) and a positive lysis reaction is a good indication of the enteropathogenicity of the organism.

The disease or infection is characterized by diarrhea, abdominal pain, mild fever, and headache. The symptoms usually appear 4–48 hours after the consumption of the food. Recovery is usually uneventful and is complete within 2–6 days. The dose required to cause such illness has been estimated from 10^6–10^9 cells (Thomson, 1972). There is also evidence that the infection does not spread from person to person (e.g. *Shigella*).

Foods implicated in outbreaks include salted cucumber, crab, prawns, cockles and raw fish and their products. Such reports also indicate that outbreaks occurred not only in Japan but also the United Kingdom and the United States.

PSEUDOMONAS

The *Pseudomonas* species are common soil and water organisms and are present as spoilage organisms on most foods. They are gram negative rods that often produce water soluble pigments. Although there are many species, the most important species with respect to pathogenicity to man is *Pseudomonas aeruginosa*. The organism is found in human feces and in wounds and urinary tract infections in man. In outbreaks of a dysentery type infection, *P. aeruginosa* has been isolated under circumstances which suggest an etiologic role. There is also evidence that *P. aeruginosa* can easily be isolated from foods (Shooter *et al.*, 1969). The symptoms of the disease are diarrhea, cramps, vomiting and nausea. The length of incubation from time of eating to the manifestation of the infection has not been established. One outbreak incriminating *P. aeruginosa* occurred 2 to 4 hours after ingestion of the food. The duration of the disease ranges from one to several days with more severe reactions among children (Bryan, 1969). It would appear that only certain strains of *P. aeruginosa*, under specific conditions, are capable of causing a food-borne outbreak. Such conditions must be rare since such illnesses are seldom reported, leading to difficulty in establishing the incidence of *P. aeruginosa* in food as a food-borne hazard.

One such species of *Pseudomonas* has been well established as a cause of food poisoning; that is *Pseudomonas cocovenenans*. This organism has been isolated from a fermented coconut product called Bongkrek. This product is very popular in Central Java and Indonesia and is normally prepared by placing fermenting grated coconut with *Rhizopus oryzae*. On occasions, this fermentation is overtaken by *P. cocovenenans* which inhibits any further development of the fungus. The production of unsaturated fatty acid (called bongkrek acid) by the *Pseudomonas* species causes hypoglycemia and death in those persons eating the infected product (Foster, 1973).

The *Pseudomonas* group are extremely adaptable and are capable of surviving under extreme conditions and emerge later as the dominant organisms, especially after antibiotic treatment. Some are also capable of producing antibacterial pigments which can eliminate competition from other organisms. Their medical history of infection encourages the idea that they could play a role in food-borne disease.

PROTEUS

The characteristics of this group like that of the *Pseudomonas*, makes them likely, but unproven candidates for causing food-borne infection. They are gram negative, glucose fermenting, proteolytic and produce H_2S. One of the major characteristics of this group is its ability to swarm, spreading rapidly over moist, solid nutritive media. The *Proteus* group is commonly associated with water, soil, sewage, manure and human feces. The most common pathogenic species to man are *Proteus vulgaris* and *Proteus morgani*. Both organisms cause infections of the gastrointestinal and genitourinary tract (Morgan, 1965).

Symptoms ascribed to food poisoning by the *Proteus* group are fever, diarrhea, vomiting and general malaise. Such symptoms occurred about 4 hours after eating the suspected food (prawn). The duration of the illness is about 2 days.

One interesting point in the cases reported by Bryan (1969) is the observation by the author of the similarity between staphylococcal intoxication and the symptoms of illness caused by the *Proteus* species.

Further experiments by several researchers with feeding studies using foods inoculated with high levels of *Proteus* organisms led to inconclusive results. It would appear that, if *Proteus* species are involved in food poisoning, very exacting conditions are required. An alternate theory is that *Proteus* species might well act as a synergist with other organisms capable of causing food poisoning. When *P. morganii* and *V. parahaemolyticus* were fed to monkeys, symptoms of diarrhea and vomiting occurred. The feeding of either organism alone, however, did not cause any illness.

YERSINIA

This group of organisms has received more than its fair share of publicity in the annals of history. Belonging to this group is *Yersinia pestis* formerly *Pasteurella pestis*, the plague organism. This organism, *Yersinia enterocolitica*, and *Yersinia pseudotuberculosis* have now been placed in the family *Enterobacteriaceae*. The latter two species have been the subject of discussions as to possible food poisoning organisms. The status of *Y. enterocolitica* and its presence in food has been reviewed by Lee (1977).

The *Yersinia* group are gram negative bacteria either motile or non motile, with variable catalase and H_2S production. The major source of *Y. pseudotuberculosis* is from animals and is the probable source of most human infection. It has been isolated from deer, otters, birds, cats and mice (Meyer, 1965). The disease caused by *Y. pseudotuberculosis* can be of two forms, one being similar to typhoid, in which the disease is generalized, or septic, and the other is the local or benign form. Death usually results from the former route. Outbreaks of food poisoning or gastroenteritis have been reported; the symptoms of headache, fever, chills and abdominal pain being manifested after 24 hours of ingestion of the food (Bryan, 1969). Contamination of the food by rodents was suspected in this particular case.

Yersinia enterocoliticus has also been reported as a cause of gastroenteritis. Like *Salmonella* and *Y. pseudotuberculosis*, *Y. enterocoliticus* can be systemic and can invade tissues outside the digestive tract. Sources of *Y. enterocoliticus* have been reported from animals such as cattle, dogs, cats and chickens, suggesting zoonosis is an important mode of transmission. Implied in such isolations is the fact that foods such as meats can be a source of infection.

Surveys by Leistner *et al.* (1975), Hanna *et al.* (1976) and others have confirmed the isolation of *Y. enterocoliticus* from chicken meat, pork, beef, lamb, oyster, milk, ice cream and fish. One important aspect of the organism is its ability to grow at refrigerated temperatures (0–4°C) and its survival during freezing (Lee, 1977). Symptoms of the disease are abdominal pain, fever, vomiting and diarrhea. These have been reported in a number of outbreaks (Gutman *et al.*, 1973). Foods implicated in such outbreaks included raw milk, and chocolate milk.

Although there is evidence to suggest that the *Yersinia* group can be a source of food-borne disease, such evidence indicates that the incidence is low (2%). Further research is needed in methodology for their isolation and in the testing of virulence of food isolates.

BRUCELLA

There are three important species in this group, *Brucella abortus*, *Brucella suis* and *Brucella melitensis*. *Brucella abortus* is associated with disease in cattle, *Brucella suis*, in swine and *Brucella melitensis* in goats and sheep. These species are characterized in part by being gram negative coccobacilli, may require CO_2 for growth (*B. abortus*), produce H_2S from proteins and produce growth on thionin- or fuchin-containing media. The resulting disease for man from these species is brucellosis. Such a disease may result from contact with an infected animal or by ingestion of a food. In technically developed countries, the disease, originally contacted by drinking raw milk, has been reduced by pasteurization. Those cases now reported in the United States and Canada usually arise as a result of close contact with an infected animal. In many other parts of the world there are still cases reported where milk and milk products have been incriminated.

The incubation period for brucellosis is anywhere from 7–28 days. The symptoms include aches, chills, fever and loss of appetite. The term "undulant" fever arises from the undulating temperature (101–105°F) of the patient during a 24 hour period, rising late in the afternoon and early morning and then declining. One result of brucellosis, in cases of serious infection, is that complete recovery may take many years. Death can result, if the *Brucella* organisms get into the lymph tissue and invade the organs leading to endocarditis and renal necrosis.

As indicated earlier, products such as cheese, cream, milk, ice cream and infected uncooked meats have been reported as a source of brucellosis. The critical point in such outbreaks is the preparation of such products without pasteurization. The *Brucella* organisms can survive in products like cheese made from unpasteurized milk for 40 days or more. Current regulations in many countries require such cheese to be stored for 60 days so that the *Brucella* die out. Many countries now have a control vaccination program whereby animals have periodic checks before milk or meat is supplied as a food source.

BACILLUS CEREUS

Although other *Bacillus* species have been suggested as possible food poisoning agents, evidence for *Bacillus cereus* has now been well established. The role of *B. cereus* as a food poisoning organism has only been demonstrated adequately in the last 10 years. This new awareness is reflected in the literature by the number of research papers and review papers (Hobbs 1969; Geopfert *et al.*, 1972; Gilbert *et al.*, 1976). Many of the current developments arose from early European reports from countries such as Hungary, Scandinavia and the Netherlands (Foster, 1973).

B. cereus is a gram positive, rod shaped organism capable of forming spores which are fairly resistant to heat. This organism is also capable of growing under anaerobic conditions. It is widely distributed in nature and is found in soil, water, air and a number of foods. *B. cereus* has also been of some concern with the spoilage of some dairy products (Davies *et al.*, 1973). It has an optimum temperature of 35°C for growth but is also capable of growth at 10°C. Most strains produce haemolysis on blood agar, when incubated aerobically.

The type of food poisoning associated with *B. cereus* is similar to *Clostridium perfringens*. The incubation period is 8–16 hours followed by such symptoms as diarrhea, nausea, abdominal pain and on occasions, an accompanying fever. The number of cells of the organism required to cause illness has been estimated at 10–100 million (Hauge, 1955). Numbers of *B. cereus* isolated from stool specimens are usually low. Various foods have been reported to be implicated in outbreaks of food poisoning by *B. cereus*. These include soups, various meats, fried rice, cooked vegetables, ice cream and milk.

Valuable information regarding *B. cereus* food poisoning has come from Hungary where reported cases are much higher than in most other countries (8–10%). In such outbreaks, meat and meat products play an important role; with the source being spores from spices used in many Hungarian dishes (Ormay *et al.*, 1969). Dried foods may also be a source of *B. cereus* since the spores may survive the drying process.

Countries reporting food-borne outbreaks incriminating *B. cereus*, apart from those already mentioned, include the United Kingdom, Canada, Rumania, USSR, USA, Germany and Australia.

Another member of the *Bacillus* group has also been suggested as an agent in causing food poisoning. This organism, *Bacillus subtilis*, is cited by Bryan (1969) and Hobbs (1969). The incubation period and symptoms are similar to *B. cereus*. Likewise, another characteristic is the low recovery of *B. subtilis* organisms from stool specimens after an outbreak. Viable spores of *B. subtilis* were isolated from cooked turkey after a heat treatment of 176.7°C for 8 hours.

STREPTOCOCCUS

When one considers or discusses food poisoning by the Streptococci one usually thinks of the enterococci or Group D streptococci. Included under this classification are *Streptococcus faecalis* and *Streptococcus faecium*, sometimes referred to as faecal streptococci. These two organisms account for most of the reported cases in the literature. These gram positive organisms are found in the intestinal tract of man and animals at level of 1–100 million per gram of feces. They are characterized by a number of properties, including the ability to grow in 6.5% sodium chloride and at a pH of 9.6. One important characteristic is their hemolytic nature on blood agar. Some strains are alpha hemolytic (i.e. able to produce a greenish zone around the colonies), while some are beta hemolytic and are capable of producing a clear zone around the colony. Little can be made of this difference, however, when discussing possible food poisoning strains. One also should be cautioned that, like many of the organisms discussed in this chapter, the role of the enterococci in food-borne infection hazards is very much under suspicion. In reported cases, implication of this group has been mainly based on the isolation of high numbers in the suspected foods (Food-borne disease in Canada, 1973).

Where the enterococci have been implicated in an illness, the symptoms have been nausea, vomiting and, depending on the severity of the disease, diarrhea. The time of incubation has been reported as an average of 6–12 hours (Bryan, 1969).

Much of the controversy surrounding the enterococci has been the inconclusive evidence in feeding studies. In human studies levels of 10^5 to 10^{10} cells were fed to volunteers with both positive and negative results. One of the most comprehensive studies was by Deibel and Silliker (1963) in which 23 enterococci strains were tested. Nine of these strains have been implicated in food-borne disease. They were unable to produce any illness in the volunteers. Filtrate of enterococci cultures have also failed to produce illness. It would appear, as suggested by Roseburg (1965), that certain conditions may be required to give rise to enterococci food poisoning. Such conditions may depend on pH, synergistic action with other bacteria, incubation temperature, and culture medium.

Foods incriminated in food poisoning by the enterococci include meat pie, fish, meat balls, chocolate pudding, cheese, and pork sausage.

There are a number of comments in the literature regarding the lack of positive evidence regarding the enterococci and food poisoning. These organisms have been isolated in a number of foods. A survey by Insalata et al., (1969) found that about 10% of the 5719 processed food survey contained enterococci. One would suspect that food poisoning cases would be more frequent than currently reported. One suggestion by Longrée (1972) is that the illness, being so mild, may result in the disease being largely overlooked.

Whatever the role the enterococci have in producing illness, one cannot help but reflect that perhaps the presence of organisms of fecal origin in food might have clouded the true position of the streptococci as an etiological agent in food-borne disease.

LISTERIA

The relation of *Listeria monocytogenes* to food-borne disease is tenuous, and is largely based on possible transmission of the organism from infected animal or animal products in the food chain. Such a transmission has not been confirmed, although there are reported cases of listeriosis, one involving raw milk (Reed, 1965).

L. monocytogenes in a gram positive rod, motile and facultative in its oxygen requirements. Once in contact with humans it can manifest itself in a number of ways such as meningitis, septicema, conjunctivitis, mononucleosis and abortion. It has been isolated from a number of animals including cows, poultry, pigs, sheep and rabbits. One source of contamination is the ingestion of contaminated silage. The nature of this organism is such that it can persist in unfavorable conditions such as direct sunlight, salting, drying and heating (Bryan, 1969).

One of the major problems in determining the role of *L. monocytogenes* in food-borne disease, is that the methodology for it's isolation is difficult. The knowledge involving carrier rates is also meager, as is knowledge on the epidemiology of listeriosis in humans.

ERYSIPELOTHRIX

Erysipelothrix infection is primarily an occupational disease as a result of handling infected meat, poultry, shellfish or fish. It takes the form of a localized skin infection and does not involve the gastrointestinal tract as do most other diseases discussed in this chapter.

The species *Erysipelothrix rhusiopathiae* is a gram positive rod and will grow aerobically or anaerobically. It has also been shown to be a hemolytic on blood agar.

It has a wide distribution in nature, animals being the main source of infection.

This infection is often referred to as erysiplloid and it can be fatal in animals.

The disease in man, as indicated, tends to localize, although septicemia and endocarditis (entering the blood stream and attacking the inner lining of the heart), can cause death. When it localizes on the skin it appears as a well defined elevated spot and then gradually spreads with the centre becoming purple. Other symptoms include headache and malaise accompanied by a throbbing in the infected area.

Food-borne infections are rare but their survival in infected food is possible since they can resist salting, pickling and smoking of foods (Reed, 1965).

MYCOBACTERIA

This group of gram positive bacteria plays a limited role in food-borne disease. However, they play a positive role in human disease, since they cause tuberculosis, a serious and dangerous illness to man. The tubercle bacilli of the human type (*Mycobacterium tuberculosis*) is transmitted by airborne droplets from an infected person. There is evidence that the bacilli of the bovine type can be transmitted to man by unpasteurized milk. Cases of milk-borne tuberculosis have been reported.

FOODBORNE INFECTION DUE TO PARASITES (see also Chapter 9)

Ingestion of raw or semicooked food by humans can lead to infection by parasites. Such foods as pork, beef, lamb, shellfish, vegetables and wild game, have been incriminated as a vehicle for infection. Food-borne parasitic infections have been known to man for hundreds of years since visible evidence such as worms can be seen in faecal material. Many such parasites have exotic life cycles and only a few of the known parasites will be dealt with here to give some idea of the nature of the beast. There is an excellent review by Healy *et al.* (1969) for those who wish further information.

The common parasite infection *Clonorchis sinensis* is well known in China, Japan and the Far East. It is found in the bile duct of man and other mammals and requires a fresh water snail and fresh water fish such as carp to complete its life cycle. Eggs of the parasite are contained in fecal material of an infected human and can be eaten by certain snails. Within the snail a reproductive cycle of the parasite takes place ending up with the release of the inactive parasite (a cercariae). The cercariae is then capable of infecting the fish by penetrating under the scales and forming a cyst in the flesh (metacercariae stage). Human infection can come about by consumption of the raw or partially cooked fresh water fish containing the metacercaral stage of the worm. The type of infection which follows will depend on the number of worms in the bile duct and the duration of the infection. If the infection is light, there may be no clinical symptoms since the worm does not multiply in man. If heavy infection takes place there may be bilary obstruction leading to inflammation of the bile duct and eventual blocking of the bile flow. Other organs such as the gall bladder and liver can be affected. Infection can be prevented by cooking, since the cysts of *Clonorchis* are not very resistant to the elevated temperatures of boiling or baking.

A similar parasite disease to *Clonoretis* is *Paragonimiasis* (lung fluke disease) caused by the trematode *Paragonimus westermani*. This trematode lives in the lung tissues and other parts of the body such as the intestinal walls or abdominal cavity of man. The infection leads to destruction of the lung tissue with the coughing up of blood; similar to pulmonary tuberculosis. Human infection is transmitted by eating infected fresh water crabs or cray fish. Such infections have been reported in Eastern countries such as China. It is not endemic in such countries as the United States or Europe.

Man acquires the disease through an intermediate host, the aquatic snail similar to *Clonorchis*. The lung fluke discharges its eggs into the sputum which then can contaminate surface waters in which it lives. The snail is infected with the first larval stage. After further developments in the snail (reproduction stage), the parasite encysts in the flesh of a fresh water crab or cray fish. The final host can be man or animals such as dogs, cats and pigs, or wild animals such as tigers or leopards. Human infection comes from eating the uncooked crab or crayfish. When infection is heavy, severe disease or death can occur, especially in lung fluke infection where migration to the liver or brain occurs.

A parasite which has influenced man for 30 centuries and is the best known is *Trichinella spiralis*, the causitive agent of trichinosis. This disease is world wide, with pigs as the major source of infection. Outbreaks in Central Europe were common where raw pork sausages were considered delicacies. Other outbreaks involving wild game are becoming more important. There are several reports both from the United States and Canada where the eating of bear meat has caused trichinosis. (Food-Borne Disease in Canada, 1973). The life cycle of *Trichinella* can be simply illustrated in Fig. 1.4. Interestingly enough, the human host is a dead end for the development of the parasite. In man, during the disease, certain symptoms have

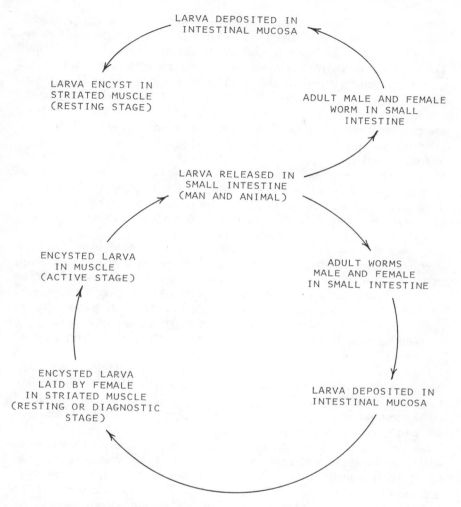

FIG. 1.4.   THE LIFE CYCLE OF *TRICHINELLA SPIRALIS.*

been noted. When the larva invades the intestinal mucosa, gastroenteritis can occur. Once the larva starts to penetrate the muscle
tissue, pain and fever can occur. If the larva encysts in the brain or
the myocardium (heart muscle), the results can be more serious.
However, most cases tend to be less severe, and once the larva are
encapsulated in the muscle tissue, symptoms become less pronounced and finally subside. The cysts of *T. spiralis* are not very

resistant to heat and can be killed by exposing them to temperatures of 55-65°C. Freezing is also used to kill the cysts using temperatures of -10C to -20C for time periods of 12–30 days (Healy *et al.*, 1969). *Taenia saginata* and *Taenia solium* are two tapeworms for which man is the sole natural final host. The beef tapeworm or *T. saginata* can be 10 metres or more and is made up of hundreds of segments called proglottids. The head or scotex from which the proglottids are produced attaches itself to the human small intestine. When the terminal proglottid matures, the gravid segment (about 2 cm long) is passed out singly or in chains from the bowels of the infected human. On occasions, the gravid segment can rupture in the intestine and the released eggs, pass out in the feces. If the eggs or proglottids are picked up by cattle, they develop into onchospheres, and penetrate the stomach wall and then spread to various parts of the animal. In places where there is no control of human faecal pollution of grazing lands, the cycle of infection of man to cattle becomes uncontrolled. It can occur in any country where such practices are tolerated and where the eating of undercooked beef is also a custom. Infections by this cestode usually produces limited symptoms, such as hunger pains or weight loss, but nausea, colic, diarrhea and vomiting have also been reported. Proper cooking or correct holding time during freezing can prevent infection with *T. saginata*.

Taenia solium is the tapeworm associated with pork. It has been reported in countries on a world wide basis. It's life cycle is similar to *T. saginata* except that pigs are the carriers. There is however an added danger with *T. solium* in man. There are cases when humans act as the host and are reinfected by regurgitation of eggs from the intestine into the stomach. The parasite can then become lodged in the stomach wall and then penetrate into other vital organs, e.g. heart, liver and brain. Cysticercosis of the latter organ is the most serious of the infections by this parasite. This can also occur if the eggs rather than the cyst are ingested by humans. Reports by Huang (1969) reveal some very interesting cases of infections by *T. saginata* and *T. solium* in countries like China and Tibet.

Broad tapeworms of the genus *Dibothriocephalus* are also involved in parasitic infection of man. *Diphyllobothrium latrium* is a common tapeworm in fish such as salmon where it acts as the intermediate host. It is found in parts of Asia, Australia, North and South America and Europe. The worm, which may be up to 10 feet in length, can exist in man and other animal hosts in the small intestine. The released proglottids contain the eggs, and can pass out of the host via the feces. If such material contaminates lakes or ponds, the eggs develop into free swimming organisms and can then be consumed by fish. After development in the fish, the larval stage at this point of the cycle is known as sparganum. Sparganosis results when the fish is eaten raw or undercooked by man. Infection is common in countries where smoked salmon, trout, caviar, frogs, snakes and even tadpoles are eaten as delicacies. The tapeworm, once in man, relies on much of its nutrition to be supplied by the host, and because of its size, its demands can lead to nutritional deficiencies and anemia similar to pernicious anemia (B_{12} deficiency). Other symptoms such as nausea, vomiting, diarrhea and weakness can also result. Like many of the other parasitic infections the host, such as man, may show no symptoms at all (asymptomatic).

PREVENTION OF FOOD-BORNE DISEASES

In much of the discussion on food-borne diseases, several concerns should be noted. One concern is the need to control, in some way, the quality of food reaching the market place. In modern society there is a range of exotic foods from all over the world in most developed countries. The obvious point of control is at the import level where either the exporting country or importing country should have some quality and safety assurance program.

One increasing concern is the amount of food being prepared away from home. Centralized food preparation is now very common in North America and Europe. A few years ago most foods were prepared in the home, and when food poisoning resulted from improper sanitation or preparation, only the immediate family or friends were involved. With a centralized system, the potential for mass food poisoning is evident, no matter how improbable it might be with current technology. There are safeguards in our system such as the various regulations, the damage due to adverse publicity and most companies' awareness and action to prevent problems of food poisoning. Such companies have programs to monitor incoming raw materials, treatment of such materials to remove any possible foci of infection, sanitation programs of final products, cleaning of equipment and regular plant check ups.

When such programs are adhered to, food-borne infection hazards are minimized. Despite these safeguards, there is always one factor which cannot be eliminated no matter how much care is taken and this is human error. Monitoring programs such as microbiological determinations is a good method to detect such errors. The problem is that such determinations (e.g. *Salmonella*) take time, and so, very often the product has been consumed before the results are available. This is true of highly perishable products with short shipping times.

Another concern is the need to recognize the opportunistic nature of food-borne disease causing organisms. Many of these organisms are poor competitors in the presence of other organisms. Once such competition has been eliminated by processes such as heating, then pathogens can grow, if contamination takes place. There are many food-borne outbreaks which can attest to this point. Outbreaks at parties, social suppers and restaurants have many common factors. The most important of these is improper holding temperatures. Most food-borne pathogens can grow at temperatures between 4°C (40°F) and 60°C (140°F) reaching their maximum doubling time around 37°C. If food is cooked, recontaminated with a pathogen and held at room or warming temperatures, then an outbreak is likely. Also, one has to remember that, after being cooked, food held at room temperature for a while, cools slowly, especially if held in bulk. During this cooling cycle the temperature can remain ideal for maximum doubling time for some time. If one could ensure all perishable food were held at less than 4°C or higher than 60°C then most food poisoning outbreaks could be eliminated.

Elimination of food-borne hazards can also be brought about by controlling those animals which are natural carriers of disease. A great deal of success has occurred with such eradication programs for *Brucella*, tuberculosis or even trichinosis. A great deal of effort is still needed to limit the problem of *Salmonella* in poultry. This particular program has attracted a great deal of publicity and many proposals. The problem is compounded by economic considerations and at what point in the production an eradication program would be successful. A program of *Salmonella* eradication in breeder flocks is certainly possible but many question whether the cost could be justified in terms of the current risk of salmonellosis to the consumer. Any large program such as *Salmonella* eradication in poultry would most certainly affect the price of the final product to the consumer.

Some attention has been given to the dangers of eating raw or undercooked meats or fish products. Recent evidence of this was demonstrated by the *Salmonella* outbreak in rare roast beef in the United States.

The important point in helping such outbreaks and those, for example, caused by parasites, is to use internal temperatures as a guide to correct cooking. Recommended temperatures around 170°F are usually considered to be sufficient to control or kill disease producing organisms. There are other measures which should also be taken, since heating should not be relied on for complete elimination. Such measures should include sanitation practices to prevent introduction or cross-contamination of the food (Longrée, 1972).

The simple truth is that man possesses the knowledge to prevent food loss by spoilage and knows a great deal about how to prevent and control food-borne diseases. It is now a matter for individual societies to determine just how much it is prepared to invest in that knowledge.

REFERENCES

ANGELOTTI, R., FOTER, M. J. and LEWIS, K. H., 1961. Time-Temperature effects on salmonellae and staphylococci in foods. 1. Behavior in refrigerated foods. Am. J. Publ. Hlth. 51.: 76–83.

AYRES, J. C., 1960. Temperature relationships and some other characteristics of the microbial flora developing on refrigerated beef. Food Res. 25: 1–18.

BARROW, G. I., MILLER, D. C., 1976. *Vibrio parahaemolyticus* and Seafoods. *In* Microbiology in Agriculture Fisheries and Food Series 4, F. A. Skinner and J. G. Carr (Editors) Academic Press, London, New York.

BENJAMIN, M. I. W., WHEATER, D. M., and SHEPHERD, P. A., 1955. Inhibition and stimulation of growth and gas production by clostridia. J. Appl. Bacteriol. 19: 159–162.

BRAY, J., 1945. Isolation of antigenically homogeneous strains of *Bac. coli neapolitanum* from summer diarrhea of infants. J. Path. Bacteriol. 57: 239–247.

BRYAN, F. L., 1969. Infections due to miscellaneous microorganisms in Food-Borne Infections and Intoxications, H. Riemann, Academic Press, New York and London.

CHO, I. C., and BRATZLER, L. J., 1970. Effect of sodium nitrite on flavour of cured pork. J. Food Sci. 35: 668–671.

CHRISTIAN, J. H. B., and STEWART, B. J., 1973. Survival of *Staphylococcus aureus* and *Salmonella newport* in dried foods as influenced by water activity and oxygen. *In* The Microbiological Safety of Food, B. C. Hobbs and J. H. B. Christian (Editors), Academic Press, London and New York.

COLLINS-THOMPSON, D. L., SØRHAUG, T., WITTER, L. D. and ORDAL, Z. J., 1971. Glycerol Ester Hydrolase Activity of *Microbacterium thermophactum*. Appl. Microbiol. 21: 9–12.

COLLINS-THOMPSON, D. L., CHANG, P. C., DAVIDSON, C. M., LARMOND, E., and PIVNICK, H., 1974. Effects of nitrite and storage temperature on the organoleptic quality and toxinogenesis by *Clostridium botulinum* in vacuum-packaged side bacon, J. Food Sci. 39: 607–609.

COMMITTEE ON SALMONELLA, NATIONAL RESEARCH COUNCIL
1969. An Evaluation of the Salmonella Problem. A report of USDA and
FDA. National Academy of Sciences Publ. 1683 Washington D.C.

DACK, G. M., 1956. Food Poisoning, University of Chicago Press, Chicago.

D'AOUST, J. Y. and PIVNICK, H., 1976. Small infectious doses of *Salmonella*. Lancet: 1: 866.

DAVIES, F. L. and WILKINSON, G., 1973. *Bacillus cereus* in milk and
dairy products. *In* The Microbiological Safety of Food. B. C. Hobbs and
J. H. B. Christian (Editors). Academic Press, London, New York.

DEIBEL, R. H., and SILLIKER, J. H., 1963. Food poisoning potential of
the enterococci. J. Bacteriol. 85: 827–832.

DUITSCHAEVER, C. L., and IRVINE, D. M., 1971. A case study: Effect
of mold on growth of coagulase positive staphylococci in cheddar cheese.
J. Milk Food Technol. 34:583.

EMARD, L. O. and VAUGHN, R. H., 1952. Selectivity of sorbic acid media
for catalase negative lactic acid bacteria and clostridia. J. Bacteriol.
63: 487–494.

FABIAN, F. W., 1946. Cheese as the cause of epidemics. J. Milk Technol.
9: 129–143.

FOOD-BORNE DISEASE IN CANADA ANNUAL SUMMARY 1973,
Health Protection Branch, Health and Welfare Canada Ottawa 1976.

FOSTER, E. M., 1973. Food poisoning attributed to controversial agents:
Bacillus cereus, Pseudomonas sp. and faecal streptococci. *In* Microbial
Food-Borne Infections and Intoxications. A. Hurst, J. M. de Mann (Editors). Can. Inst. Food Sci. Technol. 6: 73–77.

FRAIZER, W. C. 1967. Spoilage of Foods. *In* Food Microbiology 2nd
Edition, McGraw-Hill Book Company, New York.

FUJINO, T., OKUNO, Y., NAKADA, D., AOYAMA, A., FUKAI, K.,
MUKAI, T., and UEHO, T., (1953). On the bacteriological examination
of shirasu food poisoning. Med. J. Osaka Univ. 4: 299–307.

GILBERT, R. J., and TAYLOR, A. J., 1976. *Bacillus cereus* Food Poisoning.
In Microbiology in Agriculture, Fisheries and Food (Series 4). F. A.
Skinner and J. G. Carr (Editors), Academic Press, London, New York.

GOEPFERT, J. M., SPIRA, W. M., and KIM, H. U., 1972. *Bacillus cereus*:
food poisoning organism. A review. J. Milk Fd. Technol. 35: 213–227.

GRAVES, R. R., and FRAIZER, W. C., 1963. Food microorganisms influencing the growth of *Staphylococcus aureus*. Appl. Microbial 11: 513–516.

GUTMAN, L. T., OTTESEN, E. A., QUAN, T. J., NOCE, P. S., KATZ,
S. L., 1973. An interfamilial outbreak of *Yersinia enterocolitica* enteritis.
New Engl. J. Med. 288: 1372–1377.

HANDBOOK ON HUMAN NUTRITIONAL REQUIREMENTS. FAO
Nutritional/WHO Monograph Series No. 61. Food and Agriculture Organization of the United Nations Rome 1974.

HANNA, M. O., ZINK, D. L., CARPENTER, Z. L., and VANDERZANT,
C. 1976. *Yersinia enterocolitica*-like organisms isolated from vacuum-packaged beef and lamb. J. Food Sci. 41: 1254–1256.

HAUGE, S. 1955. Food poisoning caused by aerobic sporeforming bacilli. J. Appl. Bact. 18: 591–595.

HEALY, G. R., GLEASON, N. N. 1969. Parasitic Infections. *In* Food-Borne Infections and Intoxications. H. Riemann (Editor). Academic Press. New York and London.

HIGGINBOTTOM, C. 1953. The effects of storage at different relative humidities on the survival of microorganisms in milk powder and pure cultures dried in milk. J. Dairy Res. 20: 65–69.

HOBBS, B. C., 1969. *Clostridium perfringens* and *Bacillus cereus* infections. *In* Food-Borne Infections and Intoxications. H. Reimann (Editor). Academic Press, New York and London.

HOWARD, D. H., 1956. The preservation of bacteria by freezing in glycerol in broth. J. Bacteriol. 71: 625–626.

HUANG, C. T., 1969. Food Customs and Microbial Food Poisoning. University of Hong Kong Gazette 17(2), 1–9.

HURST, A., 1972. The antagonism between *Streptococcus lactis* and sporeforming microbes. J. Milk Food Tech. 35: 418–420.

HURST, A., 1973. Microbial Antagonism in Food. *In* Microbial Food-Borne Infections and Intoxications. A. Hurst and J. M. de Mann (Editors). Can. Inst. Food Sci. Technol. J. 6: 33–43.

HURST, A., 1977. Bacterial injury: a review. Can. J. Microbiol. 23: 936–944.

IENISTEA, C., 1973. Significance and detection of histamine in food. *In* The Microbiological Safety of Foods. B. C. Hobbs and J. H. B. Christian (Editors). Academic Press, London and New York.

INGRAM, M., 1962. Microbiological principles in prepacking meats. J. Appl. Bacteriol. 25: 259–281.

INGRAM, M., 1976. The microbiological role of nitrite in meat production. In Microbiology in Agriculture, Fisheries and Food. F. A. Skinner and J. G. Carr (Editors). Academic Press, London, New York.

INSALATA, N. F., WITZEMAN, J. S., and SUNGA, F. C. A., 1969. Streptococci in Industrial Processed Foods—An Incidental Study. Food Technol. 23: 1316–1318.

INSALATA, N. F., 1973. Enteropathogenic *E. coli*—A New Problem for the Food Industry. Food Technol. 27: 56–58.

LEE, W. H., 1977. An Assessment of *Yersinia enterocolitica* and its presence in Foods. J. Food Protect. 40: 486–489.

LEISTNER, L., HECHELMANN, H., KASHIWAZAKI, M., and ALBERTZ, R., 1975. Nachweis von *Yersinia enterocolitica* in Faeces und Fleisch von Schweinen, Rindern und Geflugel. Fleischwirtschaft 55: 1599–1602.

LEVINE, A. S., and FELLERS, C. R., 1940. Action of acetic acid on food spoilage micro-organisms. J. Bacteriol. 39: 499–515.

LONGRÉE, K., 1972. Quantity Food Sanitation. 2nd Edition, John Wiley and Sons, New York.

MARIER, R., WELLS, J. G., SWANSON, R. C., CALLAHAN, W., and MEHLMAN, I. J., 1973. An outbreak of Enteropathogenic *Escherichia coli* food-borne disease traced to imported French cheese. Lancet. 2: 1376–1378.

MARTH, E. H., 1969. Salmonellae and Salmonellosis associated with milk and milk products. J. Dairy Sci. 52: 283–315.

MEDICAL RESEARCH COUNCIL REPORT. 1947. Bacteriology of Spray-dried Egg with particular reference to Food Poisoning. Spec. Rep. Ser. 260.1, HMSO, London.

MEHLMAN, I. J., SANDERS, A. C., SIMON, N. T., and OLSON, J. C., 1974. Methodology for Recovery and Identification of Enteropathogenic *Escherichia coli*. J. AOAC: 57.101–110.

MEYER, K. F., 1965. *Pasteurella* and *Francisella*. *In* Bacterial and Mycotic Infections of Man. R. J. Dubos and J. G. Hirsch (Editors), J. B. Lippincott Company, Philadelphia.

MORGAN, H. R., 1965. The Enteric Bacteria. *In* Bacterial and Mycotic Infections of Man. R. J. Dubos and J. G. Hirsch (Editors). J. B. Lippencott Company, Philadelphia.

MOSSEL, D. A. A., and INGRAM, M., 1955. The physiology of the microbial spoilage of foods. J. Appl. Bacteriol. 18: 233–268.

NICKERSON, J. T., SINSKEY, A. J., 1972. Microbiology of Foods and Food Processing. American Elsevier Publishing Company.

ORMAY, L., and NOVOTNY, T., 1969. The significance of *Bacillus cereus* food poisoning in Hungary. *In* The Microbiology of Dried Foods. E. H. Kampelmacher, M. Ingram and D. A. A. Mossel. International Association of Microbiological Societies.

PIERSON, M. D., COLLINS-THOMPSON, D. L. and ORDAL, Z. J. 1970. Microbiological, Sensory and Pigment changes of aerobically and anaerobically packaged beef Food Technol. 24: 129–133.

REED, R. W., 1965. *Listeria* and *Erysipelothrix*. *In* Bacterial and Mycotic Infections of Man., R. J. Dubos and J. G. Hirsch (Editors). J. B. Lippincott Company, Philadelphia.

RIEMANN, H., 1969. Food Processing and Preservation Effects. *In* Food-Borne Infections and intoxications. H. Riemann (Editor). Academic Press, London and New York.

ROSEBURG, T., 1965. Bacteria Indigenous to Man. *In* Bacterial and Mycotic Infections of Man. R. J. Dubos and J. G. Hirsch (Editors). J. B. Lippincott Company, Philadelphia.

SAKAZAKI, R., 1969. Halophilic Vibrio Infections. *In* Food-Borne Infections and Intoxications. H. Riemann (Editor). Academic Press, New York and London.

SAKAZAKI, R., 1971. Recent trends in *Salmonella* and enteropathogenic *Escherichia coli* infection in man. 6th Joint Meeting U.S. Japanese Panel on Toxic Microorganisms, Tokyo.

SCHIEMANN, D. A., 1976. Occurance of *Klebsiella pneumoniae* in Dairy Products. J. Milk Food Technol. 39: 467–469.

SCOTT, W. J., 1957. Water relation of food spoilage microorganisms. Adv. Food Res. 7: 83–127.

SEMINIANO, E. N. and FRAIZER, W. C., 1966. Effects of *Pseudomonas* and *Achromobacter* on growth of *Staphylococcus aureus*. J. Milk Food Tech. 29: 161–164.

SHOOTER, R. A., COOKE, E. M., GAYA, H., KUMAR, P., PATEL, N., PARKER, M. T., THOM, B. T., and FRANCE, D. R., 1969. Food and medicaments as possible sources of hospital strains of *Pseudomonas aeruginosa*. Lancet (June 12), 1227.

SHOOTER, R. A., COOKE, E. M., ROUSSEAU, S. A., and BREADEN, A. L., 1970. Animal sources of common serotypes of *Escherichia coli* in the food of hospital patients. Lancet 2: 226–227.

SOJKA, W. J., 1973. Enteropathogenic *Escherichia coli* in Man and Farm Animals. *In* Microbial Food-Borne Infections and Intoxications. A. Hurst and J. M. de Mann (Editors). Can. Inst. Food Sci. Technol. J. 6: 5–16.

SUZUKI, A., KAWANISKI, T., KONUMA, T., TAKAYAMA, S., 1973. A survey of *Salmonella* contamination of imported poultry meats. Bulletin of the National Institute of Hygienic Sciences (Eisei Skikinjo Hokoku) 91: 88–93.

TAYLOR, J., and McCOY, J. H., 1969. *Salmonella* and *Arizona* infections. *In* Food-Borne Infections and Intoxications. H. Riemann (Editor). Academic Press, New York, London.

THIMANN, K. V., 1963. The Life of the Bacteria. 2nd Edition. MacMillan Co., New York.

THOMPSON, W. K., PIVNICK, H., 1972. Food Poisoning due to *Vibrio paraheomolyticus*. Can. J. Pub. Health 63: 515–516.

VANDERZANT, C., 1968. Inactive phenomena among bacteria of dairy origin. J. Dairy Sci. 51: 991–995.

WEISER, H. M., MOUNTNEY, G. J., GOULD, W. A., 1971. Food Spoilage. *In* Practical Food Microbiology and Technology 2nd Edition. AVI Publishing Co., Westport, Conn.

WEISER, R. S., and OSTERUD, C. M., 1945. Studies on the death of bacteria. I. Influence of the intensity of the freezing temperature, repeating fluctuations of temperature and periods of exposure to freezing temperatures on the mortality of *Escherichia coli*. J. Bacteriol. 50: 413–439.

WILDER, A. N., and MacCREADY, R. A., 1966. Isolation of *Salmonella* from poultry. New England J. of Med. 274: 1453–1460.

YANG, H., JONES, G. A., 1969. Physiological characteristics of Enteropathogenic and non-pathogenic coliform bacteria isolated from Canadian pasteurized dairy products. J. Milk and Food Technol. 32: 102–109.

2

Sources of Food Spoilage Microorganisms

Scott E. Martin and Z. John Ordal (Deceased)

The control of microorganisms is necessary throughout the entire food production and distribution system. Microbial spoilage (the decay or decomposition of a food to an undesirable state as caused by bacteria or fungi) is an important factor influencing both the cost and availability of food. One step in alleviation of global food shortages would be achieved if food crops were handled, stored and processed in such a way that spoilage was reduced. Inherent to this goal is an awareness of the origins of microorganisms responsible for the spoilage. The outline presented in Table 2.1 divides the sources of microbial contamination into two major areas. This classification is arbitrary, and any listed category may have the same potential importance as a source of contamination.

PRIMARY SOURCES

The sources of microbial contamination as outlined in Table 2.1 have themselves been contaminated by one or more of four primary sources of contamination. The four sources arbitrarily chosen are soil, water, air, and finally, animal, which includes, but is not limited to, contamination introduced by insects, rodents, and man. Soil is the single richest source of bacteria and fungi. It contains a wide variety of microorganisms, as well as large total numbers. Frazier (1967) states that nearly every microorganism important in food spoilage can originate from the soil. Estimates of the number of bacteria per gram of soil vary widely, influenced by numerous

TABLE 2.1.   SOURCES OF MICROBIAL CONTAMINATION

I. Prior to Processing
    Raw Materials
        Field and orchard
        Animal, poultry, and fish
    Harvest, storage

II. During and After Processing
    Processing equipment and added ingredients
    Water
    Processed food
        Packaging
        Storage
        Distribution

environmental and physical factors, but the total number in fertile surface soil is usually between 10^8 to 10^{10} bacteria per gram (Alexander, 1967; Carpenter, 1967). The genera of bacteria most commonly found in soil are: *Pseudomonas, Rhizobium, Agrobacterium, Alcaligenes, Chromobacterium, Micrococcus, Sarcinia, Corynebacterium, Arthrobacter, Bacillus, Clostridium* and *Mycobacterium* (Alexander, 1967; Buchanan and Gibbons, 1974). Other genera found in soil and important to the food microbiologist include: *Escherichia, Proteus, Erwinia, Streptococcus* and *Leuconostoc* (Frazier, 1967). Psychrotropic bacteria found in soil are of special concern to those involved in the prevention or delay of spoilage, as refrigeration in processing plants and in the household has been extended. Psychrotrophic bacteria ranged from 10^5 to 10^7 organisms per gram of soil, with Gram positive and Gram variable, nonsporeforming rods and Gram negative rods comprising 63% and 32%, respectively, of the total psychrotrophs (Druce and Thomas, 1970). These authors found that Gram positive or Gram variable, nonsporeforming rods resembled *Corynebacterium* and *Arthrobacter*, while the Gram negative rods resembled *Pseudomonas, Enterobacter* and *Acinetobacter*. Psychrotrophic aerobic sporeformers have also been isolated from soil (Michels and Visser, 1976). These bacteria were present in the range of 10^3 to 10^5 per gram of soil, and their spores had low heat resistance ($D_{90°C}$ values from 1 to 11 min). Larkin and Stokes (1967) identified four species as members of the *Bacillus* genus. Yeasts found in soil include *Candida, Cryptococcus, Debaryomyces, Hansenula, Lipomyces, Pichia, Pullalaria, Rhodotorula, Saccharomyces, Schizoblastosporion, Torula, Torulaspora, Torulopsis, Trichosporon,* and *Zygosaccharomyces*, while some of the most commonly encountered molds are: *Aspergillus, Botrytis, Gliocladium, Penicillium, Scopulariopsis, Trichoderma, Trichothecium, Alternaria, Clado-*

sporium, Fusarium, Mucor, Rhizopus and *Chaetomium* (Alexander, 1967).

Water

Microorganisms in water represent a combination of those naturally occurring and those originating from soil and pollution. The source of water influences both the number and types of microorganisms present. Rain water is generally free of microorganisms (<100 per ml), except during the early period of a storm (Carpenter, 1967). Ground waters have passed through successive layers of soil and rock, removing most of the bacteria, leaving only a few hundred per milliliter (Frazier, 1967). Surface water varies in the number and types of bacteria present, due to numerous physical and chemical variables. Freshwater which was obtained from rivers in England contained an average of 3×10^3 bacteria per ml (Jones, 1970). Druce and Thomas (1970), studying psychrotrophic bacteria from various sources in Wales, found that approximately half of the total bacteria isolated from farm water supplies were psychrotrophic. Seawater, collected from waters around Woods Hole, Massachusetts, contained approximately 1×10^6 bacteria per ml (Watson, *et al.*, 1977). Examples of genera normally found in freshwater include: *Pseudomonas, Escherichia, Proteus, Chromobacterium, Micrococcus, Streptococcus* and *Bacillus* (Frazier, 1967). Fungi found in water include *Geotrichum, Monilia, Oidiodendron, Epicoccum, Monosporium, Trichothecium, Cladosporium, Cephalosporium, Fusarium, Aspergillus* and *Penicillium* (Cooke, 1963).

Air

Air is the third primary source of contamination. Microorganisms are present in air due to numerous causes such as dust generation, aerosols produced by solid waste and waste treatment facilities, and by talking, coughing or sneezing (Lighthart and Frisch, 1976). These authors also cite various nonhuman sources of contamination such as bubble bursting of microbial-laden surface films of surface waters and the action of rain drops, wind or thermal convection. Therefore, the air around the field being harvested, as well as the air around a manufacturing plant, may be high in number and types of airborne contaminants. As an example of the number of bacteria found in air, Elliott *et al.*, (1976) determined the colony-forming particles (CFP) found in one cubic meter of air in a housed swine unit. At 30.5 cm from the floor, they found, on the average, 3.4

$\times 10^5/m^3$ total CFP, $2.4 \times 10^4/m^3$ fecal coliforms and 30 CFP/m^3 of *Staphylococcus aureus*. Mold spores of numerous types are present in air, and yeasts, especially asporogenous chromogenic types, are also found (Frazier, 1967).

Carrions

The final primary source of contamination described above is animal, by which we mean contamination carried by insects, birds, rodents, the surface of meat animals, humans, sewage and numerous others. Generally, these sources carry spoilage organisms from soil or water, and thereby contaminate the food product with microorganisms from these sources, as well as with microorganisms which are typical of their own flora. Sewage is a widely recognized source of contamination under this heading, and carries pathogens, coliform bacteria, anaerobes, enterococci and other intestinal bacteria. Sewage organisms can be found on crops when untreated domestic sewage is used as fertilizer. When untreated sewage is dumped into surface waters, it can contaminate shellfish, fish and other seafood, as well as crops, if the freshwater subsequently is used in irrigation. Food handlers represent yet another primary source of microbial contamination. The flora of human skin has been shown to contain members of the following groups or genera of bacteria: *Micrococcaceae*, streptococci, *Mimae*, nonlipophilic and lipophilic diphteroids or coryneform bacteria, propionebacteria, enterics and other gram negative bacilli, *Bacillus*, *Neisseria*, *Streptomyces* and mycobacteria (Kloos and Musselwhite, 1975).

The degree of microbial contamination of a harvested product is variable and is related to numerous conditions. In the case of field and orchard crops, the microbial load on the product will be a function of the soil and climatic conditions, the husbandry practices used in producing the crop, and the cleanliness and quality of the harvesting equipment. Susceptibility to microbial spoilage will be related to crop maturity, freedom from adhering soil or deleterious material, degree of harvest injury, and handling practices.

It is, therefore, apparent that almost all perishable foods become contaminated with a wide variety and by large numbers of microorganisms. Listed in Table 2.2 are the number and principle types of spoilage bacteria found on a variety of foods. Only a small fraction of this initial population, however, takes part in spoilage, under any given condition of storage (Mossel, 1971). Four parameters have been identified as influencing which microorganisms are able to survive and grow, thereby causing spoilage of the product

TABLE 2.2.   NUMBER AND TYPES OF BACTERIA ASSOCIATED WITH VARIOUS FOOD PRODUCTS

Product	Average bacterial count of raw produce	Sample size*	Genera dominating when spoilage occurs during standard conditions of storage	Reference
Milk	2.9×10^4/ml $<10^2 - 10^8$/ml	15	*Bacillus, Arthrobacter, Microbacterium, Streptococcus, Corynebacterium, Pseudomonas*	Washam *et al.*, 1977 Donnelly *et al.*, 1976 Patel and Blankenagel, 1972
Whole beef	10^2/in^2	106	*Pseudomonas, Microbacterium, Lactobacillus, Moraxella-Acinetobacter, Bacillus Streptococcus, Achromobacter*	Seideman *et al.*, 1976 Frazier, 1967
Ground beef	10^6/g	955	*Pseudomonas, Achromobacter, Acinetobacter*	Jay, 1972 Goepfert, 1976 Enswiler *et al.*, 1976
Chicken	10^5/carcass cavity	10	*Pseudomonas, Flavobacterium, Moraxella*	Blankenship *et al.*, 1975 Freeman *et al.*, 1976
Atlantic salmon	2.9×10^2/cm^2 of skin	56	*Pseudomonas, Flavobacterium, Moraxella*	Horsley, 1973
Shrimp	2.2×10^5/g	1,462	*Pseudomonas, Flavobacterium, Achromobacter Moraxella, Acinetobacter*	Lee and Pfeifer, 1977 Baer *et al.*, 1976 Foster *et al.*, 1977
Lettuce	6.6×10^5/g	120	*Pseudomonas, Achromobacter*	Frazier, 1967 Wright *et al.*, 1976 Ercolani, 1976

*The number of samples used to calculate the average.

(Patterson, 1968; Mossel, 1971). These are: (1) the intrinsic factors: the physical and chemical properties of the food; (2) changes in the microbial flora as a result of processing procedures: (3) the extrinsic factors: temperature, humidity and oxygen tension; and finally, (4) the implicit factors: the rate of growth of the microorganism, symbiosis between contaminants, and antagonism between contaminants.

FRUITS AND VEGETABLES

Fruits and vegetables are continually exposed to three of the four primary sources of contamination (soil, water and air) as they grow and mature. Animal contamination may also occur. Both vegetables and fruits have protective coverings which limit or exclude microorganisms from fleshy inner tissues. It should be noted, however, that bacteria are often present in apparently healthy inner plant tissues. Bacteria were isolated from 30% of surface-sterilized unbroken ovules and from 15% of surface-sterilized unbroken seeds of herbaceous plants, indicating that these microorganisms were present when the ovules and seeds were formed (Mundt and Hinkle, 1976). In addition to the microorganisms found within plant tissues, fruit and vegetable inner tissues are subject to contamination from microorganisms entering due to the presence of natural openings (stomata, lenticels, etc.) and from wounds caused by mechanical (animal) damage. The mechanical damage can occur during harvesting or during transportation to market or the processing plant. Both cooling and washing can affect the storage of fruits and vegetables. During both processes, spoilage organisms can be spread from localized sites of spoilage to the whole batch of food. In addition, recirculated or untreated, reused washwater may be a source of high numbers of spoilage bacteria. Removal of spoiled fruits and vegetables, as well as trimming of spoiled parts, may be of benefit in the retardation of spoilage, but these processes also represent additional opportunities for mechanical damage to occur. Finally, washwater left on the food surfaces may encourage the growth of spoilage microorganisms.

Every piece of equipment that a food product comes in contact with during processing operations is a potential source of microbial contamination. When pieces of fruit and vegetable are permitted to accumulate on equipment, these sites become sources for the growth of a large number of contaminants. In canning, as in other food processing operations, not only must consideration be given to the control of the total microflora, but the operation should be such that there is little or no opportunity for the development of organisms capable of producing spoilage in the processed product. The "machinery mold" *Geotrichum candidum* is widely distributed in nature, and is one of the most persistent of molds in the canning factory (Eisenberg and Chichowicz, 1977). The presence of this mold in masses is an indication of unsanitary processing facilities. Cameron and Esty (1940), Bohrer (1949), and others have demonstrated that some of the most important sources of contamination by the thermophilic "flat-sour" spoilage organisms come from within the cannery. These thermophilic organisms are able to grow in heated equipment, such as blanchers, mixing or storage tanks, fillers, etc. The control of such contamination is dependent on the correct design and operation of the processing equipment, coupled with adequate and efficient cleaning procedures.

The bacteria which are responsible for most of the soft rots of vegetables during transport and storage are the soft rot coliform bacterium, *Erwinia carotovora* and the members of the pseudomonads (Lund, 1971; Lelliot *et al.*, 1966). Other bacteria associated with spoilage of these foods are members of the following genera: *Achromobacter, Flavobacterium, Lactobacillus, Bacillus,* and *Acetobacter* (Mossel and Ingram, 1955). Fruits are generally resistant to bacterial spoilage, due, in part, to their pH. Fruits which are susceptible to microbial spoilage include peppers, cucumbers and tomatoes. Some of the genera of fungi associated with fruit and vegetable spoilage and the type of spoilage they cause are: *Alternaria* (alternaria rot, black rot); *Aspergillus* (black mold rot); *Botrytis* (gray mold rot); *Sphaerotheca* (powdery mildew); *Coryneum* (Coryneum blight); *Colletotrichum* (anthracnose); *Diplodia* (stem-end rot); *Fusarium* (stem end rot, Fusarium rot); *Penicillium* (blue mold rot); *Pythium* (pythium rot); *Monilinia* (brown rot); *Phytophthora* (downy mildew); *Rhizopus* (Rhizopus soft rot); and *Sclerotinia* (watery soft rot, brown rot; U. S. Dept. of Agriculture, 1966; 1968; 1972). Using a Most Probable Number Technique, Koburger and Norden (1975) found 2.4×10^2 and 1.5×10^5 fungal organisms per gram on fresh lettuce and string beans, respectively.

Several of the factors influencing spoilage during the storage of vegetables were outlined by Lund (1971). The first is temperature. Many vegetables can be stored at 0°C, but some are damaged by such low temperatures and must be stored at 7–10°C. The correct relative humidity is also important. When the humidity is too high, or when water is found on the surface of vegetables, microbial spoilage is favored. The recommended relative humidity is between 90–95%. Controlled atmosphere storage can also increase the useful life of some fruits and vegetables, both by directly affecting microbial growth, and by improving the physiological condition of the product.

Under ordinary storage conditions, the molds described above are responsible for the microbial spoilage of many fruits and vegetables. Molds are aerobic organisms, so they can be controlled, if the product is stored in the absence of oxygen. However, when fresh produce is stored under anaerobic conditions, the anaerobic respiration of the plant tissue produces undesirable changes in the product, thus making the anaerobic storage of fruits and vegetables impractical. Experience and extensive research have demonstrated that, if the gaseous storage environment for such products is modified and controlled, then the storage life of many products is markedly extended. Optimum composition of controlled atmosphere (CA) storage varies with the particular product, but, in general, the carbon dioxide concentration of the CA should be markedly increased and the oxygen concentration reduced. Such CA also affects the physiological changes taking place in the product and controls the rate at which physiological decomposition occurs. When CA is used with refrigerated storage, the quality and storage life of the product is markedly extended.

While the more dramatic effect of CA is reflected in controlling mold spoilage, such atmospheres also affect bacterial growth. Some data substantiating this are presented in Tables 2.3 and 2.4. The data for Table 2.3 were obtained by inoculating the surface of synthetic agar with a suspension sufficiently dilute to give a maximum of several hundred colonies per agar plate. Duplicate plates were incubated in an air incubator, or in an incubator modified so

TABLE 2.3. EFFECT OF INCUBATION TEMPERATURE AND ATMOSPHERE ON MICROBIAL GROWTH

Organism	Temperature °C.	Time for Appearance of Colonies (Days)[1]	
		Air	CA[2]
Pseudomonas fluorescens	21.1—23.9	<2	7
	7.2	<6	14
	1.1	15	>25
Erwinia amylovora	21.1—23.9	<2	>22
	7.2	22	>22
Escherichia coli	21.1—23.9	1	1
	7.2	11	>21

Source: C. W. Hastings (1967).

[1]The surface of a synthetic medium was inoculated with a suspension of the test organism, incubated under the indicated conditions, and observed periodically for the appearance of recognizable colonies.

[2]Controlled atmosphere, 3.5% O_2; 9.8% CO_2.

TABLE 2.4. EFFECT OF INCUBATION TEMPERATURE AND ATMOSPHERE OF STORAGE ON MICROBIAL GROWTH ON FRESH CUT CORN

Microorganism	Temperature °C.	Days of Storage	Plate count/Gm.[1]	
			Air	CA[2]
Control, uninoculated		0	6	6
	21.1—23.9	1	3,000	3,000
	7.2	1	75	25
	7.2	2	1,000	50
Pseudomonas fluorescens	21.1—23.9	1	1,700	1,700
	7.2	1	150	40
	7.2	2	1,000	95
Erwinia amylovora	21.1—23.9	1	4,200	2,600
	7.2	1	110	30
	7.2	2	5,400	90
Lactobacillus plantarum	21.1—23.9	1	1,300	2,000
	7.2	1	130	25
	7.2	2	4,500	95

Source: C. W. Hastings (1967).
[1]Plate count organisms per gram x 10^4.
[2]Controlled atmosphere, 1.8% O_2; 10% CO_2.

that it was continuously flushed with the indicated oxygen, carbon dioxide, and nitrogen mixture. The plates were examined at intervals to determine the incubation time for the appearance of recognizable colonies. When the plates were incubated at room temperature and *Escherichia coli* was the test organism, there was no apparent effect of the CA. When the incubation temperature was reduced to 7.2°C and the inoculated plates incubated in CA, there was no apparent growth after 21 days. However, when such plates were removed from CA and placed in air at room temperature, normal colonies appeared after overnight incubation. In the case of the other two test organisms, *Pseudomonas fluorescens* and *Erwinia amylovora*, the CA delayed the appearance of the colonies at the several test temperatures. The data presented in Table 2.4 were obtained by inoculating fresh cut corn with the test organism and incubating under the indicated conditions. At room temperature the CA was without apparent effect, but when the storage temperature was reduced to 7.2°C, the effect of the CA in controlling cell multiplication of the three test organisms became evident.

The inoculum used in these studies was large. It would be expected that a more effective suppression of growth would be demonstrable with a smaller inoculum. It is the combination of the proper controlled atmosphere with reduced storage temperature that produces the desired control of the potential spoilage organisms. Equally important as the control of the microorganism is the control of the physiological changes occurring in the plant material. As previously mentioned, each product has its own specification for the composition of the CA for optimum results. The controlled atmosphere storage of fruits, particularly apples, is now widely used.

ANIMAL PRODUCTS

In the case of animal products, contamination is a constant problem from the time of bleeding until the product is consumed or processing is complete. In the slaughterhouse there are a number of potential sources of infection, such as the hide, soil adhering to the hide, intestinal contents (if released unintentionally during dressing operations), airborne contamination, the water supply, equipment (knives, saws, cleavers, and hooks), the receptacles and containers, and the personnel.

Empey and Scott (1939), in their classical paper, identified the magnitude of microbial contamination in an abattoir. Part of their data is abstracted and presented in Table 2.5. These data reflect some of the microbial counts that were obtained when the plates were incubated at 20°C. The magnitude of these counts clearly demonstrates the ease of contamination of the flesh of the animals during slaughter and evisceration. The composition of the flora is varied and complex, but will contain organisms that are capable of growth and multiplication during subsequent storage and handling operations.

Several genera of bacteria usually found on noncomminuted beef include: *Pseudomonas, Achromobacter, Proteus, Micrococcus, Flavobacterium, Aeromonas, Streptococcus, Bacillus, Clostridium, Lactobacillus,* and *Alcaligenes* (Frazier, 1967; Jay, 1970). Six genera of molds isolated from spoiled whole beef were *Thamnidium, Mucor, Rhizopus* (all cause "whiskers" on beef), *Cladosporium* (black spot), *Penicillium* (green patches) and *Sporotrichum* (white spot). Three gen-

TABLE 2.5.   TYPICAL MICROBIAL COUNTS IN SOURCES OF MICROBIAL CONTAMINATION IN AN ABATTOIR

Sources and Methods of Calculation	Bacteria	Yeasts	Molds
Hides (no./cm² surface)	3.3×10^6	580	850
Surface soils (no./g dry wt)	1.1×10^8	5×10^4	1.2×10^5
Faeces (no./g dry wt)	9.0×10^7	2×10^5	6×10^4
Rumen (no./g dry wt)	5.3×10^7	1.8×10^5	1600
Air-borne contamination (no. deposited from air/cm²/hr)	140	—	2
Water used on slaughterhouse floors (max.no./ml)	1.6×10^5	30	480

Source: Empey and Scott (1939).

era of yeast often found on refrigerator-spoiled beef are: *Candida*, *Torulopsis*, and *Rhodotorula*. At the day of slaughter, a total plate count of 4.5×10^3 bacteria/6.45 cm^2 was found on the surface of beef carcasses (Lazarus *et al.*, 1977). Most of the coliform bacteria on red meat probably originate from contamination on the hides of the animal (Shooter *et al.*, 1974) although the intestinal tract is another source. When 85 hides and 75 surface meat samples were tested for coliforms, all 160 gave positive presumptive and confirmed tests, of which 144 were positive for fecal coliforms (Newton *et al.*, 1977). Coliforms were found to be: *Escherichia coli*, *Enterobacter cloacae*, *Klebsiella pneumonia*, *Enterobacter aerogenes*, *Enterobacter liquefaciens*, and *Serratia*.

Strains of *Pseudomonas* are thought to be the organism primarily responsible for the spoilage of meat at refrigerator temperatures when stored under humid conditions (Ingram and Dainty, 1971). When the number of bacteria exceed $10^7/cm^2$, the meat develops a distinct "off" odor, and the first stage in "slime formation" begins when numbers reach $10^8/cm^2$. The meat continues to spoil as the levels of bacteria increase to 10^9 or more/cm^2. It has been shown that *Pseudomonas* group I and II types develop most rapidly on spoiled chicken stored at 2°C, and eventually dominated the spoilage association after 12 days. (McMeekin, 1975; 1977). *Acinetobacter/Moraxella* strains also grew rapidly, but after 8 days decreased in proportion in relation to the pseudomonads.

Controlled atmosphere of a type can also be used to preserve the useful shelf life of meat. Three cuts of beef (knuckles, ribs and chucks) were vacuum packaged in three types of packages with differing oxygen transmission rates and stored at 1–3°C (Seideman *et al.*, 1976). Psychrotrophic and mesophilic counts (*Pseudomonas*, *Moraxella-Acinetobacter*) remained low during the first 7 to 14 days of storage, whereas there was a rapid increase in these counts in defective packages. After 28–35 days of storage, *Lactobacillus* species were predominant when the packages were properly sealed. Oxygen permeability was not the only factor inhibiting the growth of spoilage microorganisms. These authors suggested that CO_2-permeability may also have been involved.

SUMMARY

Microbial spoilage is one cause of the global problem of food shortages. Four primary sources of spoilage microorganisms are soil, water, air and animal. The numbers and types of microorganisms from these sources vary due to numerous environmental and physical factors. These factors, and others, influence which small part of the initial population will take part in spoilage of the food product.

REFERENCES

ALEXANDER, M. 1967. Introduction to Soil Microbiology. John Wiley and Sons, New York.

BAER, E. F., DURAN, A. P., LEININGER, H. V., READ, R. B., SCHWAB, A. H., and SWARTZENTRUBER, A. 1976. Microbiological quality of frozen breaded fish and shellfish products. Appl. Environ. Microbiol. *31*, 337–341.

BLANKENSHIP, L. C., COX, N. A., CRAVEN, S. E. and RICHARDSON, R. L. 1975. Total rinse method for microbiological sampling of the internal cavity of eviscerated broiler carcasses. Appl. Microbiol. *30*, 290–292.

BOHRER, C. W. 1949. A bacteriologist inspects pea and corn canning plants. Canner *109*, No. 10, 12–14, 24, 30.

BUCHANAN, R. and GIBBONS, N. 1974. Bergey's Manual of Determinative Bacteriology. Williams and Wilkins Co., Baltimore, Maryland.

CAMERON, E. J. and ESTY, J. R. 1940. Comments on the microbiology of spoilage in canned foods. Food Res. *5*, 549–577.

CARPENTER, P. 1967. Microbiology. Saunders Co., Philadelphia, Pennsylvania.

COOKE, W. B. 1963. A Laboratory Guide to Fungi in Polluted Waters, Sewage, and Sewage Treatment Systems: Their Identification and Culture. U.S. Dept. of Health, Education, and Welfare, Cincinnati, Ohio.

DONNELLY, C. B., GILCHRIST, J. E., PEELER, J. T., and CAMPBELL, J. E. 1976. Spiral plate count method for the enumeration of raw and pasteurized milk. Appl. Environ. Microbiol. *32*, 21–27.

DRUCE, R. G. and THOMAS, S. B. 1970. An ecological study of psychrotrophic bacteria of soil, water, grass and hay. J. Appl. Bacteriol. *33*, 420–435.

EISENBERG, W. and CICHOWICZ, S. 1977. Machinery mold-indicator organism in food. Food Science and Technol. *31*, 52–56.

ELLIOTT, L., McCALLA, T. and DESHAGER, J. 1976. Bacteria in the air of housed swined units. Appl. Environ. Microbiol. *32*:270–273.

EMPEY, W. A. and SCOTT, W. J. 1939. Investigation on chilled beef. Part I. Microbial contamination acquired in the meat works. Australian Council Sci. and Ind. Res. Bull. *126*.

ENSWILER, B. S., PIERSON, C. J. and KOTULA, A. W. 1976. Bacteriological quality and shelf life of ground beef. Appl. Environ. Microbiol. *31*. 826–830.

ERCOLANI, G. L. 1976. Bacteriological quality assessment of fresh marketed lettuce and fennel. Appl. Environ. Microbiol. *31*, 847–852.

FOSTER, J. F., FOWLER, J. L. and DACEY, J. 1977. A microbial survey of various fresh and frozen seafood products. J. Food Protect. *40*, 300–303.

FRAZIER, W. 1967. Food Microbiology. McGraw-Hill Book Co., New York.

FREEMAN, L. R., SILVERMAN, G. J., ANGELINI, P., MERRITT, C. and ESSELEN, W. B. 1976. Volatiles produced by microorganisms isolated from refrigerated chicken at spoilage. Appl. Environ. Microbiol. *32*, 222–231.

GOEPFERT, J. M. 1976. The aerobic plate count, coliform and *Escherichia coli* content of raw ground beef at the retail level. J. Milk Food Technol. *39*, 175–178.

HASTINGS, C. W. 1967. Personal communication. Res. Sta., Campden, 25.

HORSLEY, R. W. 1973. The bacterial flora of the Atlantic Salmon (*Salmo salar* L.) in relation to its environment. J. Appl. Bact. *36*, 377–386.

INGRAM, M. and DAINTY, R. 1971. Changes caused by microbes in spoilage of meats. J. appl. Bacteriol. *34*, 21–39.

JAY, J. M. 1970. Modern Food Microbiology. Reinhold Book Corporation, New York.

JAY, J. M. 1972. Mechanism and detection of microbial spoilage in meats at low temperatures: A status report. J. Milk Food Technol. *35*, 467–471.

JONES, J. G. 1970. Studies on freshwater bacteria: Effect of medium composition and method on estimates of bacterial population. J. appl. Bacteriol. *33*, 679–686.

KLOOS, W. and MUSSELWHITE, M. 1975. Distribution and persistence of *Staphylococcus* and *Micrococcus* species and other aerobic bacteria on human skin. Appl. Microbiol. *30*, 381–395.

KOBURGER, J. A. and NORDEN, A. R. 1975. Fungi in Foods. VII. A comparison of the surface, pour plate, and most probable number methods for enumeration of yeasts and molds. J. Milk Food Technol. *38*, 745–746.

LARKIN, J. M. and STOKES, J. L. 1967. Taxonomy of psychrophilic strains of *Bacillus*. J. Bacteriol. *94*, 889–895.

LAZARUS, C., ABU-BAKAR, A., WEST, R. and OBLINGER, J. L. 1977. Comparison of microbial counts on beef carcasses by using the moist-swab contact method and secondary tissue removal technique. Appl. Environ. Microbiol. *33*, 217–218.

LEE, J. S. and PFEIFER, D. K. 1977. Microbiological characteristics of Pacific shrimp (*Pandalus jordani*). Appl. Environ. Microbiol. *33*, 853–859.

LELLIOTT, R., BILLING, E. and HAYWARD, A. 1966. A determinative scheme for the fluorescent, plant pathogenic pseudomonads. J. appl. Bacteriol. *29*, 470–489.

LIGHTHART, B. and FRISCH, A. 1976. Estimate of viable airborne microbes downwind from a point source. Appl. Environ. Microbiol. *31*, 700–704.

LUND, B. 1971. Bacterial spoilage of vegetables and certain fruits. J. appl. Bacteriol. *34*, 9–20.

McMEEKIN, T. A. 1975. Spoilage association of chicken breast muscle. Appl. Microbiol. *29*, 44–47.

McMEEKIN, T. A. 1977. Spoilage association of chicken leg muscle. Appl. Environ. Microbiol. *33*, 1244–1246.

MICHELS, M. J. M. and VISSER, F. M. W. 1976. Occurrence and thermo-resistance of spores of psychrophilic and psychrotrophic aerobic spore-formers in soil and foods. J. appl. Bacteriol. *41*, 1–11.

MOSSEL, D. A. A. 1971. Physiological and metabolic attributes of microbial groups associated with foods. J. appl. Bacteriol. *34*, 95–118.

MOSSEL, D. A. A. and INGRAM, M. 1953. The physiology of the microbial spoilage of foods. J. appl. Bacteriol. *18*, 232–268.

MUNDT, J. O. and HINKLE, N. F. 1976. Bacteria within ovules and seeds. Appl. Environ. Microbiol. *32*, 694–698.

NEWTON, K., HARRISON, J. C. and SMITH, K. 1977. Coliforms from hides and meat. Appl. Environ. Microbiol. *33*, 199–200.

PATEL, G. B. and BLANKENAGEL, G. 1972. Bacterial counts of raw milk and flavor of the milk after pasteurization and storage. J. Milk Food Technol. *35*, 203–206.

PATTERSON, J. T. 1968. Bacterial flora of chicken carcasses treated with high concentrations of chlorine. J. appl. Bacteriol. *31*, 544–550.

SEIDEMAN, S. C., VANDERZANT, C., HANNA, M. O., CARPENTER, Z. L. and SMITH, G. C. 1976. Effect of various types of vacuum packages and length of storage on the microbial flora of wholesale and retail cuts of beef. J. Milk Food Technol. *39*, 745–753.

SHOOTER, R. A., COOKE, E., O'FARRELL, S., BETTELHEIM, K., CHANDLER, M. and BUSHROD, F. 1974. The isolation of *Escherichia coli* from a poultry packing station and an abattoir. J. Hyg. *73*:245–247.

UNITED STATES DEPARTMENT OF AGRICULTURE. 1966–1972. Market diseases of fruits and vegetables (series). U.S. Govt. Printing Office, Washington, D.C.

AGRICULTURE HANDBOOK NO. 28. Tomatoes, peppers and egg-plants, McColloch, L. P., Cook, H. T., and Wright, W. R., 1968.

AGRICULTURE HANDBOOK NO. 303. Asparagus, onions, beans, peas, carrots, celery and related vegetables. Smith, M. A., *et al.*, 1966.

AGRICULTURE HANDBOOK NO. 414. Stone fruits: cherries, peaches, nectarines, apricots and plums. Harvey, J. M., Wilson, L. S. and Kaufman, J. 1972.

WASHAN, C. J., OLSON, H. C. and VEDAMUTHU, E. R. 1977. Heat-resistant psychrotrophic bacteria isolated from pasteurized milk. J. Food Protect. *40*:101–108.

WATSON, S., NOVITSKY, J., QUINBY, H. and VALOIS, F. 1977. Determination of bacterial number and biomass in the marine environment. Appl. Environ. Microbiol. *33*, 940–946.

WRIGHT, C., KOMINOS, S. D. and YEE, R. B. 1976. *Enterobacteriaceae* and *Pseudomonas aeruginosa* recovered from vegetable salads. Appl. Environ. Microbiol. *31*, 453–454.

3

Microbial Problems in Food Safety With Particular Reference to *Clostridium Botulinum*

A. H. W. Hauschild

THE DIFFERENT KINDS OF BOTULISM

Clostridium botulinum causes three distinct types of illnesses: the "classical" form of botulism, wound botulism and infant botulism. The classical form results from ingestion of food with a pre-formed neurotoxin. It is characterized by an incubation period of about one to two days, severe neurological impairments such as disturbance of vision, poor pupillary reaction, difficulty to speak and swallow, dry mucous membrane, depressed reflexes and respiratory difficulties, and a high fatality rate. Before antisera and intensive respiratory care became readily accessible, the fatality rate was over 50% in the U.S. and Canada (CDC, 1974; Dolman, 1954). Since then, it has decreased significantly and is now about 20% in North America (Black and Arnon, 1977; Bowmer and Wilkinson, 1976; CDC, 1975; 1976, 1974; and Health and Welfare Canada, 1975, 1976, 1977). Recent large outbreaks in the United States (CDC, 1977; CDC, 1978) suggest that, with prompt treatment, the fatality rate may be reduced further. Early diagnosis and treatment in a recent Canadian outbreak (Health and Welfare Canada, 1978) led to full recovery of a patient with a total of about 3×10^5 mouse lethal doses in the blood stream. Wound botulism is a rare disease (CDC, 1975; Merson and Dowell, 1973) and develops essentially like other clostridial wound infections

(Smith and Holdeman, 1968). The incubation period is approximately one week, but the neurological signs are similar to those of botulism from toxin ingestion. Infant botulism was not recognized as a distinct infectious disease until 1976 (CDC, 1976; CDC, 1978). Although the neurological impairments are similar to the other two forms of botulism, the disease has certain unique characteristics: (1) only infants under half a year seem to be affected, (2) the illness develops very slowly, and (3) it results from ingestion of *C. botulinum* spores and subsequent germination, multiplication and toxin production in the intestine (Arnon *et al*, 1977; Hauschild *et al*, 1978). Since botulinal spores are common in dust and soil, it seems impossible to keep these entirely away from the infant. The only food source of spores suspected of causing infant botulism was honey (Arnon *et al*, 1977; CDC, 1978), and it may be possible to lower the spore content by eliminating some of the less hygienic practices in the honey industry. Sugiyama *et al.* (1978) found *C. botulinum* spores in 18 out of 241 25–g honey samples tested.

It is possible that the *in vivo* production of botulinal toxin may occasionally contribute to botulism from toxin ingestion, namely by prolonging the illness or causing relapses through continued toxigenesis in the intestine (CDC, 1973 and Smith, 1977). However, it is unlikely that small numbers of spores alone would give rise to significant numbers of botulinal cells in the human intestine other than that of the infant. The overriding concern of the food microbiologist, therefore, remains freedom of the food from botulinal toxin, rather than from spores.

From here on, this chapter will deal only with botulism from toxin ingestion and the terms "botulism" and "botulism from toxin ingestion" will be used synonymously.

HISTORICAL DEVELOPMENTS IN THE RECOGNITION OF BOTULISM AND ITS CAUSES

The first documented cases of botulism date back to the early 18th century in Southern Germany (Meyer, 1928), but the association of the illness with food has probably been known for many more centuries. The earliest known edict that may have helped to contain the disease was promulgated by the Byzantine Emperor Leo VI (886–911); it specifically made the manufacture of blood sausage illegal. From the meticulous records of Kerner (Kerner, 1817, 1820) on over 200 cases of botulism in Württenberg, S. Germany, in the early 19th century we know that blood sausage was a common cause of botulism, and the term "Botulismus" (botulism), proposed by Müller in 1870, is derived from the Latin *botulus* (sausage). Kerner also experimented and described conditions, favorable to anaerobiosis, under which sausages were likely to become toxic. However, the association of botulism with a specific microorganism was not known until 1896 when van Ermengem investigated a large outbreak of botulism from ham in Ellezelles, Belgium (Van Ermengem, 1897). Van Ermengem isolated the causative microorganism which he called *Bacillus botulinus*, demonstrated that botulism was an intoxication rather than an infection, and characterized the toxin with respect to its sensitivity to chemical agents and heat. At the time, it was believed that the causative foods were limited to meats, fish and related products, but a few years later Landmann (1904) discovered *C. botulinum* in canned white beans that had caused a large outbreak in Darmstadt, Germany. Since then, the relative number of outbreaks from vegetables has generally increased and now by far exceeds the outbreaks from meat products in the United States (Black and Arnon, 1977; CDC, 1974, 1975, 1976). For details on the early history of botulism, see the following references: Dickson, 1918; Dolman, 1964; Meyer, 1928; Smith, 1977.

CLOSTRIDIUM BOTULINUM TYPES INVOLVED IN BOTULISM OUTBREAKS

The strains of the *C. botulinum* species are divided into seven types, A to G, depending on the neurotoxins that they produce. The types commonly responsible for botulism in man are A, B and E. Only two type F outbreaks, both from home-prepared foods, are known, one from liver paste in Denmark (Dolman and Murakami, 1961; Moeller and Scheibel, 1960), the other from venison jerky in California (CDC, 1966; Midura *et al*, 1972). A single type D outbreak, from raw salted ham in Tchad, has been documented (Demarchi *et al*, 1958). Two reports have dealt with the possible involvement of type C in small outbreaks, but without conclusive evidence (Meyer and Eddie, 1953; Prévot *et al*, 1955). Matveev *et al*. (1967) mentioned two cases of botulism from type C in the U.S.S.R. Type G was isolated from soil in Argentina (Ciccarelli *et al*, 1977; Giménez and Ciccarelli, 1970) but has never been implicated in an outbreak.

In contrast to their effects in humans, types C and D are frequent causes of botulism in some animal species. The animals most severely affected in terms of economic and ecological losses are listed in Table 3.1. Smaller outbreaks of botulism have also been reported for swine, turtle, fox and other species (Smith, 1977).

With the exception of the type E and F toxins, the botulinal toxins are serologically distinct proteins. Types Cα and D produce more than a single toxin (Jansen, 1971) (Table 3.2). A few strains of type A (subtype Af) are known that produce small quantities of type F toxin, with a ratio A:F of 100:1 (Giménez and Ciccarelli, 1978; Sugiyama et al, 1972). Type E produces only a single toxin, but its antitoxin neutralizes small amounts of type F toxin (Rymkiewicz, 1970).

GEOGRAPHIC DISTRIBUTION OF BOTULISM OUTBREAKS, AND C. BOTULINUM IN THE ENVIRONMENT

Although human botulism is worldwide, the frequency of outbreaks, the botulinal types responsible for outbreaks, and the causative foods differ widely between countries and even between geographical or cultural regions within countries.

The most frequently reported outbreaks occur in the northern part of the globe and roughly north of 30° latitude: in North America, European countries, U.S.S.R., northern Iran, some parts of China and Japan. Undoubtedly, some of the reasons for this uneven distribution of outbreaks are differences in the distribution of botulinal spores, climate, and traditions in the preparation of foods. Another likely factor is the surveillance and reporting system for food-poisoning diseases in individual countries. For example, no botulism was recorded in Japan until 1951. Since then, several hundred cases have become known. Since most of the peccant foods

TABLE 3.1.   SPECIES AFFECTED IN BOTULISM OUTBREAKS OF ECONOMICAL OR ECOLOGICAL IMPORTANCE.

Species	Causative types of *C. botulinum*[a]	Main areas affected
Man	A, B, E, (F), (D)	Worldwide
Cattle	Cβ, D, (B)	S. Africa, Australia, U.S.A., S. America, Turkey
Sheep	Cβ	Australia, Africa
Horse	Cβ, D, (B)	Australia, S. Africa, Europe, U.S.A.
Mink	Cβ, E, (A)	Europe, N. America
Aquatic wild birds	Cα (Cβ)	Worldwide
Domest. birds[b]	Cα, Cβ, (A)	U.S.A., Europe

[a]Uncommon causes in brackets
[b]Chicken and pheasant

TABLE 3.2.   TOXINS PRODUCED BY *C. BOTULINUM*, TYPES A-G.

	A	B	Cα	Cβ	D	E	F	G
			C. botulinum type					
Toxins—major	A	B	C_1	C_2	D	E	F	G
—minor	F[a]	—	C_2,D	—	C_1	—	—	—

[a]Only few type A strains known to produce F toxin.

were traditionally Japanese, one must assume that botulism outbreaks in that country were common long before the first reported outbreak; they were either not recognized as such or were not reported. Some confirmed outbreaks have been reported from the southern hemisphere, namely Australia, India, Africa and South America (Centro Panamericano de Zoonosis, 1975; Smith, 1977).

The conditions leading to outbreaks of botulism from types A and B are usually distinctly different from conditions leading to type E outbreaks. This is due mainly to differences in (1) the natural habitat of their spores, (2) their resistance to heat, hydrogen ion concentration and curing salts, and (3) their temperature requirements for growth and toxigenesis. These characteristics are listed in Table 3.3. Some of the data are approximations only. For example, the D_{100} values differ considerably between strains of the same type and depend on the environment in which the spores have been formed, and on the food in which they are heated.

Due to their low heat resistance, type E spores are quickly killed in any heat process for low-acid foods, so that botulism outbreaks from underprocessed foods are commonly caused by types A and B only. Similarly, these two types develop in foods that are traditionally preserved mainly by a low pH (marinated foods) or by salt such as raw ham when the vinegar or salt concentrations are accidentally below the critical level. Raw, home-preserved ham is a frequent vehicle of botulism in European countries, particularly Poland and France where the causative type is usually B, probably due to its predominance in the environment in these countries. A correlation between the frequency at which *C. botulinum* types are involved in outbreaks, and their distribution in the environment is apparent from American data (CDC, 1974; Meyer and Eddie, 1965; Smith, 1978); type A predominates in soil between the Rocky Mountains and the Pacific, and the ratio of type A to type B outbreaks in the three Pacific states from 1899–1973 was nearly 12:1.

Type E botulism is encountered mainly in the more northern regions, i.e., Alaska, the Canadian Pacific coast, the Northwest Territories, northern Quebec and Labrador, Greenland, Scandinavian countries, the Soviet Union, northern Iran and northern Japan. These and adjacent regions have in common an abundance of type E spores in their environment, particularly in the coastal areas and around inland lakes where spores can be readily isolated from marine deposits, soil and fish. This was shown in a number of surveys conducted in states along the U. S. Pacific coast (Craig *et al.*, 1968; Craig and Pilcher, 1967; Eklund and Poyski, 1967), British Columbia, particularly the coastal inlets (Dolman and Iida, 1963), Alaska, (Miller *et al.*, 1972), the Great Lakes area (Bott, *et al.*, 1968; Foster *et al.*, 1965; and Pace *et al.*, 1967) the Gulf of St. Lawrence (Laycock and Loring, 1972), along the Canadian Atlantic coast (Laycock and Longard, 1972), the New England (Nickerson *et al.*, 1967) and Norwegian coasts (Cann, *et al.*, 1967), the Skagerrak, Kattegat, Baltic Sea and Swedish soils and rivers (Johannsen, 1963 and 1965), soils and waters of the U.S.S.R. (Kravchenko and Shishulina, 1967), the Caspian Sea (Rouhbakhsh-Khaleghdoust, 1973; Rouhbakhsh-Khaleghdoust and Portaghva, 1977) and along the coast and in inland waters of northern Japan (Kanasawa, *et al*; 1968; Nakamura, 1956). Almost invariably, type E spores outnumber the spores of the other clostridial types in fish; in some areas, the incidence of type E may be as high as 75 to 100% (Foster, *et al.* 1965 and Johannsen, 1963). Around the British Isles, the incidence of type E is surprisingly low (Cann *et al.*, 1967 and 1968). Cann *et al.* (1968) detected *C. botulinum* in only 15 out of 429 samples of bottom deposits, and all of these were of type B.

Spores of *C. botulinum* type E have been traced from coastal areas, along river beds, to inland areas, but are seldom found in the deep seas. This would indicate that *C. botulinum* in the marine environment is of terrestrial origin (Bott, *et al.*, 1968; Dolman and Iida, 1963; Foster, *et al.*, 1965; Johannsen, 1965; and Laycock and Loring, 1972. However, it appears certain also, that the microorganism multiplies in the marine environment, namely in the organic matter of the sediments. The ability of type E to grow at low temperatures is certainly an important contributing factor to its predominance in coastal areas (Table 3.3).

After a type E botulism outbreak from smoked trout fillets in

TABLE 3.3.   CHARACTERISTICS OF *C. BOTULINUM*, TYPES A, B AND E[a].

Characteristic	Type	
	A B	E
Habitat	Soil	Soil & marine env.
D$_{100}$	25 min.	<10 sec.
Inhibitory pH	4.6	5.0
Inhibitory NaCl conc.	9%	5%
Min. temp. for growth	10°C	3.5°C

[a]excluding non-proteolytic type B strains.

northern Germany in 1970 in which three people died (Baumgart, 1970), several trout farms were surveyed for botulinal spores. Spores were isolated from a number of ponds, particularly from pond sediments and from trout (Baumgart, 1970; Burns and Williams, 1975; Cann, et al., 1975; and Huss, et al., 1974). Type E predominated among the isolates, except in one case; Burns and Williams (1975) only isolated nonproteolytic type B from a trout farm in Scotland. This group of type B, along with non-proteolytic type F is closely related to type E, namely in its growth and toxin production below 5°C (Eklund, et al., 1967; Midura, et al., 1972 and Roberts and Hobbs, 1968).

With few exceptions, the foods responsible for type E outbreaks are either fish or fish products, or marine mammals. Generally, such foods have been (1) stored for some time above 3.5°C and consumed raw or inadequately heated, (2) poorly fermented, (3) preserved with little salt and/or at a relatively high pH (see Table 3.3) or (4), more rarely, canned.

Raw or mildly cooked meats are a common cause of botulism in Canada and Alaska. From 1974 to 1977, 20 confirmed botulism outbreaks occurred in Canada (Health and Welfare Canada, 1975, 1976 and 1977). All of these were caused by type E, and 18 took place in Eskimo communities. The incriminated foods, mostly meats from marine mammals (seal, walrus, whale) and fish, were raw or under-cooked and had invariably been kept for extended periods without adequate temperature control. By tradition, some meats are also hung up for slow drying, or may undergo an uncontrolled fermentation (Dolman, 1964). The hazards from large numbers of botulinal spores in the environment and from the ability of type E to grow at low temperatures (Schmidt, et al., 1961) are thus compounded by traditional eating habits. It was estimated that the chance of Eskimos contracting botulism was 500 times greater than for the remaining population in North America (Stuart, et al., 1970). The data on Canadian cases over the last 10 years (Bowmer and Wilkinson, 1976; Dolman, 1974 and Health and Welfare Canada, 1975, 1976 and 1977) suggest a comparable ratio of about 3,000:1; for the Eskimo vs the non-native population of Canada the ratio is 35,000:1.

From 1947 to 1976, 21 botulism outbreaks were recorded in Alaska. One was caused by type B; the other 20 were caused by type E (Eisenberg and Bender, 1976).

Fermented foods involved in type E outbreaks are salmon eggs (stink eggs) prepared traditionally in Indian communities of British Columbia, the Yukon (Bowmer and Wilkinson, 1976 and Dolman,

1964) and Alaska (CDC, 1974 and Eisenberg and Bender, 1976), "izushi", a common Japanese fermentation product of fish, rice and vegetables (Nakamura, *et al.*, 1956 and Nakano and Kodama, 1968) and "kirikomi" which is prepared similarly (Sakaguchi, 1969). Since smaller fish used for izushi are often not eviscerated (Dolman, 1964), type E spores have actually been demonstrated in apparently whole-some lots of izushi, both home-prepared and commercial (Saka-guchi, 1969). "Rakefish", lightly salted and fermented trout, have been the cause of botulism outbreaks in Norway; the *C. botulinum* types involved were B and E (Dolman and Iida, 1963 and Smith, 1977).

In the third category of foods causing type E botulism, smoked and salted fish or fish products have been the main causes of outbreaks in the United States (Pace, *et al.*, 1967), Europe (Baum-gart, 1970), Iran (Pourtaghva, *et al.*, 1975) and the U.S.S.R. (Matveev, *et al.*, 1967). Apart from the heating and drying effects of smoking, the smoke *per se* seems to have little or no inhibitory effect on the outgrowth of *C. botulinum*. Pickled (sour) herrings have also been the cause of outbreaks (Dolman, *et al.*, 1950 and Sakaguchi, 1969), possibly because *C. botulinum* developed before the acetic acid lowered the pH throughout the product.

In view of the high sensitivity of type E spores to heat, type E botulism from canned products would seem unlikely. One such outbreak, from canned tuna, occurred in Detroit in 1963, but the organism seems to have entered the cans via defective seams, after the heat process (Foster, *et al.*, 1965 and Johnston *et al.*, 1963). Post-process leakage may also account for some of the other rare type E outbreaks from canned products (Sakaguchi, 1969). Alaska salmon from a rim-damaged can was the cause of a recent type E outbreak in England.

FOOD PRESERVATION

In the preservation of foods, *C botulinum* is controlled either by killing the spores, inhibiting their outgrowth or a combination of both as in canned cured meats. Many inhibitors perform satis-factorily only if the spores are severely damaged, usually by heat.

Table 3.4 lists the major factors in the control of *C. botulinum*, as well as of other hazardous or spoilage microorganisms in foods. Since Appert's discovery in the early 19th century that foods could be preserved by heating them in sealed containers, the heat process

TABLE 3.4.   MAIN FACTORS IN THE CONTROL OF *C. BOTULINUM* IN FOODS[a].

Category	Heat process	Other factors	*C. botulinum* spores killed	*C. botulinum* spores inhib.	Foods
1	Fo $\geq$ 2.5	—	+	—	Low acid (pH >4.6), canned, non-cured
2	Pasteurization	pH	—	+	Acid (pH 4.0 – 4.6) and high-acid (pH <4.0), canned: acidic or acidified fruit and vegetable juice; fruits and vegetables; marinated vegetables; pickled fish
3	Pasteurization	Refrigeration	—	+	Canned crab meat; smoked fish[b]
4	Fo = 0.2 – 0.6	Salt, nitrite	—	+	Shelf-stable canned, cured meats
5	Pasteurization[c]	Salt, nitrite, refrigeration	—	+	Perishable canned, cured meats; perishable cured meat; smoked fish[b]
6	—	pH, nitrite, salt, drying	—	+	Raw, fermented, dry meats
7	Pasteurization	pH, salt	—	+	Fish, caviar; meats; process cheese
8	—	salt	—	+	Raw meats; raw fish and fish products

[a]Foods of marginal concern as potential hazards from *C. botulinum* are not listed.
[b]Preservation with nitrite allowed in some countries only.
[c]Except for raw fermented meats.

has played an increasing role in food preservation. In the following pages, the various types of food preservation will be discussed.

Category 1—preservation by heat.

Most of the canned foods, namely meats, fish and vegetables, are of the low-acid type, with a pH of over 4.6. The heat processing of such foods has been described in detail (Stumbo, 1973). Briefly, the centre of each can should receive a heat treatment sufficient to decrease a hypothetically large population of the most heat-resistant *C. botulinum* spores (of types A and B; see Table 3.3) by at least 12 $\log_{10}$. The viable count decreases logarithmically with time. At 121°C (or 250°F, the reference temperature of the food industry) the time needed to reduce the viable count of the most resistant botulinal spores by one log (or by 90%) is 0.21 minutes. This is the $D_{250°F}$ value for *C. botulinum*. For a 12 D treatment, or the reduction of the viable spore count by 12 logs, a period of 2.5 minutes at 250°F would be needed. If we assume an average of one botulinal spore in each can before processing, we would expect one toxic can in a thousand billion cans produced with a 12 D treatment or, at the present production rate of 30–35 billion per annum in North America, one toxic can in 30 years.

The D value decreases logarithmically with increasing temperature. Although there is some variation between strains, the accepted temperature increase or decrease needed to lower or raise the D value by one log is 10°C (18°F). This is the Z value for *C. botulinum*. Accordingly, 10 minutes at 111°C (232°F) or 0.1 minute at 131°C (268°F) are both equivalent to 1 minute at 121°C (250°F). By convention, 1 minute at 250°F has been given the "lethal value" (F) of 1. On the basis of recorded temperatures and the accepted Z value of 18°F, the total lethal value ($F^{Z=18}$ or Fo) can be calculated for any heat process. An Fo value of 2.5 is thus equivalent to a 12 D process, also referred to as the "minimum botulinum cook". Several spoilage organisms of the mesophilic and thermophilic groups, namely *B. stearothermophilus*, *C. thermosaccharolyticum* and *C. nigrificans*, produce spores with a far greater heat resistance than that of *C. botulinum*. For the control of spoilage organisms in low-acid foods, the Fo value of most industrial canning processes is, therefore, considerably higher than 2.5.

It is obvious that the safety of canned products is largely determined by the original number of viable *C. botulinum* spores in the food. Surveys in the U.K. indicate about 2 spores per Kg of bacon (Roberts and Ingram, 1976; Roberts and Smart, 1976 and Roberts and Smart, 1977), while American data showed slightly lower spore contents in ham, poultry, frankfurters and luncheon meats (Abrahamsson and Riemann, 1971; Greenberg, *et al.*, 1966; Insalata, *et al.*, 1969; Riemann, *et al.*, 1972 and Taclindo, *et al.*, 1967). Recent surveys in Canada (Hauschild and Hilsheimer, unpublished data) showed 0.20 botulinal spores per Kg of processed meats, <0.064 spores per Kg of bacon and 0.064 total *C. botulinum* per Kg of bacon. The botulinal spore content of vegetables may be considerably higher. A Canadian survey showed over 100 spores of *C. botulinum* type B per Kg of fresh commercial mushrooms (Hauschild, *et al.*, 1975).

Most incidents of botulism from canned foods are due to underprocessing in the home. As early as 1917, Dickson (1917) demonstrated the inadequacy of preserving vegetables in jars by heat-processing them for 2–3 hours in boiling water, yet this procedure was still in use several decades later. Although this practice has become less common in recent years, possibly through educational efforts of Government agencies (Agriculture Canada, 1976 and Ministry of Agriculture and Food of Ontario), the incidence of botulism associated with home-canned foods is still high. In 1974 and 1975 alone, 24 such outbreaks were recorded in the United States (Acton and Dick, 1977 and CDC, 1976). The last such outbreak in Canada was recorded in 1965 (Dolman, 1974).

Compared to home-canned foods, the number of botulism outbreaks from commercially canned products is small (Lynt, *et al.*, 1975; Merson, *et al.*, 1974 and Meyer, 1956). Most of these were underprocessed; not because of shortcomings in established procedures, but as a result of human failure or faulty equipment. Defective seams or corroded cans occasionally result in food poisoning from *S. aureus*, but they have also been the cause of botulism outbreaks (Johnston, *et al.*, 1963).

Of 84 confirmed outbreaks in Canada (since 1919), four have been caused by commercial products, and only one of these (beets) was a non-cured, low-acid canned food.

It should be pointed out that the canned products associated with botulism are likely to represent only a small fraction of canned foods that become toxic. For example, in 1973 and 1974 alone, the U.S. Food & Drug Administration found 30 canned mushroom products from nine different U.S. and foreign plants to contain botulinal toxin; none of these had caused illness (Lynt, *et al.*, 1975). There are two main reasons for this: (1) most meats and vegetables are heated before being consumed, and botulinal toxin is destroyed in the process; (2) viable *C. botulinum* strains in underprocessed cans are likely to be proteolytic and to produce offensive odors in the advanced stages of development; toxic canned foods, therefore, are likely to be consumed only in the early stages of botulinal growth.

For some foods, the cans are being replaced by heat-resistant, flexible plastic pouches (Rouhala and Clegg, 1978).

Category 2—preservation by pH and pasteurization.

The minimum pH requirement for growth and toxigenesis of *C. botulinum* is above 4.6 (Baird-Parker and Freame, 1967; Hauschild, *et al.*, 1975; Ingram and Robinson, 1951; Ito, *et al.*, 1976; Ito and Chen, 1978 and Townsend *et al.*, 1954). Therefore, the products listed in Category 2 of Table 3.4, would not allow *C. botulinum* to develop. The heat treatment is applied in order to eliminate viable cocci and non-sporing bacteria as well as molds and yeast that may even grow in high-acid foods (Odlaug and Pflug, 1978). Heating to about 82°C (180°F) at the centre of the can is adequate for the destruction of these microorganisms.

Most fruits and a few vegetables (rhubarb) have a naturally low pH; other products, namely vegetables, fish and some fruits (figs, banana puree) require addition of acidulants, usually in the form of lemon juice or organic acids (citric, adipic, malic, acetic and others) (Ito, *et al.*, 1978 and Sapers, *et al.*, 1978). Muskmelons which have a high natural pH are also canned in mixture with acidic fruits (Nath and Ranganna, 1977).

With adequate pH control and sufficient heating, these products should be safe. However, they have been implicated in a number of botulism outbreaks (CDC, 1974). Involvement of fruit juices and of fruits appears to be due mainly to inadequate heating and survival of fungal spores (Huhtanen, *et al.*, 1976). Fungal growth may be accompanied by a sufficient pH increase to allow *C. botulinum* to develop, mainly in a narrow layer underneath the fungal mat. Huhtanen *et al.* (1976) have shown that significant increases in pH can usually be measured only in undisturbed products. Dickson (1917) investigated an outbreak with 5 fatal cases from canned apricots in 1915. "The food was all apparently of good quality except the apricots which were moulded on the surface. . . .".

The common cause of botulinal growth in marinated vegetables, (mushrooms, pimientos, artichokes etc.) and fish (Bismarck herring, German "brathering" and "rollmops") is insufficient vinegar in the marinade. Most of such foods responsible for outbreaks have been prepared in the home (CDC, 1978); a rare outbreak caused by a commercial marinated product has been described in detail (Todd, *et al.*, 1974).

In recent years, a large number of commercial marinated products has been recalled, or barred from entry into the U.S. (Food Chem. News, March 3, 1975; Feb. 16, 1976) because their pH exceeded the critical level of 4.6. A Canadian survey showed some of the products to be very close to this level (Hauschild, *et al.*, 1975). Most tomato varieties are sufficiently acidic to prevent growth of *C. botulinum* in tomato juice or canned tomatoes, but some low-acid varieties, among them golden-yellow tomatoes, have been incriminated in outbreaks, and the most likely causes were low acid contents (CDC, 1974).

Dolman *et al.* (1950) described an outbreak from home-pickled herring. Also, he found that *C. botulinum* type E could grow and produce toxin in the fish, while the pH of the marinade did not exceed the range of 4.5 to 4.8. Since type E does not grow at this pH (Table 3.3), it is likely that the microorganism developed before complete equilibration of the pH.

Category 3—preservation by pasteurization and refrigeration.

(a) **Canned crab meat.**—By recommendation of a Tri-State Seafood Committee, canned crab meat is pasteurized at 85°C (185°F) for 1 minute in the center (Lynt, *et al.*, 1977). A more severe heat process would result in discoloration and off-flavor. The pasteurization process reduces the number of viable type E spores by about 1 to 4 logs, depending on the strain (Lynt, *et al.*, 1977), and has essentially no effect on the spores of proteolytic strains. The product, therefore, has to be kept under refrigeration, and for a limited period (6 months) only.

Cans inoculated with *C. botulinum* type E spores and processed at 85°C gave no indication of outgrowth at 4°C (Cockey and Tatro, 1974 and Solomon, *et al.*, 1977), but in view of the ability of type E strains to grow below 4°C and the relatively low reduction rate of viable spores in the heat process, it is conceivable that occasionally a can might become toxic during storage. However, the temperature requirements for growth from type E spores in crab meat are much greater than in broth (Solomon, *et al.*, 1977).

Potential hazards may arise from temperature abuse during storage, particularly since the non-proteolytic strains of *C. botulinum* produce very little off-odour in the product. However, pasteurized canned crab meat has never been implicated as the cause of botulism since it was first produced about 25 years ago (Cockey and Tatro, 1974).

(b) **Smoked fish.**—Not all smoked fish belongs to this category. In the United States, some fish are cured with nitrite which provides an additional safety factor (see below). Kautter and Lilly (1970) detected nitrite in commercial smoked sable and salmon locks but not in smoked whitefish or carp; the Canadian Food and Drugs Act does not allow the addition of nitrite to fish.

Two outbreaks of type E botulism from commercial vacuum-packed smoked fish (CDC, 1960 and 1963) have led to considerable work on the conditions of survival and outgrowth of *C. botulinum* in smoked fish and to strict regulations governing processing and storage of the product. The essential provisions in the United States are (Olson, 1968 and Pace and Krumbiegel, 1973): hot smoking of each portion of fish to a minimum of 82.2°C (180°F) for at least 30 minutes, and prompt cooling and packaging after the process; labelling the packages to the effect that they must be kept refrigerated; the product must not be sealed in air-impermeable bags; storage must be at ≤4.4°C (40°F) and must not exceed 7 days from the date of smoking.

The hot smoking process may reduce the incidence of viable *C. botulinum* type E spores by a few logs (Alderman, *et al.*, 1972; Christiansen, *et al.*, 1968 and Pace and Krumbiegel, 1973), but due to the high incidence of type E spores in raw fish, they can still be detected in the smoked product (Abrahamsson, 1967 and Pace and Krumbiegel, 1973). The destruction of spores during the smoking process depends largely on the water activity of the product; the heat resistance of type E spores was found to increase with decreasing water activity of the fish (Alderman, *et al.*, 1972). The possible role of bactericidal ingredients of the smoke in the control of *C. botulinum* will be discussed below.

The use of air-permeable packaging is a most important safety factor. Growth and toxigenesis of *C. botulinum* in smoked fish proceeds faster in evacuated air-impermeable pouches than in air-permeable bags, while spoilage is delayed due to the lack of oxygen. A critical level of toxin may, therefore, be reached without any overt signs of a change in organoleptic quality. Since, unlike proteolytic forms of *C. botulinum*, type E produces relatively little off-odor itself, vacuum-packed smoked fish may be toxic for a long time before the consumer is warned by a noticeable deterioration of the product. Thatcher *et al.* (1962) incubated smoked fish with *C. botulinum* spores of types A and E in either air or hydrogen at 30°C. Although all samples became toxic within 24 hours, only those inoculated with type A, and those inoculated with type E and stored in air had off-odors; the type E samples incubated anaerobically remained organoleptically acceptable. Similarly, Kautter (1964) reported that type E-inoculated smoked ciscoes, packed in air-tight pouches and stored at 10°C, did not spoil for at least 18 days but became toxic after 5 days. The Canadian Food and Drugs Act disallows the packaging of smoked fish in sealed, air-impermeable bags, unless (1) the product receives the full botulinum cook, (2) contains 9% salt, or (3) is customarily cooked before eating, or (4) is kept frozen; each of these conditions would make the product unpalatable.

In addition to the botulism outbreaks from smoked fish, there have been several reports of other forms of food poisonings from such products (Gangarosa, 1968).

Category 4—preservation by heat, salt and nitrite.

When cured meats are heated to Fo of 2.5, their organoleptic quality deteriorates to the extent that they become essentially inedible (Silliker, et al., 1958), and many cured meats show excessive shrinkage. In practice, the heat process of shelf-stable canned cured meats ranges from approximately 0.2 to 0.6 Fo (Duncan, 1970; Greenberg, et al., 1965 and Riemann, 1966). This process merely reduces the number of viable C. botulinum spores by about 1 to 3 logs and heat-damages most of the surviving spores. These are then inhibited by a combination of salt and nitrite but remain viable for considerable periods and may be cultured in broth without difficulty (Greenberg, 1972; Pivnick, et al., 1969 and Silliker, et al., 1958).

The interaction between heat, salt and nitrite in the control of C. botulinum in meats has been shown in a number of publications (Lechowich, et al., 1978; Nordin, et al., 1975; Pivnick and Thacker, 1970 and Roberts, et al., 1976). As one of the three factors is increased, the others may be reduced. However, each of the factors is limited; heat and salt by organoleptic changes, and nitrite by the concern over potential adverse effects on the consumer and, consequently, government regulations. The hazards from nitrite and nitrosamines in foods are discussed in this book by Sen (Chapter 10). The requirement of the three variable factors for the control of C. botulinum in a product such as canned, cured meat can be plotted 3-dimensionally in the form of individual cubes (Roberts, 1973 and Roberts and Ingram, 1973). Any condition below the surface of any cube would be expected to allow outgrowth of C. botulinum, while conditions above the surface would afford stability.

Pivnick et al. (1969) processed canned pork luncheon meat, with a total of 3.4×10^8 botulinal spores per 340–g can, to Fo = 0.6 and determined the amounts of salt and/or nitrite needed to inhibit outgrowth. A brine concentration of 6.1% completely inhibited the spores, but a decrease in the brine concentration to 4.5% necessitated a linear increase in nitrite from 0 ppm to 300 ppm. At the industrial nitrite level of about 150 ppm, the required salt content of the brine was 5.3%. No outgrowth in these experiments indicated death plus inhibition of spores equivalent to a heat process of somewhat over 8 D for a non-cured low-acid product. In practice, the minimum requirements indicated by Pivnick et al. (1969) are frequently not met; for example, at 150 ppm of nitrite, the heat process may be well below Fo = 0.6 and the brine concentration below 5.3. Therefore, the good safety record of shelf-stable canned cured meats is likely due to the low number of botulinal spores in meats. It has also been suggested that spores as natural contaminants of meats may have a lower heat resistance than laboratory-prepared spores (Riemann, 1963).

The reactive form of nitrite in meat is the undissociated nitrous acid. Consequently, the effect of nitrite is strongly dependent upon the pH and increases approximately ten-fold with each fall of one pH unit in the range of pH 7.0 to 5.5 (Ingram, 1973 and Nordin, et al., 1975). However, since cured meats undergo no significant pH changes before processing, the role of pH in the safety of these products receives relatively little attention; cured canned meats have a pH of around 6.0.

The control of C. botulinum in canned cured products, attributable to the destruction and inhibition of spores, has been expressed in the following formula (Pivnick and Petrasovits, 1973): Pr (protection) = Ds (destruction) + In (inhibition) or: $\log_{10}$ reduction of spores capable of outgrowth in the cured product = $\log_{10}$ destruction by heat + $\log_{10}$ inhibition of the surviving spores. At Fo = 0.6, 5% brine and 150 ppm of nitrite, Pr was estimated at approximately 8 $\log_{10}$ units (Ds = 3; In = 5) (Pivnick and Petrasovits, 1973).

Ingram and Roberts (1971) expressed In as n′ – n or $\log_{10}$ reduction of spores growing out in the cured meat minus $\log_{10}$ destruction (Ds), and calculated In as:

$$n' - n = n \left(\frac{D}{D'} - 1\right),$$

where D′ is the "pseudo-D" value, the time required to effect a $\log_{10}$ reduction in outgrowth, rather than death (D). While D′ is determined by the amount of curing agents, these ingredients have little or no effect on D (Ingram and Roberts, 1971; Nordin et al., 1975; Pivnick et al., 1970 and Roberts et al., 1966).

During the thermal process of cured meats, most of the nitrite, as measured in the Griess reaction, disappears. During subsequent storage, the nitrite content declines further to essentially zero. The half life of nitrite in cured ham, pH 6.0, has been expressed by Nordin (1969) in the following formula:

$$\log_{10} \text{ (half life in hours)} = 0.65 - 0.025 \times {}^{\circ}\text{C} + 0.35 \times \text{pH}.$$

Some reactions of nitrite with meat components, i.e., the formation of nitrosyl myoglobin and nitrosyl hemochrome (pigments of cured non-heated and heated meats) and the role of reductants such as ascorbic and isoascorbic acids in the reactions, are well understood (Kemp, 1974 and Wolff and Wasserman, 1972). Nitrite also reacts, via nitrous acid, with amino and sulfhydryl groups of protein, is partially converted to nitrate and gas in the form of various oxides, and some nitrite nitrogen is found in lipid fractions.

Little is known about the pathway(s) relevant to the inhibition of *C. botulinum*, but recent publications show significant progress in this field. Johnston and Loynes (1971) and Ashworth and Spencer (1972) found that ascorbate and other reducing agents (cysteine, thioglycolate) enhanced the inhibitory effect of nitrite on *C. botulinum* in comminuted meats. The work of Crowther *et al.* (1976) gave some evidence for such an effect of ascorbate in bacon, while Bowen *et al.* (1974) failed to find any enhancing effect of ascorbate on the action of nitrite in wieners. Tompkin *et al.* (1978) recently demonstrated a dramatic enhancement of the inhibitory effect of nitrite by isoascorbate in perishable canned cured meats; a combination of 50 ppm of nitrite and 0.02% of isoascorbate had essentially the same inhibitory effect on *C. botulinum* as 156 ppm of nitrite alone. Tompkin *et al.* (1978 and 1979) demonstrated that, by enhancing the activity of nitrite, erythorbate (isoascorbate), ascorbate and cysteine acted as chelating agents, rather than as reductants or antioxidants; the relevant ions sequestered were ferrous and ferric ions. Addition of iron salts or iron powder to cured pork resulted in complete neutralization of the inhibitory effect of nitrite (Tompkin, *et al.*, 1978). The authors offered the following hypothesis (Tompkin *et al.*, 1978): nitrous oxide (from nitrous acid) reacts with an essential iron compound, such as ferredoxin, in the germinated botulinal cell and thus prevents it from outgrowth. Iron compounds in the meat capable of binding nitrous oxide will act as competitors. Addition of chelating agents to the meat, therefore, increases the amount of nitrous oxide available for the cellular reaction. This hypothesis is consistent with the observation (Tompkin *et al.*, 1979) that nitrite has little effect on outgrowth of *C. botulinum* in pork heart or beef heart, both of which have a high iron content.

Nitrite has little or no effect on the death rate of *C. botulinum* during heating. It appeared for a while that it might control *C. botulinum* by stimulating the rate of germination (Duncan and Foster, 1968); germinated cells in turn are highly susceptible to curing salts. However, at the concentrations applied in cured meats, nitrite does not significantly affect the germination of *C. botulinum* spores. Instead, it inhibits outgrowth of the germinated cell (Ingram, 1973 and Pivnick, *et al.*, 1970).

There is some evidence that the action of nitrite on *C. botulinum* is supplemented by an inhibitor(s), produced by reactions of nitrite with meat components. An anticlostridial compound, also known as Perigo factor, is produced when certain bacteriological media are heated in the presence of nitrite (Perigo, *et al.*, 1967 and 1968). However, the Perigo factor is neutralized when meat is added to these media (Johnston *et al.*, 1969). It is unlikely, therefore, that the Perigo factor plays a role in the control of *C. botulinum* in cured meats. However, it appears that an inhibitor(s) of a different nature, often referred to as Perigo type factor (PTF) is produced by an interaction between nitrite and meat. Johnston *et al.* (1969) heat-processed meat suspensions with and without nitrite and subsequently adjusted each batch to the same nitrite level (120 ppm). Despite equal nitrite contents, outgrowth of *C. botulinum* in batches heat-processed in the presence of nitrite was significantly lower than in batches processed without nitrite. Similar results with pork (Ashworth, *et al.*, 1972 and 1973 and Chang and Akhtar, 1974) and lean beef (Hargreaves and Ashworth, 1973) have been reported, while little or no inhibitor seemed to be produced in corned beef (Hargreaves and Ashworth, 1973). One might argue that the two systems, heated originally with different concentrations of nitrite and then made up to the same nitrite level, are not strictly comparable because nitrite might disappear somewhat faster in the system originally heated with the least nitrogen. Johnston and Loynes (1971) have shown that the nitrite requirement for complete inhibition of *C. botulinum* was dependent upon the preceding treatment of the meat. This criticism would not apply to the work of Chang *et al.* (1974) and Pivnick *et al.* (1973) who processed luncheon meat with different concentrations of nitrite, incubated the cans until the nitrite content was essentially zero, and challenged them with *C. botulinum* spores. Meat containing 200 ppm of nitrite at the time of processing inhibited 1.4 $\log_{10}$ spores more than meat processed without nitrite; inhibition (In) attributable to PTF was thus 1.4 units. With less nitrite (100–150 ppm) In from PTF was only 0.3 units, while about 2.2 units were attributable to the salt (4.5% in the brine). Lee *et al.* (1978) confirmed the inhibition of *C. botulinum* in cured pork after nitrite depletion. It appears that in some products such as luncheon meats and ham, PTF may play some accessory role in the control of *C. botulinum*.

The safety record of commercial canned, cured, shelf-stable meat is exceptionally good (CDC, 1974). A single outbreak from such a product (liverpaste) involving two cases with one death occurred in Canada in 1963. The product had a brine concentration of only 3.3%, a low concentration of nitrite and had apparently been underprocessed (Erdman and Idziak, 1965; Pivnick, *et al.*, 1969 and Thatcher, *et al.*, 1967).

Category 5—preservation by pasteurization, salt, nitrite and refrigeration.

(a) Perishable canned cured meats.—For organoleptic reasons, some canned cured meats receive only a very mild heat process, adequate to kill the vegetative microflora but insufficient to even damage *C. botulinum* spores of types A and B; the internal temperature may barely reach 70°C (158°F). The USDA requires a minimal internal temperature of 66°C for perishable canned cured meats. Such meats are mainly comminuted pork, ham of 3 lb and above, and bacon. The need for refrigeration is indicated on the label.

As long as the products are refrigerated, they are safe from botulinal hazards, even without nitrite. Christiansen *et al.* (1973) found no toxin in any of 80 cans of comminuted cured meat that had been processed with *C. botulinum* and kept at 7°C for 6 months, but 7 out of 20 cans without nitrite swelled within a month and became sour and discolored. This spoilage could be prevented by as little as 50 ppm of nitrite.

It is a common experience that perishable cured meats are frequently temperature-abused, both at the retail and the consumer level. When this occurs, nitrite plays an important role in delaying the development of *C. botulinum* (Christiansen, *et al.*, 1973; Greenberg, 1972 and Tompkin, *et al.*, 1977). The safe period of storage at elevated temperatures is limited; some cans of comminuted pork with a large spore load (400,000/can) were found to become toxic even with 300 and 400 ppm of nitrite when they were kept for some weeks at 27°C (Christiansen, *et al.*, 1973). Tompkin *et al.* (1977) determined the predicted swelling time at 27°C for perishable canned comminuted pork containing about 25,000 botulinal spores per can; for cans processed with 0, 50, 100 and 156 ppm of nitrite, the swelling times were about 7, 30, 83 and 94 days, respectively.

After formulation of the raw meat, some nitrite (25%–30% at 150 ppm) is no longer detectable by the Griess reaction. Further loss during the mild heat process is minimal, but during storage at 27°C, the nitrite level declines to almost zero within two months (Christiansen, *et al.*, 1973 and Greenberg, 1972).

During storage, the decline in the nitrite content is accompanied by a decrease in viable *C. botulinum* spores. The decrease in inhibition from nitrite is, therefore, balanced by a diminishing probability of outgrowth, due to smaller spore numbers. Christian *et al.* (1978) likened this balance to a race between death of the germinated cells and nitrite depletion. However, if temperature abuse occurs after a prolonged refrigerated storage, this balance could be off-set because the decrease in residual nitrite at lower temperatures proceeds at a faster rate than the decrease in viable *C. botulinum* spores (Tompkin, *et al.*, 1978).

Despite frequent temperature abuses and only limited protection from the curing salts, perishable canned cured meats have an unblemished safety record. Christiansen *et al.* (1973) noticed that all their toxic cans had also become putrid. It is obvious, therefore, that the warning odors in these products are an important safety factor.

(b) Perishable cured meats (uncanned).—Of the various factors controlling *C. botulinum* in these products, only those common to all of them are listed in Table 3.4. For example, the pH which has a major role in the preservation of fermented meats is not listed in the table. It also affects the conditions for spore outgrowth in nonfermented meats, but these foods are generally prepared without concerted attempts to control the pH which varies considerably between products, even of the same kind (Riemann, *et al.*, 1972). Other controlling factors are reducing compounds (ascorbate and isoascorbate), starter cultures, fermentable carbohydrates, acidulants and smoke; these will be discussed below.

The important role of nitrite in the control of *C. botulinum* has been demonstrated for fermented and unfermented sausages (Ala-Huikku, *et al.*, 1977; Christiansen, *et al.*, 1975; Dethmers, *et al.*, 1975 and Kueper and Trelease, 1974), wieners (Hustad, *et al.*, 1973), ham (Pivnick, *et al.*, 1967), jellied tongue (Pivnick, *et al.*, 1967) and bacon (Christiansen, *et al.*, 1974; Collins-Thompson, 1974 and Ivey, *et al.*, 1978).

An attempt to produce liver sausage without added nitrite failed, due to subsequent spoilage from anaerobes (Ala-Huikku, *et al.*, 1977). However, considerable reduction in the amount of nitrite added from the currently common concentration of 150–200 ppm (μg/g) may be feasible for a number of products without compromising on safety or keeping quality. For example, Hustad, *et al.* (1973) found that 50 ppm of nitrite in wieners sufficed to control *C. botulinum* at an abusive temperature of 27°C for a minimum of 28 days, long after putrefaction occurred. This concentration is also adequate to fully develop the cured-meat color and flavor. Similarly, 50 ppm of nitrite were sufficient to control *C. botulinum* in Thuringer sausage, provided it also contained fermentable carbohydrate (Kueper and Trelease, 1974). The nitrite content of liver sausage can probably be reduced to 100 ppm or lower (Ala-Huikku, *et al.*, 1977).

Since the frying of bacon is linked with formation of nitrosamines, considerations for reduction of nitrite in bacon are particularly urgent. Three publications (Christiansen, *et al.*, 1974; Collins-Thompson, *et al.*, 1974 and Ivey, *et al.*, 1978) and an unpublished report (M.D. Pierson, pers. comm.) deal with the outgrowth of *C. botulinum* in bacon prepared with different concentrations of nitrite. No outgrowth occurred with proper refrigeration, even without nitrite. However, bacon must be sufficiently inhibitory to *C. botulinum* to allow for some temperature abuse during transport or in the home. On the assumptions that botulinal spores introduced experimentally into bacon are equivalent to naturally contaminating spores, and that the spores capable of outgrowth and toxigenesis follow the Poisson distribution in bacon, one may calculate the probability of a spore to give rise to toxin production in temperature-abused bacon, i.e., by (1) applying the equation of Halvorson and Ziegler (1933) to the relative number of packages becoming toxic, and (2) relating the MPN to the number of spores introduced per package. With 40–60 ppm of nitrite and a temperature abuse equivalent to 27–30°C for about one week, the values were of the order of 10^{-6} (one spore per million).

British surveys indicated about two botulinal spores per Kg of bacon (Roberts and Ingram, 1976; Roberts and Smart, 1976 and 1977). On the basis of these data, the probability of a one-lb package of bacon to become toxic would also be of the order of 10^{-6}, provided that every package received a minimum abuse. In a recent Canadian survey (Hauschild and Hilsheimer, unpublished date) 104 one-lb packages were analyzed. Four 75-g samples from each package were cultured, two for spores and two for total *C. botulinum*. Of 208 samples in each group, none were positive for spores, and only one was positive for total *C. botulinum*. These results correspond to < 0.064 spores/Kg and 0.064 total *C. botulinum*/Kg, and to a 10^{-7} – 10^{-8} probability for toxicity in abused packages.

The proportion of packages that are accidentally, or through ignorance, kept sufficiently long at abusive temperatures to allow outgrowth of *C. botulinum* has been frequently speculated upon. The specification for refrigerated storage is likely to reduce the probability of a package becoming toxic by another 3 logs. Likewise, the fact that essentially all bacon is fried or cooked before consumption probably increases the safety level again by about 3 logs. Therefore, the protection of low-nitrite bacon may even surpass the equivalent of the 12 D value. By comparison, the D equivalent of shelf-stable canned cured meat has been estimated as about 6 (Riemann, *et al.*, 1972). The potential benefits from reducing nitrite addition from currently permissible levels (120 ppm in the U.S., 150 ppm in Canada) to 50 ppm would, therefore, seem to outweigh any potential risk from botulinal hazards.

Ivey *et al.* (1978) have shown that the combination of 40 ppm of nitrite with 0.26% of potassium sorbate in bacon is at least as effective in its antibotulinal activity as 120 ppm of nitrite alone. However, the supplementation of low-level nitrite with sorbate, particularly at such a high concentration, seems unnecessary.

The use of nitrite in the curing of meats is relatively recent, but nitrite has nearly completely replaced nitrate (saltpeter) which has served as a curing agent for at least 2000 years (Binkerd and Kolari, 1975). Compared to nitrite, the effect of nitrate as an organoleptic and preservative agent is low (Christiansen, *et al.*, 1975 and Hustad, *et al.*, 1973) because its preservative action depends on its prior conversion to nitrite. Consequently, the present commercial use of nitrate as a curing agent is essentially limited to products for which a slow generation of nitrite may be desirable, i.e., in the slow fermentation of meats. Data gathered by the American Meat Institue Foundation show a drastic decline in the use of nitrate between 1970 and 1974, while the use of nitrite remained almost unchanged during that period (Binkerd and Kolari, 1975).

The effect of salt in curing has been known since time immemorial. The critical levels needed to inhibit *C. botulinum* in meats vary between individual products (Pivnick and Barnett, 1965 and Pivnick and Bird, 1965) and depend largely on the pH (Pivnick and Barnett, 1965) which may vary from <5.0 to 6.5 (Riemann, *et al.*, 1972). From a number of publications, Riemann *et al.* (1972) plotted the salt content in the brine of various cured meats against the pH, and related the data to inhibition or non-inhibition of *C. botulinum* at the 6 D level. For types A and B, the salt required in the brine increased linearly from 2% at pH 4.9 to 7% at pH 5.7. For type E the salt required was <1% below pH 5.3 and about 5% at pH 6.6. The initial nitrite concentrations were not known, but they were likely in the 150–200 ppm range.

In recent years, water activity (a_w) has received increasing attention in the evaluation of food safety. The water activity of a food may be expressed as $a_w = RH/100$ or as $a_w = p/p_0$ where RH is the relative humidity, p the vapor pressure of the food and p_0 the vapor pressure of water. The major factors contributing to the a_w of a food are the water and salt contents, but carbohydrates, fats and other ingredients also contribute. In liquid cultures the minimum a_w values for growth of *C. botulinum* types A, B and E are about 0.95, 0.94 and 0.97, respectively (Ohye, *et al.*, 1967 and Troller, 1973); these levels correspond to salt concentrations of approximately 8.0%, 9.4% and 5.1%. In media adjusted with glycerol instead of salt, the minimum a_w levels may be slightly lower (Troller, 1973).

On the basis of potential outgrowth of food-borne pathogens and the need for refrigeration, Leistner and Rödel (1976) and Rödel *et al.* (1976) grouped meats into three categories (A-C).

Group A.—(highly perishable) includes all meats with a_w >0.95 *and* pH >5.2. These are to be refrigerated at ≤5°C.

Group B.—(perishable) includes meats with (i) pH 5.0 – 5.2 *and* a_w >0.95 or (ii) a_w = 0.91 – 0.95 *and* pH >5.2. These are to be refrigerated at ≤10°C.

Group C.—(shelf-stable) includes meats with (i) pH <5.0; (ii) a_w <0.91 or (iii) pH ≤5.2 *and* a_w ≤0.95. These require no refrigeration.

Fresh meats, blood and bologna-type sausages and cooked ham fall into category A, wieners and liverwurst into categories A and B, and raw, dry fermented sausages into category C. Raw fermented hams, raw smoked meats and various sausages (metwurst, teewurst) show considerable variation in pH and a_w and are divided among each of the three categories (Leistner, *et al.*, 1971 and Rödel, *et al.*, 1976). For the control of *C. botulinum* alone the requirements for refrigeration would be somewhat less stringent.

Home-prepared cured meats, particularly cured ham in Europe (Sebald, 1970), have frequently been involved in outbreaks of botulism. The common condition that allowed toxigenesis in these products was inadequate control of the nitrite and salt contents. In many instances nitrite may have been missing altogether in the curing process (Roberts, *et al.*, 1976). Commercial cured meats are frequently involved in various kinds of food poisoning, mostly staphylococcal, but their record with respect to botulinal poisoning is exceptionally good. Commercial vacuum-packed luncheon meat was implicated in one botulism outbreak involving three people in Idaho (CDC, 1965).

As discussed in connection with smoked fish, vacuum-packaging of cured meats in air-impermeable pouches adds another potentially hazardous factor to the product. While the common spoilage microorganisms are inhibited in the absence of oxygen, development and toxigenesis of *C.botulinum* is not. Due to the delay in spoilage, cured meats may be organoleptically acceptable for some time after critical amounts of toxin have been produced (Pivnick and Barnett, 1965 and Pivnick and Bird, 1965). Toxic meats in vacuum-packed pouches are, therefore, more likely to be eaten than meats kept under less rigid exclusion of air.

As mentioned above, nitrite may provide an additional safety factor in the preservation of smoked fish in some countries (Kautter and Lilly, 1970 and Pace and Krumbiegel, 1973).

Supplementary factors in the safety of cured meats.

Ascorbate and isoascorbate (erythorbate).—The original purpose of adding ascorbate or isoascorbate to cured meats was to improve color formation and color stability (Rust and Olson, 1973). More recently, ascorbate has been shown to suppress the formation of nitrosamines (Fan and Tannenbaum, 1973; Mirvish, *et al.*, 1972 and Mottram, *et al.*, 1975). The role of these compounds in the control of *C. botulinum* has been discussed above. Although they increase the efficacy of nitrite at a level of about 0.02%, both have been found to reduce the inhibitory effect of nitrite in canned cured meat when applied at excessive levels. (Tompkin, *et al.*, 1979). This effect is attributable to accelerated depletion of residual nitrite. Similarly, Bowen and Deibel (1974) reported reduced efficacy of nitrite in the presence of large amounts of ascorbate in bacon.

Starter cultures, fermentable carbohydrates, glucone-delta-lactone, polyphosphates.— Starter cultures of *Lactobacillus* and/or *Pediococcus* species are important in assuring desirable and reproducible flavor and texture of fermented meats, but their main purpose is a rapid reduction of the pH to levels at which the growth of food-borne pathogens is progressively inhibited (Genigeorgis, 1976). The function of starter cultures is dependent upon fermentable carbohydrates which are usually added in the form of 1–2% dextrose. Christiansen *et al.* (1975) showed that 50 ppm of nitrite in combination with dextrose (with or without starter culture) were more effective in controlling *C. botulinum* in summer-style sausage than 150 ppm of nitrite alone. The effect of dextrose on *C. botulinum* in fermented sausages has also been shown by others (Kueper and Trelease, 1974). Glucono-delta-lactone (GDL) is added to cured meats such as frankfurters, bolognas, luncheon meats and fermented sausages to a concentration of 0.4%. It reduces the pH of the meat only by about 0.5 units (Acton and Dick, 1977) but has the advantage over other acidulants in that it acidifies the product slowly, through hydrolysis to gluconic acid, and this allows heating and smoking at elevated temperatures in the early stages of processing without the separation of fat which occurs when the pH is reduced rapidly. GDL also has some accelerating effect on color development (Acton and Dick, 1977 and Rust and Olson, 1973).

Polyphosphates are added to ham, bacon and poultry to reduce shrinkage during heating (Neer and Mandigo, 1977). They are unlikely to affect the control of *C. botulinum* other than indirectly by their effect on the pH.

Smoke.—There are three principal smoking processes: cold smoking (below 40°C), hot smoking (about 60–85°C) and treatment with liquid smoke which may be mixed with comminuted meats, injected into meats, or surface-applied. The main purpose of smoking is flavor development; this is actually the only purpose of liquid smoking because liquid smoke contains none of the bactericidal ingredients of the original pyrolysate (Gorbatov, *et al.* 1971).

The antibacterial effect of smoke, in particular on micrococci and staphylococci, has been shown in several investigations, but its effect on clostridia appears to be slight (Niinivara and Pohja, 1956 and Polymenides, 1977); in the centre of larger products it has probably no effect. The bactericidal effect has been attributed to phenolic compounds (Draudt, 1963; Handford and Gibbs, 1964 and Roberts, 1975) and to formaldehyde (Jensen, 1943 and Nielsen and Pedersen, 1967). In view of these findings it is of interest that, according to Sebald (1970), some French producers of commercial ham injected small quantities of formaldehyde into the deeper portions of the meat.

The main effect of the smoking process in the control of *C. botulinum* seems to be linked with the heating, particularly in the hot smoking of fish (Pace and Krumbiegel, 1973), and a reduction in the water activity due to the drying effect (Roberts, 1975). The bactericidal effects of smoke components and the slight reduction in pH from smoking (Handford and Gibbs, 1964) are probably marginal.

Cagetory 6—Preservation by pH, salt, nitrite and drying.

Products of this category include fermented dried ham and raw, fermented dry sausages: Italian salamis, cervelats, mordatella, pepperoni, sopressata, German plockwurst and Landjäger, and others (American Meat Institute, 1953). They are characterized by low water activity and pH which are usually below 0.91 and 5.0, respectively (Rödel, et al., 1976). The roles of starter cultures and fermentable carbohydrates in achieving proper pH levels have been discussed above; the low water activities are obtained by adequate salt and a prolonged drying process of 1–6 months at about 10–14°C (American Meat Institute, 1953). Improperly processed commercial dry meats are occasionally involved in staphylococcal outbreaks, but they have not been implicated in botulism on this continent.

Category 7—Preservation by pasteurization, pH and salt.

The foods in this category are predominantly fish and fish products. Their safety record is relatively good. However, imported caviar was the cause of a major botulinal type B outbreak in Japan (Fukuda, et al., 1970); Sebald (1970) listed a single case of type E botulism from caviar in France, and an incident of suspected botulism from lumpfish caviar was reported in Canada (Health and Welfare Canada, 1977). Hauschild and Hilsheimer (1979) found that outgrowth and toxigenesis of C. botulinum types A and B in lumpfish caviar was inhibited at salt concentrations in the brine of ⩾5.6% or at pH ⩾5.0 and at combinations of >4.0% salt at pH <5.2 or of >4.7% salt at pH 5.6. A survey of commercial caviar products in Canada indicated that nearly all of these would effectively control C. botulinum, but a few might conceivably produce toxin, if temperature-abused. The survey revealed also that, in a number of products, the control was on the basis of the salt content alone; the pH values ranged from 4.6 to 6.8. In some cases of botulism from home-prepared products such as pickled pigs feet (CDC, 1974) or pickled fish (Craig, et al., 1968) it is often difficult to determine whether the intended preservation was based on a combination of the salt content and pH, or on one of the two factors alone.

Process cheese which is commonly merely pasteurized (between 70 and 90°C) should be mentioned here also. The pH of these products is rarely below 5.2 (Kosikowski, 1977); the most critical factor in the control of clostridia, therefore, is the water activity. Occasional outbreaks from process cheese have been reported (Kosikowski, 1977); a detailed account of a fatal incident from Liederkranz cheese has been given by Meyer and Eddie (1951).

Category 8—Preservation by salt.

This is probably the oldest method of preserving meat and fish. The salting of meat has been largely replaced by nitrite/salt curing, but fish, such as raw salt herring, and fish eggs are still being preserved by salt alone. Botulism outbreaks from salted products have occurred frequently in the past; the causes may be (1) lack of adequate refrigeration immediately after salting (a few weeks may be required between salting and salt penetration into the core of large hams to a concentration that would inhibit *C. botulinum*); (2) faulty washing-out of the salt with respect to duration and temperature, or (3) insufficient salt in products such as fish eggs that are consumed without prior washing-out.

One of the only four botulism outbreaks from commercial products in Canada was caused by imported salted herring, but the source of the botulinal spores or the type of abuse that led to toxigenesis are unknown (Dolman, 1961). Currently, salted fish eggs in the Caspian Sea area, known in Iran as "ashbal", are a frequent cause of botulism type E outbreaks (Pourtaghva, *et al.*, 1975 and Rouhbakhsh-Khaleghdoust and Portaghva, 1977).

Table 3.4 lists only the products of major concern with respect to potential hazards from botulism. Omitted from the table are locally produced foods such as fermented salmon eggs or "izushi" etc. (see above) that are prepared under faulty concepts of food preservation. Of the various products of marginal concern, only one shall be mentioned here because a problem with it arose relatively recently, as a result of a new form of packaging. Sugiyama and Yang (1975) reported that mushrooms, packaged in cardboard containers and covered in a sealed polyvinylchloride film for extension of shelf life, allowed toxigenesis by *C. botulinum* of types A and B within 3 to 4 days at 20°C. Consumption of oxygen by product respiration exceeded oxygen diffusion through the film resulting in a rapid internal drop in the redox potential. So far, mushrooms packed in sealed plastic film have not caused illness from botulinal toxin, but this work certainly revealed a potential problem, particularly in view of the large numbers of *C. botulinum* spores in mushrooms and the occasional use of raw mushrooms in salads. The problem may be overcome by one or two small perforations in the plastic film (Kautter, *et al.*, 1978 and Sugiyama, *et al.*, 1975 and 1978) which allows adequate air passage to maintain a sufficiently high redox potential, but barely affects the loss of moisture and the shelf life of the product.

The early recognition of similar potential problems will be a major factor in preventing large-scale outbreaks of botulism in the future.

BIBLIOGRAPHY

ABRAHAMSSON, K. 1967. Occurrence of type E *Cl. botulinum* in smoked eel. *In* M. Ingram and T. A. Roberts (ed): Botulism 1966. Chapman & Hall Ltd., London; pp. 73–75.

ABRAHAMSSON, K. and RIEMANN, H. 1971. Prevalence of *Clostridium botulinum* in semipreserved meat products. Appl. Microbiol. *21*: 543–544.

ACTON, J. C. and DICK, R. L. 1977. Cured pigment and colour development in fermented sausage containing glucono-delta-lactone. J. Food Protec. *40*: 398–401.

AGRICULTURE CANADA. 1976. Canning Canadian fruits and vegetables. Public. No. 1560. Food Advisory Services, Ottawa.

ALA-HUIKKU, K., NURMI, E., PAJULAHTI, H. and RAEVUORI, M. 1977. Effect of nitrite, storage temperature and time on *Clostridium botulinum* type A toxin formation in liver sausage. Europ. J. Appl. Microbiol. *4*: 145–149.

ALDERMAN, G. G., KING, G. J. and SUGIYAMA, H. 1972. Factors in survival of *Clostridium botulinum* type E spores through the fish smoking process. J. Milk Food Technol. *35*: 163–166.

AMERICAN MEAT INSTITUTE. 1953. Sausage and ready-to-serve Meats. Institute of Meat Packing, Univ. of Chicago, Chicago, Ill.

ARNON, S. S., MIDURA, T. F., CLAY, S. A., WOOD, R. M. and CHIN, J. 1977. Infant botulism. Epidemiological, clinical, and laboratory aspects. J. Am. Med. Assoc. *237*: 1946–1951.

ASHWORTH, J., HARGREAVES, L. L., and JARVIS, B. 1973. The production of an antimicrobial effect in pork heated with sodium nitrite under simulated commercial pasteurization conditions. J. Food Technol. *8*: 477–484.

ASHWORTH, J. and SPENCER, R. 1972. The Perigo effect in pork. J. Food Technol. *7*: 111–124.

BAIRD-PARKER, A. C. and FREAME, B. 1967. Combined effect of water activity, pH and temperature on the growth of *Clostridium botulinum* from spore and vegetative cell inocula. J. Appl. Bacteriol. *30*: 420–429.

BAUMGART, J. 1970. Nachweis von *Clostridium botulinum* Typ E bei handelsfertigen Forellen. Fleischwirtsch. *50*: 1545–1546.

BINKERD, E. F. and KOLARI, O. E. 1975. The history and use of nitrate and nitrite in the curing of meat. Food Cosmet. Toxicol. *13*: 655–661.

BLACK, R. E. and ARNON, S. S. 1977. Botulism in the United States, 1976. J. Infect. Dis. *136*: 829–832.

BOWEN, V. G., CERVENY, J. G. and DEIBEL, R. H. 1974. Effect of sodium ascorbate and sodium nitrite on toxin formation of *Clostridium botulinum* in Wieners. Appl. Microbiol. *27*: 605–606.

BOWEN, V. G., and DEIBEL, R. H. 1974. Effects of nitrite and ascorbate on botulinal toxin formation in wieners and bacon. *In*: Proc. Meat Ind. Res. Conf., American Meat Institute, Chicago; pp. 63–68.

BOWMER, E. J. and WILKINSON, D. A. 1976. Botulism in Canada, 1971–74. Can. Med. Assoc. J. *115*: 1085–1086.

BOTT, T. L., JOHNSON, J., FOSTER, E. M. and SUGIYAMA, H. 1968. Possible origin of the high incidence of *Clostridium botulinum* type E in an inland bay (Green Bay of Lake Michigan). J. Bacteriol. *95*: 1542–1547.

BURNS, G. F. and WILLIAMS, H. 1975. *Clostridium botulinum* in Scottish fish farms and farmed trout. J. Hyg., Camb. *74*: 1–6.

CANN, D. C., TAYLOR, L. Y. and HOBBS, G. 1975. The incidence of *Clostridium botulinum* in farmed trout raised in Great Britain. J. Appl. Bacteriol. *39*: 331–336.

CANN, D. C., WILSON, B. B. and HOBBS, G. 1968. Incidence of *Clostridium botulinum* in bottom deposits in British coastal waters. J. Appl. Bacteriol. *31*: 511–514.

CANN, D. C., WILSON, B. B., HOBBS, G. and SHEWAN, J. M. 1967. *Cl. botulinum* type E in the marine environment of Great Britain. *In* M. Ingram and T. A. Roberts (Ed): Botulism 1966. Chapman & Hall Ltd., London; pp. 62–65.

CDC (CENTER FOR DISEASE CONTROL). 1960. Botulism. Morbid. Mortal. W. Rep. *9*(38): 2.

CDC. 1963. Epidemic botulism related to smoked fish ingestion. Morbid. Mortal. W. Rep. *12*(40): 329–330.

CDC. 1965. Botulism - Idaho. Morbid. Mortal. W. Rep. *14*(27): 225–226.

CDC. 1966. Botulism - California. Morbid. Mortal. W. Rep. *15*(41): 349, 356.

CDC. 1966. Botulism type F - California. Morbid. Mortal. W. Rep. *15*(42): 359.

CDC. 1973. Botulism - Kentucky. Morbid. Mortal. W. Rep. *22*(50): 417–418.

CDC. 1974. Botulism - Alabama. Morbid. Mortal. W. Rep. *23*(10): 90,95.

CDC. 1974. Botulism - Idaho, Utah. Morbid. Mortal. W. Rep. *23*(27): 241–242.

CDC. 1974. Botulism in the United States, 1899–1973. DHEW. Publ. 74–8279. Atlanta, Georgia.

CDC. 1975. Botulism - United States, 1974. Morbid. Mortal. W. Rep. *24*(5): 39.

CDC. 1976. Botulism in 1975 - United States. Morbid. Mortal. W. Rep. *25*(9): 75.

CDC. 1976. Botulism in infants - California. Morbid. Mortal. W. Rep. *25*(34): 269.

CDC. 1978. Follow-up on infant botulism - United States. Morbid. Mortal. W. Rep. *27*(3): 17–18, 23.

CDC. 1977. Follow-up on botulism - Michigan. Morbid. Mortal. W. Rep. *26*(16): 135.

CDC. 1978. Follow-up on botulism - New Mexico. Morbid. Mortal. W. Rep. *27*(17): 145.

CDC. 1978. Botulism - Puerto Rico. Morbid. Mortal. W. Rep. 27(38): 356–357.

CENTRO PANAMERICANO DE ZOONOSIS. 1975. Food-borne diseases. Boletin Informativo, Engl. ed., Buenos Aires, Argentina.

CHANG, P.-C. and AKHTAR, S. M. 1974. The Perigo effect in luncheon meat. Can. Inst. Food Sci. Technol. J. 7: 117–119.

CHANG, P.-C., AKHTAR, S. M., BURKE, T. and PIVNICK, H. 1974. Effect of sodium nitrite on Clostridium botulinum in canned luncheon meat: evidence for a Perigo-type factor in the absence of nitrite. Can. Inst. Food Sci. Technol. J. 7: 209–212.

CHRISTIANSEN, L. N., DEFFNER, J., FOSTER, E. M. and SUGIYAMA, H. 1968. Survival and outgrowth of Clostridium botulinum type E spores in smoked fish. Appl. Microbiol. 16: 133–137.

CHRISTIANSEN, L. N., JOHNSTON, R. W., KAUTTER, D. A., HOWARD, J. W. and AUNAN, W. J. 1973. Effect of nitrite and nitrate on toxin production by Clostridium botulinum and on nitrosamine formation in perishable canned comminuted cured meat. Appl. Microbiol. 25: 357–362.

CHRISTIANSEN, L. N., TOMPKIN, R. B., SHAPARIS, A. B., JOHNSTON, R. W. and KAUTTER, D. A. 1975. Effect of sodium nitrite and nitrate on Clostridium botulinum growth and toxin production in a summer style sausage. J. Food Sci. 40: 488–490.

CHRISTIANSEN, L. N., TOMPKIN, R. B., SHAPARIS, A. B., KUEPER, T. V., JOHNSTON, R. W., KAUTTER, D. A. and KOLARI, O. J. 1974. Effect of sodium nitrite on toxin production by Clostridium botulinum in bacon. Appl. Microbiol. 27: 733–737.

CHRISTIANSEN, L. N., TOMPKIN, R. B. and SHAPARIS, A. B. 1978. Fate of Clostridium botulinum in perishable canned cured meat at abuse temperature. J. Food Protec. 41: 354–355.

CICCARELLI, A. S., WHALEY, D. N., McCROSKEY, L. M., GIMENEZ, D. F., DOWELL, V. R. and HATHEWAY, C. L. 1977. Cultural and physiological characteristics of Clostridium botulinum type G and the susceptibility of certain animals to its toxin. Appl. Env. Microbiol. 34: 843–848.

COCKEY, R. R. and TATRO, M. C. 1974. Survival studies with spores of Clostridium botulinum type E in pasteurized meat of the blue crab Callinectes sapidus. Appl. Microbiol. 27: 629–633.

COLLINS-THOMPSON, D. L., CHANG, P. C., DAVIDSON, C. M., LARMOND, E. and PIVNICK, H. 1974. Effect of nitrite and storage temperature on the organoleptic quality and toxinogenesis by Clostridium botulinum in vacuum-packed side bacon. J. Food Sci. 39: 607–609.

CRAIG, J. M., HAYES, S. and PILCHER, K. S. 1968. Incidence of Clostridium botulinum type E in salmon and other marine fish in the Pacific northwest. Appl. Microbiol. 16: 553–557.

CRAIG, J. M. and PILCHER, K. S. 1967. The natural distribution of Cl. botulinum type E in the Pacific coast areas of the United States. In M. Ingram and T. A. Roberts (Ed): Botulism 1966. Chapman & Hall, Ltd., London; pp. 56–61.

CROWTHER, J. S., HOLBROOK, R., BAIRD-PARKER, A. C. and AUSTIN, B. L. 1976. Role of nitrite and ascorbate in the microbiological safety of vacuum-packed sliced bacon. Proc. 2nd. Int. Symp. Nitrite Meat Prod., Zeist. Pudoc. Wageningen; pp. 13–20.

DEMARCHI, J., MOURGUES, C., ORIO, J. and PRÉVOT, A.-R. 1958. Existence du botulisme humain de type D. Acad. Natl. Méd. Bull. *142*: 580–582.

DETHMERS, A. E., ROCK, H., FAZIO, T. and JOHNSTON, R. W. 1975. Effect of added sodium nitrite and sodium nitrate on sensory quality and nitrosamine formation in Thuringer sausage. J. Food Sci. *40*: 491–495.

DICKSON, E. C. 1917. Botulism. J. Am. Med. Assoc. *69*: 966–968.

DICKSON, E. C. 1918. Botulism. A clinical and experimental study. Rockefeller Inst. Med. Res., Monog. 8. New York; pp. 1–114.

DOLMAN, C. E. 1954. Additional botulism episodes in Canada. Can. Med. Assoc. J. *71*: 245–249.

DOLMAN, C. E. 1961. Further outbreaks of botulism in Canada. Can. Med. Assoc. J. *84*: 191–200.

DOLMAN, C. E. 1964. Botulism as a world health problem. Proc. Symp. R. A. Taft Sanit. Eng. Center, Cincinnati, Ohio. P. H. S. Publ. No. 999–FP-1; pp. 5–32.

DOLMAN, C. E. 1974. Human botulism in Canada (1919–1973). Can. Med. Assoc. J. *110*: 191–200.

DOLMAN, C. E., CHANG, H., KERR, D. E. and SHEARER, A. R. 1950. Fish-borne and type E botulism: two cases due to home-pickled herring. Can. J. Publ. Health *41*: 215–229.

DOLMAN, C. E. and IIDA, H. 1963. Type E botulism: its epidemiology, prevention and specific treatment. Can. J. Publ. Health *54*: 293–308.

DOLMAN, C. E. and MURAKAMI, L. 1961. *Clostridium botulinum* type F with recent observations on other types. J. Infect. Dis. *109*: 107–128.

DUNCAN, C. L. 1970. Arrest of growth from spores in semi-preserved foods. J. Appl. Bacteriol. *33*: 60–73.

DUNCAN, C. L. and FOSTER, E. M. 1968. Nitrite-induced germination of putrefactive anaerobe 3679h spores. Appl. Microbiol. *16*: 412–416.

DRAUDT, H. N. 1963. The meat smoking process: a review. Food Technol. *17*: 85–90.

EISENBERG, M. S. and BENDER, T. R. 1976. Botulism in Alaska, 1947 through 1974. J. Am. Med. Assoc. *235*: 35–38.

EKLUND, M. W. and POYSKY, F. 1967. Incidence of *Cl. botulinum* type E from the Pacific coast of the United States. *In* M. Ingram and T. A. Roberts (Ed): Botulism 1966. Chapman & Hall Ltd., London; pp. 49–55.

EKLUND, M. W., POYSKY, F. T. and WIELER, D. I. 1967. Characteristics of *Clostridium botulinum* type F isolated from the Pacific coast of the United States. Appl. Microbiol. *15*: 1316–1323.

ERDMAN, I. E. and IDZIAK, E. S. 1965. Underprocessing—a major factor leading to a recent botulism incident. Can. J. Publ. Health *56*: 26.

FAN, T. Y. and TANNENBAUM, S. R. 1973. Natural inhibitors of nitrosation reactions: the concept of available nitrite. J. Food. Sci. *38*: 1067–1069.

FOSTER, E. M., DEFFNER, J. S., BOTT, T. L. and McCOY, E. 1965. *Clostridium botulinum* food poisoning. J. Milk Food Technol. *28*: 86–91.

FUKUDA, T., KITAO, T., TANIKAWA, H. and SAKAGUCHI, G. 1970. An outbreak of type B botulism occurring in Miyazaki Prefecture. Jap. J. Med. Sci. Biol. *23*: 243–248.

GANGAROSA, E. J. 1968. Epidemic of febrile gastroenteritis due to *Salmonella java* traced to smoked whitefish. Am. J. Publ. Health *58*: 114–121.

GENIGEORGIS, C. A. 1976. Quality control of fermented meats. J. Am. Vet. Med. Assoc. *169*: 1220–1228.

GIMÉNEZ, D. F. and CICCARELLI, A. S. 1970. Another type of *Clostridium botulinum*. Zbl. Bakteriol. I, Abt. Orig. *215*: 221–224.

GIMÉNEZ, D. F. and CICCARELLI, A. S. 1978. New strains of *Clostridium botulinum* subtype Af. Zbl. Bakteriol. I. Orig. A *240*: 215–220.

GORBATOV, V. M., KRYLOVA, N. N., VOLOVINSKAYA, V. P., LYASKOVSKAYA, Y. N., BAZAROVA, K. I., KHLAMOVA, R. I. and YAKOVLEVA, G. Y. 1971. Liquid smoke for use in cured meats. Food Technol. *25*: 71–77.

GREENBERG, R. A. 1972. Nitrite in the control of *Clostridium botulinum*. Proc. Meat Ind. Res. Conf., Chicago; pp. 25–34.

GREENBERG, R. A., BLADEL, B. O. and ZINGELMANN, W. J. 1965. Radiation injury of *Clostridium botulinum* spores in cured meats. Appl. Microbiol. *13*: 743–748.

GREENBERG, R. A., TOMPKIN, R. B., BLADEL, B. O., KITTAKA, R. S. and ANELLIS, A. 1966. Incidence of mesophilic *Clostridium* spores in raw pork, beef, and chicken in processing plants in the United States and Canada. Appl. Microbiol. *14*: 789–793.

HALVORSON, H. O. and ZIEGLER, N. R. 1933. Application of statistics to problems in bacteriology. J. Bacteriol. *25*: 101–121.

HANDFORD, P. M. and GIBBS, B. M. 1964. Antibacterial effects of smoke constituents on bacteria isolated from bacon. *In* N. Molin (Ed) Microbial Inhibitors in Food. Almqvist & Wiksell, Stockholm; pp. 333–346.

HARGREAVES, L. L. and ASHWORTH, J. 1973. The Perigo effect: the inhibition of clostridia in meat and in culture media which have been heated with sodium nitrite. VI. The Perigo effect in beef and in pasteurized pork. Brit. Food Manuf. Ind. Res. Assoc. *193*: 1–22.

HAUSCHILD, A. H. W., ARIS, B. J. and HILSHEIMER, R. 1975. *Clostridium botulinum* in marinated products. Can. Inst. Food Sci. Technol. J. *8*: 84–87.

HAUSCHILD, A. H. W., BOWMER, E. J. and GAUVREAU, L. 1978. Infant botulism. Can. Med. Assoc. J. *118*: 484.

HAUSCHILD, A. H. W. and HILSHEIMER, R. 1979. Effect of salt content and pH on toxigenesis by *Clostridium botulinum* in caviar. J. Food Protec. *42*: 245–248.

HEALTH AND WELFARE CANADA. 1975, 1976 and 1977. Botulism

in Canada—summary for 1974; 1975; 1976. Can. Dis. W. Rep. *1*(3); *2*(13); *3*(12).

HEALTH AND WELFARE CANADA. 1978. Botulism at Cape Dorset, N. W. T. Can. Dis. W. Rep. *4*(20): 77–79.

HUSS, H. H., PEDERSEN, A. and CANN, D. C. 1974. The incidence of *Clostridium botulinum* in Danish trout farms. I. Distribution in fish and their environment. J. Food Technol. *9*: 445–450.

HUHTANEN, C. N., NAGHSKI, J., CUSTER, C. S. and RUSSELL, R. W. 1976. Growth and toxin production by *Clostridium botulinum* in moldy tomato juice. Appl. Env. Microbiol. *32*: 711–715.

HUSTAD, G. O., CERVENY, J. G., TRENK, H., DEIBEL, R. H., KAUTTER, D. A., FAZIO, T., JOHNSTON, R. W. and KOLARI, O. E. 1973. Effect of sodium nitrite and sodium nitrate on botulinal toxin production and nitrosamine formation in wieners. Appl. Microbiol. *26*: 22–26.

INGRAM, M. 1973. The microbiological effects of nitrite. Proc. 2nd. Int. Symp. Nitrite Meat Prod., Zeist. Pudoc. Wageningen; pp. 63–75.

INGRAM, M. and ROBINSON, R. H. M. 1951. The growth of *Clostridium botulinum* in acid bread media. J. Appl. Bacteriol. *14*: 62–72.

INGRAM, M. and ROBERTS, T. A. 1971. Application of the 'D-concept' to heat treatments involving curing salts. J. Food Technol. *6*: 21–28.

INSALATA, N. F., WITZEMAN, S. J., FREDERICKS, G. J. and SUNGA, F. C. A. 1969. Incidence study of spores of *Clostridium botulinum* in convenience foods. Appl. Microbiol. *17*: 542–544.

ITO, K. A., CHEN, J. K., LERKE, P. A., SEEGER, M. L. and UNVERFERTH, J. A. 1976. Effect of acid and salt concentration in fresh-pack pickles on the growth of *Clostridium botulinum* spores. Appl. Env. Microbiol. *32*: 121–124.

ITO, K. A. and CHEN, J. K. 1978. Effect of pH on growth of *Clostridium botulinum* in foods. Food Technol. *32*: 71–76.

ITO, K. A., CHEN, J. K., SEEGER, M. L., UNVERFERTH, J. A. and KIMBALL, R. N. 1978. Effect of pH on the growth of *Clostridium botulinum* in canned figs. J. Food Sci., *43*: 1634–1635.

IVEY, F. J., SHAVER, K. J., CHRISTIANSEN, L. N. and TOMPKIN, R. B. 1978. Effect of potassium sorbate on toxinogenesis by *Clostridium botulinum* in bacon. J. Food Protec. *41*: 621–625.

JANSEN, B. C. 1971. The toxic antigenic factors produced by *Clostridium botulinum* types C and D. Onderstepoort J. Vet. Res. *38*: 93–98.

JENSEN, L. B. 1943. Action of hardwood smoke on bacteria in cured meats. Food Res. *8*: 377–387.

JOHANNSEN, A. 1963. *Clostridium botulinum* in Sweden and the adjacent waters. J. Appl. Bacteriol. *26*: 43–47.

JOHANNSEN, A. 1965. *Clostridium botulinum* type E in foods and the environment generally. J. Appl. Microbiol. *28*: 90–94.

JOHNSTON, R. W., FELDMAN, J. and SULLIVAN, R. 1963. Botulism from canned tuna fish. Publ. Health Rep. *78*: 561–564.

JOHNSTON, M. A. and LOYNES, R. 1971. Inhibition of *Clostridium botulinum* by sodium nitrite as affected by bacteriological media and meat suspensions. Can. Inst. Food Technol. J. *4*: 179–184.

JOHNSTON, M. A., PIVNICK, H. and SAMSON, J. M. 1969. Inhibition of *Clostridium botulinum* by sodium nitrite in a bacteriological medium and in meat. Can. Inst. Food Technol. J. *2*: 52–55.

KANASAWA, K., ONO, T., KARASHIMADA, T. and IIDA, H. 1968. Distribution of *Clostridium botulinum* type E in Hokkaido, Japan. *In* M. Herzberg (ed): Toxic Microorganisms. Proc. 1st. U.S.-Japan Conf., Honolulu, Hawaii, pp. 299–303.

KAUTTER, D. A. 1964. *Clostridium botulinum* type E in smoked fish. J. Food Sci. *29*: 843–849.

KAUTTER, D. A. and LILLY, T. 1970. Detection of *Clostridium botulinum* and its toxin in food. I. Detection of *Clostridium botulinum* type E in smoked fish. J.A.O.A.C. *53*: 710–712.

KAUTTER, D. A., LILLY, T. and LYNT, R. 1978. Evaluation of the botulism hazard in fresh mushrooms wrapped in commercial polyvinyl-chloride film. J. Food Protec. *41*: 120–121.

KEMP, J. D. 1974. Nitrate and nitrite substitutes in meat curing. Food Prod. Dev. *8*: 64–70.

KERNER, J. 1817. Vergiftung durch verdorbene Würste. Tübinger Blätter für Naturwissenschaften und Arzneykunde *3*: 1–25.

KERNER, J. 1820. Neue Beobachtungen über die in Würtemberg so häufig vorfallenden tödtlichen Vergiftungen durch den Genuss geräucherter Würste. C. F. Osiander, Tübingen, Germany.

KOSIKOWSKI, F. V. 1977. Cheese and Fermented Milk Foods, 2nd. ed. Edward Brothers Inc., Ann Arbor, Michigan.

KRAVCHENKO, A. T. and SHISHULINA, L. M. 1967. Distribution of *Cl. botulinum* in soil and water in the U.S.S.R. *In* M. Ingram and T. A. Roberts (Ed): Botulism 1966. Chapman & Hall Ltd., London; pp. 13–19.

KUEPER, T. V. and TRELEASE, R. D. 1974. Variables affecting botulinum toxin development and nitrosamine formation in fermented sausages. Proc. Meat Indust. Res. Conf., American Meat Institute Foundation, Washington; pp. 69–74.

LANDMANN, G. 1904. Ueber die Ursache der Darmstädter Bohnenvergif-tung. Hyg. Rundschau 14: 449–452.

LAYCOCK, R. A. and LONGARD, A. A. 1972. *Clostridium botulinum* in sediments from the Canadian Atlantic seaboard. J. Fish. Res. Bd. Can. 29: 443–446.

LAYCOCK, R. A. and LORING, D. H. 1972. Distribution of *Clostridium botulinum* type E in the Gulf of St. Lawrence in relation to the physical environment. Can. J. Microbiol. *18*: 763–773.

LECHOWICH, R. V., BROWN, W. L., DEIBEL, R. H. and SOMERS, I. I. 1978. The role of nitrite in the production of canned cured meat products. Food Technol. *32*: 45–58.

LEE, S. H., CASSENS, R. G. and SUGIYAMA, H. 1978. Factors affecting inhibition of *Clostridium botulinum* in cured meats. J. Food Sci. *43*: 1371–1374.

LEISTNER, L., HERZOG, H. and WIRTH, F. 1971. Untersuchungen über die Wasseraktivität (a_w – Wert) von Rohwurst. Fleischwirtsch. *51*: 213–216.

LEISTNER, L. and RÖDEL, W. 1976. Inhibition of microorganisms in food by water activity. In F. A. Skinner (Ed): Inhibition and Inactivation of Vegetative Microbes. Academic Press, London; pp. 219–237.

LYNT, R. K., KAUTTER, D. A. and READ, R. B. 1975. Botulism in commercially canned foods. J. Milk Food Technol. 38: 546–550.

LYNT, R. K., SOLOMON, H. M., LILLY, T. and KAUTTER, D. A. 1977. Thermal death time of Clostridium botulinum type E in meat of the blue crab. J. Food Sci. 42: 1022–1025.

MATVEEV, K. I., NEFEDJEVA, N. P., BULATOVA, T. I. and SOKO-LOV, I. S. 1967. Epidemiology of botulism in the U.S.S.R. In M. Ingram and T. A. Roberts (Ed): Botulism 1966. Chapman & Hall Ltd., London; pp. 1–10.

MERSON, M. H. and DOWELL, V. R. 1973. The epidemiologic, clinical and laboratory aspects of wound infection. New Engl. J. Med. 289: 1005–1010.

MERSON, M. H., HUGHES, J. M., DOWELL, V. R., TAYLOR, A., BARKER, W. H. and GANGAROSA, E. J. 1974. Current trends in botulism in the United States. J. Am. Med. Assoc. 229: 1305–1308.

MEYER, K. F. 1928. Botulismus. In Handbuch per Pathogenen Mikro-organismen, 3rd. Ed., Vol. 4, Part 2. Fischer and Schwarzenberg, Jena; pp. 1269–1364.

MEYER, K. F. 1956. The status of botulism as a world health problem. Bull. World Hlth. Org. 15: 281–298.

MEYER, K. F. and EDDIE, B. 1951. Perspectives concerning botulism. Zeitschr. Hyg. 133: 255–263.

MEYER, K. F. and EDDIE, B. 1953. Clostridium botulinum type C and human botulism. Att. del 6. Congr. Intern. di Microbiol., Rome. 2: 123–124.

MEYER, K. F. and EDDIE, B. 1965. Sixty-five years of human botulism in the United States and Canada. Univ. of California Printing Dept., San Francisco, Calif.

MIDURA, T. F., NYGAARD, G. S., WOOD, R. M. and BODILY, H. L. 1972. Clostridium botulinum type F: isolation from venison jerky. Appl. Microbiol. 24: 165–167.

MILLER, L. G., CLARK, P. S. and KUNKLE, G. A. 1972. Possible origin of Clostridium botulinum contamination of Eskimo foods in Northwestern Alaska. Appl. Microbiol. 23: 427–428.

MINISTRY OF AGRICULTURE AND FOOD OF ONTARIO. Home Canning Ontario Fruits and Vegetables. Publ. No. 468. Toronto, Ontario.

MIRVISH, S. S., WALLCAVE, L., EAGEN, M. and SHUBIK, P. 1972. Ascorbate-nitrite reaction: possible means of blocking the formation of carcinogenic N-nitroso compounds. Science 177: 65–68.

MOELLER, V. and SCHEIBEL, I. 1960. Preliminary report on the isolation of an apparently new type of Cl. botulinum. Acta Pathol. Microbiol. Scand. 48: 80.

MOTTRAM, D. S., PATTERSON, R. L. S., RHODES, D. N. and GOUGH, T. A. 1975. Influence of ascorbic acid and pH on the formation of N-nitrosodimethylamine in cured pork containing added dimethylamine. J. Food Sci. Agric. *26*: 47–53.

NAKAMURA, Y., IIDA, H., SAEKI, K., KANZAWA, K. and KARASHI-MADA, T. 1956. Type E botulism in Hokkaido, Japan. Jap. J. Med. Sci. Biol. *9*: 45–58.

NAKANO, W. and KODAMA, E. 1968. On the reality of "izushi", the causal food of botulism, and on its folkloric meaning. *In* M. Herzberg (Ed): Toxic Microorganisms. Proc. 1st. U.S.-Japan Conf., Honolulu, Hawaii; pp. 388–392.

NATH, N. and RANGANNA, S. 1977. Evaluation of the thermal process for acidified canned muskmelon. J. Food Sci. *42*: 1306–1310.

NEER, K. L. and MANDIGO, R. W. 1977. Effects of salt, sodium tripolyphosphate and frozen storage time on properties of a flaked, cured pork product. J. Food Sci. *42*: 738–742.

NICKERSON, J. T. R., GOLDBLITH, S. A., DiGIOIA, G. and BISHOP, W. W. 1967. The presence of *Cl. botulinum*, type E in fish and mud taken from the Gulf of Maine. *In* M. Ingram and T. A. Roberts (Ed): Botulism 1966. Chapman & Hall Ltd., London; pp. 25–33.

NIELSEN, S. F. and PEDERSEN, H. O. 1967. Studies on the occurrence and germination of *Cl. botulinum* in smoked salmon. *In* M. Ingram and T. A. Roberts (Ed): Botulism 1966. Chapman & Hall Ltd., London; pp. 66–72.

NIINIVARA, F. P. and POHJA, M. S. 1956. Über die Reifung der Rohwurst. I. Mitteilung. Die Veränderungen der Bakterienflora während der Reife. Z. Lebensmittelunters. Forsch. *104*: 413–422.

NORDIN, H. R. 1969. The depletion of added sodium nitrite in ham. Can. Inst. Food Technol. J. *2*: 79–85.

NORDIN, H. R., BURKE, T., WEBB, G., RUBIN, L. J. and VAN BIN-NENDYK, D. 1975. Effect of pH, salt and nitrite in heat processed meat on destruction and out-growth of P.A. 3679. Can. Inst. Food Sci. Technol. J. *8*: 58–66.

ODLAUG, T. E. and PFLUG, I. J. 1978. *Clostridium botulinum* and acid foods. J. Food Protec. *41*: 566–573.

OHYE, D. F. and CHRISTIAN, J. H. B. 1967. Combined effects of temperature, pH and water activity on growth and toxin production by *Cl. botulinum* types A, B and E. *In* M. Ingram and T. A. Roberts (Ed): Botulism 1966. Chapman & Hall Ltd., London; pp. 217–223.

OHYE, D. F., CHRISTIAN, J. H. B. and SCOTT, W. J. 1967. Influence of temperature on the water relations of growth of *Cl. botulinum* type E. *In* M. Ingram and T. A. Roberts (Ed): Botulism 1966. Chapman & Hall Ltd., London; pp. 136–143.

OLSON, J. C. 1968. U.S. regulatory administration for control of microbiological health hazards in foods. *In* M. Herzberg (Ed): Toxic Microorganisms. Proc. 1st. U.S.-Japan Conf., Honolulu, Hawaii; pp. 398–403.

PACE, P. J. and KRUMBIEGEL, E. R. 1973. *Clostridium botulinum* and smoked fish production: 1963–1972. J. Milk Food Technol. *36*: 42–49.

PACE, P. J., KRUMBIEGEL, E. R., ANGELOTTI, R. and WISNIEWSKI,

H. J. 1967. Demonstration and isolation of *Clostridium botulinum* types from whitefish chubs collected at fish smoking plants of the Milwaukee area. Appl. Microbiol. *15*: 877–884.

PERIGO, J. A. and ROBERTS, T. A. 1968. Inhibition of clostridia by nitrite. J. Food Technol. *3*: 91–94.

PERIGO, J. A., WHITING, E. and BASHFORD, T. E. 1967. Observations on the inhibition of vegetative cells of *Clostridium sporogenes* by nitrite which has been autoclaved in a laboratory medium discussed in the context of sub-lethally processed cured meats. J. Food Technol. *2*: 377–397.

PIVNICK, H. and BARNETT, H. 1965. Effect of salt and temperature on toxinogenesis by *Clostridium botulinum* in perishable cooked meats vacuum-packed in air-impermeable plastic pouches. Food Technol. *19*: 140–143.

PIVNICK, H., BARNETT, H. W., NORDIN, H. R. and RUBIN, L. J. 1969. Factors affecting the safety of canned, cured, shelf-stable luncheon meat inoculated with *Clostridium botulinum*. Can. Inst. Food Technol. J. *2*: 141–148.

PIVNICK, H. and BIRD, H. 1965. Toxinogenesis by *Clostridium botulinum* types A and E in perishable cooked meats vacuum-packed in plastic pouches. Food Technol. *19*: 132–140.

PIVNICK, H. and CHANG, P.-C. 1973. Perigo effect in pork. Proc. Int. Symp. Nitrite Meat Prod., Zeist. Pudoc, Wageningen; pp. 111–116.

PIVNICK, H., JOHNSTON, M. A., THACKER, C. and LOYNES, R. 1970. Effect of nitrite on destruction and germination of *Clostridium botulinum* and putrefactive anaerobes 3679 and 3679h in meat and in buffer. Can. Inst. Food Technol. J. *3*: 103–109.

PIVNICK, H. and PETRASOVITS, A. 1973. A rationale for the safety of canned shelf-stable cured meat. Protection = Destruction + Inhibition. 19 Reunion Europ. Cherch. Viande. Paris, France; pp. 1086–1094.

PIVNICK, H., RUBIN, L. J., BARNETT, H. W., NORDIN, H. R., FERGUSON, P. A. and PERRIN, C. H. 1967. Effect of sodium nitrite and temperature on toxinogenesis by *Clostridium botulinum* in perishable cooked meats vacuum-packed in air-impermeable pouches. Food Technol. *21*: 100–102.

PIVNICK, H. and THACKER, C. 1970. Effect of sodium chloride and pH on initiation of growth by heat-damaged spores of *Clostridium botulinum*. Can. Inst. Food Technol. J. *3*: 70–75.

POLYMENIDES, A. 1977. Räuchern von Fleisch und Fleischerzeugnissen. Fleischwirtsch. *57*: 1787–1793.

POURTAGHVA, M., MACHOUN, A., FATOLLAH-ZADEH, KHODADOUST, A., TAEB, H., FARZAM, H. and FARHANGUI. 1975. Le botulisme en Iran. Med. Malad. Infect. *5*: 536–539.

PRÉVOT, A.-R., TERRASSE, J., DAUMAIL, J., CAVAROC, M., RIOL, J. and SILLIOC, R. 1955. Existence en France du botulisme humain de type C. Bull. Acad. Nat. Méd. *139*: 355–358.

RIEMANN, H. 1963. Safe heat processing of canned cured meats with regard to bacterial spores. Food Technol. *17*: 39–49.

RIEMANN, H. 1966. Botulism, its dangers and prevention in the meat industry. Proc. Res. Conf. AMIF, Chicago. pp. 27–41.

RIEMANN, H., LEE, W. H. and GENIGEORGIS, C. 1972. Control of *Clostridium botulinum* and *Staphylococcus aureus* in semi-preserved meat products. J. Milk Food Technol. *35*: 514–523.

ROBERTS, T. A. 1973. Inhibition of bacterial growth in model systems in relation to the stability and safety of cured meats. Proc. Int. Symp. Nitrite Meat Prod., Zeist. Pudoc, Wageningen; pp. 91–101.

ROBERTS, T. A. 1975. The microbiological role of nitrite and nitrate. J. Sci. Food Agric. *26*: 1755–1760.

ROBERTS, T. A., GILBERT, R. J. and INGRAM, M. 1966. The effect of sodium chloride on heat resistance and recovery of heated spores of *Clostridium sporogenes* (PA 3679/S$_2$). J. Appl. Bacteriol. *29*: 549–555.

ROBERTS, T. A. and HOBBS, G. 1968. Low temperature growth characteristics of clostridia. J. Appl. Bacteriol. *31*: 75–88.

ROBERTS, T. A. and INGRAM, M. 1973. Inhibition of growth of *Cl. botulinum* at different pH values by sodium chloride and sodium nitrite. J. Food Technol. *8*: 467–475.

ROBERTS, T. A. and INGRAM. M. 1976. Nitrite and nitrate in the control of *Clostridium botulinum* in cured meats. Proc. Int. Symp. Nitrite Meat Prod., Zeist. Pudoc, Wageningen; pp. 28–38.

ROBERTS, T. A., JARVIS, B. and RHODES, A. C. 1976. Inhibition of *Clostridium botulinum* by curing salts in pasteurized pork slurry. J. Food Technol. *11*: 25–40.

ROBERTS, T. A. and SMART, J. L. 1976. The occurrence and growth of *Clostridium* spp. in vacuum-packed bacon with particular reference to *Cl. perfringens* (welchii) and *Cl. botulinum*. J. Food Technol. *11*: 229–244.

ROBERTS, T. A. and SMART, J. L. 1977. The occurrence of clostridia, particularly *Clostridium botulinum*, in bacon and pork. *In* A. N. Barker, J. Wolf, D. J. Ellar, G. J. Dring and G. W. Gould (Ed): Spore Research 1976 II. Academic Press, New York; pp. 911–915.

RÖDEL, W., PONERT, H. and LEISTNER, L. 1976. Einstufung von Fleischerzeugnissen in leicht verderbliche, verderbliche und lagerfähige Produkte. Fleischwirtsch. *56*: 417–418.

ROUHALA, L. M. and CLEGG, L. F. L. 1978. A model system for testing the microbiological stability of foods processed in laminated flexible pouches. J. Appl. Bacteriol. *44*: 75–90.

ROUHBAKHSH-KHALEGHDOUST, A. 1973. The incidence of *Clostridium botulinum* type E in fish and bottom deposits in Caspian coastal waters. Iran J. Publ. Health *1*: 153–154.

ROUHBAKHSH-KHALEGHDOUST, A. and PORTAGHVA, M. 1977. A large outbreak of type E botulism in Iran. Trans. Royal Soc. Trop. Med. Hyg. *71*: 444.

RUST, R. E. and OLSON, D. G. 1973. Meat Curing Principles and Modern Practice. Koch Supplies Inc., Kansas City, Mo.

RYMKIEWICZ, D. 1970. Studies on type F *Clostridium botulinum*. Exp. Med. Microbiol. *22*: 13–20.

SAKAGUCHI, G. 1969. Botulism-type E. *In* H. Riemann (Ed): Food-borne

Infections and Intoxications. Academic Press, New York. pp. 329–358.

SAPERS, G. M., PHILLIPS, J. G., TALLEY, F. B., PANASIUK, O. and CARRE, J. 1978. Acidulation of home canned tomatoes. J. Food Sci. *43*: 1049–1052.

SCHMIDT, C. F., LECHOWICH, R. V. and FOLINAZZO, J. F. 1961. Growth and toxin production by type E *Clostridium botulinum* below 40°F. J. Food Sci. *26*: 626–630.

SEBALD, M. 1970. Sur le botulisme en France de 1956 à 1970. Bull. Acad. Nat. Méd. *154*: 703–707.

SILLIKER, J. H., GREENBERG, R. A. and SCHACK, W. R. 1958. Effect of individual curing ingredients on the shelf stability of canned comminuted meats. Food Technol. *12*: 551–554.

SOLOMON, H. M., LYNT, R. K., LILLY, T. and KAUTTER, D. A. 1977. Effect of low temperatures on growth of *Clostridium botulinum* spores in meat of the blue crab. J. Food Protec. *40*: 5–7.

SMITH, L. DS. 1977. Botulism. The Organism, its Toxins, the Disease. C. C. Thomas, Publ., Springfield, Illinois.

SMITH, L. DS. 1978. The occurrence of *Clostridium botulinum* and *Clostridium tetani* in the soil of the United States. Health Lab. Sci. *15*: 74–80.

SMITH, L. DS. and HOLDEMAN, L. V. 1968. The Pathogenic Anaerobic Bacteria. C. C. Thomas, Publ., Springfield, Illinois.

STUART, P. F., WIEBE, E. J., McELROY, R., CAMERON, D. G., TODD, E. C. D., ERDMAN, I. E., ALBALAS, B. and PIVNICK, H. 1970. Botulism among Cape Dorset eskimos and suspected botulism at Frobisher Bay and Wakeham Bay. Can J. Publ. Health *61*: 509–517.

STUMBO, C. R. 1973. Thermobacteriology in Food Processing, 2nd. ed. Academic Press, New York.

SUGIYAMA, H., MILLS, D. C. and KUO, L.-J. C. 1978. Number of *Clostridium botulinum* spores in honey. J. Food Protec. *41*: 848–850.

SUGIYAMA, H., MIZUTANI, K. and YANG, K. H. 1972. Basis of type A and F toxicities of *Clostridium botulinum* strain 84 (36933). Proc. Soc. Exp. Biol. Med. *141*: 1063–1067.

SUGIYAMA, H. and RUTLEDGE, K. S. 1978. Failure of *Clostridium botulinum* to grow in fresh mushrooms packaged in plastic film overwraps with holes. J. Food Protec. *41*: 348–350.

SUGIYAMA, H. and YANG, K. H. 1975. Growth potential of *Clostridium botulinum* in fresh mushrooms packaged in semipermeable plastic film. Appl. Microbiol. *30*: 964–969.

TACLINDO, C., MIDURA, T., NYGAARD, G. S. and BODILY, H. L. 1967. Examination of prepared foods in plastic packages for *Clostridium botulinum*. Appl. Microbiol. *15*: 426–430.

THATCHER, F. S., ERDMAN, I. E. and PONTEFRACT, R. D. 1967. Some laboratory and regulatory aspects of the control of *Cl. botulinum* in processed foods. *In* M. Ingram and T. A. Roberts (Ed): Botulism 1966. Chapman & Hall Ltd., London; pp. 511–521.

THATCHER, F. S., ROBINSON, J. and ERDMAN, I. 1962. The "vacuum pack" method of packaging foods in relation to the formation of the botulinum and staphylococcal toxins. J. Appl. Bacteriol. *25*: 120–124.

TODD, E., CHANG, P. C., HAUSCHILD, A., SHARPE, A., PARK, C. and PIVNICK, H. 1974. Botulism from marinated mushrooms. Proc. IV Int. Congr. Food Sci. Technol. *3*: 182–188.

TOMPKIN, R. B., CHRISTIANSEN, L. N. and SHAPARIS, A. B. 1977. Variation in inhibition of *C. botulinum* by nitrite in perishable canned comminuted cured meat. J. Food Sci. *42*: 1046–1048.

TOMPKIN, R. B., CHRISTIANSEN, L. N. and SHAPARIS, A. B. 1978. Antibotulinal role of isoascorbate in cured meat. J. Food Sci. *43*: 1368–1370.

TOMPKIN, R. B., CHRISTIANSEN, L. N. and SHAPARIS, A. B. 1978. The effect of iron on botulinal inhibition in perishable canned cured meat. J. Food Technol. *13*: 521–527.

TOMPKIN, R. B., CHRISTIANSEN, L. N. and SHAPARIS, A. B. 1978. Enhancing nitrite inhibition of *Clostridium botulinum* with isoascorbate in perishable canned cured meat. Appl. Env. Microbiol. *35*: 59–61.

TOMPKIN, R. B., CHRISTIANSEN, L. N. and SHAPARIS, A. B. 1978. Effect of prior refrigeration on botulinal outgrowth in perishable canned cured meat when temperature abused. Appl. Env. Microbiol. *35*: 863–866.

TOMPKIN, R. B., CHRISTIANSEN, L. N. and SHAPARIS, A. B. 1978. Causes of variation in botulinal inhibition in perishable canned cured meat. Appl. Env. Microbiol. *35*: 886–889.

TOMPKIN, R. B., CHRISTIANSEN, L. N. and SHAPARIS, A. B. 1979. Iron and the antibotulinal efficacy of nitrite. Appl. Env. Microbiol. *37*: 351–353.

TOWNSEND, C. T., YEE, L. and MERCER, W. A. 1954. Inhibition of the growth of *Clostridium botulinum* by acidification. Food Res. *19*: 536–542.

TROLLER, J. A. 1973. The water relations of food-borne bacterial pathogens. A review. J. Milk Food Technol. *36*: 276–288.

VAN ERMENGEM, E. 1897. Ueber einen neuen anaëroben Bacillus und seine Beziehungen zum Botulismus. Z. Hyg. *26*: 1–56.

WOLFF, I. A. and Wasserman, A. E. 1972. Nitrates, nitrites, and nitrosamines. Science *177*: 15–19.

4

Staphylococcal Food Poisoning

Merlin S. Bergdoll

One of the more unpleasant experiences of man is staphylococcal poisoning. This disease results from the consumption of food in which certain strains of *Staphylococcus aureus* have grown and produced a substance known as enterotoxin. Symptoms usually appear about two or three hours after eating, with salivation followed by nasuea, vomiting, abdominal cramping and diarrhea. Prostration may occur in severe cases. Most patients return to normal in a day or two, and death is rare. The time of onset and the severity of symptoms depend on the amount of enterotoxin consumed as well as individual susceptibility to the enterotoxin. It is not uncommon for some individuals to escape illness in outbreaks involving several people.

Staphylococcal food poisoning is not a reportable disease and its true incidence is unknown. Owing to the relatively short duration, most cases are never seen by a physician and go unrecognized. Only those outbreaks involving large numbers of people, as may occur at picnics, group dinners or public institutions are likely to come to the attention of public health authorities, who then may try to establish the cause of illness.

Most of the foods incriminated in staphylococcal poisoning have been heated to some degree and then recontaminated with *S. aureus*. Thus baked ham, roast fowl, potato salad, chicken salad, custards and cream-filled bakery products are common vehicles. Dried milk and cheese have been incriminated on occasions.

THE CAUSE OF STAPHYLOCOCCAL FOOD POISONING.

Staphylococcus aureus is a common inhabitant of the animal body. It causes a variety of infections of man ranging in severity from

pimples, boils and abscesses to osteomyelitis, pneumonia and meningitis. It is also a frequent cause of mastitis in cattle. Surveys have
shown that 30 to 50% of the human population carry S. aureus in the
nose and throat; thus, there is ample opportunity for contamination
of food, although the chances are greatest with food handlers who
have skin infections on the hands, arms or face.

Research on staphylococcal food poisoning was hampered for
many years by lack of a suitable laboratory test for the enterotoxins.
Excepting man, the only satisfactory test animals for the toxin are
monkeys and cats. The usual laboratory animals such as mice,
guinea pigs and rabbits do not have a vomiting mechanism. Monkeys are expensive and soon become refractory to the enterotoxins.
Cats often show nonspecific vomiting, especially with certain food
extracts. The use of human volunteers is practically impossible
under present day restrictions, hence, it is necessary to use either
cats or monkeys when working with unidentified enterotoxins.

Much effort has been and still is being expended on the development of methods for the detection of the enterotoxins. Although
attempts have been made to relate enterotoxin to some biological
reaction of the staphylococci the only usable methods for the detection of the enterotoxins are based on their reactions with specific
antibodies for each of the five identified enterotoxins [enterotoxins
A, B, C, D, and E (SEA, etc.)]. These methods will be discussed later
but none of them are necessarily easy to perform, and as a result,
investigators are constantly seeking markers which can be identified with enterotoxin production. Even though the staphylococci
produce a variety of lethal, hemolytic or dermonecrotic toxins, so
far, no correlation of these with enterotoxin production has been
discovered. Differences in susceptibility to bacteriophage have made
it possible to develop a staphylococcal typing system (Blair and
Williams, 1961). It has been shown that certain phage types usually
are associated with hospital-acquired infections while other phage
types are the main causes of bovine mastitis. Attempts have been
made to associate phage type with enterotoxigenicity but no such
correlation exists as S. aureus strains in each phage group have
been shown to be enterotoxigenic. It is true that a majority of the
strains involved in food poisoning belong to phage group III but this
is not a usable criterion for assessing the importance of any given S.
aureus strain. Phage-typing is a useful technique for establishing
the source of a food poisoning outbreak. It is often possible, for
example, to demonstrate staphylococci of the same phage type in an
incriminated food and in a food handler who prepared the food.

Coagulase production has been used over the years for assessing
the importance of staphylococcal strains as early studies indicated

all enterotoxigenic strains were coagulase-positive. An exception to this generalization was reported by Bergdoll *et al.* in 1967 when they discovered that some enterotoxigenic strains were coagulase-negative. Because such strains are relatively rare, the coagulase test is still used to determine whether a strain should be examined for enterotoxin production. Heat stable nuclease (DNase) is also an important characteristic of *S. aureus* and can be used in the place of coagulase to determine the importance of staphylococci as possible enterotoxin producers. While practically all enterotoxigenic *S. aureus* strains produce coagulase, DNase or both, not all *S. aureus* strains produce enterotoxin. The percentage that do depends to a certain extent on the source of the organisms, as a higher percentage of those isolated from human sources produce enterotoxin than those isolated from animal sources. Overall, probably less than 50% of *S. aureus* strains are enterotoxigenic.

DNase has become a tool in determining whether staphylococci have grown in a particular food product, for example, in milk before drying, a process which usually kills these organisms. Normally, but not always, DNase will be found in a food product in which staphylococci have grown. There are exceptions, because it is not produced in at least one or two foods, and a few strains are DNase-negative, but it is recommended where other tests are not avilable. DNase is relatively easy to detect (Tatini *et al.*, 1976) and a number of studies have been done on its detection in foods.

The staphylococci produce a number of enterotoxins, six of which have been identified, purified and characterized (SEA, SEB, SEC$_1$, SEC$_2$, SED and SEE) (Bergdoll, 1979). They are identified on the basis of the production of specific antibodies (Casman *et al.*, 1963), each of which, with the exception of SEC$_1$ and SEC$_2$, produce a major antibody which does not cross-react with the other enterotoxins. In the case of the two SEC's, they both react with the same major antibody but each has a second antibody which reacts only with its homologous toxin. There are unidentified enterotoxins (how many is not known) but they appear to be of minor importance in food poisoning, causing less than 5% of the outbreaks. Enterotoxin A is the type most commonly implicated in food poisoning outbreaks, with SED next in order of frequency (Casman *et al.*, 1967). Individual cultures of *S. aureus* may produce either a single enterotoxin or a mixture. In a survey of staphylococci isolated from 75 different food poisoning incidents, Casman *et al.* (1967) found that 49% produced SEA alone, and 29% produced a mixture of SEA and some other type, usually SED. Ten percent of the cultures produced SED alone. Only four percent produced SEB and a similar number SEC. Results from other sources indicate that SEC is a more frequent cause of food poisoning than was indicated above.

Identification of new enterotoxins is not being pursued at the present time because it is time consuming and expensive, primarily due to the need for animal testing. Each of the enterotoxins produce minor antibodies which in some cases cross-react with some of the other enterotoxins. It is possible that these antibodies may be used to detect more than one of the identified enterotoxins as well as some of the unidentified ones. This would be an important development in the enterotoxin field.

Physicochemical studies of the enterotoxins show them to be simple proteins with molecular weights of 25,000 to 30,000. Their isoelectric points vary from pH 7.0 to 8.6 with some variation in each of the enterotoxins because of a difference in the number of amide groups in the same enterotoxin. The amino acid composition of the enterotoxins are quite similar. However, on the basis of the differences in their composition they can be divided roughly into two groups, with SEA, SED and SEE in one group and SEB and the SEC's in the other group. This is reflected also in the cross-reacting antibodies produced by the enterotoxins, because there are cross-reactions between SEA, SED and SEE and cross-reactions between SEB and the SEC's but apparently none between the two groups.

Each of the enterotoxins has one cystine residue which forms a loop (cystine loop) of 20 amino acid residues near the center of the single chain molecule. The amino acid sequence of the cystine loop in the different enterotoxins is not the same but a sequence of seven residues which includes the half-cystine in the C-terminal end of the loop is the same for SEA and SEB. It is expected that this is a common area to all of the enterotoxins and is the toxic site. Confirmatory evidence is not yet available. The conclusion that the toxic site for the enterotoxins is the same is based primarily on the fact that the action of the enterotoxins is identical.

The amount of enterotoxin necessary to cause illness in humans is not known but evidence from food poisoning outbreaks indicate that as little as 0.1 μg or less might be sufficient for some individuals. One study with human volunteers (Dangerfield, 1973) required much larger amounts (3.5 μg) to produce illness. The results are not indicative of what might happen in a food poisoning outbreak, because a selected population was used in the study. The emetic dose for rhesus monkeys is reported to be about 1 μg/kg for both SEA and SEB (Schantz et al., 1965; Schantz et al., 1972) although the results from feedings in the Food Research Institute indicate a higher level, 5 to 10 μg per animal (approximately 3 kg) except for SED which, in a limited number of feedings, required 15 to 20 μg.

The pharmacological action of the enterotoxins is not clearly understood, even though much work has been done in the area. It is concluded that the main observable symptoms, nausea, abdominal pain, vomiting and diarrhea, are due to the direct action of the enterotoxin in the abdominal viscera (Sugiyama and Hayama, 1965). It differs from some of the other food-borne illnesses in that the toxin involved does not give a positive ileal loop test. The site of action has not been determined but it has been shown that enterotoxin can cause enterocolitis in man and certain animals. A number of other effects have been noted when the enterotoxin was injected into monkeys intravenously in relatively large doses (Bergdoll, 1972). While these effects could be important, they probably are inconsequential in food poisoning because of the small amounts of toxin usually ingested in the latter case.

The enterotoxins are much more stable in many respects than most proteins. In the active state, they are resistant to proteolytic enzymes such as trypsin, chymotrypsin, rennin and papain. Pepsin destroys the activity of SEB at a pH of 2.5 or below (Bergdoll, 1970; Schantz et al., 1965) but is ineffective at higher pH values. This resistance to proteolytic enzymes may be the explanation why the toxins are active when ingested. Pepsin would not inactivate them because the pH of the stomach is usually higher than 2.5 after the ingestion of food.

The enterotoxins are considered to be quite heat stable because their activity in humans (Jordan et al., 1931), in monkeys (Davidson et al., 1938), and in kittens (Dolman and Wilson, 1938) was only partially destroyed when solutions of the crude toxin were boiled for 30 minutes. Since this earlier work, a number of studies have been conducted to obtain more specific information about the heat stability of the enterotoxins. Most of the results are reported in terms of loss of reaction with the specific antibodies, primarily with the single gel diffusion tube method which has a minimal sensitivity of 1 μg/ml. This method is useful in determining the thermal inactivation curves (Humber et al., 1975) but it is not possible to determine the small amount of enterotoxin that may remain and is adequate to cause food poisoning. A review of this work is included in a chapter on staphylococcal intoxication by Bergdoll (1979) and also was reviewed by Tatini (1976). In general, all information available indicates that enterotoxin in food is not easily inactivated by heat and the larger the amount present the more heat is required to reduce the quantity to undetectable levels. The higher the temperature the more rapidly the enterotoxin is denatured, with the times and temperatures used in normal processing of canned foods (for example, 118°C for 86 minutes) sufficient to destroy the quantity of enterotoxin present in foods involved in food poisoning outbreaks (0.5 to 10 µg per 100 grams of food). Ordinary cooking procedures, pasteurization and drying are of too short duration or done at too low temperatures to inactivate the toxin. Baking of foods which is done at higher temperatures and for longer periods of time usually will result in the destruction of the toxin. If relatively large amounts of toxin are present in the food from mishandling before processing, it is possible that not all of the toxin would be completely inactivated by any of the heat treatments that might be applied.

PREVENTION OF STAPHYLOCOCCAL FOOD POISONING.

Three conditions must be met for staphylococcal food poisoning to occur: (1) enterotoxigenic staphylococci must be present; (2) the food must support growth and toxin production; and (3) the food must remain at a suitable temperature for sufficient time.

1. Contamination with enterotoxigenic staphylococci.—Food handlers are the most likely sources of contamination. Especially hazardous are personnel with suppurating infections, such as acne or boils. Contamination may also result from coughing or sneezing. Milk from mastitic udders may carry enterotoxigenic staphylococci.

S. *aureus* is not usually resistant to heat. Ordinary cooking or pasteurization (as applied to milk) will give effective protection. Zottola *et al.* (1965) killed 97% of 226 strains of S. *aureus* by heating at 152°F for 21 seconds. Walker and Harmon (1966) achieved 99.9% destruction by heating S. *aureus* in whole milk at 140°F for 12 minutes. Thus, the control of contamination is largely a matter of sanitation. Special care must be taken with cooked or pasteurized foods.

2. Composition of food.—Many foods will support growth and toxin formation by S. *aureus*. Notable exceptions are foods with pH below 5.0 or equilibrium relative humidity below 86%. S. *aureus* is not a strong competitor and its growth is suppressed by other microorganisms in most raw or fermented foods. For example, S. *aureus* has been found to grow and produce enterotoxin in meat only if the meat were collected aseptically or if it were cooked before inoculation (Casman *et al.*, 1963). Likewise, enterotoxin is apt to form in cheese only if the starter culture is inhibited by bacteriophage attack or bacteriostatic chemicals. Thus, staphylococcal food poisoning is usually associated with foods that have been heated to destroy the natural microflora and then recontaminated with enterotoxigenic staphylococci.

3. Temperature and time for growth.—The temperature range for growth of *S. aureus* is approximately 5° to 45°C, with best growth at 35° to 37°C. The rate of multiplication naturally declines as the temperature approaches the minimum. For example, Walker and Harmon (1965) observed these generation times in milk: 30° to 37°C, 0.5 to 1 hr.; 21°C, 1.7 to 2 hr.; and 10°C, 18 to 28 hr. Several studies have been made on the effect of temperature on the production of enterotoxins. Production does occur over a range of temperatures, 10° to 45°C, with some variation within the enterotoxin types. Maximum production appears to occur around 40°C for most of the enterotoxins with less produced at higher and lower temperatures, with the amount decreasing as the temperature is decreased. Because of the slower growth at the lower temperatures a longer time is required for the enterotoxin to be produced. The time required for enterotoxin to be produced in a food under ideal conditions is not known specifically but natural outbreaks have occurred in which the incubation time was no more than 4 to 5 hr. (Dack, 1956). Little information is available from laboratory studies concerning this matter but Ghosh and Laxminarayana (1973) were able to detect enterotoxin in milk after 3 to 6 hr. of incubation at 35°C when a large inoculum was used. Donnelly *et al.*, (1968) reported detecting enterotoxin in milk after incubation at 35°C for 6 to 9 hr.

The question about the relationship of numbers of staphylococci and the production of enterotoxin often arises. In general, one should be concerned when around 10^6 organisms per gram of food are present. In some instances enterotoxin may be detectable with smaller numbers but this is relatively rare, except in cases where the staphylococci were killed during processing such as in drying operations or where the organisms die off during storage such as in the case of cheese. The number of organisms associated with enterotoxin production varies with the food product and the conditions under which the food is processed and subsequently held. In the case of cheese, enterotoxin production was associated with anywhere from 7 to 50 million organisms per gram depending on the type of cheese (Tatini *et al.*, 1970). No enterotoxin was observable in some types of cheese even though good growth was obtained. Although staphylococci will grow under anaerobic conditions, enterotoxin is not always produced under these conditions (Barber and Deibel, 1972). The water activity also affects enterotoxin production as in some cases good growth of the staphylococci occurs without concurrent enterotoxin production (Troller and Stinson, 1975; Troller, 1976).

INCRIMINATION OF SPECIFIC FOODS IN OUTBREAKS OF STAPHYLOCOCCAL FOOD POISONING.

To protect the public health it is often necessary to identify the food responsible for an outbreak of staphylococcal poisoning. Sometimes the food can be identified tentatively by epidemiological means, but for medico-legal purposes direct proof, based on laboratory analysis, is needed.

Counts of coagulase-positive staphylococci in a suspect food may provide useful information, but in themselves such counts are rarely enough. Selective culture procedures have been devised which take advantage of the high salt tolerance and mannitol fermenting capacity of *S. aureus*. Typical staphylococcal colonies from a selective medium can be isolated and tested for coagulase. Large numbers of coagulase-positive staphylococci, for example, millions per gram, certainly are suggestive. Yet one still has the problem that some coagulase-positive staphylococci do not produce enterotoxin.

On the other hand, small numbers of staphylococci do not necessarily exonerate a suspect food. *S. aureus* may grow in a product and then die during storage, or it may be killed by a subsequent heat treatment, leaving the enterotoxin in the food. This apparently was the case in 1956 when non fat dry milk was the vehicle in an outbreak of staphylococcal poisoning among school children in Puerto Rico. Bacteriological tests revealed few if any viable *S. aureus* cells in the milk powder, yet epidemiological evidence clearly incriminated the product. Later, samples of cheese have been shown to contain SEA without viable enterotoxigenic staphylococci (Casman *et al.*, 1967). Thus, there has long been a serious need for a laboratory test that will reveal minute quantities of enterotoxin.

The lack of a suitable and inexpensive test animal led to the search for a practical laboratory test for the detection of enterotoxin in food. All of the tests described so far are dependent on the reaction of the enterotoxins with their specific antibodies and require the testing for each enterotoxin separately. Tests involving the use of reversed passive hemagglutination (RPHA) (Silverman *et al.*, 1968), fluorescent antibodies (Genigeorgis and Sadler, 1966), immunoelectrophoresis (Gasper *et al.*, 1973), radioimmunoassay (RIA) (Johnson *et al.*, 1971; Collins *et al.*, 1972) and enzyme-linked immunosorbent assay (ELISA) (Saunders & Bartlett, 1977; Stiffler-Rosenberg and Fey, 1978) have been proposed for detection of the enterotoxin in food extracts. The first methods proposed for this purpose were those employing the microslide as the detection method (Casman and Bennett, 1965) and are still the ones used most frequently. These methods involve the concentration of the extract from 100 g of food to approximately 0.2 ml for use with the microslides. This involves extraction of the food, centrifugations, pH adjustments, ion-exchange chromatography, chloroform extractions and freeze drying. The Casman and Bennett method requires from 6 to 10 days while the method of Reiser *et al.* (1974) requires three days. The results vary with different food materials but, in general, a competent operator can detect as little as 0.1 μg of enterotoxin per 100 g of food. The microslide is not an easy method to use as many people have difficulty achieving the sensitivity required for this work (0.05 to 0.1 μg/ml). The RPHA method initially appeared to be a useful one as its sensitivity (1.5 ng/ml) allowed for a simple extraction of the food materials, but nonspecific agglutination by substances extracted from the foods made interpretation of results difficult. The RIA method has received a lot of attention because of its sensitivity (1 ng/ml). Of the several methods proposed the one reported by Miller *et al.* (1978) appears to be the method of choice for the detection of enterotoxin in foods. The drawbacks to RIA are: the handling of radioactive materials, the need for rather expensive counting equipment and the requirement for purified enterotoxins. The latter is a problem which is not easily resolvable as some of the enterotoxins are difficult to purify and are not available for distribution. The ELISA method is one which is approximately equal in sensitivity to RIA but is a more practical one because very little in the way of equipment is required. Much more work is needed before it can be recommended as a usable method for the detection of enterotoxins in foods. One drawback is that purified enterotoxins are required for this method also. Tests for enterotoxin in foods by the RIA and ELISA tests can be accomplished within one working day. Detection of enterotoxins in foods is being done at the present time in many laboratories but, unfortunately, no commercial laboratories offer this service. It is expected that it will be provided as the methods become simpler and less expensive to perform.

REFERENCES

BARBER, L. E., and DEIBEL, R. H. 1972. Effect of pH and oxygen tension on staphylococcal growth and enterotoxin formation in fermented sausage. Appl. Microbiol. *24*, 891–898.

BERGDOLL, M. S. 1970. The staphylococcal enterotoxins. *In* Microbial Toxins, Vol. III, S. J. Ajl, T. C. Montie, and S. Kadis (Editors). Academic Press, Inc., New York.

BERGDOLL, M. S. 1972. The staphylococcal enterotoxins. *In* The Staphylococci, J. O. Cohen (Editor). Wiley Interscience, New York.

BERGDOLL, M. S. 1979. Staphylococcal intoxication. *In* Food-Borne Infections and Intoxications, Second Edition, H. Rieman and F. L. Bryan (Editors). Academic Press, Inc., New York.

BERGDOLL, M. S., WEISS, K. F., and MUSTER, M. J. 1967. The production of staphylococcal enterotoxin by a coagulase-negative organism. Bacteriol. Proc. *1967*, 12.

BLAIR, J. E., and WILLIAMS, R. E. O. 1961. Phage typing of staphylococci. Bull. World Health Org. *24*, 771–784.

CASMAN, E. P., and BENNETT, R. W. 1965. Detection of staphylococcal enterotoxin in food. Appl. Microbiol. *13*, 181–189.

CASMAN, E. P., BERGDOLL, M. S., and ROBINSON, J. 1963. Designation of staphylococcal enterotoxins. J. Bacteriol. *85*, 715–716.

CASMAN, E. P., McCOY, D. W., and BRANDLY, P. J. 1963. Staphylococcal growth and enterotoxin production in meat. Appl. Microbiol. *11*, 498–500.

CASMAN, E. P., BENNETT, R. W., DORSEY, A. E., and ISSA, J. A. 1967. Identification of a fourth staphylococcal enterotoxin, enterotoxin D. J. Bacteriol. *94*, 1875–1882.

COLLINS, W. S., II, METZGER, J. F., and JOHNSON, A. D. 1972. A rapid solid phase radioimmunoassay for staphylococcal B enterotoxin. J. Immunol. *108*, 852–856.

DACK, G. M. 1956. Food Poisoning. 3rd Edition. University of Chicago Press, Chicago, Ill.

DANGERFIELD, H. G. 1973. Effects of enterotoxins after ingestion by humans. 73rd Annual Meeting of the American Society for Microbiology, Miami, Fl. May 6–11.

DAVISON, E., DACK, G. M., and CARY, W. E. 1938. Attempts to assay the enterotoxic substance produced by staphylococci by parental injection of monkeys and kittens. J. Infect. Dis. *62*, 219–223.

DOLMAN, C. E., and WILSON, R. J. 1938. Experiments with staphylococcal enterotoxin. J. Immunol. *35*, 13–30.

DONNELLY, C. B., LESLIE, J. E., and BLACK, L. A. 1968. Production of enterotoxin A in milk. Appl. Microbiol. *16*, 917–924.

GASPER, E., HEIMSCH, R. C., and ANDERSON, A. W. 1973. Quantitative detection of type A staphylococcal enterotoxin by Laurell electroimmunodiffusion. Appl. Microbiol. *25*, 421–426.

GENIGEORGIS, C., and SADLER, W. W. 1966. Immunofluorescent detection of staphylococcal enterotoxin B II. Detection in foods. J. Food Sci. *31*, 605–609.

GHOSH, S. S., and LAXMINARAYANA, H. 1973. Growth of staphylococci in relation to normal microflora and enterotoxin production in milk. Indian J. Animal Sci. *43*, 981–986.

HUMBER, J. Y., DENNY, C. B., and BOHRER, C. W. 1975. Influence of pH on the heat inactivation of staphylococcal enterotoxin A as determined by monkey feeding and serological assay. Appl. Microbiol. *30*, 755–758.

JOHNSON, H. M., BUKOVIC, J. A., KAUFFMAN, P. E., and PEELER, J. T. 1971. Staphylococcal enterotoxin B: Solid-phase radioimmunoassay. Appl. Microbiol. *22*, 837–841.

JORDAN, E. O., DACK, G. M., and WOOLPERT, O. 1931. The effect of heat, storage and chlorination on the toxicity of staphylococcus filtrates. J. Prev. Med. *5*, 383–386.

MILLER, B. A., REISER, R. F., and BERGDOLL, M. S. 1978. Detection of staphylococcal enterotoxins A, B, C, D, and E in foods by radioimmunoassay, using staphylococcal cells containing protein A as immunoadsorbent. Appl. Environ. Microbiol. *36*, 421–426.

REISER, R., CONAWAY, D., and BERGDOLL, M. S. 1974. Detection of staphylococcal enterotoxin in foods. Appl. Microbiol. *27*, 83–85.

SAUNDERS, G. C., and BARTLETT, M. L. 1977. Double-antibody solid-phase immunoassay for the detection of staphylococcal enterotoxin A. Appl. Environ. Microbiol. *34*, 518–522.

SCHANTZ, E. J., ROESSLER, W. G., WAGMAN, J., SPERO, L., DUNNERY, D. A., and BERGDOLL, M. S. 1965. Purification of staphyloccal enterotoxin B. Biochemistry *4*, 1011–1016.

SCHANTZ, E. J., ROESSLER, W. G., WOODBURN, M. J., LYNCH, J. M., JACOBY, H. M., SILVERMAN, S. J., GORMAN, J. C., and SPERO, L. Purification and some chemical and physical properties of staphylococcal enteortoxin A. Biochemistry *11*, 360–366.

SILVERMAN, S. J., KNOTT, A. R., and HOWARD, M. 1968. Rapid, sensitive assay for staphylococcal enterotoxin and a comparison of serological methods. Appl. Microbiol. *16*, 1019–1023.

STIFFLER-ROSENBERG, G., and FEY, H. 1978. Simple assay for staphylococcal enterotoxins A, B, and C: Modification of enzyme linked immunosorbent assay. J. Clin. Microbiol. *8*, 473–479.

SUGIYAMA, H., and HAYAMA, T. 1965. Abdominal viscera as site of emetic action for staphylococcal enterotoxin in the monkey. J. Infect. Dis. *115*, 330–336.

TATINI, S. R. 1976. Thermal stability of enterotoxins in food. J. Milk Food Technol. *39*, 432–438.

TATINI, S. R., JEZESKI, J. J., MORRIS, H. A., OLSON, J. C., JR., and CASMAN, E. P. 1971. Production of staphylococcal enterotoxin A in cheddar and colby cheeses. J. Dairy Sci. *54*, 815–825.

TATINI, S. R., CORDS, B. R., and GRAMOLI, J. 1976. Screening for staphylococcal enterotoxins in food. Food Technol. *30*, 64–74.

TROLLER, J. A. 1976. Staphylococcal growth and enterotoxin production-factors for control. J. Milk Food Technol. *39*, 499–503.

TROLLER, J. A. and STINSON, J. V. 1975. Influence of water activity on growth and enterotoxin formation by *Staphylococcus aureus* in foods. J. Food Sci. *40*, 802–804.

WALKER, G. C., and HARMON, L. G. 1965. The growth and persistence of *Staphylococcus aureus* in milk and broth substrates. J. Food Sci. *30*, 351–358.

WALKER, G. C., and HARMON, L. G. 1966. Thermal resistance of *Staphyloccus aureus* in milk, whey, and phosphate buffer. Appl. Microbiol. *14*, 584–590.

ZOTTOLA, E. A., AL-DULAIME, A. N., and JEZESKI, J. J. 1965. Heat resistance of *Staphylococcus aureus* isolated from milk and cheese. J. Dairy Sci. *48*, 774.

5

Salmonella Food Poisoning

John H. Litchfield

Salmonella food poisoning is a term used to describe food-borne infections caused by bacteria of the genus *Salmonella*. This chapter will emphasize conventional salmonella food poisoning although food-borne outbreaks of typhoid fever caused by *Salmonella typhi* will also be mentioned. This discussion will cover the nature of the disease, incidence in the United States and worldwide, the type of food affected, the growth survival and control of *Salmonella* in foods, and methods for their isolation, identification and enumeration in foods. There are numerous published reviews on the *Salmonellae* and on their role in food-borne diseases that should be consulted for further information (Bowmer, 1965; Bryan, 1975, 1978, - Communicable Disease Center, 1965; Dack, 1956; Edwards and Ewing, 1972; Fagerberg and Avens, 1976; Foster, 1969; Hinshaw and McNeil, 1951; Hobbs, 1962, International Commission on Microbiological Specifications for Foods, 1974, 1978; Litchfield, 1973; Marth, 1969; National Academy of Sciences, 1969, 1975; Prost and Riemann, 1967; Riemann, 1969; Taylor and McCoy, 1969; Slanetz, *et al.*, 1963; Speck, 1976; Troller, 1976; Van Oye, 1964; Weiner, 1974).

NATURE OF THE DISEASE

Causative Organisms

Bacteria of the genus *Salmonella* constitute the major group in the family *Enterobacteriaceae* that causes food-borne infections in humans or feed-borne infections in domestic livestock which in turn, can be transmitted to humans.

120

The 8th Edition of Bergey's Manual of Determinative Bacteriology classifies the genus *Salmonella* as Genus IV in Tribe I *Esherichieae* of the family *Enterobacteriaceae*.

The *Salmonellae* are Gram-negative, faculatitively anaerobic, nonsporeforming, rod-shaped bacteria. Four subgenera are distinguished in Genus IV based on the following biochemical reactions; dulcitol, lactose β-galactosidase, d-tartrate, mucate, malonate, gelatin and KCN (Buchanan and Gibbons, 1974). Table 5.1 summarizes the general morphological and biochemical characteristics of the *Salmonellae*. Edwards and Ewing (1972), provide an extensive discussion of biochemical reactions of the divisions and groups of the *Enterobacteriaceae*.

TABLE 5.1. MORPHOLOGICAL AND BIOCHEMICAL CHARACTERISTICS OF THE SALMONELLAE[a]

Characteristic	Description or Reaction[b]
Shape	rod
Gram Stain	+
Oxygen relationships	Facultatively anaerobic
Motility	Usually by peritrichous flagella— (*Salmonella gallinarum, Salmonella pullorum* are non-motile)
Colony Size	1–4 mm
Catalase	+
Oxidase	–
Glucose	+ (No gas produced by *S. typhi*)
H$_2$S	+[c]
Indole	–
Voges-Proskauer	–
Methyl red	+
Nitrate reduction	+
Simmons citrate	D
Citrate	+
Urease	–
KCN	D
Mucate	D
Gluconate	–
Malonate	D
D-Tartrate	D
Lactose	– (Except for *S. arizonae*, +)
Dulcitol	+[d]
β-Galactosidase	D
Gelatin hydrolysis	D
Arginine dihydrolase	– or + (S)
Lysine decarboxylase	+[e]
Ornithine decarboxylase	+
Phenylalanine deaminase	–

[a] Buchanan and Gibbons, (1974); Edwards and Ewing (1972).
[b] +, positive m 24–48 hours; – negative; s, slow reaction, greater than 24 hr.; D, different biochemical types (+ or –).
[c] *Salmonella cholerae* – suis –D.
[d] *Salmonella typhi, S. enteriditis, S. cholerae-suis* and *S. pullorum*-delayed or negative.
[e] *S. enteritidis* bioserotype *paratyphi* A-negative.

It should be noted that, although members of the genus *Salmonella* do not produce a positive Voges Proskauer reaction for acetoin (acetylmethylcarbinol) in the usual MRVP medium, various *Salmonella* serotypes can produce acetoin, diacetyl or both in glucose minimal broth supplemented as needed by yeast extract or by amino acids required by the strains (Garibaldi and Bayne, 1970).

The members of the genus *Salmonella* cannot be distinguished by biochemical and morphological characteristics alone. Serological methods for classification such as the Kauffmann - White scheme based on somatic (O) and flagellar (H) antigens have enabled the identification of over 1700 distinct serotypes (Edwards and Ewing, 1972; Kauffmann, 1966). Over the years, approximately ten percent of these serotypes have been encountered in food-borne outbreaks and *Salmonella typhimurium, S. heidelberg, S. enteritidis, S. blockley; S. muenchen* and *S. saint paul* are typical serotypes that have been isolated in connection with recent food-borne outbreaks of salmonellosis (Center for Disease Control, 1977).

Symptoms and Clinical Course

The onset of symptoms of *Salmonella* food poisoning is usually sudden with nausea, vomiting, abdominal pain, headache, chills, fever and diarrhea. The incubation period before symptoms occur ranges from 12 to 24 hours after ingesting food contaminated with *Salmonella*. However, it may be as short as 3 hours or as long as 30 hours (Dack, 1956).

In general, the patient may experience a temperature rise to 38 C or higher and the acute stage may last for one to two days. Diarrhea and vomiting may lead to dehydration. Recovery is generally complete in seven days (Taylor and McCoy, 1969). Confirmation of salmonella food poisoning requires isolation of salmonellae from the implicated food and from the stools of all persons showing clinical symptoms of the disease followed by identification using biochemical and serological methods. It is not always possible to meet both of those criteria and the Center for Disease Control (1977b) uses either of them as a laboratory confirmation of an outbreak of salmonella food poisoning.

It should be borne in mind that humans may excrete salmonella for a considerable period of time after infection. Most persons cease excreting *Salmonella* after 3 to 4 weeks following infection but some may continue to excrete for weeks, months or several years. Food handlers and workers in meat slaughtering plants are more likely to be asymptomatic carriers than the general population (National

Academy of Sciences, 1968). In the case of infection by *S. typhi*, the causative organism of typhoid fever, the affected person becomes a permanent carrier and continues to excrete the organism in the feces or urine.

An important consideration in salmonella food poisoning is that intact salmonella cells are the causative agents and must be consumed with food or water to cause illness. In contrast, botulism caused by *Clostridium botulinium*, *Clostridium perfringens* food poisoning, and staphylococcus food poisoning caused by *Staphylococcus aureus* are food-borne intoxications resulting from the ingestion of foods containing toxins produced by these organisms rather than the organisms themselves.

In studies with human volunteers, McCullough and Eisele (1951, a,b,c,d) investigated the infective dosages required to produce food poisoning of three strains each of *S. anatum* and *S. meleagridis*, one strain each of *S. bareilly*, *S. derby* and *S. newport* and four strains of *S. pullorum* isolated from spray dried, powdered whole egg. Minimal infective doses were not established for *S. anatum* and *S. meleagridis*, Numbers of cells required to produce symptoms of illness ranged from as low as 12,000 for *S. anatum* and *S. Melegridis*, and 125,000 for *S. bareilly* to 16×10^9 for one strain of *S. pullorum*. It is important to note that the salmonella cultures used in these studies were isolated from commercial samples of spray dried whole egg powder and prepared in a standardized suspension in eggnog for administration to healthy adult male human volunteers. Although initial viable counts of the suspensions were determined, no counts were made of the number of salmonella excreted. Also, blood cultures were not taken to determine possible blood stream invasion.

Actual experience in human outbreaks of salmonellosis indicates that considerably smaller numbers of organisms may cause illness than would be indicated by studies with suspensions of salmonella cultures administered to adult human volunteers. In an outbreak of salmonellosis in Cincinnati, Ohio, a count of 23,000 *S. infantis* was obtained on brilliant green sulfadiazine agar (Angelotti *et al*, 1961). In other series of outbreaks caused by consumption of frozen dessert containing contaminated frozen egg yolk, an average count of *S. typhimurium* of 65 per 100 g was obtained from samples of the dessert and a count of 1,100 per 100 g was obtained from the egg yolk (National Academy of Sciences, 1969).

The mortality associated with salmonella food poisoning is quite low. Table 5.2 shows case fatality ratios in the range of 0.02 to 0.28 percent over the five year period 1972–1976. The major hazards are

TABLE 5.2.   CONFIRMED OUTBREAKS OF SALMONELLA FOOD POISONING
1971–1976

Year	Outbreaks	Number of Cases	Deaths	Case Fatality Ratio, (percent)
1972	36	1,880	4	0.21
1973	33	2,462	7	0.28
1974	35	5,499	1	0.02
1975	38	1,573	2	0.12
1976	28	1,169	3	0.26

Source: Black et al, 1978; Center for Disease Control, U.S. Department of Health, Education and Welfare, 1976c, 1977b.

to aged and severely ill persons and to infants. Patients in mental institutions, hospitals and nursing homes face a greater risk of infection with *Salmonella* than the general population.

INCIDENCE AND SOURCES OF INFECTION

Incidence

United States.—Prior to 1963, food-borne outbreaks of Salmonella food poisoning investigated by state and local health departments were reported to the National Office of Vital Statistics. However, since that time, the Center for Disease Control and the Association of State and Territorial Epidemiologists and Laboratory Directors have conducted a Salmonella Surveillance Program involving a more thorough investigation and reporting of local, state, and interstate outbreaks than in the past. Over the period 1963–1975, poultry, meat, and eggs accounted for 99 (21%), 69 (15%) and 53 (11%), respectively, of 463 outbreaks of non-typhoid salmonellosis in which the food vehicle of transmission was identified (Cohen and Blake, 1977).

Table 5.3 summarizes food-borne outbreaks of salmonellosis reported to the Center for Disease Control in 1976.

The places where the foods were mishandled and number of outbreaks at each type of location were food processing establishments (3), food service establishments (18), homes (4) and unknown or unspecified (3) (Center for Disease Control, 1977b).

Table 5.4 summarizes food-borne outbreaks of salmonellosis caused by mishandling of food in food processing establishments from 1974–1976. Note that commercially prepared roast beef products were prominent as a vehicle for transmission and that cheese was implicated in one large outbreak.

The largest outbreak of food-borne salmonellosis ever reported to the Center for Disease Control affected approximately 3,400 persons at a free outdoor barbecue in September, 1974 at a fair at Windrow, Arizona. *Salmonella newport* was the causative organism and the vehicle of transmission was potato salad that had been held for up to 16 hours at improper temperatures. The Center for Disease Control investigators determined from rates of diarrheal illness over two weeks following the outbreak in persons who did not attend the barbecue and by examining stool cultures obtained after the outbreak, that secondary person-to-person transmission did not occur (Horwitz *et al.*, 1977).

It is apparent that salmonella food poisoning occurs in all regions of the United States. The greatest numbers of outbreaks occur from May to September, the warmer months, when large outdoor gatherings are likely to be held, and foods are likely to be held at temperatures favorable to the growth of the *Salmonellae* (Center for Disease Control, 1977b).

According to Black *et al*, (1978), the outbreaks and cases of food-borne disease reported to the Center for Disease Control are probably less than one percent of the total incidence. It can be concluded that the incidence of salmonella food poisoning is probably far greater than the number of cases cited in Table 5.3 would indicate.

Worldwide.—Worldwide statistics on the incidence of Salmonellosis in Western and Eastern Europe, Africa, Asia, and Latin America are given in Van Oye's monograph (1964). A comparison of data on incidence of salmonella food poisoning in six countries in the 1973–75 period showed 1,685 outbreaks (21,428 cases) for England and Wales, 106 outbreaks (9,534 cases) in the United States and 58 outbreaks (2,253 cases) in Canada. There were 412 outbreaks in Japan from 1968–72, and 9 in Australia from 1967–71. Deaths and case fatality ratios for the 1973–75 period were 131 (0.6%) in England and Wales, 10 (0.1%) in the United States, and 1 (0.04%) in Canada (Todd, 1978). These widely differing statistics on incidences in different countries may reflect, in part, the extent and completeness of reporting food-borne outbreaks of salmonellosis.

Over the years, outbreaks of typhoid fever caused by *S. typhi* have been associated with food. Food products implicated in some of these outbreaks include cheese (Foley and Poisson, 1945; Menzies, 1944), corned beef (Ash *et al*, 1964); and canned ox-tongue (Couper *et al*,

TABLE 5.3.   FOOD-BORNE OUTBREAKS OF SALMONELLOSIS REPORTED TO THE CENTER FOR DISEASE CONTROL IN 1976[a]

Vehicle of Infection	Serotype	No. Persons Affected	Location Where Food Mishandled and Eaten[a]	State
Turkey, dressing, squash	Salmonella, Group E Species unknown	73	Banquet room (B)	California
Turkey	S. saint-paul	54	Camp (B)	New York
Chicken casserole	S. typhimurium	18	Church (C)	New Jersey
Roast beef	S. typhimurium	119	School (B)	Wisconsin
	S. newport	9	Home (D)	Massachusetts
Precooked roast beef	S. bovis morbificans	21	Restaurant and deli (A)	Connecticut, Delaware, Massachusetts, New Jersey, Pennsylvania
Prime rib, roast beef, ham	S. london	37	Restaurant (B)	Minnesota
Corned beef	S. san diego	2	Home (B)	New Jersey
Cheese	S. heidelberg	339	Restaurant (A)	Colorado
Ice cream	S. typhimurium	7	Home (C)	Michigan
Peruvian cheese and tomato dish	S. typhi	8	Dress factory (C)	Florida
Nutrient supplement	S. infantis	4	Hospital (A)	California, Colorado
Bread stuffing, gravy, corn, Apple Betty topping	S. heidelberg	78	Resort Inn (B)	Maine

Tuna & Macaroni salads	*S. thompson*	15	Hospital (B)	Missouri
Potato salad	*S. saint paul,* *S. typhimurium*	42	Restaurant (B)	Pennsylvania
Potato salad, macaroni salad	*S. typhimurium*	27	Picnic (C)	Washington
Salad dressing	*S. copenhagen*	29	Restaurant (B)	Oklahoma
Japanese food	*S. enteritidis*	15	Restaurant (B)	New York City
Mexican food	*S. heidelberg*	7	Restaurant (B)	California
Unknown	*S. typhimurium*	12	Social hall (B) (Mob. home)	California
Unknown	*S. typhimurium*	24	School (B)	Maine
Unknown	*S. typhimurium*	48	Nursing home (B)	Massachusetts
Unknown	*S. typhimurium*	44	Nursing home, jail (D)	New Hampshire
Unknown	*S. typhimurium,* *S. give*	3	Restaurant (B)	Connecticut
Unknown	*S. muenchen*	35	Wedding reception (D)	Massachusetts
Unknown	*S. heidelberg*	24	Cafeteria (B)	Washington
Unknown	*S. heidelberg* *S. schwarzengrund*	17	Mental Institution (B)	Pennsylvania
Unknown	*S. blockley*	58	Nursing home (B)	Washington

[a]Center for Disease Control. 1977b
[b]Place where food is mishandled (A) food processing establishment, (B) food service establishment, (C) home, (D) unknown.

TABLE 5.4. FOOD-BORNE OUTBREAKS OF SALMONELLOSIS CAUSED BY MISHANDLING OF FOOD IN FOOD PROCESSING ESTABLISHMENTS 1974–1976

Salmonella Serotype	Vehicle	No. of Cases	Year
Salmonella dublin	Certified raw milk	3	1974
Salmonella typhimurium	Cider	1,500	1974
Salmonella saint paul	Pre-cooked roast beef	54	1975
Salmonella singapore	roast-beef sandwiches	13	1975
Salmonella heidelberg	cheese	339	1976
Salmonella bovis-morbificans	pre-cooked roast beef	21	1976
Salmonella infantis	nutrient supplement	4	1976

(a)Black et al, 1978, Center for Disease Control, 1976c, 1977b.

1956). An outbreak was also associated with food served at a wedding reception (Caraway and Bruce, 1961). In the case of the canned ox-tongue outbreak, circumstantial evidence indicated that product contamination probably occurred as a result of polluted river water for cooling the cans after processing.

Sources of Infection

The intestinal tracts of animals are the major source of the Salmonellae in nature (Fox, 1974; McCoy, 1976; Wiener, 1974). Approximately 1 to 3% of all domestic animals are infected with these organisms (National Academy of Sciences, 1969).

Domestic livestock reservoirs of salmonellae include chickens (Galton et al., 1955; Hobbs, 1961, 1962, 1974; Kaufmann and Feeley, 1968; Lee, 1973; Osborne, 1976; Taylor and McCoy, 1969; Van Schothorst et al., 1974), turkeys (Bryan et al., 1968a; Kumar et al., 1972; McBride et al., 1978; Taylor and McCoy, 1969); cattle (Galton et al., 1954; Mackel et al., 1965; McCoy, 1976; Osborne, 1976; Taylor and McCoy, 1969), swine (Galton, 1954); Mackel et al., 1965; Osborne, 1976; Newell and Williams, 1971; Taylor and McCoy, 1969; Tsai et al., 1971), horses (Anderson and Lee, 1976; Morse et al., 1978), and sheep (Osborne, 1976).

Household pets including cats and dogs (Morse et al., 1978; Osborne, 1976); and turtles (Osborne, 1976) are also sources of salmonellae. Salmonellae have been isolated from such diverse sources as Moroccan food snails and frogs legs (Andrews et al., 1975, 1977), frogs (Sharma et al., 1974), and a wide range of wild and zoo animals including bats, birds, rats, mice, squirrels, rabbits, mink, elephants, lions, lizards and snakes (National Academy of Sciences, 1969; Osborne 1976). A wide variety of insects including flies, fleas, roaches, lice and ticks may also carry salmonellae (National Academy of Sciences, 1969).

In contrast to terrestrial animals, fish and shellfish do not carry members of the Enterobacteriae unless they encounter polluted water. Fish and shellfish caught in the open sea do not contain coliforms or Salmonellae while those caught close to shore in polluted waters may contain these organisms (Buttiaux, 1962; Gulasekharam et al., 1956).

Salmonella typhi and S. enteritidis serotypes Paratyphi A, B and C are primarily adapted to human hosts and rarely cause disease in other species. Secondary hosts for the Paratyphi B serotype are poultry, swine, cattle and dogs. Other host-adapted Salmonellae are S. pullorum and S. gallinarum, poultry; S. dublin, cattle; S. choleraesuis, swine; S. abortus-ovis, sheep; and S. abortus-equi, horses (McCoy, 1976; National Academy of Sciences, 1969).

Host-adapted serotypes cause severe illness including invasion of the bloodstream, high morbidity and mortality and produce a permanent carrier state. In contrast, non-adapted salmonellae cause a less severe illness characterized by gastroenteritis and short duration with low morbidity and mortality. These non-host adapted organisms can produce illness in both animals and humans.

Salmonellae can be transmitted by the routes animal to man, man to man, man to animal, and animal to animal. The man to animal route is the least important, but it can occur through polluted water, sewage, polluted soils or grazing areas, or contaminated feedstuffs (Bryan, 1977; Taylor and McCoy, 1969; Wiener, 1974). Routes of infection may involve several species. For example, Morse et al., (1978) reported possible transmission of S. typhimurium var. Copenhagen from horse to human to dog.

Apparently healthy pigs may be Salmonella carriers, and excrete these organisms in their feces. Increases in excretion of salmonellae during transportation from the farm to the meat packing plant may result from animals infecting each other under crowded conditions in trucks or holding pens, or from excretion of cecal contents into the rectum, exposing organisms previously masked from detection (Williams and Newell, 1970). Tsai et al., 1973 observed that the stress of truck transportation did not alter the excretion of salmonellae by pigs. Significantly, high incidences of salmonellae were found in the cecum and colon contents but not in samples of the spleen, liver, kidney or various muscles. After transportation and slaughter, salmonellae were detected in the cecal and colon contents of two Chester White and two Poland China pigs that were not excreting salmonellae in the feces.

Hobbs, 1974 describes the cycle of transmission of salmonellae as feedstuffs — animal — foodstuffs — man with a cycle of perpetuation through use of contaminated animal processing by-products as a component of feeds. This points out that host-specific serotypes will spread more rapidly than non-host-specific types as a result of large increases in number in the infected animal and extensive environmental scattering of the heavily contaminated liquid feces.

Bains and MacKenzie (1974) correlated high mortality in infected broiler flocks with incidence of *Salmonella* in grain component of rations. The chain of transmission was through feed, breeder birds, and day-old birds to dressed broilers.

Genetic Characteristics and Transferrable Drug Resistance in Salmonellae

At this point, the genetic characteristics of the *Salmonellae* that are important in connection with human and animal salmonellosis should be mentioned. Genetic recombination by the mechanisms of conjugation, lysogeny by bacteriophage, or transduction occurs among the *Enterobacteriaceae* including the *Salmonellae*. Consequently, the *Salmonellae* have been used extensively in microbial genetics investigations.

Lysogeny can result in antigenic changes that are important in serological typing of *Salmonella*. Also, the occurrence of biochemically atypical *Salmonella* in outbreaks while not frequent, may cause problems in identification.

Transferrable drug resistance in the *Salmonellae* has received considerable attention in recent years in view of the use of antibiotics as feed additives for poultry and swine. Resistance to antibiotics is transferred from a resistant (donor) strain to a sensitive (recipient) strain through the mediation of plasmids called R-factors.

Pocurull *et al.* (1971) investigated multiple drug resistance in 1251 *Salmonella* cultures representing 61 serotypes obtained from disease outbreaks in chickens, turkeys, geese, ducks, pigeons, quail, cattle, sheep, swine, horses, cats and dogs in 37 states and one territory of the United States. They demonstrated resistance to one or more antimicrobial drugs in 935 cultures, resistance transfer in 267 multiply-resistant cultures, and ability to transfer all or part of the resistant pattern to antibiotic-sensitive recipients in 181 cultures. Three cultures of *S. typhimurium* and one culture of *S. saint paul* were resistant to more than 5 drugs.

Anderson *et al.* (1977) found three types of resistant plasmids in *S. typhimurium* strains of Middle Eastern origin distributed from England to India. They concluded that these strains probably represented a clone of *S. typhimurium* that had been distributed over a 2,500 to 4,000 mile distance.

It is apparent that drug resistant strains of *Salmonella* are widely distributed not only in the United States but also elsewhere in the world. These organisms have been involved in disease outbreaks in domestic livestock, pets and humans. It is clear that regulatory agencies will be placing increasing restrictions on using antibiotics in animal feeds in view of these findings (see also Chapter 15).

SALMONELLA IN FOODS

Growth Characteristics of the Salmonellae

As a prelude to a discussion of salmonellae in foods, some of the factors that affect the growth of these organisms will be reviewed briefly. A detailed discussion of methods for isolating, identifying and enumerating salmonellae will be presented later in this chapter.

Table 5.5. summarizes some of the important growth characteristics of the *Salmonellae*. Minimum growth temperatures of the *Salmonellae* in foods tends to increase with decreasing pH (National Academy of Sciences, 1969). The effects of pH on growth are a function of temperature, nutrients, salt or sugar concentrations (water activity) and presence of natural inhibitors or food preservatives. There is a need for further studies to define the interactions among these variables in actual food systems. (Chung and Goepfert, 1970; Matches and Liston, 1968b, 1972a, b).

As mentioned previously, the *Salmonellae* are facultative anaerobes, but growth is more rapid under aerobic conditions. There is a need for further studies to determine the effects of oxidation-reduction potential on the growth of salmonellae in foods. *Salmonella* strains grow on a simple medium consisting of glucose as the carbon and energy source, ammonium salts as the nitrogen source,

TABLE 5.5.   SUMMARY OF GROWTH CHARACTERISTICS OF SALMONELLAE

Characteristics	Range of Values	References
Temperature, °C	Minimum—5.3—*S. heidelberg* Maximum—46.2—*S. bareilly* Optimum—35–37	Matches and Liston, 1968 Elliott and Heiniger, 1965 National Academy of Sciences, 1969
	Range in foods—7–45—mixed culture in chicken a la King and custard	Angelotti *et al.*, 1961a, b, c
pH	Range—4.1–9.0 Optimum—6.5–7.5	Banwart and Ayres, 1957 National Academy of Sciences, 1969
Salt tolerance	8–9% NaCl	Troller, 1973, 1976 Christian and Scott, 1953
Water activity, a_w	Minimum—0.94 Range—0.94–0.999 Optimum—0.995	Christian, 1955 Christian and Scott, 1953 Christian and Scott, 1953

and mineral salts. Strains of a given serotype may require specific amino acids or vitamins, or both. According to Stokes and Bayne (1958), various *Salmonella* strains required the amino acids arginine, aspartic acid, cystine, histidine, leucine, methionine and threonine, the purines adenine and guanine, and the vitamins thiamine and nicotinic acid.

Effects of Heating, Drying, Freezing, Irradiation and Chemical Treatment

Effects of Heating.—Most *Salmonella* strains are readily destroyed at 60–65°C at high water activities. Table 5.6 summarizes the results of some studies of heat resistance of salmonellae in laboratory media. Thermal destruction of salmonellae in foods is covered subsequently in connection with discussions on various foods products.

Ng *et al.* (1969) claimed that *S. senftenberg* 775W was unique in its heat resistance among the *Salmonellae*. However, Baird-Parker, *et al.* (1970) found that a strain of *S. bedford* had a thermal resistance similar to that of *S. senftenberg* 775W. Heat resistance of the 775W strain could not be related to water activity over the a_w range 0.85–0.98, but increased markedly in the presence of sucrose reaching a maximum $D_{60°}$ value of 75.2 min at a 75.6% (W/W) sucrose concentration ($a_w = 0.90$).

Goepfert *et al.* (1970) did not find a direct relationship between

TABLE 5.6.   HEAT RESISTANCE OF SELECTED *SALMONELLA* STRAINS IN LABORATORY MEDIA

Serotype	Medium	Temperature	pH	D Value[a] (minutes)	Reference
Salmonella anatum	15.4% sucrose (W/W)	57.2	6.9	0.8–1.3	Goepfert, *et al.*, 1970
S. bedford 286	Heart infusion broth	60	7.4	4.3	Baird-Parker, *et al.*, 1970
S. blockley 2004	Trypticase soy-1% yeast extract broth	57	6.8	5.8	N.g., 1966
S. montevideo	15.4% sucrose (W/W)	57.2	6.9	0.7–1.4	Goepfert, *et al.*, 1970
S. senftenberg 775W	(a) 15.4% sucrose (W/W)	57.2	6.9	13.5–16.0	Goepfert, *et al.*, 1970
	(b) Heart infusion broth	(1) 60	7.4	6.3	Baird-Parker, *et al.*, 1970
		(2) 60	7.4	37	Davidson, *et al.*, 1966
	(c) Trypticase soy-1% yeast extract broth	57	6.8	31	Ng, 1966
S. senftenberg 436	Heart infusion broth	60	7.4	0.62	Baird-Parker, *et al.*, 1970
S. tennessee	15.4% sucrose (W/W)	57.2	6.9	0.6–0.9	Goepfert, *et al.*, 1970
S. typhimurium TM-1	(a) Trypticase soy-1% yeast extract broth	57	6.8	1.2	Ng. 1966
	(b) Trypticase soy broth	55	—	8.5	Elliott and Heiniger, 1965
S. typhimurium	15.4% sucrose (W/W)	57.2	6.9	0.7–1.5	Goepfert, *et al.*, 1970

[a] time in minutes required for 90% destruction

water activity and heat resistance of five *Salmonella* serotypes. Troller, (1973) concludes that there is a large strain to strain variability in heat resistance of salmonellae at different water activities and that factors other than water activity influence the heat sensitivity of those organisms.

Although some rough strains of 775W have up to two-fold higher heat resistance than smooth strains, studies with a large number of strains show that heat resistance among strains cannot be related to the degree of roughness (Thomas *et al.*, 1966; Liu *et al.*, 1968). It is important to note that thermal resistance of the 775W strain is a function of temperature and that at temperatures below 55° the Z value (slope of the thermal death curve) decreases markedly to values similar to those of other salmonellae (Elliott and Heiniger, 1965; Licciardello *et al.*, 1965; Baird-Parker, *et al.*, 1970).

Survivor curves for heated salmonellae may show deviations from a logarithmic relationship. For example, the shape of survivor curves of *S. anatum* heated at 55°C and the heat resistance of this organism changed with increasing age of the culture (Moats *et al.*, 1971a, b).

According to Corry and Roberts (1970) the heat resistance of *S. typhimurium* could be increased by repeated heat treatment at 55°C in phosphate buffer at pH 7.0. The D value increased from 3.7 min to 11.7 min after heat cycling for 39 times. This thermal resistance approached that of *S. senftenburg* 775W used as a comparative culture (D=13.0 min, 55C, pH 7.0). This selection for heat resistance did not result in any increase in pathogenicity or change in the usual biochemical characteristics. Sublethal heating of salmonellae may result in thermal injury followed by repair and recovery. Cells of *S. typhimurium* that were heat-injured at 48°C for 30 minutes were unable to grow in various selective media for enteric bacteria such as Levine Eosine Methylene Blue Agar, but recovered and grew in an enriched medium such as Trypticase Soy Broth (Clark and Ordal, 1969). The recovery process was dependent upon synthesis of ribsosomal ribonucleic acid (RNA), adenosine triphosphate and new protein (Tomlins and Ordal, 1971).

In a study of the effects of various solutes on thermally induced death and injury in *S. typhimurium*, Lee and Goepfert, (1975) observed that heating cells in phosphate buffer was more delterious than heating in distilled water. They concluded that RNA degradation was not the primary cause of cell death in this organism.

D'Aoust, (1978) found that lactate, mannitol and α-glycerophosphate (in order of decreasing effectiveness) brought about 90% or greater recovery of heat injured *S. typhimurium* when these compounds were added to Levine Eosine Methylene Blue salts agar. He observed that susceptibility to heat injury was affected markedly by growth and storage conditions prior to heat treatment.

Effects of Drying and Freezing.—It has been mentioned previously that the minimum water activity for the growth of salmonellae is approximately 0.93–0.94 (Troller, 1973). The extent of destruction of salmonellae as a result of drying is extremely variable. Drying may reduce the number of viable cells, but the survivors may persist for extended periods and die at a very low rate (National Academy of Sciences, 1969).

The same conclusion can be reached on the effects of freezing on the viability of salmonellae. In addition, the resistance of a variety of *Salmonella* serotypes to drying and freezing may be related to the frequency of isolation of these organisms in salmonellosis. (Enkiri and Alford, 1971).

S. typhimurium did not survive freeze-drying in 2% gelatin at a platen temperature of 160°F, but did survive in a 2% gelatin containing 5% dextrose (glucose) at this temperature according to Sinskey *et al.* (1964). Survival in such model systems at a platen temperature of 120 F was greater than at 160 F, but at both temperatures depended upon the composition of the model system. Cell damage during storage depended upon composition of the model system, presence of air and the relative humidity, with greatest damage at low relative humidity values (Sinskey *et al.*, 1967).

Ray *et al.* (1971 a, b) found that 70–90% of *S. anatum* cells that survived freeze drying in non-fat milk solids were injured as measured by failure to grow on a selective plating medium containing deoxycholate. Within 1 hour after rehydration, repair of this injury took place. The rate of repair was independent of pH between 6.0 and 7.0, and was reduced by lowering the temperature from 35°C to 1°C. They concluded that the lipopolysaccharide portion of the cell wall may be involved in freeze drying injury and that the repair of this injury requires synthesis of adenosine triphosphate but not synthesis of protein, ribonucleic acid or cell wall mucopeptide.

Subsequently, Ray *et al.* (1972) reported that fast freezing in a dry-ice acetone bath and slow thawing of *S. anatum* suspended in water injured more than 90% of the surviving cells. Again, the injury appeared to involve the cell wall lipopolysaccharides and the repair process required adenosine triphosphate synthesis.

In addition to *S. anatum*, Janssen and Busta (1973) demonstrated freezing injury and repair in 10 other *Salmonella* serotypes. Rapid freezing of. *S. gallinarium* at −40°C at a rate of 18°C per min. resulted in a 5-fold increase in recovery over that obtained with a freezing rate of 0.5°C per min. (Raccach and Juven, 1973). Furthermore, metabolic injury to *S. gallinarium* by freezing at −75°C did not alter pathogenicity as measured by mortality in chicks injected with injured or uninjured cells (Sorrells *et al.*, 1970).

Effects of Radiation

Ionizing Radiation.—Table 5.7 summarizes the resistance of selected *Salmonella* serotypes to ionizing radiation in the form of beta and gamma radiation. The radiation sensitivity of salmonellae is generally greater in the presence of oxygen than in its absence (Ley *et al.*, 1963). There is relatively little effect of pH on radiation sensitivity of salmonellae in contrast to the effect on heat sensitivity.

The lower radiation sensitivity of the heat-resistant *S. senftenberg* 775W as compared to other serotypes indicated that the mechanism of radiation resistance differs from that of heat resistance (Lineweaver, 1966).

Radiation resistance can develop in *Salmonella* cultures subjected to repeated cycles of gamma radiation. For example, the frequency of radio-resistant cells in a population of *S. newport* was in the order of 1 in. 8.9×10^6 (Licciardello *et al*, 1969a). In addition, *Salmonella* cultures subjected to repeated radiation (3 - 20 cycles) may show changes in morphological and serological characteristics, but do not show major changes in biochemical characteristics, increases in antibiotic resistance or changes in virulence (Licciardello *et al.*, 1969b; Epps and Idziak 1970; Previte *et al.*, 1971 a, b.). Radiation resistant cultures showed decreased pathogenicity for day-old chicks and this was not increased by five serial passages in chicks (Epps and Idziak, 1970).

Ultraviolet Radiation.—Ijichi, (1966) investigated the effects of ultraviolet radiation of liquid egg white inoculated with *S. typhimurium* or *S. senftenberg* 775W using a rotating bowl device fitted with ultraviolet lamps at 2537A. At maximum conditions of exposure, reductions of 10^6 to 10^7 salmonellae were obtained with both serotypes. Again, radiation sensitivity did not follow heat sensitivity patterns in these cultures.

Effects of Chemical Treatment.—A wide range of chemical compounds have been evaluated for their anti-microbial effects on salmonellae. Typical examples include hydrogen peroxide (Rogers *et al.*, 1966), ethylene oxide (Sair, 1966), and sorbic acid (Park and

TABLE 5.7.   RESISTANCE OF SELECTED *SALMONELLA* STRAINS TO IONIZING RADIATION

Serotype	Medium	D_{10} Value [a] Megareds	Reference
Salmonella anatum	Nutrient broth— 0.3% yeast extract	0.050–0.054	Epps and Idziak, 1970
S. blockley (wild type)	Nutrient broth— 0.3% yeast extract	0.045–0.050	Epps and Idziak, 1970
S. enteritidis *S. gallinarum*	Brain heart infusion Phosphate buffer pH7 (aerated)	0.042 0.013	Previte *et al.*, 1971b Ley *et al.*, 1963
S. give (wild type)	Nutrient broth— 0.3% yeast extract	0.040–0.043	Epps and Idziak, 1970
S. heidelberg	Brain heart infusion	0.078	Previte *et al.*, 1971b
S. infantis (wild type)	Nutrient broth— 0.3% yeast extract	0.026–0.030	Epps and Idziak, 1970
S. manhattan (wild type)	Nutrient broth— 0.3% yeast extract	0.018–0.020	Epps and Idziak, 1970
S. newport	Brain heart infusion	0.060	Previte *et al.*, 1971b
S. senftenberg	Phosphate buffer, pH7 (aerated)	0.013	Ley *et al.*, 1963
S. typhi	crabmeat, diluted 1:10 with water	0.027	Dyer *et al.*, 1966
S. thompson	Brain heart infusion	0.084	Previte *et al.*, 1971b
S. typhimurium	(1) phosphate buffer (aerated)	0.048	Ley *et al.*, 1963
	(2) brain heart infusion	0.048	Previte *et al.*, 1971b
S. wichita	crabmeat diluted 1:10 in water	0.026	Dyer *et al.*, 1966

[a]Radiation dose required to inactivate 90% of the initial numbers present.

Marth, 1972a), volatile and non-volatile fatty acids ranging from formic to stearic acids and citric, lactic and phosphoric acids (Goepfert and Hicks, 1969, Khan and Katamay, 1969). In addition, natural products and their constituents may exert inhibitory effects on salmonellae. Examples are onion and garlic, and certain volatile aliphatic disulfide constituents (Johnson and Vaughn, 1969, Wilson and Andrews, 1976), spices, particularly allspice, cassia, cinnamon, clove, and oregano, (Julseth and Deibel, 1974, Wilson and Andrews, 1976) and cocoa (Busta and Speck, 1968). The identities of the inhibitory constituents in the spices and in cocoa have not been established.

Salmonellae are inhibited by volatile compounds in milk (Kul-shrestha and Marth, 1970, 1972, 1975 a,b). Also, culture filtrates of starter cultures for cultured dairy products, for example, *Leuconostoc citrovorum* may also be inhibitory (Sorrells and Speck, 1970).

SALMONELLAE IN SELECTED FOOD AND FEED PRODUCTS

Feedstuffs—Animal By-Products and Fishmeal

Animal By-Products.—It has been pointed out previously that feedstuffs are a major source of transmission of salmonellae to domestic livestock. The incidence of salmonellae in animal by-products and wastes from meatpacking plants in many parts of the world has been documented extensively (Clise and Snecker, 1965; Morehouse and Wedman, 1961; Harvey and Price, 1967, 1968; Hobbs, 1974; Patterson, 1969; Riley, 1969; Smyser *et al.*, 1970; Van-derpost and Bell, 1977).

Environmental conditions in rendering plants are an important factor in the transmission of salmonella (Magwood *et al*, 1965). For example, a variety of serotypes have been isolated from environmental swabs taken during various stages of processing and from flies in the plant as well as from the finished product (Loken *et al.*, 1968).

The resistance of salmonellae to dry heat apparently was greater in naturally contaminated than in artificially contaminated products. Heating at 180 F for 63 min. was required to reduce salmonellae to undectable numbers in the naturally contaminated product but only 15 min. heating at 155 F was required for the artificially contaminated product (Rasmussen *et al.*, 1964).

Increasing the water activity of meat and bone meal prior to heating from a normal 0.6 to 0.9 reduced the time required at 90°C to achieve a 10^6 reduction in *S. typhimurium* added to the meal. The heat of pelleting the meal can also cause significant destruction of salmonellae. (Riemann, 1968; Scott *et al.*, 1975).

Die off of salmonellae inoculated into animal by-product and fishmeals is related to water activity. *Salmonella* populations were stable at a_w values in the range of 0.405 to 0.415. However, the numbers decreased markedly when moisture levels were increased to a_w values as high as 0.94 which did not permit growth (Carlson and Snoeyenbos, 1969).

S. senftenberg 775W populations inoculated into meat and bone meal remained stable at temperatures up to 50 C at 50% moisture (a_w = 0.25) but, a 5-log. die off was observed at 50 C in a 15% moisture meal (a_w = 0.82) in a 72 hour period. (Liu *et al.*, 1969).

It should be emphasized that salmonellae in naturally contaminated animal by-products may be more difficult to inactivate by heating as compared with salmonellae added to those products in experimental studies. Individual cells and clumps of cells in naturally contaminated products may be coated with proteins and fats which retard heat transfer considerably and require heating for longer times at a given temperature to achieve lethal effects. Current U.S. Department of Agriculture regulations require heating at 170°F for 30 min. during rendering of carcass and parts passed for cooking into rendered pork fat or tallow (Code of Federal Regulations, 1978b).

Fish Meal.—Fishmeals, particularly menhaden fishmeals that are widely used as animal feed ingredients; may be contaminated with salmonellae (Garrett and Hamilton 1971; Jacobs, 1963).

Morris *et al.*, (1970) were unable to isolate salmonellae from fish freshly caught in the Gulf of Mexico. However, these organisms were isolated frequently from the boats and from dockside water at the processing plants. The raw product areas of fish meal plants were the source of more isolations when the plants were operating and less when the plants were idle, with the reverse being the case in the processing areas. It appeared that salmonellae survived and multiplied when the plants were idle resulting in contamination of the first portion of incoming raw materials.

Reducing water activity of fish meals inoculated with salmonellae did not reduce initial death rates during storage (Doesburg *et al.*, 1970a). Heat treatment of inoculated fish meal for 10 min at 80°C, at a water activity of 0.72 reduced counts of *S. senftenberg* to undetectable levels (Doesburg *et al.*, 1970b).

The U.S. National Marine Fisheries Service has developed sanitation guidelines for *Salmonella* control in fish meal production. (Garrett and Hamilton, 1971). Recommended procedures for minimizing *Salmonella* isolation from fish meal include:

(1) Heating the product for 10 min at 190°F

(2) Recycling the first 45 minutes of production back through the cookers

(3) Preventing uncontrolled transfer of personnel and equipment from the pre-dry to the post-dry areas, and preventing excessive moisture and dust accumulation in the post-dry area

(4) Maintaining suitable cleaning in pre-dry and post-dry areas
(5) Maintaining proper insect, rodent and bird control
(6) Maintaining proper control over storage and shipping of the final product.

Poultry and Eggs

Poultry.—Chickens, turkeys and ducks and their eggs are important sources of *Salmonella* that have been implicated in foodborne outbreaks of salmonellosis. Considerable attention has been given to the control of *Salmonella* in these products starting from the source of production, through subsequent processing to the final consumer.

Poultry and Poultry Products.—A wide variety of *Salmonella* serotypes has been isolated from chickens and turkeys prior to and during processing and from the environments in processing plants. (Glagen *et al.*, 1966; Jarolmen *et al.*, 1976; Knivett, 1971; Ladiges and Foster, 1974; Morris *et al.*, 1968; Patterson, 1969; Patrick *et al.*, 1973; Pivnick, 1970; Surkiewicz *et al.*, 1969; Wilson *et al.*, 1962; Zottola *et al.*, 1970). In addition, numerous surveys have been made of *Salmonella* and related enteric bacteria in dressed chickens, turkeys, and further processed chickens and turkey products down to the retail level (Bryan *et al.*, 1968; Cotterill, *et al.*, 1977; Duitschaever, 1977; Hussemann and Wallace, 1951; Mercuri *et al.*, 1978; Morris and Ayres, 1960; Sadler *et al.*, 1960; Sadler and Corstvet, 1965; Swaminathan *et al.*, 1978c; Woodburn, 1964).

There has been a substantial effort in recent years to effect control of *Salmonella* in poultry during processing and to provide information on proper time-temperature conditions for thermal destruction during cooking. Typical chemical treatments of broiler and fryer chickens to reduce salmonellae that have been tried experimentally include spraying with 200ppm chlorine solutions, (Dixon and Pooley, 1961; Thomson *et al.*, 1976) sodium hypochlorite (Wabeck *et al.*, 1968), immersion in hot 3% succinic acid (Juven *et al.*, 1974) and immersion in 0.5% glutaraldehyde, pH 8.6 (Thomson *et al.*, 1977). In general, these treatments have led to adverse changes in the appearance and odor in the finished product. In the United Kingdom, poultry carcasses are often treated with polyphosphates for moisture loss control and this treatment apparently increases the susceptibility of salmonellae to death during cold storage (Foster and Mead, 1976). At the present, emphasis is on controlling salmonella during poultry processing by improving inplant processing conditions.

Immersion chilling with simultaneous agitation (spin chilling) is widely practiced in the poultry industry. Reddy *et al.*, (1978) found that spin chilling of turkeys, when practiced under proper operating conditions and with properly operated and maintained equipment, was highly effective in reducing microbial counts. No salmonellae were recovered in any of the samples taken. Present U.S. Department of Agriculture regulations specify chilling all carcasses to an internal temperature of 5°C (U.S. Department of Agriculture, 1978a).

There is also considerable interest in water conservation in poultry processing. Lillard (1978a,b) evaluated the feasibility of recycling broiler chicken chiller water or gizzard splitter water. Filtering bird chiller water with diatomaceous earth using a vertical tank leaf filter followed by treatment with 26-29ppm chlorine gas eliminated fecal coliforms and salmonellae.

There have been a number of studies on the effects of heating on and the thermal resistance of various *Salmonella* serotypes in chicken and turkey products (Angelotti *et al.*, 1960b; Bayne *et al.*, 1965; Bryan *et al.*, 1968; Bryan and McKinley, 1974; Dawson, 1966; Hussemann and Buyske, 1954; Klose *et al.*, 1971, Malone and Watson, 1970; Rogers and Gunderson, 1958; Woodburn *et al.*, 1962). Typical thermal death times or decimal reduction times are as follows:

(1) for *S. oranienburg* inoculated into turkey meat. F=3.5 min, at 60°C (10^4 cells/g. of meat) (Cotterill *et al.*, 1977);

(2) for *S. typhimurium*, and *S. senftenberg* in chicken meat, D= 5 min at 60°C and 10–15 min at 65°C respectively (3×10^8 cells/g. of meat) (Bayne *et al.*, 1965)

(3) for *S. senftenberg* and *S. manhattan* in chicken a la king, D=9.61 min and 0.40 min, at 140°F (10^7 cells/tube) (Angelotti *et al.*, 1960c)

Studies have also been conducted on the treatment of poultry with ionizing radiation for *Salmonella* control (Licciardello *et al.*, 1970; Previte *et al.*, 1970). Typical doses for 90% reduction in gamma radiation treatment of chicken meat, for *S. oranienburg* and *S. typhimurium*, were 24 and 20 Kilorads, respectively, in air pack and 39 and 31 Kilorads, respectively, in vacuum pack, (Licciardello *et al.*, 1970).

Eggs.—The role of eggs and egg products contaminated with *Salmonella* in food-borne illness has been recognized for many years. (Dack, 1956; Hobbs, 1961, 1962; Taylor and McCoy, 1969; Thatcher and Montford, 1962). Eggs may become contaminated with salmonellae while in the oviduct, or more frequently, by penetration of these organisms into the egg after laying. (Stokes *et al.*, 1956). The importance of *Salmonella* contamination of processed egg products was first recognized in the course of producing dehydrated eggs for military rations during World War II. (See Hinshaw and McNiel, 1951 for a review of the earlier literature). It was recognized that heat-treatment of egg products was necessary for *Salmonella* control (Schneider, 1951; Solowey and Calesnick, 1948; Solowey *et al.*, 1947, 1948; Winter, 1946).

However, no amount of control during processing can take the place of control of sanitary conditions and incidence of salmonella infections on farms where eggs are produced. The U.S. Department of Agriculture in cooperation with the states has developed a National Poultry Improvement Plan which includes salmonella control.

Washing eggs to remove adhering soil and organic matter can lead to *Salmonella* infection (March, 1969). Use of hot water in washing (150°F for 3 min) may reduce counts markedly (Bierer and Barnett, 1965). Also, treating shell eggs with disinfectants has been investigated, but did not eliminate salmonellae from the membrane after penetration of the shell. (Bierer and Barnett, 1962; Rizk *et al.*, 1966).

Methods that have been investigated for treating egg products during processing for *Salmonella* control include use of salmonella antagonists during egg white fermentation (Flippin and Mickelson, 1960; Mickelson and Flippin, 1960), and treatment with acetic acid (Garibaldi, 1968); hydrogen peroxide (Rogers, 1966) ethylene oxide (Sair, 1966); or ionizing radiation (Comar *et al.*, 1963; Lineweaver, 1966), ultraviolet radiation (Ijichi, 1966; Ijichi *et al.*, 1964) and microwave cooking (Baldwin *et al.*, 1968).

There are many published reports on processes and conditions for thermal destruction of salmonellae in egg products (Angelloti *et al.*, 1960 a, b, c; Anellis *et al.*, 1954; Ayres and Slosberg, 1949; Beloin and Schlosser, 1963; Bergquist, 1966; Clarenburg and Burger, 1950; Corry, 1974; Dabbah *et al.*, 1971; Garibaldi *et al.*, 1969 a, b; McBee and Cotterill, 1971; Putnam, 1966). Table 5.8 summarizes the results of some of these studies.

One of the problems in thermal processing of egg products, particularly egg whites is alteration of the functional properties of the heated product (Banwart and Ayres, 1956, 1957; Forsythe, 1966). Attempts have been made to reduce the thermal processing times required to destroy salmonellae in eggs by the use of various chemical treatments including adjustment of pH and adding chelating agents (Garibaldi *et al.*, 1969a) and addition of β-propionolactone, ethylene oxide and butadiene dioxide (Lategan and Vaughn, 1964). The pH of egg products is an important factor in thermal resistance

TABLE 5.8. SUMMARY OF SELECTED STUDIES ON THERMAL RESISTANCE OF SALMONELLAE IN EGG PRODUCTS

Serotype	Product	Temperature	D value (min)	Reference
Salmonella typhimurium	Egg white, pH 9.2	54.8°C	0.94	Garibaldi, et al., 1969
	Egg yolk	140°F	0.4	"
	Whole egg, pH 7.5	140°F	0.28	"
S. senftenberg775W	Whole egg, pH 8.0	140°F	1.5	Anelis et al., 1954
S. pullorum	Whole egg, pH 8.0	140°F	0.18	"
S. senftenberg775W	albumen, pH 9.1	57.8°C	2.1–2.4	Corry and Barnes, 1968
S. typhimurium	albumen, pH 9.1	57.8°C	0.13	"

of salmonellae (Cotterill, 1968, Corry and Barnes, 1968). Sugar and salt tend to increase the resistance of salmonellae in eggs, and thermal processes for the destruction of egg products containing these additives must be adjusted accordingly (Cotterill and Glauert, 1967, 1969, 1971).

U.S. Department of Agriculture regulations require the pasteurization of liquid egg albumen at 134°F for 3.5 min. or 132°F for 6.2 min; whole egg at 140°F for 3.5 min, plain yolk at 142°F for 3.5 min or 140°F for 6.2 min and yolk with 2% or more sugar or 2–12% salt added, 146°F for 3.5 min or 144°F for 6.2 min. Spray-dried albumen must be heated to 130°F or greater and held at that temperature not less than 7 days and until it is negative for *Salmonella*.

Meat

Fresh meat and improperly processed meat products have long been recognized as a source of *Salmonella* infections in humans. Both beef and pork products can be contaminated with salmonellae at the processing plant or at retail and food service outlets (Akman and Park, 1974; Bailey et al., 1972; Bryan, 1978b, Childers and Keahey, 1970; Childers et al., 1973; Davidson and Webb, 1973; Field et al., 1977; Galton et al., 1954; Goepfert and Chung, 1970a, b; Goepfert and Kim, 1975; Jordan et al., 1973; Palumbo and Alford, 1970; Patterson, 1969; Rey et al., 1971; Takacs and Nagy, 1975; Takacs and Simonffy, 1970; Tiwari and Maxcy, 1972; Weissman and Carpenter, 1969; Williams and Spencer, 1973). In the past few years, there have been a number of food-borne outbreaks of salmonellosis attributed to inadequately precooked roasts of beef (Center for Disease Control, 1976, a, b, 1977a).

Salmonellae may survive for extended periods in luncheon meats or fermented or non-fermented sausage products for extended periods in refrigerated storage (Goepfert and Chung, 1970, a, b; Smith et al., 1975 a, b). Heating to an internal temperature of 57.8 to 58.9 C for 3.5 hours followed by drying at 21C at 50–55% relative humidity for four days was required to reduce salmonellae in an inoculated nonfermented snack sausage to below detectable levels (Smith et al., 1977).

Cooking beef roasts contaminated with S. typhimurium to an internal temperature of 141.5°F did not result in a Salmonella-free product (Blankenship, 1978). U.S. Department of Agriculture regulations require the cooking of commercial cooked beef and roast beef to a minimum temperature of 145°F in all parts of the meat or at times and temperatures giving an equivalent thermal process (U.S. Department of Agriculture, 1978b).

Milk and Dairy Products

There have been many food-borne outbreaks of salmonellosis attributed to milk and dairy products such as cheese and ice cream (Armstrong, 1970; Julseth and Deibel, 1969; Marth, 1969). Salmonella have been isolated on many occasions from naturally contaminated dry milk products (Ray et al., 1971 c, d) and these organisms grow readily in milk and related dairy products (Subramanian and Marth, 1968, Marth 1969).

S. typhi and other salmonellae may survive for extended periods in cheese (Goepfert et al., 1970; Moquot et al., 1963; Park, et al., 1970 a, b; Wade and Lewis, 1928; White and Custer, 1976). Also, survival of salmonellae has been demonstrated in cultured dairy products (McDonough et al., 1970; Park and Marth, 1972 b, c). However, S. typhimurium did not survive in cottage cheese whey at pH 4.5 - 4.6 at 25–35°C longer than 3 days (Westhoff and Engler, 1973).

Heat treatment of skim milk at 80 to 121°C stimulated the growth of S. typhimurium over that obtained in non-heated skim milk (Singh and Mikolajcik, 1971). Because of the problem of survival of salmonellae in dry milk products, there has been considerable attention given to the thermal resistance of salmonellae isolated from these products and on the survival of salmonellae during spray drying (Dabbah et al., 1971a; Dega et al., 1972; Li Cari and Potter, 1970 a, b, c, d; McDonough and Hargrove, 1969; Read et al., 1968, Thomas, et al., 1966).

Typical thermal resistance (D) values of salmonellae in sterile whole milk are 3.6 to 5.7 sec at 62.8 C; 1.1 to 1.8 sec at 65.6 C and 0.28 to 0.52 sec at 68.3 C. *S. senftenberg* 775W had higher D values of 34.0 sec at 65.6 C and 10.0 sec at 68.3 C than other serotypes (Read *et al*, 1968). *S. tennessee* was considerably more resistant to spray drying in skim milk and showed a D value of 5.3 sec at 65.6 C which was 2 to 5 times greater than the values obtained with *S. typhimurium*, *S. thompson* or *S. kentucky* (LiCari and Potter, 1970d).

Salmonellae may be aerosolized during milk processing operations. The survival of *S. new brunswick* aerosolized from skim milk was greater than that after aerosolization from distilled water which indicated a possible protective effect of skim milk solids (Starsky *et al.*, 1972 a, b).

Fish

It was pointed previously that *Salmonella* contamination of fish takes place during processing and handling. Smoked fish has been implicated as a vehicle for salmonellosis (Olitzky *et al.*, 1956) and shellfish have been associated with many outbreaks over the years.

In addition to thermal processing, ionizing radiations (Matches and Liston, 1968a, Shiflet *et al.*, 1968) and microwave heating (Baldwin *et al.*, 1971) have been investigated for use in destroying salmonellae in fishery products. In the case of microwave cooking of fish, including tuna pies and casseroles and fish fillets dishes and sticks, treatment was required for longer than usual times to achieve lethal temperatures (Baldwin *et al.*, 1971).

Miscellaneous Foods

Confectionery.—Salmonellosis outbreaks have been attributed to confectionery (Harvey *et al.*, 1961) and confectionery ingredients such as chocolate (D'Aoust, 1978, D'Aoust *et al.*, 1975, Raines and Swanson, 1975) and desicated coconut (Wilson and McKenzie, 1955, Seiler, 1960). The survival and heat resistance of salmonellae in milk chocolate has been investigated (Barille and Cone, 1970; Barille *et al.*, 1970, Goepfert and Biggie, 1968).

Typical D values for *S. typhimurium* and *S. senftenberg* ranged from 720 to 1050 min and 360 to 480 min, respectively, at 70°C to 72–78 min and 36 to 42 min, respectively, at 90°C. under dry heat conditions. (Goepfert and Biggie, 1968).

Grain and Bakery Products.—Salmonellae have been isolated from wheat as a result of transmission by stored grain insects

(Crumbine *et al.*, 1971). Salmonellae can survive in wheat stored at 25C at relative humidities from 7 to 22% for extended periods (Crumbine and Foltz, 1969).

Isolations of salmonellae have also been made from pasta products such as spaghetti and macaroni. Extrusion, drying and storage conditions used in spaghetti manufacture result in a loss in viability of *Salmonella* contaminants, but the organisms will survive in the product for extended periods (Walsh *et al.*, 1974). D values for *S anatum* in semolina dough decrease to a minimum at a water activity around 0.8 at 50–65°C and higher resistances were observed at lower and higher a_w values. Consequently, it is desirable to pasterize the dough at high a_w values (Hsieh *et al.*, 1976).

Food Service Products, and other Miscellaneous Foods.—Commercially wrapped sandwiches, prepared salads and barbecued food sold in restaurants and vending machines may contain salmonellae (Christiansen and King, 1971; Khan and McCaskey, 1973; Maxcy, 1978; McCroan *et al.*, 1964). These foods have also been implicated in food-borne salmonellosis. Health foods may also contain salmonellae (Thompson *et al.*, 1977). Also, salmonellae may survive in contaminated chip dips (Wekell and Martinsen, 1977), in fruit juices (Mossel and deBruin, 1960) and in a variety of dehydrated foods (Mossel and Shennan, 1976), frozen foods (Chou and Marth, 1969; Georgala and Hurst, 1963; Splittstoesser and Segen, 1970; Woodburn and Strong, 1960) and prepared and packaged foods (Adinarayanan *et al.*, 1965).

Salmonella Control

The World Health Organization has published a report on recommended methods of sampling and surveillance programs for food-borne diseases including salmonellosis (WHO, 1974) Watson, (1976) reviewed *Salmonella* control in Western European countries and concluded that improvements were needed in reducing *Salmonella* contamination of foodstuffs and in reducing the potential hazards associated with livestock slaughter.

In the United States the Food and Drug Administration considers foods containing salmonellae to be adulterated and subject to seizure under the provisions of the Food Drug and Cosmetic Act (Olson, 1975). Bryan (1978a,b) concluded that inadequate cooling of foods was the major factor in food-borne diseases including salmonellosis followed by inadequate cooking and reheating, and contamination of food by workers, utensils, equipment or food contact surfaces. The role of human fingers (Pether and Gilbert, 1971) and slicing machines

(Gilbert, 1969) must always be considered in any investigation of salmonella food poisoning.

METHODS FOR ISOLATION, IDENTIFICATION AND ENUMERATION OF SALMONELLAE

General Methods

There is an extensive number of publications on laboratory procedures for isolating, identifying and enumerating salmonellae in foods (AOAC, 1975, Food and Drug Administration, 1976; Galton et al., 1968; Harvey and Price, 1974; ICMSF, 1974, 1978; Lewis and Angelotti, 1964; Poelma and Silliker, 1976; Silliker and Greenberg, 1969; U.S. Department of Agriculture 1974).

The International Commission on Microbiological Specifications for Foods has sponsored extensive studies on analytical methods for detecting salmonellae in various types of foods (Gabis and Silliker, 1974 a, b, 1977; Erdman, 1974; Idziak et al., 1974; Silliker and Gabis, 1973, 1974).

The major problem in detecting salmonellae in foods is to enrich the small numbers of salmonellae to detectable levels while suppressing the growth of far larger numbers of other enteric bacteria such as coliforms and food spoilage microorganisms that may be present.

Conventional schemes for detecting salmonellae involve preenrichment of the food in a medium such as lactose broth (Hall et al., 1964; North, 1961) followed by selective enrichment in such media as selenite or tetrathionate broths (North and Bartram, 1953; Carlson and Snoeyenbos, 1974; Greenfield and Bigland, 1970; Kafel and Bryan, 1971; Palumbo and Alford, 1970; Whitehill and Gardner, 1970), plating on selective media such as brilliant green agar or other selective agars (Banwart and Ayres, 1953; Bisciello and Schrade, 1974; Fagerberg and Avens, 1976; Goo et al., 1975; Hoben et al., 1973, Littell, 1977; Moats, 1978; Mohr and Kinner, 1976; Reamer et al., 1974; Restaino et al., 1977; Wells et al., 1958). Suspect colonies on the selective plating media are isolated and confirmed by biochemical reactions (Cox and Williams, 1976; Eskenazi and Littell, 1978; Litchfield, 1973) and by serological typing with polyvalent, group, and type specific antisera. (See Litchfield 1973 for a more detailed review). Phage typing techniques can also be used to identify salmonellae. However, there are large numbers of phage types for each salmonella serotype. A large stock of bacteriophages must be maintained to make this system workable (Anderson, 1964;

Gershman, 1970, 1974 a, b, 1976 a, b, 1977; Guinee *et al.*, 1974; Sechter and Gerichter, 1968, 1969). In general, a wide range of modified methods has been developed for isolating and identifying salmonellae in specific foods (Andrews *et al.*, 1976, 1977; Edel and Kempelmacher, 1973; Hall, 1969; Hargrove *et al.*, 1971; Hoben *et al.*, 1973a,b; Iveson *et al.*, 1964; Iveson and MacKay-Scollay, 1972; Sadivski, 1977; VanSchothorst and VanLeusden, 1972, 1975a,b; VanSchothorst *et al.*, 1977; Wells *et al.*, 1958, 1971; Wilson *et al.*, 1975).

Sampling procedures are also very important in the detection of *Salmonella* (Cox and Blakenship, 1975; Cox *et al.*, 1978; Huhtanen *et al.*, 1972) The ICMSF publications (1974, 1978) provide an extensive discussion of sampling plans for detecting salmonellae in foods.

Enumeration of salmonellae in foods is difficult because of the small numbers present. Most probable number methods are useful, provided that the sample is prepared properly to eliminate clumps and disperse salmonellae (Hall *et al.*, 1964). Direct plating is not too useful since the biochemical reactions of salmonellae on selective media are usually obscured by the other microorganisms present.

Improved and Rapid Methods

In recent years a number of commercial test "kits" containing a battery of biochemical reaction tubes have been developed (Cox and Mercuri, 1976, Cox *et al.*, 1977; Guthertz and Okoluk, 1978; Hansen *et al.*, 1974). These "kits" enable the biochemical characterization of *Salmonella* isolates without preparing and maintaining a stock of the numerous media that would be required ordinarily. These products vary in their ability to consistently give suitable biochemical responses. A careful check of any suspect cultures using conventional biochemical procedures should be made where the "kit" test result is variable.

Fluorescent antibody (FA) techniques offer a rapid and reliable means for identifying salmonellae isolated from foods. (Insalata *et al.*, 1972, 1975; Fantasia *et al.*, 1975; Ladiges *et al.*, 1974; Laramore and Moritz, 1969; Mohr *et al.*, 1974). The FA staining procedure is performed on a sample taken directly from the enrichment broth, eliminating the selective plating and biochemical test steps in the conventional procedure.

Thomason, (1976) and Thomason and Dodd (1976) provide a discussion of this procedure and its limitations. For example, non-specific staining may be a problem (Tharrington *et al.*, 1978). Preparation of a specially purified conjugate from the immunoglobulin G fraction of *Salmonella* polyvalent flagellar antiserum is a means for controlling this problem (Swaminathan *et al.*, 1978a).

Rapid FA methods have been developed for detecting salmonellae in milk (Reamer and Hargrove, 1972; Reamer et al., 1969, 1974). Recently, automated fluorescent antibody procedures have been developed and evaluated (Munson et al., 1976; Thomason et al., 1975).

Another rapid method that has received considerable attention is enrichment serology. In this procedure a sample from an enrichment broth is used to inoculate "M" broth followed after 6 to 8 hr. incubation at 35°C by serological identification using a pooled poly "H" antiserum. (Sperber and Deibel, 1969; Fantasia et al., 1969; Hilker, 1975; Hilker and Solberg, 1973).

Other recent novel procedures include pyrolysis gas liquid chromatography (Emswiler and Kotula, 1978), enzyme-labelled antibodies (Krysinski and Meinsch, 1977), a timed release capsule procedure (Sveum and Hartman, 1977) and membrane filter disc immunoimmobilization (Swaminathan et al., 1978b). None of these procedures has been in use for a sufficient period to permit a definitive evaluation.

Any new procedure for detecting salmonellae in food products must be compared on a statistical basis with conventional methods such as the Official AOAC method (Association of Official Analytical Chemists, 1975) before any conclusions can be made regarding its usefulness. Rapid methods are useful in screening large numbers of food samples. However, conventional cultural and serological methods must be used to confirm suspect samples.

REFERENCES

ADINARAYANAN, N., V. D. FOLTZ and F. McKINLEY. 1965. Incidence of salmonellae in prepared and packaged foods. J. Infect. Dis. 115:19–26.

AKMAN, M. and R. W. A. PARK. 1974. The growth of salmonellae on cooked cured pork. J. Hyg. (Camb.) 72:369–378.

ALFORD, J. A., and S. A. PALUMBO. 1969. Interaction of salt, pH, and temperature on the growth and survival of salmonellae in ground pork. Appl. Microbiol. 17:528–532.

ANDERSON, E. S. 1964. Phase typing of Salmonella other than S. typhi. p. 89–110. In E. Van Oye (ed.), the world problem of salmonellosis. Junk, the Hague.

ANDERSON, E. S., E. J. THRELFALL, J. M. CARR, M. M. McCONNELL, and H. R. SMITH. 1977. Clonal distribution of resistance plasmid-carrying Salmonella typhimurium mainly in the Middle East. J. Hyg. (Camb.) 79:425–461.

ANDERSON, G. D. and D. R. LEE. 1976. Salmonella in horses: a source of contamination of horsemeat in a packing plant under Federal inspection. Appl. Environ. Microbiol. 31:661–663.

ANDREWS, W. H., C. D. DIGGS, J. J. MIESCHIER, C. R. WILSON, W. N. ADAMS, S. A., FURFERI, and J. F. MUSSELMAN. 1976. Validity of members of the total caliform and fecal caliform groups for indicating the presence of *Salmonella* in the Quahaug, *Mercenaria mercenaria* J. Milk Food Technol. 39:322–324.

ANDREWS, W. H., C. R. WILSON, A. ROMERO, and P. L. POELMA. 1975. The Moroccan food snail, *Helix aspersa*, as a source of *Salmonella* Appl. Microbiol. 29:328–330.

ANDREWS, W. H., C. R. WILSON, P. L. POELMA, and A. ROMERO. 1977. Comparison of methods for the isolation of *Salmonella* from imported frog legs. Appl. Environ. Microbiol. 33:65–68.

ANGELOTTI, R., M. J. FOTER, and K. H. LEWIS. 1961. Time-temperature effects on salmonellae and staphylococci in foods. I. Behavior in refrigerated foods. Amer. J. Public Health 51:76–83.

ANGELOTTI, R., M. J. FOTER, and K. H. LEWIS. 1961. Time-temperature effects on salmonellae and staphylococci in foods. II. Behavior of warm holding temperatures. Amer. J. Public Health. 51:83–88.

ANGELOTTI, R., M. J. FOTER, and K. H. LEWIS. 1961. Time-temperature effects on salmonellae and staphylococci in foods. III. Thermal death time studies. Appl. Microbiol., 9:308–315.

ANGELOTTI, R., G. C. BAILEY, M. J. FOTER and K. H. LEWIS. 1961. Salmonella infantis isolated from ham in food poisoning incident. Public Health Rep. 76:771–776.

ANELLIS, A., J. LUBAS, and M. M. RAYMAN. 1954. Heat resistance in liquid eggs of some strains of the genus Salmonella. Food Res. 19:337–395.

ARMSTRONG, R. W., T. FODOR, G. T. CURTIS, A. B. COHOEN, W. T. MARTIN and J. FELDMAN 1970. Endemic *Salmonella* gastroenteritis due to contaminated imitation ice cream. Amer. J. Epidemiol. 91:300–307.

ASH, I., G. D. W. McKENDRICK, M. H. ROBERTSON, and H. L. HUGHES. 1964. Outbreak at typhoid fever connected with corned beef. Brit. Med. J. (1), 1474–1478.

ASSOCIATION OF OFFICIAL ANALYTICAL CHEMISTS. 1975. Official methods of analysis 12th ed., Assoc. Offic. Anal. Chem., Washington, DC secs. 46.013–46.026.

AYRES, J. C. and H. M. SLOSBERG 1949. Destruction of salmonella in egg albumen. Food Technol. 3:180–183.

BAILEY, G. K., P. K. FRASER, C. P. WARD, G. BOUTTELL, and E. KINNEAR. 1972. Enteritis due to *S. panama* from infected ham. J. Hyg. (Camb.) 70:113–119.

BAINS, B. S. and M. A. MACKENZIE. 1974. Transmission of Salmonella through an integrated poultry organization. Poultry Sci. 53:1114–1118.

BAIRD-PARKER, A. C., M. BOOTHROYD, and E. JONES. 1970. The

effect of water activity on the heat resistance of heat sensitive and heat resistant strains of salmonellae. J. Appl. Bacteriol. 33:515–522.

BAKER, R. C. 1974. Microbiology of eggs. J. Milk Food Technol. 37:265–268.

BALDWIN, R. E., M. CLONIGER, and M. L. FIELDS. 1968. Growth and destruction of *Salmonella typhimurium* in white foam products cooked by microwaves. Appl. Microbiol. 16:1929–1934.

BALDWIN, R. E., M. L. FIELDS, W. C. POON, and B. KORSCHGEN. 1971. Destruction of *Salmonella* by microwave heating of fish with implications for fish products. J. Milk Food Technol. 34:467–470.

BANWART, G. J., and J. C. AYRES. 1953. Effect of various enrichment broths and selective agars upon the growth of several species of *Salmonella*. Appl. Microbiol. 1:296–301.

BANWART, G. J., and J. C. AYRES. 1956. The effect of high temperature storage on the content of Salmonella and on the functional properties of dried egg white. Food Technol. 10:68–73.

BARNWART, G. J. and J. C. AYRES, 1957. The effect of pH on the growth of Salmonella and functional properties of liquid egg white. Food Technol. 11:244–246.

BARRILE, J. C., and J. F. CONE. 1970. Effect of added moisture on the heat resistance of Salmonella anatum in milk chocolate. Appl. Microbiol. 19:177–178.

BARRILE, J. C., J. F. CONE, and P. G. KEENEY. 1970. A study of salmonellae survival in milk chocolate. Manufacturing Confectioner 50:34–39.

BAYNE, H. G., J. A. GARIBALDI and H. LINEWEAVER. 1965. Heat resistance of *Salmonella tphimurium* and *Salmonella senftenberg* 775W in chicken meat. Poultry Sci. 44:1281–1284.

BELOIN, A., and G. C. SCHLOSSER. 1963. Adequacy of cooking procedures for the reduction of salmonellae. Amer. J. Public Health. 53:782–791.

BERQUIST, D. H. 1966. Heat pasteurization of plain egg white and egg white powder. pp. 65–67 In: The Destruction of Salmonellae. U.S. Department of Agriculture ARS 74–37 Albany, California.

BEUCHAT, L. R., and E. K. HEATON. 1975. *Salmonella* survival on pecans as influenced by processing and storage conditions. Appl. Microbiol. 29:795–801.

BIERER, B. W. and B. D. BARNETT. 1962. Killing *Salmonella* on egg shells with disinfectants. J. Amer. Vet. Med. Assoc. 140–159–161.

BIERER, B. W. and B. D. BARNETT. 1965. Killing *Salmonella* on egg shells by increasing wash temperature. J. Amer. Vet. Med. Assoc. 146:735–736.

BISCIELLO, N. B., and T. P. SCHRADE. 1974. Evaluation of Hektoen enteric agar for the detection of *Salmonella* in foods and feeds. J. Assoc. Off. Anal. Chem. 57:992–996.

BLACK, R. E., R. C. COX and M. A. HORWITZ. 1978. Outbreaks of food-borne disease in the United States, 1975. J. Infectious Diseases. 137:213–218.

BLANKENSHIP, L. C. 1978. Survival of *Salmonella typhimurium* experimental contaminant during cooking of beef roasts. Appl. Environ. Microbiol. 35:1160–1165.

BLANKENSHIP, L. C. and N. A. COX. 1976. Modified water rinse sampling for sensitive non-adulterating salmonellae detection on eviscerated broiler carcasses. J. Milk Food Technol. 39:680–681.

BOWMER, E. J. 1965. Salmonellae in foods-a review. J. Milk Food Technol. 28:74–86.

BRYAN, F. L. 1975. Status of food-borne disease in the United States. J. Environ. Health. 38:74–83.

BRYAN, F. L. 1977. Diseases transmitted by foods contaminated by wastewater. J. Food Prot. 40:45–56.

BRYAN, F. L. 1978a. Impact of food-borne diseases and methods of evaluating control programs. J. Environ. Health. 40:315–323.

BRYAN, F. L. 1978b. Factors that contribute to outbreaks of food-borne disease. *J. Food Protection.* 41:816–827.

BRYAN, F. L., J. C. AYRES, and A. A. KRAFT, 1968a. Contributory sources of salmonellae on turkey products. Amer. J. Epidemiol. 87:578–591.

BRYAN, F. L., J. C. AYRES, and A. A. KRAFT. 1968b. Salmonellae associated with further-processed turkey products. Appl. Microbiol. 16:1–9.

BRYAN, F. L., J. C. AYRES, and A. A. KRAFT. 1968c. Destruction of salmonellae and indicator organisms during thermal processing of turkey rolls. Poultry Sci. 47:1966–1978.

BRYAN, F. L. and T. W. McKINLEY. 1974. Prevention of food-borne illness by time-temperature control of thawing cooking, chilling and reheating turkeys in school lunch kitchens. J. Milk Food Technol. 37:420–429.

BUCHANAN, R. E. and N. E. GIBBONS, eds. 1974. Bergey's Manual of Determinative Bacteriology. 8th Ed. Williams & Wilkins, Baltimore, Md.

BUSTA, F. F., and M. L. SPECK. 1968. Antimicrobial effect of cocoa on *Salmonella*. Appl. Microbiol. 16:424–425.

BUTTIAUX, R. 1963. Salmonella problems in the sea. *In*: Fish as Foods. Vol. 2, pp. 503–519, G. Borgstrom (ed.) Academic Press, New York.

CAMPBELL, A. G. and J. GIBBARD. 1944. The survival of *E. typhosa* in cheddar cheese manufactured from infected raw milk. Can. J. Public Health 35:158–164.

CARAWAY, C. T., and J. M. BRUCE. 1961. Typhoid fever epidemic following a wedding reception. Public Health Rep. 76:427–430.

CARLSON, V. L. and G. M. SNOEYENBOS. 1970. Effect of moisture on salmonella populations in animal feeds. Poultry Sci. 49:717–725.

CARLSON, V. L., and G. H. SNOEYENBOS. 1974. Comparative efficacies of selenite and tetrathionate enrichment broths for the isolation of salmonella serotypes. Am. J. Vet. Res. 35:711–718.

CENTER FOR DISEASE CONTROL. 1976a. *Salmonella saint-paul* in precooked roasts of beef. Morbid. Mortal. Weekly Rep. 25:34–39.

CENTER FOR DISEASE CONTROL. 1976b. *Salmonella boris-morbificans* in precooked roasts of beef. Morbid. Mortal. Weekly Rep. 25–333.

CENTER FOR DISEASE CONTROL. 1976c. Food-borne and water-borne disease outbreaks. U.S. Department of Health Education and Welfare. Pub. CDC 76–8185, Atlanta, Georgia.

CENTER FOR DISEASE CONTROL. 1977a. Multi-state outbreak of *Salmonella newport* transmitted by pre-cooked roasts of beef. Morbid. Mortal. Weekly Rep. 26:277–278.

CENTER FOR DISEASE CONTROL. 1977b. Food-borne and water-borne disease outbreaks. Annual Summary U.S. Department of Health, Education and Welfare, Pub. CDC 78–8185, Atlanta, Georgia.

CHILDERS, A. B., and E. E. KEAHEY. 1970. Sources of *Salmonella* contamination of meat following approved livestock slaughtering procedures. J. Milk Food Technol. 33:10–12.

CHILDERS, A. B., E. E. KEAHEY and P. G. VINCENT. 1973. Sources of salmonellae contamination of meat following approved livestock slaughtering procedures II. J. Milk Food Technol. 36:635–638.

CHRISTIAN, J. H. B., and W. J. SCOTT. 1953. Water relations of salmonellae at 30°C. Australian J. Biol. Sci. 6:565–573.

CHRISTIAN, J. H. B. 1955. The influence of nutrition on the water relations of *Salmonella oranienburg*. Austral. J. Biol. Sci. 8:75–82.

CHOU, C. C. and E. H. MARTH. 1969. Microbiology of some frozen and dried foodstuffs. J. Milk Food Technol. 32:372–378.

CHRISTIANSEN, L. N. and N. S. KING. 1971. The microbial content of some salads and sandwiches in retail outlets. J. Milk Food Technol. 34:289–293.

CHUNG, K. C., and J. M. GOEPFERT. 1970. Growth of *Salmonella* at low pH. J. Food Sci. 35:326–328.

CLARK, C. W., and Z. J. ORDAL. 1969. Thermal injury and recovery of *Salmonella typhimurium* and its effect on enumeration procedures. Appl. Microbiol. 18:332–336.

CLARENBURG, A. and H. C. BURGER. 1950. Survival of *Salmonellae* in boiled duck's eggs. Food Res. 15:340–342.

CLISE, J. D., and E. E. SEVECKER. 1965. Salmonellae from animal byproducts. Public Health Rep. 80:899–905.

CODE OF FEDERAL REGULATIONS 1978a. Title 7. Chapter 1, Part 59, Inspection of Eggs and Egg Products. 59.500–59, 575, U.S. Government Printing Office, Washington, D.C.

CODE OF FEDERAL REGULATIONS 1978b. Title 9. Animal and Animal Products. Part 381, Poultry Products Inspection Regulations. U.S. Government Printing Office, Washington, D.C.

COHEN, M., and P. BLAKE. 1977. Trends in food-borne salmonellosis outbreaks 1963–1975. J. Food Prot. 40:798–800.

COMMUNICABLE DISEASE CENTER, 1965. Proceedings, National Conference on Salmonellosis. Public Health Service Pub. No. 1262 U.S. Department of Health, Education and Welfare, Washington, D.C.

CORRY, J. E. L. 1974. The effect of sugar and polyols on the heat resistance of salmonellae. J. Appl. Bacteriol. 37:31–43.

CORRY, J. E. L., and E. M. BARNES. 1968. The heat resistance of salmonellae in egg albumen. Brit. Poultry Sci. 9:253–260.

CORRY, J. E. L., A. G. KITCHELL, and T. A. ROBERTS. 1969. Interactions with recovery of Salmonella typhimurium damaged by heat or gamma radiation. J. Appl. Bacteriol. 32:415.

CORRY, J. E. L. and T. A. ROBERTS. 1970. A note on the development of resistance to heat and gamma radiation in Salmonella. J. Appl. Bacteriol. 33:733–737.

COTTERILL, O. J. 1968. Equivalent pasteurization temperatures to kill salmonellae in liquid egg white at various pH levels. Poultry Sci. 47:354–65.

COTTERILL, O. J., and J. GLAUERT. 1967. Pasteurization requirements to destroy Salmonella in egg products containing sugar and salt. Poultry Sci. 46:1248.

COTTERILL, O. J., and J. GLAUERT. 1969. Thermal resistance of salmonellae in egg yolk products containing sugar or salt. Poultry Sci. 48:1156–1166.

COTTERILL, O. J., and J. GLAUERT. 1971. Thermal resistance of salmonellae in egg yolk containing 10% sugar or salt after storage at various temperatures. Poultry Sci. 50:109–115.

COTTERILL, O. J., J. GLAUERT, and G. F. KRAUSE. 1973. Thermal destruction curves for Salmonella oranienberg in egg products. Poultry Sci. 52:568–577.

COTTERILL, O. J., H. P. GLAUERT, and W. D. RUSSELL. 1977. Microbial counts and thermal resistance of Salmonella oranienberg in ground turkey meat. Poultry Sci. 56:1889–1892.

COUPER, W. R. M., K. W. NEWELL, and D. J. H. PAYNE. 1956. An outbreak of typhoid fever associated with canned ox-tongue. The Lancet, June 30, 1057–1059.

COX, N. A., and L. C. BLANKENSHIP. 1975. Comparison of rinse sampling methods for detection of salmonellae on eviscerated broiler carcasses, J. Food Sci. 40:1333–1334.

COX, N. A., F. McMAN, and D. Y. C. FUNG. 1977. Commercially available mini kits for identification of Enterobacteriaceae: A review, J. Food Protection. 40:866–872.

COX, N. A., and A. J. MERCURI. 1976. Rapid confirmation of suspect Salmonella colonies by use of the Minitek system in conjunction with serological tests. J. Appl. Bacteriol. 41:389–394.

COX, N. A., A. J. MERCURI, D. A. TANNER, M. O. CARSON, J. E. THOMSON, and J. S. BAILEY. 1978. Effectiveness of sampling methods for Salmonella detection on processed broilers. J. Food Protection 41:341–343.

COX, N. A., and J. E. WILLIAMS. 1976. A simplified biochemical system to screen salmonella isolates for serotyping. Poultry Sci. 55:1968–1971.

COMER, A. G., A. W. ANDERSON, and E. H. GARRAD. 1963. Gamma irradiation of Salmonella species in frozen whole eggs. Can. J. Microbiol. 9:321–327.

CRUMBINE, M. H., and V. D. FOLTZ. 1969. Survival of *Salmonella montevideo* on wheat stored at constant relative humidity. Appl. Microbiol. 18:911–914.

CRUMBINE, M. H., V. D. FOLTZ and J. O. MORRIS. 1971. Transmission of *Salmonella montevideo* in wheat by stored-product insects. Appl. Microbiol. 22:578–580.

CRAVEN, P. C., D. C. MACKEL, W. B. BAINE, W. H. BARKER, E. J. GANGAROSA, M. GOLDFIELD, H. ROSENFELD, R. ALTMAN, G. LACHAPELLE, J. W. DAVIS and R. C. SWANSON. 1975. International outbreak of *Salmonella eastbourne* infection traced to contaminated chocolate. Lancet 1:788–793.

DABBAH, R., W. A. MOATS and V. M. EDWARDS. 1971a. Heat survivor curves of food-borne bacteria suspended in commercially sterilized whole milk. I. Salmonellae. J. Dairy Sci. 54:1583–1588.

DABBAH, R., W. A. MOATS, and V. M. EDWARDS. 1971b. Survivor curves of selected *Salmonella enteritidis* serotypes in liquid whole egg concentrates at 60°C. Poultry Sci. 50:1772–1776.

DACK, G. M. 1956. Food Poisoning, 3rd Ed. The University of Chicago Press, Chicago, Ill. pp. 150–194.

DACK, G. M., and G. LIPPITZ. 1962. Fate of staphylococci and enteric microorganisms introduced into a slurry of frozen pot pies. Appl. Microbiol. 10:472–479.

D'AOUST, J. Y. 1977. *Salmonella* and the chocolate industry. A review. J. Food Protection. 40:718–727.

D'AOUST, J. Y. 1978. Recovery of sublethally heat-injured *Salmonella typhimurium* on supplemented plating media. Appl. Environ. Microbiol. 35:483–486.

D'AOUST, J. Y., B. J. ARIS, P. THISDELE, A. DURANTE, N. BRISSON, D. DRAGON, G. LACHAPELLE, M. JOHNSTON, and R. LAIDLEY. 1975. *Salmonella eastbourne* outbreak associated with chocolate. Can. Inst. Food Sci. Technol. J.8:181–184.

DAVIDSON, C. M., M. BOOTHROYD, and D. L. GEORGALA. 1966. Thermal resistance of *Salmonella senftenberg*. Nature 212:1060–1061.

DAVIDSON, C. M., and G. WEBB. 1973. The behavior of salmonellae in vacuum-packaged cooked cured meat products. Can. Inst. Food Sci. Technol. J. 6:41–44.

DAWSON, L. E. 1966. Destruction of salmonellae in turkey rolls and other meat. pp. 72–78 *In*: The Destruction of Salmonellae. U.S. Department of Agriculture ARS 74–37. Albany, California.

DEGA, C. A., J. M. GOEPFERT, and C. M. AMUNDSON. 1972. Heat resistance of salmonellae in concentrated milk. Appl. Microbiol. 23:415–420.

DIXON, J. M. S., and F. E. POOLEY. 1961. The effect of chlorination on chicken carcasses infected with salmonellae. J. Hyg. (Camb.) 59:343–348.

DOESBURG, J. J., E. C. LAMPRECHT, and M. ELLIOTT. 1970a. Death rates of salmonellae in fishmeals with different water activities. I. During storage. J. Sci. Food Agric. 21:632–635.

DOESBURG, J. J., E. C. LAMPRECHT, and M. ELLIOTT. 1970b. Death rates of salmonellae in fishmeals with different water activities. II. During heat treatment. J. Sci. Food Agric. 21:636–640.

DOUGHERTY, T. J. 1974. *Salmonella* contamination in a commercial poultry (broiler) processing operation. Poultry Sci. 53:814–821.

DUITSCHAEVER, C. L. 1977. Incidence of *Salmonella* in retailed raw cut-up chicken. J. Food Protection. 40:191–192.

DYER, J. K., A. W. ANDERSON, and P. DUTIYABODHI. 1969. Radiation survival of food pathogens in complex media. Appl. Microbiol. 14: 92–97.

EDWARDS, P. R., and W. H. EWING. 1972. Identification of *Enterobacteriaceae*, 3rd ed. Burgess Publishing Co., Minneapolis.

ELLIOTT, R. P., and P. K. HEINIGER. 1965. Improved temperature-gradient incubator and the maximal growth temperature and heat resistance of *Salmonella*. Appl. Microbiol. 13:73–76.

ENKIRI, N. K., and J. A. ALFORD. 1971. Relationship of the frequency of isolation of *Salmonella* to their resistance to drying and freezing. Appl. Microbiol. 21:381–382.

EMSWILER, B. J., and A. W. KOTULA. 1978. Differentiation of *Salmonella* serotypes by pyrolysis-gas-liquid chromatography of cell fragments. Appl. Environ. Microbiol. 35:97–104.

EPPS, N. A., and E. S. IDZIAK. 1970. Radiation treatment of foods. II. Public health significance of irradiation-recycled *Salmonella*. Appl. Microbiol. 19:338–344.

ERDMAN, I. E. 1974. ICMSF methods studies. IV. International collaborative assay for the detection of *Salmonella* in raw meat. Can. J. Microbiol. 20:715–720.

ESKENAZI, S., and A. M. LITTELL. 1978. Dulcitol-malonate-phenylalanine agar for the identification of *Salmonella* and other *Enterobacteriaceae*. Appl. Environ. Microbiol. 35:199–201.

FAGERBERG, D. J., and J. S. AVENS. 1976. Enrichment and plating methodology for *Salmonella* detection in food. A review. J. Milk Food Technol. 39:628–646.

FANTASIA, L. D., J. P. SCHRADE, J. F. YAGER, and D. DEBLER. 1975. Fluorescent antibody method for the detection of *Salmonella*: development, evaluation and collaborative study. J. Assoc. Off. Anal. Chem. 58:828–844.

FANTASIA, L. D., W. H. SPERBER, and R. H. DEIBEL. 1969. Comparison of two procedures for detection of salmonella in food, feed, and pharmaceutical products. Appl. Microbiol. 17:540–541.

FIELD, R. A., F. C. SMITH, D. D. DEANE, G. M. THOMAS, and A. W. KOTULA. 1977. Sources of variation at the retail level in bacteriological condition of ground beef. J. Food Protection 40:385–388.

FLIPPIN, R. S., and M. N. MICKELSON. 1960. Use of *Salmonellae* antagonists in fermenting egg white. I. Microbial antagonists of *Salmonellae*. Appl. Microbiol. 8:366–370.

FOLEY, A. R., and E. POISSON. 1945. A cheese-borne outbreak of typhoid fever. 1944. Can. J. Public Health. 35:116–118.

FOOD AND DRUG ADMINISTRATION. 1976. Bacteriological analytical manual, 4th ed. Food and Drug Administration, Rockville, Md.

FORSYTHE, R. H. 1966. Performance and stability of heat-pasteurized whole egg and yolk. pp. 52–55. *In*: The Destruction of Salmonellae. U.S. Department of Agriculture, Albany, California.

FOSTER, E. M. 1969. The problem of salmonellae in foods. Food Technol. 23:75–78.

FOSTER, R. D., and G. C. MEACH. 1976. Effect of temperature and added polyphosphate on the survival of salmonellae in poultry meat during cold storage. J. Appl. Bacteriol. 41:505–510.

FOX, M. D. 1974. Recent trends in salmonellosis epidemiology. J. Am. Vet. Med. Assoc. 165:990–993.

GABIS, D. A., and J. H. SILLIKER. 1974a. ICMSF methods studies. II. Comparison of analytical schemes for detection of *Salmonella* in high-moisture foods. Can. J. Microbiol. 20:663–669.

GABIS, D. A., and J. H. SILLIKER. 1974b. ICMSF methods studies. VI. The influence of selective enrichment media and incubation temperatures on the detection of *Salmonella* in dried foods and feeds. Can. J. Microbiol. 20:1509–1511.

GABIS, D. A., and J. H. SILLIKER. 1977. ICMSF methods studies. IX. The influence of selective enrichment broths, differential plating media, and incubation temperatures on the detection of *Salmonella* in dried foods and feed ingredients. Can. J. Microbiol. 23:1225–1231.

GALTON, M. M., W. D. LOWERY, and A. V. HARDY. 1954. Salmonella in fresh and smoked pork sausage. J. Infect. Dis. 95:232–235.

GALTON, M. M., D. C. MACKEL, A. L. LEWIS, W. C. HAIRE, and A. V. HARDY. 1955. Salmonellosis in poultry and poultry processing plants in Florida. Amer. J. Vet. Res. 16:132–137.

GALTON, M. M., G. K. MORRIS, and W. T. MARTIN. 1968. Salmonellae in foods and feeds. Review of isolation methods and recommended procedures. U.S. Dept. of Health, Education and Welfare/Public Health Service. Communicable Disease Center, Atlanta, Georgia.

GALTON, M. M., W. V. SMITH, H. B. McELRATH and A. B. HARDY. 1954. *Salmonella* in swine, cattle, and the environment of abbatoirs. J. Infectious Dis. 95:236–245.

GARIBALDI, J. A. 1966. Factors affecting the heat sensitivity of salmonellae. pp. 34–37. *In*: The Destruction of Salmonellae. U.S. Department of Agriculture ARS 74–37, Albany, California.

GARIBALDI, J. A. 1968. Acetic acid as a means of lowering heat resistance of salmonella in yolk products. Food Technol. 22:1031–1033.

GARIBALDI, J. A., and H. G. BAYNE. 1970. Production of acetoin and diacetyl by the genus *Salmonella*. Appl. Microbiol. 20:855–856.

GARIBALDI, J. A., K. IJICHI and H. G. BAYNE. 1969a. Effect of pH and chelating agents on the heat resistance and viability of *Salmonella typhimurium* Tm-1 and *Salmonella senftenberg* 775W in egg white. Appl. Microbiol. 18:318–322.

GARIBALDI, J. A., R. P. STRAKA, and K. IJICHI. 1969b. Heat resistance of *Salmonella* in various egg products. Appl. Microbiol. 17:491–496.

GARRETT, E. S., and R. HAMILTON. 1971. Sanitation guidelines for the control of *Salmonella* in the production of fish meal. U.S. Department of Commerce, NOAA Technical Report NMFS CIRC-354. U.S. Government Printing Office, Washington, D.C.

GEORGALA, D. L., and M. A. BOOTHROYD. 1965. A system for detecting salmonellae in meat and meat products. J. Appl. Bacteriol. 28:206–212.

GEORGALA, D. L., and A. HURST. 1963. The survival of food poisoning bacteria in frozen foods. J. Appl. Bacteriol. 26:346–358.

GERSHMAN, M. 1972. Preliminary report: A system for typing *Salmonella thompson*. Appl. Microbiol. 23:831–832.

GERSHMAN, M. 1974a. A phage typing system for *Salmonella anatum*. Avian Dis. 18:565–568.

GERSHMAN, M. 1974b. A phage typing system for *Salmonella newport*. Can. J. Microbiol. 20:769–771.

GERSHMAN, M. 1976a. A phage typing system for salmonellae: *Salmonella senftenberg*. J. Milk Food Technol. 39:682–683.

GERSHMAN, M. 1976b. Phage typing system for *Salmonella enteritidis*. Appl. Environ. Microbiol. 32:190–191.

GERSHMAN, M. 1977. A phage typing system for salmonellae: *Salmonella heidelberg*. J. Food Protection. 40:43–44.

GILBERT, R. J. 1969. Cross-contamination by cooked-meat slicing machines and cleaning cloths. J. Hyg. (Camb.) 67:249–254.

GLEGEN, W. P., M. P. HINES, M. KERBOUGH, M. E. GREEN, and J. KOOMAN. 1966. *Salmonella* in two processing plants. J. Amer. Vet. Med. Ass. 148:550–552.

GOEPFERT, J. M. 1969. The interrelationship between pH, temperature, oxygen supply and available water on growth and survival of *Salmonella* organisms with special reference to confectionary products. Manufacturing Confectioner 49:68–72.

GOEPFERT, J. M., and R. A. BIGGIE. 1968. Heat resistance of *Salmonella typhimurium* and *Salmonella senftenberg* 775W in milk chocolate. Appl. Microbiol. 16:1939–1940.

GOEPFERT, J. M. and K. C. CHUNG. 1970a. Behavior of salmonellae in sliced luncheon meats. Appl. Microbiol. 19:190–192.

GOEPFERT, J. M., and K. C. CHUNG. 1970b. Behavior of *Salmonella* during the manufacture and storage of a fermented sausage product. J. Milk Food Technol. 33:185–191.

GOEPFERT, J. M., and R. HICKS. 1969. Effect of volatile fatty acids on *Salmonella typhimurium*. J. Bacteriol. 97:956–958.

GOEPFERT, J. M., I. K. ISKANDER, and C. H. AMUNDSON. 1970. Relation of the heat resistance of salmonellae to the water activity of the environment. Appl. Microbiol. 19:429–433.

GOEPFERT, J. M. and H. U. KIM. 1975. Behavior of selected food borne pathogens in raw ground beef. J. Milk Food Technol. 38:449–452.

GOEPFERT, J. M., N. F. OLSON, and E. H. MARTH. 1968. Behavior of *Salmonella typhimurium* during manufacture and curing of Cheddar cheese. Appl. Microbiol. 16:862–866.

GOO, V. Y. L., G. Q. L. CHING, and J. M. GOOCH. 1975. Comparison of brilliant green agar and Hektoen enteric agar media in the isolation of salmonellae from food products. Appl. Microbiol. 26:288–292.

GREENBLATT, A. P., V. C. DALEY, and M. C. BRESLOW. 1947. Salmonella epidemic from commercially prepared sandwiches. Bull. U.S. Army Med. Dept. 71:345–348.

GREENFIELD, J., and C. H. BIGLAND. 1970. Selective inhibition of certain enteric bacteria by selenite media incubated at 35° and 43°C. Can. J. Microbiol. 16:1267–1271.

GUINEE, P. A. M., W. J. VAN LEEUWEN, and D. PRUYS. 1974. Phage typing of S. typhimurium in the Netherlands. 1. The phage typing system. Zbl. Bakt. Hyg., I. Abt. Orig. A.226:194–200.

GULASEKHARAM, J. T.: VELAUDAPILLAI and G. R. NILES. 1956. The isolation of Salmonella organisms from fresh fish sold in a Colombo fish market. J. Hyg. (Camb.). 54:581–584.

GUTHERTZ, L. S. and R. L. OKOLUK. 1978. Comparison of miniaturized multifast systems with conventional methodology for identification of Enterobacteriaceae from foods. Appl. Environ. Microbiol. 35:109–112.

HAGBERG, M. M., F. F. BUSTA, E. A. ZOTTOLA, and E. A. ARNOLD. 1973. Incidence of potentially pathogenic microorganisms in further-processed turkey products. J. Milk Food Technol. 36:625–634.

HALL, H. E., D. F. BROWN, and R. ANGELOTTI. 1964. The quantification of salmonellae in foods by using the lactose pre-enrichment method of North. J. Milk Food Technol. 27:235–240.

HALL, J. R. 1969. Enhanced recovery of Salmonella montevideo from onion powder by the addition of potassium sulfite to lactose broth. J. Assoc. Offic. Anal. Chem. 52:940–942.

HANSEN, S. L., D. R. HARDESTY, and B. M. MYERS. 1974. Evaluation of the BBL Minitek system for the identification of Enterobacteriaceae. Appl. Microbiol. 28:798–801.

HARGROVE, R. E., F. E. McDONOUGH and W. A. MATTINGLY. 1969. Factors affecting survival of Salmonella in cheddar and colby cheese. J. Milk Food Technol. 32:480–484.

HARGROVE, R. E., F. E. McDONOUGH, and R. H. REAMER. 1971. A selective medium and presumptive procedure for the detection of Salmonella in dairy products. J. Milk Food Technol. 34:6–11.

HARVEY, R. W. S. and T. H. PRICE. 1967. The isolation of salmonella from animal feeding stuffs. J. Hyg. (Camb) 65:237–244).

HARVEY, R. W. S. and T. H. PRICE. 1974. Isolation of salmonellas. Public Health Laboratory Service. Monograph Series No. 8. London. Her Majesty's Stationery Office.

HARVEY, R. W. S. and T. H. PRICE. 1978. A survey of Argentine and Lebanese bone for salmonellas with particular reference to the isolation of Salmonella typhimurium. J. Appl. Bacteriol. 44:241–247.

HARVEY, R. W. S., T. H. PRICE, A. R. DAVIS and R. B. MORLEY-DAVIES 1961. An outbreak of salmonella food poisoning attributed to bakers' confectionery. J. Hyg. (Camb) 59:105–108.

HILKER, J. S. 1975. Enrichment serology and fluorescent antibody procedures to detect salmonellae in foods. J. Milk Food Technol. 38:227–231.

HILKER, J. S., and M. SOLBERG. 1973. Evaluation of a fluorescent antibody-enrichment serology combination procedure for the detection of salmonellae in condiments, food products, food by-products and animal feeds. Appl. Microbiol. 26–751–756.

HINSHAW, W. R. and E. McNEIL. 1951. Salmonella infection as a food industry problem. Adv. Food Res. 3:209–240. Academic Press, New York.

HOBBS, B. C. 1961. Public Health significance of salmonella carriers in livestock and birds. J. Appl. Bacteriol 24:340–352.

HOBBS, B. C. 1962. Salmonellae, p. 224–248. In J. C. Ayres, A. A. Kraft, H. E. Snyder and H. W. Walker (ed.). Chemical and biological hazards in food, Iowa State University Press, Ames.

HOBBS, B. C. 1974. Microbiological hazards of meat production. Food Manufacture 49:29–34, 54.

HOBEN, D. A., D. H. ASHTON, and A. C. PETERSON. 1973a. A rapid presumptive procedure for the detection of *Salmonella* in foods and food ingredients. Appl. Microbiol. 25:123–129.

HOBEN, D. A., D. H. ASHTON and A. C. PETERSON. 1973b. Some observations on the incorporation of novobiocin into Hektoen enteric agar for improved *Salmonella* isolation. App. Microbiol. 26:126–127.

HORWITZ, M. A., R. A. POLLARD, M. H. MERSON, and S. M. MARTIN. 1977. A large outbreak of food-borne salmonellosis on the Navajo nation Indian reservation, epidemology and secondary transmission. Amer. J. Public Health 67:1071–1079.

HSIEH, F. H., K. ACOTT, and T. P. LABUZA. 1976. Death kinetics of pathogens in a pasta product. J. Food Sci. 41:516–519.

HUHTANEN, C. N., J. NAGHSKI, and E. S. DELLAMONICA. 1972. Efficiency of Salmonella isolation from meat and bone meal of one 300 g sample versus ten 30 g samples. Appl. Microbiol. 23:688–692.

HUSSEMANN, D. L. and J. K. BUYSKE. 1954. Thermal death time-temperature relationships of *Salmonella typhimurium* in chicken muscle. Food Res. 19:351–355.

HUSSEMANN, D. L. and M. A. WALLACE. 1951. Studies on the transmission of Salmonella by cooked fowl. Food Res. 16:89–96.

IDZIAK, E. S., J. M. A. AIRTH, and I. E. ERDMAN. 1974. ICMSF Methods Studies III. An appraisal of 16 contemporary methods used for detection of *Salmonella* in meringue powder. Can. J. Microbiol. 20:703–714.

IJICHI, K. 1966. Sensitivity of salmonellae to uv energy. pp. 50–52 In: The Destruction of Salmonellae. U.S. Department of Agriculture ARS 74–37. Albany, California.

IJICHI, K., O. A. HAMMERLE, H. LINENEAVER and L. KLINE. 1964. Effects of UV irradiation of egg liquids on Salmonella destruction and performance quality with emphasis on egg white. Food Technol. 18:124.

INTERNATIONAL COMMISSION ON MICROBIOLOGICAL SPECIFICATIONS FOR FOODS. 1974. Microorganisms in food. 2. Sampling for microbiological analysis: Principles and specific applications. University of Toronto Press. Toronto, Ontario, Canada. 213 pp.

INTERNATIONAL COMMISSION ON MICROBIOLOGICAL SPECIFICATIONS FOR FOODS. 1978. Microorganisms in Food 1: Their significance and methods of enumeration. University of Toronto Press. Toronto, Ontario, Canada. 434 pp.

INSALATA, N. F., W. G. DUNLAP, and C. W. MAHNKE. 1975. A comparison of cultural methods used with microcolony and direct fluorescent-antibody techniques to detect salmonellae. J. Milk Food Technol. 38:201–203.

INSALATA, N. F., C. W. MAHNKE, and W. G. DUNLAP. 1972. Rapid, direct fluorescent-antibody method for the detection of salmonellae in food and feeds. Appl. Microbiol. 24:645–649.

IVESON, J. B., N. KOVACS, and W. LAURIE. 1964. An improved method of isolating salmonellae from contaminated desiccated coconut. J. Clin. Pathol. 17:75–78.

IVESON, H. B., and E. M. MACKAY-SCOLLAY. 1972. An evaluation of strontium chloride. Rappaport and strontium selenite enrichment for the isolation of salmonellas from man, animals, meat products and abattoir effluents J. Hyg. (Camb.) 70:367–384.

JACOBS, J., P. A. M. GUINEE, E. H. KAMPELMACHER, and A. VAN KEULEN. 1963. Studies on the incidence of salmonella in imported fish meal (in English). Zentralb. fur Veterinarmedizen 10:542–550.

JANSSEN, D. W. and F. F. BUSTA. 1973. Injury and repair of several *Salmonella* serotypes after freezing and thawing in milk solids. J. Milk Food Technol. 36:520–522.

JAROLMEN, H., R. J. SAIRK and B. F. LANGWORTH. 1976. Effect of chlortetracycline feeding on the *Salmonella* reservoirs in chickens. J. Appl. Bacteriol. 40:153–161.

JOHNSON, M. G. and R. H. VAUGHN. 1969. Death of *Salmonella typhimurium* and *Escherichia coli* in the presence of freshly reconstituted garlic and onion. Appl. Microbiol. 17:903–905.

JORDAN, M. C., K. E. POWELL, T. E. COROTHERS, and R. J. MURRAY. 1973. Salmonellosis among restaurant patrons: The incisive role of a meat slicer. Amer. J. Public Health. 63:982–985.

JULSETH, R. M. and R. H. DEIBEL. 1969. Effect of temperature on growth of *Salmonella* in rehydrated skim milk from a food-poisoning outbreak. Appl. Microbiol. 17:767–768.

JULSETH, R. M., and R. H. DEIBEL. 1974. Microbial profile of selected spices and herbs at import. J. Milk Food Technol. 37:414–419.

JUVEN, B. J., N. A. COX, A. J. MERCURI, and J. E. THOMPSON. 1974. A hot acid treatment for eliminating *Salmonella* from chicken meat. J. Milk Food Technol. 37:237–239.

KAFEL, S. and F. L. BRYAN. 1977. Effects of enrichment media and incubation conditions on isolating salmonellae from ground meat filtrade. Appl. Environ. Microbiol. 34:285–291.

KAUFMANN, A. F. and J. C. FEELEY. 1968. Culture survey of *Salmonella* at a broiler-raising plant. Public Health Rep. 83:417–422.

KAUFFMANN, F. 1966. The Bacteriology of Enterobacteriaceae. Williams & Wilkins, Baltimore, Md.

KHAN, M., and M. KATAMAY. 1969. Antagonistic effect of fatty acids against *Salmonella* in meat and bone meal. Appl. Microbiol. 17:402–404.

KHAN, N. A. and J. A. McCASKEY. 1973. Incidence of salmonellae in commercially prepared sandwiches for the vending trade. J. Milk Food Technol. 36:315–316.

KLOSE, A. A., V. F. KAUFMAN, H. G. BAYNE, and M. F. POOL. 1971. Pasteurization of poultry meat by steam under reduced pressure. Poultry Sci. 50:1156–1160.

KNIVETT, V. A. 1971. *Salmonella typhimurium* contamination of processed broiler chickens after subclinical infection. J. Hyg. (Camb.). 69: 497–505.

KRYSINSKI, E. P. and R. C. HIMSCH. 1977. Use of enzyme-labeled antibodies to detect *Salmonella* in foods. Appl. Environ. Microbiol. 33: 947–954.

KULSHRESTHA, D. C., and E. H. MARTH. 1970. Inhibition of lactic streptococci and some pathogenic bacteria by certain milk associated volatile compounds as measured by disc assay. J. Milk Food Technol. 33:305–310.

KULSHRESTHA, D. C., and E. H. MARTH. 1972. Inhibition of *Staphylococcus* aureus and *Salmonella typhimurium* by some volatile compounds associated with milk. J. Dairy Sci. 55:670.

KULSHRESTHA, D. C., and E. H. MARTH. 1975a. Inhibition of *Streptococcus lactis* and *Salmonella typhimurium* by mixtures of some volatile and non-volatile compounds associated with milk. J. Milk Food Technol. 38:138–141.

KULSHRESTHA, D. C., and E. H. MARTH. 1975b. Some volatile and non-volatile compounds associated with milk and their effects on certain bacteria. A review. J. Milk Food Technol. 38:604–620.

KUMAR, M. C., H. R. OLSON, L. T. AUSHERMAN, W. B. THURBER, M. FIELD, W. H. HOHLSTEIN, and B. S. POMEROY. 1972. Evaluation of monitoring programs for *Salmonella* infection in turkey breeding flocks. Avian Dis. 16:644–648.

LADIGES, W. C., and J. F. FOSTER. 1974. Incidence of *Salmonella* in beef and chicken. J. Milk Food Technol. 37:213–214.

LADIGES, W. C., J. F. FOSTER, and W. M. GANZ. 1974. Comparison of *Salmonella* polyvalent H antisera, direct fluorescent antibody, and culture procedures in detecting salmonellae from experimentally contaminated ground beef under frozen storage. J. Milk Food Technol. 37:369–371.

LATEGAN, P. M. and R. H. VAUGHN. 1964. The influence of chemical additives on the heat resistance of *Salmonella typhimurium* in liquid whole egg. J. Food Sci. 29:339–344.

LARAMORE, C. R., and C. W. MORTIZ. 1969. Fluorescent-antibody technique in detection of salmonellae in animal feed and feed ingredients. Appl. Microbiol. 17:352–354.

LEE, J. A. 1973. Salmonellae in poultry in Great Britain: p. 197 In: The Microbiological Safety of Food, B. C. Hobbs and J. H. B. Christian, eds. Academic Press, New York.

LEE, A. C., and J. M. GOEPFERT. 1975. Influence of selected solutes on the thermally induced death and injury of *Salmonella typhimurium*. J. Milk Food Technol. 38:195–200.

LEWIS, K. H. and R. ANGELOTTI, EDS. 1964. Examination of foods for enteropathogenic and indicator bacteria. Public Health Service Pub. No. 1142. U.S. Department of Health, Education, and Welfare. 1964.

LEY, F. J., B. J. FREEMAN, and B. C. HOBBS. 1963. The use of gamma radiation for the elimination of salmonellae from various foods. J. Hyg. (Camb.) 61:515–529.

LICARI, J. J., and N. N. POTTER. 1970a. *Salmonella* survival during spray drying and subsequent handling of skim milk powder. I. Salmonella enumeration. J. Dairy Sci. 53:865–876.

LICARI, J. J. and N. N. POTTER. 1970b. *Salmonella* survival during spray drying and subsequent handling of skim milk powder. II Effects of drying conditions. J. Dairy Sci. 53:871–876.

LICARI, J. J., and N. N. POTTER. 1970c. *Salmonella* survival during spray drying and subsequent handling of skim milk powder. III. Effects of storage temperature on *Salmonella* and dried milk properties. J. Dairy Sci. 53:877–882.

LICARI, J. J. and N. N. POTTER. 1970d. *Salmonella* survival differences in heated skim milk and in spray drying of evaporated milk. J. Dairy Sci. 53:1287–1289.

LICCIARDELLO, J. J., J. T. R. NICKERSON, and S. A. GOLDBLITH. 1965. Destruction of salmonellae in hard boiled eggs. Amer. J. Public Health, 55:1622–1628.

LICCIARDELLO, J. J., J. T. R. NICKERSON, and S. A. GOLDBLITH. 1970. Inactivation of *Salmonella* in poultry with gamma radiation. Poultry Sci. 49:663–675.

LICCIARDELLO, J. J., J. T. R. NICKERSON, S. A. GOLDBLITH, W. W. BISHOP and C. A. SHANNON. 1969a. Effect of repeated irradiation on various characteristics of *Salmonella*. Appl. Microbiol. 18:636–640.

LICCIARDELLO, J. J., J. T. R. NICKERSON, S. A. GOLDBLITH, C. A. SHANNON and W. W. BISHOP. 1969b. Development of radiation resistance in *Salmonella* cultures. Appl. Microbiol. 18:24–30.

LILLARD, H. S. 1978a. Improving quality of bird chiller water for recycling by diatomaceous earth filtration and chlorination. J. Food Sci. 43:1528–1531.

LILLARD, H. S. 1978b. Evaluation of bird chiller water for recycling in giblet flumes. J. Food Sci. 43:401–403.

LINEWEAVER, H. 1966. Sensitivity of salmonellae to beta and gamma energy, pp. 47–50. In: The Destruction of Salmonellae. U.S. Department of Agriculture ARS 74–37. Albany, California.

LITCHFIELD, J. H. 1973. *Salmonella* and the food industry—Methods for isolation, identification and enumeration. Crit. Rev. Food Technol. 3: 415–456.

LITTELL, A. M. 1977. Plating medium for differentiating Salmonella arizonae from other Salmonellae. Appl. Environ. Microbiol. 33:485–487.

LIU, T. S., G. H. SNOEYENBOS, and V. L. CARLSON. 1969. The effect of moisture and storage temperature on a *Salmonella senftenberg* 665W population in meat and bone meal. Poultry Sci. 48:1628–1633.

LOKEN, K. I., K. H. CULBERT, R. E. SOLEE, and B. S. POMEROY. 1968. Microbiological quality of protein food supplements products by rendering plants. Appl. Microbiol. 16:1002–1005.

McBEE, I. E. and O. J. COTTERILL. 1971. High temperature storage of spray-dried egg white. B. Thermal resistance of *Salmonella oranienburg*. Poultry Sci. 50:452–458.

McBRIDE, G. B., B. BROWN, and B. J. SKURA. 1978. Effect of bird type, growers and season on the incidence of salmonellae in turkeys. *J. Food Sci.* 43:323–326.

McCOY, J. H. 1976. Salmonella infections of man derived from animals. J. Roy. Soc. Health 96(1) i25–30.

McCROAN, J. E., T. W. McKINLEY, A. BRIN, and W. C. HENNING. 1964. *Staphylococcus* and salmonellae in commercial wrapped sandwiches Public Health Reports. 79:997–1004.

McCULLOUGH, N. B. and C. W. EISELE. 1951.a Experimental human salmonellosis. I. Pathogenicity of strains of *Salmonella meleagridis* and *Salmonella anatum* obtained from spray-dried whole egg. J. Infectious Dis. 88:278–289.

McCULLOUGH, N. B. and C. W. EISELE. 1951.b Experimental human salmonellosis, II. Immunity studies following experimental illness with *Salmonella meleagridis* and *Salmonella anatum*. J. Immunol. 66:595.

McCULLOUGH, N. B. and C. W. EISELE. 1951.c Experimental human salmonellosis. III. Pathogenicity of strains of *Salmonella newport, Salmonella derby*, and *Salmonella bareily* obtained from spray-dried whole egg. J. Infectious Dis. 89:209–213.

McCULLOUGH, N. B. and C. W. EISELE. 1951.d Experimental human salmonellosis. IV. Pathogenicity of strains of *Salmonella pullorum* obtained from spraydried whole egg. J. Infectious Dis. 89:259–265.

McDONOUGH, F. E., and R. E. HARGROVE. 1968. Heat resistance of *Salmonella* in dried milk. J. Dairy Sci. 51:1587–1591.

McDONOUGH, F. E., R. E. HARGROVE, and R. P. TITSLER. 1967. The fate of salmonellae in the manufacture of cottage cheese. J. Milk Food Technol. 30:354–356.

MACKEL, D. C., L. F. LANGLEY, and C. J. PIRCHAL. 1965. Occurrence in swine of *Salmonella* and serotypes of *Escherichia Coli* pathogenic to man. J. Bacteriol. 89:1434–1435.

MAGWOOD, S. E., J. FUNG, and J. L. BYRNE. 1965. Studies on *Salmonella* contamination of environment and product of rendering plants. Avian Dis. 9:302–308.

MARCH, B. E. 1969. Bacterial infection of washed and unwashed eggs with reference to salmonellae. Appl. Microbiol. 17:98–101.

MARTH, E. H. 1969. Salmonellae and salmonellosis associated with milk and milk products. A review. J. Dairy Sci. 52:283–315.

MATCHES, J. R., and J. LISTON. 1968a. Growth of salmonellae on irradiated and nonirradiated seafoods. J. Food Sci. 33:406.

MATCHES, J. R., and J. LISTON. 1968b. Low temperature growth of Salmonella. J. Food Sci. 33:641–645.

MATCHES, J. R., and J. LISTON. 1972a. Effects of incubation temperature on the salt tolerance of Salmonella. J. Milk Food Technol. 35:39–44.

MATCHES, J. R., and J. LISTON. 1972b. Effect of pH on low temperature growth of Salmonella. J. Milk Food Technol. 35:49–52.

MAXCY, R. B. 1978. Lettuce salad as a carrier of microorganisms of public health significance. J. Food Protection. 41:435–438.

MENZIES, D. B. 1944. An outbreak of typhoid fever in Alberta traceable to infected Cheddar cheese. Can. J. Public Health. 35:431–438.

MERCURI, A. R., N. COX, M. O. CARSON, and D. A. TANNER. 1978. Relation of Enterobacteriaceae counts to Salmonella contamination of market broilers. J. Food Protection. 41:427–428.

MICHENER, H. D., and R. P. ELLIOT. 1964. Minimum growth temperatures for food-poisoning, fecal-indicator and psychrophilic microorganisms. Adv. Food Res. 13:349–396.

MICKELSON, M. N. and R. S. FLIPPIN. 1960. Use of Salmonellae antagonists in fermenting egg white. Appl. Microbiol. 8:371–377.

MILONE, N. A. and J. A. WATSON. 1970. Thermal inactivation of Salmonella senftenberg 775W in poultry meat. Health Lab. Sci. 7:199–225.

MOATS, W. A. 1978. Comparison of four agar plating media with and without added novobiocin for isolation of salmonellae from beef and deboned poultry meat. Appl. Environ. Microbiol. 36:747–751.

MOATS, W. A., R. DABBAH, and V. M. EDWARDS. 1971a. Survival of Salmonella anatum heated in various media. Appl. Microbiol. 21:476–481.

MOATS, W. A., R. DABBAH, and V. M. EDWARDS. 1971b. Interpretation of nonlogarithmic survivor curves of heated bacteria. J. Food Sci. 36:523–526.

MOATS, W. A, and J. A. KINNER. 1976. Observations on brilliant green agar with an H₂S indicator. Appl. Environ. Microbiol. 31:380–384.

MOHR, H. K., H. L. TRENK, and M. YETERIAN. 1974. Comparison of fluorescent-antibody methods and enrichment serology for the detection of Salmonella. Appl. Microbiol. 27:324–328.

MOQUOT, G., P. LAFONT, and L. VASSAL. 1963. Nouvelles observations concernant la survie des Salmonella dans les fromages. Ann Inst. Pasteur. 104–570.

MOREHOUSE, L. G. and E. E. WEDMAN. 1961. Salmonella and other disease producing organisms in animal by-products - a survey. J. Amer. Vet. Med. Assoc. 139:989–995.

MORRIS, G. K., B. L. McMURRAY, M. M. GALTON, and J. G. WELLS.

1968. A study of the dissemination of Salmonellosis in a commercial broiler chicken operation. Amer. J. Vet. Res. 30:1413–1421.

MORRIS, G. K., W. T. MARTIN, W. H. SHELTON, J. G. WELLS, and P. S. BRACHMAN. 1970. Salmonella in fish meal plants: relative amounts of contamination at various stages of processing and a method of control. Appl. Microbiol. 19:401–408.

MORRIS, G. K., and J. G. WELLS. 1970. *Salmonella* contamination in a poultry processing plant. Appl. Microbiol. 19:795–799.

MORRIS, J. G. and J. C. AYRES. 1960. Incidence of *Salmonellae* on commercially processed poultry. Poultry Sci. 39:1139–1145.

MORSE, E. V., K. W. KERSTING, L. E. SMITH, JR., E. P. MYHROM, and D. E. GREENWOOD. 1978. Salmonellosis: Possible transmission from horse to human to dog of infection. Amer. J. Public Health. 68:497–499.

MOSSEL, D. A. A. 1963. Le survie des Salmonellae dans des differents products alimentaires. Ann. Inst. Pasteur. 104:547–569.

MOSSEL, D. A. A. and A. S. DeBRUIN. 1960. The survival of *Enterobacteriaceae* in acid liquid foods stored at different temperatures. Ann. Inst. Pasteur deLille. 11:65–72.

MOSSEL. D. A. A., and J. L. SHENNAN. 1976. Microorganisms in dried foods: their significance, limitation and enumeration. *J. Food Technol.* 11:205–220.

MUNSON, T. E., J. P. SCHRADE, N. B. BISCIELLO, JR., L. D. FANTASIA, W. H. HARTUNG, and J. J. O'CONNOR. 1976. Evaluation of an automated fluorescent antibody procedure for detection of *Salmonella* in foods and feeds. Appl. Environ. Microbiol. 31:514–521.

NATIONAL ACADEMY OF SCIENCES. 1969. An evaluation of the *Salmonella* problem. National Academy of Sciences, Washington, D.C., 207p.

NATIONAL ACADEMY OF SCIENCES. 1975. Prevention of microbial and parasitic hazards associated with processed foods. A guide for the food processor. National Research Council. National Academy of Sciences, Washington.

NEWELL, K. W., and L. P. WILLIAMS. 1971. Control of *Salmonella* affecting swine and man. J. Amer. Vet. Med. Ass. 158:89–98.

NG, H. 1966. Heat sensitivity of 300 *Salmonella* isolates. pp 39–41 In: the Destruction of *Salmonellae*. U.S. Department of Agriculture ARS 74–37 Albany, California.

NG, H., H. G. BAYNE, and J. A. GARIBALDI. 1969. Heat resistance of *Salmonella*: the uniqueness of *Salmonella senftenberg* 775W. Appl. Microbiol. 17:78–82.

NORTH, W. R., JR. 1961. Lactose preenrichment method for isolation of *Salmonella* from dried egg albumin. Appl. Microbiol. 9:188–195.

NORTH, W. R., and M. T. BARTRAM. 1953. The efficiency of selenite broth of different compositions in the isolation of *Salmonella*. Appl. Microbiol. 1:130–134.

OLITZKY, I., A. M. PERRI. M. A. SHIFFMAN, and M. WERRIN. 1956. Smoked fish as a vehicle of *Salmonellosis*. Public Health Rep. 71:773–779.

OLSON, J. C., JR. 1975. Development and present status of FDA *Salmonella* sampling and testing plans. J. Milk Food Technol. 38:369–371.

OSBORNE, A. D. 1976. *Salmonella* infections in animals and birds. J. Ray. Soc. Health. 96 (1):30–33.

OSBORNE, W. W., R. P. STRAKA, and H. LINEWEAVER. 1954. Heat resistance of strains of *Salmonella* in liquid whole egg, egg yolk, and egg white. Food Res. 19:451–463.

PACE, P. J., K. J. SILVER, and H. J. WISNIEWSKI. 1977. *Salmonella* in commercially produced dried dog food: possible relationship to a human infection caused by *Salmonella entertidis* serotype Havana. J. Food Protection. 40:317–321.

PALUMBO, S. A., and J. A. ALFORD. 1970. Inhibitory action of tetrathionate enrichment broth. Appl. Microbiol. 20:970–976.

PALUMBO, S. A., C. N. HUHTANEN, and J. L. SMITH. 1974. Microbiology of the frankfurter process: *Salmonella* and natural aerobic flora. Appl. Microbiol. 27:724–732.

PARK, H. S. and E. H. MARTH. 1972a. Inactiation of *Salmonella typhimurium* by sorbic acid. J. Milk Food Technol. 35:532–539.

PARK, H. S. and E. H. MARTH. 1972b. Behavior of *Salmonella typhimurium* in skim milk during fermentation by lactic acid bacteria. J. Milk Food Technol 35:482–488.

PARK, H. S. and E. H. MARTH. 1972c. Survival of *Salmonella typhimurium* in refrigerated cultured milks. J. Milk Food Techno. 35:489–495.

PARK, H. S., E. H. MARTH, J. M. GOEPFERT, and N. F. OLSON. 1970a. The fate of *Salmonella typhimurium* in the manufacture and ripening of low-acid Cheddar cheese, J. Milk Food Technol. 33:280–284.

PARK, H. S., E. H. MARTH, and N. F. OLSON. 1970b. Survival of *Salmonella typhimurium* in cold pack cheese food during refrigerated storage. J. Milk Food Technol. 33:383–388.

PATTERSON, J. T. 1969. *Salmonella* in meat and poultry. Poultry plant cooling waters and effluents and animal feeding stuffs. J. Appl. Bacteriol. 32:329–337.

PATRICK, T. E., J. A. COLLINS, and T. L. GOODWIN. 1973. Isolation of *Salmonella* from carcass of steam-and water-scalded poultry. J. Milk Food Technol. 36:34–36.

PETHER, J. V. S., and R. J. GILBERT. 1971. Survival of *Salmonella* on fingertips and transfer of organisms to foods. J. Hyg. (Camb.) 69:673–681.

PIVNICK, H. 1970. *Salmonella* in poultry products. Can Inst. Food Technol. J. 3:A50–A51.

POELMA, P. L. and J. H. SILLIKER. 1966. *Salmonella*, pp 301–328. In: Compendium of Methods for the Microbiological Examination of Foods. M. L. Speck, ed. American Public Health Association, Washington, D. C.

POCURULL, D. W., S. A. GAINES, and H. D. MERCER. 1971. of infectious multiple drug resistance among *Salmonella* isolated from animals in the United States. Appl. Microbiol. 21:358–362.

PREVITE, J. J., Y. CHANG, and H. M. EL BISI. 1970. Effect of radiation pasteurization on *Salmonella*. I. Parameters affecting survival and recovery from chicken. Can J. Microbiol. 16:465–471.

PREVITE, J. J., Y. CHANG, W. SCRUTCHFIELD and H. M. EL BISI. 1971. Effects of radiation pasteurization on *Salmonella* 11. Influence of repeated radiation growth cycles on virulence and resistance to radiation and antibiotics. Can. J. Microbiol. 17:105–110.

PREVITE, J. J., Y. CHANG, and H. M. EL BISI. 1971. Effects of radiation pasteurization on *Salmonella*. 111. Radiation lethality and the frequency of mutation to antibiotic resistance. Can. J. Microbiol. 17: 385–389.

PROST. E., and H. RIEMANN. 1967. Food-borne *Salmonellosis*. Ann. Rev. Microbiol. 21:495–528.

PUTNAM, G. W. 1966. Engineering aspects of egg pasteurization, pp. 56–57. In: The Destruction of *Salmonellae*. U.S. Department of Agriculture ARS 74–37., Albany, California.

RACCACH, M. and B. J. JUVEN. 1976. Effect of suspending and plating media on the recovery of *Salmonella gallinarum* following freezing and thawing. J. Food Technol. 11:221–228.

RAINES, J. W., and R. C. SWANSON. 1975. International outbreak of *Salmonella eastbourne* infection traced to contaminated chocolate. Lancet 1:788–793.

RASMUSSEN, O. G., R. HANSEN, J. J. JACOBS, and O. H. M. WILDER. 1964. Dry heat resistance of *Salmonellae* in rendered animal by-products. Poultry Sci. 43:1151–1157.

RAY, B., D. W. JANSSEN, and F. F. BUSTA. 1972. Characterization of the repair of injury induced by freezing *Salmonella anatum*. Appl. Microbiol. 23:803–809.

RAY, B., J. J. JEZESKI, and F. F. BUSTA. 1971a. Effect of rehydration on recovery, repair, and growth of injured freeze-dried *Salmonella anatum*. Appl. Microbiol. 22:181–189.

RAY, B., J. J. JEZESKI, and F. F. BUSTA. 1971b. Repair of injury in freeze-dried *Salmonella anatum*. Appl. Microbiol. 22:401–407.

RAY, B., J. J. JEZESKI, and F. F. BUSTA. 1971c. Isolation of *Salmonella* from naturally contaminated dried milk products. J. Milk Food Technol. 34:389–393.

RAY, B., J. J. JEZESKI, and F. F. BUSTA. 1971d. Isolation of *Salmonella* from naturally contaminated dried milk products. J. Milk Food Technol. 34:423–427.

READ, R. B., JR., J. G. BRADSHAW, R. W. DICKERSON, JR., and J. T. PEELER. 1968. Thermal resistance of Salmonellae isolated from dry milk. Appl. Microbiol. 16:998–1001.

REAMER, R. H., R. E. HARGROVE, and F. E. McDONOUGH. 1969. Increased sensitivity of immunofluorescent assay for *Salmonella* in non-fat dry milk. Appl. Microbiol. 18:328–331.

REAMER, R. H., and R. E. HARGROVE. 1972. Twenty four hour immunofluorescence technique for the detection of *Salmonella* in non-fat dry milk. Appl. Microbiol. 23:78–81.

REAMER, R. H., R. E. HARGROVE, and F. E. McDONOUGH. 1974. A selective plating agar for direct enumeration of *Salmonella* in artificially contaminated dairy products. J. Milk Food Technol. 37:441–444.

REDDY, K. V., A. A. KRAFT, R. J. HASIAK, W. W. MARION, and D. K. HOTCHKISS. 1978. Effect of spin chilling and freezing on bacteria on commercially processed turkeys. J. Food Sci. 43:334–336.

RESTAINO, L., G. S. GRAUMAN, W. A. McCALL, and W. M. HILL. 1977. Effects of varying concentrations of novobiocin incorporated into two *Salmonella* plating media on the recovery of four *Enterobacteriaceae*. Appl. Environ. Microbiol. 33:585–589.

REY, C. R., A. A. KRAFT, and R. E. RUST. 1971. Microbiology of beef. shell frozen with liquid nitrogen. J. Food Sci. 36:955–958.

RIEMANN, H. 1968. Effect of water activity on the heat resistance of *Salmonella* in "dry" material. Appl. Microbiol. 16:1621–1622.

RIEMANN, H. D. 1969. Food-Borne Infections and Intoxications. Academic Press, New York.

RIEMANN, H. 1969. Food processing and preservation effects, pp 490–541. In: Food-Borne Infections and Intoxications. H. Riemann (ed). Academic Press, New York.

RILEY, P. B. 1969. *Salmonella* infection: the position of animal food and its manufacturing process, p. 101. In: Bacterial Food Poisoning, J. Taylor, ed. The Royal Society of Health, London.

RIZK, S. S., J. C. AYRES, and A. A. KRAFT. 1966. Disinfection of eggs artificially inoculated with *Salmonella*. I. Application of several disinfectants. Poultry Sci. 45:764–769.

ROGERS, A. B., M. SEBRING and R. W. KLINE. 1966. Hydrogen peroxide pasteurization process for egg white, pp 68–72. In: The Destruction of Salmonellae U.S. Department of Agriculture, ARS 74–37, Albany, California.

ROGERS, R. E., and M. F. GUNDERSON. 1958. Roasting of frozen stuffed turkeys. I. Survival of *Salmonella pullorum* in inoculated stuffing. Food Res. 23:87–95.

SADLER, W. W., R. YAMAMOTO, H. E. ADLER, and G. F. STEWART. 1960. Survey of market poultry for *Salmonella* infection. Appl. Microbiol. 9:72–76.

SADLER, W. W., and R. E. CORSTVET. 1965. Second survey of market poultry for *Salmonella* infection. Appl. Microbiol. 13:348–351.

SADOVSKI, A. Y. 1977. Technical note: Acid sensitivity of freeze injured salmonellae in relation to their isolation from frozen vegetables by preenrichment procedure. *J. Food Technol.* 12:85–91.

SAIR, L. 1966. Sensitivity of salmonellae to expoxides. pp 41–46. In: The Destruction of Salmonellae. U.S. Department of Agriculture ARS 74–37, Albany, California.

SCHAFFNER, C. P., K. MOSBACH, V. C. BIBIT, and C. H. WATSON. 1967. Coconut and *Salmonella* infection. Appl. Microbiol. 15:471–475.

SCHNEIDER, M. D., 1951. Investigation of *Salmonella* content of powdered whole egg with not more than two percent moisture content. IV Bactericidal action of the preheather on *Salmonella* in liquid whole egg. Food Technol. 5:349–352.

SECHTER, I., and C. B. GERICHTER. 1968. Phage typing scheme for *Salmonella braendercup*. Appl. Microbiol. 16:1708–1712.

SECHTER, I., and C. B. GERICHTER. 1969. Phage typing scheme for *Salmonella blockley*. Ann L'Inst. Pasteur. 116:190–199.

SEGALOVE, M. and G. M. DACK. 1944. Growth of a stain of *Salmonella enteritidis* experimentally inoculated into canned foods. Food Res. 9:1–5.

SEILER, D. A. L. 1960. The effect of time and temperature on the survival of salmonellae in desiccated coconut. Month. Bull. Ministry Health (Gt. Brit.), 19:211–212.

SHARMA, V. K., Y. K. KAVTA, and I. P. SINGH. 1974. Frogs as carriers of *Salmonella and Edwardsiella*; Antonie van Leevwenhoek J. Microbiol. Seral. 40:171–175.

SHIFLETT, M. A., J. S. LEE, and R. O. SINNHUBER. 1967. Effect of food additives and irridation on survival of *Salmonella* in oysters. Appl. Microbiol. 15:476–479.

SILLIKER, J. H., and D. A. GABIS. 1973. ICMSF methods studies I. Comparison of analytical schemes for detection of *Salmonella* in dried foods. Can J. Microbiol. 19:475–479.

SILLIKER, J. H., and D. A. GABIS. 1974. ICMSF methods studies V. The influence of selective enrichment media and incubator temperatures on the detection of sallmonellae in raw frozen meats. Can J. Microbiol. 20:813–816.

SILLIKER, J. H., and R. A. GREENBERG. 1969. Laboratory methods, analysis of foods for specific food poisoning organisms - *Salmonella*. pp. 467–475 in Food-Borne Infections and Intoxications. ed. Academic Press, New York.

SINGH, V. K., and E. M. MIKOLAJCIK. 1971. Influence of heat treatment of skim milk upon growth of enteropathogenic and lactic bacteria. J. Milk Food Technol. 34:204–208.

SINSKEY, T. J., A. H. McINTOSH, I. S. PABLO, G. J. SILVERMAN, and and S. A. GOLDBLITH. 1964. Considerations on the recovery of microorganisms from freeze-dried foods. Health. Lab. Sci. 1:297–306.

SINSKEY, T. J., G. J. SILVERMAN, and S. A. GOLDBLITH. 1967. Influence of platen temperature and relative humidity during storage on the survival of freeze-dried *Salmonella typhimurium*. Appl. Microbiol. 15:22–30.

SKOLL, S. L., and H. O. DILLENBERG. 1963. *Salmonella thompson* in cake mix. Can J. Public Health. 54:325–329.

SMYSER, C. F., G. H. SNOEYENBOS, and B. McKIE. 1970. Isolation of salmonellae from rendered by-products and poultry litter cultured in enrichment media incubated at elevated temperatures. Avian Dis. 14: 248–254.

SOLOWEY, M., and E. J. CALESNICK. 1948. Survival of salmonella in reconstituted egg powder subjected to holding and scrambling. Food Res. 13:216–226.

SOLOWEY, M., R. R. SUTTON, and E. J. CALESNICK. 1948. Heat resistance of *Salmonella* organisms isolated from spray-dried whole-egg powder. Food Technol. 2:9–14.

SOLOWEY, M., V. H. McFARLANE, E. H. SPAULDING, and C. CHE-

MERDA. 1947. Microbiology of spray-dried whole egg II. Incidence and types of salmonella. Amer. J. Public Health. 37:971–982.

SORRELLS, K. M., and M. L. SPECK. 1970. Inhibition of *Salmonella gallinarum* by culture filtrates of *Leuconostoc çitrovorum*. J. Dairy Sci. 53:239–241.

SORRELLS, K. M., M. L. SPECK, and J. A. WARREN. 1970. Pathogenicity of *Salmonella gallinarum* after metabolic injury by freezing. Appl. Microbiol. 19:39–43.

SPECK, M. L. (ED.) 1976. Compendium of methods for the microbiological examination of foods. American Public Health Association, Washington, D.C.

SPERBER, W. H., and R. H. DEIBEL. 1969. Accelerated procedure for *Salmonella* detection in dried foods and feeds involving only broth cultures and serological reactions. Appl. Microbiol. 17:533–539.

SLANTEZ, L. W., C. O. CHICHESTER, A. R. GAUFIN, and Z. J. ORDAL, EDS. 1963. Microbiological Quality of Foods. Academic Press, New York, pp. 84–101.

SMITH, J. L., S. A. PALUMBO, J. C. KISSINGER, and C. N. HUHTANEN. 1975a. Survival of *Salmonella dublin* and *Salmonella typhimurium* in Lebanon bologna. J. Milk Food Technol. 38:150–154.

SMITH, J. L., C. N. HUHTANEN, J. C. KISSINGER, and S. A. PALUMBO. 1975b. Survival of salmonellae during pepperoni manufacture. Appl. Microbiol. 30:759–763.

SMITH, J. L., C. N. HUHTANEN, J. C. KISSINGER, and S. A. PALUMBO. 1977. Destruction of *Salmonella* and *Staphylococcus* during processing of a non-fermented snack sausage. J. Food Protection. 40:465–467.

SPLITTSTOESSER, D. F., and B. SEGEN. 1970. Examination of frozen vegetables for salmonellae. J. Milk Food Technol. 33:111–113.

STERSKY, A. K., D. R. HELDMAN, and T. I. HEDRICK. 1972a. Viability of airborne *S. newbrunswick* under various conditions. J. Dairy Sci. 55: 14–18.

STERSKY, A. K., D. R. HELDMAN, and T. I. HEDRICK. 1972b. Effects of aerosolization on *Salmonella newbrunswick*. J. Milk Food Technol. 35: 528–531.

STOKES, J. L., and H. G. BAYNE. 1958. Growth-factor dependent strains of salmonellae. J. Bacteriol. 76:417–421.

STOKES, J. L., W. W. OSBORNE, and H. G. BAYNE. 1956. Penetration and growth of *Salmonella* in shell eggs. Food Res. 21:510–518.

SCOTT, J. A., J. E. HODGSON, and J. C. CHANEY. 1978. Incidence of salmonellae in animal feed and the effect of pelleting on content of *enterobacteriaceae*. J. Appl. Bacteriol. 39:41–46.

SUBRAMANIAN, C. S., and E. H. MARTH. 1968. Multiplication of *Salmonella typhimurium* in skim milk with and without added hydrochloric, lactic, and citric acids. J. Milk Food Technol. 31:323–326.

SURKIEWICZ, B. F., R. W. JOHNSTON, A. B. MORAN, and G. W. KRUM. 1969. A bacteriological survey of chicken eviscerating plants. Food Technol. 23:1066–1069.

SVEUM, W. H., and P. A. HARTMAN. 1977. Timed-release capsule method for the detection of Salmonellae in foods and feeds. Appl. Environ. Microbiol. 33:630-634.

SWAMINATHAN, B., J. C. AYRES, and J. E. WILLIAMS, 1978a. Control of nonspecific staining in the fluorescent antibody technique for the detection of salmonellae in foods. Appl. Environ. Microbiol. 35:911-919.

SWAMINATHAN, B., J. M. DENNER, and J. C. AYRES. 1978b. Rapid detection of Salmonellae in foods by membrane filter-disc immunoimmobilization technique. J. Food Sci. 43:1444-1447.

SWAMINATHAN, B., M. A. B. LINK, and J. C. AYRES. 1978c. Incidence of salmonellae in raw meat and poultry samples in retail stores. J. Food Protection. 41:518-520.

TAKACS, J., and G. B. NAGY. 1975. Demonstration of salmonellae in sausages of high fat content. Acta Veterina Academiae Scientiarum Hungaricae. 25:303-312.

TAKACS, J., and Z. SIMONFFY. 1970. Das Salmonellen-problem bei Dauerwursten. Die Fleischwirtschaft. 50:1200-1202.

TAMMINGA, S. K., R. R. BEUMER, and E. H. KAMPELMACHER. 1978. The hygienic quality of vegetables grown in or imported into the Netherlands: a tentative survey. J. Hyg. (Camb.) 80:143-154.

TAYLOR, J., and J. H. McCOY. 1969. Salmonella and Arizona infections, pp. 3-72. In: H. Riemann (ed.), Food-borne infections and intoxications. Academic Press, New York, New York.

THARRINGTON, G., JR., D. H. ASHTON, J. R. HATFIELD, and F. H. FRY. 1978. Nonspecific staining of a Lactobacillus by Salmonella fluorescent antibodies. J. Food Sci. 43:548-552.

THATCHER, F. S., and J. MONTFORD. 1962. Egg products as a source of salmonellae in processed foods. Can. J. Public Health. 53:61-69.

THOMAS, C. T., J. C. WHITE, and K. LONGREE. 1966. Thermal resistance of salmonellae and staphylococci in foods. Appl. Microbiol. 14:815-820.

THOMASON, B. M. 1976. Fluorescent antibody detection of salmonellae. pp. 329-343. In: Compendium of Methods for the Microbiological Examination of Foods. M. L. Speck, ed. American Public Health Association. Washington, D.C.

THOMASON, B. M., J. W. BIDDLE, and W. B. CHERRY. 1975. Detection of Salmonellae in the environment. Appl. Microbiol. 30:764-767.

THOMASON, B. M., G. A. HEBERT, and W. B. CHERRY. 1975. Evaluation of a semiautomated system for direct fluorescent antibody detection of salmonellae. Appl. Microbiol. 30:557-564.

THOMASON, B. M., W. B. CHERRY, and D. J. DODD. 1977. Salmonellae in health foods. Appl. Environ. Microbiol. 34:602-603.

THOMASON, B. M., and D. J. DODD. 1976. Comparison of enrichment procedures for fluorescent antibody and cultural detection of salmonellae in raw meat and poultry. Appl. Environ. Microbiol. 31:787-788.

THOMASON, J. E., N. A. COX, and J. S. BAILEY. 1977. Control of

Salmonella and extension of shelf life of broiler carcasses with a glutaraldehyde product. J. Food Sci. 42:1353–1355.

TIWARI, N. P., and R. B. MAXCY. 1972. Comparative growth of salmonellae, coliforms and other members of the microflora of raw and radorized ground beef. J. Milk Food Technol. 35:455–460.

TODD, E. C. D. 1978. Food-borne disease in six countries - A comparison. J. Food Protection. 41:559–565.

TODD, E., and H. PIVNICK. 1973. Public health problems associated with barbecued foods: A review. J. Milk Food Technol. 36:1–18.

TOMLINS, R. I., and Z. J. ORDAL. 1971. Requirements of *Salmonella typhimurium* for recovery from thermal injury. J. Bacteriol. 105:512–518.

TROLLER, J. A. 1973. The water relations of food-borne bacterial pathogens. A review. J. Milk Food Technol. 36:276–288.

TROLLER, J. A. 1976. *Salmonella* and *Shigella*. *In*: M. deFigueiredo and D. F. Splittstoesser (eds.), Food Microbiology: Public Health and Spoilage Aspects. AVI. Westport, Conn.

TSAI, R. Y. T., J. M. GOEPFERT, R. G. CASSENS, and E. J. BRISKEY. 1971. A comparison of *Salmonella* excretion by stress-susceptible and stress-resistant pigs. J. Food Sci. 36:889–891.

U.S. DEPARTMENT OF AGRICULTURE, APHIS. 1974. Microbiology laboratory guidebook. Washington, D.C.

U.S. DEPARTMENT OF AGRICULTURE. 1978a. Water in poultry chillers. Federal Register. 43:14043–14044.

U.S. DEPARTMENT OF AGRICULTURE. 1978b. Cooking requirements for cooked beef and roast beef. Federal Register. 43:30791–30793.

U.S. GENERAL ACCOUNTING OFFICE. 1974. *Salmonella* in raw meat and poultry: An assessment of the problem. Washington, D.C.

VANDERPOST, J. M., and J. B. BELL. 1977. Bacteriological investigation of Alberta Meat Packing Plant wastes with emphasis on *Salmonella* isolation. Appl. Environ. Microbiol. 33:538–545.

VAN OYE, E., ED. 1964. The World Problem of Salmonellosis. Dr. W. Junk. Publishers, The Hague, Netherlands.

VAN SCHOTHORST, M., and F. M. VAN LEUSDEN. 1972. Studies on the isolation of injured salmonellae from foods. Zbl. Bakt. Hyg. I. Abt. Orig. A. 221:19–29.

VAN SCHOTHORST, M., and F. M. VAN LEUSDEN. 1975. Comparison of several methods for the isolation of *salmonella* from egg products. Can. J. Microbiol. 21:1041–1045.

VAN SCHOTHORST, M., F. M. VAN LEUSDEN, W. EDEL, and E. H. KAMPELMACHER. 1974. Further studies on the presence of *Salmonella* in chickens and hens in the Netherlands. Zbl. Vet. Med. B. 21:723–728.

VAN SCHOTHORST, M., F. M. VAN LEUSDEN, J. JEUNINK, and J. DEDREW. 1977. Studies on the multiplication of *Salmonellae* in various enrichment media at different incubation temperatures. J. Appl. Bacteriol. 42:157–163.

WABECK, C. J., D. V. SCHWALL, G. M. EVANCHO, J. G. HECK and
A. B. ROGERS. 1968. *Salmonella* and total count reduction in poultry
treated with sodium hypochlorite solution. Poultry Sci. 47:1090–1094.

WADE, E. M., and S. LEWIS. 1928. Longevity of typhoid bacilli in cheddar
cheese. Am. J. Pub. Health. 18:1480–1488.

WALSH, D. E., B. R. FUNKE and K. R. GRAALUM. 1974. Influence of
spaghetti extruding conditions drying and storage on the survival of
salmonella typhimurium. J. Food Sci. 39:1105–1106.

WATSON. W. A. 1976. *Salmonella* control in certain European countries.
J. Roy. Soc. Health. 96 (1):21–25.

WEISS, K. F., J. C. AYRES, and A. A. KRAFT. 1965. Inhibitory action of
selenite on *E. coli. P. vulgaris. S. thompson.* J. Bacteriol. 90:857–862.

WEISSMAN, M. A., and J. A. CARPENTER. 1969. Incidence of *salmonellae* in meat and meat products. Appl. Microbiol. 17:899–902.

WEKELL, M. M., and C. S. MARTINSEN. 1975. Growth of *Salmonella
heidelberg* at room temperature in irridiated and nonirridated potato
chip dip. J. Milk Food Technol. 38:259–261.

WELLS, F., D. BERQUIST, and R. H. FORSYTHE. 1958. A comparison
of selective media for the isolation of *Salmonella* from commercial egg
white solids. Appl. Microbiol. 6:198–200.

WELLS, J. G., G. K. MORRIS, and P. S. BRACHMAN. 1971. New method
of isolating *salmonellae* from milk. Appl. Microbiol. 21:235–239.

WESTHOFF, D. C., and T. ENGLER. 1973. The fate of *Salmonella typhimurium* and *Staphylococcus aureus* in cottage cheese whey. J. Milk
Technol. 36:19–22.

WETHINGTON M. C., and F. W. FABIAN. 1950. Viability of food poisoning *staphylococci* and *salmonellae* in salad dressing and mayonnaise.
Food Res. 15:125–133.

WHITE, C. H., and E. W. CUSTER. 1976. Survival of *Salmonella* in
Cheddar Cheese. J. Milk Food Technol. 39:328–331.

WHITEHILL, A. R., and B. L. GARDNER. 1970. Evaluation of selenite-F
broth as a selective medium for *Salmonella.* Dev. Ind. Microbiol. 12:
249–252.

WHO STUDY GROUP. 1974. Food-borne disease: Methods of sampling and
examination in surveillance programmes. World Health Organization.
Technical Report Series No. 543, Geneva Switzerland.

WIENER, H. 1974. The origins of salmonellosis. Animal Health Institute,
Suite 1009, 1717 K Street, N.W., Washington, D.C. 20006.

WILLIAMS, E. F., and R. SPENCER. 1973. Abattoir practices and their
effect on the incidence of *salmonella* in meat. p. 41 In: The Microbiological Safety of Foods. B. C. Hobbs and J. H. B. Christian, eds. Academic Press, New York.

WILLIAMS, L. P., and K. W. NEWELL. 1970. *Salmonella* excretion in
joy-riding pigs. Amer. J. Public Health. 60:926–929.

WILSON, C. R., and W. H. ANDREWS 1976. Sulfite compounds as neu-
tralizers of spice toxicity for *Salmonella.* J. Milk Food Technol. 39:
464–466.

WILSON, C. R., W. H. ANDREWS, and P. L. POELMA. 1975. Evaluation of cultural methods to isolate *Salmonella* from pressed yeast and dried inactive yeast. J. Milk Food Technol. 38:383–385.

WILSON, E., R. S. PAFFERNBARGER, M. J. FOTER, and K. H. ITO. 1962. Prevalence of *salmonellae* in meat and poultry products. J. Infect. Dis. 109:166–171.

WILSON, J. M., and R. DAVIES. 1976. Minimal medium recovery of thermally injured *Salmonella senftenberg* 4969. J. Appl. Bacteriol. 40:365–374.

WILSON, M. M., and E. F. MACKENZIE. 1955. Typhoid fever and Salmonellosis due to consumption of infected desiccated coconut. J. Appl. Bacteriol. 18:510–521.

WILSON, V. R., G. J. HERMAN, and J. BALOWS. 1971. Report of a new system for typing *Salmonella typhimurium* in the United States. Appl. Microbiol. 21:774–776.

WINTER, A. R., G. F. STEWART, V. H. McFARLANE, and M. SO-LOWEY. 1946. Pasteurization of liquid egg products III. Destruction of *Salmonella* in liquid whole egg. Amer. J. Public Health. 36:451–460.

WOODBURN, M. 1964. Incidence of Salmonellae in dressed broiler-fryer chickens. Appl. Microbiol. 12:492–495.

WOODBURN, M., M. BENION, and G. E. VAIL. 1962. Destruction of Salmonellae and staphylocci in precooked poultry products by heat treatment before freezing. Food Technol. 16 (6):98–100.

WOODBURN, M. J., and D. H. STRONG. 1960. Survival of *Salmonella typhimurium, Staphylococcus aureus* and *Streptococcus faecalis* frozen in simplified food substrates. Appl. Microbiol. 8:100–113.

YAMAMOTO, R., W. W. SADLER, H. E. ADLER, and G. F. STEWART. 1961. Comparison of media and methods for recovering *Salmonella typhimurium* from turkeys. Appl. Microbiol. 9:76–80.

ZOTTOLA, E. A., D. L. SCHMELTZ, and J. J. JEZESKI. 1970. Isolation of Salmonellae and other airborne microorganisms in turkey processing plants. J. Milk Food Technol. 33:395–399.

6

Viruses in Foods

R. Di Girolamo

It is a well documented fact that foods, if improperly handled or processed, can harbor a number of pathogenic, or potentially pathogenic microorganisms (Bjnoe and Yurack, 1964; Dolman, 1964; Hall and Angelotti, 1965). Among this latter group we must include the viruses, which are rather unique entities. The viruses are extremely small, ranging from 25 to 250 mm in size. Although they share some characteristics with living organisms, they are not truly alive. They replicate inside a living cell using this host cell's metabolism. In fact, viruses, by definition, are considered as obligate intracellular pathogens that do contain a nucleic acid but which use the host cell's synthetic machinery to produce infectious particles, the virions. Structurally, virions are composed of a central core of DNA or either single stranded or double stranded RNA. This central core, termed a genome, is surrounded by a protein coat, the capsid. The capsid is constructed of repeating protein molecules called capsomers. The entire structure, genome and capsid coat, is termed the nucleocapsid. Some viruses, in addition to the capsid, possess a second coat, the peplos or envelope. This envelope, rich in lipids, originates from the host cell as the virus "leaks out."

There are a number of ways by which viruses may be transmitted. Probably the most common of these is via "direct" or "distance contact", i.e. from host to host by touching or through short distances in air. Indirect transmission may occur also via contact with inanimate objects (formites), by the bite of certain infected insects (vectors), or by water and food (vehicles).

There are a variety of ways in which viruses may enter foods. Viral contamination may be considered *primary* when the food

176

product already contains virus at the time of slaughter or harvest or *secondary* when it occurs during processing, storage, or distribution of a food. This distinction into primary and secondary contamination may seem irrelevant, but is important since the method(s) of prevention or elimination of viruses will vary according to type or method of contamination.

Examples of products which may show primary contamination include meats, (especially beef), milk, vegetables, and shellfish (Cliver, 1971). Potentially, any food free of virus at time of harvest or slaughter may become contaminated secondarily sometime before reaching the consumer. There are many ways in which this can occur; via insects (zoonosis), contact with contaminated surfaces, contact with polluted water, or via handling by an infected food handler. Epidemiological evidence has incriminated each of these possibilities in food-borne cases of infectious hepatitis and poliomyelitis (Cliver, 1969; Dingman, 1916, Eisenstein *et al.*, 1963; Joseph *et al.*, 1965; Lipari, 1951; Roos, 1956).

TYPES IN FOODS

The general types of viruses recovered from foodstuffs will depend also upon the nature of the product and to some extent on whether the viral species has entered as a result of primary or secondary contamination. Foods of animal origin become contaminated with a variety of viruses, some of which can survive processing and occur in the final product. These viruses can be grouped into three categories; those that infect animals but not humans, those of animal origin (zoonosis) capable of infecting humans, and those of human origin capable of infecting animals.

Examples of viruses in the first category include the agents of Newcastle's disease, hog cholera, foot and mouth disease, and rinderpest virus. Although representing no direct public health hazard to humans they are of great economic importance to farmers and cattlemen. In addition, data obtained from research with these animal diseases may prove applicable to food-borne viruses capable of infecting humans.

The second category of viruses include the enteroviruses, the adenoviruses, and the reoviruses. Enteroviruses of bovine origin can replicate in human cell cultures and there is serological evidence showing their infectivity for humans (Cliver and Bohl, 1962; Kasel *et al.*, 1963). These agents, however, are not considered to represent any great threat. A far more significant threat is presented by the agent of tick-borne encephalitis, which is a true zoonotic virus. Subtypes of this virus are found in eastern Europe where this agent is transmitted via ticks to dairy goats which shed the virus in their milk during infection (Ernek *et al.*, 1968). Laboratory studies have indicated that cattle and sheep experimentally become infected and likewise shed virus in their milk (Gresikova, 1958). To date, however, only goats have been shown to eliminate the virus under actual field conditions.

The reoviruses, as a group, can actually be considered as overlapping between categories two and three. Similar types of reovirus have been isolated from humans, cattle, fowl, and swine and cross infection of reoviruses from cattle to human and vice-versa have occurred under controlled experimental conditions. Evidence indicating recovery of a bovine strain of Reovirus Type 1 from humans is shown in Table 6.1. However, such cross transmissions have not, as yet, been shown to occur in nature (Kasel *et al.*, 1963).

CONTROLS

In the case of vegetables, "primary contamination" occurs rather passively as a result of using sewage effluent and sewage sludge for irrigation and fertilization (Bagdassar' Yan, 1964; Mazur and Paciorkiewicz, 1973). Recovery of poliovirus in vegetables has been recorded under conditions in which the vegetables (lettuce and radishes) were grown in soil previously flooded with sewage sludge and from vegetables grown in soil irrigated with inoculated sludge (Larkin *et al.*, 1976). The results of these studies are summarized in Tables 6.2–6.4. Land disposal of sewage sludge and effluent is

TABLE 6.1.   RECTAL VIRUS ISOLATION FROM ADULT VOLUNTEERS FOLLOWING INTRANASAL INOCULATION WITH A BOVINE STRAIN OF REOVIRUS TYPE 1. (KASEL *ET AL.*, 1963)

Volunteer	Virus recovery (days after inoculation)					
	1–7		8–14		15–21*	
	No.	%	No.	%	No.	%
T.M.	6/7+	86	2/7	29	0/4	0
C.D.	7/7	100	3/7	43	0/4	0
C.B.	6/7	86	4/7	57	0/4	0
Totals	19/21	90	9/21	43	0/12	0

*Specimens were not tested on days 18, 19 and 20.
+Numerator indicates No. of virus-positive specimens; denominator, No. of specimens tested.

TABLE 6.2. PERSISTENCE OF POLIOVIRUS 1 ON VEGETABLES SPRAY-IRRI-GATED WITH INOCULATED SEWAGE SLUDGE AND EFFLUENT, JUNE-JULY, 1974 (PFU/g)* (LARKIN ET AL., 1976)

Day tested	Sludge irrigated			Effluent irrigated		
	Lettuce	Radish leaves	Radish	Lettuce	Radish leaves	Radish
1	180*	4,400	950	20	500	172
3	4	60	300	3	7.8	80
6	1.4	6	46	0.5	4	10
8	0.2	0.2	14	0.4	0.2	7
10	0	0.2	0.4	0.2	2	0.2
14	0.2	0	0.6	0	0.1	0

*Average of three 100-g samples.

currently being done by over 1,000 communities in the continental United States and is being contemplated by many municipalities as a method of tertiary sewage treatment (Sullivan et al., 1973).

Present procedures used to eliminate pathogens from sewage effluents and sludge are not particularly effecacious in removing or inactivating viruses. Enteroviruses and reoviruses can be consistently recovered from sewage sludge and treated sewage plant discharge (Clarke and Kobler, 1964; Kelly and Sanderson, 1959; Grinstein et al., 1970). Hence, a potential public health hazard could ensue from hasty extensive use of sewage sludge and/or fluid discharge for fertilization or irrigation, unless proper sterilization procedures are carried out. To date, however, this does not appear to be a major public health problem.

Shellfish (oysters, mussels, clams and cockles), in contrast to vegetables, become actively contaminated as a result of feeding. These bivalvular molluscs feed by selectively ingesting potential food particles strained from large volumes of water. The food

TABLE 6.3. RECOVERY OF VIRUS IN THE DRAINAGE FROM VEGETABLE PLOTS FLOODED WITH INOCULATED SEWAGE SLUDGE OR NONCHLORINATED SEC-ONDARY EFFLUENT, JUNE-JULY, 1975. (LARKIN ET AL., 1976)

Days after flooding	Virus recovered (PFU/ml)	Liters collected
0	10,000	11
1	10,000	23
3	500	23
6	0	30
8	10	20
10	0	24
14	2	40
17	0	129

TABLE 6.4.   RECOVERY OF POLIOVIRUS 1 FROM VEGETABLES HARVESTED FROM EFFLUENT FLOODED SOIL (PFU/100g). (LARKIN *ET AL.*, 1976)

Item	August-October 1973	June-July 1974
Day 13:*		
Lettuce	0	0
Radish tops+	20	20
Day 17:		
Lettuce	0	0
Radish tops	20	80
Radish	20	0
Day 23:		
Lettuce	60	60
Radish tops	0	20
Radish	0	20

*First day that plants were of sufficient size for sampling.
+Radish not yet matured (i.e., 1 in. or 2–3 cm).

particles become enmeshed on mucus sheaths secreted and then ingested by these animals during the filtering process (Galtsoff, 1964). This method of feeding permits the entrapment of bacteria and viruses. Recently, it has been demonstrated that this bonding of viruses to the mucus sheaths is ionic in nature (Di Girolamo, *et al.*, 1977). Consequently, only a brief period of exposure to contaminated seawater would be necessary for these creatures to become highly polluted. (Tables 6.5, 6.6 and Fig. 6.1).

All species of commercially valuable shellfish have been shown capable of accumulating high titers of virus under experimental conditions (Atwood *et al.*, 1964; Canzonier, 1971; Crovari, 1958; Di Girolamo, 1975; Liu, 1971). The major site of this uptake is the digestive area, but there is no apparent replication (Liu *et al.*, 1966, 1968). Shellfish contaminated with viruses have been isolated also from the marine environment. As seen in Tables 6.7 and 6.8, viruses recovered from these animals have included poliovirus types 1 and

TABLE 6.5.   UPTAKE OF POLIOVIRUS BY SHELLFISH MUCUS. (DI GIROLAMO *ET AL.*, 1977)

Amt. of poliovirus added to mucus homogenate (PFU/ml)	Amt. of poliovirus bound by shellfish mucus (PFU/g)	Virus uptake by shellfish mucus (%)
5.8×10^5	3.8×10^5	65
4.8×10^4	2.8×10^4	60
9.0×10^3	5.8×10^3	64
9.0×10^2	6.6×10^2	72

TABLE 6.6.   ACCUMULATION OF FOUR REPRESENTATIVE ENTERIC VIRUSES BY SHELLFISH MUCUS. (DI GIROLAMO *ET AL.*, 1977)

Virus sample	Control virus titer	Amt. of virus bound by mucus (PFU/g)	Virus uptake by shellfish mucus (%)
Poliovirus 1 (1sc–2ab)	1.8×10^4	1.2×10^4	68.0
Coxsackie A–9	1.6×10^4	1.0×10^4	53.0
Coxsackie B–4	6.3×10^4	3.5×10^4	53.0
Echo 1	1.5×10^4	5.7×10^3	38.0

3, echovirus types 4 and 9, and coxsackie virus types A and B (Bendinelli and Ruski, 1969; Fugate *et al.*, 1975; Metcalf and Stiles, 1968).

Bivalves may not be the only "shellfish" capable of ingesting viruses. In a recent study crabs were also shown to acquire viruses under laboratory conditions. The viruses were found to survive in chilled, frozen, or cooked crabs (Di Girolamo *et al.*, 1972; Di Girolamo and Daly, 1973). However, the actual presence of viruses in market crabs has not been demonstrated. Consequently, the actual hazard presented by contaminated crabs remains unresolved.

As stated previously, any product may become contaminated after harvesting either through contact with contaminated surfaces, polluted water, insects, or an infected food handler. Of all these possibilities, the one most frequently responsible for food contamination is the latter. The asymptomatic, or actively infected food handler has been responsible for such bacterial diseases as salmonellosis (Bowmer, 1964, 1965) as well as some milk-borne poliomyelitis outbreaks (Knapp *et al.*, 1926; Lipari, 1951) and certain food-borne incidences of hepatitis (Seddon, 1961). In fact, in a 1976 survey of 497 outbreaks of food-borne diseases, 275 of these were found to be a direct result of food mishandling (U.S. DHEW Publ., 1976).

VIRUSES TRANSMITTED VIA FOOD

The one human virus disease first known to be transmitted via food is poliomyelitis. The earliest recorded incident seems to have taken place in 1914. The vehicle in this outbreak was thought to be raw milk (Jubb, 1915). In fact, in every food associated outbreak of poliomyelitis the vehicle has proven to be milk (generally unpasteurized) which may have been contaminated either by an infected food handler or flies which had first come in contact with human

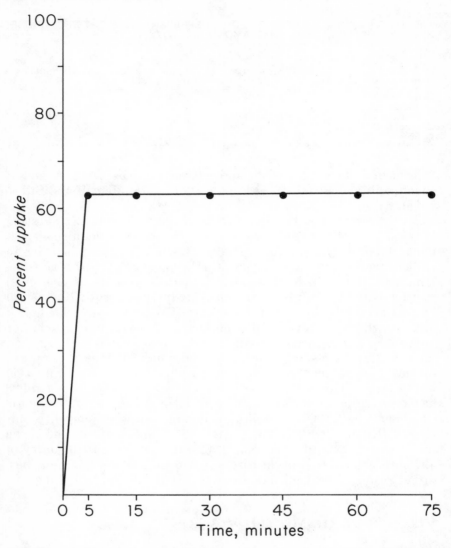

FIG. 6.1.   UPTAKE OF VIRUS BY MUCUS DURING INCUBATION AT 5°C FOR 75 MINUTES. (DI GIROLAMO *ET AL.*, 1977)

feces. An excellent and thorough review of food associated polio-myelitis has been published elsewhere (Cliver, 1967). One signifi-cant point is that these outbreaks are no longer seen in the con-tinental U.S. or other nations which use improved standards in the pasteurization of milk and production of foods; and/or where there is an active vaccination program against this disease.

TABLE 6.7. ENTEROVIRUS ISOLATIONS FROM PARALLEL TESTS OF SEWAGE EFFLUENTS, RIVER WATER, AND SHELLFISH IN RIVER. (METCALF AND STILES, 1968)

Day	Sewage Effluent	River Water*	Shellfish⁺
1	Reovirus I‡	Poliovirus III	0
4	Poliovirus III	Poliovirus III	Poliovirus III
5	Poliovirus III	Poliovirus III	0
6	Poliovirus I	0	Poliovirus III
7	Poliovirus I	Poliovirus I	0
8	Poliovirus I	Poliovirus I	Poliovirus I
9	Reovirus I	Poliovirus I	Poliovirus I
10	Poliovirus I	Poliovirus I	Poliovirus I
11	Poliovirus I	Poliovirus I	0
14	ECHO-9	Poliovirus I	Poliovirus I

*River water collections made one-fourth mile downstream from sewage outlet.
⁺Shellfish collections made at point of river water collections.
‡Virus isolations made in AG or LLCMK₂ cultures.

During the past two decades, the virus which has replaced poliomyelitis as a problem in foods is the agent of infectious hepatitis. Until quite recently this virus remained unidentified, although a number of viruses were suspect as the probable agent of the disease (Ludberg, 1965; Krugman et al., 1963; Hartwell et al., 1968). There is as yet no generally accepted laboratory host for this virus. Hence, its association with foods must still be based upon recognized common-source outbreaks.

The virus is spread via the fecal-oral route in a manner similar to the enteric viruses. The incubation period for infectious hepatitis varies from 10 to 50 days with the average between 28 and 30 days. Prodromal symptoms have been reported as being similar to those of poliomyelitis, with fever, loss of appetite, nausea, and occasionally vomiting plus aches in the head and joints (Di Marco, 1963; Davis et al., 1968). As the disease progresses there is hepatic involvement, with the liver first becoming tender then stonelike on palpation. The urine becomes dark and the feces clay colored. Typically, there is jaundicing of the eyes and skin, caused by hepatic insufficiency and subsequent leakage of bile salts into the blood. The presence of these bile salts can lead to a feeling of severe prurience. Resolution of the disease is slow, six weeks to six months. Fortunately, the mortality rate is low, less than 2%. The epidemiological aspects of both infectious and "serum hepatitis", i.e. hepatitis A & B, are summarized in Table 6.9 (Guidon and Pierach, 1973).

To date there have been approximately 47 outbreaks of infectious hepatitis in which foods of one type or another have been implicated. These outbreaks have been reviewed extensively (Cliver, 1966, 1967, 1969, 1971; Portnoy et al., 1976). The vehicles implicated included milk, pastries, salad vegetables, and meats, with shellfish being the one source named most frequently. In each incident the food was

TABLE 6.8. RESULTS OF VIROLOGICAL AND BACTERIOLOGICAL TESTS ON POOLS OF MUSSELS. (BENDINELLI AND RUSCHI, 1969)

Source	Pools examined	Cytopathic agents isolated on			Virus identified	Bacteria at 37°C (per ml)[a]	Coliform (per ml)
		HeLa	HEp-2	Human amnion			
Mussel farm	32	0	0	0		5,100 ± 2,500[b]	231 ± 177[b]
Heavily polluted sea	36	0	0	5	1 Echovirus 5 2 Echovirus 6 1 Echovirus 8 1 Echovirus 12 1 Coxsackie- virus A18	49,800 ± 14,240	5,240 ± 2,820

[a]Bacteriological tests were performed on the initial homogenate.
[b]Mean ± SD.

TABLE 6.9. DIFFERENT FEATURES OF INFECTIOUS HEPATITIS AND "SERUM HEPATITIS" (GUIDON AND PIERACH, 1973)

Feature	Infectious Hepatitis	"Serum Hepatitis"
Route of Infection	Mainly fecal-oral, but also parenteral	Mainly parenteral, but also fecal-oral
Incubation Period	29–36 days, no difference in oral or parenteral route of infection	35–200 days 65 days after parenteral exposure 98 days after oral infection
Clinical Features:		
Onset	Usually abrupt	Usually slow and insidious
Anorexia	Present	Usually absent
Abdominal pain	Infrequent	Frequently appearing after 1–2 weeks
Ambulation	Usually non-ambulant	Usually ambulant
Arthralgia	Usually insignificant or absent	As early as 3 to 4 weeks before the onset of jaundice, particularly in small joints and at night
Ratio of icteric to aniceteric	About 1:1 in adults About 1:12 in children	About 1:100
Fever over 38°C	Common	Less Common
Mortality rate	0.1.0.3% in young adults, higher in older individuals, especially in women after menopause	Up to 1%
Duration of infectivity and carrier state	Blood: for days Feces: for weeks to months	Blood: for months to years Feces: undetermined
Prophylactic value of gamma globulin	Good	Variable results
Diagnostic biochemical and serological tests:		
SGOT	Spiking rise and short duration of elevated activity (1–3 weeks)	Gradual rise and long duration (up to six months)
Thymol turbidity	Consistently abnormal High increase 3–4 days before abnormal SGOT, return to normal 5–35 days later	Relatively normal Slight increase during acute stage
Au Antigen/Antibody	Negative	Positive 30–80 days after exposure in blood, urine & feces, only transient in 65%
EHAA (Milan)	Positive	Negative

either eaten raw or after insufficient heating. The immediate source of contamination proved to be either water polluted with sewage, human sewage itself, or an infected food handler.

Another viral agent, or possibly agents, transmitted via foods are those responsible for viral gastroenteritis. This is a disease characterized usually, but not always, by diarrhea and vomiting with no obvious bacterial agent being recovered from the fecal specimens of patients. Such an agent may have been responsible for an outbreak of food poisoning involving 35 persons who had eaten cockles. Virus-like particles were recovered from fecal specimens of three victims, but not from the cockles themselves (Anon., 1977).

DESTROYING VIRUSES

By their very nature, viruses must replicate inside living cells. The cells of vertebrate animals normally die soon after the creature is slaughtered for food; those for vegetables and shellfish survive longer, but are unsuitable for the replication of viruses producing diseases in humans. Consequently, the amount of virus present in the food at the time of contamination represents the total viral load of the product. Thereafter, the viruses either persists or becomes inactivated. A virus is considered inactivated when it has lost its ability to produce infection. This loss of infectivity is usually permanent. It results from alteration or degredation of either the capsid coat proteins, the lipid envelope (if present), or the viral genome.

There are a number of methods by which inactivation of viruses may be effected. These include, use of heat (thermal inactivation), freezing, drying, and exposure to ionizing radiations. Even the chemical composition of the food medium can have an effect.

Cooking is the most obvious method of inducing thermal inactivation. Any cooking process should prove quite effective provided the portion of the food containing the viruses is heated for a sufficient length of time. This is especially important with shellfish because any viruses present are apt to be in the digestive area which is the innermost portion of these animals. Experimentally, Di Girolamo et al. (1970), demonstrated that poliovirus can survive in shellfish cooked according to standard household procedures (Tables 6.10–6.12) including steaming. Steaming, especially of clams in the shell, is a popular cooking method. This practice has led to several sporadic outbreaks of infectious hepatitis (Koff et al., 1967; U.S. Public Health Service, 1972).

TABLE 6.10. RECOVERY OF POLIOVIRUS FROM OYSTERS PROCESSED BY STEWING (DI GIROLAMO ET AL., 1970)

Sample	Processing time (min)	Virus PFU/g or M1	Per Cent of survival	Internal temp (C)
Oyster	0	1.0×10^4	100	18.5
	2	2.7×10^2	27	33.5
	4	1.4×10^2	14	57.0
	8	1.0×10^2	10	75.0
Milk	0	0	0	
	2	1.4×10^2	14	
	4	1.2×10^1	12	
	8	7.0×10^1	7	

Another highly effective method of thermal inactivation is pasteurization. This is significant since a variety of viruses can occur and survive in milk or milk products, especially under storage conditions. Proper pasteurization practices are important also since improperly pasteurized milk has been responsible for at least one outbreak of infectious hepatitis (Raska et al., 1966). However, any viruses present in milk will be inactivated, if pasteurization is carried out according to U.S. Public Health Service recommendations (Gresikova et al., 1961; Sullivan et al., 1971B).

Freezing may, or may not, be effective as a means of inactivation. In some cases, survival of viruses has been shown to occur in refrigerated and frozen foods which have been held beyond the usable shelf life of the product, as can be seen in Tables 6.13 and 6.14 (Di Girolamo et al., 1972; Di Girolamo and Daley, 1973; Lynt, 1966). Where freezing has produced inactivation, this was due to denaturation of the nucleic acid core (Dimmock, 1967). The nature of the food itself may aid also in this process of denaturation (Cliver et al., 1970).

Drying, depending upon the process used, can also affect viruses. Freeze drying of experimentally contaminated food has proven highly effective in inactivating them (Heidelbaugh and Giron, 1969). However, experimentally, residual viruses were shown to survive

TABLE 6.11. RECOVERY OF POLIOVIRUS FROM FRIED OYSTERS PROCESSED FOR 8 MIN. (DI GIROLAMO ET AL., 1970)

Processing time (min)	Virus PFU/g	Per cent of survival	Internal temp (C)
0	1.2×10^4	100	23.0
3	7.6×10^3	61	40.0
6	4.4×10^2	36	72.5
8	1.7×10^1	13	100.0

TABLE 6.12.   RECOVERY OF POLIOVIRUS FROM BAKED OYSTERS PROCESSED FOR 20 MIN. (DI GIROLAMO *ET AL.*, 1970)

Processing time (min)	Virus PFU/g	Per cent of survival	Internal temp (C)
0	1.9×10^4	100	25
5	4.6×10^3	24.2	41
15	3.9×10^2	20.5	68
20	2.7×10^1	12.7	90

during storage of foods at refrigerator temperatures (Cliver *et al.*, 1970).

In recent years use of ionizing radiation, especially gamma radiation from a cobalt 60 source, has been proposed as a means of eliminating spoilage and potentially pathogenic microorganisms from foods (Liuzzo *et al.*, 1967, Mac Lean and Welander, 1960). The ability of gamma radiation to inactivate viruses in foods has been reported (Sullivan *et al.*, 1971A). However, the doses used (0.4 to 0.45 M) may not inactivate all the virus present in some foods or may cause organoleptic changes in the product rendering it unpalatable and unmarketable (Di Girolamo *et al.*, 1972).

The nature and chemical composition of food products can affect the stability of viruses. This subject has been reviewed extensively by Cliver (1971). Of particular interest has been the discovery that extracts of certain foods (e.g. ground beef) may contain viral inhibitors. These inhibitors can form chemical bonds with a virus at 4°C rendering the virus inactive. However, this reaction can be reversed by reaction with acid (Konowalchuk and Speirs, 1973).

TESTING FOR VIRUSES

The only method available at present to test for viruses is to expose a living host to a contaminated suspension and look for signs of infection in the host. The usual host systems employed for test purposes include embryonated eggs, laboratory animals, and tissue monolayers. The latter, which are probably the most widely used, are generally cells of monkey or human origin grown in glass or plastic vessels. Signs of "infection" in these tissue monolayers consist of small areas of tissue destruction due to viral activity. These small ragged holes are referred to as plaques or cytopathogenic effects (C.P.E.).

At present there is no standard method available for testing foods for viruses. Current assay procedures can be divided into three

TABLE 6.13. RECOVERY OF COLIPHAGE T4 FROM CONTAMINATED SAMPLES OF UNPROCESSED AND PROCESSED EDIBLE CRABS STORED FOR 30 DAYS AT -20°C (DI GIROLAMO AND DALEY, 1973)

Sample	Storage time (days)	No. of virus in crab tissue (PFU/g[a])	Survival (%)
Unprocessed crab	0	3.6×10^4	100
	1	3.6×10^4	100
	20	1.5×10^4	42
	30	1.2×10^4	35
Processed crab	0	2.6×10^2	100
	1	2.6×10^2	100
	20	1.4×10^2	55
	30	45	17

[a]PFU, plaque forming unit.

general types: dilution techniques, concentration techniques, and extraction techniques.

In the dilution technique the food sample is suspended in fluid then homogenized to a fine consistency. This procedure facilitates handling of the sample and dilutes down any food component which might prove harmful to tissues. Usually, serial decimal dilutions of the homogenized sample are prepared and assayed for virus. The dilution method has several distinct advantages: 1) It permits isolation and quantal determination of viruses; 2) It does not require a great amount of costly equipment; 3) It is usually less time consuming compared to other techniques. Unfortunately, these dilution methods may not be quite as sensitive as other procedures.

A great degree of sensitivity can be obtained by using concentration techniques. All these procedures are based upon removing as much fluid from assay samples as possible, but not the viruses which become "concentrated" in the remaining fluid medium. Many techniques have been suggested for purposes of concentration. Some of the most efficient have proven to be those using either filtration or ultrafiltration (Konowolchuk and Speirs, 1973; Turney et al., 1973). However, these methods also have their problems and shortcomings (Cliver, 1971).

An extraction procedure differs from the above in that the bulk of the food substance, contaminating bacteria, etc. are excluded from the sample suspension as a first step. Extraction may be performed by using a fluid to disassociate viruses from food particles (Sullivan et al., 1970) or food solids may be separated via centrifugation and/or filtration (Larkin et al., 1975). Extraction procedures tend to be more selective than dilution methods, but there still remains some risk of losing viruses during suspension or separation.

TABLE 6.14.   RECOVERY OF ADDED TYPE B6 COXSACKIEVIRUS FROM FOOD PREPARATIONS AFTER STORAGE (LYNT, 1966)

Preparation	Zero Time	Storage conditions							
		Room temp			10 C		-20 C		
		3 to 4 days	7 days	1 week	2 weeks	1 month	1 month	2 months	5 months
Blank	9	0.16	0.5	0.5	0.16	0.16	0.5	0.5	0.5
Hashed brown potatoes	3.2	0.5	0.5	1.6	0.16	<.05	1.6	1.6	1.6
Breaded shrimp	0.5	0.16	0.16	1.6	0.16	0.16	0.5	0.5	1.6
Shrimp roll	5	0.16	0.5	1.6	0.5	0.16	0.16	0.5	0.5
Cole slaw	5	0.005	<0.005	0.016	<0.005	0.005	0.016	0.005	0.016

Each of the procedures described has its own advantages and disadvantages. One shortcoming common to all techniques is the time element involved. It still requires two to seven days before the results of viral activity (C.P.E.) become apparent. Other techniques, such as the use of serological or chemical means of detection, are not presently available, either because they are not perfected due to technical difficulties, or because of cost. Therefore, determination of viral infectivity remains the only reliable means of detection at present.

PREVENTING TRANSMISSION OF VIRUSES

The most obvious way to prevent foods from serving as the vehicles for transmission of viral diseases is to prevent the viruses from entering in the first place. How this will be accomplished will depend to some extent on how the particular food is usually contaminated i.e. by primary or secondary means. Where the source of contamination is water polluted with raw sewage, human feces, or insects, these must be prevented from contacting foods. This may be accomplished by being absolutely certain of the purity of the water supply, by avoiding exposure of the product to insects, and by strict adherence of food handlers to sanitary practices. This task is not an easily accomplished one.

Where diseases are transmitted via the flesh of sick animals, their milk or eggs, then use of these should be prevented. In this situation, the foods should be tested for viral contamination. Testing is absolutely essential in the case of certain "high risk" foods, such as shellfish which have a long record as vectors in the transmission of infectious hepatitis.

Ultimately, the consumer must also assume the responsibility of using proper procedures in handling, storing, and processing foods and, as in the case of shellfish, making absolutely certain the food product comes from a safe or reputable source.

SUMMARY

Among the pathogenic microorganisms which find their way into foods are the viruses. The viruses are small entities which must replicate inside a living host cell. They enter foods as a result of either primary or secondary contamination. Food products which may show primary contamination include meats (especially beef), milk, vegetables, and shellfish. Any product may become contaminated by secondary means, especially if exposed to an infected food handler, contaminated insects, or food preparation surfaces. Virus diseases known to be spread via foods include infectious hepatitis, poliomyelitis, and tick-borne encephalitis. This latter disease may be spread not only by the bite of a tick but also by drinking infected raw milk or inhaling infected material. The symptoms usually consist of severe headaches, acute central nervous involvement plus the symptoms of encephalitis. Hospitalization, when required, may last from 3 to 6 weeks with a prolonged period of convalescence. The mortality rate varies from 1% to 4.5% (Blaskovic, 1967).

Viruses which have lost their ability to produce infection are said to be "inactivated". Inactivation of viruses in foods may be caused by heating (thermal inactivation), by freezing, drying, or the use of gamma radiation. Of these methods, the most reliable is heating, provided all portions of the food are heated for a sufficient length of time. Pasteurization is also a highly efficient method of inactivating viruses in milk and/or milk products. Freezing may, or may not, inactivate viruses. When freezing is effective, this is probably due to denaturation of the nucleic acid core. As with freezing, the efficacy of drying will depend upon the method used. Freeze drying seems to affect viruses, but residual viruses may survive subsequent storage of the product.

Gamma radiation, using cobalt 60 source, has been proposed as a means of eliminating harmful or food spoilage microorganism from foods. The suitability of this method as a means of destroying viruses is questionable. Doses of radiation sufficiently high to produce 90% or better inactivation of virus may also induce organoleptic changes rendering the product unpalatable.

Methods used to assay foods for viruses depend upon determining the cytopathogenic effects produced by these pathogens in host cells. Test methods may be divided into three basic types: dilution techniques, extraction techniques and concentration techniques. Each of these assay procedures has its own advantages and disadvantages. A major disadvantage common to all is the time required to obtain results (2–7 days).

The most obvious way to prevent foods serving as vehicles for the spread of viral diseases is to prevent entrance of viruses into food stuffs. The method(s) of prevention will depend, to some extent, upon the nature of the food. Those foods subject to primary contamination should be assayed for viral content and/or protected from exposure to raw sewage or sewage polluted waters. Products subject to secondary contamination should be handled using the most rigid sanitary practices. These foods must be protected from contaminated surfaces, insects or food handlers. Ultimately the consumer must also play an important role through proper handling and storage of foods and by purchasing foods from reputable sources.

ACKNOWLEDGEMENTS

The author wishes to express his sincere appreciation to the following individuals, societies, and journals for granting permission to reproduce the data presented in this chapter.

The Academic Press and The Proceedings of The Society for Experimental Biology and Medicine.

The American Journal of Epidemiology and Dr. T. Metcalf.

The American Society for Microbiology and Applied and Environmental Microbiology.

Dr. L. B. Baldwin, Institute for Food and Agricultural Science, University of Florida, Gainesville.

E. P. Larkin, Symposium Proceedings, 1976. Viral Aspects of Applying Municipal Waste on Land.

The Minnesota Medical Society and Minnesota Medicine.

The author also wishes to express special thanks to Mrs. Lucy Allen for the care and long hours spent typing this manuscript.

BIBLIOGRAPHY

ANON. 1977. Abstract on Hygiene. 52:612.

ATWOOD, R. P., CHERRY, J. D., and KLEIN, J. O. 1964. "Clams and viruses, studies with coxsackie B-5 virus." Comm. Dis. Center Hepatitis Survey Rpt. 20.

BAGDASAR 'YAN, G. A. 1964. "Sanitary examination of soil and vegetables from irrigation fields for the presence of viruses." Gigiene i Sanitariya 11:37–39.

BENDINELLI, M., and RUSKI, A. 1969. "Isolation of human enteroviruses from mussels." Appl. Microbiol. 18:531–532.

BLASKOVIC, D. et al. 1967. "Studies on tick-borne encephalitis." Bull. W.H.O. 36 suppl. 1:5–13.

BOWMER, E. 1964. "The challenge of salmonellosis, major public health problem." Amer. J. Med. Sci. 247:467–470.

BOWMER, E. 1965. "Salmonellae in food, a review." J. Milk Food Techno. 28:74–80.

BJNOE, E. T., and YURACK, J. A. 1964. "Salmonellosis in Canada." in

The World Problem of Salmonellosis. Monographiae Biological, Vol. XIII, Dr. J. W. Junk, publisher. The Hague.

CANZONIER, W. J. 1971. "Accumulation and elimination of coliphage S-13 by the hard clam *Mercenaria mercenaria.*" Appl. Micriobio. *24*: 1024–1031.

CLARKE, N. A., and KOBLEER, S. W. 1964. "Human enteric viruses in sewage." Health Lab. Sc. *1*:44–50.

CLIVER, D. O., and BOHL, E. H. 1962. "Isolation of enteroviruses from a herd of dairy cattle." J. Dairy Sci. *45*:921–924.

CLIVER, D. O. 1966. "Implications of food-borne infectious hepatitis." Public Health Rpt. *81*:159–165.

CLIVER, D. O. 1967. "Food associated viruses." Health Lab. Sci. *4*:213–221.

CLIVER, D. O. 1969. "Viral infection" in *Food Borne Infections and Intoxications.* Riemann, H. (Editor) Acad. Press. N.Y. p. 73.

CLIVER, D. O., KOSTENBADER, K. D., JR. and VALLINOS, M. R. 1970. "Stability of viruses in low moisture foods." J. Milk Food Technol. *33*: 484–491.

CLIVER, D. O. 1971. "Transmission of Viruses Through Foods" in *Critical Reviews in Environmental Control.* *1*:551–579.

CROVARI, P. 1958. "Some observations on the depuration of mussels infected with poliomyelitis virus." Ig. Mod. *5*:22–32.

DI GIROLAMO, R., LISTON, J., and MATCHES, J. 1970. "Survival of virus in chilled, frozen, and processed oysters." Appl. Microbiol. *20*:58–63.

DI GIROLAMO, R., LISTON, J., and MATCHES, J. 1972. "Effects of irradiation on the survival of virus in West Coast oyster." Appl. Microbiol. *24*:1005–1006.

DI GIROLAMO, R., WICZYNSKI, L., DALEY, M., and MIRANDA, F. 1972. "Preliminary observations on the uptake of poliovirus by West Coast shore crabs." Appl. Microbiol. *23*:170–171.

DI GIROLAMO, R., WICZYNSKI, L., DALEY, M., MIRANDA, F., and VIEHWEGER, C. 1972. "Uptake of bacteriophage and their subsequent survival in edible West Coast crabs after processing." Appl. Microbio. *23*:1073–1076.

DI GIROLAMO, R., and DALEY, M. 1973. "Recovery of bacteriophage from contaminated chilled and frozen samples of edible West Coast crabs." Appl. Microbiol. *25*:1020–1022.

DI GIROLAMO, R., LISTON, J., and MATCHES, J. 1975. "Uptake and elimination of poliovirus by West Coast oysters." Appl. Microbiol. *29*:260–264.

DIMMOCK, N. J. 1967. "Differences between the thermal inactivation of picorna viruses at "high" and "low" temperatures." Virology *31*:338–353.

DINGMAN, J. C. 1916. "Report of a possible milk borne epidemic of infantile paralysis." N.Y. State J. Med. *16*:589.

DOLMAN, C. E. 1964. "Botulism as a World Health Problem." in *Botulism Proceeding Of A Symposium.* U.S. Public Health Serv. Publ. 999-f-p. *1*.

EISENSTEIN, A. B., AACH, R. D., JACOBSOHN, W., and GOLDMAN,

A. 1963. "An epidemic of hepatitis in a general hospital. Probable transmission by contaminated orange juice." J.A.M.A. *185*:171–174.

ERNEK, E., KOZUCK, O., and NOSEK, J. 1968. "Isolation of tick-borne encephalitis virus from blood and milk of goats grazing in the Tribec focus zone." J. Hyg. Epidemiol. Microbiol. Immunol *12*:32–36.

FUGATE, K. J., CLIVER, D. O., and HATCH, M. T. 1975. "Enteroviruses and potential bacteriological indicators in Gulf Coast oysters." J. Milk Food Technol. 38:100–104.

GALTSOFF, P. 1964. The American Oyster, *Crassostrea virginica.* U.S. Fish and Wildlife Serv. Fish Bulletin *64*:27–35.

GRESIKOVA, M. 1958. "Isolation of the tick-borne encephalitis virus from the blood and milk of subcutaneously infected sheep." Acta Virol. *2*:113.

GRESIKOVA, M. 1958. "Excretion of the tick-borne encephalitis virus in the milk of subcutaneously infected cows." Acta Virol. *2*:188.

GUIDON, LUKAS, M. D., and PIERACH, CLAUS, A., M. D. 1973. "Infectious hepatitis after ingestion of raw clams." Minn. Med. *56*(1)15–19.

HALL, H. E., and ANGELOTTI, R. 1965. "*Clostridium perfringens* in meat and meat products." Appl. Microbiol. *13*:353–357.

HARTWELL, W. V., AVERNHEIMER, A. H., and PEARCE, G. W. 1968. "Examination by chromatography and immunodiffusion of an adenovirus-3 isolated from humans with infectious hepatitis." Appl. Microbiol. *16*:1859–1864.

HEIDELBAUGH, N. D., and GIRON, D. J. 1969. "Effect of processing on recovery of poliovirus from inoculated foods." J. Food Sci. *34*:239–241.

JOSEPH, P. R., MILLOR, J. D., and HENDERSON, Da. A. 1965. "An outbreak of hepatitis traced to food contamination." New England J. Med. *273*:188–194.

KASEL, J. A., ROSEN, L., and EVANS, H. E. 1963. "Infection of human volunteers with a reovirus of bovine origin." Proc. Soc. Exp. Biol. Med. *112*:979.

KELLY, S. M., and SANDERSON, W. W. 1959. "Viruses in sewage." New York Dept. of Health News *36*:18.

KNAPP, A. C., GODFREY, E. S., JR., and AYCOCK, J. L. 1926. "An outbreak of poliomyelitis." J.A.M.A. *87*:635.

KOFF, J. S., GRADY, G. F., CHALMERS, T. C., METCALF, J. W., and SWARTZ, B. L. 1967. "Viral hepatitis in a group of Boston hospitals. III. Importance of exposure to shellfish in a non-epidemic period." New England J. Med. *276*:703–710.

KONOWALCHUK, J., and SPEIRS, J. I. 1973. "Identification of a viral inhibitor in ground beef." Canad. J. Microbiol. *19*:177–181.

KONOWALCHUK, J., and SPEIRS, J. I. 1973. "An efficient ultrafiltration method for enterovirus recovery from ground beef." Canad. J. Microbiol. *19*:1054–1056.

KRUGMAN, S., WARD, R. W., and GILES, P. 1963. "Studies on the natural history of infectious hepatitis." Prospects in Virol. *3*:159–169.

LARKIN, E. P., TIERNEY, J. T., SULLIVAN, R., and PEELER, J. T.

1975. "Collaborative study of the glass wool filtration method for the recovery of virus inoculated into ground beef." Journal of the AOAC 58:576–578.

LARKIN, E. P., TIERNEY, J. T., and SULLIVAN, R. 1976. "Persistence of virus on sewage irrigated vegetables." J. Env. Eng. Div. ASCF 102: 29–35.

LIPARI, M. 1951. "A milk borne poliomyelitis episode." N.Y. State J. Med. 16:589.

LIU, O. C., SERAICHEKAS, H.R., and MURPHY, B. L. 1966. "Fate of poliovirus in Northern Quahaugs." Proc. Soc. Exp. Biol. Med. 121:601–607.

LIU, O. C., SERAICHEKAS, H. R., and MURPHY, B. L. 1968. "Viral depuration of the Northern Quahaug." Appl. Microbiol. 15:307–315.

LIU, O. C. 1971. "Viral Pollution and Depuration of Shellfish" in Proc. Nat. Spec. Conf. Disenfect. ASCE, New York p. 397–428.

LIUZZO, J. A., FARAG, K., and NOVAK, A. F. 1967. "Effect of low level raiation on the proteolytic activity of bacteria in oysters." J. Food Sci. 32:104–112.

LUDBERG, R. 1956. "On the nature of the viral agent of hepatitis." Amer. J. Publ. Health 53:1406–1410.

LYNT, R. K., JR. 1966. "Survival and recovery of enteroviruses from foods." Appl. Microbiol. 14:218–222.

MAC LEAN, D. F., and WELANDER, C. 1960. "The preservation of fish with ionizing radiation: bacterial studies." Food Technol. 14:251–254.

MAZUR, B., and PACIORKIEWICZ, W. 1973. "The spread of enteroviruses in man's environment." Medical Microbiol. Exps. (Trans) 23:83–98.

METCALF, T. G., and STILES, W. C. 1968. "Enteroviruses within an estuarine environment." Amer. J. Epidemical. 88:379–391.

PORTNOY, B. L., MACKOWIAK, P. A., GARAWAY, C. T., WALKER, J. A., McKINLEY, T. W. and KLEIN, C. A., JR. 1975. "Oyster associated hepatitis. Failure of shellfish certification programs to prevent outbreaks." JAMA 233:1065–1068.

RASKA, K., HELEL, J., JEZEK, J., KUBELKA, Z., LITOV, M., NOVAK, K., RADOVSKY, J., SERY, V., ZEJDL, J., and ZIKMUND, V. 1966. "A milk borne infectious hepatitis epidemic." J. Hyg. Epidemiol. Microbiol. Immunol. 10:413.

ROOS, B. 1956. "Hepatitis epidemic conveyed by oysters." Svenska Lakartidningen 53:989–1003.

SEDDON, J. H. 1961. "An epidemiological survey of infectious hepatitis in a country town." N.Z. Med. J. 60:55.

SULLIVAN, R., FASSOLITIS, A. C. and READ, R. B., JR. 1970. "Method for isolating viruses from ground beef." J. Food Sci. 35:624–626.

SULLIVAN, R., FASSOLITIS, A. C., LARKIN, E. P., READ, R. B., JR., PEELER, J. T. 1971A. "Inactivation of thirty viruses by gamma radiation." Appl. Microbiol. 22:61–65.

SULLIVAN, R., FASSOLITIS, A. C., and READ, R. B., JR. 1971B.

"Thermal resistance of certain oncogenic viruses suspended in milk and milk products." Appl. Microbiol. 22:315–320.

SULLIVAN, R. H., COHN, M. M. and BAXTER, S. S. 1973. "Survey of facilities using land application of wastewater." EPA–430/9–73–006 Office of Water Programs Oper.

TURNEY, J. T., SULLIVAN, R., LARKIN, E. P., and PEELER, J. T. 1973. "Comparison of methods for the recovery of virus inoculated into ground beef." Appl. Microbiol. 26:497–501.

U.S. PUBLIC HEALTH SERVICE 1972. Morbidity, Mortality Weekly Rpt. 21(2):20.

U.S. DEPT. OF HEALTH, ED. AND WELFARE, 1976. "Food-borne and Water-borne Disease Outbreaks." Annual Summary DHEW Publ. 76–8185 p. 75.

7

Mycotoxins in Foods and Feeds

D. K. Salunkhe, M. T. Wu, J. Y. Do and Melanie R. Maas

Although fungal contamination of foods and feeds is a constantly recurring phenomenon, the attitude towards it has been far from consistent. For many years, a tendency to regard fungi as harmless was common; mild fungal growth was treated as a regrettable nuisance, detracting from the appearance of the product and perhaps causing degrees of losses or spoilages. In severe cases commodities thus deemed unfit for human consumption might still be diverted for animal feeds. On the contrary, some people seem actually to have preferred, on the basis of flavor, foods with a touch of moldiness, while fungal cultures have frequently been used as essential agents in the preparation of various traditional fermented food products. An incident in 1960 changed the general attitude toward the fungal contamination of food and prompted an awareness of the scope of the problem. The discovery of aflatoxin in 1961 has led to a vast amount of research in chemistry, biochemistry, mycology, toxicology, nutrition, and food technology. An equally striking development has been the focusing of attention on the wider problem of mycotoxins as a class.

Mycotoxins are secondary fungal metabolites that produce toxic reactions in animals or humans exposed to them. They comprise a group of chemical compounds widely diverse in their nature and biological effects. The term "mycotoxicosis" introduced in 1962 by Forgacs and Carll is defined as "poisoning of the host following entrance into the body of toxic substances of fungal origin." This term is usually limited to those pathological entities associated with ingestion of food or feed, on which the fungi proliferate and give rise to toxic products. Although poisoning due to ingestion of certain basidiomycetes should be included under this heading, mushrooms as a class, are usually handled separately. Ergot, the oldest form of mycotoxicosis known, is caused by ingestion of rye and other cereals infected with sclerotia of the fungus *Claviceps purpurea*. The etiology of ergot was determined in the sixteenth century, but the alkaloids produced by the mold, which are all derivatives of lysergic acid were not chemically identified until the 1930's (Rothlin and Cerletti, 1954). The first recorded mycotoxicosis is probably the bread staggering syndrome reported by Woronin in 1891 and attributed to *Gibberella zeae*. At the turn of the last century, equine encephalitis was reported in the state of Maryland (Buckley, 1901). Several years later, a fetal hemorrhagic afebrile disease of cattle was shown to be a mycotoxicosis (Schofield, 1924). In 1954, Forgacs and Carll isolated a toxigenic *Aspergillus clavatus* which produced hemorrhage and hyperkeratosis in calves. Likewise, during the years 1942–1955, a hemorrhagic syndrome caused significant losses of chicken in Delaware and West Virginia. Forgacs and Carll (1955) using grains inoculated with selected strains of the genera *Aspergillus*, *Penicillium*, and *Alternaria*, reproduced the toxic syndrome in chickens. Facial eczema of ruminants was shown to be of fungal origin (Percival, 1959). In Russia, mycotoxicoses were also reported. One of these is a human disease known as 'aseptic angina' and later as 'Alimentary Toxic Aleukia (ATA)'. The main causative agent was reported to be *Fusarium sporotrichoides*. ATA may also affect cattle, sheep, chickens, horses, and swine (Mayer, 1953). The second mycotoxicosis is equine stachybotryotoxicosis, which also affects humans, calves, sheep, dog, and swine.

During the last two decades, research on mycotoxins has revealed the seriousness of mycotoxin contamination of many important agricultural products such as aflatoxins on corn, peanuts, grain sorghum, cottonseeds, Brazil nuts, pistachio nuts, almonds, walnuts, pecans, filberts, copra, rice, legumes, peppers, potatoes, dried fruits and dairy products; zearalenone on corn; ochratoxins on corn, wheat, barley, oats, rye, and coffee beans; citrinin on rice, barley; penicillic acid on corn, dried beans; and patulin on apples, peaches, apricots, pears, cherries, etc. A report by Wilson and Nuovo (1973) showed up to 45 ppm patulin has been detected from "Organic Apple Cider". An extensive survey of patulin contamination of apple juice in terms of geography and type of pack by U.S. Food & Drug Administration showed levels of from 40 to 150 $\mu g/l$ of juice in 46 of the samples (Bureau of Foods, FDA, 1974). Three samples had 270 $\mu g/l$ and 1 sample 440 $\mu g/l$. In the Washington, D.C. area, levels of 49 to 309 ppb patulin were found in 8 of 13 commercial apple cider samples analyzed in another study (Ware *et al.*, 1974). Thus fungal infection of foods and feeds not only results in economic losses but also poses a potential health hazard to man and animals.

FUNGAL CONTAMINATION OF FOODS AND FEEDS

Fungal spores are ubiquitously distributed, and foodstuff may be contaminated with a variety of fungi. For example, Christensen (1957) reported a wide variety of fungi that invade the seeds of cereal grains and their products. These microorganisms have been grouped into three categories, field fungi, storage fungi, and advanced decay fungi, according to the stage at which invasion and growth occur. If an equally complex fungal flora is assumed to be present in other commodities, it is apparent that the majority of food raw materials are liable, in varying degrees, to the growth of contaminating fungi at some stage during their harvest, storage, transport, or processing. The possibility for spore germination and growth on a given product is governed by several factors including moisture content, relative humidity, and temperature. During the entire postharvest period, food crops are essentially in a state of storage and mold growth on them is avoidable only by careful regulation of moisture content, temperature and other environmental conditions.

Mature fruits and vegetables are highly susceptible to invasion by specific pathogenic microorganisms because they are high in moisture and nutrient content and are no longer protected by the intrinsic factors which conferred resistance during their development. In addition, many fruits become more easily injured as they approach full maturity and, therefore, are more vulnerable to pathogens which enter through wounds. Postharvest losses of fresh fruits and vegetables due to microbial decay have been reported as follows: 10–15% for lettuce (Ceponis, 1970), 15–30% for orange (Smoot, 1969), 15–24% for peach (Wells and Harvey, 1970), 14–18% for strawberry (Harris et al., 1967), 40–50% for pineapple (Ramsey, 1938), 20–33% for mango (Singh, 1960), 50% for yam (Burton, 1970).

Fruits on the plant may be infected by direct penetration of certain fungi through the intact cuticle or through wounds and natural openings in the surface of the fruits. Alternatively, many postharvest diseases are initiated through injuries created during and subsequent to harvest, such as cut stem and mechanical damage to the surface of the product during handling and transporting. Preharvest infection by fungi also plays a role in post harvest decay. Some genera of pathogenic fungi sporulate abundantly in lesions on stem, leaves, and flower parts of infected plants as well as dead plant materials. Rain and wind transport spores of these fungi to flowers and fruits at every stage of their development. These spores germinate when free water is present and develop to a limited extent before growth is halted by the resistance of the cells of the immature fruits. Infection of stone fruits by *Monilinia fructicola* (Jenkins and Reinganum, 1965; Kable, 1971) and strawberries and grapes by *Botrytis cinerea* (Powelson, 1960) while these fruits are developing on the plant is a major consideration in the epidemiology of postharvest disease initiated by these fungi. Major storage diseases of apples in Europe are initiated by penetration of spores of *Phylyctaenea vagabunda* and *Cryptosporiopsis curvispora* into the lenticular cavities of the fruit during periods of relatively high temperature and humidity late in the summer (Bompei *et al.*, 1969). These fungi develop to a very limited extent in the lenticular cavity and then become quiescent until the fruit begins to ripen in storage. At that time, the fungi become active once again and invade the flesh of the apple, producing a lesion around the lenticel.

Propagules of pathogenic fungi are abundant in the atmosphere and on the surface of fruits and vegetables as they approach maturity in the field (Lukezic *et al.*, 1967). Fungi such as *Penicillium*, *Rhizopus* and *Geotrichum* are not capable of directly penetrating the cuticle and epidermis of the host but if they gain entry through an injury or natural opening, these fungi may cause devastating rots of mature produce. Fresh fruits and vegetables cannot be harvested without creating some injuries through which these and other pathogens may infect their hosts.

High and low temperatures which physically damage the surface cells of fresh produce invariably increase infection by pathogens which invade through wounds. According to Freidman (1960), a major portion of the decay found in the terminal markets has been attributed to predisposition of the produce by high and low temperatures. Several investigators have observed that tomatoes were more prone to attack by weak pathogens after exposure to low temperatures which did not obviously damage the produce (McColloch and Worthington, 1952). Storage of produce in an atmosphere of greater than 90% relative humidity favors the development of postharvest disease by maintaining injuries in a moist condition which permits the development of pathogenic fungi. In Florida, it has been recognized for years that degreening citrus fruits with low concentrations (ca. 50 ppm) of ethylene gas produces a substantial increase in stem end rot during storage (McCornack, 1972). This treatment accelerates senescence and abscission of the button of the fruit, thereby stimulating the development of quiescent infection of *Diplodia* in this organ and providing a means for this fungus to invade the fruit (Brown and Wilson, 1968). Grain, oil seeds, and prepared foods and feeds are also prone to fungal damage. This constitutes a major factor in losses of food or raw and processed material which have a significant impact on the total world food supplies.

CONTAMINATIONS OF FOODS AND FEEDS WITH MYCOTOXINS AND THEIR HEALTH HAZARDS

A. Aflatoxins

More information is available about aflatoxins than any other single group of mycotoxins. The aflatoxins are a closely related group of secondary fungal metabolites that have been shown to be mycotoxins. A renaissance in mycotoxin research developed as a direct result of concurrent outbreaks of disease in poultry and fish during 1960's in diverse geographic locations. The most prominent development was the report of severe losses of turkey poult in Britain (Blount, 1961). Since the etiological agent involved in the disease was not known, the disorder was named "Turkey X Disease." According to Spensley (1963), the acute forms of disease were characterized by loss of appetite, lethargy, weakness of wings and a distinctive attitude of the head and neck at the time of death. Blount (1961) conducted the histological examinations of the diseased birds and found an acute hepatic necrosis associated with bile duct proliferation. Examination of the feed source showed that a common factor in disease outbreaks was the utilization of a Brazilian groundnut in the rations (Blount, 1961; Sargeant *et al.*, 1961). At the same period, symptoms analogous to those of the "Turkey X" disorder were reported in outbreaks of disease in other farm animals, particularly ducklings, following ingestion of feeds containing groundnut meal (Allcorft *et al.*, 1961; Asplin and Carnaghan, 1961; Loosmore and Harding, 1961; Loosmore and Markson, 1961; Sargeant *et al.*, 1961). Allcroft and Carnaghan (1963) subsequently studied the groundnut meal from various producing countries and found that 13 of the samples from diverse locations contained the toxin. Researchers at the Tropical Products Institute in London, the Central Veterinary Laboratory at Weybridge, and Unilever Laboratories in Holland and Britain are credited with the initial successes in narrowing the etiological agent of Turkey X Disease to groundnut meal and the isolation of the toxic material (Carnaghan and Allcroft, 1962;

Sargeant *et al.*, 1961; Van der Zidjen *et al.*, 1962). Feeding a ration containing 20% of the toxic groundnut meal to rats for 30 weeks increased the incidence of hepatic carcinoma in the animals (Lancaster *et al.*, 1961).

Sinnhuber *et al.*, (1965) and Wolf and Jackson (1963) demonstrated an interesting parallel between the developments associated with the identification of the etiological agent involved in Turkey X Disease and that of an epizootic of liver cancer in hatchery-reared rainbow trout. The outbreaks of trout hepatomas were associated with the ingestion of toxic factors in the cottonseed meals (Sinnhuber *et al.*, 1965).

Isolation of the toxins was facilitated by the discovery that a characteristic blue fluorescence parallels the toxicity observed in duckling tests (Sargeant *et al.*, 1961; Smith and McKernan, 1962). The toxic factors were later separated into distinct compounds (Nesbitt *et al.*, 1962; Sargeant *et al.*, 1961; Van der Zidjen *et al.*, 1962). These components have been given the name aflatoxins, identifying their generic origin. Chemically, the aflatoxins are highly substituted coumarin and contain a fused dihydrofurofuran configuration, which is peculiar to a limited number of compounds of natural origin. The compounds occur in two series, aflatoxin B_1 and derivatives and aflatoxin G_1 and derivatives. These include B_1, B_2, G_1, G_2, M_1, M_2, GM_1, B_{2a}, G_{2a}, R_0, B_3, $1-OCH_3B_2$, $1-OCH_3B_2$, and $1-CH_3G_2$

Investigations of feed contaminated with *A. flavus* established that a typical toxicity syndrome involved liver damage in many mammals, fish and birds (Kraybill and Shimkin, 1964; Lancaster, 1968). Aflatoxin initiation of tumors has been clearly described in ducklings, rats, ferrets, trout, guinea pigs, mice, and sheep. Histological examinations of the injured livers following aflatoxin treatment has demonstrated a routine pattern of cellular changes, including peripheral zone necrosis and biliary proliferation (Kraybill and Shimkin, 1964). In addition to hepatomas, aflatoxin has also been implicated in the induction of neoplasms in the glandular stomach, kidney, lung, salivary gland, lachrymal gland, colon, and skin (Butler and Barnes, 1964, 1966; Lancaster, 1968; Wogan and Newberne, 1967). Recently, Howarth and Wyatt (1976) demonstrated that feeding a low level of aflatoxin (10–15 μg/g diet) significantly decreased the egg production of hens. Hatchability of fertile eggs was also significantly decreased.

Since the initial implication of *A. flavus* in Turkey X Disease, aflatoxin production by this organism and by *A. parasiticus* has been well documented. Not all strains of *A. flavus* produced aflatoxins. Other fungi have also been reported to produce aflatoxins, but confirmatory studies are lacking. Those organisms that are capable of toxin production generally synthesize only two or three aflatoxins under a given set of conditions. When they occur as food contaminants, aflatoxin B_1 is always present. Although aflatoxin B_2, G_1 and G_2 have also been reported in contaminated products, they generally occur less frequently than B_1 and have never been reported in the absence of B_1. This is an important point, because B_1 has the highest potency of the group as a toxin and as a carcinogen.

Since the first report in 1961, the natural occurrence of aflatoxin in peanuts and peanut meals has been reported by several investigators (Allcroft and Carnaghan, 1962; Allcroft and Lewis, 1963). Samples of fermented peanut press cake from Indonesia have been shown to contain high levels of aflatoxin (Van Veen et al., 1967). Taber and Schroeder (1967) analyzed 78 samples of farmer stock peanuts grown in nine different geographical areas in the United States during 1964 and found that concentrations of aflatoxins in peanuts from all areas varied from 0–91 ppb and only two samples contained more than 50 ppb. Tung and Ling (1968) studied aflatoxin contamination of the peanut in Taiwan. They found preparations, 4 of 12 sources of peanut cakes and 8 of 17 peanut oils, contained aflatoxin. *Aspergillus flavus* and aflatoxins have been shown to occur in California cottonseed prior to harvest (McMeans and Ashworth, 1966). The *A. flavus* group is commonly found on rough rice immediately after harvest, but infection of kernels at this point appears to be unusual. Tung and Ling (1968) did not detect aflatoxin in rice collected from the market in Taiwan. However they found *A. flavus* and high concentration of toxin in rice that may have been involved in an intoxication of 25 people. Shotwell *et al.* (1969a) have carried out extensive surveys of cereal grains and soybean from commercial channels for the presence of aflatoxins. Very low levels of what appeared to be aflatoxin were detected in a total of 9 out of 1368 samples of wheat, sorghum and oats. In Taiwan, dried sweet potato was found to be heavily contaminated with *A. flavus* and contained 0.01–0.18 ppb aflatoxin (Tung & Ling, 1968). In an investigation of mixed commercial feeds, aflatoxin was found in one or more components of the feed (Schultz and Motz, 1967). Aflatoxin-positive components included pea, wheat, rye, soy, peanut, and rapeseed meal. Even grape wine has also been found to contain aflatoxins (Schuller *et al.*, 1967). Kiermeir *et al.* (1977) recently reported that 69% of cheese samples tested contained detectable amounts of aflatoxin M_1.

Contamination of aflatoxins in stored field corn has drawn more attention recently. According to Shotwell *et al.* (1969a, 1969b) on samples mainly from the corn belt, 9 out of 1,368 samples of wheat, grain sorghum and oats, and 35 out of 1,311 corn samples contained small amounts of aflatoxin at levels up to 19 ng/kg. Their second survey of U.S. corn showed a similarly low incidence of aflatoxins at levels up to 25 ng/kg (Shotwell *et al.*, 1970). Most of the aflatoxin contamination was in the lower grades of corn. Incidence of contamination of southern corn of all grades except U.S. No. 1 was near 30% and ranged from 5 µg/kg to 308 µg/kg with only 16% of the contaminated samples containing more than 100 µg/kg (Shotwell *et al.*, 1973). Generally, low incidence and low levels of aflatoxins in grains and grain products were observed (Scott, 1973). Lillehoj *et al.* (1975a) studied the stored corn in southeast Missouri bins and found heavy contaminations of *A. flavus* of stored corn with 0.4 ppm aflatoxin B_1 in one sample. Hesseltine *et al.*, (1975 & 1976) studied white corn under a government loan in 1972 and 1973 to the Agricultural Stabilization and Conservation Service at Diehlstadt, Missouri. They found that of the 1283 samples, each representing a truck load of corn, 394 contained aflatoxin and of these *A. flavus* group was present in all but seven. The occurrence of *A. flavus* closely paralleled the aflatoxin content of the corn samples. Lillehoj *et al.* (1975b) also reported that corn samples from 5% of the test ear exhibited blue-green-yellow fluorescence while those from 2.5% of the test ears contained aflatoxin B_1 in excess of 20 ppb. Aflatoxin was detected in significantly more earworm damaged samples than in those with no insect damage. On aflatoxin contamination of corn in the field, Anderson *et al.* (1975) reported that aflatoxin has been found in corn samples at all stages of development and maturity from the late milk stage until harvest. Insect damage was also associated with aflatoxin contamination of grains. A recent report by Lillehoj *et al.* (1976) showed that some Iowa corn on plants before harvest in 1975 contained an average of 430 ppb aflatoxins. Shotwell and her associates (1976) even detected aflatoxin M_1 in stored corn as well as freshly harvested corn. Recently, LaPrade and Manwiller (1977) reported that 39% of naturally infested corn ears had detectable levels of aflatoxin B_1.

Dry figs imported into the United States are sometimes rejected because of contamination with aflatoxin (Anonymous, 1974a). Its presence in foreign produced figs suggests that aflatoxin could, under certain conditions, be present in domestic figs and pose a threat to domestic production. The Food and Drug Administration instituted a sampling program for 13 domestic food products including figs (Anonymous, 1974b). A recent study by Buchanan *et al.* (1975) showed that immature fig fruits did not support colonization and aflatoxin production by *Aspergillus flavus*, but become susceptible when ripe. While sun-drying on the tree, fruits were particularly vulnerable to fungal infection and colonization. Fig fruits accumulated fairly high levels of aflatoxins. Dickens and Welty (1975) indicated that bright greenish-yellow fluorescence under long wave ultraviolet light was observed on the shells of 7% of the nuts in samples from 46 aflatoxin contaminated commercial lots of Iranian pistachio nuts. Kernels from the fluorescent nuts contained 50% of the aflatoxin in the samples. Colonization by *Aspergillus flavus* and aflatoxin contamination in harvested almonds in California has been recently reported (Phillips *et al.*, 1976).

B. Sterigmatocystins

The sterigmatocystins are a group of closely related fungal metabolites characterized by a xanthone nucleus fused to a dihydrodifurano or tetrahydrodifurano moiety. The most economically important member of the group is sterigmatocystin from *Aspergillus versicolor* (Bullock *et al.* 1962). Other members include asperotoxin (3-hydroxy-6,7-dimethoxydifuroxanthone) (Rodricks *et al.*, 1968a and 1968b; Waiss *et al.*, 1968), o-methylsterigmatocystin (Burkhardt and Forgacs, 1968) and dihydro-o-methylsterigmatocystin (Cole and Kirksey, 1970) from *Aspergillus flavus*; 5-methoxysterigmatocystin (Holker and Kagal, 1968), 6-demethylsterigmatocystin (Elsworthy *et al.*, 1970), dihydrosterigmatocystin (Hatsuda *et al.*, 1972) and dihydrodemethylsterigmatocystin (Hatsuda *et al.*, 1972) from *Aspergillus versicolor*. The major differences among the various sterigmatocystins are the presence or absence of unsaturation in the difurano ring system (similar to aflatoxin B_1 and B_2) and in the substitution pattern on positions 6, 7, and 10 of the xanthone ring system and/or position 3 of the difurano system.

The toxicity of sterigmatocystin analogs on primary cell culture was greater for the compounds containing the $\Delta^{1,2}$-furobenzofuran-ring system than those containing a saturated furobenzofuran-ring system (Englebrecht and Altenkirk, 1972). The carcinogenicity of sterigmatocystin has been well documented (Dickens *et al.*, 1966; Purchase and Van Der Watt, 1968 and 1970). A carbonyl group unsaturated in the a,b position and an unsaturated bond in the $\Delta^{1,2}$-position are required for carcinogenicity according to Englebrecht and Altenkirk (1972). Also, a methoxy group at position 6 enchanced toxicity of these compounds, and a methoxy group at position 7 decreased toxicity.

Since sterigmatocystins, versicolorins, and aflatoxins all contain the

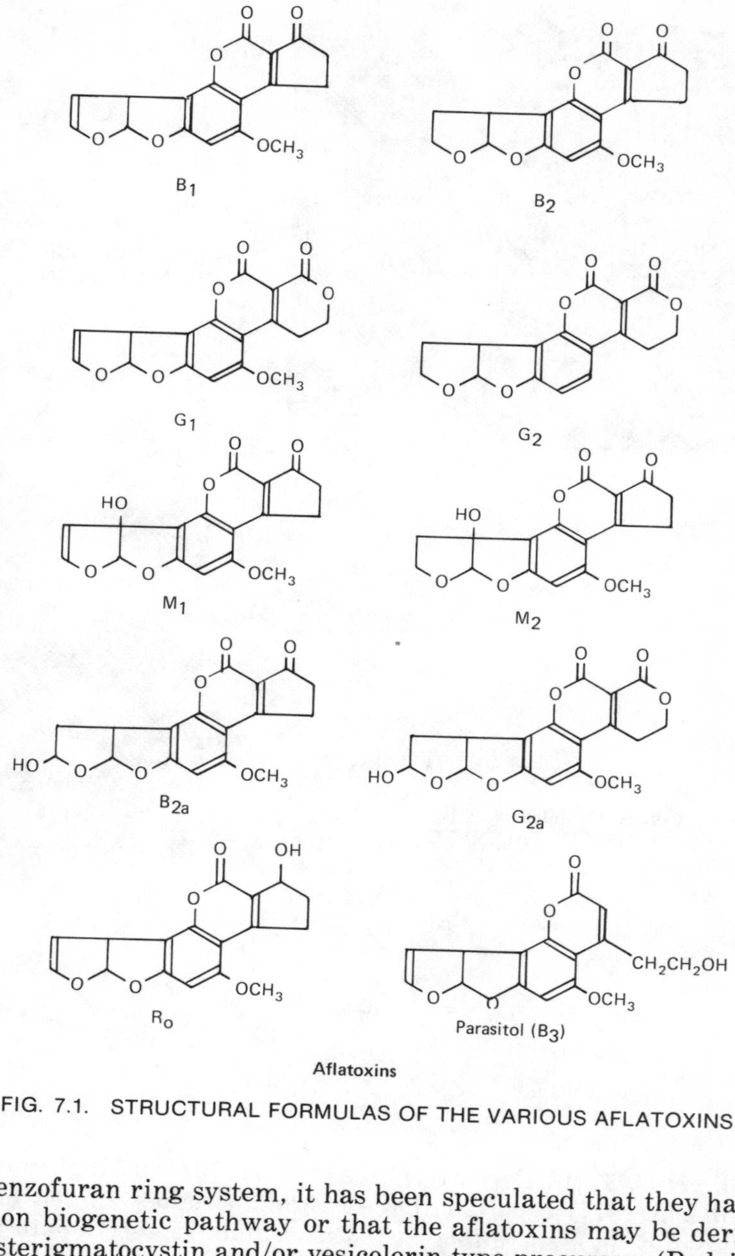

Aflatoxins

FIG. 7.1.   STRUCTURAL FORMULAS OF THE VARIOUS AFLATOXINS

furobenzofuran ring system, it has been speculated that they have a common biogenetic pathway or that the aflatoxins may be derived from sterigmatocystin and/or vesicolorin type precursors (Rodricks, 1969; Holker and Mulheirn, 1968; Holker and Underwood, 1964;

FIG. 7.2.   EXTENSIVE MOLD GROWTH ON CORN

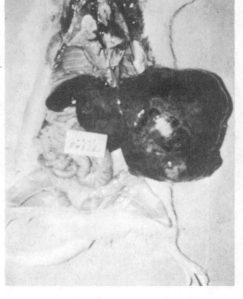

FIG. 7.3.   LIVER TUMOR IN A RAT FED   AFLATOXIN-CONTAMINATED PEANUT   MEAL

Biollaz *et al.* 1970). Recently, Hsieh *et al.*, (1973) and Singh and Hsieh (1976 and 1977) reported that sterigmatocystin was converted to aflatoxin B_1 by *Aspergillus parasiticus*, and proposed a pathway for the biosynthesis of aflatoxin B_1 to be acetate $\rightarrow\rightarrow\rightarrow$ averufin $\rightarrow$ versiconal acetate $\rightarrow$ versicolorin A $\rightarrow$ sterigmatocystin $\rightarrow$ aflatoxin B_1.

FIG. 7.4.   CONVULSIVE SEIZURE IN MOUSE DUE TO TREMORGEN OF *AS-PERGILLUS FLAVUS*

C. Ochratoxins

In a screening program to determine possible causes of certain animal diseases in South Africa, Scott (1965) showed that three of five strains of *Aspergillus ochraceus* isolated from local domestic legumes and cereal products were toxigenic. The toxic metabolites

FIG. 7.5.   LIVER OF DOG WITH SUBACUTE HEPATITIS X SHOWING EXTENSIVE FATTY CHANGE, HEPATIC CELL NECROSIS, AND GROUPS OF REGENERATING HEPATOCYTES

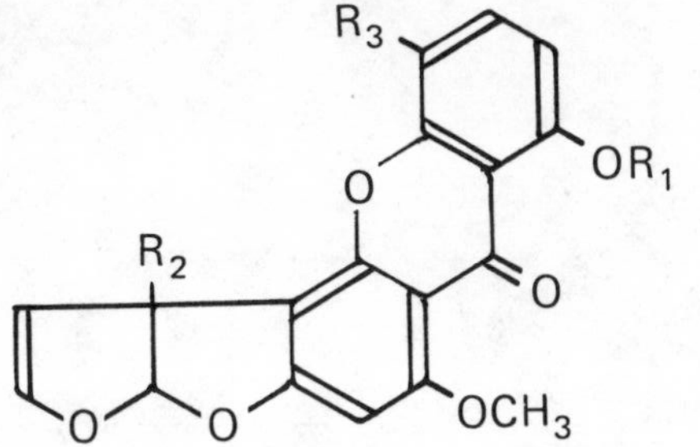

	R_1	R_2	R_3
Sterigmatocystin	H	H	H
O-Methyl sterigmatocystin	CH_3	H	H
Aspertoxin	CH_3	OH	H
5-Methoxy sterigmatocystin	H	H	OCH_3

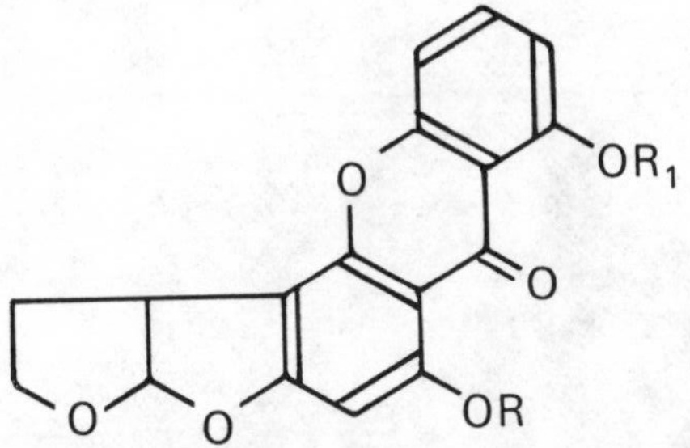

	R	R_1
Dihydro-o-methylsterigmatocystin	CH_3	CH_3
Dihydrosterigmatocystin	CH_3	H
Dihydrodemethylsterigmatocystin	H	H

Sterigmatocystins

FIG. 7.6.   STRUCTURAL FORMULAS OF THE STERIGMATOCYSTINS

were identified as ochratoxins. Ochratoxin A, the 7-carboxy-5-chloro-8-hydroxy-3,4-dihydro-3-R-methyl isocoumarinamide of L-β-phenylalanine is the most toxic of the ochratoxins and is produced in highest yield (Nesheim, 1969). Several investigators studied the toxicological and pathological effects produced by the ochratoxins. The LD_{50} for ochratoxin A administered per os varies from 2.1–4.67 mg/kg for the chick, swine, and trout to 22 mg/kg for the female rat (Van Walbeek, 1973; Doster and Sinnhuber, 1972; Meisner and Chan, 1974; Szezech et al., 1973).

   Penicillium viridicatum has also been identified as a producer of ochratoxin (Van Walbeek, et al., 1969; Krogh et al., 1973) and has been associated with the natural occurrence of ochratoxin A in most situations where an association could be made. Both species are widely distributed and frequently encountered on grain, legumes and other commodities usually protected by a reduction of water activity. According to Harwig and Chen (1974), a water activity that favors the growth of *P. viridicatum* on wheat and barley is about 0.90, a level far higher than that recommended for safe storage of grains and a higher rung in the water activity ladder than that favorable for the growth of *A. flavus* (0.84–0.86). Shotwell and her associates (1969c, 1970) made a survey of natural contamination of corn with some mycotoxins. Ochratoxin was found at 130 mg/kg in one Sample Grade sample of various grades of corn received from commercial markets in 1967. Three more samples with ochratoxin (83, 119, and 166 mg/kg were encountered in a later survey of 293 samples of corn intended for export (Shotwell et al., 1971). In a Danish survey of barley and oats (Krogh et al., 1973), ochratoxin was found in 58% of 33 samples of feed grains, mostly barley, taken in districts experiencing a high incidence of swine nephropathy; the average level was 3 mg/kg. Hald and Krogh (1972) reported that ochratoxin was found in 6% of 50 samples of barley selected as high quality grain at 9, 44, and 189 mg/kg. Subsequent to these findings, a farm was located at which the grain being fed to pigs was contaminated with ochratoxin. Residues of ochratoxin were found at slaughter in the kidneys of 18 of 19 pigs examined. In a Canadian survey of moldy feedstuffs, ochratoxin was found in 18 of 29 samples of heated grain (wheat, oats, and rye) at 0.03–27 mg/kg and in three of four samples of dried white beans at 0.02, 0.03, and 1.9 mg/kg (Scott et al., 1972). *Penicillium viridicatum* was consistently found to be the mold related to the presence of ochratoxin, as was also the case in the Danish studies. Ochratoxin has also been found in green coffee beans, initially in four of five samples of heavily molded beans at 20–400 mg/kg (Levi et al., 1974). The study continued with an examination of samples from 267 bags of beans originating in six countries. Ochratoxin was found in 19 samples at an average of 47 mg/kg. In another study, 68 samples were analyzed. Ochratoxin was found in two samples at about 20 mg/kg and in one sample at 80 mg/kg. *A. ochraceus* and unidentified species of *Penicillia* were found in the samples.

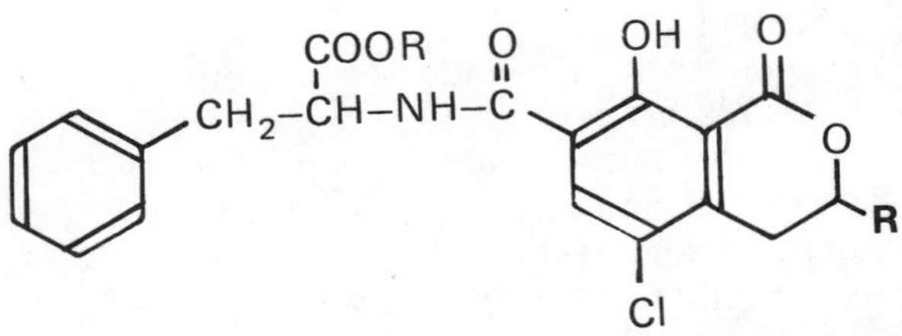

	R
Ochratoxin A	H
Ochratoxin A methyl ester	CH_3
Ochratoxin A ethyl ester	CH_2CH_3

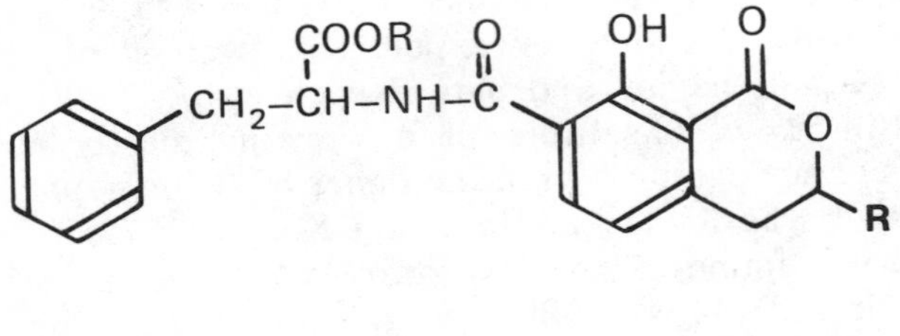

	R
Ochratoxin B	H
Ochratoxin B methyl ester	CH_3
Ochratoxin B ethyl ester	CH_2CH_3

Ochratoxins

FIG. 7.7. STRUCTURAL FORMULAS OF THE OCHRATOXINS AND DERIVATIVES

D. Patulin

Patulin, 4-hydroxy-4H-furo (3,2C) pyran-2 (6H) one is a lactone metabolite of several species of *Penicillium, Aspergillus, Gymnoascua,* and *Byssochlamys*. It is causing increasing concern as a potential toxic fungal contaminant of food and feed products. This compound has carcinogenic, mutagenic as well as teratogenic properties (Dickens and Jones, 1961; Mayer and Legator, 1969; Ciegler *et al.*, 1976). The chemical nature of patulin has been thoroughly investigated. Patulin is toxic to many biological systems. Intravenus injection of patulin into mice and rats give LD_{50} values varying from 0.3–0.7 mg patulin/20 g body weight, and 1 mg per mouse was always lethal (Broom *et al.*, 1944; Stansfeld *et al.*, 1944). The LD_{50} by oral administration to mice is 0.7 mg patulin (Broom *et al.*, 1944). Moule and Hatey (1977) reported that the toxic effect of patulin was due to the inhibition of the transcription process.

Scott (1965) isolated patulin producing molds *A. clavatus, A. terreus, P. expansum, P. urticae* from cereal grains and legumes. *P. expansum* has also been isolated from inshell pecans (Schindler *et al.*, 1974); from apricots, crab apples, and persimmons (Sommer *et al.*, 1974); from pears and grapes (Sommer *et al.*, 1974); and from apples (Harwig *et al.*, 1973a; Sommer *et al.*, 1974; Wilson and Nuovo, 1973). Patulin producing penicillia come mainly from fruits and vegetables (Atkinson, 1943). Contamination of apple and grape juices by *Byssochlamys* species may also be a hazard since *Byssochlamys nivea* has been identified as the *Gymnoascus* species able to produce patulin (Yates, 1974; Kuehn, 1958). Patulin-producing strains of fungi isolated from foods generally make up a low percentage of the total isolates. Only 1% of the penicillia isolated from flour and bread by Bullerman and Hartung (1973) were able to produce patulin *in vitro*. Six of 422 of the total fungi isolated from European style dry sausages were observed by Mintzlaff *et al.* (1972) to be capable of producing 0.02 to 0.48 mg patulin/ml of laboratory media. Of the 116 fungi isolated from corn meal only 0.9% were reported to produce patulin (Bullerman *et al.*, 1975). However 42 out of 61 naturally rotted apples yielded *Penicillium expansum* isolates which were observed to produce patulin (Harwig *et al.*, 1973a). Thus apple products appear to be the foods most likely to be contaminated with patulin, whereas a lesser chance of contamination exists for grain products.

Several fungi isolated from foods have been reported to produce patulin at refrigerated temperature in laboratory media (Lovett and Thompson, 1973). Three strains of *A. clavatus* and two each of

P. expansum and *P. patulum* produced in excess of 400 μg patulin/ml media after 100 days of incubation at 1.7 C. Patulin production in excess of 70μg/ml was observed from *P. claviforme* at 7.2 C and *P. griseofulvin* at 12.8C after 55 days incubation. *P. urticae* has been reported capable of producing up to 630 μg patulin/ml liquid medium in 3 weeks at 10 C and up to 250 μg/ml at 5 C. Levels of 60 μg and 43 μg of the toxin were observed per ml of medium within 4 days at these temperatures, respectively (Norstadt & McCalla, 1971). Using *P. expansum* in liquid medium, maximum patulin production of 0.6 mg/ml media was observed after 5–7 weeks incubation at 10 C with 0.15 mg/ml produced within 2 weeks. At 0°C approximately 0.02 mg patulin/ml of media was noted after 2 weeks, 0.6 mg/ml after 12 weeks, and a maximum amount of 0.7 mg/ml media after 18 weeks incubation (Sommer *et al.*, 1974).

Brian *et al.* (1956) discovered patulin in excess of 1000 ppm and Harwig *et al.* (1973a) found 0.02 to 17.7 mg per apple from the sap of apples rotted by *P. expansum*. Apple tissue was found to contain up to 125 μg patulin/g of tissue infected by *P. expansum*, but this was substantially lowered when incubated in atmospheres modified to 2% O_2 or 7.5% CO_2 (Sommer *et al.*, 1974). Existence of patulin in apple juice has also been reported by Scott *et al.* (1972). They detected patulin at a concentration of 1 ppm in a sample of sweet apple cider. Patulin has also been found in moldy fruits such as apples, plums, peaches, pears, apricots, and sweet cherries (Walker, 1969; Harwig *et al.*, 1973a; Buchanan *et al.*, 1974).

Wilson and Nuovo (1973) reported finding up to 45 ppm patulin in "Organic Apple Cider". An extensive survey in terms of geography and type of pack for the presence of patulin in apple juice was carried out by the U.S. Food and Drug Administration in 1973 (Bureau of Foods, FDA, 1974). Patulin was detected in 50 of 136 consumer packs of apple juice taken in the open market from all sections of the country. Levels from 40 to 150 μg/l of juice were found in 46 of the samples. Three samples had 270 μg/l and 1 sample 440 μg/l.

FIG. 7.8.   STRUCTURAL FORMULA OF PATULIN

Patulin

In the Washington, D.C. area, levels of 49 to 309 ppb patulin were found in eight of 13 commercial apple cider samples analyzed in another study (Ware *et al.*, 1974). The presence of patulin in apple juice is evidence that rotted apples are being used. Scott and Sommers (1968) showed that patulin was relatively stable in grape and apple juice, but not in orange juice, possibly due to the presence of thiols.

E. Citrinin

Citrinin, an antibiotic and nephrotoxin is produced by many species of penicillia and aspergilli (Betina *et al.*, 1964; Pollock, 1947). It was first recognized as a powerful antibiotic, but later was found to damage the kidneys of test animals, retard growth and eventually cause death (Ambrose and deEds, 1964; Carlton and Tuite, 1969 and 1970). Citrinin-producing isolates of *P. citrinum* have been obtained from yellow-colored rice of the type associated with toxic symptoms in Japan (Saito *et al.*, 1971). However, the occurrence of citrinin as a contaminant of feedstuffs has been associated with *P. viridicatum* and always as a co-contaminant with ochratoxin. In the Canadian survey of moldy feedstuffs, citrinin was found at 0.07–80 mg/kg levels in 13 of the 18 samples in which ochratoxin had been found (Scott *et al.*, 1972). In the Danish survey of grains associated with swine nephropathy, citrinin was determined at 0.16, 1.0, and 2.0 mg/kg in three of the 22 samples of barley in which ochratoxin had been detected (Krogh *et al.*, 1973). Wu *et al.* (1974a) reported that seven strains of *Penicillium viridicatum* isolated from country cured ham produced citrinin on artificial media and country-cured ham. Citrinin caused kidney damage in experimental animals (Krogh *et al.*, 1970), comparable to that caused by feeding barley contaminated with *P. viridicatum*. Thus, it is possible that both ochratoxin A and citrinin are involved in the mycotoxicosis which has affected up to 7% of the pigs in Denmark (Krogh, 1969). Hesseltine (1976) reported that citrinin occurs naturally in wheat, rye, barley, and oats in both Canada and Denmark.

F. Zearalenone

Zearalenone, a compound exhibiting uterotrophic and anabolic activity was first isolated from maize infected with the fungus *Gibberella zeae*, the peritheral stage of *Fusarium graminearum* (Stob *et al.*, 1962). Lasztity and Woller (1975) reported that zearalenone and some of its derivatives have estrogenic effect on pigs, rabbits, and deer. It is also related to the invasion of grain by other

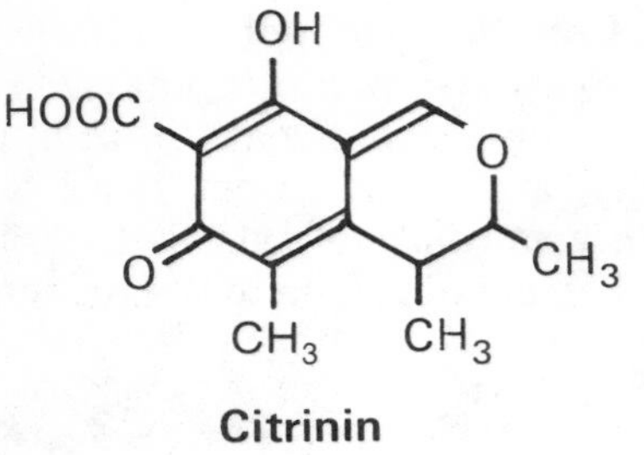

FIG. 7.9.   STRUCTURAL FORMULA OF CITRININ

Citrinin

species such as *F. tricinctum* and *F. moniliforme* (Mirocha *et al.*, 1967, 1969; Caldwell *et al.*, 1970) but particularly by *F. roseum* var *graminearum*. These organisms invade developing corn at the silking stage in period of heavy rainfall and proliferate on mature grains that have not dried because of wet weather at harvest or on grains that are stored wet (Tuite *et al.*, 1974; Caldwell and Tuite, 1970, 1974). Field observation of a pink discoloration of kernels signals the presence of *F. roseum* in corn; scab is associated with *F. roseum* in small grains. Since low temperature is needed to initiate and maintain the production of zearalenone by fungus, the presence of fungus is insufficient evidence for the presence of the metabolite. In the USDA surveys of corn from 1967 crops (Shotwell *et al.*, 1970) and of corn for export during 1968–1969 (Shotwell *et al.*, 1971), zearalenone was found in six of the 576 samples at an average level of 624 mg/kg (range 450–800 mg/kg). In some years conditions are conducive for *Fusarium* ear rot in epidemic proportions (Tuite *et al.*, 1974). Eppley *et al.* (1974) reported that corn samples collected in the spring of 1973 at terminal elevators servicing areas were found to suffer *F. roseum* damage. Zearalenone was found in 17% of 223 samples assayed, at an average level of 0.9 mg/kg with no relation to grade or intended use, including food use. Within the area surveyed there was a geographical concentration of the contamination inside a 150-mile radius of the southern tip of Lake Michigan. Zearalenone contamination of feed grains is not confined to the United States; instances of feed contamination have been reported from Finland (Roine *et al.*, 1971), Denmark (Erickson, 1968) and France (Jemmali, 1973), England, Mexico, and Yugoslavia (Stoloff, 1976; Hesseltine, 1976), and Zambia (Lovelace and Nyathi, 1977).

G. Penicillic acid

Penicillic acid (γ-Keto-β-methyl-δ-methylene-A-hexonic acid), a fungal metabolite, was first isolated from a mold culture in 1913 as

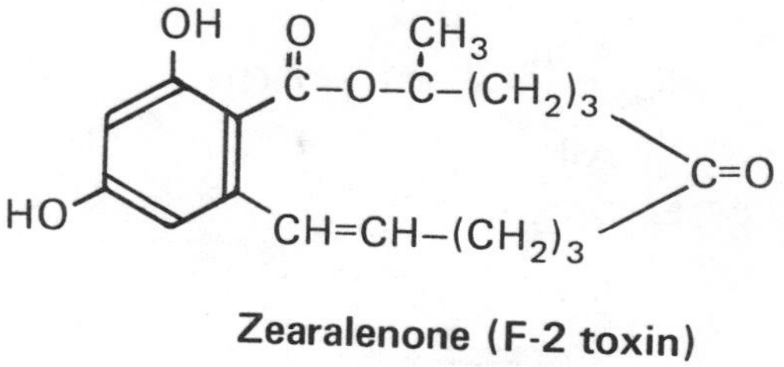

FIG. 7.10.   STRUCTURAL FOR-
MULA OF ZEARALENONE

Zearalenone (F-2 toxin)

part of a study of corn deterioration (Alsberg and Black, 1913). Penicillic acid is toxic to animals with a LD_{50} for mice of 100 mg/kg (Spector, 1957). In a survey of naturally contaminated corn, penicillic acid was found in seven of 20 random samples assayed at 5–231 mg/kg (Thorpe and Johnson, 1974). Fungi associated with blue-green discoloration of corn known as blue eye were demonstrated to be producers of penicillic acid (Ciegler and Kurtzman, 1970). Study of 48 samples of corn selected by USDA's Agricultural Marketing Service graders on those having blue-green discoloration showed that all samples contained 5 to 184 mg penicillic acid/kg. Dry bean samples, selected from a nationwide survey because of a portion of seeds contained viable *Penicillium cyclopium*, and penicillic acid was found in five of 20 samples tested in a range of 11–179 mg/kg. Ciegler *et al.*, (1972) studied the fungus species isolated from fermented sausage. Penicillic acid was produced in synthetic media by 44 of 221 isolates and by at least one isolate in seven of 18 identified

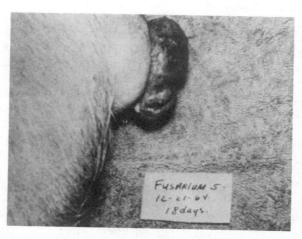

FIG. 7.11.   VAGINAL PROLAPSE IN SOW FED *FUSAR-IUM* INFECTED CORN

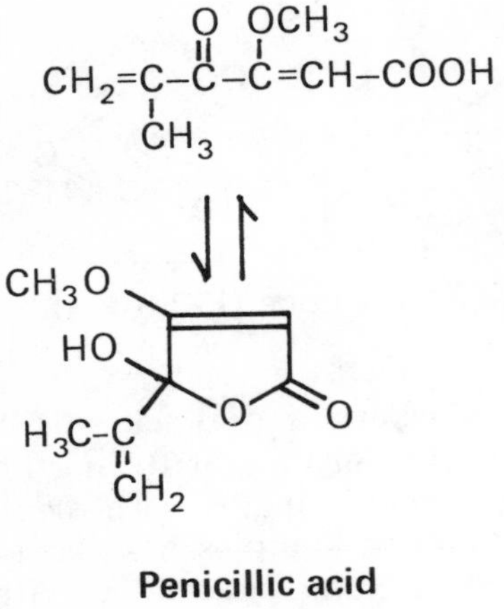

$$CH_2=\overset{\overset{\displaystyle O}{\|}}{\underset{\underset{\displaystyle CH_3}{|}}{C}}-\overset{\overset{\displaystyle OCH_3}{|}}{C}-C=CH-COOH$$

FIG. 7.12.   STRUCTURAL FORMULA OF PEN-
ICILLIC ACID

Penicillic acid

species, including some not previously recorded as producers of penicillic acid.

H. Rubratoxins

Burnside *et al.* (1957) reported a disease of pigs and cattle which followed the consumption of moldy corn. Of 13 cultures of fungi isolated from such toxic corn, only two were shown to cause illness and death when fed to experimental animals. One of the cultures was found to be *Penicillium rubrum* which produced a mycotoxin. Forgacs *et al.* (1958) fed grains with *P. rubrum* and *P. purpurogenum* to chicks and observed a hemorrhagic syndrome. In 1962, Wilson and Wilson reported preliminary results of chemical and pathological studies on the toxic metabolite isolated from *P. rubrum*. Townsend *et al.* (1966) carried out further chemical studies of the culture broth of *P. rubrum* and post mortem examinations of inspected mice, guinea pigs, and dogs, and subsequently isolated two toxic metabolites designated as rubratoxin A and B. Rubratoxin A and Rubratoxin B have the same structure but with A, one of the anhydride groups is reduced to the lactol. Rubratoxin B is the major metabolite, and it is produced by the head-to-tail, head-to-tail coupling of two C_{13} units.

The LD_{50} for rubratoxin A and B are 6.6 and 3.0 mg/kg, respectively. Oral toxicity is perhaps the most important parameter when considering the immediate implications of a toxin in food material. Carlton *et al.* (1958) investigated toxic effects in mice fed with certain species of fungus belonging to *Penicillium*. They discovered marked liver lesions in mice fed with the *P. purpurogenum* diet. Rubratoxin B was shown to be mutagenic as well as teratogenic (Evans *et al.*, 1975; Hood *et al.*, 1973).

I. Griseofulvin

Griseofulvin was isolated from *Penicillium janczewski* (*P. nigricans*) by Brian *et al.* (1946) and McGowan (1946) who termed it "curling factor" because of a peculiar inhibitory action on hyphal development in other fungi. The identity of the compound was not recognized at that time.

Griseofulvin has also been obtained from *P. patulum* (Paget and Walpole, 1958), *P. albidum*, *P. raciborskii*, *P. melinii*, *P. urticae*, *P. raistrickii*, *P. brefeldianum*, *P. viridi-cyclopium*, and *P. brunneostoloniferum* (Rhodes, 1963).

This unusual antimicrobial substance has unique value as a systemic therapeutic agent for cutaneous fungal infections. In spite of the clinical effectiveness of griseofulvin and the ability of man and animals to tolerate relatively large doses given daily for several weeks, the serious nature of toxic responses observed in recent years has served to restrict its therapeutic use somewhat. Numerous toxic manifestations for griseofulvin have been reported, included are the following symptoms. I. In the skin-angio-neurotic edema, eryth-

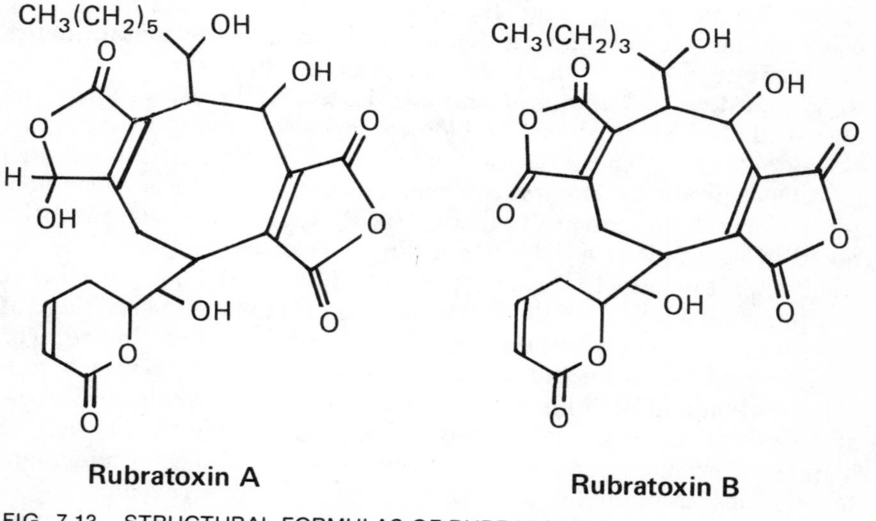

Rubratoxin A

Rubratoxin B

FIG. 7.13. STRUCTURAL FORMULAS OF RUBRATOXINS

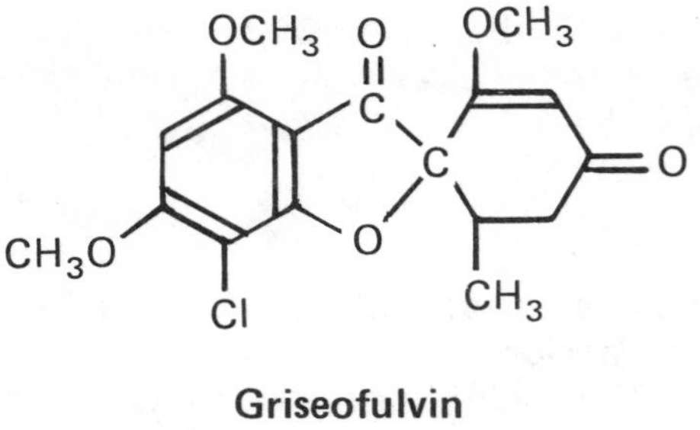

Griseofulvin

FIG. 7.14.   STRUCTURAL FORMULA OF GRISEOFULVIN

ema, urticaria, vesicular and macular eruptions, photosensitivity, and lichen-planus eruptions; II. Hematopoietic disturbances—transient leukopenia, granulocytopenia, punctate basophilia, and monocytosis; III. Neurological manifestations—blurred vision, headaches, disorientation, and vertigo; IV. Gastrointestinal signs and symptoms—anorexia, nausea, vomiting, and diarrhea; V. Oral changes— dry mouth, black hairy tongue, glossodynia, and angular stomatitis; VI. Miscellaneous effects—thirst, fatigue, malaise, fainting, and potentiation of alcohol effects (Anderson, 1965).

J. Luteoskyrin and Islanditoxin

At high humidity, stored rice is quite susceptible to contamination by many fungi, especially *Penicillium* and *Aspergillus* species. Heavy fungal growth and accumulation of fungal pigments may color the rice yellow and render it bitter. Most of the contaminated rice contained toxic materials and may produce toxic symptoms upon ingestion by humans or animals. Extensive investigation of the chemical, biochemical, and biological aspects of these toxins has been done by Japanese researchers. Strains of *Penicillium islandicum* have produced hepatotoxic metabolites that induced liver neoplasm in rat and mice (Tsunoda, 1951). Liver cirrhosis has been noted in rats fed with diets containing cereals contaminated by the fungus. Saito *et al.* (1971) and Miyake and Saito (1965) reported that continuous feeding of diets containing a small to large percentage of the fungi produced a wide range of pathologic changes in the liver of rats and mice, including acute atrophy at high levels of administration and nodular hyperplasia, bile duct hyperplasic fibrosis and cirrhosis at lower levels. Rats and mice surviving the lower level of intake for more than 1 year showed low incidences of adenomatous lesions and hepatomas. Fractionation of toxins revealed a lipophilic toxin and a hydrophilic one. The yellow lipophilic toxin, luteoskyrin is a substituted bis-polyhydroxydihydroanthraquinone and the hydrophilic toxin is islanditoxin, a toxic cyclic peptide. Luteoskyrin is slow acting and, for a 10 g mouse, the LD_{50} is 1.5 mg, subcutaneously, and 2.2 mg/10 g, orally. Liver damage caused by luteoskyrin is characterized by centrilobular necrosis of the liver and diffuse fatty metamorphosis of the liver cells. Luteoskyrin is thought to bind to DNA, to cause pigment changes and to alter activity of DNA-dependent RNA polymerase. Islanditoxin is extremely hepatotoxic, causing severe liver damage, hemorrhaging and death. LD_{50} values for 10 g mice range from 4.75 μg, subcutaneously, to 65.5 μg, orally. It is suggested that this toxin interferes with carbohydrate metabolism by causing the disappearance of glycogen granules in the injured liver.

A chlorine-containing peptide (cyclochlorotine) was also one of the yellowed rice toxins isolated from the culture filtrate, cultured fungal mat of *Penicillium islandicum* Sopp, and cultured moldy rice grain (Tatsuno *et al.* 1955). The peptide is a quick-acting hepatotoxin, causing the disappearance of glycogen granules in injured liver (Saito, 1959). The peptide accelerates glycogen catabolism, inhibits glycogen neogenesis *in vivo* (Yamazoe *et al.*, 1963), and interferes directly with carbohydrate metabolism of liver cells. The chlorine-containing peptide causes hepatic damage of the so-called acinus peripheral cytotoxic pattern characterized by vacuolation of liver and endothelial cells in the perilobular region, followed by the

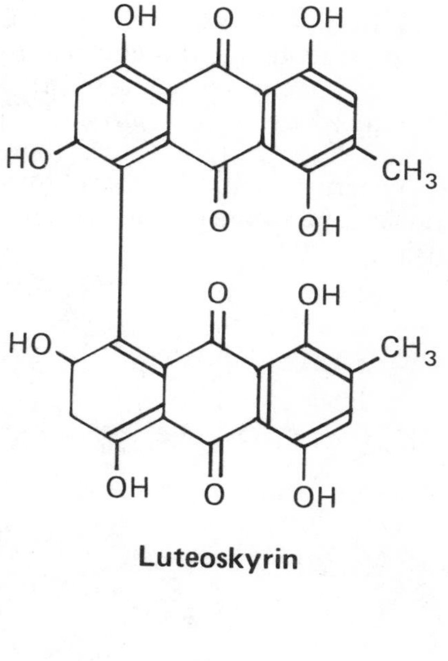

FIG. 7.15. STRUCTURAL FORMULA OF LUTEOSKYRIN

Luteoskyrin

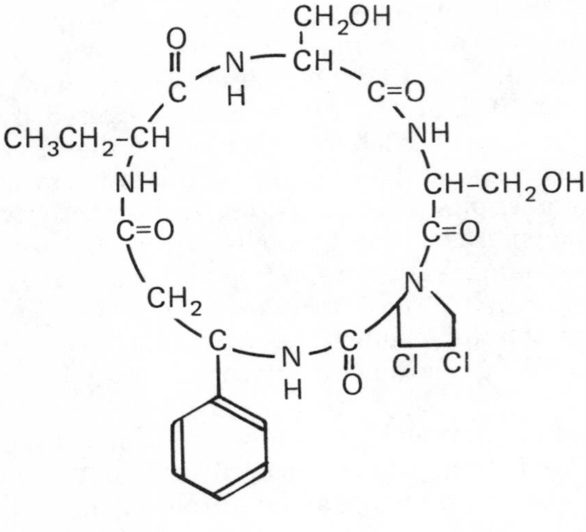

FIG. 7.16. STRUCTURAL FORMULA OF ISLANDI-TOXIN

Islanditoxin

appearance of hyaline droplets (Saito *et al.*, 1971). Results of long-term feeding experiments with this peptide revealed that it acts as a cirrhogenic agent. The carcinogenic action of this chlorine-containing peptide needs further study.

K. Fumigatin

Fumigatin was first noted as a substance responsible for a pH color changes in the culture broth of a strain of *Aspergillus flavus* isolated from Indian soil (Anslow and Raistrick, 1938). It is also produced by *A. fumigatus* (Anslow and Raistrick, 1938; Pettersson, 1963). Fumigatin has been shown to have marked *in vitro* inhibitory properties against several gram-positive and gram-negative bacteria. The range of effectiveness is approximately the same for both types.

FIG. 7.17. STRUCTURAL FORMULA OF FUMIGATIN

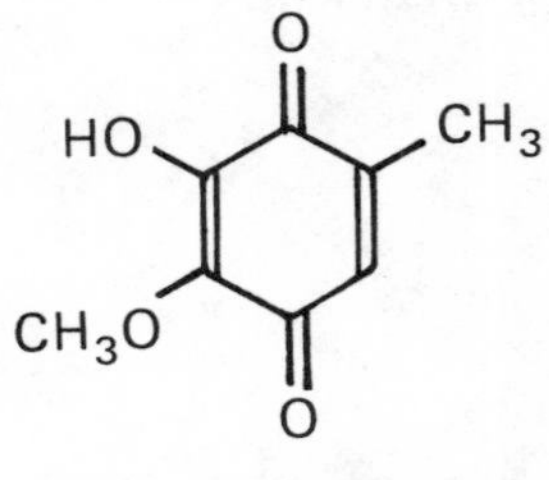

Fumigatin

L. Trichothecenes

Mycotoxicoses caused by the 12,13-epoxy-Δ^9-trichothecenes in humans and animals include alimentary toxic aleukia, stachybotryotoxicosis, moldy corn toxicosis, and the refusal-vomition phenomenon (Ciegler, 1975). Hsu *et al.* (1972) identified T-2 toxin as the cause of a lethal toxicosis in Wisconsin dairy cattle. These cattle had consumed feed containing 60% corn molded with *F. tricinctum*. The cows had extensive hemorrhaging on the serosal surface of all internal viscera, typical of previously reported cases of moldy corn poisoning (Smalley, 1973). Vomiting in animals and humans caused by the consumption of moldy wheat, barley, and flour had been reported in the early 1900's (Ciegler, 1975; Hesseltine, 1976). The extracts of corn naturally infected with *G. zeae* caused emesis in pigs and also possibly caused feed refusal (Curtin and Tuite, 1966). Both effects have been observed naturally in the midwestern U.S. However, the causative agent was not detected until a new trichothecene, vomitoxin, was isolated from corn infected in the field with *F. graminearum* by Vesonder *et al.* (1973). From barley invaded with *Fusaria*, Yoshizawa and Morooka (1973) isolated deoxynivalenol which is structurally the same as vomitoxin. Vomitoxin does not

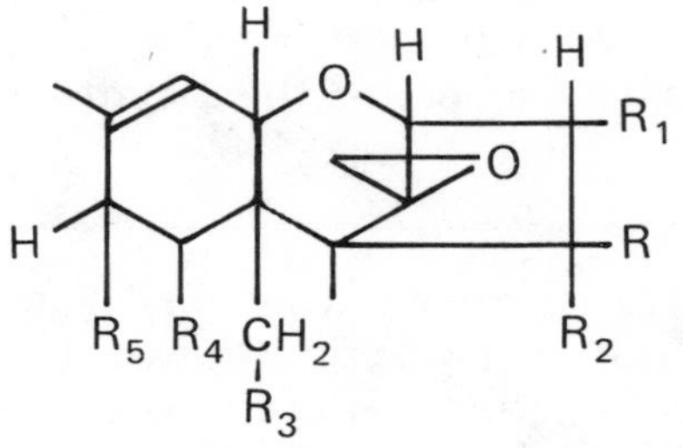

Trichothecenes

	R_1	R_2	R_3	R_4	R_5
Trichothecene	H	H	H	H	H
Trichodermin	H	OAc	H	H	H
Trichodermol	H	OH	H	H	H
T-2 toxin	OH	OAc	OAc	H	$OOCCH_2CHMe_2$
HT-2 toxin	OH	OH	OAc	H	$OOCCH_2CHMe_2$

FIG. 7.18. STRUCTURAL FORMULAS OF TRICHOTHECENES AND DERIVATIVES

appear to cause hemorrhaging and is less potent in causing dermal necrosis than T-2 toxin.

Feeding of *Fursarium tricinctum* molded corn or pure T-2 toxin at 8 and 16 ppm resulted in reduced feed intake, reduction in weight gain and decreased egg production of laying hens (Speers *et al.*, 1977). A recent report showed that trichothecenes inhibited initiation, elongation and termination of eukaryotic protein synthesis (Cundliffe and Davies, 1977).

M. Aspergillic acid

Aspergillic acid was discovered and named by White (1940) and White and Hill (1943). It is the first of a number of closely related pyrazine fungal metabolites reported. Aspergillic acid and its analogs are major metabolites of certain strains of *A. flavus* and other *Aspergillus* species. It is acutely toxic to mice (100–150 mg/kg, ip) but has no chronic effects at sublethal dosages (White and Hill, 1943). For the analogs of aspergillic acid, ranges of toxicity from near zero to toxicity equaling that of aspergillic acid have been reported (McDonald, 1973; Sasaki *et al.*, 1968; Yokotsuka *et al.*, 1968). Toxicity appeared to be related to the hydroxamic acid functionality, and little effect on toxicity was observed for differences in the 3 and 6 positioned side-chain substitutes. Aspergillic acid and most analogs can exist in either the hydroxamic acid form (2-hydroxy-pyrazine-1-oxide) or the 1-hydroxy-2-pyrazinone form.

N. Kojic acid

Kojic acid is a relatively common metabolite of *Aspergillus* species, particularly of *Aspergillus flavus*. Its chemical structure, (5-hydroxy-2-(hydroxymethyl)-4H-pyran-4-one), was elucidated mainly

FIG. 7.19.   STRUCTURAL FORMULA
OF ASPERGILLIC ACID

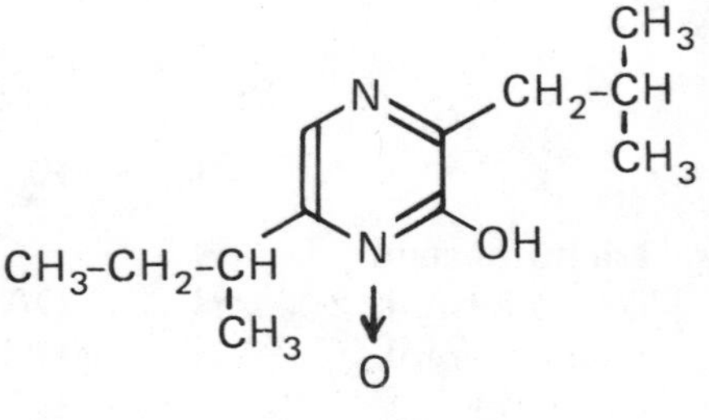

Aspergillic acid

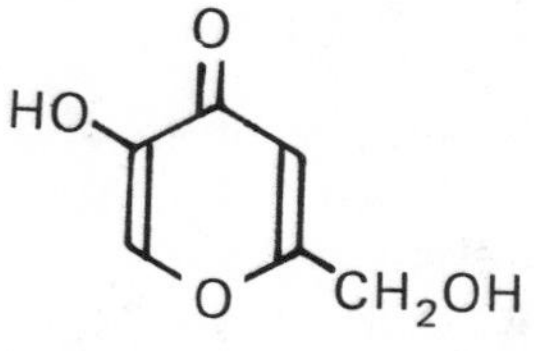

Kojic acid

FIG. 7.20. STRUCTURAL FORMULA OF KOJIC ACID

by Yubato (1924). It is characterized by a γ-pyrone nucleus substituted on positions 2 and 5 with a hydroxymethyl and a hydroxy group respectively. Classified as a convulsant, a relatively large quantity required to produce severe intoxication or death in animals. Although it has not been directly implicated in natural outbreaks of mycotoxicosis, it remains a potential problem in view of the large number of microorganisms capable of producing large amounts of it. The LD_{50} in 17 g mice was 30 mg ip injection (Morton *et al.*, 1945). Kojic acid also showed toxicity in plant cells at 10^{-1} M (Gaumann and Von Arx, 1947).

O. Terreic acid

Terreic acid was first discovered by Wilkins and Harris (1942). It is produced by *Aspergillus terreus*. It was highly toxic to mammals and showed a LD_{50} of 71–119 mg/kg to mice by intravenous injection (Kaplan *et al.* 1954). Terreic acid was found to be inhibitory to several bacteria and fungi in concentrations varying from 1.0 to 3.1 mg/ml. It was also inhibitory to *Trichomonas vaginalis*, a protozoan infecting the human vaginal tract. The chemical structure of terreic acid was proposed by Sheehan *et al.* (1958) as 2,3-epoxy-6-hydroxytoluquinone.

FIG. 7.21. STRUCTURAL FORMULA OF TERREIC ACID

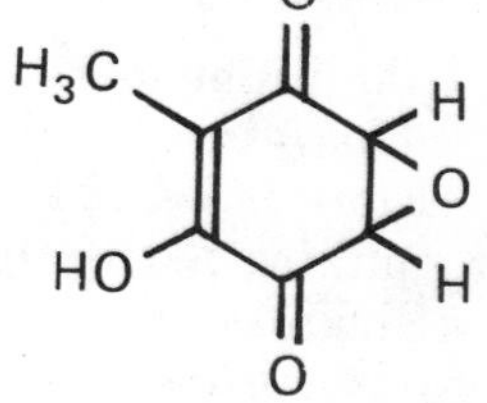

Terreic acid

P. Helvolic acid

Helvolic acid is a toxic antibiotic produced by some isolates of *A. fumigatus* and was reported by Waksman *et al.* (1943) and Chain *et al.* (1943) at almost the same time. The correct structure of helvolic acid was determined by Okuda *et al.* (1964). Twenty gram mice were found to tolerate up to 5 mg of sodium helvolate given intravenously, and 20 mg was tolerated by intragastric administration. Repeated intraperitoneal injections caused peritonitis and superficial liver lesions. With minimal effects seen in the kidneys. Helvolic acid was absorbed from the subcutaneous tissues and gastrointestinal tract and was excreted in an active form in the urine and bile. Leucocytes were not damaged by a 1:2500 dilution for 48 hours without undergoing irreversible changes. A concentration of 1:1500 suppressed outgrowth from a tissue culture explant for 24 hours, and the cells were noted to undergo vacuoler degeneration (Wilson, 1971).

FIG. 7.22.   STRUCTURAL FORMULA OF HELVOLIC ACID

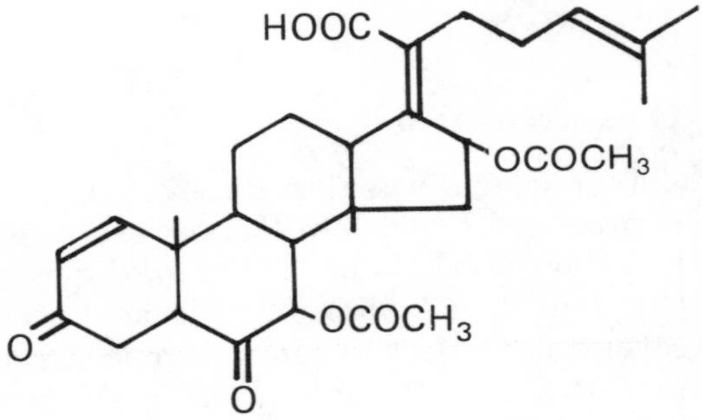

Helvolic acid

Q. Mycophenolic acid

Mycophenolic acid, a product of several *Penicillium* species, *P. stoloniferum* Thom (Alsberg and Black, 1913), *P. brevi-compactum* (Florey *et al.* 1946), *P. viridicatum* and *P. bialowiezense* (Burton, 1949), contains a five-membered lactone ring fused to a benzene moiety (Birkinshaw *et al.* 1952; Shibata *et al.*, 1964). The compound exhibits toxic properties toward bacteria, fungi, and viruses, and is an effective antitumor agent (Cline *et al.* 1969; Gilliver, 1946).

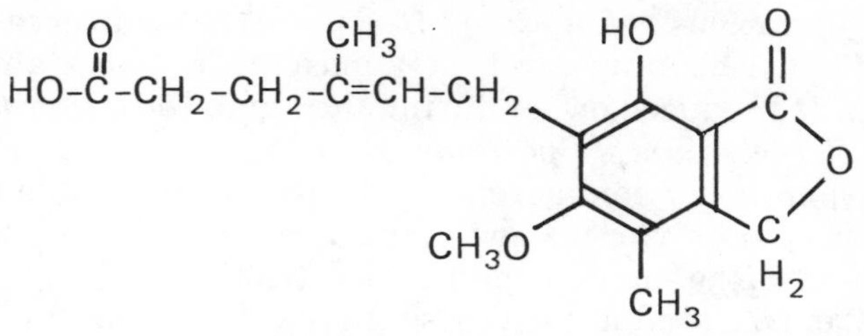

Mycophenolic acid

FIG. 7.23. STRUCTURAL FORMULA OF MYCOPHENOLIC ACID

R. Tremorgenic mycotoxins

Tremorgenic mycotoxins are one of two classes of substances that have been isolated and which appear to act at the level of the central nervous system. These toxins include several compounds which cause sustained trembling in infected animals. It is unusual for secondary fungal metabolites to cause a sustained tremoring response in animals. Only 10 tremorgenic compounds have been reported in the literature, but structures have been determined for five of these. They are verruculogens, TR-1 and TR-2, fumitremorgen B, tryptoquivaline, and tryptoquivalone. The tremorgens can be separated into three groups based on their nitrogen content: Group A consisting of penitrems A, B, and C contain only one nitrogen atom per molecule; Group B consisting of fumitremorgens A and B and verruculogens TR-1 and TR-2 contain three nitrogens per molecule (Cole and Kirksey, 1973); Group C consisting of tryptoquivaline and tryptoquivalone contain four nitrogens per molecule. *Group A.* Penitrem A was first extracted from two strains of *Penicillium cyclopium* that were the principal contaminants of feedstuffs causing disease outbreaks among sheep and horses (Wilson *et al.*, 1968). Subsequently, Ciegler (1969) isolated the same toxin from *P. palitans* NRRL 3468, a culture found in almost monotypic growth on a sample of moldy commercial feed suspected of being implicated in deaths of dairy cows. Tremorgen production among the Penicillia was confined to several species in the subsection Fasiculata, section Asymetrica. Other subsections in Asymetrica or sections other than Asymetrica tested, showed no production of tremorgens (Ciegler and Pitt, 1970). A good producer of tremorgen, *P. crustosum*, is a contaminant of various refrigerated foods, grains, and cereal products and causes a soft brown rot in apples. It is uncertain whether tremorgen is produced or not during the apple-rotting process. Four cultures: *P. cyclopium*, *P. palitans*, *P. crustosum*, and *P. puberulum*, isolated from moldy commercial feedstuffs were capable of producing Penitrem A on various agricultural commodities with low temperatures favoring toxin accumulation (Hou *et al.* 1971a). Most of the Penitrem A producers also synthesized two additional closely related tremorgens: Penitrem B and Penitrem C (Hou *et al.*, 1971b).

Penitrem A caused perceptible tremors in mice at 250 μg/kg, whereas Penitrem B required 1.3 mg/kg for a similar response. The mechanism of tremor produced by Penitrem A was speculated to be inhibiting the interneurons which inhibit the α-motor cells of the anterior horn of the spinal columns (Stern, 1971). The single dose LD_{50} of Penitrem A for mice was 1.05 mg/kg with a 95% confidence interval of 0.51–2.17 mg/kg, whereas the LD_{50} for Penitrem B was 5.84 mg/kg with a 95% confidence interval of 8.26–41.13 mg/kg (Hou *et al.*, 1971b). Penitrem C is much less toxic than Penitrem A or Penitrem B but information on its properties and toxicity are still meager (Ciegler *et al.*, 1976). Further toxicological effects of Penitrems have been described by Wilson *et al.*, (1968, 1972) and Cysewski (1973). The structures of the Penitrems have not yet been determined (Ciegler *et al.*, 1976).

Group B. Verruculogen TR-1 was produced by *Penicillium verruculosum* isolated from moldy peanuts as a result of improper storage conditions (Cole *et al.*, 1972). This tremorgenic mycotoxin causes severe tremors in mice and one-day old cockerels. It has an LD_{50} of 2.4 mg/kg in mice and 15.2 mg/kg in cockerels, i.p., and the LD_{50} of 126.7 mg/kg in mice and 365.5 mg/kg in cockerels when administered orally.

Since the conversion of verruculogen TR-1 to TR-2 was observed only under the specific conditions reported (Cole and Kirksey, 1973) and since TR-2 was not a photodegradation product, it appears that TR-2 is also of natural origin and may be a biosynthetic precursor to TR-1 (Ciegler *et al.*, 1976).

Fumitremorgens A and B are two closely related toxins produced by *Aspergillus fumigatus* (Yamazaki *et al.*, 1971 and 1974). The structure of fumitremorgen B was established and shown to contain proline and 6-methoxyindole groups (Yamazaki *et al.*, 1974). It appears identical to lanasulin, the major metabolite of *Penicillium lanosum* (Dix *et al.*, 1972). The structure of fumitremorgen A has not been determined.

I.p. injection of fumitremorgen A at 1 mg/kg mice caused more severe sustained trembling than with fumitremorgen B at a similar dose. A dose of 5 mg of either toxin (i.p.) caused 7% death within 96 hr. The fumitremorgens have not been reported in any natural occurring mycotoxicosis (Ciegler *et al.*, 1976).

Group C. Tryptoquivaline and tryptoquivalone were two tremorgenic toxins isolated from *Aspergillus clavatus*. This toxin producing fungus had been originally isolated from a sample of moldy rice collected in a Thai household where a young boy had died of an unidentified toxicosis. Two new nontoxic metabolites, kotanin and desmethylkotanin (Buchi *et al.*, 1971), and a small amount of highly toxic cytochalasin E (Buchi *et al.* 1973) were produced by the isolated strain.

Tryptoquivaline and tryptoquivalone apparently are not as toxic as those tremorgenic compounds previously reported. The two compounds when injected into weanling rats at 500 mg/kg caused tremoring which persisted for five days, and the animals died by the eight day with no observable histopathologic changes (Glinsukon *et al.*, 1974).

S. Fumagillin

Fumagillin is a metabolite of *A. fumigatus*. It is known for its antibacteriophage and amebicidal properties (McCowen *et al.*, 1951). However, because of toxic side effects encountered in human clinical trials (Schindel, 1954), medical applications of fumagillin are restricted to its use by apiarists (Katznelson and Jamieson, 1952; Bailey, 1953) and to the veterinary profession (Field *et al.*, 1956). An isolate of *A. fumigatus* from country cured ham has been shown to be highly toxic to chick embryos (Wu *et al.*, 1974b). Yokota *et al.* (1977) isolated a hemolytic toxin from mycelia and culture filtrates of *A. fumigatus*.

T. Psoralens

The psoralens are a group of furocoumarins characterized by a linearly annulated structure formed by the fusion of a furan ring in the 2 and 3 positions with the carbon atoms 6 and 7, respectively, of the coumarin structure.

Many psoralens posses potent phototoxic properties (Kuske, 1938, 1940). Pathak *et al.* (1962) showed that photosensitizing psoralens occur as natural constituents of many plants belonging principally to the botanical families *Umbelliferae, Rutaceae, Leguminosae,* and *Moraceae.* In addition, the growth of the fungus *Sclerotinia sclero-*

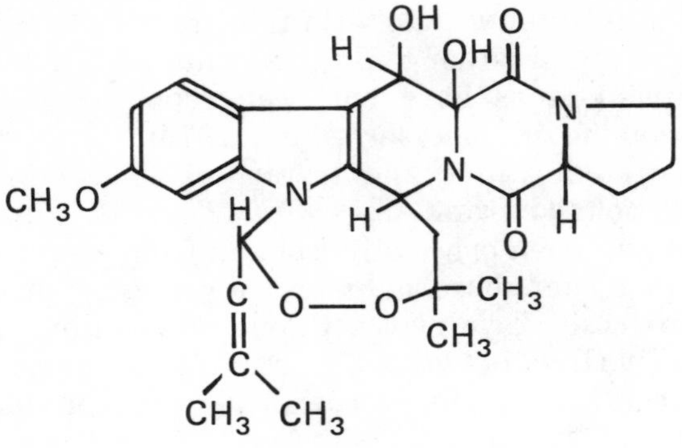

Verruculogen TR-1

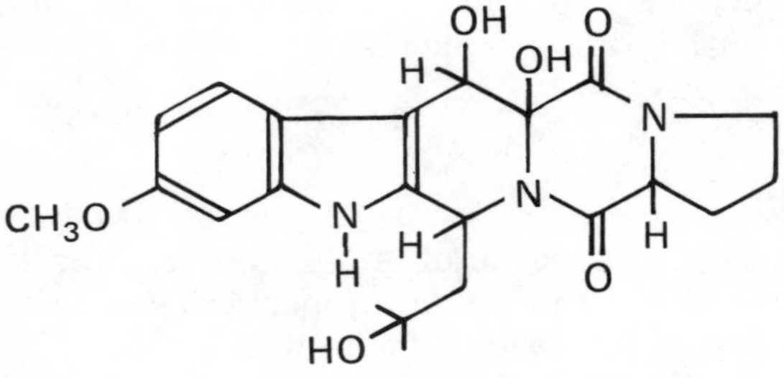

Verruculogen TR-2

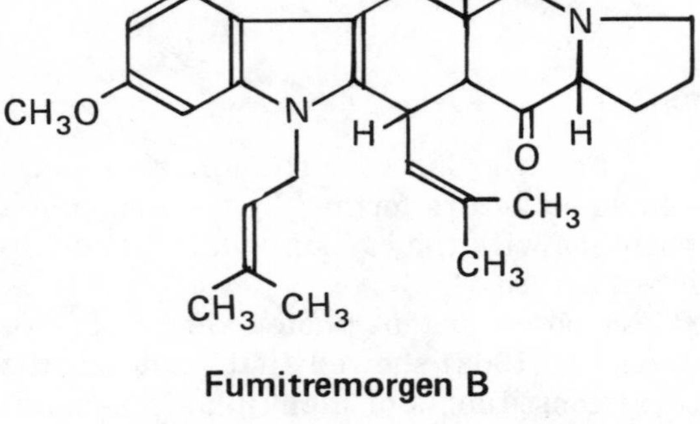

Fumitremorgen B

FIG. 7.24.   STRUCTURAL FORMULAS OF TREMOR-
GENIC TOXINS

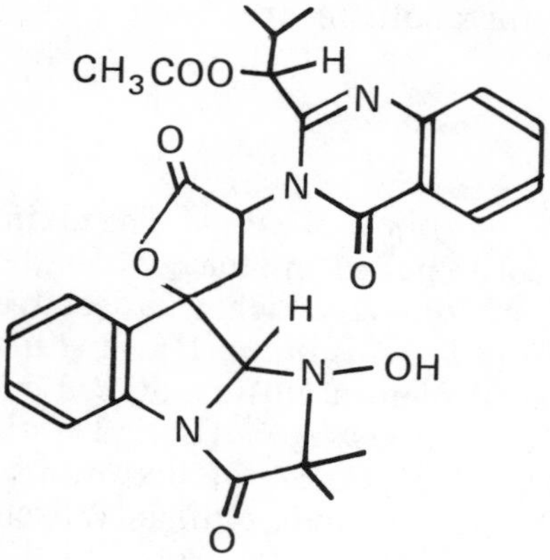

Tryptoquivaline 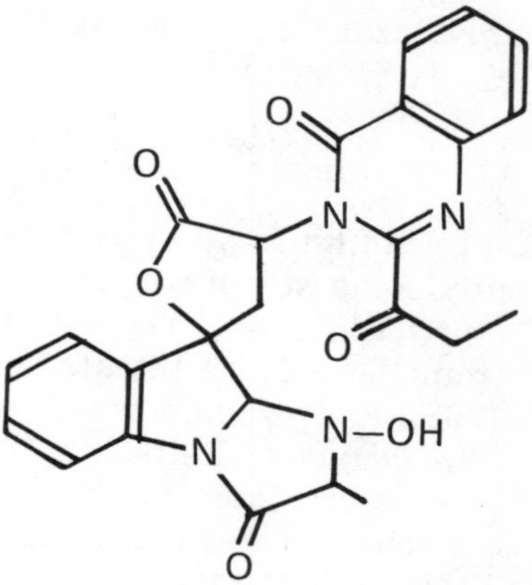   **Tryptoquivalone**

FIG. 7.25.   STRUCTURAL FORMULAS OF TREMORGENIC TOXINS (CONTINUED)

tiorum on the celery plant has been reported to result in the biosynthesis of two phototoxic psoralen derivatives, 8-methoxypsoralen and 4,5′,8-trimethylpsoralen, in the affected area of this plant (Scheel *et al.*, 1963). These compounds were shown to be the casual agents of a bullous dermatitis suffered by many workers engaged in handling celery parasitized by the fungus. The outbreak of the cutaneous lesions occurred only when exposure to sunlight or ultra-violet irradiation having wavelength in the 320 to 370 nm region followed contact with the diseased plant. It has also been shown that

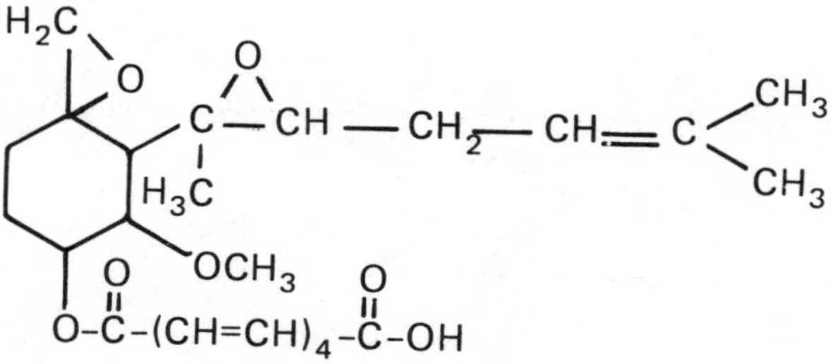

Fumagillin

FIG. 7.26.   STRUCTURAL FORMULA OF FUMAGILLIN

this hazard exists only with actively metabolizing fresh celery (Wu et al., 1972).

U. Decumbin

Decumbin is a toxic metabolite of *Penicillium decumbens*. The toxin-producing strain was isolated from corn spoiled in storage (Singleton *et al.*, 1958). The compounds, brefeldin A which was isolated from the culture filtrate of *Penicillium brefeldianum* (Harri *et al.*, 1963), and cyanein which was obtained from a culture of *Penicillium cyaneum* (Betina *et al.*, 1965), *P. simplicissimum* (Betina *et al.*, 1966) were later found to be identical substances of decumbin.

Decumbin was quite toxic and lethal to rats and goldfish (Wilson, 1971). Administration of the toxin at levels of 400, 300 and 250 mg/kg with food to fasted rats resulted in 37 deaths among 38 fed, 8 deaths among 10 fed, and no deaths among 5 fed rats, respectively. The young rats showed signs of illness in about 2 hours in the form of loss of appetite and was followed by diarrhea, lethargy, labored breathing, cyanosis, stupor, and death in about 24 hours. During au-

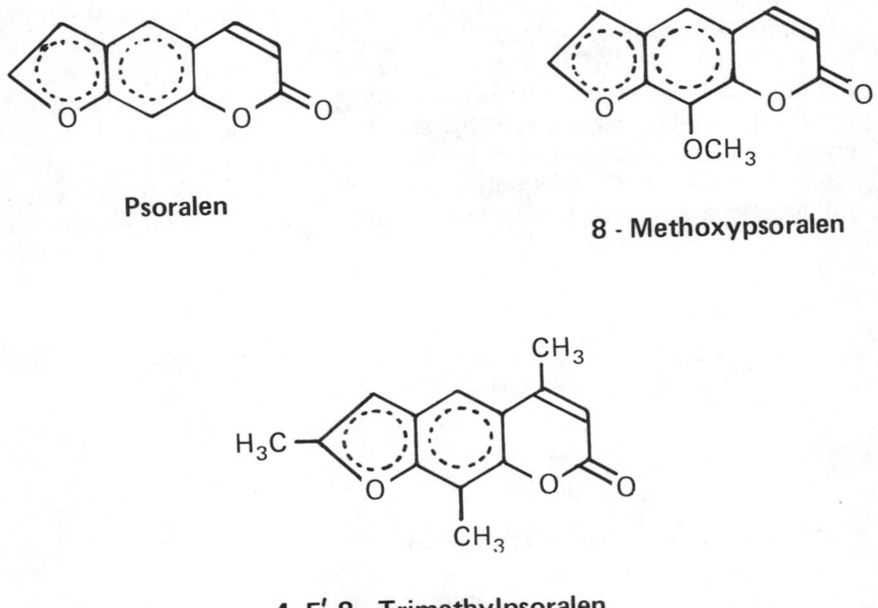

Psoralen

8 - Methoxypsoralen

4, 5', 8 - Trimethylpsoralen

FIG. 7.27.   STRUCTURAL FORMULAS OF PSORALEN AND DERIVATIVES

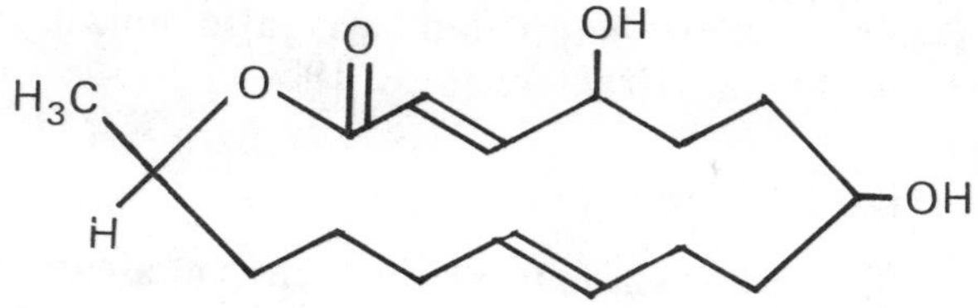

Decumbin (Brefeldin A)

FIG. 7.28.   STRUCTURAL FORMULA OF DECUMBIN

topsy there was evidence of slight nasal bleeding and congestion of the auricles and associated blood vessels. Goldfish (*Carassius auratus*) of about 1.7 gm body weight were killed in water containing 0.6 mg/liter. Seed germination of wheat (*Triticum aestivum*) was reduced by contact with the toxin.

. Cyanein was found to be effective in inhibiting growth of *Candida albicans, C. pseudotropicalis, Aspergillus fumigatus,* and *Saccharomyces cerevisiae,* but ineffective with bacteria and protozoa. It has also been reported as cytostatic for HeLa cells (Betina *et al.*, 1964). No inhibition of glycolysis was obtained in yeasts or animal cells exposed to cyanein. However, inhibition of synthesis of nucleic acids and proteins in both microbial cells and animal cells has been demonstrated (Betina *et al.*, 1966).

V. Viridicatin

Viridicatin was first isolated from the mycelium of *Penicillium viridicatum* by Cunningham and Freeman (1953). An isolate of this fungus was obtained from Argentine corn causing a poisoning of

FIG. 7.29.   STRUCTURAL FORMULA OF VIRIDICATIN

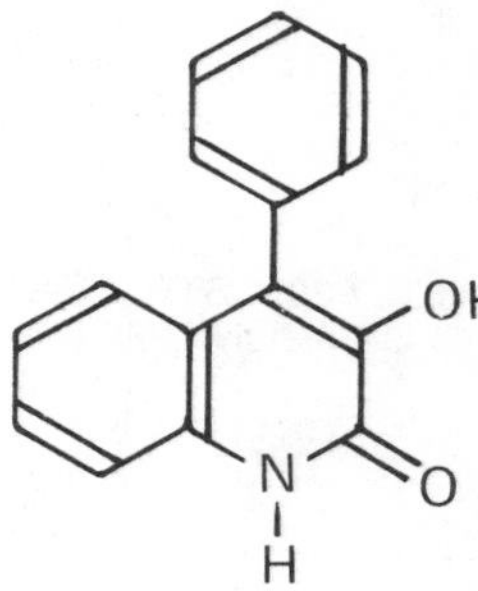

Viridicatin

swine and horses. This compound was also obtained from five strains of *P. cyclopium* (Bracken *et al.*, 1954).

W. Cyclopiazonic acid

A strain of *P. cyclopium* caused acute mycotoxicosis in ducklings and rats. The toxic substance involved is cyclopiazonic acid, a hydrophilic, colorless compound (Holzapfel, 1968). Acute toxicity of this toxin to rats has been reported (Purchase, 1971). Recently, production of cyclopiazonic acid by *Aspergillus flavus* was reported for the first time (Luk *et al.*, 1977).

MISCELLANEOUS MYCOTOXINS AND THEIR HEALTH HAZARDS

A. Ergotoxins

Ergotism is one of the oldest known mycotoxicosis. In the 1500–1600's it was known as "St. Anthony's fire." It had been determined by 1658 that fungi growing on rye and other grains were responsible for gangrenous and convulsive ergotism.

Species of *Claviceps*, particularly *Claviceps purpurea* are the causative fungus of ergotism. By 1808, ergot infested grains were used for medicinal purposes. Toxic substances from ergot grains were not isolated and purified until 1934 (Groger, 1972) when six derivatives of lysergic acid were isolated and identified.

The effects of ergot can be divided into three groups depending on the site of action: (1) peripheral effects, (2) neurohumoral effects, and

FIG. 7.30. STRUCTURAL FORMULA OF CY-CLOPIAZONIC ACID

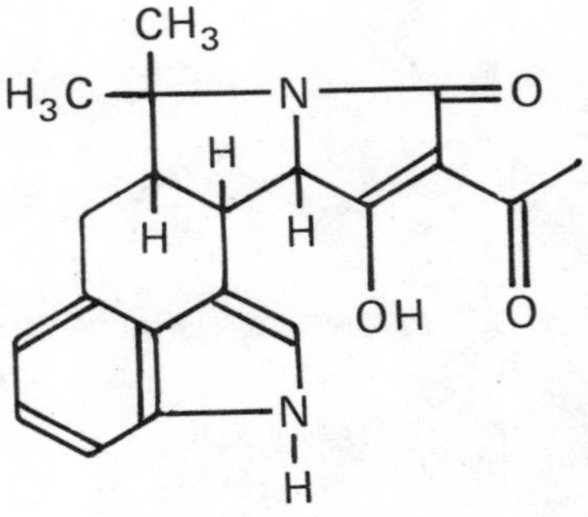

Cyclopiazonic acid

(3) effects on the central nervous system. The most pronounced peripheral effect is the constricting of the smooth muscles in the uterus. Ergot causes reproductive failure in herbivorous mammals, particularly cattle, sheep, and pigs. Feeding of milled ergot sclerotia containing ergotamine caused severe illness of pregnant sheep. Feeding at later stages in pregnancy resulted in less acute illness in the ewe but caused fetal death (Greatorex and Mantle, 1974). The production of hallucinations is one of the most famous effects of ergot.

The last outbreak of epidemic proportions occurred in the U.S. in 1825. Russia, in 1926 and 1927, and England in 1928, experienced large outbreaks of ergotism. These two incidences have been the last reported outbreaks, although small occurrences still take place. Commercial cereal dealers must continually take precautions against ergot infected grains. Ergotism is presently a problem in animals, because people feel it is acceptable to feed animals moldy grains or grasses.

B. Mushroom toxins

Amanita toxins: This group of toxins are produced in the poisonous mushrooms of *Amanita* species, including green *Amanita phalloides* (Vaill. ex Fr.) Secr., white *Amanita verna* (Bull, ex Fr.) Pres. ex Vitt, *Amanita virosa*, *Amanita tenuifolia*, *Amanita bisporigera*. Amanita toxins have also been found in some species of *Galerina* (Tyler et al., 1963).

The white or green *Amanita* causes more than 95% of fatal mushroom poisonings. The high percentage of fatalities occurs from *A. phalloides* and related mushrooms because the first symptoms of intoxication—vomiting and diarrhea, which are not caused by the lethal toxins—do not become apparent until several hours after ingestion, and during this time the deadly toxins have already reached the liver and kidneys. The slow-acting amatoxins cause death by specifically destroying the liver cells. Their fatal action, however, starts shortly after the toxins reach the organ and is irreversible.

The investigation of the poisonous substances of *A. phalloides* by Wieland and Wieland (1959), Wieland (1967), Lynen and Wieland (1937), led to grouping the toxins into two families: amatoxins and phallotoxins. The phallotoxins consist of phalloidin, phallacidin, phalloin, phallisin, and phallin B. The amatoxins include α-, β-, γ-, and ϵ-amanitin, and amanin.

The phallotoxins are fast acting; mice or rats die within 1 or 2

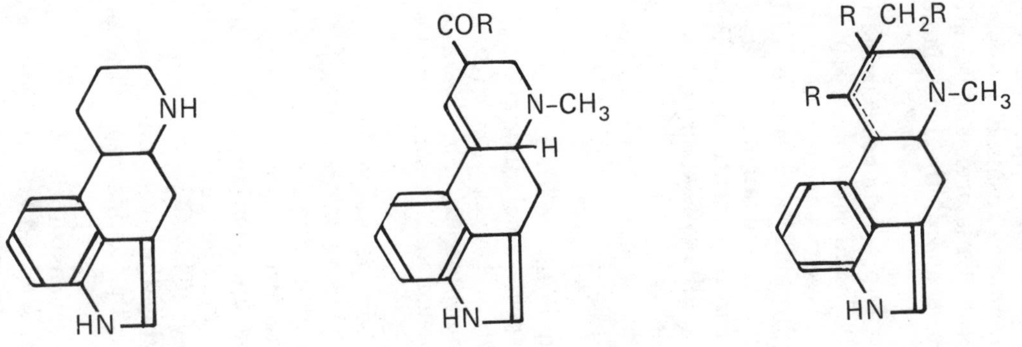

FIG. 7.31. STRUCTURAL FORMULAS OF ERGOT ALKALOIDS

Clavine alkaloids
R=H, OH

Lysergic acid derivative
R=Tripeptide or
smaller units

Ergoline

hours at higher dose levels, whereas amatoxins are slow-acting. The toxicity of amatoxins is 10 to 20 times stronger than the phallotoxins. Different animal species show different sensitivity to the toxins. Rats can survive five-fold amanitin doses as compared with mice, but are more sensitive to phallotoxins ($LD_{50} = 0.75$ mg/kg). The LD_{50} of phalloidin for guinea pigs is 3.5 mg/kg. The phallotoxins and amatoxins are cyclopeptides which have in common a sulfur-containing bridge formed by the coupling of cysteine sulfur with the indole nucleus of a tryptophan, and a γ-hydroxylated amino acid.

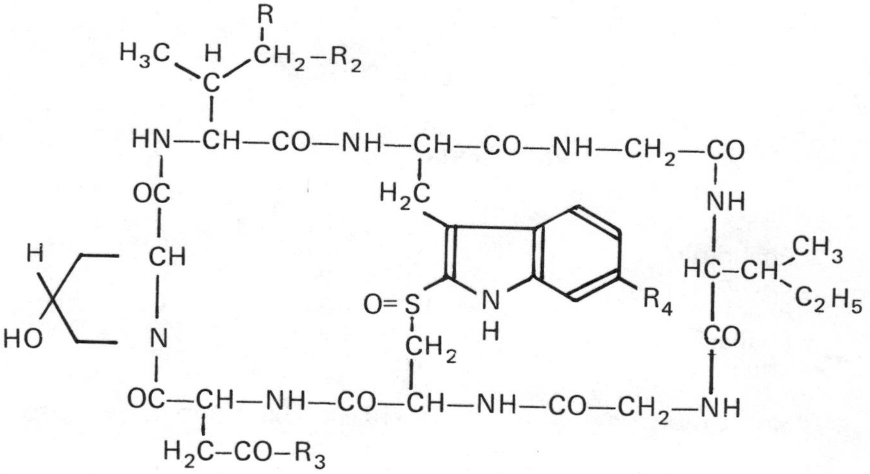

Amatoxins

	R_1	R_2	R_3	R_4
α - Amanitin	OH	OH	NH_2	OH
β - Amanitin	OH	OH	OH	OH
γ - Amanitin	OH	H	NH_2	OH
ϵ - Amanitin	OH	H	OH	OH
Amanin	OH	OH	OH	H
Amanullin (nontoxic)	H	H	NH_2	OH

FIG. 7.32.   STRUCTURAL FORMULAS OF AMATOXINS AND DERIVATIVES

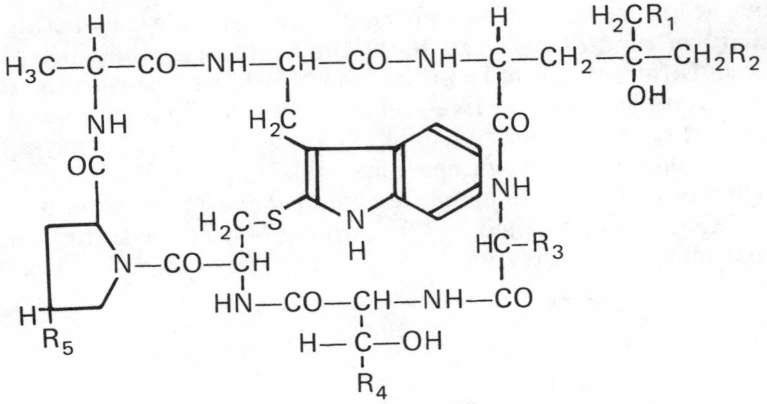

Phallotoxins

	R_1	R_2	R_3	R_4	R_5
Phalloidin	OH	H	CH_3	CH_3	OH
Phalloin	H	H	CH_3	CH_3	OH
Phallisin	OH	OH	CH_3	CH_3	OH
Phallacidin	OH	H	$CH(CH_3)_2$	COOH	OH
Phallin (tentatively)	H	H	$CH_2C_6H_5$	CH_3	H

FIG. 7.33. STRUCTURAL FORMULAS OF PHALLOTOXINS AND DERIVATIVES

C. Others

Four new tryptoquivaline related mycotoxins have been isolated from *Aspergillus clavatus* (Buchi *et al.*, 1977). This included hydroxylamine nortryptoquivaline, and the three secondary amines, deoxytryptoquivaline, deoxynortryptoquivaline, and deoxynortryptoquivalone.

Brevianamide A, a mycotoxin, has been shown to be produced by *Penicillium brevi-compactum*, *P. ochraceum*, and *P. viridicatum* (Birch and Wright, 1969 and 1970; Robbers *et al.*, 1973 and 1975; Wilson *et al.*, 1973). This mycotoxin caused renal and hepatic lesions (Carlton *et al.*, 1972).

Monoacetoxyscirpenol, a new mycotoxin was produced by *Fusarium roseum* Gibbonsum which was isolated from corn suspected of causing illness and death in several dairy cattle (Pathre *et al.*, 1976).

When this isolate was grown on corn or rice and consumed by rats, turkey poults, swine or young chickens, it caused illness and death. Two toxic metabolites were isolated from corn on which this isolate was grown. Monoacetoxysirpenol, the major toxic component, has been shown to be 15-acetoxy, 3α, 4β-dihydroxy-12,13-epoxy-trichothec-9ene.

CONTROL, INACTIVATION OR ELIMINATION OF SOME MYCOTOXINS IN FOODS AND FEEDS

Aflatoxins are very stable compounds. Autoclaving may destroy or inactivate small amounts. Gamma rays show no appreciable effects on the toxins, but ultraviolet light seems to have a slight photo-destructive effect. Chemicals such as ammonia, methylamine, and H_2O_2 will reduce the toxic activity, but means of complete detoxification, other than combustion, have not been found.

There is increasing interest in agents that could be used to control the growth and aflatoxin production by *A. flavus* and *A. parasiticus*. The effects of sodium acetate and propionate on the growth and aflatoxin production by *A. parasiticus* NRRL 2999 was studied in AMU medium (modified AM medium) + 2% yeast extract to determine the possible use of this compound as a means of controlling aflatoxin production (Buchanan and Ayres, 1976). At pH 4.5, sodium acetate at a concentration of 1.0g/100ml completely inhibited growth and prevented aflatoxin production. Propionic acid was found to be more inhibitory than acetate. Surface application of acetic acid, propionic acid, or other suitable fatty acids by spraying, dipping or flotation prior to storage may be a potential means of preventing the growth and aflatoxin production by *A. flavus* and *A. parasiticus* in the product. The prevention of aflatoxin formation on foods and feeds is essentially a problem of preventing the growth of aflatoxin-producing fungi. This can be achieved by at least three means: control of the environment; use of chemical antifungal agents; and utilization of natural resistance factors in agricultural commodities.

The use of chemical antifungal agents to control aflatoxin production has been extensively investigated (Majundar *et al.*, 1965; Malla *et al.*, 1967; Parpia and Sreenivasamurthy, 1970; Bean *et al.*, 1971; Rao and Harein, 1972; Hsieh, 1973; Buchanan and Ayres, 1976). However, the use of certain antifungal agents must be viewed with reservations since there is a possibility of ecological problems developing later. Reports on the natural resistance of certain crops or of certain varieties of crops to aflatoxin contamination have been documented from time to time. The elucidation of the precise chemical nature of factors responsible for varietal differences in susceptibility will help plant geneticists to breed strains with such characteristics. The cause of a markedly low aflatoxin production on the opaque-2 variety of maize as compared to other varieties was reported to be due to the high concentration of a low molecular weight protein in the resistant variety (Nagarajan and Bhat, 1972). One peanut variety, among 60 different varieties screened, has been found to be resistant to aflatoxin contamination (Rao and Tulpule, 1967).

A compound with high inhibitory activity towards *A. parasiticus* was isolated from the potatoes by Swaminathan and Koehler (1976). They suggested the compound to be a hydroxy-cinnamic acid derivative structurally similar to caffeic acid but lacking the ortho-dihydroxy structure of caffeic acid.

The concentration of sodium hypochlorite and pH are important factors in reducing the concentration of aflatoxins in the protein isolates to nondetectable levels during its preparation from raw peanuts and defatted peanut meal. At pH 9, 0.3% sodium hypochlorite reduced the aflatoxin B_1 content in the protein isolate from 300 ppb to below detectable quantities and the aflatoxin B_2 content from 52 ppb to 2 ppb. For defatted peanut meal, at pH 8, 0.25% sodium hypochlorite reduced the content of aflatoxin B_1 and B_2 (136 and 36 ppb, respectively) to below detectable quantities in the protein isolate (Natarajan et al., 1975).

Destruction by chemical means shows the greatest potential for reducing aflatoxin content. Several chemical agents have been suggested as effective in destroying aflatoxin (Trager and Stoloff, 1967; Mann et al., 1970). Some of these reagents have been used experimentally for destruction of aflatoxin in peanut meal (Sreenivasamurthy et al., 1967; Dwarakanath et al., 1968; Dollear et al., 1968) but have not been demonstrated to be commercially feasible, with the exception of ammoniation which is being used to salvage aflatoxin contaminated cottonseed and cottonseed meal. The most common and effective reagent reported in the literature is sodium hypochlorite (NaOCl). This has been recommended as a safety measure for disposal of contaminated materials in laboratories doing aflatoxin research (Trager and Stoloff, 1967). Treatment of peanut meal with formaldehyde and calcium hydroxide also effectively inactivated aflatoxin (Codifer et al., 1976).

It is known that *Aspergillus flavus* will grow on soybeans, but no aflatoxin is formed on that substrate. The reason is not known. Soybeans are high in trypsin inhibitor but it is not known if this is important. If an inhibitor gene can be found and bred into the genetic complement of corn and other commodities, toxin formation could probably be prevented, regardless of harvesting, storage, and processing techniques. This solution would, therefore, be the most practical one for developing countries (Weinberg, 1975).

Foods that contain cinnamon may not readily support mold growth and/or mycotoxin production. Growth of *Aspergillus parasiticus* (NRRL 2999 and NRRL 3000) on raisin bread containing cinnamon was very limited and aflatoxin production was absent or greatly reduced compared to that observed on the other breads. The inhibitory effect was greater on aflatoxin production than on mold growth, with cinnamon levels of 2.0% inhibiting aflatoxin production by 97–99%. An alcoholic extract of cinnamon, in similar amounts, was also inhibitory to mold growth and aflatoxin production (Bullerman, 1974). It is generally accepted that the essential oils of certain spices have antimicrobial activities (Frazier, 1967). However, their actions against specific microbial metabolic processes have not been elucidated.

Diener and Davis (1972) investigated the effect of various levels of carbon dioxide (CO_2), oxygen (O_2), and nitrogen (N_2) in combination with several temperatures and relative humidities (RH) on aflatoxin production and fungal growth in peanuts. Aflatoxin production was limited by 40% CO_2 and above at 77°F and by 20% CO_2 and above at 86% RH and 63°F.

Aflatoxin production in sound mature peanut kernels decreased with increasing concentrations of CO_2 from 0.03% to 100%. Reducing O_2 concentrations generally reduced aflatoxin production. Noticeable decreases in aflatoxin resulted when O_2 was reduced from 5% to 1%, regardless of CO_2 concentration. Lowering temperature or relative humidity from the optimum also reduced aflatoxin production (Diener and Davis, 1972).

To control, reduce or eliminate ochratoxin contamination, factors which can modify mold development (including moisture, temperature, aeration, time, and substrate) have been investigated (Levi *et al.*, 1974; Lillehoj, 1973; Harwig and Chen, 1974). Ochratoxins are largely destroyed in coffee by heating to 200°C for 5 min or more under simulated coffee roasting conditions (Levi *et al.*, 1974), but 30% remained intact in oatmeal or rice cereals after 3 hr of autoclaving (Trenk *et al.*, 1971). A 2–7% transmission of ochratoxin A from contaminated barley (1–5000 mg/kg) to the beer (6–20 mg/l) was observed on malting and brewing the beer (Krogh et al., 1974). Wu and Ayres (1974) reported that the insectide dichlorovos could be used to inhibit growth and ochratoxin production of A. ochraceus.

Patulin disappeared from apple juice when the juice was fermented by 3 different yeasts (Harwig et al., 1973). This effect has also been tested with 2 strains of yeast used commercially in English cider making. The affinity of patulin for sulfur dioxide is of little significance at the SO_2 concentration (below 200 ppm) used in the processing of apple juice and cider (Burroughs, 1977). However, at a concentration of 2000 ppm SO_2 and 14 ppm patulin, combination was 90% complete in 2 days. Removal of SO_2 liberated only part of the patulin. It was suggested that two mechanism were involved: one reversible (opening the hemiacetal ring) and one irreversible (SO_2 addition at the double bond). Sands et al. (1976) recently reported the successful use of activated charcoal for the removal of patulin from cider. Activated charcoal at 5 mg/ml reduced patulin in naturally contaminated cider to nondetectable levels. Gamma irradiation at 200 Krad has been reported to inhibit both growth and fungal production of Penicillium expansum and Penicillium patulum (Adams et al., 1976b). Subatmospheric pressure lower than 160 mm Hg also inhibited growth and patulin production of both fungi (Adams et al., 1976a).

Inhibition of zearalenone biosynthesis by the insectide dichlorvos has been reported by Wolf et al. (1972) and Wolf and Mirocha (1973). Berisford and Ayres (1976) found that the insecticide naled (1,2-dibromo,2,2-dichloroethyl dimethyl phosphate), applied as a fumigant or in a liquid preparation at levels of 30 and 100 ml/l, completely inhibited zearalenone production by Fusarium graminearum in liquid media and on corn. They suggested that naled was a potential protectant only when applied prior to initial growth of the fungus.

Wieland et al. (1968, 1969) discovered an antitoxin, antamanide, in extracts of Amanita phalloides. When the antitoxin was administered simultaneously with phalloidin to the white mouse, it counteracted a 100% lethal dose of 5 mg/kg of this toxin. The protective dose is 0.5 mg/kg. Antamanide is a cyclic decapeptide consisting of only four different L-amino acids: alanine, valine, phenylalanine, and proline, in a molar ratio of 1:1:4:4 (Prox et al., 1969; Koenig and Geiger, 1969). The therapeutic use of this cyclic decapeptide, which can also be synthesized, is rather restricted, since it must be given before or at least at the time of toxin administration. The therapeutic value of these compounds in Amanita poisoning requires further investigation.

DETECTION, DETERMINATION, AND DETOXIFICATION OF SOME MYCOTOXINS

All mycotoxins should be handled as very toxic substances. All manipulations should be performed under a hood whenever possible and precautions, such as a glove box, should be utilized when toxins are in a dry form due to their electrostatic nature and resultant tendency to disperse into working areas. Extraction of all samples should be carried out in a explosion proof blender.

After extraction in the proper solvent, the crude extract is reduced in volume by using a continuous feed rotary evaporator. In most cases the concentrated crude extract is applied to a chromatographic column in order to clean up the extract, i. e. remove high molecular weight compounds, lipids, and/or pigments. An alternate cleanup procedure is described by Roberts and Patterson (1975), which involves a membrane system to remove high molecular weight substances. Following cleanup, thin layer chromatography (TLC) methods are utilized for detection and separation of mycotoxins present in the sample. In most mycotoxin methods, TLC is a critical step in mycotoxin detection and separation so care must be taken in preparation, spotting, and developing of spots. Due to batch to batch variation in particle size and site activity of adsorbents and variations such as, the temperature and humidity to which plates are exposed during handling, spotting, and developing, adjustments of layer thickness and solvent polarity may be required. Since reactive vapors, e.g., O_3, SO_2, and HCl can affect adsorbents and the stability of adsorbed spots, TLC should be carried out in a laboratory free of volatile reagents. In order to protect the adsorbent layer and the developed spots, TLC plates should be handled in a box with an inert atmosphere or a clean glass plate can be placed over the adsorbent layer while spotting and after development and drying. After development of plates, always dry plates completely before exposure to visualizing agents, such as UV light. Since UV light from sunlight or fluorscent lamps can catalyze changes in the compounds being examined, exposure to UV light should be avoided, especially in the presence of solvent. Exposure of developed spots to UV light should be held to the minimum time needed for visualization.

Quantitative determination of mycotoxins can be carried out using a number of physical methods including: 1) quanitative thin layer chromatography; 2) spectrophotometry; 3) gas liquid chromatography; and/or 4) high pressure liquid chromatography. Of these methods the most commonly utilized is quantitative TLC. This is described in more detail below. Quantitative TLC can be performed by visual or densitometric analysis. When using visual analysis one should spot 3, 5, and 7 μl samples of clean extract on the plate along with 2, 3, 4, and 5 μl samples of a standard mycotoxin solution. After development of the chromatogram, examine the plate and compare the intensities of the sample spots with those of the standards. From this a rough quantitative estimation of the amount of mycotoxin present in the extract can be made. Densitometric analysis is carried out by spotting sample aliquots on the plate with duplicate aliquots of the mycotoxin standard solution adjusted to provide 5 ng of toxin. During spotting, care should be taken not to damage the gel layer since significant errors can be introduced by such damage. After development of the plate, scan the plate with a densitometer in accordance with the instructions of the manufacturer. Irradiate the mycotoxin spots with the proper wavelength and observe the emission of the absorbed light at the optimum wavelength of the mycotoxin being quantified. For example, aflatoxin samples are irradiated with a 365 nm source and emission of light is measured at 420–460 nm. Quanitity of mycotoxin present in the samples can be calculated by using the average area of the emission peaks. Details on the utilization of other analytical techniques for the screening of samples containing more than one mycotoxin are described by Roberts and Patterson (1975); Wilson *et al*, (1976); and Engstrom *et al*, (1977). The physical methods are utilizable for analysis of any mycotoxin that has a characteristic absorption spectra in the UV, visible, or infrared regions. One large group of mycotoxins, the toxic trichothecenes, lack characteristic absorption patterns so the development of spectral methods for determining trichothecenes has thus far been prevented. At the present time qualitative and quantitative analysis of samples for trichothecenes has been carried out on the basis of biological screening tests. The bioassays include: 1) oral or parentenal administration of sample extract to rats, mice, and guinea pigs; 2) dermatotoxicity tests; 3) cytotoxicity tests; 4) phytotoxicity tests; and/or 5) inhibition of protein synthesis. These methods are reviewed in detail by Epply (1975).

Detoxification of spillage, containers, work surface areas, and personnel can be accomplished by applying one of several chemicals including undiluted bleach, concentrated acids, and strong bases. Areas of spillage should be completely covered with 5–6% NaOCl (household bleach). The bleach should be at least onetenth of the volume of the spill. Contact time should be at least thirty seconds.

All glassware, including TLC plates, should be soaked in a solution of bleach diluted tenfold with water for at least 30 seconds and care should be taken to scrape off all adhering material. If chromic acid cleaning solution is to be used, no other treatment is necessary. Work surfaces should be wiped with towels soaked in a tenfold dilution of bleach. Hoods are detoxified by mixing equal volumes of undiluted bleach with 6N HCl. This procedure will generate chlorine gas which will diffuse throughout the hood. Allow the gas to remain in the hood several minutes before turning on the exhaust fan.

All workers who handle mycotoxins should wear protective gloves routinely. If hands should become contaminated, they should be washed immediately with undiluted bleach followed by washing with soap. If the skin is too sensitive for this treatment, sodium perborate with a detergent can be utilized. If toxic material should enter the mouth, a gargle containing 1% sodium perborate and 1% sodium bicarbonate should be employed. Details in the detoxification especially of aflatoxins, procedures can be found in a paper by Stoloff and Trager (1965).

In general, detoxification of mycotoxins in biological systems occurs in the liver. The liver microsomal enzymes carry out many oxidative reactions including aliphatic oxidation, aromatic hydroxylation, N-dealkylation, O-dealkylation, S-demethylation, deamination, sulfoxide formation, desulfuration, N-oxidation, and N-hydroxylation. Metabolism of most of the major groups of mycotoxins has not been studied in detail. Some information is available on the metabolism of ochratoxins and aflatoxins. In both cases, hydroxylation appears to be the major metabolic route with ochratoxin A being converted to 4-hydroxy-ochratoxin A and aflatoxin B_1 is converted to aflatoxin M_1, which is a hydroxylated form of aflatoxin B_1.

More general information on these topics is given by Howitz (1975) and Purchase (1974).

SIGNIFICANCE TO THE PUBLIC HEALTH AND ECONOMY

The deleterious effects of fungal spoilage of many types of crops, especially grains, oilseeds, fruits, and vegetables which are prone to fungal damage, result in reducing the usefulness of the commodity for food and feed purposes and cause serious economic consequences. In addition to this, contamination of the affected commodities with mycotoxins produced during the fungal growth is a serious health problem. Ingestion of contaminated foodstuffs and feedstuffs causes toxicity syndromes known as mycotoxicoses. All the mycotoxicoses have posed direct health hazards to human consumers, or have caused significant losses in domestic animals used for food. The net result in either case has been a reduction in the utilizable food supply, either by directly limiting human consumption or by decreasing available animal protein sources.

REFERENCES

ADAMS, K. B., M. T. WU and D. K. SALUNKHE. 1976a. Effects of subatmospheric pressures on the growth and patulin production of *Penicillium expansum* and *Penicillium patulum*. Lebensm.-Wiss. u.-Technol 9:153.

ADAMS, K. B., M. T. WU and D. K. SALUNKHE. 1976b. Effects of gamma radiation on the growth and patulin production of *Penicillium expansum* and *Penicillium patulum*. J. Environ. Exptl. Bot. 16:189.

ALLCROFT, R. and R. B. A. CARNAGHAN. 1962. Groundnut *Aspergillus flavus* toxin (aflatoxin) in animal. Vet. Record 74:863.

ALLCROFT, R. and R. B. A. CARNAGHAN. 1963. Groundnut toxicity: an examination for toxin in human food products and animal fed toxic groundnut meal. Vet. Record 75:259.

ALLCROFT, R., R. B. A. CARNAGHAN, K. SARGEANT, and J. O'KELLY. 1961. A toxic factor in Brazilian groundnut meal. Vet. Record 73:428.

ALLCROFT, R. and G. LEWIS. 1963. Groundnut toxicity in cattle: Experimental poisoning of calves and a report on clinical effect in older cattle. Vet. Record 75:487.

ALSBERG, C. L. and O. F. BLACK. 1913. Contributions to the study of maize deterioration. U.S. Dept. Agr. Bull. Bur. Plant Ind. 270:7.

AMBROSE, A. M., F. DeEDS. 1964. Some toxicological and pharmacological properties of citrinin. J. Pharmacol. Exp. Ther. 88:173.

ANONYMOUS, 1974a. Imported figs recalled because of aflatoxin. Food Chem. News 16:17.

ANONYMOUS. 1974b. FDA to sample 13 domestic food products for mycotoxins. Food Chem. News 16:4.

ANDERSON, D. W. 1965. Griseofulvin, biology and clinical usefulness. Ann. Allergy 23:103.

ANDERSON, H. W., E. W. NEHRING, and W. R. WICHSER. 1975. Aflatoxin contamination of corn in the field. J. Agr. Food Chem. 23:775.

ANSLOW, W. K., and H. RAISTRICK. 1938. Studies in the biochemistry of microorganisms. LVII. Fumigatin (3-hydroxy-4-methoxy-2:5-toluquinone) and spinulosin (3:6-dihydroxy-4-methoxy-2:5-toluquinone), metabolic products respectively of *Aspergillus fumigatus* Fresenius and *Penicillium spinulosum* Thom. Biochem. J. 32:687.

ASPLIN, F. D. and R. B. A. CARNAGHAN. 1961. The toxicity of certain groundnut meals for poultry with special reference to their effect on ducklings and chickens. Vet. Record. 73:1215.

ATKINSON, N. 1943. Antibacterial substances produced by some common Penicillia Aust. J. Expt. Biol. and Med. Sci. 21:15.

BAILEY, L. 1953. Effects of fumagillin upon *Nosema* pigs (Zander). Nature 171:212.

BEAN, G. A., W. L. KLARMAN, G. W. RAMBO and J. B. SANFORD. 1971. Dimethyl sulfoxide inhibition of aflatoxin synthesis by *A. flavus*. Phytopathol. 61:380.

BERISFORD, Y. C. and J. C. AYRES. 1976. Use of the insecticide naled to control zearalenone production. J. Agr. Food Chem. 24:973.

BETINA, V., L. DROBNICA, P. NEMEC, and M. ZEMANOVA. 1964. The antifungal activity of the antibiotic, cyanein. J. Antibiotics (Tokyo) 17:93.

BETINA, V., J. FUSKA, A. KJAER, M. KUTKOVA, P. NEMEC, and R. H. SHAPIRO. 1966. Production of cyanein by *Penicillium simplicissimum*. J. Antibiotics (Tokyo). 19:115.

BETINA, V., P. NEMEC, S. KOVAK, A. KJAER, and R. H. SHAPIRO. 1965. The identity of cyanein and brefeldin A. Acta. Chem. Scand. 19:519.

BETINA, V., P. NEMEC, M. KUTKOVA, J. BALAN, and S. KOVAC. 1964. The isolation of citrinin from *Penicillium notatum*. Chem. Zyesti 18:128.

BIOLLAZ, M., G. BUCHI, and G. MILNE. 1970. The biosynthesis of the alfatoxins. J. Am. Chem. Soc. 92:1035.

BIRCH, A. J. and J. J. WRIGHT. 1969. The breveanamides: a new class of fungal alkaloid. Chem. Commun. 1969:644.

BIRCH, A. J. and J. J. WRIGHT. 1970. Studies in relation to biosynthesis XLII. The structural elucidation and some aspects of the biosynthesis of the brevianamides A and E. Tetrahedron 26:2329.

BIRKINSHAW, J. H., H. RAISTRICK, and D. J. ROSS. 1952. Studies in the biochemistry of microorganisms. 86. The molecular constitution of mycophenolic acid, a metabolic product of *Penicillium brevi-compactum* Dierekx. Part 3. Further observations on the structural formula for mycophenolic acid. Biochem. J. 50:630.

BLOUNT, W. T. 1961. Turkey "X" disease. Turkeys 9:53, 58, 61, 77.

BOMPEIZ, G., F. MORGAT, B. POIRET, and J. P. Planque. 1969. Chemical control of apple rot in storage: efficacy of benomyl and thiabendazole. Acad. Agr. France, Paris, Compt. Rend. Hebd. Seances 55:776.

BRACKEN, A., A. POCKER, and H. RAISTRICK. 1954. Studies in the biochemistry of microorganisms. 93. Cyclopenin, a nitrogen-containing metabolic product of *Penicillium cyclopium* Westling. Biochem. J. 57:587.

BRIAN, P. W., P. J. CURTIS, and H. J. HEMMING. 1946. A substance causing abnormal development of fungal hyphae produced by *Penicillium janczewskii* Zal. I. Biological assay, production and isolation of "curling factor." Brit. Mycol. Soc. Trans. 29:173.

BRIAN, P. W., G. W. ELSON and D. LOWE. 1956. Production of patulin in apple fruits by *Penicillium expansum*. Nature 178:263.

BROOM, W. A., E. BULBRING, C. J. CHAPMAN, J. W. F. HAMPTON, A. M. THOMPSON, J. UNGAR, R. WIEN and G. WOOLFE. 1944. The pharmacology of patulin. Brit. J. Exp. Pathol. 25:195.

BROWN, G. E., and W. C. WILSON. 1968. Mode of entry of *Diplodia natalensis* and *Phomopsis citri* into Florida oranges. Phytopathol. 58:736.

BUCHANAN, JR., R. L., and J. C. AYRES. 1976. Effects of sodium acetate on growth and aflatoxin production by *Aspergillus parasiticus* NRRL-2999. J. Food Sci. 41:128.

BUCHANAN, J. R., N. F. SOMMER, and R. J. FORTLAGE. 1975. *Aspergillus flavus* infection and aflatoxin production in fig fruits. Appl. Microbiol. 30:238.

BUCHANAN, J. R., N. F. SOMMER, R. J. FORTLAGE, E. C. MAXIE, F. G. MITCHELL, and D. P. H. HSIEH. 1974. Patulin form *Penicillium expansum* in stone fruits and pears. J. Am. Soc. Hort. Sci. 99:262.

BUCHI, G., D. H. KLAUBERT, R. C. SHANK, S. M. WEINREB, and G. N. WOGAN. 1971. Structure and synthesis of kotanin and desmethyl-kotanin, metabolites of *Aspergillus glaucus*. J. Org. Chem. 36:1143.

BUCHI, G., Y. KITAURA, S. S. YUAN, H. E. WRIGHT, J. CLARDY, A. L. DEMAIN, T. GLINSUKON, N. HUNT, and G. N. WOGAN. 1973. Structure of cytochalasin E, a toxic metabolite of *Aspergillus clavatus*. J. Am. Chem. Soc. 95:5423.

BUCHI, G., K. C. LUK, B. KOBBE, and J. M. TOWNSEND. 1977. Four new mycotoxins of *Aspergillus clavatus* related to tryptoquivaline. J. Org. Chem. 42:244.

BUCKLEY, S. S. 1901. Acute hemorrhagic encephalitis prevalent among horses in Maryland. Am. Vet. Review 25:99.

BULLERMAN, L. B. 1974. Inhibition of aflatoxin production by cinnamon. J. Food Sci. 39:1163.

BULLERMAN, L. B., J. BACA, and W. T. STOTT. 1975. An evaluation of potential mycotoxin producing molds in corn meal. Cereal Foods World 20:248.

BULLERMAN, L. B., H. M. BARNHART, and T. E. HARTUNG. 1973. Use of γ-irradiation to prevent aflatoxin production in bread. J. Food Sci. 38:1238.

BULLERMAN, L. B., and T. E. HARTUNG. 1973. Mycotoxin producing potential of molds from flour and bread. Cereal Sci. Today 18:346.

BULLOCK, E., J. C. ROBERTS, and J. G. UNDERWOOD. 1962. Studies in mycological chemistry. Part XI. The structure of isosterigmatocystin and an amended structure for sterigmatocystin. J. Chem. Soc. p. 4179.

BUREAU OF FOODS. FDA. 1974. Evaluation of results from FY-73 Compliance Program, Mycotoxins in Foods.

BURKHARDT, H. J., and J. FORGACS. 1968. o-Methylsterigmatocystin, a new metabolite from *Aspergillus flavus*, Link Ex Fries. Tetrahedron 24:717.

BURNSIDE, J. E., W. L. SIPPEL, J. FORGACS, W. T. CARLL, M. B. ATWOOD, and E. R. DOLL. 1957. A disease of swine and cattle caused by eating moldy corn. II. Experimental production with pure cultures of molds. Am. J. Vet Res. 18:817.

BURROUGHS, L. F. 1977. Stability of patulin to sulfur dioxide and to yeast fermentation. J. A.O.A.C. 60:100.

BURTON, C. L. 1970. Diseases of tropical vegetables on the Chicago market. Trop. Agr. (Trinidad) 47:303.

BURTON, H. S. 1949. Antibiotics from *Penicillia*. Brit. J. Exptl. Pathol. 30:151.

BUTLER, W. H. and J. M. BARNES. 1964. Toxic effect of groundnut meal containing aflatoxin to rats and guinea pigs. Brit. J. Cancer 17:699.

BUTLER, W. H. and J. M. BARNES. 1966. Carcinoma of the glandular stomach in rats given diets containing aflatoxin. Nature 209:90.

CALDWELL, R. W. and J. TUITE. 1970. Zearalenone production in field corn in Indiana. Phytopathol. 60:1696.

CALDWELL, R. W. and J. TUITE. 1974. Zearalenone in freshly harvested corn. Phytopathol. 64:752.

CALDWELL, R. W., J. TUITE, M. STOBB and R. BALDWIN. 1970. Zearalenone production by *Fusarium* species. Appl. Microbiol. 20:31.

CARLTON, W. W., J. TUITE and R. W. CALDWELL. 1972. Mycotoxicosis induced in mice by *Penicillium ochraceum*. Toxicol. Appl. Pharmacol. 21:130.

CARLTON, W. W., and J. TUITE. 1969. Toxicosis in miniature swine induced by corn cultures of *Penicillium viridicatum*. Toxicol. Appl. Pharmacol. 14:636.

CARLTON, W. W., and J. TUITE. 1970. Mycotoxicosis induced in guinea pigs and rats by corn cultures of *Penicillium viridicatum*. Toxicol. Appl. Pharmacol. 16:345.

CARLTON, W. W., J. TUITE and P. MISLIVEC. 1958. Investigations of the toxic effects of mice of certain species of *Penicillium*. Toxicol. Appl. Pharmacol. 13:372.

CARNAGHAN, R. B. A. and R. ALLCROFT. 1962. Groundnut toxicity. Vet. Record 74:925.

CEPONIS, M. J. 1970. Diseases of California head lettuce on the New York market during the spring and summer months. Plant Disease Reptr. 54:964.

CHAIN, E., H. W. FLOREY, M. A. JENNINGS, and T. I. WILLIAMS. 1943. Helvolic acid, an antibiotic produced by *Aspergillus fumigatus*, mut. helvola Yuill. Brit. J. Exp. Pathol. 24:108.

CHRISTENSEN, C. M. 1957. Deterioration of stored grains by fungi. Bot. Rev. 23:108.

CHRISTENSEN, C. M. 1965. Fungi in cereal grains and their products. In 'Mycotoxins in Foodstuffs' (G. N. Wogan, ed). pp. 9, M.I.T. Press, Cambridge, MA.

CHRISTENSEN, C. M., G. H. NELSON, and C. J. MIROCHA. 1965. Effect on the white rat uterus of a toxic substances isolated from *Fusarium*. Appl. Microbiol. 13:653.

CIEGLER, A. 1969. Tremorgenic toxin from *Penicillium palitans*. Appl. Microbiol. 18:128.

CIEGLER, A. 1975. Mycotoxins: Occurrence, chemistry, biological activity. Lloydia 38:21.

CIEGLER, A., A. C. BECKWITH and L. K. JACKSON. 1976. Teratogenecity of patulin and patulin adducts formed with cysteine. Appl. Environ. Microbiol. 31:644.

CIEGLER, A. and C. P. KURTZMAN. 1970. Penicillic acid production by blue-eye fungi on various agricultural commodities. Appl. Microbiol. 20:761.

CIEGLER, A., H. J. MINTZLAFF, W. MACHNIK, and L. LEISTNER. 1972. Untersuchungen uber das Toxinbidungsvermogen von Rohwursten Isolierter Sohimmelpilze der Gattung *Penicillium*. Fleischwirtschaft 52: 1311.

CIEGLER, A., and J. I PITT. 1970. Survey of the genus *Penicillium* for tremorgenic toxin production. Mycopathol. Mycol. Appl. 42:119.

CIEGLER, A., R. F. VESONDER, and R. J. COLE. 1976. Tremorgenic mycotoxins. In "Mycotoxins and other fungal related food problems." Ed. J. V. Rodricks. Advances in chemistry series 149. Am. Chem. Soc., Washington, D.C.

CLINE, J. C., J. D. NELSON, K. GERZON, R. H. WILLIAMS, and D. C. DELONG. 1969. In vitro antiviral activity of mycophenolic acid and its reversal by guanine-type compounds. Appl. Microbiol. 18:14.

CLUTTERBUCK, P. W., A. E. OXFORD, H. RAISTRICK, and G. SMITH. 1932. Biochemistry of microorganisms. XXIV. The metabolic products of the *Penicillium brevi-compactum* series. Biochem. J. 26:1441.

CODIFER, L. P., G. E. MANN, and F. G. DOLLEAR. 1976. Aflatoxin inactivation: treatment of peanut meal with formaldehyde and calcium hydroxide. J. Am. Oil Chem. Soc. 53:204.

COLE, R. J. and J. W. KIRKSEY. 1970. Dihydro-o-methylsterigmatocystin, a new metabolite from *Aspergillus flavus*. Tetrahedron Lett. 35:3109.

COLE, R. J., and J. W. KIRKSEY. 1973. The mycotoxin verruculogen: a 6-0-methylindole. J. Agr. Food Chem. 21:927.

COLE, R. J., J. W. KIRKSEY, J. H. MOORE, B. R. BLANKENSHIP, U. L. DIENER, and N. D. DAVIS. 1972. Tremorgenic toxin from *Penicillium verruculosum*. Appl. Microbiol. 24:248.

CUNDLIFFE, E. and J. E. DAVIES. 1977. Inhibition of initiation, elongation, and termination of eukaryotic protein synthesis by trichothecene fungal toxins. Antimicrobiol Agent & Chemotherapy 11:491.

CURTIN, T. M., and J. TUITE. 1966. Emesis and refusal of feed in swine associated with *Gibberella zeae*-infected corn. Life Sci. 5:1937.

CYSEWSKI, S. J., JR. 1973. A tremorgenic mycotoxin from *Penicillium puberulum*, its isolation and neuropathic effects. Ph.D. Thesis. Iowa State University, Ames, IA.

DICKENS, F. and H. E. H. JONES. 1961. Carcinogenic activity of a series of reactive lactones and related substances. Brit. J. Cancer 15:85.

DICKENS, F., H. E. H. JONES, and H. B. WAYNFORTH. 1966. Oral, sucutaneous, and intratracheal administration of carcinogenic lactones and related substances: The intratracheal administration of cigarette tar in the rat. Brit. J. Cancer 20:134.

DICKENS, J. W. and R. E. WELTY. 1975. Fluorescence in pistachio nuts contaminated with aflatoxin. J. Am. Oil Chem. Soc. 52:448.

DIENER, U. L. and N. D. DAVIS. 1966. Aflatoxin production by isolates of *Aspergillus flavus*. Phytopathol. 56:1390.

DIENER, U. L. and N. D. DAVIS. 1972. Atmospheric gases and aflatoxin production of peanuts. Highlights of Agr. Res. Agr. Expt. Stn. of Auburn Univ., Auburn, AL.

DIX, D. T., J. MARTIN, and C. E. MOPPETT. 1972. Molecular structure of the metabolite lanosulin. Chem. Commun. 1168.

DOLLEAR, F. G., G. E. MANN, L. P. CODIFER, JR., H. K. GARDNER, JR., S. P. KOLTUN, and H. L. E. VIX. 1968. Elimination of aflatoxin from peanut meal. J. Am. Oil Chem. Soc. 45:862.

DOSTER, R. C. and R. O. SINNHUBER. 1972. Comparative rates of hydrolysis of ochratoxins A and B *in vitro*. Food Cosmet. Toxicol. 10:389.

DWARAKANATH, C. T., E. T. RAYNER, G. E. MANN, and F. G. DOLLEAR. 1968. Reduction of aflatoxin levels in cottonseed and peanut meals by ozonization. J. Am. Oil Chem. Soc. 45:93.

ELSWORTHY, G. C., J. S. E. HOLKER, J. M. McKEOWN, J. B. ROBINSON, and L. J. MULHEIRN. 1970. The biosynthesis of the aflatoxins. Chem. Commun. 26:1069.

ENGLEBRECHT, J. C., and B. ATTENKIRK. 1972. Comparison of some biological effects of sterigmatocystin and aflatoxin analogues on primary cell cultures. J. Nat. Cancer Inst. 48:1647.

ENGSTROM, G. W., J. L. RICHARD, and S. J. CYSEWSKI. 1977. High pressure liquid chromatographic method for detection and resolution of rubratoxin, aflatoxin, and other mycotoxins. J. Agr. Food Chem. 25:833.

EPPLEY, R. M., L. STOLOFF, M. W. TRUCKSESS and C. W. CHUNG. 1974. Survey of corn for *Fusarium* toxins. J. AOAC. 47:632.

EPPLEY, R. M. 1975. Methods for detection of trichothecenes. *J. AOAC.* 5:906.

EPSTEIN, E., M. P. STEINBERG, A. I. NELSON, and L. S. WEI. 1970. Aflatoxin production as affected by environmental contributions. J. Food Sci. 35:389.

ERICKSON, E. 1968. Estrogenic factors in moldy grain. Nord. Vet. Med. 20:396.

EVANS, M. A., B. J. WILSON and R. D. HARBISON. 1975. Toxicity and mutagenic effects of rubratoxin B. Pharmacologists 17:399.

FIELD, J. B., D. A. FILLER, L. T. BASCOY, F. COSTA, and A. BORYCZKA. 1956. Antibiotics as sources of anticarcinogens. Fed. Proc. 15:250. 15:250.

FLOREY, H. W., M. A. JENNINGS, K. GILLIVER, and A. G. SANDERS. 1946. Mycophenolic acid, an antibiotic from *Penicillium brevi-compactum* Dierchx. Lancet. I:46.

FLOSS, H. G., H. GUENTHER, and L. A. HADWIGER. 1969. Biosynthesis of furanocoumarins in diseased celery. Phytochem. 8:585.

FORGACS, J. and W. T. CARLL. 1962. Mycotoxicosis. Adv. Vet. Sci. 7:274. hemorrhagic disease in poultry. Vet. Med. 50:172.

FORGACS, J. and W. T. CARLL. 1962. Mycotoxicosis. Adv. in Vet. Sci. 7:274.

FORGACS, J., H. KOCH, W. T. CARLL and R. H. WHITE-STEVENS.

1958. Additional studies on the relation ship of mycotoxicoses to the poultry hemorrhagic syndrome. Am. J. Vet. Res. 19:744.

FRAZIER, W. C. 1967. Food Microbiology, 2nd ed. McGraw-Hill Book Company, New York, N.Y.

FRIEDMAN, B. A. 1960. Market diseases of fresh fruits and vegetables. Econ. Bot. 14:145.

GAUMANN, E., and A. VON ARX. 1947. Antibiotica als pflanzliche plasmagifte. II. Ber. Schweiz. Botan. Ges. 57:174.

GILLIVER, K. 1946. Inhibitory action of antibiotics on plant pathogenic bacteria and fungi. Ann. Botany (London) 10:271.

GLINSUKON, T., S. S. YUAN, R. WIGHTMAN, Y. KITAURA, G. BUCHI, R. C. SHANK, G. N. WOGAN, and C. M. CHRISTENSEN. 1974. Isolation and purification of cytochalasin E and two tremorgens from *Aspergillus clavatus*. Plant Foods Man 1:113.

GREATOREX, J. C. and P. G. MANTLE. 1974. Effect of rye ergot on the pregnant sheep. J. Reprod. Fert. 37:33.

GROGER, D. 1972. Ergot. In "Microbial Toxins," VIII. Ed., Kadis, S., Ciegler, A. and Ajl, S. J. Academic Press, New York, N.Y.

HALD, B. and P. KROGH. 1972. Ochratoxin residues in bacon pigs. Abstr. IUPAC-sponsored symp. "Control of mycotoxins", Goeteborg, Sweden, 18.

HARRI, E., W. LOEFFLER, H. P. SIGG, H. STAHELIN, and C. TAMM. 1963. Uber die isolierung neuer stoffwechselprodukte aus *Penicillium brefeldianum* Dodge. Helv. Chim. Acta 46:1235.

HARRIS, C. M., F. M. PORTER, and J. M. HARVEY. 1967. Freight shipment of California strawberries in mechanically refrigerated cars, test shipments. 1967. U.S. Dept. of Agr., ARS 51.

HARWIG, J. and Y. K. CHEN. 1974. Some conditions favoring production of ochratoxin A and citrinin by *Penicillium viridicatum* in wheat and barley. Can J. Plant Sci. 54:17.

HARWIG, J., Y. K. CHEN, B. P. C. KENNEDY, and P. M. SCOTT. 1973a. Occurrence of patulin and patulin-producing strains of *Penicillium expansum* in natural rots of apple in Canada. Can. Inst. Food Sci. Technol. J. 6:22.

HARWIG, J., P. M. SCOTT, B. P. C. KENNEDY, and Y. K. CHEN. 1973b. Disappearance of patulin from apple juice fermented by Saccharomyces sp. Can. Inst. Food Sci. Technol. J. 6:45.

HATSUDA, Y., T. HAMASAKI, M. ISHIDA, K. MATSUI, and S. HARA. 1972. Dihydrosterigmatocystin and dihydrodemethyl-sterigmatocystin, new metabolites from *Aspergillus versicolor*. Agr. Biol. Chem. 36:521.

HESSELTINE, C. W. 1976. Mycotoxins other than aflatoxins. In Proc. 3rd Int. Biodegradation Symp. Eds. J. M. Sharpley and A. M. Kaplan. Applied Science Publishers Ltd., London.

HESSELTINE, C. W., R. J. BOTHAST, and O. L. SHOTWELL. 1975. Aflatoxin occurence in some white corn under loan, 1971: IV. Mold flora. Mycologia 67:392.

HESSELTINE, C. W., O. L. SHOTWELL, W. F. KWOLEK, E. B. LILLEHOJ, W. K. JACKSON, and R. J. BOTHAST. 1976. Aflatoxin occurrence in 1973 corn at harvest. II. Mycological studies. Mycologia 68:341.

HOLKER, J. S. E., and S. A. KAGAL. 1968. 5-Methoxysterigmatocystin, a metabolite from a mutant strain of *Aspergillus versicolor*. Chem. Commun. 24:1574.

HOLKER, J. S. E., and L. J. MULHEIRN. 1968. The biosynthesis of sterigmatocystin. Chem. Commun. 24:1576.

HOLKER, J. S. E., and J. G. UNDERWOOD. 1964. A synthesis of a cyclopentenocoumarin structurally related to aflatoxin B. Chem. Ind. 1865.

HOLZAPFEL, C. W. 1968. The isolation and structure of cyclopiazonic acid, a toxic metabolite of *Penicillium cyclopium* westling. Tetrahedron 24:2101.

HOOD, R. D., J. E. INNES, and A. W. HAYES. 1973. Effects of rubratoxin B on prenatal development in mice. Bull. Env. Contam. Toxicol. 10:200.

HOU, C. T., A. CIEGLER, and C. W. HESSELTINE. 1971a. Tremorgenic toxins from Penicillia. III. Tremortin production by *Penicillium* species on various agricultural commodities. Appl. Microbiol. 21:1101.

HOU, C. T., A. CIEGLER, and C. W. HESSELTINE. 1971b. Tremorgenic toxins from penicillia. II. A new tremorgenic toxins, tremortin B, from *Penicillium palitans*. Can. J. Microbiol. 17:599.

HOWARTH, B. JR., and R. D. WYATT. 1976. Effect of dietary aflatoxin on fertility, hatchability and progeny performance of broiler breeder hens. Appl. Environ. Microbiol. 31:680.

HOWITZ, W. (Editor). 1975. Official Methods for Analysis. Association of Official Analytical Chemists, Washington, D.C.

HSIEH, D. P. H. 1973. Inhibition of aflatoxin biosynthesis by dichlorovos. J. Agr. Food Chem. 21:468.

HSIEH, D. P. H., M. T. LIN, and R. C. YAO. 1973. Conversion of sterigmatocystin to aflatoxin B1. Biochem. Biphys. Res. Commun. 52:992.

HSU, I. C., E. B. SMALLEY, F. M. STRONG, and W. B. RIBELIN. 1972. Identification of T-2 toxin in moldy corn associated with a lethal toxicosis in dairy cattle. Appl. Microbiol. 24:684.

JEMMALI, M. 1973. Presence d'nn facteur Oestrogenique d'origine fongique la zearalenone ou F-2 comme contaminant Naturel, dans du Mais. Ann. Microbiol. (Inst. Pasteur). 124B:109.

JENKINS, P. T., and C. REINGANUM. 1965. The occurrence of a quiescent infection of stone fruits caused by *Sclerotinia fructicola* (Wint.) Rehm. Aust. J. Agr. Res. 16:131.

KABLE, P. F. 1971. Significance of short term latent infections in control of brown rot of peach fruits. Phytopathol. Z. 70:173.

KAPLAN, M. A., I. R. HOOPER, and B. HEINEMANN. 1954. Antibiotic-producing properties of *Aspergillus terreus*. Antibiot. Chemotherapy 4:746.

KATZNELSON, H. and C. A. JAMIESON. 1952. Control of Nosema disease of honeybees with fumagillin. Science 115:70.

KIERMEIER, F., G. WEISS, G. BEHRINGER, and M. MILLER. 1977.

Uber das Vorkommen und den Gehalt von Aflatoxin M_1 in Kasen des Handels. Z. Lebensm. Unters-Forsch. 163:268

KOENIG, W. and G. GEIGER. 1969. Novel synthesis of antamanide. Ann. Chem. 727:125.

KRAYBILL, H. F. and M. B. SHIMKIN. 1964. Carcinogenesis related to foods contaminated by processing and fungal metabolites. Adv. Cancer Res. 8:191.

KROGH, P. 1969. The pathology of mycotoxicoses. J. Stored Prod. Res. 5:259.

KROGH, P., B. HALD, P. GJERTSEN, and F. MYKEN. 1974. Fate of ochratoxin A and citrinin during malting and brewing experiments. Appl. Microbiol. 28:31.

KROGH, P., B. HALD and J. PEDERSEN. 1973. Occurrence of ochratoxin A and citrinin in cereals associated with mycotoxic porcine nephropathy. Acta. Path. Microbiol. Scand. B. 81:689.

KROGH, P., E. HASSELAGER, and P. FRIIS. 1970. Studies on fungal nephrotoxicity. 2. Isolation of two nephrotoxic compounds from *Penicillium viridicatum* Westling: Citrinin and oxalic acid. Acta Pathol. Microbiol. Scand. B. 78:401.

KUEHN, H. H. 1958. A preliminary survey of the Gymnoascaceae. Mycologia 50:417.

KUSKE, H. 1938. Photosensitization of the skin by plant substances. I. Light sensitization by furocoumarin as a cause of various plant dermatoses. Arch. Dermatol. Syphilis. 178:112.

KUSKE, H. 1940. Percutaneous photosensitization by active substances and extracts from plants. Dermatologica 82:273.

LANCASTER, M. C. 1968. Comparative aspects of aflatoxin induced hepatic tumors. Cancer Res. 28:2288.

LANCASTER, M. C., F. P. JENKINS, and J. M. PHILP, K. SARGEANT, A. SHERIDAN and J. O'KELLY. 1961. Toxicity associated with certain samples of ground nuts. Nature 192:1095.

LAPRADE, J. C. and A. MANWILLER. 1977. Relation of insect damage, vector, and hybrid reaction to aflatoxin B1 recovery from field corn. Phytopathol. 67:544.

LASZTITY, R. and L. WOLLER. 1975. Effects of zearalenone and some derivatives on animals fed on contaminated fodder. Acta Alimentaria 4:189.

LEVI, C. P., H. L. TRENK, and H. K. MOHR. 1974. Study of the occurrence of ochratoxin A in green coffee beans. J. AOAC. 57:866. 57:866.

LILLEHOJ, E. B. 1973. Feed sources and conditions conducive to production of aflatoxin, ochratoxin, Fusarium toxins, and zearalenone. J. Am. Vet. Med. Ass. 163:1281.

LILLEHOJ, E. B., D. I. FENNELL and S. HARA. 1975a. Fungi and aflatoxin in a bin of stored white maize. J. Stored Prod. Res. 11:47.

LILLEHOJ, E. B., D. I. FENNELL, and W. F. KWOLEK. 1976. *Aspergillus flavus* and aflatoxin in Iowa corn before harvest. Science 193:495.

LILLEHOJ, E. B., W. F. KWOLEK, D. I. FENNELL and M. S. MIL-

BURN. 1975b. Aflatoxin incidence and association with brigh greenish-yellow fluorescence and insect damage in a limited survey of freshly harvested high moisture corn. Cereal Chem. 52:403.

LILLEHOJ, E. B., M. S. MILBURN, and A. CIEGLER. 1972. Control of *Penicillium martensii* development and penicillic acid production by atmospheric gases and temperatures. Appl. Microbiol. 24:198.

LOOSMORE, R. M. and L. M. MARKSON. 1961. Poisoning of cattle by Brazilian groundnut meal. Vet. Record 73:813.

LOOSMORE, R. M. and J. D. J. HARDING. 1961. A toxic factor in Brazilian groundnut causing liver damage in pigs. Vet. Record 73:1362.

LOVELACE, C. E. A. and C. B. NYATHI. 1977. Estimation of the fungal toxins, zearalenone and aflatoxin, contaminating opaque maize beer in Zambia. J. Sci. Food Agr. 28:288.

LOVETT, J. and R. G. THOMPSON. 1973. Low temperature patulin production by *Aspergillus* and *Penicillium* species. Bacteriol Proc. 1973, E71 (Abstract).

LUK, K. C., B. KOBBE, and J. M. TOWNSEND. 1977. Production of cyclopiazonic acid by *Aspergillus flavus* Link. Appl. Environ. Microbiol. 33:211.

LUKEZIC, F. L., W. J. KAISER, and M. M. MARTINEZ. 1967. The incidence of crown rot of boxed bananas in relation to microbial populations of the crown tissue. Can. J. Bot. 45:413.

LYNEN, F. and U. WIELAND. 1937. Toxins of *Amanita* species. IV. Ann. Chem. 533:93.

MACDONALD, J. C. 1973. Toxicity, analysis, and production of Aspergillic acid and its analogues. Can. J. Biochem. 51:1311.

MAJUMDAR, S. K., K. S. NARASIMHAN, and H. A. B. PARPIA. 1965. Microecological factors of microbial spoilage and the occurrence of mycotoxins in stored grains. IN "Mycotoxins in Foodstuffs," Ed., Wogan, G. N. M.I.T. Press, Cambridge, MA.

MALLA, D. A., J. F. DIEHL, and D. K. SALUNKHE. 1967. In vitro susceptibility of strains of *Penicillium viridicatum* and *Aspergillus flavus* to β-radiation. Experientia 23:492.

MANN, G. E., L. P. CODIFER, JR., H. K. GARDNER, JR., S. P. KOLTUN, and F. G. DOLLEAR. 1970. Chemical inactivation of aflatoxins in peanut and cottonseed meals. J. Am. Oil Chem. Soc. 47:173.

MAYER, C. F. 1953. Endemic panmyelotoxicosis in the Russian grain belt. Part I. The clinical aspects of alimentary toxic aleukia (ATA). A comprehensive review. Military Surgeon 113:173.

MAYER, V. W. and M. S. LEGATOR. 1969. Production of petite mutants of *Saccharomyces cerevisiae* by patulin. J. Agr. Food Chem. 17:454.

McCOLLOCH, L. P. and J. T. WORTHINGTON. 1952. Low temperatures as a factor in the susceptibility of mature-green tomatoes to *Alternaria* rot. Phytopathol. 42:425.

McCORNACK, A. A. 1972. Effect of ethylene degreening on decay of Florida citrus fruit. Proc. Florida State Hort. Soc. 84:270.

McGOWAN, J. C. 1946. A substance causing abnormal development of fungal hyphae produced by *Penicillium janczewskii* Zal. II. Preliminary notes on the chemical and physical properties of "curling factor." Brit. Mycol. Soc. Trans. 29:188.

McCOWEN, M. C., M. E. CALLENDER, and J. F. LAWLIS, JR. 1951. Fumagillin (H-3), a new antibiotic with amebicidal properties. Science 113:202.

McKINNEY, F. E. 1971. Food Irradiation: Potato sprout inhibition by radiation. Part I. Isotopes Radiation Technol. 8:187.

McMEANS, J. L. and L. J. ASHWORTH, JR. 1966. Preharvest occurrence of *Aspergillus flavus* and aflatoxins in California cotton seed. Phytopathol. 56:889.

MEISNER, H. and S. CHAN. 1974. Ochratoxin A, an inhibitor of mitochondrial transport systems. Biochemistry 13:2795.

MINTZLAFF, J. H., A. CIEGLER, and L. LEISTNER. 1972. Potential mycotoxin problems in mold-fermented sausage. Zeit. fur Leben. Unter. und-Forsch. 150:133.

MIROCHA, C. J., C. M. CHRISTENSEN, and G. H. NELSON. 1967. Estrogenic metabolite produced by *Fusarium graminearum* in stored corn. Appl. Microbiol. 15:497.

MIROCHA, C. J., C. M. CHRISTENSEN, and G. H. NELSON. 1969. Biosynthesis of the fungal estrogen F-2 and a naturally occurring derivative (F-3) by *Fusarium moniliforme*. Appl. Microbiol. 17:482.

MIYAKE, M. and M. SAITO. 1965. Liver injury and liver tumors induced by toxins of *Penicillium islandicum* growing on yellowed rice. In "Mycotoxins in foodstuffs." G. N. Wogan (ed.) M.I.T. Press, Cambridge, MA.

MORTON, H. E., W. KOCHOLATY, R. JUNOWICZ-KOCHOLATY, and A. KELNER. 1945. Toxicity and antibiotic activity of kojic acid produced by *Aspergillus luteo-virescens*. J. Bacteriol. 50:579.

MOULE, Y. and F. HATEY. 1977. Mechanism of the *in vitro* inhibition of transcription by patulin, A mycotoxin from *Byssochlamys nivea*. FEBS Letters 74:121.

NAGARAJAN, V. and R. B. BHAT. 1972. Factors responsible for varietal differences in aflatoxin production in maize. J. Agr. Food Chem. 20:911.

NATARAJAN, K. R., K. C. RHEE, C. M. CATER, and K. F. MATTIL. 1975. Destruction of aflatoxins in peanut protein isolates by sodium hypochlorite. J. Am. Oil Chem. Soc. 52:160.

NESBITT, B. F., J. O'KELLY, K. SARGEANT and A. SHERIDAN. 1962. Toxic metabolites of *Aspergillus flavus*. Nature 195:1062.

NESHEIM, S. 1969. Isolation and purification of ochratoxins A and B and preparation of their methyl and ethyl esters. J. AOAC. 52:975.

NORSTADT, F. A. and T. M. McCALLA. 1971. Growth and patulin formation by *Penicillium urticae* Banier in pure and mixed cultures. Plant and Soil 34:97.

OKUDA, S., N. IWASAKI, K. TSUDA, Y. SANO, T. HATA, S. UDAGAWA, Y. NAKAYAMA, and H. YAMAGUCHI. 1964. Helvolic acid. Chem. Pharm. Bull. 12:121.

PAGET, G. E. and A. I. WALPOLE. 1958. Some cytological effects of griseofulvin. Nature 182:1320.

PARPIA, H. A. B. and V. SREENIVASAMURTHY. 1970. Importance of aflatoxins in foods with respect to India. In "Third International Congress-Food Science and Technology; Symposium Proceedings," p. 701. Institute of Food Technologists, Chicago, IL.

PATHAK, M. A., F. DANIELS, JR., and T. B. FITZPATRICK. 1962. The presently known distribution of furocoumarins (psoralens) in plants. J. Invest. Dermatol. 39:225.

PATHRE, S. V., C. J. MIROCHA, C. M. CHRISTENSEN and J. BEHRENS. 1976. Monoacetoxysirpenol, a new mycotoxin produced by *Fusarium roseum* Gibbosum. J. Agr. Food Chem. 24:97.

PERCIVAL, J. C. 1959. Photosensitivity diseases in New Zealand. XVII. The association of *Sporidesmium bakeri* with facial eczema. New Zealand J. Agr. Res. 2:1040.

PETTERSSON, G. 1963. Toluquinones from *Aspergillus fumigatus*. Acta Chem. Scand. 17:1771.

PHILLIPS, D. J., M. UOTA, D. MONTICELLI and C. CURTIS. 1976. Colonization of almond by *Aspergillus flavus*. J. Am. Soc. Hort. Sci. 101:19.

POLLOCK, A. V. 1947. Production of citrinin by five species of *Penicillium*. Nature 160:331.

POWELSON, R. L. 1960. Initiation of strawberry fruit rot caused by *Botrytis cinerea*. Phytopathol. 50:491.

PROX, A., J. SCHMID, and H. OTTENHEYM. 1969. Sequence analysis by combined gas chromatography and mass spectrometry. II. Structure of antamanide. Ann. Chem. 722:179.

PURCHASE, I. F. H. 1971. The acute toxicity of the mycotoxin cyclopiazonic acid to rats. Toxicol. Appl. Pharmacol. 18:114.

PURCHASE, I. F. H. (Editor). 1974. Mycotoxins. Elsevier Scientific Publishing Company, Amsterdam, Holland.

PURCHASE, I. F. H., and J. J. VAN DER WATT. 1968. Carcinogenicity of sterigmatocystin. Food Cosmet. Toxicol. 6:555.

PURCHASE, I. F. H., and J. J. VAN DER WATT. 1970. Carcinogenicity of sterigmatocystin. Food Cosmet. Toxicol. 8:289.

RAMSEY, G. B. 1938. Fruit and vegetable diseases on the Chicago market in 1937. Plant Disease Reptr. Suppl. 106:62.

RAO, H. R. G. and P. K. HAREIN. 1972. Inhibition of aflatoxin and zearalenone biosynthesis by dichlorovos. Bull. Envir. Contam. and Toxicol. 10:112.

RAO, K. S. and P. G. TULPULE. 1967. Varietal differences of groundnut in the production of aflatoxin. Nature 214:738.

RHODES, A. 1963. Griseofulvin: production and biosynthesis. Progr. Ind. Microbiol. 4:165.

ROBERTS, B. A. and D. S. P. PATTERSON. 1975. Detection of twelve mycotoxins in mixed animal feedstuffs, using a novel membrane cleanup procedure. *J. AOAC*. 58:1178.

ROBBERS, J. E., M. G. MARCELLO, W. W. CARLTON, and J. F. TUITE. 1973. The isolation and identification of brevianamide A from corn cultures of *Penicillium viridicatum* (Abstract). Lloydia 36:440.

ROBBERS, J. E., J. W. STRAUS, and J. TUITE. 1975. The isolation of brevianamide A from *Penicillium ochraceum*. Lloydia 38:355.

RODRICKS, J. V. 1969. Fungal metabolites which contain substituted 7,8-dihydrofuro-(2,3-b)furan (DHFF) and 2,3,7,8,-tetrahydrofuro (2,3-b),furans (THFF). J. Agr. Food Chem. 17:457.

RODRICKS, J. V., E. LUSTIG, A. D. CAMPBELL, and L. STOLOFF. 1968a. Aspertoxin, a hydroxy derivative of o-methylsterigmatocystin from aflatoxin-producing cultures of *Aspergillus flavus*. Tetrahedron Lett. 25:2975.

RODRICKS, J. V., K. R. HENERY-LOGAN, A. D. CAMPBELL, L. STOLOFF, and M. J. VERETT. 1968b. Isolation of a new toxin from cultures of *Aspergillus flavus*. Nature 217:668.

ROINE, K., E. L. KORPINEN and K. KALLELA. 1971. Mycotoxicosis as a probable cause of infertility in dairy cows. Nord. Vet. Med. 23:628.

ROTHLIN, E. and A. CERLETTI. 1954. Pharmacological principles of therapy with *Veratrum* alkaloids. Schweiz. Med. Wochenschr 84:137.

SAITO, M. 1959. Liver cirrhosis induced by metabolites of *Penicillium islandicum*. Acta Pathol. (Japan) 9:785.

SAITO, M., M. ENOMOTO and T. TATSUNO. 1971. Yellow rice toxins. Luteoskyrin and related compounds, chlorine-containing compounds and citrinin. In "Microbial toxins." Vol. VI. Fungal Toxins. A. Ciegler. S. Kadis, and S. J. Ajl (eds.). Acad. Press. New York, N.Y.

SANDS, D. C., J. L. McINTYRE, and G. S. WALTON. 1976. Use of activated charcoal for the removal of patulin from cider. Appl. Environ. Microbiol. 32:388.

SARGEANT, K., A. SHERIDAN, J. O'KELLY and R. B. A. CARNAGHAN. 1961. Toxicity associated with certain samples of groundnuts. Nature 192:1096.

SASAKI, M., Y. ASAO, and T. YOKOTSUKA. 1968. Compounds produced by molds. V. Isolation of nonfluorescent pyrazines. 2. Nippon Nogei Kagaku Kaishi. 42:351.

SCHEEL, L. D., V. B. PERONE, R. L. LARKIN, and R. E. KUPEL. 1963. The isolation and characterization of two phototoxic furanocoumarins (psoralens) from diseased celery. Biochemistry 2:1127.

SCHINDEL, L. 1954. Treatment of amebiasis with fumagillin. J. Am. Med. Assoc. 155:903.

SCHINDLER, A. F., A. N. ABADIE, J. S. GECAN, P. B. MISLIVEC and P. M. BRICKEY. 1974. Mycotoxins produced by fungi isolated from inshell pecans. J. Food Sci. 29:213.

SCHOFIELD, F. W. 1924. Damaged sweet clover: the cause of a new disease in cattle simulating hemorrhagic septicemia and blackleg. Am. Vet. Med. Assoc. J. 64:553.

SCHULLER, P. L., T. OCKHUIZEN, J. WERRINGLOER, and P. MARQUARDT. 1967. Aflatoxin B1 und Histamin in Wein. Arzneimettel-Forsch. 17:888.

SCHULTZ, J. and R. MOTZ. 1967. Properties of aflatoxins and the possibility of their detection. Arch. Exptl. Veterinaermed 21:129.

SCOTT, DE. B. 1965. Toxigenic fungi isolated from cereal and legume products. Mycopathol. Mycol. Appl. 25:213.

SCOTT, P. M. 1973. Mycotoxins in stored grain, feeds, and other cereal products. IN "Grain Storage: Part of a system." Eds., R. N. Sinha and W. E. Muir. Avi Publishing Co., Westport, CT.

SCOTT, P. M., W. F. MILES, P. TOFT and J. G. DUBE. 1972. Occurrence of patulin in apple juice. J. Agr. Food Chem. 20:450.

SCOTT, P. M. and E. SOMMERS. 1968. Stability of patulin and penicillic acid in fruit juice and flour. J. Agr. Food Chem. 16:483.

SCOTT, P. M., W. VAN WALBEEK, J. HARWIG, and D. I. FENNELL. 1971. Occurrence of a mycotoxin, ochratoxin A, in wheat and isolation of ochratoxin A and citrinin-producing strains of *Penicillium viridicatum*. Can. J. Plant Sci. 50:583.

SCOTT, P. M., W. VAN WALBEEK, B. KENNEDY, and D. ANYETI. 1972. Mycotoxins (ochratoxin A, citrinin, and sterigmatocystin) and toxigenic fungi in grains and other agricultural products. J. Agr. Food Chem. 20:1103.

SHANK, R. C. and G. N. WOGAN. 1972. Dietary aflatoxins and human liver cancer. 1. Toxigenic molds in foods and foodstuffs of tropical Southeast Asia. Food Cosmet. Toxicol. 10:51.

SHANK, R. C., G. N. WOGAN, J. B. GIBBSON, and A. NONDASUTA. 1972. Dietary aflatoxins and human liver cancer. 2. Aflatoxins in market foods and foodstuffs of Thailand and Hong Kong. Food Cosmet. Toxicol. 10:61.

SHANK, R. C., J. E. GORDON, G. N. WOGAN, A. NONDASUTA, and B. SUBHAMANI. 1972. Dietary aflatoxins and human liver cancer. 3. Field survey on rural Thai families for ingested aflatoxins. Food Cosmet. Toxicol. 10:71.

SHANK, R. C., N. BHAMARAPRAVATI, J. E. GORDON, and G. N. WOGAN. 1972. Dietary aflatoxins and human liver cancer. 4. Incidence of primary liver cancer in two municipal populations of Thailand. Food Cosmet. Toxicol. 10:171.

SHANK, R. C., P. SIDDHICHAI, B. SUBHAMAI, N. BHAMARAPRAVATI, J. E. GORDON, and G. N. WOGAN. 1972. Dietary aflatoxins and human liver cancer. 5. Duration of primary liver cancer and prevalence of hepotomegaly in Thailand. Food Cosmet. Toxicol. 10:181.

SHEEHAN, J. C., W. B. LAWSON, and R. J. GAUL. 1958. The structure of terreic acid. J. Am. Chem. Soc. 80:5536.

SHIBATA, S., S. NATORI, and S. UDAGANA. 1964. List of Fungal Products. Univ. of Tokyo Press, Tokyo, Japan.

SHOTWELL, O. L., C. W. HESSELTINE, and M. L. GOULDEN. 1973. Incidence of aflatoxin in southern corn 1969–1970. Cereal Sci. Today 18:192.

SHOTWELL, O. L., M. L. GOULDEN, and C. W. HESSELTINE. 1976.

Aflatoxin M₁. Occurrence in stored and freshly harvested corn. J. Agr. Food Chem. 24:683.

SHOTWELL, O. L., C. W. HESSELTINE, H. R. BURMEISTER, W. F. KWOLEK, G. M. SHANNON, H. H. HALL, G. M. GOULDEN, E. B. VANDERGRAFT, and M. S. MILBURN. 1969a. Survey of cereal grains and soybeans for the presence of aflatoxin: I. Wheat, grain sorghum, and oats. Cereal Chem. 46:446.

SHOTWELL, O. L., C. W. HESSELTINE, H. R. BURMEISTER, W. F. KWOLEK, G. M. SHANNON, H. H. HALL, G. M. GOULDEN, E. B. VANDERGRAFT, and M. S. MILBURN. 1969b. Survey of cereal grains and soybeans for the presence of aflatoxins. II. Corn and soybean. Cereal Chem. 46:454.

SHOTWELL, O. L., C. W. HESSELTINE and M. L. GOULDEN. 1969c. Ochratoxin A: occurrence as a natural contaminant of a corn sample. Appl. Microbiol. 17:765.

SHOTWELL, O. L., C. W. HESSELTINE, M. L. GOULDEN and E. E. VANDEGRAFT. 1970. Survey of corn for aflatoxin, zearalenone and ochratoxin. Cereal Chem. 47:700.

SHOTWELL, O. L., C. W. HESSELTINE, E. E. VANDEGRAFT, and M. L. GOULDEN. 1971. Survey of corn-from different regions for aflatoxin, ochratoxin and zearalenone. Cereal Sci. Today 16:266.

SHOTWELL, O. L., G. M. SHANNON, M. L. GOULDEN, M. S. MILBURN, and H. H. HALL. 1968. Factors in oats that could be mistaken for aflatoxin. Cereal Chem. 45:236.

SINGH, L. B. 1960. The Mango. Interscience Publishers, New York. N.Y.

SINGH, R. and D. P. H. HSIEH. 1976. Enzymatic conversion of sterigmatocystin into aflatoxin B₁ by cell-free extracts of *Aspergillus parasiticus*. Appl. Environ. Microbiol. 31:743.

SINGH, R. and D. P. H. HSIEH. 1977. Aflatoxin biosynthetic pathway: Elucidation by using blocked mutants of *Aspergillus parasiticus*. Arch. Biochem. Biphys. 178:285.

SINGLETON, V. L., N. BOHONOS, and A. J. ULLSTRUP. 1958. Decumbin, a new compound from a species of *Penicillium*. Nature 181:1072.

SINNHUBER, R. O., J. H. WALES, R. H. ENGELBRECHT, D. E. AMEND, J. L. AYRES, W. F. ASHTON, and W. KRAY. 1965. Aflatoxins in cottonseed meal and hepatoma rainbow trout. Fed. Proc. 24:627.

SMALLEY, E. B. 1973. T-2 toxins. J. Am. Vet. Med. Assoc. 163:1278.

SMITH, R. H. and W. McKERNAN. 1962. Hepatotoxic action of chromatographically separated fractions of *Aspergillus flavus* extracts. Nature 195:1301.

SMOOT, J. J. 1969. Decay of Florida citrus fruits stored in controlled atmospheres and in air. Proc. 1st Intern. Citrus Symp., Univ. Calif. (Riverside) 3:1285.

SOMMER, N. F., J. R. BUCHANAN, and R. J. FORTLAGE. 1974. Production of patulin by *Penicillium expansum*. Appl. Microbiol. 28:589.

SPECTOR, W. S. 1957. Handbook of Toxicology. Vol. II. Sanders, Philadelphia, PA.

SPEERS, G. M., C. J. MIROCHA, C. M. CHRISTENSEN, and J. C.

BEHRENS. 1977. Effects on laying hens of feeding corn invaded by two species of *Fusarium* and pure T-2 mycotoxin. Poultry Sci. 56:98.

SPENSLEY, P. C. 1963. Aflatoxins, the active principle in turkey "X" disease. Endeavour 22:75.

SREENIVASAMURTHY. V., H. A. B. PARPIA, S. SRIKANTA, and A. SHANKER. 1967. Detoxification of aflatoxin in peanut meal by hydrogen peroxide. J. AOAC. 50:350.

STANSFELD, J. M., A. E. FRANCIS, and C. H. STUART-HARRIS. 1944. Laboratory and clinic trials of patulin. Lancet 247:370.

STERN, P. 1971. Pharmacological analysis of the tremor induced by Cyclopium toxin. Yugoslav. Physiol. Pharmacol. Acta. 7:187.

STOB, M., R. S. BALDWIN, J. TUITE, F. N. ANDERSON and K. G. GILLETTE. 1962. Isolation of an anabolic, uterotrophic compound from corn infested with *Gibberella zeae*. Nature 196:1318.

STOLOFF, L. 1976. Occurrence of mycotoxins in foods and feeds. In "Mycotoxins and Other Fungal Related Food Problems," Ed., J. V. Rodricks. Advances in chemistry series 149, Amer. Chem. Soc., Washington, D.C.

STOLOFF, L. and W. TRAGER. 1965. Recommended decontamination procedures for aflatoxins. *J. AOAC*. 48:681.

SWAMINATHAN, B. and P. E. KOEHLER. 1976. Isolation of an inhibitor of *Aspergillus parasiticus* from white potatoes (*Solanum tuberosum*). J. Food Sci. 41:313.

SZCZECH, G. M., W. W. CARLTON and W. W. TUITE. 1973. Ochratoxicosis in beagle dogs. II. Pathology J. Vet. Pathol. 10:219.

TABER, R. A. and H. W. SCHROEDER. 1967. Aflatoxin producing potential of isolates of the *Aspergillus-flavus-oryzae* group from peanuts (*Arachis hypohaea*). Appl. Microbiol. 15:140.

TATSUNO, T., M. TSUKIOKA, Y. SAKAI, Y. SUZUKI, and Y. ASAMI. 1955. The toxic substance in yellowed rice. Pharm. Bull. (Japan) 3:476.

THORPE, C. W. and R. L. JOHNSON. 1974. Analysis of penicillic acid by gas-liquid chromatography. J. AOAC. 57:861.

TOWNSEND, R. J., M. O. MOSS, and H. M. PECK. 1966. Isolation and characterization of hepatatotoxins from *Penicillium rubrum*. J. Pharm. Pharmacol. 18:471.

TRAGER, W. and L. STOLOFF. 1967. Possible reactions for aflatoxin detoxification. J. Agr. Food Chem. 15:679.

TRENK, H. L., M. E. BUTZ, and F. S. CHU. 1971. Production of ochratoxins in different cereal products by *Aspergillus ochraceus*. Appl. Microbiol. 21:1032.

TSUNODA, H. 1951. Studies on a poisonous substance produced on the cereals by a *Penicillium* Spp. under storage. Jap. J. Nutr. 8:185.

TUITE, J., G. SHANER, G. RAMBO, J. FOSTER, and R. W. CALDWELL. 1974. The *Gibberella* ear rot epidemics of corn in Indiana in 1965 and 1972. Cereal Sci. Today 19:238.

TUNG, T. C. and K. H. LING. 1968. Study on aflatoxin of foodstuffs in Taiwan. J. Vitaminol. 14:48.

TYLER, V. E., JR., L. R. BRADY, R. G. BENEDICT, J. M. KHANNA, and M. H. MALONE. 1963. Chromatographic and pharmacologic evaluation of some toxic *Galerina* species. Lloydia 26:154.

VAN DER ZIDJEN, A. S. M., W. A. A. B. KOELENSMID, J. BOLDINGH, C. B. VARETT, W. Q. ORD, and J. PHILP. 1962. *Aspergillus flavus* and turkey X disease. Nature 195:1060.

VAN VEEN, A. G., D. C. W. GRAHAM and K. H. STEINKRAUS. 1967. Fermented peanut presscake. Cereal Sci. Today 12:109.

VAN WALBEEK, W. 1973. Fungal toxins in foods. Can. Inst. Food Sci. Technol. J. 6:96.

VAN WALBEEK, W., P. M. SCOTT, J. HARWIG, and J. W. LAWRENCE. 1969. *Penicillium viridicatum* Westling: a new source of ochratoxin A. Can. J. Microbiol. 15:1281.

VESONDER, R. F., A. CIEGLER, and A. H. JENSEN. 1973. Isolation of the emetic principle from *Fusarium*-infected corn. Appl. Microbiol. 26: 1008.

WAISS, A. C., JR., M. WILEY, D. R. BLACK, and R. E. LUNDIN. 1968. 3-Hydroxy-6,7-dimethoxydifuroxanthone-a new metabolite from *Aspergillus flavus*. Tetrahedron Letters 28:3207.

WAKSMAN, S. A., E. S. HORNING, and E. L. SPENCER. 1943. Two antagonistic fungi, *Aspergillus fumigatus* and *Aspergillus clavatus*, and their antibiotic substances. J. Bacteriol. 45:233.

WALKER, J. R. L. 1969. Inhibition of the apple phenolase systems through infection by *Penicillium expansum*. Phytochem. 8:561.

WARE, G. M., C. W. THORPE, and A. E. POHLAND. 1974. A liquid chromatographic method for the determination of patulin in apple juice. J. AOAC. 57:1111.

WEINBERG, J. H. 1975. Blighted bounty: a twist in the staff of life. Sci. News 107:12.

WELLS, J. M. and J. M. HARVEY. 1970. Combination heat and 2,6-dichloro-4-nitroaniline treatments for control of *Rhizopus* and brown rot of peaches, plums, and nectarines. Phytopathol. 60:116.

WHITE, E. C. 1940. Bactericidal filtrate from a mold culture. Science 92:127.

WHITE, E. C. and J. H. HILL. 1943. Studies on antibacterial products formed by molds. I. Aspergillic acid, a product of a strain of *Aspergillus flavus*. J. Bacteriol. 45:433.

WIELAND, T. 1967. The toxic peptide of Amanita phalloides. Fortschr. Chem. Org. Naturestoffe. 25:214.

WIELAND, T., G. LUEBEN, H. OTTENHEYM, and H. SCHIEFER. 1969. Constituents of *Amanita phalloides*. XXXIX. Isolation and characterization of an antitoxic cyclopeptide, antamanide, from the lipophilic fraction. Ann. Chem. 722:173.

WIELAND, T., G. LEUBEN, H. OTTENHEYM, J. FAESEL, J. X. DE VRIES, W. KONZ, A. PROX. and J. SCHMID. 1968. The discovery, isolation, elucidation of structure and synthesis of antaminide. Angew. Chem. Intern. Ed. Engl. 7:204.

WIELAND, T., and O. WIELAND. 1959. Chemistry and toxicology of the toxins of *Amanita phalloides*. Pharmacol. Rev. 11:87.

WILKINS, W. H., and G. C. M. HARRIS. 1942. Investigation into the production of bacteriostatic substances by fungi. I. Preliminary examination of 100 fungal species. Brit. J. Exp. Pathol. 23:166.

WILSON, B. J. 1971. Miscellaneous Aspergillus toxins. In "Microbial Toxins." VI. Eds., A. Ciegler, S. Kadis, and S. J. Ajl. Academic Press, New York, N.Y.

WILSON, B. J., T. HOEKMAN, and W. D. DETTBARN. 1972. Effects of a fungus tremorgenic toxin (penitrem A) on transmission in rat phrenic nerve-diaphragm preparation. Brain Res. 40:540.

WILSON, B. J. and C. H. WILSON. 1962. Extraction and preliminary characterization of a hepatotoxic substance from cultures of *Penicillium rubrum*. J. Bacteriol. 84:283.

WILSON, B. J., C. H. WILSON, and A. W. HAYES. 1968. Tremorgenic toxin from *Penicillium cyclopium* grown on food materials. Nature 220: 77.

WILSON, B. J., D. T. C. YANG and T. M. HARRIS. 1973. Production, isolation and preliminary toxicity studies of brevianamide A from cultures of *Penicillium viridicatum*. Appl. Microbiol. 26:633.

WILSON, D. M. and E. J. NUOVO. 1973. Patulin production in apples decayed in *Penicillium expansum*. Appl. Microbiol. 26:124.

WILSON, D. M., W. H. TABOR, and M. W. TRUCKESS. 1976. Screening method for the detection of aflatoxin, ochratoxin, zearalenone, penicillic acid, and citrinin. *J. AOAC*. 59:125.

WOGAN, G. N. and P. M. NEWBERNE. 1967. Dose-response characteristics of aflatoxin B1 carcinogenesis in the rats. Cancer Res. 27:2370.

WOLF, H. and E. W. JACKSON. 1963. Hepatomas in rainbow trout: descriptive and experimental epidemiology. Science 142:676.

WOLF, J. C., J. R. LIEBERMAN, and C. J. MIROCHA. 1972. Inhibition of F-2 (zearalenone) biosynthesis and perithecium production in *Fusarium roseum* Graminearum. Phytopathol 62:937.

WOLF, J. C., and C. J. MIROCHA. 1973. Regulation of sexual reproduction in *Gibberella zeae* (*Fusarium roseum* Graminearum) by F-2 (Zearalenone). Can. J. Microbiol. 19:725.

WU, M. T., P. E. KOEHLER, and J. C. AYRES. 1972. Isolation and identification of xanthotoxin (8-methoxypsoralen) and bergapten (5-methoxy-psoralen) from celery infected with *Sclerotinia sclerotiorum*. Appl. Microbiol. 23:852.

WU, M. T. and J. C. AYRES. 1974. Effects of dichlorvos on ochratoxin production by *Aspergillus ochraceus*. J. Agr. Food Chem. 22:536.

WU, M. T., J. C. AYRES, and P. E. KOEHLER. 1974a. Production of citrinin by *Penicillium viridicatum* on country-cured ham. Appl. Microbiol. 27:427.

WU, M. T., J. C. AYRES, and P. E. KOEHER. 1974b. Toxigenic aspergilli and penicillia isolated from aged cured meats. Appl. Microbiol. 28:1094.

WU, M. T., J. C. AYRES, P. E. KOEHLER, and G. CHASSIS. 1974. Toxic metabolite produced by *Aspergillus wentii*. Appl. Microbiol. 27:337.

YABUTA, T. J. 1924. The constitution of kojic acid, a γ-pyrone derivative formed by *Aspergillus oryzae* from carbohydrates. J. Chem. Soc. 125:575.

YAMAZAKI, M., K. SASAGO, and K. MIYAKI. 1974. The structure of fumitremorgin B (FTB), a tremorgenic toxin from *Aspergillus fumigatus* Fres. Chem. Commun. 408.

YAMAZAKI, M., S. SUZUKI, and K. MIYAKI. 1971. Tremorgenic toxins from *Aspergillus fumigatus*. Chem. Pharm. Bull. 19:1739.

YAMAZOE, S., M. NAKAO, K. HAYASHI, K. NAGANO, T. MOTEGI, S. UESUNGI, and T. KANO. 1963. Effects of poisons of *Penicillium islandicum* on metabolism in isolated livers from animals. Gunma J. Med. Sci. 12:73.

YATES, A. R. 1974. The occurrence of *Byssochlamys* sp. molds in Ontario Can. Inst. Food Sci. Technol. J. 7:148.

YOKOTA, K., H. SHIMADA, A. KAMAGUCHI, and O. SAKAGUCHI. 1977. Studies on the toxin of *Aspergillus fumigatus*. Purification and some properties of hemolytic toxin (asp—hemolysin) from culture filtrates and mycelia. Microbiol. Immunol. 21:11.

YOSHIZAWA, T. and N. MOROOKA. 1973. Deoxynivalenol and its mono-acetate: New mycotoxins from *Fusarium roseum* and moldy barley. Agr. Biol. Chem. 37:2933.

8

Control of Foodborne Diseases

Frank L. Bryan

Factors that frequently contribute to outbreaks of foodborne disease are: improper cooling of foods, lapse of a day or more between preparing and serving, infected persons having touched foods which are not subsequently heat-processed, inadequate time or temperature or both during heat processing of foods, insufficiently high temperature during hot storage of foods, inadequate time or temperature or both during reheating of previously cooked foods, ingesting contaminated raw foods or raw ingredients. Therefore, control of foodborne diseases must be based on preventing these situations from occurring.

Activities for the prevention and control of foodborne diseases should include: epidemiologic surveillance of foodborne diseases, microbiological surveillance of foods, hazard analyses, critical control point designation and control action, training of professional and industrial personnel, and education of the public.

FACTORS THAT CONTRIBUTE TO FOODBORNE DISEASE OUTBREAKS

Epidemiologic and research data collected over the past century have demonstrated that the following sequence of events must occur for persons to get a foodborne disease: (1) the etiologic agent must be present either in citizens of a community, in food-source animals, or in the environment in which foods are grown, harvested, processed,

265

or stored; (2) the agent itself (or the organism that produces it, if it is one of several toxins) must contaminate a food during the growing period or during harvesting, processing, storage, or preparation; (3) then, one of the following events must happen: (a) the agents must be present on or in the contaminated food in sufficient numbers or concentration to survive the remainder of the growing period, storage, and processing and still cause illness: (b) bacteria on or in foods in insufficient numbers to cause illness must multiply and reach quantities or produce toxins in sufficient quantities to cause illness; (c) microorganisms, particularly bacteria, enter food preparation areas on or in raw foods, where they are transferred to workers' hands or to equipment surfaces, which if inadequately washed, will then contaminate other foods that they subsequently touch (and hence, if bacteria, multiply as described in 3b); (4) sufficient quantities of the contaminated food that contain enough of the agent to exceed a person's resistance-susceptibility threshold must be ingested. Ingestion of foods contaminated to this level can result in sporadic cases of illness as well as outbreaks. Whether or not outbreaks are detected depends on the number of persons who ingested the contaminated food and on the socio-cultural attitudes of the populace to report illness and the efficiency of a health agency to determine that the illness is foodborne and epidemiologically related to other cases. When numbers of pathogens insufficient to cause illness are ingested, an infected individual may become a carrier and may contaminate other foods that he touches.

In the United States, factors that have been frequently shown to contribute to outbreaks of foodborne disease are   listed in Table 8.1 (Bryan, 1978a).

FACTORS

The factors that most often contribute to outbreaks of foodborne disease vary, depending on the type of establishment in which foods are handled. In food service establishments during 1973 through 1976, the factors that have contributed to foodborne outbreaks are (in order of frequency of occurrence): inadequate cooling of foods, lapse of a day or more between preparing and serving foods, insufficiently high temperature during hot storage of foods, infected person ving touched foods which are not subsequently heat-processed,

TABLE 8.1.   THE MOST IMPORTANT FACTORS CONTRIBUTING TO THE OCCUR-RENCE OF 427 OUTBREAKS OF FOODBORNE DISEASE OCCURRING IN FOOD PROCESSING PLANTS, FOOD SERVICE ESTABLISHMENTS, AND HOMES, 1973 TO 1976, AND IN 1,152 OUTBREAKS OF FOODBORNE DISEASE FROM ALL PLACES OF MISHANDLING, 1961 TO 1976, BY RANK AND PERCENT

Contributing factor	Rank and Percent				
	Food service (235 outbreaks)	Homes (122 outbreaks)	1973–1976 Food processing (32 outbreaks)	Total* (427 outbreaks)	1961–1976 Total (1,152 outbreaks)
Improper cooling	1 (63)	1 (30)	3 (16)	1 (46)	1 (46)
Lapse of day or more between preparing and serving	2 (29)	5 (11)	(all)	2 (20)	2 (21)
Infected person	4 (26)	6 (8)	5 (9)	3 (18)	3 (20)
Inadequate thermal process, canning, or cooking	9 (5)	3 (21)	1 (25)	6 (11)	4 (16)
Improper hot storage	3 (27)	7 (6)		4 (16)	5 (16)
Inadequate reheating	5 (25)	9 (5)		5 (16)	6 (12)
Ingesting contaminated raw food or ingredient	11 (2)	2 (22)	1 (25)	7 (11)	7 (11)
Cross contamination	8 (6)	12 (2)		11 (4)	8 (7)
Inadequate cleaning of equipment	6 (9)	13 (1)		8 (6)	9 (7)
Obtaining foods from unsafe sources	13 (1)	5 (11)	6 (6)	11 (4)	10 (5)
Using leftovers	7 (7)	10 (4)		9 (5)	11 (4)
Toxic species mistaken for edible varieties		4 (13)		11 (4)	
Faulty fermentations		7 (6)	3 (16)	16 (2)	
Incidental additives				15 (2)	
Intentional additives	11 (2)				

Source: (Bryan, 1978a).
*Includes other outbreaks which do not fall in the other three classes.

inadequate time or temperature, or both, during reheating of pre-viously cooked foods, and other factors listed in Table 8.1.

The factors that occasioned outbreaks of foodborne diseases when foods were mishandled in homes during the same period are (in order of frequency of occurrence): inadequate cooling, inadequate time or temperature, or both, during canning or cooking, mistaking toxic species of mushrooms and other plants for edible varieties, obtaining foods from unsafe sources, lapse of a day or more between preparing and serving, and other factors listed in Table 8.1.

Factors that have contributed to the occurrence of outbreaks of foodborne disease which resulted from foods mishandled in food processing plants are: contaminated raw ingredient, inadequate heat processing, inadequate cooling and faulty fermentations. A detailed review has been prepared previously (Bryan, 1974, 1978a).

Foodborne disease outbreaks can be prevented by eliminating any of the factors that are necessary for outbreaks to occur. Those factors that epidemiologic evidence has indicated as the most frequent contributors to these outbreaks merit special efforts. Elimination of these factors calls for minimizing contamination, preventing growth of bacteria and molds, and killing microorganisms or destroying their toxins.

PROCEDURES THAT MINIMIZE CONTAMINATION

Measures for preventing and limiting contamination include attempts to keep the sources and reservoirs of pathogenic microorganisms or poisonous substances away from the environment where food is produced, processed, or prepared, or, if this is not possible, to prevent or minimize the dissemination of these microorganisms or substances from their source to foods. Accomplishment of these measures, however, is oftentimes difficult to achieve. This is so because agents that cause microbial foodborne illness have their natural reservoir in humans who process or prepare foods, in animals which serve as the source of food, in water and soil from which food is harvested, and because agents that cause chemical foodborne illnesses are sometimes used to promote growth of animals or plants, to protect them from insects and fungi, to hold and transport foods, and to clean equipment in food processing and preparation establishments. Nevertheless, there are some effective remedies available.

Reduce Incidence of Disease in Animals

Control of diseases in animals that serve as a source of food aids in preventing dissemination of infectious agents on and in meat and meat products. For example, immunizations can protect animals from such diseases as brucellosis, tuberculosis and Q fever. Testing and slaughtering of reactors are parts of animal health programs aimed at control of tuberculosis and brucellosis in cattle and at control of *Salmonella pullorum* and *Salmonella gallinarum* in chickens. Eradication of foodborne salmonellosis would have to start with elimination of *Salmonella* in animal feed and prevention of *Salmonella* infection in animals. Prevention and treatment of mastitis in dairy cows, along with rapid cooling of milk, are essential if milkborne staphylococcal intoxications are to be prevented.

Processing Foods of Animal Origin

Animals or incoming raw foods are the primary sources of the pathogenic microorganisms for the processing plant and for products processed therein. Washing the raw products during processing can sometimes reduce the number of microorganisms on them or reduce the quantity of units that have pathogens on them. But, with the introduction of water, microorganisms that are in the drip water are readily spread over units of food and to equipment surfaces and back to other units of foods. Thus, more items become contaminated than were contaminated originally. Certain processes, such as defeathering poultry, force microbes into feather follicles and the outer layer of the skin, and the flagella and fimbriae of these organisms aid in attaching to these surfaces. Subsequent washing removes few of these microorganisms. In some processes, chlorination has been used to reduce surface contamination of products such as poultry. Such treatment will only lower counts by one log (Tomkin, 1977). Its main effect is to kill microorganisms that are in chill water that can be spread during chilling; it has little effect on organisms that are attached. Washing shell eggs removes over 90% of the microorganisms that are on the shells and disinfecting (with 100 to 200 ppm available chlorine or equivalent) kills many of the remaining organisms. Dirt in wash water, however, can nullify a disinfectant's effectiveness. Some problems are encountered when egg shells are washed. For instance, washing destroys the cuticle and presence of water on shells enhances microbial penetration of the shell.

Cleaning and Disinfection of Equipment

To minimize transfer of contamination from one raw item to another or to cooked foods, utensils and equipment should, at the end of each work day, if not more frequently, be flushed with water, washed with warm water and a detergent, rinsed, and disinfected. Utensils and equipment that have been used for raw foods should be given this treatment before they are permitted to touch cooked foods or other foods which are not subsequently cooked. Microorganisms are more effectively removed from equipment surfaces by physical action of detergents and the mechanical action of scrubbing than by disinfection, but a sufficient concentration of a suitable disinfectant kills many of the few organisms that remain on cleaned and rinsed surfaces. Equipment should be of such design and made of such materials that will facilitate cleaning.

Planning Plant Layout

The operational flow and plant layout and kitchen layout should segregate raw food operations from food which will not be heated before consumption. Whenever possible, raw foods should not be processed by the same equipment or in the same room that is used for foods which will not be heated before consumption. Backtracking of the designed sequence of food flow must be avoided to prevent cross-contamination.

Hygiene of Workers

Workers in food establishments are often sources of foodborne pathogens. They usually carry *Clostridium perfringens* in their intestinal tracts and shed it in their stools. They are sometimes infected with shigellae, *Salmonella typhi*, one of a variety of other serotypes of *Salmonella*, hepatitis A virus, and other enteric microorganisms, or they sometimes have beta-hemolytic streptococci in their throats. They frequently carry coagulase-positive staphylococci in their anterior nares and on their skin. Staphylococci are resident on the skin and are frequently a source of infection following cuts, scrapes, and burns. *Salmonella, C. perfringens,* and *Staphylococcus aureus* and other microorganisms can also become transients on hands as a result of touching contaminated raw foods of animal origin or a contaminated surface. If any of these microorganisms reach the hands of food workers, and they frequently do, they can be transferred readily to the foods that are touched by the worker.

Persons who have open lesions, boils, sinus infections, or diarrhea should not handle cooked foods or uncooked foods that readily support growth of pathogenic foodborne bacteria. Management should establish a procedure by which workers with such infections will report their condition to supervisors; supervisors should suspend such persons until the condition has cleared up, instruct them to get treatment, or arrange for them to do tasks that do not require contact with food. Regulations that prohibit ill persons from handling food are difficult to enforce and are not very effective because healthy persons are just as likely to carry *S. aureus* and *C. perfringens* and to contaminate food. Water-tight dressings can prevent dissemination of pathogens from small lesions, but they must be kept dry and replaced when they become soiled. This, however, is difficult to ensure during food processing and preparation. Hands must be thoroughly washed after the dressing has been replaced.

Health Examinations

There is sporadic clattering from citizens for food workers to be tested and certified to be healthy by a health authority. As a result, particularly in the past in developed countries and currently in developing countries where enteric diseases are endemic, regulations sometimes require or have required that food establishment workers have medical examinations or screening tests before they are hired or periodically (i.e., annually) while they are employed as food workers. Whenever such tests are a part of a food protection program, it must be decided upon as to the action to take when a person is found to be infected or otherwise shows evidence of probably being infected. The action could be to either exclude this person from employment or to treat this person until a test or a series of tests produces no indication of infection. If neither course is practicable nor is going to be enforced, there is no point to the examination or test. They become a waste of time and public or company funds.

Tests are sometimes given to detect cases of syphilis and tuberculosis. Neither, however, is transmitted by infected persons while handling foods (Bryan, 1979). Screening tests for these diseases can be of value in case finding but not in food protection.

It would not be practicable to treat or exclude from food processing and preparation persons who are nasal carriers of *S. aureus* or intestinal carriers of *C. perfringens*. Therefore, no effort should be made to identify such persons. Even if a treatment did remove these organisms, they would reestablish in their niche in a few days or weeks.

Finding *Salmonella* in stool specimens of persons is more significant because it is not normally carried by healthy persons. But, other than in areas where salmonellosis is highly prevalant, such examinations would be costly and frequently negative. The usefulness of this test appears even more dubious in light of the fact that raw foods of animal origin are often sources of salmonellae. Such foods, if improperly cooked or foods that have been subjected to cross-contamination from such foods after cooking, are found to be vehicles in outbreaks of salmonellosis more frequently than foods touched by carriers. Also, workers are occupationally at risk of becoming carriers of *Salmonella* after they touch raw foods of animal origin and then touch their mouths or after they eat contaminated foods prepared where they work. So, they can be free of the organism the day of examination and infected the next. Screening tests cannot be done frequently enough to be effective at detecting persons continually exposed to such a risk of infection. Carriers of *Salmonella* and other enteric pathogens often have negative stools during periods they are not experiencing diarrhea. To detect infectious agents from stools of such carriers, multiple cultures have to be taken after saline pugation, which would not be practicable as a routine procedure. If a food handler is found positive for *Salmonella*, antimicrobial treatment would not be recommended because such treatment sometimes prolongs the carrier state.

Testing stools can detect carriers of *S. typhi* or shigellae, but these tests are too costly for use even in developed countries except in the case of certain institutions in which one of these diseases is known to be endemic.

Negative cultures often convey to food handlers the erroneous concept that they are incapable of transmitting foodborne pathogens, so that if they touch foods the outcome will be insignificant. A program of a public health agency or an industry that seldom reveals a positive culture, often gives administrators the unjustified feeling of accomplishment that they are protecting the public or safeguarding the food supply.

Handwashing

Handwashing can effectively remove transient foodborne pathogens that are acquired by fingering the nose, brushing the hair, touching dirty equipment, eliminating wastes, or handling raw foods (Lowbury *et al.*, 1964). Resident bacteria, as *S. aureus*, are not readily removed by this means (Lowbury *et al.*, 1963). Routine handwashing tends to bring staphylococci to the surface of the skin, and more may be found on hands after washing than before (Brodie, 1965). Bacteria are removed in the course of handwashing, not by bactericidal action of the soap, but by the combination of the soap's emulsifying action on lipids and other oils and grease, the abrasive effect of rubbing, and the effect of wash water in carrying away the loosened, dispersed, entrained particles and organisms. The use of disinfectants after handwashing does not kill enough *S. aureus* to make its use worthwhile (Lowbury, 1961; Brodie, 1965). After removal, by whatever means, resident bacteria reestablish their previous numbers in time (Price, 1938).

Use of Gloves

Gloves can be a barrier between contaminated hands or infected lesions and food. However, gloves can become contaminated with transient bacteria just as hands do if they contact contaminated foods. Bacteria, such as S. aureus, can accumulate as hands perspire and even multiply on hands that are encased in gloves. Then, even greater numbers than would otherwise have been on hands can contaminate foods that are handled if the gloves are ruptured or punctured. Rubber gloves have the further disadvantage that they are seldom washed as frequently as hands or as well as other pieces of equipment at the end of a day's operation. So, bacteria find harborage in the build-up on the gloves and may multiply thereon. Disposable gloves can prevent the transfer of contamination from hands or from lesions on hands to food. They are valuable in preventing transfer of microorganisms which are associated with hands, when handling cooked foods or foods which will not be subsequently heated. They should be worn only for the task for which they are provided and discarded when they become soiled, punctured, or torn, and after having been used in performing other tasks.

Other Measures

The chances of contaminating foods can also be reduced by using either clean utensils or automated equipment for tasks that are sometimes done by hand. Clean work clothes or coats worn either over or instead of workers' personal clothes will reduce somewhat the chances for contamination of foods, but they will not prevent transfer of organisms from raw foods to these clothes and then to other foods that touch them. They also will not prevent airborne transmission of staphylococci from workers' bodies to foods.

PROCESSES THAT AFFECT SURVIVAL AND GROWTH OF MICROORGANISMS

In a favorable environment, populations of pathogenic foodborne bacteria, as well as other bacteria, grow in the following fashion. Initially, there is a lag phase of an hour or more during which the organisms adjust to their new environment and there is little change in number of cells. At the end of this phase, multiplication begins and accelerates to a logarithmic constant rate. This logarithmic phase continues until nutrients are nearly depleted and toxic metabolic products accumulate. At this time, and because of these reasons, the rate of growth slows and the population remains constant; this is referred to as the maximum stationary phase. This phase is followed in time by a progressive die-off of cells. Foodborne pathogens can be controlled by processing or storing foods so that the microorganisms associated with them remain in their lag phase and by processing foods so that these microorganisms are killed.

Heat Processing

When bacterial cells or spores are heated, they will either be injured in varying degrees or be killed. Whatever viability comes about is a result of differences among individual cells and of the growth phase the population is in when subjected to heat. Bacterial populations that are in the exponential (logarithmic) phase of growth, for instance, generally are less resistant to heat than cells that are in any other stage of growth. Microorganisms grown at temperatures near their maximum are found to be more heat resistant than those grown at optimal or lower temperatures.

Injured survivors of a heat process either die or recover, depending on the degree of injury and the further conditions to which they are subjected. Heat-damaged cells behave in one medium as if they are dead, but in another medium (usually one that is enriched) or under different conditions or time of incubation, multiply, sometimes after a considerable lag. Sublethal heat treatment can cause a stimulated germination effect on spores. If time-temperature exposure is more severe than necessary for the stimulatory effect required, but not lethal, germination can be delayed.

Injury or death of cells is random among cells of a population and is the result of a single injurious or lethal event. In general, when a population of bacterial cells or spores is exposed to high temperatures, it decreases exponentially. A straight line is formed when the number of survivors is plotted against the times of exposure at a given temperature on semi-log paper. The time in minutes to kill 90% of the cells or spores (or to destroy 90% of a toxin) at any given temperature is called the D value for that temperature. Survival time decreases when the heating temperature increases. When time (or D value units) in minutes is plotted on a log scale against temperature plotted on a linear scale, the degree of temperature required for the thermal destruction curve (or death time curve) to traverse one log cycle is known as the z value. Thus, z values reflect the relative resistance of a microorganism to different temperatures. D and z values are used to determine safe time-temperature relationships to ensure destruction of specific organisms in the heat processes of foods. (Examples of D and z values for common foodborne pathogens are given in Table 8.2).

The minimum safe heat processing for canned, low-acid foods (pH above 4.5) is designed to reduce a population of 10^{12} *Clostridium botulinum* spores to a population of 10^0. In another situation, if there is one spore in each of 10^{12} containers of low-acid food before processing, only one container will contain a live spore after processing. Thus, there is one chance in 10^{12} that any particular container will harbor a live spore. This is called the 12 D concept. A process that provides this effect for *C. botulinum* type A and B spores is equivalent to or higher than 121.1°C for 2.5 minutes. (This is calculated by multiplying the $D_{121.1}$ value of 0.204 minutes by 12 = 2.45 minutes.) Because *C. botulinum* has a z value of 10°C (18°F), 24.5 minutes would be required if a temperature of 111.1°C were used, and 245 minutes would be required if a temperature of 101.1°C were used. (See National Canners Association Research Laboratories, 1968; Stumbo, 1973, for methods of calculating processes.)

Vegetative cells are far less heat resistant than spores, but they are likely to show a greater range of heat resistance among the individuals of a populations. A process that provides a 5 D to 12 D for a specific vegetative pathogenic organism (as *Salmonella* is sometimes used to design heat processes that are intended to kill vegetative bacteria, depending on the anticipated number of organisms in the product. For instance, regulations in the United States require heat treatment of 60°C (140°F) for 3.5 minutes to pasteurize liquid eggs. Such treatment results in a 10^1-fold to 10^2-fold reduction in numbers of viable *S. senftenberg* 775W and a 10^5-fold reduction of other serotypes of *Salmonella*. Regulations in the United Kingdom require a temperature of 64.4°C (148°F) for 2.5 minutes, which results in a 10^5-fold reduction of *S. senftenberg* 775W and a 10^{12}-fold reduction of other serotypes of *Salmonella* (Genigeorgis and Riemann, 1979). From a practical standpoint, even the lowest requirement would be likely to render the eggs free of salmonellae, because eggs are seldom contaminated either with *S. senftenberg* 775W or 10^5 or more salmonellae. Current milk pasteurization standards (145°F for 30 minutes; 161°F for 15 seconds) result in more than a 10^{12}-fold reduction of viable salmonellae in milk (Read

TABLE 8.2. D AND z VALUES FOR IMPORTANT FOODBORNE PATHOGENS IN VARIOUS FOOD SUBSTRATES OR MEDIA, DATA FROM SELECTED STUDIES.

Organism	Substrate	D value (°C)	z value 10C (18F)	Reference
Clostridium botulinum Type A	Phosphate buffer	$D_{121.1} = 0.204$		Esty and Meyer, 1922
	Phosphate buffer	$D_{121.1} = 0.1$ to 0.2	14.7–16.3F	Stumbo et al., 1950
	Phosphate buffer	$D_{115.5} = 0.24$ to 0.77		Tsuji and Perkins, 1962
	Phosphate buffer	$D_{112.8} = 1.09$ to 1.23		Ito et al., 1967
	Phosphate buffer	$D_{112.8} = 0.15$ to 1.32		Ito et al., 1967
Clostridium botulinum Type B	Media and phosphate buffer	$D_{80} = 0.33$ to 1.25 $D_{80} = 1.78$ to 2.3	7.4–10.8F 14–15F	Roberts and Ingram, 1965 Schmidt, 1964
Clostridium botulinum Type E	White fish	$D_{80} = 1.6$ to 4.3	10.3 to 13.6F	Crisley et al., 1967
	Phosphate buffer	$D_{77} = 0.77$ to 1.95		Ito et al., 1967
	Media	$D_{90} = 3$ to 5	6 to 8C	Roberts, 1968
Clostridium perfringens Heat-sensitive strain	Media	$D_{90} = 15$ to 145 $D_{100} = 6$ to 17	9 to 24 C 9 to 24 C	Roberts, 1968 Roberts, 1968
Clostridium perfringens Heat-resistant strain	Chicken a la king	$D_{60} = 5.17$ to 5.37	9.3–10.5F	Angelotti et al., 1961
Staphylococcus aureus	Custard	$D_{60} = 7.68$ to 7.82	9.95 to 10.5F	Angelotti et al., 1961
	Beef bouillon	$D_{60} = 2.2$ to 2.6	10F (5.6C)	Thomas et al., 1966
	Green pea soup	$D_{60} = 6.7$ to 10.4	10F (5.6C)	Thomas et al., 1966
	Skim milk	$D_{60} = 1.0$ to 6.5	10F (5.6C)	Thomas et al., 1966
	0.5% NaCl	$D_{60} = 2.0$ to 2.5	10F (5.6C)	Thomas et al., 1966

Food	D-value	F value	Reference
Turkey stuffing	$D_{60} = 15.4$	12.3F	Webster and Esselen, 1956
	$D_{73.9} = 0.15$	12.3F	Webster and Esselen, 1956
Chicken a la king	$D_{60} = 0.39$ to 0.40	8.7 to 9.2F	Angelotti et al., 1961
Custard	$D_{60} = 2.39$ to 2.49	13 to 15.8F	Angelotti et al., 1961
Milk chocolate	$D_{70} = 720$ to 1050	34.2F	Goepfert and Biggie, 1968
	$D_{80} = 222$		
	$D_{90} = 72$ to 78		
Roast beef	$D_{51.7} = 61$ to 62	10	Goodfellow and Brown, 1978
	$D_{57.2} = 3.8$ to 4.2	10	Goodfellow and Brown, 1978
	$D_{57.2} = 0.6$ to 0.7	10	Goodfellow and Brown, 1978
Chicken a la king	$D_{60} = 9.22$ to 9.99	11.50 to 12.15	Angelotti et al., 1961
Custard	$D_{60} = 9.36$ to 12.05	11.86 to 12.25	Angelotti et al., 1961
Beef bouillon	$D_{60} = 5.8$	10F	Thomas et al., 1966
Greenpea soup	$D_{60} = 5.2$ to 10.6	10F	Thomas et al., 1966
Skim milk	$D_{60} = 6$ to 10.8	10F	Thomas et al., 1966
Milk chocolate	$D_{70} = 360$ to 480	32.4F	Goepfert and Biggie, 1968
	$D_{80} = 96$ to 144		Goepfert and Biggie, 1968
	$D_{90} = 30$ to 42		Goepfert and Biggie, 1968

Salmonella, serotypes other than *S. senftenberg* 775W

Salmonella senftenberg 775W

et al., 1968). It results also in a greater than 10^6 reduction of viable *Mycobacterium tuberculosis* cells (Kells and Lear, 1960).

Certain intrinsic properties of the food affect the heat resistance of microorganisms. The pH has a pronounced effect on heat resistance of bacterial spores and vegetative cells. The heat resistance of an organism has a narrow pH range outside of which the resistance falls rapidly (Hansen and Riemann, 1963). Generally, the heat resistance is at a maximum at a pH close to neutral. Spores of *C. botulinum*, for instance, have maximum resistance from pH 6.3 to 6.9, and there is a marked reduction of heat resistance at pH 4.9 or below (Esty and Meyer, 1922). The pH of the food also determines which of the survivors of a heat treatment will grow.

The water activity of a food also has a pronounced effect on the heat resistance of microorganisms. At a_w 0.2 to 0.4, for instance, the heat resistance of spores is much higher than at a_w 0.85; and at that a_w it is higher than at a_w 0.99 (See Table 8.2.)

Fats and oils have a protective effect on microorganisms, probably because of the localized absence of moisture, but most bacteria are unable to grow in the oil phase. They can, however, migrate from the oil phase into the aqueous phase where they can multiply. High concentrations of sugars also result in increased resistance for microorganisms. Proteins also have a protective effect on microorganisms.

Chemical preservatives (e.g. sulfur dioxide, hydrogen peroxide, certain essential oils, and nitrites) generally decrease the heat resistance of vegetative bacteria. Certain concentrations of NaCl, nitrites, and other curing ingredients have an inhibitory effect on the germination of spores which have survived the heat process. Several antibiotics have an inhibitory action on mildly heat-treated spores of *C. botulinum*.

By using demonstrated D and z values for vegetative bacteria, it is possible to stipulate or check conditions of cooking or reheating processes in food service and catering operations to see if these result in a safe product (Bryan, 1978b). These values indicate that a temperature of 73.9°C (165°F) for a few seconds should result in death of most vegetative bacteria, and a temperature of 71.1°C (160°F) will result in the death of most of these organisms within a minute. To kill 10^7 salmonellae and staphylococci in custard and chicken a la king, a temperature of 65.6°C (150°F) must be imposed over a period ranging from 0.3 to 11.3 minutes, or a temperature of 60°C (140°F) must be imposed over a period of 3 to 81.5 minutes, depending on the type species and strain and the substrate (Angelotti *et al.*, 1961).

Any temperature above the maximum for growth of a particular organism is lethal to it. The higher the temperature, the more rapid the destruction. Salmonellae will die in dried eggs, for instance, when stored at 52°C for 7 to 10 days. In food service operations, these organisms will die in most foods when these foods are stored in hot-holding devices at temperatures at or above 55°C, if the storage time is long enough.

Cooling

Bacterial foodborne pathogens multiply at temperatures ranging from around 3 to 50°C. Each species has its characteristic temperature range and an optimal temperature for growth. The lag before multiplication is least and the multiplication rates are most rapid for each at or near its specific optimal temperature. In general, the growth rate for microorganisms increases as temperature increases, until the optimal temperature for growth is reached. Growth rates decrease as the temperature approaches either limit of their growth range and finally stops. These organisms remain viable for a long period at temperatures too low for growth. When the temperatures climb above the minimum, growth resumes. As temperatures approach the minimum at which growth can occur, a progressively longer lag period occurs before growth commences. Temperatures below optimum also cause lower than maximal rates of growth, selection for cold-tolerant species and strains, sometimes a smaller yield of cells, and may affect the nutritional requirements or the composition of the cell. When water activity, pH, oxidation-reduction potential and a supply of essential nutrients are near the optimum for a particular kind of microorganism, the lag phase is short (perhaps an hour) near the optimal temperature. Cells in the exponential phase of growth show a minimal lag when transferred to a similar substrate at the same temperature; whereas, those in the lag or stationary phases of growth have extended lag periods. The lag phase lengthens as the temperature decreases from the optimum until it reaches infinity, just beyond the minimal temperature for growth.

Storage at a particular temperature has a considerable impact on species which can grow best in a diversified microbial population. Some species tend to predominate numerically over others because the storage temperature has been favorable to their growth. Table 8.3 lists data on minimal, optimal, and maximal temperatures for growth of bacterial pathogens that cause most foodborne diseases. High temperatures (45–50°C) select for *Bacillus cereus, C. per-*

TABLE 8.3.   MINIMAL, OPTIMAL, AND MAXIMAL TEMPERATURES (°C) FOR GROWTH OF IMPORTANT FOODBORNE PATHOGENS

Organism	Minimum	Optimum	Maximum
Aspergillus flavus (Aflatoxin production)	11	18–35	42
Bacillus cereus	7	30	49
Clostridium botulinum types A and B	10–12	37	48–50
Clostridium botulinum type E	3.3	30	45
Clostridium perfringens	15–20	43–45	50
Salmonella	5.2	37–43	44–47
Staphylococcus aureus	6.7	37–40	45
Vibrio parahaemolyticus	3–13	37	42–44
Yersinia enterocolytica	3	30–37	43

fringens, and sometimes *C. botulinum* types A and B; low storage temperatures (3–5°C) select for *C. botulinum* type E and *Yersinia*.

Growth of foodborne disease bacteria can be prevented by completing the processing of food within an hour or two so that the bacterial contaminants remain in their lag phase of growth. Growth can also be prevented by cooling foods rapidly. Many foods, because of their very nature or their bulky size or large volumes, cool slowly when they are stored in refrigerators. Cooling time can be shortened by increasing the ratio of surface to volume, as by slicing meat, by cutting large roasts into 1.25-pound chunks, or by cutting poultry into parts, providing the resulting pieces are stacked so that the surfaces are exposed to circulating air, liquid, or cold surfaces. In food service operations, solid and semisolid foods should be put into pans that are no more than 4 inches high. Sliced, solid food can also be cooled faster by putting it in contact with ice, circulating cold water, or cold surfaces (Bryan and Kilpatrick, 1971; Bryan and McKinley, 1974). Several large cuts of meat or carcasses of poultry cannot be cooled with sufficient rapidity in a covered pan, in a cooling unit without circulating air, or in any container more than 4 inches high. Liquid foods can be cooled faster than otherwise by putting them into pans surrounded by ice or by water baths, by adding ice to a slightly more concentrated mixture than ordinarily used, or by stirring in a mixer for 30 to 60 minutes before refrigeration. To be assured of preventing bacterial foodborne disease outbreaks, cooked foods which are to be served cold or reheated on a day subsequent to cooking should be cooled (from 55°C or higher) to

21°C within 2 hours and then to 7°C or lower to delay or prevent growth of pathogenic foodborne bacteria.

Freezing

Freezing and frozen storage have effects on microorganisms that are detrimental in ways other than those produced by cooling and cold storage. Freezing kills a percentage of the microbial flora as well as arrests multiplication of surviving microorganisms. A 10- to 100-fold decrease from freezing is not uncommon. Freezing also injures cells and they are unable to multiply until they have recovered. Freezing prevents multiplication by the freezing of water outside the cell and a corresponding lowering of water activity as well as by the low temperature typical of this state. This lowering of water activity is why some yeasts and molds predominate in foods stored at low temperatures. Low temperatures slow bacterial growth and low water activity prevents their growth entirely. Yeasts and molds are capable of growing at much lower water activity values than bacteria. Freezing removes liquid water from both the external environment and the interior of the cell itself, causing dehydration of the surface as water migrates from within the cell through its cell wall and freezes outside, and thus increases the concentration of solutes both inside and outside the cell. Ice crystals formed inside or outside the cell during freezing and during formation of larger ice crystals from smaller ice crystals damage cell membranes. Cells are injured by one or more of these events.

During frozen storage, the death rate is substantially reduced. Death at this time is caused by either recrystallization of small unstable intracellular ice crystals or by prolonged contact with concentrated solutes in the residual unfrozen intracellular solution (Ray and Speck, 1973). The lower the temperature of frozen storage and the more favorable the properties of the substrate (such as low-acid food), the better are the chances for survival of the remaining microorganisms. More vegetative microorganisms are killed or injured at –2°C to –10°C than at –15°C, and more at the latter temperature than at –30°C.

Thawing creates additional detrimental effects. The faster foods thaw, the greater the number of bacteria that survives within them. *Salmonella, Vibrio parahaemolyticus,* and other Gram-negative bacteria are more sensitive to cold shock caused by freezing, frozen storage, and thawing than staphylococci and streptococci. The characteristic arrangement of the latter organisms in clusters or in chains no doubt influences survival, because a few individual cells in the aggregate are oftentimes randomly spared. These survivors can subsequently grow into colonies on media that are used to enumerate cells. Vegetative cells of *C. botulinum* and *B. cereus* are only slightly more resistant to freezing than are the Gram-negative organisms. Vegetative cells of *C. perfringens* are rather sensitive to frozen storage. Spores, bacterial toxins, and viruses are quite resistant to freezing. Protozoa and helminths succumb to freezing temperatures within a few days. Survival of parasites is longer at higher freezing temperatures and in thicker pieces of meat or fish.

Reduced Water Activity

Growth and metabolism of microorganisms depend on the presence of water in an available form. Reducing available water in the aqueous phase of a food by either removing some of the water (dehydration), or by adding solutes (as in curing, salting, syruping, or sugaring) slows the growth rate of microorganisms. When solutes (e.g. salts or sugars) are added to foods, some water molecules become oriented about solute molecules; others become adsorbed on.to insoluble constitutents of the food. Small solute molecules (e.g. NaCl) can, weight for weight, attract and hold more water molecules in a pattern than can large solute molecules (e.g. sucrose).

The most useful measure of the amount of available water in a food is the food's water activity (a_w). The a_w is the ratio of the food's water vapor pressure to the vapor pressure of pure water at the same temperature. As a solution becomes more concentrated, its vapor pressure decreases, and the corresponding a_w decreases from a maximum of 1, the a_w of pure water. Relative values for a_w to various concentrations of sodium chloride, inverted sugar, glucose, and sucrose are listed in Table 8.4.

TABLE 8.4.   RELATION OF a_w AT 25°C TO APPROXIMATE CONCENTRATIONS OF SODIUM CHLORIDE, INVERTED SUGAR, GLUCOSE, AND SUCROSE (% w/w).

a_w	NaCl	Inverted sugar	Glucose	Sucrose
1.000	0	0	0	0
.99	1.7	4.1	8.9	15.5
.98	3.4	8.2	15.7	26.1
.96	6.6	16.4	28.5	39.7
.94	9.4	24.7	37.8	48.2
.92	11.9	32.9	43.7	54.4
.90	14.2	41.1	48.5	58.5
.86	18.2	57.5	58.5	65.6
.80	23.1			
.75	26.5			

From ICMSF, unpublished.

Within the a_w range from 0.98 to 0.999, most foodborne pathogens and bacteria causing spoilage grow rapidly. Only the growth of *V. parahaemolyticus* is restricted at an a_w above 0.994. Growth of these pathogens is progressively slowed in foods as water activity is reduced. Most pathogenic foodborne bacteria stop growing when the a_w becomes as low as 0.95 to 0.96. At an a_w of 0.93 or lower, only *S. aureus* and some toxigenic molds multiply. *Staphylococcus aureus* can multiply at a_w values as low as 0.83. Molds grow at lower a_w values than bacteria, but there are no reports of mycotoxin production below a_w 0.83 (Northolt *et al.*, 1976). Table 8.5 lists minimal a_w values at which common foodborne pathogens can grow.

In foods with a_w above 0.95, *S. aureus* grows poorly when in competition with other bacteria that are also present. But, at a_w 0.94, which would slow down or completely stop the growth of most of its bacterial competitors, the growth rate of *S. aureus* is only about one half of that at its optimum a_w (Scott, 1953). So, at a_w 0.94, *S. aureus* can grow without much competition from other microorganisms. However, production of enterotoxin has not been shown in foods having an a_w of 0.92 or below (Troller and Stinson, 1975).

Foodborne pathogenic microorganisms remain viable for long periods in many dried foods with low a_w, but die rapidly in heavily salted foods that have a low a_w. The a_w of a food is a determinant of the time-temperature combination required to kill foodborne pathogens. For example, the lower the a_w, the longer the time needed at any given lethal temperature.

TABLE 8.5. MINIMUM a_w AT WHICH COMMON FOODBORNE PATHOGENS GROW AT NEAR OPTIMAL TEMPERATURES.

Organism	Minimum a_w
Aspergillus flavus	0.71–0.78
Bacillus cereus	0.93–0.95
Clostridium botulinim type A	0.95
Clostridium botulinum type B	0.94
Clostridium botulinum type E	0.97
Clostridium perfringens	0.95–0.96
Escherichia coli	0.95
Salmonella spp.	0.95
Staphylococcus aureus	0.83–0.86
Vibrio parahaemolyticus	0.94

..cidifying

The hydrogen ion concentration or certain undissociated organic acids (such as acetic, lactic, and sorbic) have a destructive or inhibitory effect on microorganisms. The pH of the food encourages or discourages the growth or survival of various kinds of microorganisms. In general, bacteria grow best when the pH is at a neutral level; yeasts and molds grow best in a food that has a pH low enough to slow or stop bacterial growth. Few pathogenic foodborne microorganisms, except the toxigenic fungi, can grow at pH 4.5 and none below pH 4.0. Table 8.6 lists information on the minimal, optimal, and maximal pH at which common foodborne pathogens grow.

Some foods have an inherently low pH; others, by fermentation, produce acids that decrease pH, and the pH of others can be decreased by the addition of acids. Certain foods (most fruits and soft drinks) contain a significant amount of acid that inhibits growth of pathogenic, foodborne bacteria. Other foods (e.g. sauerkraut, pickles, green olives, buttermilk, cheese, and fermented sausage) have their pH decreased when lactic acid bacteria metabolize carbohydrates and produce lactic acid. The acids formed by these bacteria not only contribute, in part, to the characteristic flavors and aromas of fermented foods, but suppress the growth of undesirable bacteria. If lactic acid bacteria do not become the dominant flora during the early stages of the fermentation process, undesirable,

TABLE 8.6. MINIMAL, OPTIMAL, AND MAXIMAL pH FOR GROWTH OF COMMON FOODBORNE PATHOGENS.

Organism	Minimum	Optimum	Maximum
Aspergillus flavus	1.7–2.4	3.4–5.5	9.3
Bacillus cereus	4.4–5.0		9.3
Clostridium botulinum type A	4.8	7	8–8.5
Clostridium botulinum type B	4.8	7	8–8.5
Clostridium botulinum type E	5.0	7	8.5–8.9
Clostridium perfringens	5.0	7	8–9
Salmonella spp.	4.1–5.5	7	8–9
Staphylococcus aureus	4.3	7	8–9
Vibrio parahaemolyticus	4.8	7.4–8.5	11

even pathogenic, bacteria can multiply to high numbers. Anything (a high ratio of pathogens to lactics; destruction of lactics by bacterophage; paucity of fermentable carbohydrates; or temperature, pH, a_w, or redox potential, favoring growth of pathogens over the lactics) that interferes with the growth of lactic acid bacteria can allow this to happen.

Other Processing

Several other processes affect the survival and growth of microorganisms, but their use is at present restricted by regulatory agencies. They either act selectively and frequently permit survival of foodborne pathogens, or they are difficult to control.

Preservatives, antibiotics, smoking, and vacuum packaging act selectively against microorganisms, and sometimes they favor survival of pathogens more than the survival of natural microbial competitors. For this reason, regulatory agencies restrict the use of antibiotics and many preservatives. Smoking is used primarily to change the flavor of meats, fish, and poultry, but smoking also dries the surface, adds certain chemical substances which inhibit growth of some microbes, and some microbes succumb to the heat that sometimes is associated with smoking. Vacuum packaging prolongs shelf life of many products by arresting the growth of aerobic bacteria and molds, but anaerobic and facultatively anaerobic bacteria can multiply, if other conditions (temperature, a_w, pH, and the substrate) are favorable.

Fermentation of foods is an effective way of preventing growth of pathogenic bacteria, but the process is complex and must be critically controlled. Foodborne disease outbreaks, particularly those caused by *S. aureus*, can occur when growth of the usual fermenting organisms is delayed or suppressed.

Irradiation can be used to either sterilize or pasteurize a product. Sterilization, referred to as radappertization, requires exposure to approximately 4.8 Mrad. Pasteurization, referred to as radicification, requires exposure to radiation in the range from 0.5 to 0.7 Mrad. Neither process has as yet been approved in the United States.

PUBLIC HEALTH OR QUALITY CONTROL ACTIVITIES

Public health activities and quality control programs should be based on the elimination of the factors that have been demonstrated to contribute to foodborne disease outbreaks. Prevention and control activities must minimize contamination, prevent growth of bacteria and molds, and kill microorganisms.

Epidemiologic Surveillance of Foodborne Diseases

Surveillance of the diseases that are associated with ingredients and products processed or prepared by the segment of the food industry in which an establishment or control agency is concerned is an essential aspect of foodborne disease control. Such surveillance programs are concerned with the incidence of disease that can be foodborne and the sources of the agents that cause them, the relative capabilities of various foods to become vehicles for these diseases, the places where foods are mishandled, and the operations that frequently lead to outbreaks. An awareness of these items can be gained by reading articles in scientific and medical journals, summaries of foodborne disease reports, and surveillance reports.

Components of a surveillance program are: notification of illness, identifying outbreaks, investigating outbreaks, interpreting investigative data, and disseminating findings.

After notification of a suspected foodborne illness is received, thorough investigation should be made (Bryan *et al.*, 1976). During the initial phase of the investigation, investigators must define a case, verify diagnosis of cases, decide whether or not an outbreak occurred, count cases in different population groups and calculate rates, and make epidemiologic (time, place, and person) associations of the cases. From this information, hypotheses about the mode of transmission of the agents that caused the illness are formed. These must be confirmed or refuted by subsequent investigation.

During the next phase of the investigation, additional cases and other persons at risk of having the same time or place of exposure or common eating or cultural habits are sought. Also, the place at which the exposed persons got or ate the suspected food must be scrutinized. At this place, investigators must determine from managers and workers the step-by-step process that foods underwent. They must critically review the likelihood of contamination of incoming products or ingredients and each step of the operation for other possible sources and modes of contamination. Each processing step, which is designed to heat or otherwise treat foods to reduce the number of bacteria present, must be evaluated to determine the likelihood of survival of pathogenic microorganisms. Also, each period of cold, warm, or room-temperature storage in the establishment and while in transit, must be evaluated for time-temperature conditions that could have been conducive to bacterial multiplication. The likelihood of subsequent mishandling upon arrival at its designation must also be determined. After the operations that contributed to the outbreak have been identified, control measures must be devised or selected from known measures and explained and demonstrated to managers and staff. If necessary, staff may have to be trained to use safe procedures. The hazardous procedures must be terminated and replaced with safe procedures before the incriminated product can again be processed or prepared in the implicated establishment.

To confirm the etiology of a disease, specimens must be collected from ill persons. Samples of suspected foods should be collected whenever possible. Samples of raw ingredients and from the environment in which suspicious foods were processed should be collected, also, if definitive typing procedures are to be used (Kazal, 1976). Specimens should also be taken from workers, if there is any indication that these persons could have contaminated the suspected foods. Isolation from ill persons of a pathogen, or demonstration of a toxin, which produces the same syndrome as occurred in the cases confirms the etiology. Finding the same definitive type of organism or large numbers ($>10^5$) of specimens which are known to be foodborne pathogens from the epidemiologically implicated food confirms the vehicle. Isolation of the same definitive type of microorganisms (which either caused the infection or that produces a toxin that causes the same syndrome as affecting the ill) from an appropriate specimen from a person who had an opportunity to contaminate the food, from a sample of raw food, or the environment can either confirm the source or detect a link in a chain of transmission. The source of infection can sometimes be established by definitively typing (serotyping, phage typing, or colicin typing) the isolates from these specimens and samples and those from foodsource animals or their feed or environment or from persons.

A report of the incident, including a description of epidemiologic patterns and a detailed account of the way the implicated food was contaminated, the reason the etiologic agents survived processes, and the circumstances that allowed pathogenic bacteria or toxigenic molds to multiply should be made and sent to persons concerned with surveillance of foodborne disease. This report, along with others, should be summarized as to the type of establishment at which mishandling occurred, the type of vehicle, and the operations that contributed to the outbreak. Cumulative data can show where priorities must be set for hazard analysis, for food protection control programs, and training of students, professional health personnel, and food industry managers and workers.

When reviewing summary data of the incidence of foodborne

illness, it must be kept in mind that reports of such illness are sporadic and incomplete. People rarely tell health or food control agencies that they have had diarrheal illnesses; they rarely seek medical help for these ailments. Even when medical attention is sought, foodborne illness is seldom identified because of the similarity of the symptoms of these illnesses with other ailments, and foodborne disease outbreaks are obscured because of the usual sporadic ways afflicted persons seek treatment. Specimens are seldom taken to confirm clinical diagnosis because supportive treatment and time usually effect a cure. When complaints are made to health agencies, priorities of other matters sometimes prevent a complete investigation. Reports are not always sent along the usual surveillance channels and so are never recorded in summary data. Also, data from individual reports are not always complete, and they are sometimes inaccurate.

Naive reliance on surveillance data can lead some persons, especially those who have a vested interest, to believe that foodborne diseases are of minor significance because few people die of the consequences or relatively few cases and few outbreaks are reported for vast population groups. Other persons believe that a particular food is generally safe because there is little reported evidence of its having been a vehicle. This can be true, if the food is of such a nature that precludes bacterial growth, but otherwise there may be just a false sense of security. The understanding of the role of a food as a vehicle can change when surveillance improves, when current control measures are relaxed because of food fads or economic pressure, when traditional foods are processed in modified ways, and when new foods are introduced. The available surveillance data, however, do indicate that foods are mishandled in certain types of establishments, that some foods are vehicles, repetitively, and that there are identifiable operations which are followed that lead to foodborne disease outbreaks. Control programs should deal primarily with these identified operations until other aspects of the foodborne disease problem have been elucidated by either epidemiologic or other scientific investigation.

Microbiological Surveillance of Foods

Foods (and the environment in which foods are processed or prepared) can be tested either for groups of microorganisms that are indicative of contamination, survival, or multiplication, or for specific species indicative of spoilage or health risk. Samples for these tests can be taken of ingredients as they are received by processing plants; they can be taken of foods at any stage of the process that could have altered the type or number of micro-organisms; they can be taken of finished products at the plant or at various stages of distribution; they can be taken of surfaces of equipment that the food touched during processing; they can be taken of other sites in the environment in which foods were exposed; or they can be taken of waste generated as a result of the processing. Each type of sample will disclose specific information about the food or process. The samples to be collected are governed by the information desired, the skill of field and laboratory personnel, and economic necessity.

It is particularly important to sample ingredients of foods that receive no subsequent heat or other treatment lethal to likely kinds of contaminating agents, and to test these samples for foodborne pathogens that are known to be associated with them. When ingredients are found to be contaminated, they should be rejected and returned to their source, treated to destroy the contaminants, or used in a product of a nature that causes the contaminants to die (such as high-acid foods), or that will receive processing that kills the contaminants.

Collecting samples of a food at each stage of processing (particularly after the food has been exposed to any chance of contamination, after it has been heated or otherwise processed, and after periods of storage) provides information on sources of contamination, whether or not organisms are surviving processing, and whether or not organisms are multiplying. Sources and modes of contamination can sometimes be traced by definitively typing isolates of certain pathogens and comparing those taken of foods at each phase of the operation with those taken of raw ingredients, from persons who have touched the food, from equipment that contacted the food during processing, from food source animals, and from the environment in which foods were exposed. Counts of total aerobic colonies and indicator groups of bacteria or specific species before and after exposure to a process can show the effectiveness of the processing in destroying or reducing numbers of the organism tested for.

The extent of microbial growth can be determined by sampling foods before and after periods of storage and by testing these foods for numbers of aerobic colonies or indicator groups of bacteria.

Samples of finished products disclose information about the combined effects of contamination, survival, and multiplication, but they do not usually reveal the exact contribution each has made to the types or numbers of microorganisms found. Expedience, however, sometimes forces food-regulatory agencies to collect samples of finished products only. When such sampling is coupled with inspection of operations, operational problems can sometimes be identified. Tests of samples from drains, process water, effluents, stabilization ponds, food-scrap piles, vacuum-cleaner collecting bags, sweepings, and air filters can show that specific organisms have passed through an operation. Tests of fecal droppings, truck beds, delivery coops, or live animals can show that specific pathogens have entered or could enter a plant.

Enough samples must be taken if results from microbiological samples are to produce reliable data. As risk of a food being contaminated with a large number of pathogens that can cause severe illness increases, the number of samples must be increased. A protocol that is based on this idea has been developed by the International Commission on Microbiological Specifications for Foods (1974).

This group classifies foods into five degrees of hazard (utility, low hazard, moderate hazard with limited spread, moderate hazard with extensive spread, severe hazard), based on the type of hazards created by the nature of the food (composition, likelihood of contamination, and potential for supporting growth of the agents) and on the risk of foodborne disease expected as demonstrated by epidemiologic investigations or microbiologic tests. It also reveals that the degree of hazard can change because the risk could be reduced by cooking the food, or the degree of hazard could remain the same, or it could be increased by storing foods at temperatures that permit growth of contaminating microorganisms or by exposing the food to contamination subsequently. These classifications form 15 situations called sampling cases. Any raw or processed food will fall into one or more of these cases. The type of organism or group of organisms to seek during the microbiological examination of the food is then determined. This determination is based to a large extent on the sampling case in which the product has been placed and the organisms expected to be found in the product. Total counts are determined if the first three cases are used, indicator organisms are counted if the next three cases are used, and pathogens are counted or sought or toxins are determined, if the remainder of the cases are used. Either a three-class (that has a level at which or below which indicates a microbiologically satisfactory product, a level which indicates microbiologically marginally acceptable product in which only a portion of the samples can fall and remain acceptable, and a level at which and above which is unacceptable) or a two-class (that has a level at which or above which indicates a microbiologically unacceptable product) sampling plan is then selected. The selection of a sampling plan is based on the case and the type or group of organisms sought. A three-class plan is selected, if counts are to be made, and a two-class plan is selected, if the presence or absence of a specific organism is to be determined. An ample number of samples must be collected to provide acceptable probability of finding defective lots. The number of samples to collect is based on statistical probability that is acceptable and the sampling case in which a food is placed. In general, the number of samples to take increases as the case number increases. The number of marginally acceptable samples that can be tolerated in a three-class sampling plan is chosen on the basis of the sampling case and the amount of risk that is acceptable. The ICMSF (1974) has suggested the number of samples to take, the number of marginally acceptable samples, levels of microorganisms above which risk is unacceptable, and the range within which some of, but not all, samples fall and the lot still remains acceptable for several classes of foods.

Hazard Analyses, Critical Control Point Designation, and Control Action

Operations should be watched to determine all possible sources of contamination and places in a process where contamination of food by a foodborne pathogen is likely to occur. These observations should include all steps and materials of processing from incoming animals or raw materials through each following operation until the finished product leaves a plant or is eaten or sold. Concern should also be given to possible mishandling after sale or shipment.

Live animals are frequently infected with or contaminated by foodborne pathogens when they enter slaughtering operations. As a consequence, meat and poultry are frequently contaminated by *Salmonella, S. aureus,* and *C. perfringens.* Egg shells are sometimes contaminated by salmonellae. Marine fish and shellfish are frequently contaminated by *V. parahaemolyticus,* and fresh-water fish are frequently contaminated by *C. botulinum* type E. Raw vegetables in areas where these products are subjected to sewage, waste-water irrigation, washing in rivers, or night-soil fertilization can introduce enteric and soil bacteria and parasites into processing plants or kitchens. Grains and legumes can be contaminated by *B. cereus* and *C. perfringens.*   These organisms can be transferred to equipment that is used to process the contaminated items and then is used, without cleaning, for other foods, or the same foods after cooking. These organisms can also be transferred to workers' hands and then to other foods that they handle. The processes or activities by which cross-contamination can occur must be determined.

Workers must be observed so that situations in which they could or do contaminate foods are spotted. Barriers to such contamination must then be instituted, if practicable. Examples of barriers are: automated equipment, utensils or disposable gloves used to handle foods instead of bare hands, and frequent handwashing. Education and training are often necessary to inform workers of recommended techniques of personal hygiene and food handling.

Each process should be evaluated for its effect on increasing or decreasing the level of contamination. Microbiological sampling, as described in previous sections, can aid in identifying or confirming sources of contamination and modes of spread. A process that is found to increase or spread contamination calls for measures to eliminate the contamination or to modify the process to reduce the contamination. Alternately, a subsequent process can be modified to lower or eliminate the contamination or a new process can be added to do this.

In certain types of processing plants, condensate, excessive use of water, or drippage can moisten dried or concentrated foods to a level that could permit microbial growth. Samples should be taken to evaluate the a_w of products suspected of excessive water content. Likewise, the pH of acidified products should be evaluated.

Design of heating (and reheating) processes or other processes intended to be lethal to microorganisms should be based on appropriate testing and calculations to ensure an appropriate reduction of microbial populations that could be in a food. The hazard analysis should include calculations or time-temperature evaluations to confirm whether a process is or is not killing enough foodborne pathogens that might be in a product to make the product safe.

The possibility of contamination after heat processing has been completed must not be neglected. Precautionary steps could be to wrap or further process heated products in an area physically separated from areas used to process raw products. Also, different persons than those who handled the raw product should work with heat-processed foods, and they should avoid touching cooked food by using utensils, waxed paper, disposable plastic gloves, or similar aids.

Evaluation of the time-temperature exposure foods receive after heat processing is crucial in hazard analyses. Cold-holding and hot-holding equipment must be of such design and capacity and used to keep foods at temperatures out of the range that permits multiplication of foodborne pathogens. Of equal importance, operational procedures should be such that foods are cooled rapidly to temperatures below those that support multiplication of bacteria at a logarithmic rate. Whenever deficiencies are found, immediate action must be taken to correct the deficiencies.

Education and Training

Education and training are the keystones of effective industrial quality control and governmental food protection programs. Unfortunately, these vital activities are often neglected or given low priority.

Essential for the development of a food protection program in a community or nation is for the populace of either (or both) to become aware that foods can be important vehicles of disease or develop aesthetic values that incline them to demand a clean food supply. This awareness is related to acceptance of the germ theory and other scientific principles. Aesthetics often do not appear until the mode of life improves beyond a bare subsistence level. After either or both of these attitudes have developed, efforts should be made to reach community or religious leaders with information about the value of a clean, nutritious, and safe food supply. Next, teachers and nutritionists need to be trained in the principles of food safety, and they also need to be encouraged to present this information in their classes and demonstrations. Every school food service operation should be a demonstration of nutritious, clean, and safe food, as well as a means to provide a balanced meal. Information about foodborne diseases and their prevention can be presented to adults through newspaper and magazine articles, in extension classes, by talks to social clubs and other gatherings. Leaders of so-called "consumer" groups must be informed of the important food-related disease problems so that their activities will have a positive rather than a negative effect on foodborne disease prevention. During periods when there is notoriety of local or national outbreaks of foodborne disease, citizens are most receptive of information about ways to protect themselves and their families from illness. Information about foodborne disease prevention should be released at these times as well as before holidays, before picnic and camping seasons, before times when produce or fish are usually canned, and at other times public warnings are deemed necessary.

Students of public health, environmental health, microbiology, and food science (many of whom will be subsequently employed in the food industry for quality control programs, in governmental regulatory agencies, and by educational institutions) must learn the principles of foodborne disease control and should become motivated to contribute to the field and to their community. They should acquire information on the components of a contemporary food protection program, and at the same time learn to question the value of outdated procedures or policies that do not have scientific and epidemiologic bases. They should also become skillful in detecting foodborne disease problem situations and devising practical preventive measures. For this to happen, university and college faculties in this field must be so educated and take an active role in applied research regarding food safety.

Managers and supervisors in the food industry should become aware of foodborne disease problems that confront their segment of the industry. They must learn the primary sources of foodborne disease agents and the other factors that contribute to foodborne disease outbreaks. So informed, they must become motivated to ensure that their operations are either free of such situations or they must be able to recognize undesirable situations and take appropriate measures to prevent them. These managers and supervisors should, in turn, train the workers under their direction in safe ways to handle food and proper sanitary maintenance of processing and preparation equipment. In the final analysis, it is the workers who make operations safe or hazardous, so they must know how to handle foods in a safe and sanitary manner.

CONCLUSION

The activities to prevent and control foodborne disease must employ the techniques that minimize contamination and the processes that kill microorganisms or prevent their multiplication. These activities are interrelated, and often the success of one is interdependent on the application of another. If a satisfactory balance of these activities is maintained in industrial quality control programs and governmental food protection programs, the probability of occurrence and the incidence of foodborne illnesses should decrease drastically.

REFERENCES

ANGELOTTI, R., FOTER, M. J., and LEWIS, K. H. 1961. Time-temperature effects on salmonellae and staphylococci in foods. III. Thermal death-time studies. Appl. Microbiol. 9:308–315.

BRODIE, J. 1965. Hand hygiene. Scot. Med. J. 10:115–125.

BRYAN, F. L. 1974. Microbiological food hazards today—Based on epidemiological information. Food Technol. 28(9):52–66.

BRYAN, F. L. 1978a. Factors that contribute to foodborne disease outbreaks. J. Food Prot. 41:816–827.

BRYAN, F. L. 1978b. Impact of foodborne disease and methods of evaluating control programs. J. Environ. Health 40:315–323.

BRYAN, F. L. 1979. Infections and intoxications caused by other bacteria. *In* Foodborne Infections and Intoxications (H. Riemann and F. L. Bryan eds.) Academic Press, New York.

BRYAN, F. L., and KILPATRICK, E. G. 1971. *Clostridium perfringens* related to roast beef cooking, storage, and contamination in a fast food service restaurant. Am. J. Public Health 61:1869–1885.

BRYAN, F. L., and McKINLEY, T. W. 1974. Prevention of foodborne illness by time-temperature control of thawing, cooking, chilling, and reheating turkeys in school lunch kitchens. J. Milk Food Technol. 37:420–429.

BRYAN, F. L., ANDERSON, H. W., ANDERSON, R. K., BAKER, K. J., MATSUURA, H., McKINLEY, T. W., SWANSON, R. C., and TODD, E.C.D. 1976. Procedures to Investigate Foodborne Illness. International Association of Milk, Food, and Environmental Sanitarians, Ames, Iowa.

CRISLEY, E. D., PELLER, J. T., ANGELOTTI, R., and HALL, H. E. 1967. Heat resistance of *Clostridium botulinum* type E spores in whitefish chubs. Bacteriol. Proc. p. 5.

ESTY, J. R., and MEYER, K. F. 1922. The heat resistance of the spores of *C. botulinum* and allied anaerobes. XI. J. Infect. Dis. 31:650–663.

GENIGEORGIS, C., and RIEMANN, H. 1979. Food processing and hygiene. *In* Food-borne Infections and Intoxications (H. Riemann and F. L. Bryan, eds.). Academic Press, New York.

GOEPFERT, J. M., and BIGGIE, R. A. 1968. Heat resistance of *Salmonella typhimurium* and *Salmonella senftenberg* 775W in milk chocolate. Appl. Microbiol. 16:1939–1940.

GOODFELLOW, S. J. and BROWN, W. L. 1978. Fate of *salmonella* inoculated into beef for cooking. J. Food Prot. 41:598–605.

HANSEN, N. H., and RIEMANN, H. 1963. Factors affecting the heat resistance of nonsporeforming organisms. J. Appl. Bacteriol. 26:314–333.

INTERNATIONAL COMMISSION ON MICROBIOLOGICAL SPECIFICATIONS FOR FOODS. 1974. Microorganisms in Foods: 2. Sampling for microorganisms: Principles and specific applications. University of Toronto Press, Toronto.

ITO, K. A., SESLAR, D. J., MERCERN, W. A., and MEYER, K. F. 1967. The thermal and chlorine resistance of *C. botulinum* types A, B, and E spores. *In* Botulism 1966. (M. Ingram and T. A. Roberts, eds.) Chapman and Hall, London. pp. 108–122.

KAZAL, H. L. 1976. Laboratory diagnosis of foodborne diseases. Ann. Clin. Lab. Sci. 6:381–399.

KELLS, H. R., and LEAR, S. A. 1960. Thermal death time curve of *Mycobacterium tuberculosis* var. *bovis* in artificially infected milk. Appl. Microbiol. 8:234–236.

LOWBURY, E. J. L. 1961. Skin disinfection. Clin. Pathol. 14:85–90.

LOWBURY, E. J. L., LILLY, H. A., and BULL, J. P. 1963. Disinfection of hands. Removal of resident bacteria. Br. Med. J. 1:1251–1256.

LOWBURY, E. J. L., LILLY, H. A., and BULL, J. P. 1964. Disinfection of hands. Removal of transient organisms. Br. Med. J. 2:230–233.

NATIONAL CANNERS ASSOCIATION RESEARCH LABORATORIES. 1968. Laboratory Manual for Food Canners and Processors. Vol. 1. Microbiology and Processing. AVI, Westport, Conn.

NORTHOLT, M. D., VERHULSDONK, C. A. H., SOENTORO, P. S. S., and PAULSCH, W. E. 1976. Effect of water activity and temperature on aflatoxin production by *Aspergillus parasiticus*. J. Milk Food Technol. 39:107–174.

PRICE, P. B. 1938. The bacteriology of normal skin. A new quantitative test applied to study of the bacterial flora and the disinfection action of mechanical cleansing. J. Infect. Dis. 63:301–318.

RAY, B., and SPECK, M. L. 1973. Freeze-injury in bacteria. CRC Crit. Rev. Clin. Lab. Sci. 4:161–213.

READ, R. B., BRADSHAW, J. G., DICKERSON, R. W., and PEELER, J. T. 1968. Thermal resistance of salmonellae isolated from dry milk. Appl. Microbiol. 16:998–1001.

ROBERTS, T. A. 1968. Heat and radiation resistance and activation of spores of *Clostridium welchii*. J. Appl. Bacteriol. 31:133–144.

ROBERTS, T. A., and INGRAM, M. 1965. The resistance of spores of *Clostridium botulinum* type E to heat and radiation. J. Appl. Bacteriol. 28:125–137.

SCHMIDT, C. F. 1964. Spores of *C. botulinum*: Formation, resistance, germination. *In* Botulism 1964. (K. H. Lewis and K. Cassel, eds.) pp. 69–79. U.S. Department of Health, Education, and Welfare, Cincinnati, Ohio.

SCOTT, W. J. 1953. Water relations of *Staphylococcus aureus* at 30°C. Austral. J. Biol. Sci. 6:549–564.

STUMBO, C. R. 1973. Thermobacteriology in Food Processing, 2nd ed. Academic Press, New York.

STUMBO, C. R., MURPHY, J. R., and COCHRAN, J. 1950. Nature of thermal death time curves for P.A. 3679 and *Clostridium botulinum*. Food Technol. 4:321–326.

THOMAS, C. T., WHITE, J. C., and LONGREE, K. 1966. Thermal resistance of salmonellae and staphylococci in foods. Appl. Microbiol. 14:815–820.

TOMKIN, R. B. 1977. Control by chlorination. *In* Proceedings of the International Symposium on *Salmonella* and Prospects for Control. University of Guelph, Guelph, Ontario, Canada.

TROLLER, J. A., and STINSON, J. V. 1975. Influence of water activity on growth and interotoxin formation by *Staphylococcus aureus* in foods. J. Food Sci. 40:802–804.

TSUJI, K., and PERKINS, W. E. 1962. Sporulation of *Clostridium botulinum*. 1. Selection of an aparticulate sporulation medium. J. Bacteriol. 84:81–85.

WEBSTER, R. G., and ESSELEN, W. B. 1956. Thermal resistance of food poisoning organisms in poultry stuffing. J. Milk Food Technol. 19:209–212.

Toxic Chemicals in Foods

9

Food-Borne Diseases of Animal Origin

P. M. Morgan, W. W. Sadler and R. M. Wood

Authoritative documents of the World Health Organization (1959) and the U.S. Public Health Service (1965) list more than 100 infections and infestations (zoonoses) that are shared by man and other vertebrate animals in the Western Hemisphere. Approximately 30 animal diseases are considered to be transmissible to man from his food-producing animals namely through mammalian meat (Dolman 1957), milk (Kaplan *et al.* 1962), or poultry meat or eggs (Galton and Arnstein 1959). These diseases are listed in Tables 9.1–9.3. Food-borne disease is an ever present hazard to man. Increasing population pressures (of man and domestic animals), increasing contamination of animals from both feeding and processing, changes in food habits and processing techniques, and other factors have in many ways increased the potential hazard posed by diseases of animal origin.

Many factors serve to increase the importance of foods as vehicles of diseases of animal origin. Epidemiological characteristics of the disease, pathogenicity of the agent, status of disease control or eradication programs, stability or viability of the agent, level of inspection and sanitary control, and food habits reflected in processing and cooking practices all must be considered in an attempt to elucidate the importance of each disease.

TABLE 9.1.   LISTING OF ANIMAL DISEASES TRANSMISSIBLE TO MAN THROUGH MEAT

Zoonoses acquired occasionally through the intestinal tract	Zoonoses acquired chiefly through the intestinal tract
Pasteurellosis	Salmonellosis
Tularemia	Shigellosis
Pseudotuberculosis	Meat-borne helminthic zoonoses
Pasteurella multocida infection	Trichinosis
Leptospirosis	Cysticercosis (Taeniasis)
Erysipelas (Swine)	Echinococcosis[1]
Listeriosis	Rare meat-borne zoonoses, possibly
Miscellaneous	acquired by ingestion
Foot and Mouth Disease	Toxoplasmosis
Q Fever	Sarcosporidiosis
Ornithosis	Intestinal Myiasis
Vibrio foetus infection	
Commoner zoonoses acquired occupationally by meat handlers	
Anthrax	
Bovine Tuberculosis	
Brucellosis	

Source: Dolman (1957).
[1]Indirect transmission only.

DISEASES OF PRIMARY IMPORTANCE

Anthrax

The role of *Bacillus anthracis* as a cause of disease in man and animals is well known. Distribution of this agent is worldwide, and herbivorous animals are particularly susceptible. Swine have, on occasion, been found to be affected, but there are no reports of this disease in poultry. Anthrax is an acute disease, readily diagnosed both in the living animal and in the carcass at postmortem inspection. Although contacts with wool, hides, and animals are the main sources of infection in man, there are recorded cases of infection from both ingesting and handling meat of diseased animals (Dolman, 1957). In acute anthrax the milk secretion is suppressed or so altered in appearance that it is not appetizing to humans. However, there is danger of anthrax bacilli entering clean milk from the discharges of infected animals in a herd. Although there are no reports of human infection from this source, milk from a herd in which there is an anthrax case ideally should not be used; at least it should be pasteurized or otherwise heat treated before consumption.

Brucellosis

Brucella abortus, *B. suis*, and *B. melitensis* are distributed worldwide and are recognized as causing one of the most common, important, and widespread of the zoonoses. All three species can

TABLE 9.2. LISTING OF ANIMAL DISEASES TRANSMISSIBLE TO MAN THROUGH MILK

Anthax[1]
Brucellosis
Coli infections (pathogenic strains of *Escherichia coli*)
Foot and mouth disease
Leptospirosis[1]
Paratyphoid fever
Q Fever
Salmonellosis (other than typhoid and paratyphoid fevers)
Staphylococcal enterotoxic gastroenteritis
Streptococcal infections
Tick-borne encephalitis
Toxoplasmosis[1]
Tuberculosis

Source: Kaplan *et al.* (1962).
[1]Not conclusively incriminated as milk-borne, but epidemiologically probable or suspect.

infect man, but with different degrees of virulence. Brucellosis causes severe economic loss in all the food-producing mammals as well as severe and prolonged disease in man. The organisms have been isolated from naturally infected poultry, and infection has been caused in fowl by feeding naturally infected milk as well as other agents (Emmel and Huddleson, 1929). There is no evidence that infection is significant in poultry, however, nor that these animals are a source of infection for man. Brucellosis is a classical example of a milkborne zoonosis, and its epidemiology was clarified in the early part of this century. There is increasing evidence that wildlife plays a significant role in its epidemiology. Improperly cooked goat meat and hams have been incriminated in the transmission of brucellosis (Dalrymple-Champneys, 1960); (Hutchings *et al.*, 1952). Dolman (1957) feels there is evidence that the handling, and particularly the processing of meat, results in significant exposure of individuals. Because there are no pathognomonic lesions by which infected carcasses may be identified at postmortem, untold numbers of such carcasses may be handled and consumed. A probable incidence of infection in animals slaughtered in the United States before 1958 was 1.25% of cattle and 1.5% of swine (Sadler, 1960), but eradication programs have drastically reduced the numbers of infected swine and cattle since then. The degree of exposure in various countries is a direct reflection of the level of eradication and control programs that exist.

Erysipeloid

Erysipelothrix insidiosa is a frequent cause of severe disease in domestic swine and turkeys. It is recognized as the cause of an

TABLE 9.3.   LISTING OF ANIMAL DISEASES TRANSMISSIBLE TO MAN THROUGH POULTRY MEAT OR EGGS

Arthropod-borne encephalitides[1]
Brucellosis[1]
Erysipelas (swine)
Listeriosis[1]
Newcastle disease
Paracolon infections
Pasteurella multocida infection[1]
Pseudotuberculosis[1]
Psittacosis-ornithosis
Salmonellosis
Staphylococcal infections or intoxications
Tuberculosis
Toxoplasmosis[1]

Source: Galton and Arnstein (1959).
[1]No proved transmission.

occupational disease of varying severity in workers handling the meat of swine, turkeys, and fish (Sneath *et al.*, 1951). Because the disease takes the form of a septicemia in the food animals, the organism is widely distributed throughout the tissues and may even be isolated from the tissues of symptomless and lesionless animals and birds. Although meat from these species may act as a vehicle for contact transmission, there are few authenticated reports of infection through ingestion (Dolman, 1957).

Salmonellosis (See also Chapters 5 and 8)

The literature is replete with reports of surveys for Salmonellae in meat animals and poultry, processed foods, animal feeds, wild fauna, ranches and feed lots, environs of slaughter and rendering plants, and almost all other environments of animals and their products. Almost all such surveys report positive results. Some member of this ubiquitous group of organisms apparently can be found in the intestinal tract of almost any animal species and, therefore, in the environment contaminated by feces. In addition to the host-specific *Salmonella typhi*, any of the other 1700 or more serotypes apparently can cause disease in man. However, these organisms also can establish themselves in the intestinal tract of man, as they can in other animals, without development of frank disease.

The recent apparently dramatic increase in salmonellosis in man and isolation of Salmonellae from his environment probably reflects more than just an increase in prevalence. The establishment of a Salmonella Surveillance Unit at the U.S. Public Health Service, Center for Disease Control and its continuing compilation and publication of summary reports, has served to focus attention on the magnitude of the *Salmonella* problem in the United States (Center for Disease Control). Increased awareness and interest on the part of Public Health Service officials, more complete reporting, and improvement and standardization of sampling and laboratory techniques have undoubtedly contributed significantly to the apparent increase in prevalence. There is no question, however, that changes in environment, animal husbandry practices, and processing and distribution of food and feed have resulted in a increased level of hazard from these organisms. In the United States the U.S. Food and Drug Administration, as well as many state and local health departments, has instituted programs of sampling a wide variety of foods and other substances for Salmonellae. The continuing reports of their findings would appear to substantiate the ubiquity of these organisms and the magnitude of the problem of preventing their contamination of human food. Such diverse commodities as dried milk, brewers yeast, candy bars, carmine dye, and candy for dogs (which children might eat) have been recalled from commercial channels because they were found to be contaminated with Salmonellae. Although the original contamination is sometimes quite remote from the final product, usually it can be successfully traced back to animal feces.

Random samplings of poultry at slaughter have revealed an intestinal carrier rate of three to five per cent (Sadler and Corstvet, 1965). Surveys have shown that, although the percentage of swine demonstrably shedding Salmonellae in feces may be low in the farm environment, it is dramatically increased during transportation and holding before slaughter (Galton *et al.*, 1954; Williams 1964). This increases the possibility of contamination during slaughtering and processing. The massive contamination that can occur during processing of cattle, swine, and poultry has been demonstrated repeatedly (Galton *et al.*, 1954; Kampelmacher *et al.* 1961; Wilder and MacCready 1966). Although fecal contamination is the major source of Salmonellae, they also have been isolated from the lymph nodes of normal swine (Kampelmacher *et al.*, 1963). Milk is easily contaminated with fecal matter and thus is frequently the vehicle for transmission of Salmonellae. Although fresh shell eggs rarely contain Salmonellae, the shell surface may be contaminated and liquid egg products then become contaminated during breaking. Dried milk and dried and frozen eggs frequently have been found to be contaminated. Any food, including water, that has been directly or indirectly contaminated with feces may act as a vehicle for these organisms, and the risk of contamination directly reflects the conditions of sanitation under which the food was processed, distributed, and prepared for consumption.

Tuberculosis

This disease has been a scourge of mankind since recorded history began. Although the human to human transmission of *Mycobacterium hominis* accounts for the vast majority of the human cases, *M. bovis* and *M. avium* also cause disease in humans. *M. avium* apparently has a low pathogenicity for humans, and only about 50 human cases of this type have been documented worldwide. Swine are subject to infection with all three types of tubercle bacilli. The incidence of these types in hogs depends upon the locality where the animals live and the character of their feed. Infection with the human type, for example, almost always occurs in swine that are fed on garbage, while the avian type is found in swine that run with chickens. *M. bovis* is the type found in the vast majority of cases of bovine tuberculosis, but *M. hominis* has been known to infect cattle, *M. avium* has been isolated from cattle and is known to cause problems with false positive tuberculin reactions. Chickens are not susceptible to mammalian tubercle bacilli. All cases are caused by avian-type organisms.

Tubercle bacilli are frequently excreted in the milk of tuberculous cows and may also be found in a contaminated environment. There appears to be little evidence that tuberculosis is transmitted through either poultry meat or mammalian meat. The bacilli are occasionally found in the eggs from tuberculous hens, and there is a report of circumstantial evidence for this as the source of human infection (Bradbury and Younger, 1946). As with brucellosis, the risk of foodborne tuberculosis is a direct reflection of the eradication and control programs that exist in a specific country or area.

DISEASES OF SECONDARY IMPORTANCE

Listeriosis

Listeria monocytogenes is distributed worldwide, and its isolation from domestic animals is reported with increasing frequency (Graham, 1963). The finding of antibody titers in apparently healthy humans and in patients of various types suggests that this organism is a more frequent invader of man than the rarely diagnosed cases would indicate (Bruner and Gillespie, 1966). Except for the intrauterine route, the transmission chain is obscure. Several reports have indicated that *Listeria* may be excreted in the milk of cows both with and without evidence of mastitis. Seeliger (1961) has summarized evidence from human cases in Europe indicating that milk was the vehicle in many instances. Although the organism is known to cause disease in poultry as well as mammals, there is no evidence of meat-borne transmission. This disease may be more important in humans than is generally recognized, and raw milk may be a significant source of infection.

Paracolon Infections

Paracolon organisms of the Arizona group are frequently the cause of disease in turkeys and may be isolated from the feces of healthy carriers. They are known to have caused outbreaks of human gastroenteritis after ingestion of poultry meat (Galton and Arnstein, 1959) and should be considered as part of the overall "Salmonella Problem."

Q Fever

The causative organism, *Coxiella burnetti*, is distributed world-wide, and cattle, sheep, and goats are the principal reservoirs for human infection. Infection does not result in frank disease in these domestic animals, and so carriers are not readily detected. Although most human infection is by inhalation of dust previously contaminated with uterine discharge of infected animals, the organisms are excreted in milk, and infection from this source has been documented (Kaplan *et al.*, 1962). Infected cattle continue to excrete *C. burnetti* in their milk for prolonged periods of time, and the organism is relatively heat resistant. U.S. PHS recommendations for pasteurization became more stringent as a result of a study of this heat resistance (Enright *et al.*, 1957).

Staphylococcal and Streptococcal Infection

Both of these organisms vary widely between their species and strains in both pathogenicity and virulence. From the standpoint of animal-borne disease, milk is the only significant source of infection. Strains of both bacteria that are pathogenic to man have, on occasion, been isolated from the milk of cows with normal as well as mastitic udders (Kaplan *et al.*, 1962). Outbreaks of disease in man have conclusively been traced to milk as the source of staphylococcal intoxication and particularly streptococcal infection. However, human sources are far more important than other animals in the epidemiology of human disease caused by both of these agents.

DISEASES POTENTIALLY FOOD-BORNE OR OF OCCUPATIONAL SIGNIFICANCE

Escherichia coli Infection

Several types of *E. coli* implicated in infantile diarrhea have been isolated from cases of white scours in calves or from mastitic udders of cows (Kaplan *et al.*, 1962). Other possible sources of infections, such as poultry and swine, have been listed by Taylor (1961). Although the epidemiological significance of these findings is not yet understood, a potential hazard is obvious in milk from these sources.

Leptospirosis

The various species of *Leptospira* are distributed worldwide and infect common domestic animals as well as rodent populations. Acute, chronic, and latent infections of man and domestic food animals are being reported with increasing frequency (Bruner and Gillespie, 1966). *Leptospira* organisms are excreted in the urine of infected animals for long periods after recovery, resulting in pollution of water, foods, and food-handling surfaces. Fish-gutters are apt to have an increased incidence of leptospirosis. Milk-producing animals may shed the organism in the milk when infected, but human infection has not been traced to this source, nor to consumption of meats (Kaplan *et al.*, 1962). Man's infection occurs either directly from urine or tissues of a diseased animal, or indirectly through contact with contaminated water, soil, or surfaces. Domestic fowl are not involved.

Ornithosis (Psittacosis)

Turkeys and pigeons have been incriminated in the epidemiology of this disease in humans, but all cases attributed to exposure to these birds resulted from the occupational exposures of rearing or slaughtering (Galton and Arnstein 1959). There is no evidence of transmission in food.

Newcastle Disease

This virus is distributed worldwide in avian species and has been identified as a cause of disease in man. Although there have been reports of systemic involvement, the majority of cases in man have been limited to a conjunctivitis. There is no evidence of food-borne transmission, but it does constitute an occupational hazard to persons raising and slaughtering poultry (Galton and Arnstein, 1959).

Pasteurelloses

Pasteurella tularensis is known to infect humans as a result of dressing wild rabbits or deer, and infection by ingestion has been recorded in Japan and Russia (Dolman, 1957). Although infected sheep and cattle have been found in the United States, there are no reports of human infection from domestic food animals.

Pasteurella multocida is widespread in populations of domestic livestock and fowl. It is a common inhabitant of the respiratory tract of healthy animals, both domestic and wild. Recent medical literature reveals a constantly increasing incidence of disease in man caused by this organism (Schwartz and Kunz, 1959). The disease caused by this organism in domestic animals is a septicemia; therefore, the organism is present in all tissues of the body during the acute stage. Although there is obviously potential food-borne risk as well as risk in occupational exposure, no transmissions by these routes have been conclusively documented.

Pasteurella pseudotuberculosis also produces a septicemia in birds, mammals, and man. Although it is not uncommon in birds and other animals, its identification as the etiological agent of human disease is extremely rare, and the mode of transmission is unknown.

DISEASES OF QUESTIONABLE SIGNIFICANCE

Foot and Mouth Disease

This is one of the most highly contagious diseases of cattle, but man appears to be only slightly susceptible to it. The few clear-cut cases in man all have been traced to occupational exposure, and none was due to ingestion. A somewhat similar disease, vesicular stomatitis, is frequently confused with foot and mouth disease. Vesicular stomatitis virus is a common human pathogen (causing a mild influenza-like syndrome) in regions where the virus is enzootic in the livestock population (Hanson and Karstad, 1956). Occurrence of the virus in man appears to be a result of occupational exposure.

Diseases Caused by Arthropod-Borne Encephalitis Viruses

The mosquito-borne equine encephalitis viruses have been isolated from domestic poultry, thus demonstrating that they play at least a minor role in the transmission chain. Although exposure may occur through contact or ingestion, there is no evidence of transmission to humans by these routes.

The tick-borne encephalitis virus (group B of the arboviruses) is on the North American continent (World Health Organization, 1961). This virus can be transmitted to man by drinking milk from infected goats, but there have been no reports of virus isolation from the milk of naturally infected cows (Kaplan *et al.*, 1962).

Paratyphoid Infection

Salmonella paratyphi B (*schottmuelleri*) has been isolated from chickens, turkeys, cattle, sheep, and swine (Edwards *et al.*, 1948). Although there appear to be no reports of transmission to man from these sources, infected animals (particularly milk cows) could prove a potential hazard. This organism, along with the paracolons, should be considered in the overall "*Salmonella* Problem."

Infections with Shigella sp., Vibrio foetus and Entamoeba histolytica

Although Dolman (1957) believes that the report of an isolation of *Shigella sonnei* from the bone marrow of a two-month old calf has public health significance, Kaplan *et al.* (1962) point out that there is no evidence that it infects adult cattle or is excreted in milk. A single report of such an anomaly should be considered just that.

Vibrio foetus is a common cause of abortion in cattle and sheep and has been isolated from the blood of pregnant women. One of these women aborted in the last trimester of pregnancy; one gave birth prematurely; and another carried to normal term. The agent has been isolated also from the blood of a man with no occupational exposure who drank raw beef serum (Dolman, 1957). These reports suggest a potential hazard, both from occupational exposure and ingestion, but the paucity of reports suggests that infection is an extremely rare occurrence.

Entamoeba histolytica, one of the etiological agents of amoebic dysentery, can be transferred to monkeys, rodents, rabbits, cats, dogs, and pigs. *E. coli*, another agent, has been transferred to monkeys, cats, and rats. Natural infections with *E. histolytica* occur in most animals in which experimental infections have been produced. *E. polecki*, common in pigs and various ruminants, occasionally infects man (Chandler and Read, 1955). Although under certain levels of environmental sanitation there might possibly be cycling of these amoebas between man and animals, this cycle plays at most a minor role in the epidemiology of this disease. A sanitation level promoting this domestic animal-to-man cycling would be conducive to the far more significant human-to-human transmission by the feces-food route.

Toxoplasmosis, Sarcosporidiosis, and Intestinal Myiasis

Toxoplasma gondii infects both man and domestic animals. This agent can be transferred between swine and rats by feeding infected tissues. Toxoplasmosis is transmitted congenitally, following maternal consumption of improperly cooked meat. Direct evidence of transmission of the organism to man via the oral route has been documented (Desmonts *et al.*, 1965).

Sarcocystis sp. are frequently found in domestic animals, but rarely in man. Little is known of the life cycle, and there is no evidence that this is a food-borne disease. Intestinal myiasis results from ingestion of foodstuffs contaminated with larvae and cannot be considered to be a disease transmitted by domestic animals.

MEAT ANIMAL PARASITES

Of the 29 parasitic infections commonly shared by animal and man, 3 are considered major zoonoses in the Americas. The diseases produced by various stages in the life cycle of these three parasites are trichinosis, hydatidosis, and cysticercosis. Infected or contaminated meat is a vital link in maintaining the cycle of these diseases.

Trichinosis

Trichinella spiralis, a nematode, may cause trichinosis in man, swine, rodents, and other mammals. This parasite completes all stages of its life cycle in one host, but infection is incurred by consuming viable encapsulated larvae in the meat of an infected animal. Man commonly becomes infected by eating improperly cooked pork. Swine and rodents are infected by eating raw or partially cooked garbage that contains infected meat scraps. When infective larvae are ingested, the cyst walls are dissolved in the hosts' stomachs, and the immature worms travel into the intestines. There they mature and reproduce. The larvae released by adult female parasites in the mucosa penetrate the intestines, enter the blood stream, and are transported throughout the body. They penetrate the musculature of the body, where they grow and become encysted. At this point, the life cycle of *T. spiralis* has been completed. If live, encysted larvae are ingested by man or another mammal, the cycle begins again.

Clinical symptoms produced in man vary with the stage of the parasites' development. When the immature worms reach the intestines, there may be nausea, vomiting, and diarrhea. This usually occurs within 24 to 48 hours after the encysted larvae are ingested. When the larvae are migrating through the muscles, there is severe myalgia, rising fever, and occasionally, prostration and collapse. When the migration through the muscles stops, the patient may exhibit severe tissue reactions, such as swelling of the face, subcutanous hemorrhage, severe headache, and difficult breathing. A few patients show central nervous system involvement, with such symptoms as convulsions or comata. Approximately five per cent of those who develop clinical symptoms of trichinosis die of the disease.

Control measures that may be used against trichinosis include: (1) proper cooking of all garbage fed to swine; (2) freezing or heating of all pork so that any larvae present are killed; and (3) public education concerning the importance of cooking or properly treating pork and pork products.

Although it is hoped that thiabendazole and other drugs under investigation will prove beneficial, currently there is no satisfactory treatment for trichinosis in man or animal.

Research is being conducted to arrive at an effective method of diagnosis. Indirect and soluble antigen fluorescent antibody tests are being favorably reviewed.

Hydatidosis

Hydatidosis in man is caused by the larval stage (or hydatid cyst) of two species of cestodes, *Echinococcus granulosus* and *E. multilocularis*. Dogs and other canines are the natural hosts of the adult parasite. They become infected by eating the flesh of other animals that contain the hydatid cysts. Man contracts the disease by ingesting the infective ova of these two parasites. Thus, it is important to maintain sanitary practices that prevent the contamination of food and water consumed by man with the ova of *E. granulosus* and *E. multilocularis*.

Pathogenicity in man varies with the species of ova ingested. Generally, disease develops when the immature embryos have penetrated the intestinal wall and migrated to the liver, lungs, or osseous tissue, and occasionally the brain. The cyst usually has a central fluid-filled cavity surrounded by a friable membrane and a host-tissue capsule. In bony tissue, a capsule is not formed and the organism grows as a protoplasmic stream, eroding the cancellous structure of the invaded bones. Lesions resulting from soft tissue invasion vary with the organ involved. The rupture of a large cyst, releasing cystic fluid into the peritoneal cavity, may produce anaphylaxis and death.

Control measures for hydatidosis include: (1) stringent control of slaughterhouse operations so that dogs do not have access to uncooked viscera; (2) destruction of feral dogs or other wild carnivores; (3) education of people who have close contact with dogs concerning the necessity of personal hygiene practices; and (4) mass treatment of dogs to destroy the adult tapeworms.

Chemotherapy is of no value with hydatid disease. If the cyst is located where surgery is feasible, removal of the cyst is standard procedure.

Taeniasis and Cysticercosis

Within the limited scope of this chapter taeniasis refers to human infections with the adult tapeworms, *Taenia saginata* or *T. solium.* Cysticercosis, unless specified otherwise, refers to human infection with *Cysticercus cellulosae*, the larval stage of *T. solium.*

Man incurs taeniasis by eating insufficiently cooked beef or pork that is infected with *Cysticercus bovis* or *C. cellulosae*. Cattle and swine become infected with these larvae by ingesting ova of *T. saginata* or *T. solium* passed in the feces of infected humans. Transmission may be direct as a result of human defecation in areas where cattle or swine are confined, or indirect by contamination of pastures with sewage or drainage water that contains viable ova.

Except for the irritation of proglottids crawling from the anus, taeniasis in man is often symptomless. There may be, however, diarrhea, abdominal pain, and weight loss. Occasionally, there may be liver damage from toxic metabolites of the tapeworms. On rare occasions, a mass of worms may cause acute intestinal obstruction. *Taenia saginata* generally produces more severe symptoms than does *T. solium.*

Control of taeniasis in man is based upon: (1) sanitary disposal of human feces so that cattle and swine do not develop animal cysticercosis; (2) thorough meat inspection procedures to reduce the chance of infected meat being used for food; and (3) adequate cooking or freezing of all beef and pork to be consumed.

Niclosamide and Dichlorophen are considered safe and effective human anthelmintics. Niclosamide has fewer side effects and is usually the drug of choice.

In addition to being the only natural host for the adult form of *T. solium*, man may also be infected with the larval form of this parasite, *Cysticercus cellulosae*. Except for rare, nonrelated cases, man has never been known to be infected with *C. bovis*. Human cysticercosis, results when the infective ova of *T. solium* reach the duodenum. This usually occurs in one of three ways: (1) ova passed in the feces of one individual are swallowed by another (heteroinfection); (2) ova may be transferred from anus to mouth of the same individual (external autoinfection); and (3) gravid proglottids in an individual infected with adult *T. solium* may be regurgitated into the stomach, return to the duodenum, and release infective ova (internal autoinfection).

Cysticercosis in man is often a serious debilitating or even fatal disease. Clinical lesions produced depend upon the number of cysticerci that develop and the tissue in which they localize. Unless nerve endings are affected, cysticerci that develop in subcutaneous or muscle tissue cause little discomfort. The eye and brain are other common sites of cyst localization. In the eye, lesions may range from mild irritation to destruction of one or more ocular structures with subsequent blindness. Cysts in the brain may produce a variety of neurological manifestations including epileptiform seizures, collapse, and death.

Control of this disease is dependent upon: (1) personal hygiene, including sanitary disposal of human feces, and (2) inspection of swine slaughtered as food animals for man. Recent studies show promise for an antemortem antigenic test that is quite accurate in detecting cattle cysticercosis. It is hoped that a similar procedure, both accurate and practical, will be developed for swine.

Surgical removal of the cyst is the only effective treatment for cysticercosis.

COMPONENTS OF AN EFFECTIVE MEAT INSPECTION SYSTEM

An effective and standardized meat inspection program is complex. There are certain general requirements for the efficient operation of any meat inspection system. These are usually considered under the headings of antemortem inspection, postmortem inspection, packing and labeling procedures, physical facilities of the slaughterhouse, and the professional ability and personal integrity of the inspectors.

Antemortem Inspection

Adequate animal holding facilities are necessary for antemortem inspection. There must be holding pens that can be easily cleaned and that have provisions for the segregation and individual examination of animals suspected to be diseased. Personnel must be supplied for handling and restraining animals.

Postmortem Inspection

An essential aspect of postmortem inspection is that each individual carcass be thoroughly examined and that the parts of each animal retain their identity. That is, the inspector must be able to determine from which carcass each head, liver, etc. came.

Sanitary Practices

Approved sanitary practices should be maintained at all times to prevent contamination of exposed meat surfaces. All personnel should be required to practice correct personal hygiene.

Packing and Labeling Procedures

Packing and labeling should be performed under the supervision of the inspector. Official stamps should be retained by the inspector and used only under his direct supervision.

Slaughterhouse Facilities

Slaughterhouse construction and equipment should be such that adequate sanitation can be maintained and satisfactory inspections performed. Screening, ventilation, special containers for condemned meat, and sufficient hand-washing and toilet facilities for personnel are some of the things that must be considered.

Inspector Qualifications

The professional ability, conscientiousness, and integrity of the inspector are of paramount importance. The senior inspector in each slaughterhouse should be a veterinarian who has had adequate instruction in specific meat inspection procedures. He or she should not be placed in a position of being coerced or intimidated by owners or employees of the slaughterhouse. Each inspector should be paid a salary commensurate with professional ability and experience. This should be sufficient to preclude the necessity of other employment.

Standardization

It is understood that equal standards should apply to all slaughterhouses under the jurisdiction of a meat inspection system.

Approval of Meat Inspection Systems

There has been a trend in recent years toward U.S. Dept. of Agriculture approval of "meat inspection systems" for a state or country rather than of individual slaughterhouses within the state or country. There are two general reasons given for the change in policy of granting approval to slaughter houses on an individual basis. The first is that those states or countries that economically benefit most from the production and processing of meat animals should bear the major cost of inspection. A second reason is that there are some slaughterhouses that never would be subject to direct federal inspection, and therefore, locally managed inspection would be better than none.

At this point one might state that some investigators are directly opposed to the approval of meat inspection systems, rather than individual slaughterhouses. Their opposition is based on the necessity of providing safe, wholesome meat for the consumer.

TABLE 9.4.   SIGNIFICANT FOOD-BORNE DISEASES OF ANIMAL ORIGIN. STATUS OF PROGRAMS OF DISEASE CONTROL, INSPECTION, SANITATION, PASTEURIZATION

Absent or Inadequate	Active and Effective
Mammalian Meat	
Anthrax	Salmonellosis
Brucellosis	Trichinosis
Cysticercosis (Taeniasis)	
Erysipeloid	
Salmonellosis	
Trichinosis	
Tuberculosis	
Hydatidosis	
Milk	
Brucellosis	
Listeriosis	
Q Fever	
Salmonellosis	
Staphylococcal intoxication	
Streptococcal infection	
Tuberculosis	
Poultry Meat	
Erysipeloid	Salmonellosis
Salmonellosis	

Several requirements were previously listed as necessary for any satisfactory meat inspection program. In the past, every one of these requirements has been grossly violated or completely ignored in slaughterhouses considered satisfactory under the meat inspection system approval method. It is probable that this same state of affairs still exists in some areas today.

SUMMARY

As pointed out earlier in this discussion, the level of significance of each disease reflects the conditions prevailing in, and perhaps peculiar to, a given area. The extent of the development of programs of disease control and meat and milk inspection will significantly influence the level of disease hazards, as will the socioeconomic status of the populace as reflected in their overall level of sanitation. No single evaluation of the safety of foods of animal origin can be applied to all areas of the Western Hemisphere.

Two categorizations may be made: These are (1) the hazards from food-borne animal diseases existing in areas lacking adequate programs of disease control, inspection, sanitation, and pasteurization; and (2) those existing in areas where disease control programs are effective, and inspection programs and levels of sanitation comply with U.S. Department of Agriculture and U.S. Public Health Service standards. Such categorizations are presented in Table 9.4. In those areas lacking adequate animal health and food hygiene programs, a potential vehicle exists in mammalian meats for at least seven significant diseases; in milk, for at least seven; and in poultry meat, for at least two. Although other diseases that may present a potential hazard have been discussed, they are of little significance. Also, in addition to the "Animal Diseases," a variety of "Human Diseases," particularly those caused by the enterics, are readily transmitted through food products of animal origin when unsanitary handling practices offer the opportunity of contamination. In sharp contrast, in those areas with adequate programs of disease control, inspection, sanitation, and pasteurization, mammalian meat provides a potential vehicle for only two significant diseases, poultry meat for only one, and milk (pasteurized) for none. Furthermore, the degree of *Salmonella* contamination is reduced in direct proportion to the level of sanitation adhered to, and prevalence of *Trichinella* in swine is being reduced by proper husbandry and feeding practices. Chicken eggs with dirty cracked shells, all duck eggs, and all unpasteurized egg products should be considered a source of salmonellosis. However, chicken eggs with clean, sound shells, and all properly pasteurized egg products present no hazard.

REFERENCES

ACHA, P. N., and AGUILAR, F. J. 1964. Studies on cysticercosis in Central American and Panama. Am. J. Trop. Med. 13, 48–53.

BRADBURY, F. C. S., and YOUNGER, J. A. 1946. Human pulmonary tuberculosis due to avian tubercle bacilli. Lancet 250, 89.

BRANDLY, P. J. 1962. The control of trichinosis. Presented, Fourth Pan American Congress of Veterinary Medicine and Zootechnics, Mexico City, Mexico.

BRUNER, D. W. and GILLESPIE, J. H. 1966. Hagans Infectious Diseases of Domestic Animals. Cornell Univ. Press, Ithaca, N.Y.

BUTCHART, D. W. 1966. A study of the possible use of the incidence of taeniases and cysticercoses as indicators of meat inspection efficiency and the economic consequences of inefficient meat hygiene practices. Doctor of Public Health Thesis. Tulane University, New Orleans, LA 8–14.

CHANDLER, A. C., and READ, C. P. 1955. Introduction to Parasitology. 10th Edition. John Wiley & Sons, N.Y.

DALRYMPLE-CHAMPNEYS, SIR WELDON. 1960. Brucellosis Infection and Undulant Fever in Man. Oxford U. London Press.

DESMONTS, G., COUVREUR, J., ALISON, P., BAUDELOT, J., GERBEAUX, J., and LELONG, M. 1965. Etude epidemiologique sur la toxoplamose: de l'influence de la cuisson des viandes de boucherie sur la frequence de l'infection humaine. Rev. franc Etud clin biol. 10:952–958.

DEWHIRST, L. W., CRAMER, J. D., and SHELDON, J. J. 1967. An analysis of current inspection procedures for detecting bovine cysticercosis. J. Am. Vet. Med. Assoc. 150, 412–417.

DOLMAN, C. E. 1957. Epidemiology of meat-borne disease. In Meat Hygiene. WHO Monograph 33.

EDWARDS, P. R., BRUNER, D. W., and MORAN, A. B. 1948. The genus Salmonella: its occurrence and distribution in the United States. Ky. Agr. Expt. Sta. Bull. 525.

EMMEL, M. W., and HUDDLESON, I. F. 1929. Abortion disease in fowl. J. Am. Vet. Med. Assoc. 75, 578–580.

ENRIGHT, J. B., SADLER, W. W., and THOMAS, R. C. 1957. Thermal inactivation of *Coxiella burnetii* and its relation to pasteurization of milk. U.S. Public Health Monograph 47.

FAUST, E. C., BEAVER, P. C., and JUNG, R. C. 1962. Animal Agents and Vectors of Human Disease, 2nd Edition. Lea and Febiger, Philadelphia, Pa.

GALTON, M. M., SMITH, W. V., McELRATH, H. B. and HARDY, A. B.

1954. Salmonella in swine, cattle and the environment of abbatoirs. J. Infect. Diseases 95, 236–245.

GALTON, M. M. and ARNSTEIN, P. 1959. Poultry diseases in public health. Public Health Rept. (U.S.).

GEMMELL, M. A. 1960. Advances in knowledge on the distribution and importance of hyatid disease as world health and economic problems during decade 1950–1959. Helminthological Abstracts, 29, 355–369.

GORDON, J. E. (EDITOR) 1965. Control of Communicable Diseases in Man, 10 Edition. Am. Public Health Assoc., New York, N.Y.

GRAHAM, R. 1963. Listeriosis. In Diseases Transmitted from Animals to Man, 5th Edition. T. G. Hull (Editor). Chas. C. Thomas, Springfield, Ill.

HANSON, R. P., and KARSTAD, L. 1956. Enzootic vesicular stomatitis. Proc. USLSA 16, 288–292.

HUTCHINGS, L. M., McCULLOUGH, N. B., DONHAM, C. R., EISELE, C. W., and RUNNELL, D. E. 1952. The viability of *Br. melitensis* in naturally infected cured hams. PHR, 66:1402–1408.

KAMPELMACHER, E. H., GUINEE, P. A. M., HOPTRA, K., and VAN KEULEN, A. 1961. Studies on Salmonella in slaughterhouses. Zbl. Vet.-Med. 8, 1025–1042.

———— 1963. Further studies on Salmonella in slaughterhouses and in normal pigs. Zbl. Vet.-Med. 10, 1–27.

KAPLAN, M. M., ABDUSSALAN, M., and BIJLENGA, G. 1962. Diseases transmitted through milk. In Milk Hygiene. WHO Monograph 48.

MASFERRER, R. 1966. Personal communication, Department of Anatomy and Pathology, Hospital Rosales, San Salvador, El Salvador.

MATA, L. J. 1966. Personal communication, Microbiology Division, Institute of Nutrition for Central America and Panama, Guatemala City, Guatemala.

MURPHY, E. A. 1962. Inspection of imported meats. Presented, Fourth Pan American Congress of Veterinary Medicine and Zootechnics. Mexico City, Mexico.

MURPHY, E. A., and MURRAY, G. L. B. 1962. Veterinary meat inspection responsibilities. Presented, British Caribbean Veterinary Association. Georgetown, British Guiana.

SADLER, W. W. 1960. Present evidence on the role of meat in the epidemiology of human brucellosis. Am. J. Public Health 50, 504–514.

SADLER, W. W., CORSTVET, R. E. 1965. Second survey of market poultry for Salmonella infections. J. Appl. Microbiol. 13, 348–351.

SALCEDO, O. G. 1966. Epidemiology of hydatidosis in Peru. Bol. P.A.S.B. 60, 144–153.

SCHILLER, E. L. 1964. Occupational Diseases Acquired from Animals. University of Michigan Press, Ann Arbor, Michigan.

SCHWARTZ, M. N., and KUNZ, L. J. 1959. *Pasteurella multocida* Infection in man. New Engl. J. Med. 261, 889–893.

SEELIGER, H. P. R. 1961. Listeriosis. Hafner Pub. Co., New York.

SNEATH, P. H. A., ABBOTT, J. D., and CUNLIFFE, A. C. 1951. The bacteriology of erysipeloid. Brit. Med. J. 2, 1063–1068.

SOULSBY, E. J. L. 1965. Textbook of Veterinary Clinical Parasitology. F. A. Davis Company, Philadelphia, Pa.

TAYLOR, J. 1961. Host specificity and enteropathogenicity of *Escherichia coli* J. Appl. Bacteriol. 24, 316–325.

U.S. PUBLIC HEALTH SERV. 1965. Epidemiological aspects of some of the zoonoses Rept.

U.S. PUBLIC HEALTH SERV. Communicable Disease Center. Salmonella surveillance Rept. Atlanta, Ga.

WILDER, A. N., and MACCREADY, R. A. 1966. Isolation of Salmonella from poultry. New Engl. J. Med. 274, 1453–1460.

WILLIAMS, L. P., JR. 1964. The Relationship of Feed and Environment to the Recovery of Salmonellae from Market Swine. Ph.D. Thesis, Tulane Univ. of Louisiana, New Orleans, La.

WORLD HEALTH ORGAN. 1959. Joint WHO/FAO Expert Committee on zoonoses. WHO Tech. Rept. Series 169.

WORLD HEALTH ORGAN. 1961. Study group on arthropod-borne viruses. WHO Tech Rept. Series 219.

10

Nitrosamines

N. P. Sen

In recent years, the occurrence of nitrosamines, which are produced by the interaction of nitrite and secondary or tertiary amines, in foods has been the subject of considerable interest and controversy. Studies carried out in various laboratories have shown clearly that traces of nitrosamines do indeed occur in certain foodstuffs. In some cases, they are produced during cooking due to the interaction of added nitrite, which is used as a preservative, and naturally-occurring amines in foods. Although nitrosamines have been studied from the early days of organic chemistry, it is only recently that they have been the subject of intensive studies. The current interest arose from the work of Barnes and Magee (1954) who showed that dimethylnitrosamine (DMN), the simplest member of the N-nitrosamines, is a strong hepatotoxic agent capable of causing acute liver damage in a wide variety of animal species and man. Subsequent studies by these workers (Magee and Barnes, 1956) indicated that DMN is also a strong carcinogen, inducing tumours mainly in the liver and, occasionally, in the kidney of rats. Soon it became apparent from other studies (Druckrey *et al.*, 1961, 1967; Schoental, 1960; Magee *et al.* 1976; Lijinsky *et al.*, 1969) that this carcinogenic property was a common characteristic of many other N-nitroso compounds such as nitrosodialkylamines, nitrosoureas and nitroso-guanidines, and nitrosaminoacids, etc. Some of these compounds can induce tumours even after a single dose and some can cross the placental barrier to produce tumours in the offspring (Magee *et al.*, 1976).

Interest in the nitrosamine problem increased significantly in the early sixties following an outbreak of severe liver disease in sheep and mink in Norway (Bohler, 1960, 1962). Systematic investigations revealed that the sheep became severely sick after consumption of a fish meal that had been preserved with nitrite (Koppang, 1964). Continued consumption of the feed sometimes proved fatal. The causative agent was shown to be DMN, presumably formed by the interaction of the nitrite preservative and the amines present in the fish (Ender *et al.*, 1964). This discovery alerted the scientific community to the possible occurrence of traces of nitrosamines in human foods, especially those preserved with nitrates and nitrites, and prompted the search for nitrosamines in various foods used for human consumption. The concern proved to be justified as evidenced by the finding of significant levels of nitrosamines in various foods, sometimes at fairly high levels (0.1–25 ppm). It was soon realized that nitrosation of amines can also occur *in vivo* in the acidic environment of the human stomach from the ingested amines and nitrites, thus adding another dimension to the problem (Sander, 1967; Sen *et al.*, 1969).

Since there are already a few good review articles (Crosby, 1976; Magee *et al.*, 1976; Fiddler, 1975; Sen, 1974) on the subject, no attempt will be made to cover the details. Instead, an attempt will be made to summarize and update some of the important findings and to discuss the possible health hazard implications.

FORMATION OF NITROSAMINES

Nitrosamines are easily formed by the interaction of nitrite and secondary or tertiary amines, preferably under acidic conditions. Quaternary ammonium compounds can also react with nitrous acid to produce nitrosamines, although the yields are much lower than those from secondary or tertiary amines (Fiddler *et al.*, 1972). The extent of formation, however, depends on a variety of factors such as basicity of the amine, concentration of the reactants, pH, temperature and the presence or absence of catalysts or inhibitors. A clear understanding of the kinetics of the nitrosation reaction is, therefore, essential for studying the formation of nitrosamines in foods as well as in the human stomach.

Kinetics

The kinetics of nitrosamine formation from secondary amines and nitrous acid has been studied in detail by Mirvish (1972). The actual nitrosating species can be one of the following depending on the reaction conditions: nitrous anhydride, nitrous acidium ion, nitrosyl halide or nitrosyl thiocyanate (Mirvish, 1975).

For most secondary amines the nitrosation reaction proceeds via equations (1) and (2), and the reaction rate is represented by either equation (3) or (4), as shown below:

$$2HNO_2 \rightleftharpoons N_2O_3 + H_2O \dots \dots \dots \dots \dots \dots \dots (1)$$

$$\frac{R_1}{R_2} > NH + N_2O_3 \rightarrow \frac{R_1}{R_2} > N-N=O + HNO_2 \dots \dots \dots (2)$$

$$rate = k_1 [R_1R_2NH] [HNO_2]^2 \dots \dots \dots \dots \dots \dots (3)$$

$$rate = k_2 [total\ amine] [nitrite]^2 \dots \dots \dots \dots \dots (4)$$

where k_1 and k_2 are the respective rate constants.

In equation (3), the concentrations expressed are those of unprotonated amine and undissociated nitrous acid (both of which are pH dependent), and k_1 is independent of pH. Whereas, in equation (4), the total concentrations are used, and k_2 varies with pH and shows a maximum value at pH 3.4 (Mirvish, 1975). In the pH range of 9–5, the rate of nitrosation of dimethylamine would be expected to increase by 10 fold for each decrease of 1 unit of pH (Mirvish, 1970).

Since the concentration of a nonionized (unprotonated) secondary amine is inversely proportional to the basicity of the amine, the rate constant (k_2) of nitrosation of weakly basic amines would be higher than that of strongly basic amines. This is, in fact, borne out by the results of Mirvish (1972) who studied the kinetics of nitrosation reactions of various secondary amines and amino acids (Table 10.1).

TABLE 10.1. RATE CONSTANTS FOR THE NITROSATION OF SOME SECONDARY AMINES AND AMINO ACIDS

Amine	pK_a	Optimum pH	Rate constant[1] k_2	Stoichiometric rate constant[1] k_1 x 10^{-6}
Piperidine	11.2	3.0	0.027	8.6
Dimethylamine	10.72	3.4	0.10	8.9
Morpholine	8.7	3.0	14.8	15.0
Mononitrosopiperazine	6.8	3.0	400	5.0
Piperazine	5.57	3.0	5,000	3.7
L-Proline	—	2.25	2.9	—
L-Hydroxyproline	—	2.25	23.0	—
Sarcosine	—	2.5	13.6	—

[1]Values at the optimum pH, in moles $^{-2}$ 1^2 min^{-1}
Source: Mirvish (1972)

At the optimum pH, the relative rate of nitrosation of the least basic secondary amine, piperazine (pKa, 5.57), is about 185,000 faster than that of piperidine (pKa, 11.2) and about 50,000 times faster than that of dimethylamine (pKa, 10.72). Similar results were also observed by Sander et al. (1968). The basicity of the amine, the concentration of nitrite and pH are, therefore, the three most important factors in determining the formation of nitrosamines in foods.

Similar kinetic studies (Mirvish, 1972) with N-nitrosatable amino acids, suggest that the rate of nitrosation of proline, hydroxyproline and sarcosine, all of which occur widely in nature (Gray, 1976) is proportional to the concentration of the amino acid and the square of the nitrite concentration. The optimum pH of nitrosation is believed to be 2.25–2.5. The nitrosation of alkylureas and alkylurethanes (Mirvish, 1972; Sander and Schweinsberg, 1972) proceeds rapidly under acidic conditions. However, the rate does not exhibit a maximum at any pH. In the pH range of 3–1, the rate of nitrosation for these compounds increases about 10 fold for each pH unit decrease. Furthermore, in contrast to the situation with the amines or amino acids, the rate is proportional to the concentration of nitrite and not its square.

As in the case with all chemical reactions temperature has a pronounced effect on the rate of nitrosation. For every 10°C rise in temperature, the reaction rate is doubled (Foreman and Goodhead, 1975). Ender et al. (1967) have studied the effect of temperature and storage on the formation of DMN from dimethylamine hydrochloride and sodium nitrite; the results are given in Tables 10.2 and 10.3, respectively. These results suggest that heating of nitrite-containing foods or prolonged storage, even at low temperatures, may substantially influence the formation of nitrosamines. As will

TABLE 10.2. EFFECTS OF TEMPERATURE ON THE FORMATION OF DMN FROM DIMETHYLAMINE HYDROCHLORIDE AND SODIUM NITRITE[1]

Temperature, °C	DMN formed ppm
50	0.2
60	0.4
70	0.75
80	1.55
90	3.0
100	8.3

[1]Each solution contained 40 millimole each of dimethylamine hydrochloride and sodium nitrite and was heated under reflux for 15 min. The solutions were buffered at pH 6.5.
Source: Ender et al., 1967

TABLE 10.3.   EFFECT OF STORAGE ON THE FORMATION OF DMN FROM DI-METHYLAMINE HYDROCHLORIDE AND SODIUM NITRITE[1]

No. of days stored at 4°C	DMN formed ppm
2	1.6
8	6.2
18	13.5
30	25
52	42
76	68.5
104	88
128	110
157	133.5

[1]The concentration of each reactant was 40 mM, pH 6.3.
Source: Ender et al., 1967.

be discussed later, the cooking of certain foods such as bacon, cured country-style ham, and fish enhances the formation of nitrosamines.

Catalysts and Inhibitors of Nitrosamine Formation

As mentioned earlier, the rate of nitrosation can be influenced by other compounds present in the reaction mixture. Generally, weak anions such as thiocyanates, chlorides, bromides and iodide ions have all been shown to be strong catalysts of the nitrosation reaction (Boyland, 1972; Fan and Tennenbaum, 1973). Since the thiocyanate ion is a normal constituent of human saliva its presence in the human stomach may markedly increase the *in vivo* formation of nitrosamines (Boyland, 1972). Similarly, the presence of chloride and iodide ions in foods may also influence the formation of nitrosamines. Formaldehyde and other carbonyl compounds have been shown to catalyze nitrosation reaction (Keefer and Roller, 1973) even under alkaline conditions. Most of the unhindered secondary amines studied such as diethylamine, pyrrolidine, piperidine, di-*n*-propyl-amine and dimethylamine were nitrosated easily at basic pH in the presence of formaldehyde. Diisopropylamine, tertiary and quaternary amines are not affected by such catalysis because of either steric hindrance or their inability to form iminium ions (Roller and Keefer, 1974). Gallic acid and tannins, on the other hand, act as catalysts at low concentrations but are effective inhibitors of nitrosamine formation at high levels (Walker et al. 1975).

The presence of compounds which destroy nitrite can act as inhibitors of nitrosamine formation. Thus various antioxidants such as ascorbic acid, glutathione, cysteine, ascorbyl palmitate, α-tocopherol (vitamin E), propyl gallate or other phenolic compounds have been shown to inhibit the formation of nitrosamines both in foods as well as *in vivo* in experimental animals (Mirvish *et al.*, 1972; Fiddler *et al.*, 1973; Sen *et al.*, 1976; Mirvish, 1975; Mergens *et al.*, 1977; Walters *et al.* 1976). The incorporation of these chemicals in foods or in drugs, as part of the formulation, may prove to be an effective way to minimize human exposure to nitrosamines.

In Vitro and In Vivo Formation

The possibility that traces of carcinogenic nitrosamines may be formed in the acidic environment of the human stomach has received a considerable amount of support from the various experiments carried out thus far. Sander *et al.* (1967, 1968) demonstrated the formation of nitrosamines from various secondary amines and nitrite in human gastric juice. Sen *et al.* (1969) studied the formation of diethylnitrosamine (DEN) from diethylamine and nitrite in gastric juices from man, rabbit, cat, dog and rat. The formation of DEN was shown to be more rapid in the gastric juices of man and rabbit (pH of 1–2), than in the gastric juice of rat (pH 4–5). Since gastric juices contain thiocyanate ions, the optimum pH of nitrosation in human gastric juice is closer to 2 than 3.4 which would be expected if no thiocyanate ions were present (Boyland and Walker, 1974; Fan and Tannenbaum, 1973). More recently, Lane *et al.* (1974) have used the gas chromatographic-mass spectrometric technique to study the formation of DMN in human gastric juice, and obtained similar results. Fine *et al.* (1977) and his co-workers (Rounbehler *et al.* 1977) have studied the *in vivo* formation of nitrosamines in man and mouse, respectively.

The induction of tumours in experimental animals after concurrent administration of various secondary amines or amides and nitrite also supports the theory that nitrosation reaction can indeed occur in the stomach. Thus, Sander and Burkle (1969) induced esophageal and liver tumours in rats after feeding N-methylbenzylamine or morpholine and nitrite. Simultaneous feeding of nitrite with other N-nitrosatable compounds such as methylaniline, piperazine, aminopyrene, methylurea and ethylurea has also resulted in the induction of cancer in experimental animals (Sander and Schweinsberg, 1972; Greenblatt *et al.*, 1971; Ivankovic and Preussmann, 1970; Lijinsky *et al.*, 1973). Negative results were obtained when these compounds (amines or amides) or nitrite were fed alone. It should be pointed out that all the amines or amides mentioned above are weak bases and, therefore, the *in vivo* yield of the various N-nitroso compounds was high enough to induce tumours in the test animals. Similar experiments with nitrite and strongly basic a-mines such as diethylamine, dimethylamine, pyrrolidine, piperidine, etc. (Druckrey *et al.*, 1963a; Greenblatt *et al.*, 1971; Garcia and Lijinsky, 1973; Sen *et al.*, 1976b) gave negative results. The basicity of the amine is probably the most important determining factor in influencing nitrosamine formation in the stomach. The feeding of even very low levels (5 ppm) of the weakly basic amine morpholine and nitrite to hamsters has produced tumours in these laboratory animals (Shank and Newberne, 1976).

Bacterial Synthesis

Considerable research has been carried out studying the role of bacteria in the formation of nitrosamines. There are two schools of thought: one group believes that certain bacteria can actually synthesize nitrosamines by enzymatic reactions, although no cell-free enzyme system has yet been isolated. The second group postulates that the nitrosation reactions are non-enzymic in nature; the bacteria only synthesize the precursors namely, various amines and nitrite (from nitrate) which then react chemically to form the nitrosamines or the bacteria produce metabolites which can catalyze nitrosamine formation.

The formation of nitrosamines from added secondary amines and nitrate has been demonstrated to occur in the stomach of human subjects with hypoacidity (Sander and Schweinsberg, 1972), in bacterial cultures or rat cecal contents (Alam *et al.*, 1971; Klubes *et al.*, 1972; Hill and Hawksworth, 1972). Collins-Thompson *et al.*, (1972) have studied the formation of nitrosamines by several bacterial species isolated from foods, and concluded that the nitrosation reaction was non-enzymatic in nature and probably involved catalysis by bacterial metabolites. More recently, Archer *et al.* (1977) studied the nitrosation of dihexylamine under mildly acidic conditions in the presence of various bacteria and yeast cells. The nitrosation rates were similar in the presence of either boiled or unheated cells but significantly higher than in the controls. A non-enzymatic mechanism involving hydrophobic interactions of the amine with the cell constituent was proposed. The occurrence of high levels of DMN in some salted and dried fish has been attributed to the bacterial conversion of nitrate (present as an impurity in the crude salt used in the preparation of these products) to nitrite and interaction of the nitrite with the amines in the fish (Fong and Chan, 1973). The addition of benzoic acid preservative or the use of pure sodium chloride, instead of the crude salt, markedly decreased the formation of DMN in these products (Fong and Chan, 1976).

PRECURSORS OF NITROSAMINE FORMATION

It is clear from the above discussions that wherever nitrate, nitrite and nitrosatable organic compounds occur together there is a possibility of the formation of N-nitroso compounds. Nitrate and nitrite in our diet can originate from a variety of sources. Nitrate occurs in high concentrations (sometimes as high as 1,000–3,000 ppm) in vegetables such as cabbage, cauliflower, carrot, celery, lettuce, radish, beets, spinach, etc. (Ashton, 1970; White, 1975). Nitrate also occurs in water, especially in well waters in some rural areas (Comly, 1945; Burden, 1961). The concentration of nitrite in vegetables and water is usually very low, although fairly high levels have been detected in storage-abused spinach and beets (Heisler *et al.*, 1974).

In addition to the above sources, nitrate and nitrite can originate in our foods from their use as intentional food additives. These chemicals are used as food additives in many countries for the preservation of fish, meat, cheese, etc. The permissible levels of these additives in different foods vary from country to country, but they usually lie in the range of 10–200 ppm for sodium nitrite and 50–1,000 ppm for sodium nitrate (U.S. Food Additive Regulations, C.F.R. 121.1063, 121.1064, 1972; Canadian Food and Drug Act and Regulations, 1965; Meester, 1973). Nitrate and nitrite are mainly used for their role in inhibiting the outgrowth of *Clostridium botulinum* spores. Nitrate *per se* does not have any inhibitory action against these bacteria but its action is manifested by the reduction of nitrate to nitrite by microorganisms present in the foods. The addition of nitrate to cheese milk is necessary to prevent the unwanted fermentation by butyric acid bacteria belonging to the species *Clostridium tyrobutyricum*, the spores of which may survive the normal pasteurization process (Goodhead *et al.* 1976). Apart from their preserving action, these chemicals are believed to play an important role in improving the color and flavour of cured meat and fish products. Because of the present concern over the possible interaction of nitrite and amines in foods and the formation of nitrosamines, the use of nitrate and nitrite as food additives has come under serious criticism. The health authorities in many countries are reexamining the situation and trying to amend or modify the current regulations so as to minimize human exposure to these chemicals.

White (1976) has estimated the average daily ingestion of nitrate and nitrite for U.S. residents, and calculated the relative significance of various dietary sources. The data in Table 10.4 show that vegetables are the major (86%) source of nitrate in the average American diet; the rest originates from salivary excretion and cured meats. Whereas the predominant portion of the ingested nitrite comes from saliva and a smaller but significant amount (21%) from cured meats, various studies have shown that the levels of nitrite in saliva can increase markedly after consumption of meals containing nitrate-rich foods such as vegetables (Spiegelhalder *et al.*, 1976; Tannenbaum *et al.*, 1976), thereby suggesting that the ingested nitrate is converted *in vivo* in the human body to nitrite and then excreted in the saliva. Since the volume of daily excretion of saliva can be quite high (up to 1,000 ml) the high concentration of nitrite (as observed after nitrate-rich diet) in saliva can be important in the formation of nitrosamines in the highly favourable acidic conditions of the stomach. Preliminary reports by Tannenbaum *et al.* (1978) suggest that both nitrate and nitrite can be synthesized in the body, from other nitrogenous components of our diets, by intestinal bacteria thus complicating the situation even further.

Data on the occurrence of various amines in foods are scarce. The food which has been studied most extensively for its amine content is fish. Fairly highly levels of dimethylamine, trimethylamine and trimethylamine oxide have been determined in various fish, particularly those of marine origin. The concentration of trimethylamine and trimethylamine oxide in such products are quite high; sometimes as high as 100–185 mg% (Shewan, 1951). Dimethylamine is usually absent in fresh water fish but present in fairly high

TABLE 10.4. RELATIVE SIGNIFICANCE OF DIETARY SOURCES OF NITRATE AND NITRITE: ESTIMATED AVERAGE DAILY INGESTION FOR U.S. RESIDENT

Source	Nitrate		Nitrite	
	mg	%	mg	%
Vegetables	86.1	86.3	0.20	1.8
Fruits, juices	1.4	1.4	0.00	0.0
Milk and products	0.2	0.2	0.00	0.0
Bread	2.0	2.0	0.02	0.2
Water	0.7	0.7	0.00	0.0
Cured meats	9.4	9.4	2.38	21.2
Saliva	30.0[1]		8.62	76.8
Total	99.8	100	11.22	100

[1]Not included in total.
Source: White, 1976.

concentrations in gadoid fish (cod, pollack, hake and haddock). Its concentration in these fish can increase significantly after prolonged storage at freezing temperatures (Castell *et al.*, 1971). Golovnya (1976) has reported the occurrence of traces of various primary, secondary and tertiary amines in sturgeon and salmon caviar.

Relatively little is known about the occurrence of amines in other foods. Low levels of simple amines such as methylamine, ethylamine, dimethylamine, diethylamine and trimethylamine have been reported occasionally in various meat products (Landmann and Batzer, 1966; Patterson and Mottram, 1974; Cantoni *et al.*, 1969). A wide variety of primary and secondary amines have also been detected in Tilsiter, Gouda and Rossiiskii cheese (Schwartz and Thomasow, 1950; Weurman and DeRooy, 1961; Ruiter, 1973; Golovnya *et al.*, 1970). Various spices such as paprika, cayenne pepper and black pepper also contain fairly high levels of amines such as pyrrolidine and piperidine (Marion, 1950; Gough and Goodhead, 1975). Since spices are used in the preparation of various foods in different countries, they may contribute significantly to the total intake of amines in our diets. More recently, Singer and Lijinsky (1976) and Neurath *et al.* (1977) have studied the amine contents of various foods such as meat, fish, fruits, vegetables, bread, tobacco, beverages and drinking water. Traces of dimethylamine, trimethylamine, diethylamine, pyrrolidine, piperidine, morpholine, N-methylbenzylamine, N-methylaniline and a few other amines were detected in some of the products analyzed. Other amine precursors which could lead to the formation of N-nitroso compounds include proline and hydroxyproline in pork belly and bacon, piperidine in fish and soybean, sarcosine in peanut and barley, and pipecolic acid in legumes (Gray *et al.*, 1977; Sen, 1974; Ishidate *et al.*, 1971). In addition, traces of pesticide and drug residues in various foods can also serve as precursors of N-nitroso compounds (Lijinsky *et al.* 1972; Elespuru and Lijinsky, 1973; Eisenbrand *et al.*, 1974; Sen *et al.*, 1974a).

OCCURRENCES IN FOODS

Traces of nitrosamines have been detected in various foods such as cured meat products including cooked bacon, raw and smoked-cured fish, various cheeses, spice-nitrite premixes (used for the commercial preparation of cured meats), mushrooms, a solanaceous fruit, and alcoholic drinks. Some of the published results have been summarized in Table 10.5. It should be pointed out that except in a

TABLE 10.5.   POSITIVE FINDINGS OF NITROSAMINES IN FOODS

Kind	Nitrosamine detected[1]	Level (ppb)	References
Meats			
Smoked sausage, ham, bacon	DMN, DEN	trace-6	Ender and Ceh (1967)
Uncooked sausage, smoked pork belly	DEN, DPN	5	Moehler and Mayrhoffer (1969)
Cured meat (Kasseler ribs)	DEN	trace-40	Freimuth and Glaeser (1970)
Dry sausage and uncooked salami	DMN	20–80	Sen (1972)
Frankfurters	DMN	11–84	Wassermann *et al.*, (1972)
Sausages, salami, luncheon meat, pork	DMN, DEN	1–4	Crosby *et al.*, (1972)
Fried bacon and cooked-out fat	DMN, DEN, NPYR, NPIP	trace-40	Crosby *et al.*, (1972)
"	DMN, NPYR	trace-44	Sen *et al.* (1973b, 1976a, 1977a)
"	DMN, DEN, NPYR, NPIP	1–18	Gough *et al.* (1977)
"	NPYR	10–207	Fazio *et al.* (1973)
"	DMN, NPYR	1–110	Gough and Walters (1976)
"	NPYR	4–100	Havery *et al.* (1977)
"	NHPYR	trace-12	Sen *et al.* (1976d)
"	NPYR	4–41	Kawabata *et al.* (1977)
"	NPYR	3–39	Pensabene *et al.* (1974)
"	NPYR	trace-41	Gray *et al.* (1977)
"	NHPYR	trace-4	Lee *et al.* (1977)
"	DMN, NPYR	1–100	Teliing *et al.* (1974)
Raw bacon	NPRO	380–1,180	Kushnir *et al.* (1975)
"	NPRO	24–44	Sen *et al.* (1977b)
Bacon, bologna, ham, meat loaf	NHPYR, NPRO, NSAR, NHPRO	1–401	Eisenbrand *et al.* (1977)
Smoked horse meat, luncheon meat, bacon, liver sausage, minced beef	NPRO	10–2,900	Dhont *et al.* (1976)

Continued

TABLE 10.5.   (Continued)

Kind	Nitrosamine detected[1]	Level (ppb)	References
Smoked horse meat, luncheon meat, bacon, liver sausage, minced beef	DMN, DEN, MEN, NPIP	trace-15	Gorenen et al. (1976b)
Ham, sausage, fried bacon, smoked meat	DMN, DEN	2–91	Gorenen et al. (1976a)
Various spiced meat products	DMN, DEN, NPYR, NPIP	3–50	Sen et al. (1976c)
Various spiced meat (both cooked and uncooked)	DMN, NPYR, NPIP	1–250	Eisenbrand et al. (1976)
Variety of cured meat products	DMN, DEN, NPYR	trace-105	Panalaks et al. (1973, 1974)
Sausage	DMN, DEN	trace-13	Kann et al. (1976)
Smoked meat, raw and fried bacon, ham, sausage	DMN, DEN, DBN, NPYR, NPIP	trace-55	Stephany et al. (1976)
Ham	DMN	trace-18	Klein et al. (1976)
Wiener, bologna, ham, hamburger containing fish	DMN	trace-10	Nakamura and Usuki (1973)
Souse and gelatin containing meat products	DMN, NPYR	3–45	Fiddler et al. (1975)
Country-style fried ham	NPYR	trace-50	Greenberg (1976)

Fish, Fish Meal and Sea Foods

Kind	Nitrosamine detected[1]	Level (ppb)	References
Herring meal (nitrite treated)	DMN	15,000– 100,000	Sakshaug et al. (1965) Ender et al. (1964)
Fish meal (untreated)	DMN	120–450	Sen et al. (1972)
"	DMN		Koppang (1974)
"	DMN, DEN	5–417	Juszkiewicz and Kowalski (1976)
Smoked fish	DMN	trace-40	Ender and Ceh (1967)
Smoked fish, canned and baked fish	DMN	6–177	Kann et al. (1976)
Fresh fish, salted fish (cooked and uncooked)	DMN	trace-9	Crosby et al. (1972)
Smoked, nitrate or nitrite treated fish	·DMN	4–26	Fazio et al. (1971)
Salted and dried fish	DMN	10–1,000	Fong and Chan (1973)

TABLE 10.5.   (Continued)

Kind	Nitrosamine detected[1]	Level (ppb)	References
Various Chinese sea foods (dried shrimps, fish sauce, oyster sauce, sausage, dried squid, etc.)	DMN, NPYR	trace-37	Fong and Chan (1977)
Fish and sea foods (uncooked and cooked)	DMN, DEN	3–18	Iyengar et al. (1976)
Smoked, nitrite-treated sable fish	DMN	trace-6	Gadbois et al. (1975)
Various fish (cooked and uncooked)	DMN	trace-10	Gough et al. (1977)
Raw and canned fish	DMN, DEN, NPYR	trace-9	Telling et al. (1974)
Fish sausage, fermented squid	DMN	trace-10	Nakamura and Usuki (1973)

Dairy Products

Kind	Nitrosamine detected[1]	Level (ppb)	References
Havarti and Gouda cheese	DPN or DIPN	5–10	Kroeller (1967)
Grinland, Edam, Tilsit and a few other varieties of cheese	DMN	1–4	Crosby et al. (1972)
Cheshire cheese	DEN, NPYR	≃1	Alliston et al. (1972)
Cheddar, Cheshire, Gouda, Edam, St. Paulin, Stilton cheese	DMN	1–13	Gough et al. (1977)
Gouda, Havarti, Camembert, Cheddar and a few other varieties of cheese	DMN, DEN, NPIP	trace-19	Sen et al. (1977b)
Wine cheese	DMN, DEN, NPIP	7–68	Sen et al. (1977b)
Emmentaler, Parmesan, Tilsiter, Camembert, Brie, Philadelphia and a few other varieties of cheese	DMN, DEN, NPYR	trace-6	Eisenbrand et al. (1977)
Tilsit cheese	DMN, NPYR	≃4	Fiddler et al. (1977)

Spice-Nitrite Premixes

Kind	Nitrosamine detected[1]	Level (ppb)	References
Premixes for Mettwurst and Thuringer sausages	DMN, NPYR, NPIP	850–25,000	Sen et al. (1973a)

Continued

TABLE 10.5.   (Continued)

Kind	Nitrosamine detected[1]	Level (ppb)	References
Premixes for ham, wiener, corned beef, etc.	DMN, NPYR, NPIP	24–2,000	Havery et al. (1976)
Various premixes containing ground spices, spice extractives, hydrolyzed vegetable proteins, nitrate, and nitrite	DMN, NPYR, NPIP	15–15,000	Gough and Goodhead (1975)
Premixes for meat loaf, frankfurter, pepperoni, bologna, pork, sausage, chicken roll, etc.	DMN, NPYR, NPIP	8–4,000	Sen et al. (1974b)
Fruits and Vegetables			
Fruit from a solanaceous (*Solanium incanum*) bush	DMN	≃200	DuPlessis et al. (1969)
Mushrooms (*Clitocybe suaveolens*)	p-methylnitrosoaminobenzaldehyde (noncarcinogenic)	not known	Herrman (1961)
Various types of edible mushrooms	DMN	1–30	Ender and Ceh (1967)
Alcoholic Beverages			
Cider distillates	DMN	trace-10	Bogovski et al. (1974)
Various brands of beer	DMN	trace-10	Walker (1978)

[1]Abbreviations: DMN (Dimethylnitrosamine), DEN (diethylnitrosamine), DPN (di-*n*-propyl-nitrosamine), DIPN (diisopropylnitrosamine), DBN (di-*n*-butyl-nitrosamine), NPYR (nitrosopyrrolidine), NPIP (nitrosopiperidine), NHPYR (nitrosohydroxypyrrolidine), NPRO (nitrosoproline), NHPRO (nitrosohydroxyproline), NSAR (nitrososarcosine).

few cases, the levels of nitrosamines detected were extremely low, and even these were detected only in a small percentage of the samples tested. One picture that clearly emerges from these data is the fact that, among the foods examined, certain types of cured sausages and salami (all of them spiced), cooked bacon, nitrate-nitrite treated smoked fish and certain types of salted and dried fish are the main contributors of nitrosamines in our diet. The major nitrosamines detected in these foods are DMN, DEN, NPYR and nitrosopiperidine (NPIP), all of which are potent carcinogens.

The early work from the author's (Sen, 1972; Panalaks et al., 1973, 1974) laboratory and that of Wasserman et al. (1972) indicated that fairly high levels of DMN, NPYR and an unidentified (later proven to be NPIP) nitrosamine were occasionally detectable in some samples of frankfurters and sausages and salami. Both the levels and the frequency of occurrence of these nitrosamines were, however, very unpredictable. The reason for this inconsistency was not clear until later when we (Sen *et al.*, 1973a, 1974b) discovered the occurrence of excessively high levels (Table 10.5) of nitrosamines in some commercial spice-nitrite premixes which were used for the preparation of a wide variety of cured meat products. These findings were later verified by other laboratories (Havery *et al.*, 1976; Gough and Goodhead, 1975). These high levels of nitrosamines were apparently formed by the interaction of nitrite and the amines in the spices or spice extractives. Subsequently, the practice of premixing spices with nitrate or nitrite was discontinued in Canada as well as other countries, and the manufacturers were asked to package the curing salts and spices in separate containers. This action has resulted in a marked decrease in the levels of nitrosamines in various meat products (Sen and McKinley, 1974; Sen *et al.*, 1976c; Eisenbrand *et al.*, 1977; Gough *et al.*, 1977), and made these foods much safer for human consumption.

Among all the cured meat products tested, cooked bacon appears to be one item which has consistently been shown to contain fairly high levels of nitrosamines, mainly NPYR and traces of DMN (Table 10.5). The nitrosamines are believed to be formed during the high heat cooking process normally employed for frying bacon. The levels of nitrosamines in the cooked product depend on many factors such as the type of bacon, the method and the temperature of cooking (Pensabene *et al.*, 1974) and the concentration of nitrite used in preparing the bacon (Table 10.6). On the average, about 50% of the total NPYR and 70% of the total DMN produced are given off in the fumes, and one-third of the remaining nitrosamines is found in the

TABLE 10.6. THE EFFECT OF NITRITE CONCENTRATION ON THE FORMATION OF DMN AND NPYR IN FRIED BACON

Sample No.	Concentration of sodium nitrite used in preparing the bacon (ppm)	Nitrosamine	Concentration of the nitrosamine in the fried sample (ppb)[1]
1–5	0	none detected	
6–9	50	DMN	N[2], N, 3, N
		NPYR	2, 2, 4, 4
10–14	100	DMN	N, N, 2, N, 3
		NPYR	8, 7, 8, 8, 8
15–19	150	DMN	3, N, 3, N, N
		NPYR	10, 20, 10, 5, 5
20–24	200	DMN	N, 3, N, 5, 3
		NPYR	20, 20, 20, 12, 12

[1]Results expressed on the basis of the weight of the uncooked bacon.
[2]N = negative (detection limit, 2 ppb)
Source: Sen *et al.* (1974c).

cooked lean and about two-thirds is retained in the cooked-out fat (Sen *et al.*, 1976e; Gough and Walters, 1976; Warthesen *et al.*, 1976).

Periodical surveillance of the levels of nitrosamines in cooked bacon over the past few years suggests that, in both Canada and the U.S., the concentration of these chemicals has been decreasing steadily (Sen *et al.*, 1977a; Greenberg, 1976; Havery *et al.*, 1977). For example, in Canada, the average level of NPYR in cooked bacon in 1972 was found to be about 29 ppb compared to 24 ppb and 12 ppb as detected in the surveys of 1974 and 1975, respectively. The corresponding average figures in the U.S. were 67 ppb in 1972 versus 17 ppb in 1976. This decline may have been due to several factors such as: (a) better control of the input of nitrate and nitrite by the industries; (b) the lowering of permissible levels of sodium nitrite from 200 ppm to 150 ppm and the complete elimination of nitrate (in Canada); and (c) the addition of inhibitors of nitrosamine formation such as sodium ascorbate or erythorbate. Recent research in various laboratories has established that the addition of these antioxidants can significantly inhibit the formation of nitrosamines during cooking of bacon (Greenberg, 1976; Mergens *et al.*, 1977; Sen *et al.*, 1976a). Controlled addition of such inhibitors may be an effective and practical way of reducing the concentration of nitrosamines in cooked bacon and minimizing the remote possibility of any health hazards arising from the consumption of such foods.

The occurrence of significant levels of nitrosamines in smoked-cured, and salted and dried fish (Table 10.5) deserves special mention. Since these foods are consumed in large quantities in certain parts of the world, and may contain high levels of nitrosamines, prolonged consumption of these products may not be advisable. The chance of nitrosamine formation is especially greater in the marine fish than in the fresh water fish, mainly because the former type contains much higher concentrations of amines than the latter. Thus, traces of DMN have been detected in nitrite-cured sable, salmon and shad - all marine fish, but none could be detected in nitrite-cured chub (a fresh water fish) (Fazio *et al.*, 1971; Howard *et al.*, 1970). Similarly, Fong and Chan (1973) have detected fairly high levels of DMN in salted and dried marine fish (white herring, yellow croaker) which are considered to be favourite dishes along the southern coast of China and in many Southeast Asian countries. Since most of these results are based on thin-layer chromatographic or gas chromatographic-low resolution mass spectrometric techniques, it will be highly desirable to confirm these results by more reliable methods such as high resolution mass spectrometry. As mentioned earlier (under Bacterial Synthesis), the replacement of crude salt (which contains nitrate as an impurity) and the addition of benzoic acid as a preservative seems to have markedly decreased the formation of nitrosamines in these products (Fong and Chan, 1976).

CARCINOGENIC PROPERTIES OF NITROSAMINES

Many nitrosamines and, in general, most N-nitroso compounds are highly potent carcinogens. Thus far, about 130 N-nitroso compounds have been tested and approximately 80% of them shown to be carcinogenic (Preussman et al., 1976). DMN has been shown to be carcinogenic in 6 species and DEN in about 20 animal species including some subhuman primates. None of the animal species tested has been able to resist the carcinogenic action of nitrosamines, and it is, therefore, generally believed that man is also at risk, although no direct evidence of carcinogenic action has been demonstrated. The carcinogenic potency of different nitroso compounds, however, varies a great deal; some (e.g. DMN, DEN) are highly potent whereas others (e.g. nitrosodiethanolamine, nitrososarcosine) are only weakly carcinogenic. Many nitroso compounds are also strong mutagens. In contrast to the nitrosamides, the nitrosamines need enzymic activation before exerting their mutagenic properties. The acute toxicity of various nitroso compounds covers a wide range of dose levels (Magee and Barnes, 1967). Low acute toxicity (LD_{50}) may not be related to the carcinogenic potential of the compounds. Thus, the LD_{50} value for DMN is $\frac{1}{8}$th of that of DEN but the latter is probably a stronger liver carcinogen. Mink appear to be the most sensitive animal to the toxic and carcinogenic action of nitrosamines (Koppang and Rimeslatten, 1976).

One of the outstanding features of nitrosamines is their capacity to induce tumours in practically any organ. Because of this, this group of compounds is considered as one of the most versatile carcinogens known to man (Preussmann, 1974). Many of these compounds are highly organ specific, i.e. causing tumours only in a specific organ. DMN, the simplest and the most widely occurring nitrosamine in foods, causes mostly liver tumours and occasionally kidney tumours, whereas methylbenzylnitrosamine and N-nitroso-n-butyl-(4-hydroxybutyl) amine causes cancer of the esophagus and bladder, respectively (Druckrey et al., 1969; Magee et al., 1976). It is believed that this organospecificity stems from the preferential ability of a particular organ to metabolize the nitrosamine and form the ultimate carcinogen. The site of action may also depend on other factors such as mode of administration, species and age of the animal, dose level, diet and nutritional status of the animal. Table 10.7 gives a general outline of the organ specificity of various nitrosamines.

Transplacental Carcinogenesis

Various studies with laboratory animals clearly indicate that many nitrosamines and nitrosamides, if administered to the pregnant mother, can cross the placental barrier and cause cancer in the offspring. Thus, DMN, DEN, N-nitrosomethylurea and N-nitrosoethylurea have been shown to cause transplacental carcinogenesis in various species such as rat, mouse, hamster, pig, etc. (Magee *et al.*, 1976). The most critical factor appears to be the time of treatment. In rats, the administration of nitrosamine to the mothers any time after the 10th day of gestation results in the induction of tumors in the offspring; the most sensitive period seems to be between the 18th day of gestation and just prior to the date of the delivery. The lack of any transplacental carcinogenic activity prior to the 10th day of gestation may be related to the inadequate metabolic system in the fetuses during the early stages of development and hence their inability to convert the nitrosamines to ultimate carcinogens. Dose-response studies in BD rats have shown that a single dose of 2

TABLE 10.7. CARCINOGENICITY OF NITROSAMINES: ORGAN SPECIFIC EFFECTS

$$\frac{R_1}{R_2} >N-N=O$$

Compound	Target organ
$R_1 = R_2$ (e.g. DMN, DEN, DBN)	Mostly liver (kidney, bladder, lung)
$R_1 \neq R_2$ (e.g. methylbenzylnitrosamine)	Mostly esophagus (forestomach, lung, liver)
Cyclic (e.g. NPYR, NPIP)	Liver, esophagus, nasal cavity
Acylalkylnitrosamines (e.g. methylnitrosourea)	Central and peripheral nervous system (lung, forestomach, pancreas)
R_1 or R_2 with functional groups (–OH, –COOH, etc.) [e.g. NSAR, *n*-buty-(4-hydroxybutyl)-nitrosamine]	Liver, bladder, esophagus

Adapted from Wishnok (1977).

mg/kg body weight of N-nitrosoethylurea (i.e. 0.8% of LD_{50}) on the 15th day of pregnancy produced carcinogenic response in the nervous system of the newborn (Ivankovic and Druckrey, 1968). This dose is about 1/50th of that necessary to induce tumours in the adult rats (Druckrey et al., 1969), thus demonstrating the extreme sensitivity of the fetal nervous system to the carcinogenic action of nitrosoureas and probably other N-nitroso compounds. This extreme fetal sensitivity emphasizes the importance of carrying out two generation studies (to expose the fetuses to the test chemical) when testing carcinogenicity of a chemical compound and, it also underscores the importance of avoiding excessive consumption of foods containing relatively high levels of carcinogens by pregnant women.

Dose Response Study

In the first study, Magee and Barnes (1956) observed that 50 ppm of DMN in the diet produced a high incidence of malignant liver tumours in rats after a very short (26–40 weeks) induction period. At higher doses kidney tumors were observed. Even a single dose of 30 mg/kg body weight induced renal tumours (Magee and Barnes, 1959).

Terracini et al. (1967) carried out a dose response study in rats using 2, 5, 10, 20 and 50 ppm DMN in the diet and observed a significant number of liver tumours in all the groups, except the lowest one. Only one rat in the lowest dose group developed liver tumour. Since no liver tumour was observed in any of the control animals, it was concluded that a dose of 2 ppm is probably carcinogenic, and a no-effect level was not established.

Druckrey et al. (1963b) have carried out a much more extensive study with DEN and observed a very interesting quantitative response on the dose-effect relationship. By carefully controlled experiments, they demonstrated that the total dose required to produce a specific carcinogenic response in rats decreases with a decrease in the daily dose, thus indicating an accelerating effect. For example, at a daily dose of 0.3 mg/kg body weight the average time of induction of liver tumours in the experimental rats was 500 days. The corresponding figures at 0.15 mg/kg and 0.075 mg/kg doses were found to be 605 days and 830 days, respectively. If there were no accelerating effect, the average time of induction at the latter two dose levels would have been expected to be much longer—possibly, 1,000 days and 2,000 days, respectively. These investigators were unable to find a threshold dose level for DEN as all the dosages tested were shown to be carcinogenic.

The median time of tumour induction (t) was found to be related to the daily dose (d) by the equation dt^n = constant, where n varies between 1.4 and 4 depending on the structure of the nitrosamine (Druckrey et al., 1963b; Archer and Wishnok, 1977). For DEN, n was found to be 2.3 and for NPYR, n = 1.4, which indicates that the carcinogenic process for DEN accelerates at a faster rate than for NPYR. The corresponding value for DMN has not been determined, but it is probably close to that of DEN. These results suggest that, while evaluating the possible human health hazard, the accelerating effects of prolonged (over 50 to 60 years) exposure should be taken into account.

More recently, Preussmann et al. (1976) have investigated the carcinogenic dose response relationship of NPYR in rats. Four different doses (10, 3, 1 and 0.3 mg/kg body weight/day) of NPYR (in the drinking water) were fed to rats. The rats in all the groups, except the lowest one, developed a significant number of liver tumours. Although no malignant liver tumours were observed in the 0.3 mg/kg dose group, 3 benign hepatocellular adenomas were observed which did not occur in the controls. About 60 rats were used in each of the control, 0.3 mg/kg and 1.0 mg/kg groups.

ASSESSMENT OF POSSIBLE HUMAN HEALTH HAZARD

Since certain foods contain traces of nitrosamines one may ask what is the significance of these findings? Do they pose any significant health hazard to humans? In evaluating the risk one should take the following points into consideration: (a) the cumulative and accelerating effect of prolonged exposure to extremely low levels of nitrosamines; (b) relative carcinogenic potency of different nitrosamines (Archer and Wishnok, 1977); (c) synergistic effect of other carcinogens in food and the environment, and finally; (d) the extent of in vivo formation. As pointed out earlier, present knowledge in many of these areas is limited and, therefore, it is difficult to say with confidence that such low levels of nitrosamines in food do or do not pose any significant health hazard. Insufficient data on the carcinogenic threshold levels of various nitrosamines also make the task of assessing the health hazard arbitrary and unscientific.

It is even more difficult to assess the significance of the in vivo formation of nitrosamines. Although the recent studies by Fine et al. (1977) suggest that the in vivo synthesis of both DMN and DEN in humans can increase significantly after a normal mid-day meal consisting of cooked bacon, cooked spinach, fresh tomatoes, bread and beer, the data should be considered very preliminary for only 3 cases were investigated. Similar studies with whole mouse indicate that significant levels of DMN can be produced *in vivo* after administration of low levels of dimethylamine and nitrite (Rounbehler *et al.*, 1977). Further research along these lines (using a larger number of human volunteers and a wider range of foodstuffs) should be continued. Such data will be very useful in assessing the relative significance of endogenous sources of nitrosamines compared to those occurring in foods.

It might be worth mentioning that food is not the only source of nitrosamines or nitrosatable amines that we are exposed to. Various other sources are: cigarettes and tobacco, certain drugs, pesticide formulations, cutting oils, cosmetics, deionized water, etc. (Fine, 1977). The relative risk of each of these items should be assessed in their proper perspectives, and an integrated effort should be made to minimize the exposure to nitrosamines from all of these sources and not just foods. It is hoped that various research programmes now underway in different laboratories will help us to gain a better understanding of the problem and suggest ways of minimizing the concentration of these carcinogens in foods as well as the other products mentioned above.

BIBLIOGRAPHY

ALAM, B. S., SAPOROSCHETZ, I. B. and EPSTEIN, S. S. 1971. Synthesis of nitrosopiperidine from nitrate and piperidine in the gastrointestinal tract of the rat. *Nature 232*, 199–200.

ALLISTON, T. G., COX, G. B. and KIRK, R. S. 1972. The determination of steam-volatile N-nitrosamines in foodstuffs by formation of electron capturing derivatives from electrochemically derived amines. Analyst *97*, 915–920.

ARCHER, M. C. and WISHNOK, J. S. 1977. Quantitative aspects of human exposure to nitrosamines. *Food Cosmet. Toxicol. 15*, 233–235.

ARCHER, M., YANG, H. S. and OKUN, J. D. 1977. Acceleration of nitrosamine formation at pH 3.5 by microorganisms. Presented at the Fifth IARC Meeting on Analysis and Formation of N-Nitroso Compounds, August 22–24, 1977, University of New Hampshire, Durham, N.H.

ASHTON, M. R. 1970. The occurrence of nitrates and nitrites in foods. *The British Food Manufacturing Industries Research Association*, Literature Survey No. 7.

BARNES, J. M. and MAGEE, P. N. 1954. Some toxic properties of dimethylnitrosamine. *Brit. J. Ind. Med. 11*, 167–174.

BOGOVSKI, P., WALKER, E. A., CASTEGNARO, M. and PIGNATELLI, B. 1974. Some evidence of the presence of traces of nitrosamines in cider distillates. In: *N-Nitroso Compounds in the Environment*, Bogovski, P. and Walker, E. A. (Editors), International Agency for Research on Cancer, Scientific Publication No. 9, Lyon, pp. 192–196.

BOHLER, N. 1960. En ondartet leversykdom hos mink og rev. *Nor. Pelsdyrbl. 34*, 104–106.

BOHLER, N. 1962. Ondartet leversykdom hos pelsdyr i Norge. *Proc. IX Nord. Vet. Congr.*, Copenhagen, Vol. II, 774–776.

BOYLAND, E. 1972. The effect of some ions of physiological interest on nitrosamine synthesis. In: *N-Nitroso Compounds Analysis and Formation*, Bogovski, P., Preussmann, R. and Walker, E. A. (Editors), International Agency for Research on Cancer, Scientific Publication No. 3, Lyon, pp. 124–126.

BOYLAND, E. and WALKER, S. A. 1974. Thiocyanate catalysis of nitrosamine formation and some dietary implications. In: *N-Nitroso Compounds in the Environment*, Bogovski, P. and Walker, E. A. (Editors), International Agency for Research on Cancer, Scientific Publication No. 9, Lyon, pp. 132–136.

BURDEN, E. H. W. J. 1961. The toxicology of nitrates and nitrites with particular reference to the potability of water supplies. *Analyst 86*, 429–433.

CANADIAN FOOD AND DRUG ACT AND REGULATIONS. 1965. Table XI. Queen's Printer and Controller of Stationery, Ottawa.

CANTONI, C., BIANCHI, M. A., RENNON, P. and D'AUBERT, S. 1969. Putrefaction of sausages. *Archiro Veterinario Italiano 20*, 245–264.

CASTELL, C. H., SMITH, B. and NEAL, W. 1971. Production of dimethylamine in muscle of several species of gadoid fish during frozen storage, especially in relation to presence of dark muscle. *J. Fish. Res. Bd. Can. 28*, 1–5.

COLLINS-THOMPSON, D. L., SEN, N. P., ARIS, B. and SCHWING-HAMER, L. 1972. Nonenzymic *in vitro* formation of nitrosamines by bacteria isolated from meat products. *Can. J. Microbiol. 18*, 1968–1971.

COMLY, H. H. 1945. Cyanosis in infants caused by nitrates in well water. *J. Am. Med. Assoc. 129*, 112–116.

CROSBY, N. T. 1976. Nitrosamines in foodstuffs. *Res. Rev. 64*, 77–135.

CROSBY, N. T., FOREMAN, J. K., PALFRAMAN, J. F. and SAWYER, R. 1972. Estimation of steam-volatile N-nitrosamines in foods at the 1 μg/kg level. *Nature 238*, 342–343.

DHONT, J. H. 1976. Development of a method of estimating N-nitrosamino acids and its use on some meat products. *Proc. 2nd int. Symp. Nitrite Meat Prod.* Zeist, pp. 221–225.

DRUCKREY, H., PREUSSMANN, R., SCHMAEHL, D. and MUELLER, M. 1961. Chemische Konstitution und carcinogene Wirkung bei Nitrosaminen. *Naturwissenschaften 48*, 134–135.

DRUCKREY, H., STEINHOFF, D., BEUTHNER, H., SCHNEIDER, H. and KLAERNER, P. 1963a. Pruefung von Nitrit auf toxische Wirkung an Ratten. *Arzneim.–Forsch. 13*, 320–323.

DRUCKREY, H., SCHILDBACH, A., SCHMAEHL, D., PREUSSMANN,

R. and IVANKOVIC, S. 1963b. Quantitative Analyse der carcinogen Wirkung von Diaethylnitrosamin. *Arzneim.–Forsch. 13*, 841–851.

DRUCKREY, H., PREUSSMANN, R., IVANKOVIC, S. and SCHMAEHL, D. 1967. Organotrope carcinogene Wirkungen bei 65 verschiedenen N-Nitroso–Verbindungen an BD–Ratten. *Z. Krebsforsch. 69*, 103–201.

DRUCKREY, H., PREUSSMANN, R. and IVANKOVIC, S. 1969. N-Nitroso compounds in organotropic and transplacental carcinogenesis. *Ann. N.Y. Acad. Sci. 163*, 676–696.

DuPLESSIS, L. S., NUNN, J. R. and ROACH, W. A. 1969. Carcinogen in a Transkeian Bantu food additive. *Nature 22*, 1198–1199.

EISENBRAND, G., UNGERER, O. and PREUSSMANN, R. 1974. Formation of N-nitroso compounds from agricultural chemicals and nitrite. In: *N-Nitroso Compounds in the Environment*, Bogovski, P. and Walker, E. A. (Editors), International Agency for Research on Cancer, Scientific Publication No. 9, Lyon, pp. 71–74.

EISENBRAND, G., JANZOWSKI, C. and PREUSSMANN, R. 1976. Analysis, formation and occurrence of volatile and nonvolatile N-nitroso compounds: recent results. *Proc. 2nd int. Symp. Nitrite Meat Prod.*, Zeist, pp. 155–169.

EISENBRAND, G., SPIEGELHALDER, B., JANZOWSKI, C., KANN, J. and PREUSSMANN, R. 1977. Volatile and non-volatile N-nitroso compounds in foods and other environmental media. Presented at the Fifth IARC Meeting on Analysis and Formation of N-Nitroso Compounds, August 22–24, 1977, University of New Hampshire, Durham, N.H.

ELESPURU, R. K. and LIJINSKY, W. 1973. The formation of carcinogenic nitroso compounds from nitrite and some types of agricultural chemicals. *Food Cosmet. Toxicol. 11*, 807–817.

ENDER, F. and CEH, L. 1967. Vorkommen und Bestimmung von Nitrosaminen in Lebensmitteln der menschlichen und tierischen Ernaehrung. In Alkylierend wirkende Verbindungen, 2nd Conf. Tobacco Res., Freiburg, pp. 83–91.

ENDER, F., HAVRE, G., HELGEBOSTAD, A., KOPPANG, N., MADSEN, R. and CEH, L. 1964. Isolation and identification of a hepatotoxic factor in herring meal produced from sodium-nitrite preserved herring. *Naturwissenschaften 51*, 637–638.

ENDER, F., HAVRE, G. N., MADSEN, R., CEH, L. and HELGEBOSTAD, A. 1967. Studies on conditions under which N-nitroso-dimethylamine is formed in herring meal produced from nitrite-preserved herring. *Z. Tierphysiol. Tierernaehr. Futtermittelk. 22*, 181–189.

FAN, T. Y. and TANNENBAUM, S. R. 1973. Factors influencing the rate of formation of nitrosomorpholine from morpholine and nitrite: acceleration by thiocyanate and other ions. *J. Agr. Food Chem. 21*, 237–240.

FAZIO, T., DAMICO, J. N., HOWARD, J. W., WHITE, R. H. and WATTS, J. W. 1971. Gas chromatographic determination and mass spectrometric confirmation of N-nitrosodimethylamine in smoke-processed marine fish. *J. Agr. Food Chem. 19*, 250–253.

FAZIO, T., WHITE, R. H., DUSOLD, L. R. and HOWARD, J. W. 1973.

Nitrosopyrrolidine in cooked bacon. *J. Assoc. Offic. Anal. Chem. 56,* 919–921.

FIDDLER, W. 1975. The occurrence and determination of N-nitroso compounds. *Toxicol. Appl. Pharmacol. 31,* 352–360.

FIDDLER, W., PENSABENE, J. W. DOERR, R. C. and WASSERMANN, A. E. 1972. Formation of N-nitrosodimethylamine from naturally occurring quaternary ammonium compounds and tertiary amines. *Nature 236,* 307.

FIDDLER, W., PENSABENE, J. W., PIOTROWSKI, E. G., DOERR, R. C. and WASSERMANN, A. E. 1973. Use of sodium ascorbate or erythorbate to inhibit formation of N-nitrosodimethylamine in frankfurters. *J. Food Sci. 38,* 1084.

FIDDLER, W., FEINBERG, J. I., PENSABENE, J. W., WILLIAMS, A. C. and DOOLEY, C. J. 1975. Dimethylnitrosamine in souse and similar jellied cured meat products. *Food Cosmet. Toxicol. 13,* 653–654.

FIDDLER, W., DOERR, R. C. and PIOTROWSKI, E. G. 1977. Observations on the use of the thermal energy analyzer as a specific detector for nitrosamines. *Presented at the Fifth IARC Meeting and Formation of N-Nitroso Compounds,* August 22–24, 1977, University of New Hampshire, Durham, N. H.

FINE, D. H. 1977. An assessment of human exposure to N-nitroso compounds. *Presented at the Fifth IARC Meeting on Analysis and Formation of N-Nitroso Compounds,* August 22–24, 1977, University of New Hampshire, Durham, N.H.

FINE, D. H., ROSS, R., ROUNBEHLER, D. P., SILVERGLEID, A. and SONG, L. 1977. Formation *in vivo* of volatile N-nitrosamines in man after ingestion of cooked bacon and spinach. *Nature 265,* 753–755.

FONG, Y. Y. and CHAN, W. C. 1973. Dimethylnitrosamine in Chinese marine salt fish. *Food Cosmet. Toxicol. 11,* 841–845.

FONG, Y. Y. and CHAN, W. C. 1976. Methods of limiting the content of dimethylnitrosamine in Chinese marine salt fish. *Food Cosmet. Toxicol. 14,* 95–98.

FONG, Y. Y. and CHAN, W. C. 1977. Nitrate, nitrite, dimethylnitrosamine and N-nitrosopyrrolidine in some Chinese food products. *Food Cosmet. Toxicol. 15,* 143–145.

FOREMAN, J. K. AND GOODHEAD, K. 1975. The formation and analysis of nitrosamines. *J. Sci. Food Agric. 26,* 1771–1783.

FREIMUTH, U. and GLAESER, E. 1970. Zum Auftreten von Nitrosamine in Lebensmitteln. *Nahrung 14,* 357–361.

GADBOIS, D., RAVESI, E. M., LUNDSTROM, R. C. and MANEY, R. S. 1975. N-Nitrosodimethylamine in cold-smoked sable fish. *J. Agric. Food Chem. 23,* 665–668.

GARCIA, H. and LIJINSKY, W. 1973. Studies of the tumorigenic effect in feeding of nitrosamino acids and of low doses of amines and nitrite to rats. *Z. Krebsforsch. 79,* 141–144.

GOLOVNYA, R. V. 1976. Analysis of volatile amines contained in foodstuffs as possible precursors of N-nitroso compounds. In: *Environmental*

N-Nitroso Compounds Analysis and Formation. Walker, E. A., Bogovski, P. and Griciute, L. (Editors), International Agency for Research on Cancer, Lyon, Scientific Publication No. 14, pp. 237–245.

GOODHEAD, K., GOUGH, T. A., WEBB, K. S., STADHOUDERS, J. and ELGERSMA, R. H. C. 1976. The use of nitrate in the manufacture of Gouda cheese. Lack of evidence of nitrosamine formation. *Neth. Milk Dairy J. 30*, 207–221.

GORENEN, P. J., JONK, R. J. G., VAN INGEN, C. and TEN NOEVER DE BRAU, M. C. 1976a. Determination of eight volatile nitrosamines in thirty cured meat products with capillary gas chromatography-high resolution mass spectrometry: The presence of nitrosodiethylamine and the absence of nitrosopyrrolidine. In: *Environmental N-Nitroso Compounds Analysis and Formation*, Walker, E. A., Bogovski, P. and Griciute, L. (Editors), International Agency for Research on Cancer, Lyon, Scientific Publication No. 14, pp. 321–331.

GORENEN, P. J., DE COCK-BETHBEDER, M. W., JONK, R. J. G. and VAN INGEN, C. 1976b. Further studies on the occurrence of volatile N-nitrosamines in meat products by combined gas chromatography and mass spectrometry. *Proc. 2nd int. Symp. Nitrite Meat Prod.*, Zeist, pp. 227–237.

GOUGH, T. A. and GOODHEAD, K. 1975. Occurrence of volatile nitrosamines in spice premixes. *J. Sci. Food Agric. 26*, 1473–1478.

GOUGH, T. A. and WALTERS, C. L. 1976. Volatile nitrosamines in fried bacon. In: *Environmental N-Nitroso Compounds Analysis and Formation*, Walker, E. A., Bogovski, P. and Griciute, L. (Editors), International Agency for Research on Cancer, Lyon, Scientific Publication No. 14, pp. 195–203.

GOUGH, T. A., McPHAIL, M. F., WEBB, K. S., WOOD, B. J. and COLEMAN, R. F. 1977. An examination of some foodstuffs for the presence of volatile nitrosamines. *J. Sci. Food Agric. 28*, 345–351.

GRAY, J. I. 1976. N-Nitrosamines and their precursors in bacon: A review. *J. Milk Food Technol. 39*, 686–692.

GRAY, J. I., COLLINS, M. E. and RUSSEL, L. F. 1977. Formation of N-nitrosohydroxypyrrolidine in model and cured meat systems. *Can. Inst. Food Sci. Technol. J. 10*, 36–39.

GREENBERG, R. A. 1976. Nitrosopyrrolidine in United States cured meat products. *Proc. 2nd int. Symp. Nitrite Meat Prod.*, Zeist, Pudoc, Wageningen, pp. 203–210.

GREENBLATT, M., MIRVISH, S. S. and SO, B. T. 1971. Nitrosamine studies: Induction of lung adenomas by concurrent administration of sodium nitrite and secondary amines in Swiss mice. *J. Nat. Cancer Inst. 46*, 1029–1034.

HAVERY, D. C., KLINE, D. A., MILETTA, E. M., JOE, F. L. JR. and FAZIO, T. 1976. Survey of food products for volatile N-nitrosamines. *J. Assoc. Offic. Anal. Chem. 59*, 540–546.

HAVERY, D. C., FAZIO, T. and HOWARD, J. W. 1977. Trends in levels of N-nitrosopyrrolidine in fried bacon. *Presented at the Fifth IARC Meeting* on *Analysis and Formation of N-Nitroso Compounds*, August 22–24, 1977, University of New Hampshire, Durham, N.H.

HEISLER, E. G., SICILIANO, J., KRULICK, S., FEINBERG, J. and SCHWARTZ, J. H. 1974. Changes in nitrate and nitrite content, and search for nitrosamines in storage-abused spinach and beets. *J. Agr. Food Chem. 22*, 1029–1032.

HERRMANN, H. 1961. Identifizierung eines Stoffwecheselproduktes von *Clitocybe suaveolens* als 4-Methyl-nitrosamino-benzaldehyd. *Hoppe-Seyler's Z. Physiol. Chem. 326*, 13–16.

HILL, M. J. and HAWKSWORTH, G. 1972. Bacterial production of nitrosamines *in vitro* and *in vivo*. In: *N-Nitroso Compounds: Analysis and Formation*, Bogovski, P., Preussmann, R. and Walker, E. A. (Editors), International Agency for Research on Cancer, Lyon, Scientific Publication No. 3, pp. 127–129.

HOWARD, J. W., FAZIO, T. and WATTS, J. O. 1970. Extraction and gas chromatographic determination of N-nitrosodimethylamine in smoked fish: Application to smoked nitrite-treated chub. *J. Assoc. Offic. Anal. Chem. 53*, 269–274.

ISHIDATE, M., TANIMURA, A., ITO, Y., SAKAI, A., SAKUTA, H., KAWAMURA, T., SAKAI, K., MIYAZAWA, F. and WADA, H. 1971. Secondary amines, nitrites and nitrosamines in Japanese foods. *Nippon Yakuzaishikai Zasshi 23*, 47–51.

IVANKOVIC, S. and DRUCKREY, H. 1968. Transplacentare Erzeugung maligner Tumoren des Nervensystems. I. Aethyl-nitroso-harnstoffan BD IX-Ratten. *Z. Krebsforsch. 71*, 320–360.

IVANKOVIC, S. and PREUSSMANN, R. 1970. Transplacentare Erzeugung maligner Tumoren nach oraler Gabe von Aethylharnstoff und Nitrit and Ratten. *Naturwiss* enshaften *57*, 460–461.

IYENGAR, J. R. PANALAKS, T., MILES, W. F. and SEN, N. P. 1976. A survey of fish products for volatile N-nitrosamines. *J. Sci. Food Agric. 27*, 527–530.

JUSZKIEWICZ, T. and KOWALSKI, B. 1976. An investigation of the possible presence or formation of nitrosamines in animal feeds. In: *Environmental N-Nitroso Compounds Analysis and Formation*, Walker, E. A., Bogovski, P. and Griciute, L. (Editors), International Agency for Research on Cancer, Lyon, Scientific Publication No. 14, pp. 375–383.

KANN, J., TAUTS, O., RAJA, K. and KALVE, R. 1976. Nitrosamines and their precursors in some Estonian foodstuffs. In: *Environmental N-Nitroso Compounds Analysis and Formation*, Walker, E. A., Bogovski, P. and Griciute, L. (Editors), International Agency for Research on Cancer, Lyon, Scientific Publication No. 14, pp. 385–394.

KAWABATA, T. 1977. Nitrite additives, with special reference to carcinogenic N-nitroso compounds. *Japanese J. Dairy Food Sci. 26*, A37–A46.

KEEFER, R. K. and ROLLER, P. P. 1973. N-Nitrosation by nitrite ion in neutral and basic medium. *Science 181*, 1245–1247.

KLEIN, D., POULLAIN, B. and DERBY, G. 1976. N-Nitroso compounds in products widely consumed in France: hams. *Proc. 2nd int. Symp. Nitrite Meat Prod.* Zeist, Pudoc, Wageningen, pp. 289–292.

KLUBES, P., CERNA, I. RABINOWITZ, A. D. and JONDROFF, W. R. 1972. Factors affecting dimethylnitrosamine formation from simple precursors by rat intestinal bacteria. *Food Cosmet. Toxicol. 10*, 757–767.

KOPPANG, N. 1964. An outbreak of toxic liver injury in ruminants. Case reports pathological-anatomical investigations, and feeding experiments. *Nord. Veterinaermed. 16*, 305–322.

KOPPANG, N. 1974. Dimethylnitrosamine-formation in fish meal and toxic effects in pigs. *Am. J. Pathol. 74*, 95–108.

KOPPANG, N. and RIMESLATTEN, H. 1976. Toxic and Carcinogenic effects of nitrosodimethylamine in mink. In: *Environmental N-Nitroso Compounds Analysis and Formation*, Walker, E. A., Bogovski, P. and Griciute, L. (Editors), International Agency for Research on Cancer, Lyon, Scientific Publication No. 14, pp. 443–452.

KUSHNIR, I., FEINBERG, J. I., PENSABENE, J. W., PIOTROWSKI, E. G., FIDDLER, W. and WASSERMANN, A. E. 1975. Isolation and identification of nitrosoproline in uncooked bacon. *J. Food Sci. 40*, 427–428.

KROELLER, E. 1967. Untersuchungen zum Nachweis von Nitrosaminen in Tabakrauch und Lebensmitteln. *Deut. Lebensm.-Rundsch. 63*, 303–305.

LANDMANN, W. A. and BATZER, O. F. 1966. Influence of processing procedures on the chemistry of meat flavors. *J. Agr. Food Chem. 14*, 210–214.

LANE, R. P., RICE, R. H. and BAILEY, M. E. 1974. Gas chromatographic-mass spectrometric determination of N-nitrosodimethylamine formed in synthetic and human gastric juice. *J. Agr. Food Chem. 22*, 1019–1023.

LEE, J. S., LIBBY, L. M., SCANLAN, R. A. and BARBOUR, J. 1977. 3-Hydroxy-N-nitrosopyrrolidine in fried bacon and fried out fat. *Presented at the Fifth IARC Meeting on Analysis and Formation of N-Nitroso Compounds*, August 22–24, 1977, University of New Hampshire, Durham N.H.

LIJINSKY, W., TOMATIS, L. and WENYON, C. E. M. 1969. Lung tumors in rats treated with N-nitrosoheptamethyleneimine and N-nitrosooctamethyleneimine. *Proc. Soc. Exp. Biol. Med. 130*, 945–949.

LIJINSKY, W., CONRAD, E. and VAN DE BOGART, R. 1972. Formation of carcinogenic nitrosamines by interaction of drugs with nitrite. In: *N-Nitroso Compounds Analysis and Formation*, Bogovski, P., Preussmann, R. and Walker, E. A. (Editors), International Agency for Research on Cancer, Lyon, Scientific Publication No. 3, pp. 130–133.

LIJINSKY, W., TAYLOR, H. W., SNYDER, C. and NESHEIM, P. 1973. Malignant tumours of liver and lung in rats fed aminopyrene or heptamethyleneimine together with nitrite. *Nature 244*, 176–178.

MAGEE, P. N. and BARNES, J. M. 1956. The production of malignant primary hepatic tumours in the rat by feeding dimethylnitrosamine. *Brit. J. Cancer 10*, 114–122.

MAGEE, P. N. and BARNES, J. M. 1959. The experimental production of tumours in the rat by dimethylnitrosamine (N-nitrosodimethylamine). *Acta Un. Int. Contra Cancrum 15*, 187–190.

MAGEE, P. N. and BARNES, J. M. 1967. Carcinogenic nitroso compounds. *Advan. Cancer Res. 10*, 163–246.

MAGEE, P. N., MONTESANO, R. and PREUSSMANN, R. 1976. N-Nitroso compounds and related carcinogens. In: *Chemical Carcinogens*, Searle, C. E. (Editor), American Chemical Society, New York, pp. 490–625.

MARION, L. 1950. The Pyridine Alkaloids. In: The Alkaloids Chemistry and Physiology, Manske, R. H. F. and Holmes, H. L. (Editors), Academic Press Inc., New York, pp. 167–258.

MEESTER, J. 1973. Nitrate and nitrite allowances in meat products. *Proc. 1st int. Symp. Nitrite Meat Prod.*, Zeist, Krol, B. and Tinbergen, B. J. (Editors), pp. 10–14.

MERGENS, W. J., KAMM, J. J., NEWMARK, H. L., FIDDLER, W. and PENSABENE, J. 1977. Alpha-tocopherol: uses in preventing nitrosamine formation. *Presented at the Fifth IARC Meeting on Analysis and Formation of N-Nitroso Compounds*, August 22–24, 1977, University of New Hampshire, Durham, N.H.

MIRVISH, S. S. 1970. Kinetics of dimethylamine nitrosation in relation to nitrosamine carcinogenesis. *J. Nat. Cancer Inst. 44*, 633–639.

MIRVISH, S. S. 1972. Kinetics of N-nitrosation reaction in relation to tumorigenesis experiments with nitrite plus amines or ureas. In: *N-Nitroso Compounds in the Enviroment*, Bogovski, P., Preussmann, R. and Walker, E. A. (Editors), International Agency for Research on Cancer, Lyon, Scientific Publication No. 3, pp. 104–108.

MIRVISH, S. S. 1975. Formation of N-nitroso compounds: chemistry, kinetics and *in vivo* occurrence. *Toxicol. Appl. Pharmacol.* 31, 325–351.

MIRVISH, S. S., WALLCAVE, L., EAGEN, M. and SHUBIK, P. 1972. Ascorbate-nitrite reaction: possible means of blocking the formation of carcinogenic N-nitroso compounds. *Science 177*, 65–68.

NAKAMURA, M. and USUKI, M. 1973. Safety of marine products. III. Distribution of dimethylnitrosamine in marine foods. *Shokuhin Eiseigaku Zasshi 14*, 264–271.

NEURATH, G. B., DUENGER, M., PEIN, F. G., AMBROSIUS, D. and SCHREIBER, O. 1976. Primary and secondary amines in the human environment. *Food Cosmet. Toxicol. 15*, 275–282.

PANALAKS, T., IYENGAR, J. R. and SEN, N. P. 1973. Nitrate, nitrite and dimethylnitrosamine in cured meat products. *J. Assoc. Offic. Anal. Chem. 56*, 621–625.

PANALAKS, T., IYENGAR, J. R. and SEN, N. P. 1974. Further survey of cured meat products for volatile N-nitrosamines. *J. Assoc. Offic. Anal. Chem. 57*, 806–812.

PATTERSON, R. L. S. and MOTTRAM, D. S. 1974. The occurrence of volatile amines in uncured and cured pork meat and their possible role in nitrosamine formation in bacon. *J. Sci. Food Agr. 25*, 1419–1425.

PREUSSMANN, R. 1974. Formation of carcinogens from precursors occurring in the enviroment: new aspects of nitrosamine-induced tumorgenesis. *Recent Results Cancer Res. 44*, 9–15.

PREUSSMANN, R., SCHMAEHL, D., EISENBRAND, G. and PORT, R. 1976. Dose-response study with N-nitrosopyrrolidine and some comments on risk evaluation of environmental N-nitroso compounds. *Proc. 2nd int. Symp. Nitrite Meat Prod.*, Zeist, Pudoc, Wageningen, pp. 261–268.

ROLLER, P. P. and KEEFER, L. K. 1974. Catalysis of nitrosation reactions by electrophilic species. In: *N-Nitroso Compounds in the Environment*, Bogovski, P. and Walker, E. A. (Editors), International Agency for Research on Cancer, Lyon, Scientific Publication No. 9, pp. 86–89.

ROUNBEHLER, D. P., ROSS, R., FINE, D. H., IQBAL, Z. M. and EPSTEIN, S. S. 1977. Quantitation of dimethylnitrosamine in the whole mouse after biosynthesis *in vivo* from trace levels of precursors. *Science 197*, 917–918.

RUITER, A. 1973. Determination of volatile amines and amine oxides in food products. *Proc. int. Symp. Nitrite Meat Products*, Zeist, Krol. B. and Tinbergen, B. J. (Editors), pp. 37–43.

SAKSHAUG, J., SOEGNEN, E., AAS HANSEN, M. and KOPPANG, N. 1965. Dimethylnitrosamine; its hepatotoxic effect in sheep and its occurrence in toxic batches of herring meal. *Nature 206*, 1261–1262.

SANDER, J. 1967. Kann Nitrit in der menschlichen Nahrung Ursache einer Krebsentstehung durch Nitrosaminbildung sein? *Arch. Hyg. Bakteriol. 151*, 22–28.

SANDER, J. and BUERKLE, G. 1969. Induktion maligner Tumoren bei Ratten durch gleichzeitige Verfuetterung von Nitrit und sekundaeren Aminen. *Z. Krebsforsch. 73*, 54–66.

SANDER, J. and SCHWEINSBERG, F. 1972. *In vivo* and *in vitro* experiments on the formation of N-nitroso compounds from amines or amides and nitrate or nitrite. In: *N-Nitroso Compounds Analysis and Formation*, Bogovski, P., Preussmann, R. and Walker, E. A. (Editors), International Agency for Research on Cancer, Lyon, Scientific Publication No. 3, pp. 97–103.

SANDER, J., SCHWEINSBERG, F. and MENZ, H. P. 1968. Untersuchungen ueber die Entstehung cancerogener Nitrosamine im Magen. *Hoppe-Seyler's Z. Physiol. Chem. 349*, 1691–1697.

SCHOENTAL, R. 1960. Carcinogenic action of diazomethane and of nitroso-N-methylurethan. *Nature 188*, 420–421.

SCHWARTZ, G. and THOMASOW, J. 1950. The aroma of Tilsiter cheese. *Milchwissenschaft. 5*, 376–379 and 412–416.

SEN, N. P. 1972. The evidence for the presence of dimethylnitrosamine in meat products. *Food Cosmet. Toxicol. 10*, 219–223.

SEN, N. P. 1974. Nitrosamines. In: *Toxic Constituents of Animal Foodstuffs*, Liener, I. E. (Editor), Academic Press, Inc., New York, pp. 131–194.

SEN, N. P. and McKINLEY, W. P. 1974. Meat curing agents as one source of nitrosamines. *Proc. IV Int. Congress Food Sci. and Technol.* Vol. III, pp. 476–482.

SEN, N. P., SMITH, D. C. and SCHWINGHAMER, L. 1969. Formation of N-nitrosamines from secondary amines and nitrite in human and animal gastric juice. *Food Cosmet. Toxicol. 7*, 301–307.

SEN, N. P., MILES, W. F., DONALDSON, B., PANALAKS, T. and

IYENGAR, J. R. 1973a. Formation of nitrosamines in a meat curing mixture. *Nature 245*, 104–105.

SEN, N. P., DONALDSON, B., IYENGAR, J. R. and PANALAKS, T. 1973b. Nitrosopyrrolidine and dimethylnitrosamine in bacon. *Nature 241*, 473–474.

SEN, N. P., DONALDSON, B. A. and CHARBONNEAU, C. 1974a. Formation of nitrosodimethylamine from the interaction of certain pesticides and nitrite. In: *N-Nitroso Compounds in the Environment*, Bogovski, P. and Walker, E. A. (Editors), International Agency for Research on Cancer, Lyon, Scientific Publication No. 9, pp. 75–79.

SEN, N. P., DONALDSON, B., CHARBONNEAU, C. and MILES, W. F. 1974b. Effect of additives on the formation of nitrosamines in meat curing mixtures containing spices and nitrite. *J. Agr. Food Chem. 22*, 1125–1130.

SEN, N. P., IYENGAR, J. R., DONALDSON, B. A. and PANALAKS, T. 1974c. Effect of sodium nitrite concentration on the formation of nitrosopyrrolidine and dimethylnitrosamine in fried bacon. *J. Agr. Food Chem. 22*, 540–541.

SEN, N. P., DONALDSON, B., SEAMAN, S., IYENGAR, J. R. and MILES, W. F. 1976a. Inhibition of nitrosamine formation in fried bacon by propyl gallate and L-ascorbyl palmitate. *J. Agr. Food Chem. 24*, 397–401.

SEN, N. P., SMITH, D. C., MOODIE, C. A. and GRICE, H. C. 1976b. Failure to induce tumours in guinea pigs after concurrent administration of nitrite and diethylamine. *Food Cosmet. Toxicol. 13*, 423–425.

SEN, N. P., IYENGAR, J. R., MILES, W. F. and PANALAKS, T. 1976c. Nitrosamines in cured meat products. In: *Environmental N-Nitroso Compounds Analysis and Formation*, Walker, E. A., Bogovski, P., Griciute, L. (Editors), International Agency for Research on Cancer, Lyon, Scientific Publication No. 14, pp. 333–342.

SEN, N. P., COFFIN, D. E., SEAMAN, S., DONALDSON, B. and MILES, W. F. 1976d. Extraction, clean-up and estimation as methyl ether of 3-hydroxy-1-nitrosopyrrolidine, a non-volatile nitrosamine in cooked bacon at mass fractions of µg/kg. *Proc. 2nd int. Symp. Nitrite Meat Prod.*, Zeist, Pudoc, Wageningen, pp. 179–185.

SEN, N. P., SEAMAN, S. and MILES, W. F. 1976e. Dimethylnitrosamine and nitrosopyrrolidine in fumes produced during the frying of bacon. *Food Cosmet. Toxicol. 14*, 167–170.

SEN, N. P., DONALDSON, B., SEAMAN, S., COLLINS, B. and IYENGAR, J. R. 1977a. Recent nitrosamine analysis in cooked bacon. *Can. Inst. Food Sci. Technol. J. 10*, A13–A15.

SEN, N. P., DONALDSON, B. A., SEAMAN, S., IYENGAR, J. R. and MILES, W. F. 1977b. Recent studies in Canada on the analysis and occurrence of volatile and non-volatile N-nitroso compounds in foods. *Presented at the Fifth IARC Meeting on Analysis and Formation of N-Nitroso Compounds*, August 22–24, 1977, University of New Hampshire, Durham, N.H.

SHANK, R. C. and NEWBERNE, P. M. 1976. Dose-response study of the carcinogenicity of dietary sodium nitrite and morpholine in rats and hamsters. *Food Cosmet. Toxicol. 14*, 1–8.

SHEWAN, J. M. 1951. The chemistry and metabolism of the nitrogenous extractives in fish. *Biochem. Soc. Symp.* (Cambridge, England), No. *6*, 28–48.

SINGER, G. M. and LIJINSKY, W. 1976. Naturally occurring nitrosatable compounds. I. Secondary amines in foodstuffs. *J. Agr. Food Chem. 24*, 550–553.

SPIEGELHALDER, B., EISENBRAND, G. and PREUSSMANN, R. 1976. Influence of dietary nitrate on nitrite content of human saliva: Possible relevance to *in vivo* formation of N-nitroso compounds. *Food Cosmet. Toxicol. 14*, 545–548.

STEPHANY, R. W., FREUDENTHAL, J. and SCHULLER, P. L. 1976. Quantitative and qualitative determination of some volatile nitrosamines in various meat products. In: *Environmental N-Nitroso Compounds Analysis and Formation*, Walker, E. A., Bogovski, P. and Griciute, L. (Editors), International Agency for Research on Cancer, Lyon, Scientific Publication No. 14, pp. 343–354.

TANNENBAUM, S. R., WEISMAN, M. and FETT, D. 1976. The effect of nitrate intake on nitrite formation in human saliva. *Food Cosmet. Toxicol. 14*, 549–552.

TANNENBAUM, S. R. 1978. Nitrite, nitrate synthesis in human body. In: *Food Chem. News 19*, January 9 issue, p. 52.

TERRACINI, B., MAGEE, P. N. and BARNES, J. M. 1967. Hepatic pathology in rats on low dietary levels of dimethylnitrosamine. *Brit. J. Cancer 21*, 559–565.

U.S. FOOD ADDITIVE REGULATIONS. 1972. Code of Federal Regulations. Office of the Federal Register, Washington, D.C.

WALKER, E. A. 1978. Personal communication.

WALKER, E. A., PIGNATELLI, B. and CASTEGNARO, M. 1975. The effects of gallic acid on nitrosamine formation. *Nature 258*, 176.

WALTERS, C. L., EDWARDS, M. W., ELSEY, T. S. and MARTIN, M. 1976. The effect of antioxidants on the production of volatile nitrosamines during the frying of bacon. *Z. Lebensm. Unters.-Forsch. 162*, 377–385.

WARTHESEN, J. J., BILLS, D. O., SCANLAN, R. A. and LIBBEY, L. M. 1976. N-Nitrosopyrrolidine collected as a volatile during heat-induced formation in nitrite-containing pork. *J. Agr. Food Chem. 24*, 892–894.

WASSERMANN, A. E., FIDDLER, W., DOERR, R. C., OSMAN, S. F. and DOOLEY, C. J. 1972. Dimethylnitrosamine in frankfurters. *Food Cosmet. Toxicol. 10*, 681–684.

WEURMAN, C. and DeROOY, C. 1961. Volatile amines in the odors of food. *J. Food Sci. 26*, 239–243.

WHITE, J. W. 1976. Relative significance of dietary sources of nitrate and nitrite. *J. Agr. Food Chem 24*, 202.

WISHNOK, J. S. 1977. Formation of nitrosamines in food and in the digestive system. *J. Chem. Ed. 54*, 440–441.

11

Mercury in Food, Feedstuffs and the Environment

Robert L. Bradley, Jr. and Alan G. Hugunin

INTRODUCTION AND HISTORICAL DEVELOPMENT

Mercury, a natural element, has been present since the formation of the earth and known to man since prehistoric times. Great concern over its toxicity developed during the past 25 years since the first encounter with the potential danger of mercury near Minamata Bay in Japan in 1953. Several families in the area became infected with a mysterious neurological illness which, by late 1956, reached epidemic proportions. Investigators showed that this disease affected the peripheral nervous system, cerebellum, hearing, vision and less frequently, pyramidal tracts. Progressive brain damage was indicated in severe cases. Of 111 reported cases, 41 were fatal, and the majority of survivors were incapacitated in varying degrees. By 1959, the cause of the disease remained unknown. However, two facts had been established: first, that a relationship existed between fish consumption and occurrence of the disease, and secondly, that the fish responsible for the disease were caught in Minamata Bay (McAlpine and Araki, 1958, 1959). Magnesium, selenium and thallium received the most attention but the similarities of symptoms to those noted by Hunter, Bomford and Russell (1940) for methyl mercury poisoning were recognized. In 1968, a complete report on the disease expressing agreement that the toxic agent was methyl mercury was published. Mercury had drained into the bay with the effluent from an acetaldehyde plant, in which it was used as a catalyst (Kutsuna, 1968). In 1962, the danger of organic mercury became more apparent when the Minamata incident was repeated in Niigata, Japan, along the lower Agano River. Methyl mercury poisoning was officially documented in 5 deaths and 26 illnesses. Again, fish and shellfish which concentrated mercury discharged by a chemical plant were the source (Irukayama, 1968; Takizawa, 1970).

In the 1960s, Swedish scientists concerned with environmental pollution noted that wildlife species contained lethal levels of mercury. Investigating causes for decreases in wild bird populations, Johnels and Westermark (1969) found that mercury levels in feathers of museum specimens markedly increased after 1940. This increase coincided with the introduction of methyl mercury seed dressings and abruptly decreased after a ban was imposed on alkyl mercury seed dressings. Borg et al. (1966) found lethal levels of mercury in a high percentage of animals found dead in the countryside in which poisoning was suspected, including pheasants, partridges, pigeons, finches, and corvine birds as well as their predators; eagles, hawks, falcons, owls, foxes, martens and polecats. Westöo (1968) stated that Westermark and Sjostrand (1965) were the first to report that   Swedish fish contained elevated levels of mercury which Westöo (1966) identified as methyl mercury. In view of the Japanese experiences, the Swedish government adopted a limit of 1.0 ppm mercury in fish which were to be sold.

Canadian scientists developed concern over the widespread use of mercury in their country in the late 1960s but concern in North America finally peaked in 1970 when Fimreite et al. (1971) reported that fish taken from lakes Erie and St. Clair contained significant quantities of mercury. In the following months, the FDA seized several lots of fish which contained high concentrations of mercury. Lakes which showed significant mercury pollution were closed to commercial and sport fishing and reports indicated high levels of mercury in wildlife.

This chapter presents an overview of information available on: the natural levels of mercury; man's uses of mercury and their impact on the environment; types of mercury compounds; biotransformation and differences in the metabolic and toxicologic effect of the various forms; level of mercury in foods; FDA regulations on mercury and regulations for other countries; and methods of analysis for mercury.

NATURAL LEVELS OF MERCURY

The average concentration of mercury in the earth's crust is about 0.5 ppm, and its abundance ranks 62nd from the top or 19th from

the bottom on the list of elements in the earth (Watts *et al.* 1976).
Most of the mercury is concentrated in geographical belts of the
earth (Jonasson and Boyle, 1972). It can exist in any of three oxidation states in nature: the metallic state $(Hg)^0$, the binuclear state
$(Hg-Hg)^{+2}$ in which there is a covalent bond between the two atoms,
or the mercuric state $(Hg)^{+2}$. The red mercuric sulfide, cinnabar, is
the principal commercial source of mercury and it contains 86.2%
mercury. Cinnabar is found within a few hundred feet of the surface
in small veins or pockets in limestone, calcareous shales, sandstone,
serpentine, chert, andesite, basalt, and rhyolite, most generally in
areas of volcanic activity (Jones, 1971).

Ehmann and Lovering (1967) reported the mercury concentration
of the earth's deep crust and upper mantel is 0.78 to 1.48 ppm. Day
(1964) stated that mercury reaches the surface by a process other
than weathering. Being particularly mobile under magmatic and
hydrothermal conditions, he theorized that mercury compounds are
reduced by ferrous ions to metallic mercury. The highly volatile
metal travels upward toward the surface where it is fixed by
combination with sulfur. Volcanic activity releases mercury into the
litosphere, atmosphere, and hydrosphere. The vapor equilibrium of
mercury at room temperature is approximately 10 mg/m^3 of air and
this doubles with a 10 degree increase in temperature (Biram 1957).
Mercury vaporizes from the earth and water into the atmosphere
where it is found both in the vapor form and as particles (Goldwater,
1973). Because of these unique properties, mercury is subject to a
natural geological cycling which can be schematically shown (Figure
11.1). Traces of mercury are, therefore, found over the entire earth.

Metallic mercury occurs in some ores (Jones, 1971); however, the
concentration is not of commercial significance. Turekian and Wedepohl (1961) reported the mercury concentration of sedimentary
rocks ranged from 0.03 ppm in sandstone to 0.04 ppm in shales. In
the igneous rock category, granites contain an average concentration of 0.08 ppm mercury (Turekian and Wedepohl, 1961). Joensuu
(1971) analyzed a number of coal samples for mercury and found
concentrations ranging from 0.07 ppm to 33 ppm. In line with these
values, Bailey, Snavely, and White (1961) reported crude petroleum
samples collected from two wells contained 1.9–21.0 ppm mercury.

Possibilities of environmental pollution limit determination of the
natural concentration of mercury in soil. A Geological Survey (Anon,
1970) estimated background levels at 0.1 ppm. However, other studies
indicated lower background values (Fleischer, 1970; Klein, 1972).
Higher concentrations of mercury would naturally occur in top soils
rich in humus or near mercury ore deposits (Goldwater, 1971).

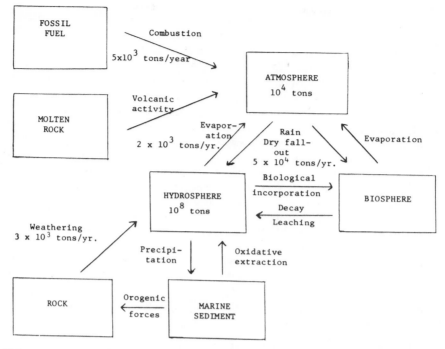

FIG. 11.1. OVERALL GLOBAL CYCLING OF MERCURY. MAN'S ANNUAL USE IS 10⁴ TONS OF WHICH ABOUT ONE HALF IS RECYCLED. (BERTINE AND GOLD-BERG, 1971; BOWEN, 1966; ERIKSSON, 1967; JOENSUU, 1971; PEAKALL AND LOVETT, 1973; NO REFERENCE NO. 331)

Atmospheric concentrations of mercury vary depending upon sampling site. Because of the relatively high vapor pressure, highest natural concentrations are found over cinnabar deposits. McCarthy *et al.* (1970) found concentrations of 0.02–20 μg Hg/m³ air over ore deposits, whereas over non-mineralized land concentrations range from 0.003–0.009 μg Hg/m³ air. Air over oceans contains the lowest concentration of mercury: 0.0006–0.0007 μg/m³ (W.H.O., 1976). Precipitation by rain or snow reduces mercury concentration in air to near zero, while other natural factors such as direction and speed of wind, temperature, and solar insulation alter the level of mercury in the atmosphere (W.H.O., 1976).

Metallic mercury is relatively insoluble in water and because of the high density (13.546 at 20 C) (Watts *et al.* 1976), it collects in low spots in waterways. Ionic forms of mercury are soluble in water, the extent depends on the acidity of the water and the presence of complexing ions. Wershaw (1970) analyzed ground water samples from 73 areas of the United States and found 83% contained less than 1 ppb mercury, whereas only 2 samples from areas near mercury ore deposits had levels in excess of 5 ppb (Wenninger, 1965). Klein (1972) analyzed 67 samples of water from ponds, lakes and rivers in Northeastern United States and reported that background levels were approximately 0.055 ppb when environmental pollution had not occurred. Joensuu (1971) estimated the amount of mercury released into waterways by natural weathering. His calculations were based on the amount of sodium and mercury present in the litosphere and on the rate at which sodium is reached by weathering; Joensuu reported a maximum of 20.9×10^4 kg of mercury may be carried to sea each year. In sea water, Burton and Leatherland (1971) found 0.014–0.021 ppb of mercury in samples collected off the coast of England. Leatherland *et al.* (1971) reported that in the Northeastern Atlantic, surface water contained 0.013-0.018 ppb mercury; the concentration decreased with depth. However, Klein and Goldberg (1970) found higher concentrations of mercury at deeper depths off the coast of Japan. Their values ranged from 0.10 ppb at surface to 0.15–0.27 ppb at lower depths (Klein and Goldberg, 1970).

MAN'S USE OF MERCURY AND THE ENVIRONMENTAL CONTAMINATION

For centuries, man's use of mercury was limited to cosmetics and medicine. In 1557, Bartolome de Medina devised a process by which silver was recovered by amalgamation (Saha, 1973). Since that time, the uses of mercury have increased and, in 1971, over 10.5×10^6 kg of mercury were produced to meet demands (Cammaroto, 1974). Increasing amounts of mercury are being recovered, repurified and sold as secondary mercury. Many processes for reclaiming mercury from mercury boilers, electrical apparatus, dental amalgams, batteries, and sludge from processes using mercury as a catalyst have been developed (Jones, 1971). In 1971 over 5.8×10^5 kg of mercury on the U.S. market were secondary mercury (Cammaroto, 1974). The difference between total mercury and secondary mercury sold is indicative of the losses of mercury which occur in the environment. Mercury pollution is unique from many other types of pollution in that the total earth concentration is not being changed; rather the problem occurs from redistribution of the mercury or conversion to a more toxic form. Thus, it is important to consider these aspects with regard to production and uses of mercury to determine present day risks.

Information on production of mercury is available (Jones, 1971). In the United States most mercury is extracted from the ore by a directly heated pyrometallurgical process. After crushing, the ore is heated in retorts or furnaces to liberate mercury as vapor which after passing through dust collectors, is cooled by a condenser system and the resultant liquid mercury collected. Although recovery from this process averages between 95–98%, only 1.24–12.4 g of mercury are obtained per kilogram of ore (Jones, 1971). The primary source of pollution in processing mercury is loss of vapor through condenser stack emissions which range from 74.2–244 mg per kilogram of ore processed (USEPA, 1973).

Uses

The largest users of mercury and trends in mercury consumption are shown in Table 11.1 New public awareness of mercury pollution has had some effects on the uses of mercury. Unfortunately, a more recent survey is not available.

The electrical industry, the largest consumer, utilizes over one-half of its consumption in the production of batteries (Greenspoon, 1970). Other electrical products utilizing significant quantities of mercury are fluorescent lamps and high intensity discharge lamps, rectifiers, oscillators, and power control switches (Greenspoon, 1970). Recycling of most of these items is impractical except for industries utilizing large quantities with volume turnovers. For this reason, much of the mercury is added to the environment with trash. Projections for mercury use in the year 2000 range from 8.6×10^5 – 13.8×10^5 kg; the broad range reflects both the continued demand for mercury batteries and a probable shift to rechargeable nickel-cadmium batteries (Greenspoon, 1970).

Use of mercury cells for production of chlorine gas accounted for over 23% of the mercury consumed in the U.S. in 1971 (Cammaroto, 1974). The chlorine production process (USEPA, 1973) involves a cell with two sections. The first section, the electrolyzer, uses mercury as a flowing cathode which flows co-currently with a salt solution such as sodium chloride. When a high current is applied between the carbon or metal anode and the mercury cathode, chlorine gas is formed at the anode and alkali amalgam at the cathode. The amalgam is separated and passed into the second section, the decomposer, where the amalgam becomes the anode to a short

TABLE 11.1.   MERCURY CONSUMED IN THE UNITED STATES, BY USE
$$(Kg \times 10^3)$$

Use	1955[1]	1965[2]	1968[3]	1970[3]	1972[3]
Agriculture[4]	255	107	118	62.4	63.3
Amalgamation	7.47	17.1	9.20	7.55	N/A
Catalysts	25.1	31.9	66.0	77.1	27.6
Dental preparations	70.3	119	106	78.8	103
Electrical apparatus	442	589	677	550	536
Electrolytic preparation of chlorine and caustic soda	107	302	602	517	397
General laboratory use	33.6	97.4	68.6	62.3	20.5
Industrial and control instruments	343	350	275	167	225
Paint:					
Antifouling	25.0	8.79	13.5	6.83	1.10
Mildew-proofing	N/A	260	351	350	282
Paper and pulp manufacture	N/A	21.3	14.4	7.79	0.34
Pharmaceuticals	54.4	112	14.6	23.8	19.9
Other[5]	609	620	285	202	147
Total known uses	1971	2635	2600	2112	1823
Total unknown uses	—	—	—	7.82	0.76
Grand total[6]	1971	2635	2600	2119.82	1823.76

N/A values not available from these references.
[1] Data from P.C.I.A.C.H. (1973) including portion listed as redistilled.
[2] Data from Erickson (1967) including portion listed as redistilled.
[3] Data from Cammaroto (1974).
[4] Includes fungicides and bactericides for industrial purposes.
[5] Includes that portion listed as miscellaneous uses for redistilled mercury P.C.I.A.C.H. (1973) and Erickson (1967).
[6] Discrepancies due to rounding off of above values.

circuited graphite cathode. Then, the amalgam is converted back to mercury and alkali metal hydroxide with the production of hydrogen gas. Caustic soda produced by this process may contain up to 7 ppm mercury (D'Itri, 1972). The Environmental Protection Agency bulletin (USEPA, 1973) identified significant losses of mercury into the environment (up to 0.91 g Hg/kilogram of Cl_2) which may occur from chloralkali production, but new processes have been developed which drastically reduce these emissions. An increasing percentage of chlorine also is being produced by the diaphragm process. Although the caustic produced by this process is less pure, limited mercury emissions are produced. In spite of improvements, pollution still is significant in chlor-alkali production. Cammaroto (1974) indicated that 27.2% of the 8.45×10^9 kg of chlorine produced was from mercury cells and 153 mg of mercury were lost in the production of a kilogram of chlorine; therefore, over 4.25×10^5 kg of mercury were lost in the production of chlorine and alkali. Greenspoon (1970) projected the need for 2.07×10^6 kg of mercury for chlor-alkali production in 2000. This appears excessive based on improved recovery and increased use of diaphragm cells.

The bactericidal and fungicidal properties of many mercury compounds have resulted in their use in paint and paper industries and agriculture. The paint industry is the third largest user of mercury.

Incorporation of phenyl mercury compounds into water-based paints inhibits mildew formation on painted surfaces as well as prolonging shelf life of paint. At a Wisconsin Department of Natural Resources hearing one manufacturer reported that the amount of phenyl mercury added per gallon ranged from 0.45 g Hg for interior water base paint to 3.41 g Hg for exterior alkyd base paint (Stewart, 1970). Moreover, measurable concentrations of mercury were found in the atmosphere of a room freshly painted with paint containing a mercurial fungicide (Sibbett et al. 1972). Health hazards from such emissions were minimal (Goldberg and Shapero, 1957). Marine paints contain mercurous compounds as antifouling agents to prevent slime buildup and plant growth on painted surfaces. Insignificant concentrations of this mercury are released into water (Hanson, 1971). Extensive research is directed to find non-mercurial preservatives which, on the whole, have not been as effective on the basis of shelf and surface life. The mercury requirements of the paint industry in the year of 2000 may reach $5.5 - 6.9 \times 10^5$ kg (Greenspoon, 1970).

Slime formation in paper and pulp machinery results in production of inferior paper. Although highly effective, the use of mercury-based slimicides has been largely discontinued in recent years as a result of the Federal Food and Drug Administration action (Byfield 1970) precluding the use for paper used in food packaging. Moreover, the concomitant development of other satisfactory slimicides has reduced this need. Replacement of mercury slimicides has aided in pollution control since approximately one-half of the mercury used passed out of the mill with the effluent whereas the other half bound to the paper fibers also polluted the environment subsequently at disposal (Byfield 1970). The pulping process utilizes large quantities of caustic soda and chlorine, both of which may contain mercury residues, if produced by the mercury cell process.

Agricultural consumption of mercury is low in comparison to other industries; however, special consideration is warranted because the mercury is applied directly to the environment. Mercurial compounds, which have broad spectrum fungal control, are unequaled by any other preparations used as seed dressings for barley, beans, corn, cotton, flax, millet, milo, oats, peanuts, peas, rice, rye, safflower, sorghum, soybeans, sugar beets and wheat. Novick (1969) listed the mercury-containing pesticides licensed by the Department of Agriculture in 1967 and also identified those subsequently restricted in use or prohibited. Use of these dressings has declined as a result of development of varieties of grain resistant to many fungal diseases (Worf, 1970). Moreover, environmental pollution by mercurial seed dressings may not be significant since common application rates would contribute only 1.8 g per hectare, which is much less than background levels. Sand *et al.* (1971) reported that of 93 soil samples analyzed, the mean concentration of mercury was 114 ppb where mercury-treated seeds had not been used, whereas a higher mean of 195 ppb mercury was found in soil where mercury-dressed seeds had been used.

Mercurial fungicides have been used also as sprays on fruit trees and tomato plants prior to development of fruit. In potato production, mercurial compounds have been used for both soil treatment and foliage sprays (Ohi *et al.* 1975). However, all such uses have been terminated by F.D.A. action. Inoue and Aomine (1969) reported that absorption of phenyl mercury acetate by clay mineral soils was subject to the law of cation exchange. Absorption was reduced greatly at or greater than pH 7 as well as below pH 5, due to the reduction of phenylmercuric ions or the increase of hydronium ions. Montmorillonite gave the highest absorption rate followed by Allophane and Kaolinite types of soils.

The highest rates of application of mercurial fungicides most likely are for turf management. Eckert (1970) stated that mercury fungicides have been used for years on golf courses. Non-mercurial fungicides are replacing the mercury compounds to control summer diseases, but no effective alternatives have been found to control winter fungus diseases referred to collectively as snow mold. An inorganic mercurial fungicide containing 60% $HgCl$ and 30% $HgCl_2$ offers broad spectrum activity and residual effectiveness; phenylmercuric acetate is used, also. Commonly applied only to the golf course greens, approximately 2.04×10^3 kg, on an elemental basis of mercury compounds, are applied annually to 182.1 hectares of greens in Wisconsin (Eckert, 1970). Of this, slightly over 2% are organic and about 98% are inorganic mercury compounds.

General laboratory uses of mercury are extensive. Cooke and Beitel (1971) stated that approximately 5.9×10^3 kg of mercury per year are lost by Canadian hospital laboratories through such uses as $HgCl_2$ for fixing tissue samples. Common uses of mercury compounds in food analysis laboratories (Bradley, 1970) are mercuric chloride to preserve milk samples, mercuric sulfate or oxide in

Kjeldahl nitrogen determinations, and mercuric sulfate in chemical oxygen demand determinations.

Medicinal uses of mercury are for pharmaceuticals, cosmetic preparations, and dental preparations. Mercury was once used to treat syphilis, as an aphrodisiac, and as a means of prolonging life (D'Itri, 1972). The use of organomercurial compounds as diuretics continues, but are being replaced by non-mercurial compounds. Organomercurials under such trade names as Mercurochrome[R], Merthiolate[R], Mercresin[R], and Metaphen[R] are still very popular antiseptics. Mercurials are frequently used in sterilizing solutions for medical instruments and are often incorporated into pharmaceuticals and cosmetics to control microbial growth. It is interesting that in 1971 as well as today, more mercury was used in dental preparations than in agriculture. Mercury is used in an amalgam with a silver-tin alloy and has proved more satisfactory with regard to relative permanence, compressive strength, abrasion resistance and ease of handling than other materials tested (Jones, 1971). With regard to pollution, dental preparations are trapped in the mouth of the patient. However, nearly all of the mercury utilized for other medical purposes eventually is added to the waterways through the sewer system.

There are almost 3,000 distinct applications of mercury, a number of which may contribute to environmental pollution (Bailey and Smith, 1964). In addition, some activities of man may inadvertantly contribute to contamination of the environment with mercury. Joensuu (1971) concluded that burning fossil fuels, which are relatively low in mercury content, released approximately 2.72×10^6 kg of mercury into the atmosphere annually, due to the large quantities burned.

Analysis of permanent snowfields in Greenland for mercury allowed evaluation of changes in atmospheric mercury concentrations over the past 2700 years (Weiss et al. 1971). Using reliable techniques to date the various strata sampled, it was concluded that activities of man have had an impact on the environment. They estimated the major sources of atmospheric pollution would be: natural runoff into waterways, 3.8×10^6 kg per year; chloroalkali production, 3×10^6 kg per year, heating of limestone and shale containing mercury in the production of cement, 1×10^5 kg per year; burning fossil fuels, 1.6×10^6 kg per year; and from natural earth degassing, $2.5 \times 10^7 - 1.5 \times 10^8$ kg per year. On the basis of these estimates, alteration of terrestrial surfaces increases exposure of crust materials, and may be the greatest source of mercury pollution. However, these levels of mercury pollution are too low to significantly affect the mercury content of mixed layers of the ocean (Weiss *et al.* 1971). Miller *et al.* (1972) gave support to this theory by their findings that museum specimens of tuna and swordfish dating back to 1878 had concentrations of mercury similar to those caught recently. Numerous other reports have concluded that ocean levels of mercury have not changed significantly.

Mercury pollution has resulted in elevated concentrations in several isolated areas. Kitamura (1968) reported that a maximum concentration of 2010 ppm mercury (wet weight) was found in sediment of Minamata Bay near the draining channel of a chemical plant and the concentration dropped sharply as a function of the distance from the effluent outlet. Konrad (1972) reported deposits of mercury in Wisconsin waterways: 684 ppm (dry weight) in bottom sediment just below the outfall of a chlorine-caustic soda plant with concentrations decreasing to 12 ppm one mile downstream; 11.5 ppm (dry weight) in bottom sediment below a city sewage treatment plant accepting waste from manufacturers of electrical batteries. Background levels ranged from 0.01–0.15 ppm Hg in bottom sediments.

Numerous incidents of fish kills have been attributed to environmental contamination by mercury. Turney (1972) reported an incident in Michigan in which the discharge of phenylmercuric acetate, used as a mildew inhibitor by a laundry, was the apparent cause of a fish kill. Derrybury (1972) reported that fish were killed in a Tennessee reservoir when steel drums which had contained phenylmercuric acetate were used on floating docks and houseboats. Johnels and Westermark (1969) reported 5 to 10-fold increases in the mercury content of pike caught below pulp mills in Sweden, when compared to the levels in pike caught upstream. Numerous other reports of localized pollution have been published.

BIOTRANSFORMATION

Dumping inorganic mercury, either in the elemental or ionic form, into waterways was considered a safe practice until recent years. It was believed that ionic mercury would readily bind to other inorganic ions, form precipitates and deposit as soil sediment, whereas, elemental mercury would remain inert and settle in low pockets in the waterway bed (Ehrlich, 1970).

Japanese scientists had suspected biological methylation of inorganic mercury in Minamata Bay (Irukayama, 1968), but the theory was not pursued after methyl mercury was discovered in the effluent of the chemical plant. Later, methyl mercury was produced enzymatically from inorganic mercury and extracts of methanogenic bacteria (Wood et al. 1968). Jensen and Jernelov (1969) found organisms in bottom sediments from fresh water aquaria that methylated inorganic mercury. Mercury bound as HgS was not methylated under anaerobic conditions, but aerobically where the sulfide was oxidized to sulfate and methylated (Joensuu, 1971). The formation of methyl mercury is not restricted to aquatic systems and was shown to function in terresterial environment (Beckert et al. 1974). When ^{203}Hg as mercuric nitrate was applied to plots of sandy loam soil having a pH of about 8.5, approximately 33% of the radioactivity was recovered as methyl mercury. In other studies, methylation by microorganisms in fresh human feces showed that maximum levels of methyl mercury were reached within 24 to 48 hours followed by a rapid decrease. Incubation of radioactive methyl mercury with feces showed a constant loss of methyl mercury over a seven-day trial (Ehmann and Lovering, 1967) whereas, Rowland et al. (1975) showed that many species of bacteria and yeasts isolated from human feces were able to form methyl mercury from $HgCl_2$. Wood et al. (1968) reported that a nonenzymatic transfer of the methyl group from Co_{+++} to Hg_{++} may occur, the process being enhanced by anaerobic conditions and increasing concentrations of alkyl cobalamine synthesizing bacteria. Landner (1971) described a process by which mercury was bound to homocysteine and methylated to form methyl mercury-homocysteine. Apparently, this reaction is linked to the methionine biosynthesis pathway and could occur under aerobic or anaerobic conditions. Methylation in lakes and rivers appears to be an aerobic process, since HgS formed under anaerobic conditions where methylation would not occur (Jernelöv, 1972). The pH of water also affects the methylation process; low pH favors formation of monomethyl mercury, high pH favors formation of dimethyl mercury (Jernelöv, 1972). The extent to which inorganic mercury is methylated in sediments of waterways has not been ascertained. Consideration of this reaction is imperative: a) methyl mercury may be absorbed by marine organisms and, therefore, represents a greater toxicological hazard than inorganic mercury in the food chain (Friberg, 1971); b) transformation of divalent inorganic mercury to mono- or dimethyl mercury can contribute to release of mercury bound to organic sediment (Jernelöv, 1969). Synthesis of methyl mercury is not the only interconversion of mercury in nature. The cycle of mercury interconversions is outlined in Figure 11.2.

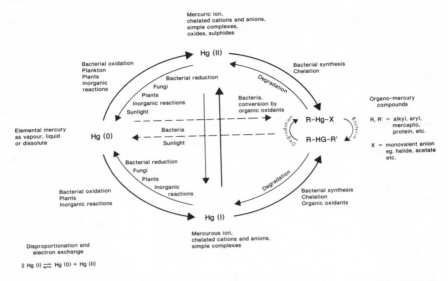

FIG. 11.2. CYCLING OF MERCURY INTERCONVERSIONS IN NATURE. (JONAS-
SON AND BOYLE, 1972)

Biochemistry

The toxicology of mercury has been the subject of numerous
investigations. Ancient authors such as Hippocrates, Pliny, Galen,
and Avicenna reported that mercury compounds were toxic, yet
differences in the metabolism and toxicology of various mercury
compounds have only recently been understood. An understanding
of the four classes of mercury compounds, namely elemental mercury, Hg^0; inorganic mercury salts, Hg^{+2} and Hg_2^{+2}; aryl mercurials,

$\boxed{O}^-Hg^+$; alkoxyalkyl mercurials, R-O-Hg^+; and alkyl mercurials,

R-Hg (R represents a short-length hydrocarbon chain) is necessary
to understand differences in their metabolism and toxicology and to
evaluate the mercury residue problem.

Absorption:

Mercury compounds may be absorbed by the body through the
gastrointestinal tract, respiratory tract or skin and is highly dependent upon the form of mercury. Reports on behavioral effects of

mercury vapor, ingested inorganic mercury, and methyl mercury were summarized (Evans *et al.* 1975). In investigations with rodents and cats, disorders of movement and posture apparently reflect damage to the cerebellum and peripheral nerves. Behavioral studies with primates revealed primarily sensory damage, especially changes in vision.

Elemental liquid mercury can be absorbed from the gastrointestinal tract. Bornmann *et al.* (1970) observed 10-fold increases in the concentration of mercury in blood and kidneys of rats orally administered metallic mercury. The extent of absorption appears relatively low and dangers of poisoning from liquid mercury by this route seem slight. Toxic signs were rarely observed in patients who received oral doses of 100 to 500 g of mercury in the treatment of bowel obstructions (Bidstrup, 1964). Goldwater (1971) stated that a person could probably swallow 454 g of liquid mercury with no adverse effects.

In the ionized form, mercury is absorbed more efficiently from the gastrointestinal tract. In studies of humans who consumed mercuric chloride and showed acute signs, an estimated 8% of the dose was absorbed (Sollmann and Schreiber, 1936). Accurate evaluation of absorption rates, following ingestion of acute doses, is difficult. The corrosive effect of mercuric chloride undoubtedly changes the permeability of the gastrointestinal wall and vomiting is almost certain to occur within one hour after ingestion. Miettinen (1973) and Rahola *et al.* (1971) fed nontoxic doses of mercuric salt to ten human volunteers and found only 85% of the dose excreted in the feces within four days. No differences were mentioned for excretion rates from patients who received mercury in a liver paste mixture or in an ionic solution. Using mice, Clarkson (1971) concluded that less than 2% of the daily intake of mercuric chloride was absorbed. Studies in which rats were dosed orally with mercuric acetate indicated that 20% of this compound was absorbed (Prickett *et al.* 1950). The solubility of inorganic mercury compounds was generally believed to be the major factor affecting absorption. Mercurous compounds are less soluble than mercuric salts and are less efficiently absorbed. The poor absorption of mercurious compounds was confirmed by autoradiographic studies (Viola and Cassano, 1968).

Organic mercurials apparently are absorbed more rapidly from the gastrointestinal tract than inorganic mercury salts. Data indicate that 50–80% of oral doses of phenyl mercury acetate were absorbed by rats (Prickett *et al.* 1950).

Alkyl mercury compounds are absorbed in the gastrointestinal tract to an even greater extent than phenyl mercury compounds.

Ekman *et al.* (1968) and Falk *et al.*(1970) fed low doses of (^{203}Hg) methyl mercury to human volunteers and reported almost complete absorption. Clarkson (1972) reported an average of 98% absorption of methyl mercury in rats fed contaminated diets. Although gastro-intestinal absorption of mercury is of primary importance with regard to mercury residue levels in foods, harmful levels of mercury may be absorbed by laboratory personnel through the respiratory tract or through the skin.

The vapor state of elemental mercury is far more dangerous to man than the liquid state because of the higher absorption rate in the Alveoli of lungs. Being relatively insoluble in water, the vapors of metallic mercury are not absorbed by the mucous membranes, but pass further down the respiratory tract (Permanent Commission and International Association on Occupational Health, 1973). Magos (1968) reported that nearly 20% of mercury vapor injected intra-venously was exhaled by rats within 30 seconds. Kudsk (1965) had human volunteers inhale concentrations of 50–350 μg mercury va-por/m^3 air and reported 71–88% absorption. Approximately 10 μg mercury/m^3 was exhaled by human volunteers who inhaled a con-centration of 200 μg/m^3 (Shepherd *et al.* 1941). Also, methyl mer-cury compounds are very volatile and readily absorbed in the respiratory tract. Mice, which Ostlund (1969) exposed to vapors of radioactive dimethyl mercury for 45 seconds, absorbed 50–80% of the total radioactivity. Since his primary concern was not the percentage of absorption, no indication was given about the extent of vaporization. Therefore, the 50–80% absorption may be lower than the actual amount possible.

Mercury compounds in aerosol or dust form also may enter the respiratory tract. Although quantitative data on absorption of mer-cury compounds in these physical states are not available, it is likely that penetration is similar to that of other heavy metals. The task group on metal accumulation (Permanent Commission and Interna-tional Association on Occupational Health, 1973) described pul-monary absorption of metals and stated that the extent of absorption was dependent upon: the size of the particles, solubility of the compound, physical characteristics of the individual, route of in-halation, tidal volume and respiratory rate.

Absorption of mercury compounds through the skin was studied by Friberg, Skog and Wahlberg (1961) who reported that 6% of an aqueous solution containing 16 mg mercuric chloride/ml was ab-sorbed through the skin of guinea pigs in 5 hours. Silberberg, Prutkin, and Leider (1969) were able to trace penetration of mer-curic ions by electron microscopy when a solution was applied to the skin. Rates of absorption of methyl mercury compounds through the skin appear similar to those of mercuric chloride (Friberg *et al.* 1961; Nordberg and Skerfving, 1972).

Transport and Transformation

Following absorption, mercury is transported in the blood to various tissues of the body. Distribution of mercury between blood components is dependent upon the form in which mercury exists (Fang *et al.* 1976).

Apparently, vapors of metallic mercury are absorbed preferentially by erythrocytes. Rabbits and monkeys exposed to mercury vapor had, immediately after exposure, 67–84% of the mercury in blood associated with the cellular fraction as opposed to the plasma fraction (Berlin *et al.* 1969). More than half the total concentration of mercuric mercury was found in plasma. Berlin and Gibson (1963) studied distribution of mercury in the blood of rabbits receiving mercuric chloride by infusion. Considerable time delay occurred between changes in serum levels and equilibration. About 50% of the mercury in blood was bound to the erythrocytes after equilibrium. Takeda *et al.* (1968) reported that four days were required for equilibration of mercuric mercury between erythrocytes and plasma in rats given subcutaneous injections. In plasma most of the mercury was apparently bound to the plasma proteins since no more than 1% of the mercury could be detected in ultrafiltrate obtained when plasma was passed through a glomular membrane (Berlin and Gibson, 1963).

The aryl and alkyl mercurials are largely bound to erythrocytes. Rabbits infused with phenylmercuric acetate and methyl mercury dicyandiamide had about 10% of phenylmercury, less than 20% of methyl mercury in plasma, and less than 1% of plasma mercury in either form was found in the ultrafiltrate (Berlin, 1963). Three human volunteers given an oral dose of radioactive methyl mercury showed radioactivity in their blood within 15 minutes (Aberg *et al.*, 1969). The ratio of mercury in erythrocytes-to-plasma was 10:1 and remained constant for 24 days following consumption. Results of numerous studies indicate that the erythrocyte-to-plasma ratio is different for various species. Reported ranges are 300:1 in rats; 20:1 in man; and 10:1 in monkeys, cats and mice (Clarkson, 1972). Suzuki *et al.* (1971) reported that blood cells collected from umbilical cords contained higher concentrations of mercury than maternal blood, whereas maternal plasma samples were higher in mercury than plasma collected from the umbilical cord. These variations were attributed to higher hemoglobin content in fetal blood and higher concentration of thiol radicals in fetal hemoglobin.

Except for the loss of the anion fraction, most mercury compounds are absorbed into the circulatory system unchanged. Vapors of elemental mercury are rapidly oxidized to mercuric ions by blood *in vitro* (Magos, 1967). An enzyme system is likely involved in the oxidation. However, the exact process could not be determined (Kudsk, 1969). Some reduction of mercuric ions to elemental mercury in the blood may occur also. Magos (1967) observed that 0.5% of added mercuric chloride was volatized from blood samples. Clarkson and Rothstein (1964) reported that following injection of mercuric chloride into rats about 4% of the amount excreted was exhaled.

Aryl and alkoxy mercury compounds are degraded to inorganic mercury in organisms. Only 20% of the mercury in the liver and 10% of the mercury in the kidney was in the organic form 48 hours after rats were injected with phenylmercuric acetate (Miller *et al.* 1960). Also, Daniels and Gage (1971) reported that when rats were administered ^{14}C-phenylmercuric acetate, about 85% of the radioactivity was detected in the urine within 96 hours, whereas only 12% of the mercury was excreted in the urine and 50–60% was excreted in the feces. Indirectly, breakdown of aryl mercury compounds was also indicated by redistribution of mercury in the animal. As was previously indicated, organic mercury compounds were found in higher concentrations in red blood cells, whereas, inorganic mercury ions were associated with the plasma fraction. Studies have shown that when phenylmercury compounds were administered, the blood distribution initially resembled that expected of organic mercurials, but later, the distribution became similar to that of inorganic mercury (Nordberg and Skerfving, 1972; Takeda *et al.* 1968). The distribution of aryl and alkoxy mercury compounds in tissues and the excretion rates are further indicators of the relatively rapid breakdown of these compounds.

Methyl mercury is apparently degraded to inorganic mercury in the body but the rate is considerably slower than reported for aryl or alkoxy mercurials. After administering methyl mercury to rats, 6–50% of the mercury excreted in the urine was inorganic mercury (Gage, 1964; Norseth and Clarkson, 1970). Methyl mercury apparently can be broken down to inorganic mercury in the liver and kidney of mice (Norseth, 1971), and liver of rats (Norseth and Clarkson, 1970). Six weeks after injection of methyl mercury in rats, the percentage of mercury remaining in the organic form was 90% in liver, spleen, and blood, 75% in plasma and brain, and 55% in kidney (Gage, 1964). The relatively slow excretion rate and constant distribution of methyl mercury among different organs are further indications of its relative stability (Nordberg and Skerfving, 1972). In guinea pigs, the extent of biotransformation is inversely proportional to the dose (Iverson and Hierlihy, 1974).

Tissue Distribution

The task group on metal accumulation discussed the factors which determine the distribution of metals among tissues (Permanent Commission and International Association on Occupational Health, 1973). They indicated that most mercury absorbed by tissues is probably found in plasma. Mercury contained in erythrocytes most likely is unavailable for direct exchange with tissues, but is one of many pools of mercury in the body. The task group stated that the exchange of mercury between the plasma and tissues undoubtedly involves a number of complex factors: a) the fraction of mercury in plasma in the diffusible form and the rate at which perfusion of tissue occurs, b) permeability of the cell membrane to the form of mercury in plasma, and c) availability and turnover rate of suitable membrane and intracellular ligands for binding mercury.

An excellent review concerning binding of mercury compounds to biological tissues and fluids is available (Clarkson, 1972). Studies covered in this review indicate that less than 1% of the mercury in plasma is in a diffusible form while binding is to thiol groups. However, ligand groups other than thiols also may be involved in binding mercury in tissues. A two-point attachment of mercury, one point through a thiol group and the other to a nonthiol ligand has been proposed for mercuric ions (Kessler *et al.* 1957) and organomercurial molecules (Weed and Ecker, 1931). A theoretical discussion of the transfer of mercury from blood to tissues was presented by Clarkson (1972). The rate of equilibration between nondiffusible bound mercury and the diffusible form will affect the tissue concentration. Availability of ligands and strength of the mercury ligand bond on each side of the cell membrane will influence the concentration gradient established across the cell membrane. Certain sulfhydryl groups show higher affinities for mercury than others; furthermore, some evidence indicates that cells may expend metabolic energy to concentrate mercury.

The mechanism which selectively permits some forms of mercury to penetrate the blood-brain barrier and the placental wall has not been completely resolved (Clarkson, 1972). The pattern of distribution of mercury among tissues depends on the chemical form.

Therefore, biotransformation of mercury to different forms results in a continuous change in distribution.

Inorganic mercury characteristically shows a nonuniform distribution among body tissues. Berlin and Ullberg (1963) injected mice with a solution of $^{203}HgCl_2$ and by autoradiographic techniques, observed its distribution over a 16-day period. Initially, high mercury content in the blood indicated intense concentrations in highly vascularized organs. In 24 hours, mercury accumulated in the liver, spleen, bone marrow, and thymus, then decreased rapidly in the blood, liver, spleen, bone marrow and myocardium. However, in the brain, renal cortex, and testes, the decline in concentration was much slower. While mercury appeared in the placenta 24 hours after injection, only traces of mercury were detected in the fetus. Throughout the entire study, the concentration of mercury was highest in the renal cortex. This agrees with data showing 85% of mercury in rats located in the kidney 15 days after injection with mercuric chloride (Rothstein and Hayes, 1960). Similar distributions were noted in guinea pigs (Nordberg and Serenius, 1969), rats (Birkhaug, 1933; Swensson and Ulfvarson, 1968), and quail (Nishimura et al. 1971). However in quail, a considerable portion of the mercury was found in the ova (Nishimura et al. 1971).

The distribution of elemental mercury, after inhaling of vapors, was similar to that observed for inorganic mercury salts. Higher concentrations of mercury were found in the brain, blood and myocardium than were noted with organic salts (Nordberg and Skerfving, 1972). This was attributed to a slight delay in oxidation of mercury vapor, permitting diffusion of vapors to occur across the blood-brain barrier (Magos, 1967, 1968). Berlin et al. (1969) reported that the concentration of mercury in the brain of mice inhaling elemental mercury vapors was 10 times higher than that following an equivalent injection of mercuric salt. Similar values were noted in guinea pigs (Nordberg and Serenius, 1969).

Phenylmercuric salts distribute themselves differently than inorganic mercury salts. Phenylmercury is retained in the blood longer, larger portions are found in the liver and alimentary tract, more mercury is retained in the skeletal muscles, and accumulation in the kidney is slower than observed for inorganic mercury salts. Similar to mercuric chloride, high concentrations of mercury were found in the placenta, but only traces could be found in the fetus (Berlin and Ullberg, 1963).

Takeda et al. (1968) reported that distribution of phenylmercury in rats, initially after administration, was similar to that observed for alkyl mercury compounds. After eight days, however, the distribution pattern more closely resembled that observed for mercuric chloride. The authors attributed the change in the distribution pattern to the rapid breakdown of phenylmercury to inorganic mercury. Rats, when injected with phenylmercury acetate, showed similar results (Gage, 1964). At the end of six weeks, the concentration of mercury in the kidney was 90 times higher than in the liver, spleen, brain, heart and blood components. Ellis and Fang (1967) studied the distribution of phenylmercury and mercuric ions in rats for 120 hours after a single oral dose. The concentration of mercury in the organs of rats treated with phenylmercury was higher, reflecting a higher gastrointestinal absorption of the organic form. Highest concentrations of mercury were found in the kidneys for both mercury compounds. Blood levels were 30–40 times higher and liver 4–5 times higher in the phenylmercury treated rats than in those receiving mercuric chloride. Distribution of mercury among tissues was dependent upon the dose of phenylmercury, indicating that saturation may occur (Ulfvarson, 1969).

Following administration of alkyl mercury compounds, a relatively even and constant distribution of mercury among the organs was reported. Berlin and Ullberg (1963) using single intravenous injections of [^{203}Hg] methyl mercury dicyandiamide in pregnant mice, studied its distribution by autoradiography. One hour after injection, the highest concentration of mercury was found in blood. However, an accumulation was noted in the kidney, liver, pancreas, mucosa of the alimentary tract, and gallbladder. After four hours, mercury was detected in the fetus, colon and muscles, while the amount of mercury in liver was at its maximum. Twenty-four hours after injection, mercury was distributed uniformly through the body and fetus, except the central nervous system that contained a lesser amount and the bony skeleton in which mercury was not evident. During the following fifteen days, the concentration of mercury decreased in all organs except the brain and spinal cord, where the concentration doubled. At day 16, the concentration of mercury in the brain was exceeded only by that in the renal cortex and colonic mucosa. A similar study (Fehling et al. 1975) concluded that in the rat, methyl mercury induces a partially reversible peripheral neuropathy with the primary target the sensory ganglion cell.

Hirota (1969), as reported by Kojima and Fujita (1973), noted similar distribution of mercury in rats given a single intramuscular injection of methyl mercury, except that the concentration of mercury in the brain was always less than that found in liver, kidney or blood within 40 days of injection. Maximum brain concentrations were reached at approximately 15 days.

Whole body retention and tissue distribution of [^{203}Hg] methyl-mercury in adult cats following a single oral dose of 78 mg of labeled methylmercuric chloride were studied by Hollins et al. (1975). Whole body half-period was 117.7 ± 1.4 days when hair was included and 76.2 ± 1.6 days with hair excluded. The latter figure is in good agreement with the 78 ±5 day half-period reported for man. Highest concentrations were found in hair, liver, gall bladder and kidney. The blood:brain ratio ranged from 0.78 to 0.98, compared with values of 0.05 to 0.8 which have been reported for humans. Thus, a higher blood level is required in cats than in humans to elicit neurological symptoms.

Friberg (1959) administered 10 daily injections of methyl mer-cury dicyandiamide and mercuric chloride to rats and studied the distribution. Blood, spleen, liver and kidney of rats receiving methyl mercury contained more uniform concentrations of mercury, while the concentration of mercury in brain was 10 times higher than found in the rats receiving mercuric chloride. However, the liver in methyl mercury treated rats had a mercury concentration 5 times higher than the brain.

Apparently, large differences exist between species with refer-ence to concentrating of methyl mercury across the blood-brain barrier. After equilibration, the ratio of concentration of mercury in blood-to-brain approached 20 for rats (Friberg, 1959; Gage, 1964; Swensson et al. 1959) 0.5–1.5 for mice (Norseth, 1971; Suzuki, 1969) and cats (Kitamura, 1968), 6.5 for pigs (Platonow, 1968), and 0.1–0.2 for man (Nordberg and Skerfving, 1972).

It is not known whether such dramatic differences exist between species in the ability to transport methyl mercury across the pla-cental membrane. Higher concentrations of mercury were found in the blood cells of umbilical cord in humans than in the maternal blood cells (Suzuki et al. 1971), while higher concentrations of mercury were found in brains of fetal rats than their mothers which had received methyl mercury chloride (Matsumoto et al. 1967).

Takeda et al (1968) studied the distribution of other alkyl mercury compounds in rats. Rats treated with ethyl and n-butyl mercury showed mercury distribution patterns similar to those noted for methyl mercury. However, increasing the chain length of the alkyl group resulted in a decrease in the brain-to-plasma ratio for mer-cury concentration. Platonow reported similar distribution for ethyl (Platonow, 1968) and methyl mercury (Platonow, 1968) given to pigs. However, in rats, transplacental transport of methyl mercury to the fetus was more rapid than mercuric nitrate (Mansour et al. 1974) or phenylmercuric acetate (Worf, 1970). These findings relate to the pathogenesis of Minamata disease.

Ellis and Fang (1967) studied the intracellular distribution of phenylmercury and mercuric ions in rat kidney and liver. Whereas concentration differences between the two organs were only slight, several differences were noted between the two forms of mercury. When the concentration of mercury in the cell fractions was based on moles of mercury per milligram of nitrogen, the differences were as follows: 1) binding of phenylmercury was greater than inorganic mercury in all fractions; nuclear, mitochondria, microsomal, and soluble, 2) the rate of accumulation was faster for phenylmercury in most fractions, while elimination was equally as fast, 3) the level of mercury in the fractions, particularly the nuclear and soluble, remained quiet constant over the 120-hour study period.

Intracellular distribution patterns for mercuric chloride, methyl mercury dicyandiamide, and methoxyethylmercury in rat liver were reported (Norseth, 1969). Concentration of mercury after injection of mercuric chloride increased during a four-day period in mito-chondrial and lysosomal fractions, and remained relatively constant in the microsomal fraction. Of three mercurials used, the smallest percentage of mercury in mitochrondria occurred with methyl mer-cury, the largest with mercuric chloride. Approximately 25.9% of methyl mercury was associated with microsomes, compared to 12.5% of mercuric chloride and 18.7% of methoxyethylmercury. Approxi-mately 25% of mercuric chloride was associated with lysosomes compared to 18% methoxy-ethylmercury and 7% methyl mercury. Concentration of mercury in nuclear and soluble fractions was not determined.

Excretion of Mercury

Inorganic mercury is excreted by the kidney, liver in bile, in-testinal mucosa, sweat glands, and salivary glands with the major routes of excretion in urine and feces (Friberg and Nordberg, 1973). Rothstein and Hayes (1960) reported that for 5–7 days after rats were injected with $Hg (NO_3)_2$, a high concentration of mercury was found in the feces. However, after 15–30% of the dose had been excreted by this route, a sharp decrease in the mercury content of the feces was seen. Remaining mercury was primarily excreted in urine. A total of 47–71% of the total dose was excreted by the two routes within 52 days. The change in the route of excretion ap-parently resulted from the change of tissue distribution of mercury. High fecal excretion occurred during the rapid decrease in all tissues with the exception of the kidney. By the fifteenth day, 86% of the remaining body burden of mercury was concentrated in the kidney, and subsequent loss resulted from clearance of mercury from the kidney. Prickett *et al.* (1950) reported similar excretion patterns from rats injected with mercuric acetate. Miettinen (1973) and Rahola *et al.* (1971) gave an oral dose of $^{203}Hg^{+2}$ to ten human volunteers and studied the retention and excretion. Eight-five percent of the dose was found in feces collected during the first five days. The biological half-life of the 15% mercury remaining after five days as determined by whole-body measurements, was 37 ± 3 days for men and 48 ± 5 days for women. These half-lives correlate with the biological half-life of mercury remaining after the first phase of excretion (Rothstein and Hayes, 1960), and therefore, the excretion rates of inorganic mercury apparently are similar for man and rat. Plummer and Bartlett (1975) investigated the distribution of mercury in the eggs, feces and carcasses of laying hens fed a ration containing 8.8 percent whalemeal containing levels of mercury up to 10 ppm. The mercury content of the diet was 0.95 ppm. About 90 percent of ingested mercury was excreted in feces. During this experimental period, mercury content of the egg albumin rose from 0.05 ppm (first week) to a range of 0.24 to 0.35 ppm (third through seventh week). Concentration in the yolks remained much lower than that in albumin. At slaughter, after 47 days on the whalemeal ration, mercury content of breast muscle ranged from 0.10 to 0.18 ppm, compared to less than 0.01 ppm in control hens. In a similar study (March *et al.* 1974) warned against processing of feathers into meal because of bioaccumulation of mercury.

Because of the varying turnover rate of inorganic mercury in different body tissues, it is generally believed that danger of repeated doses of mercury may be underestimated using calculations based on whole body biological half-life of mercury. To evaluate danger for critical organs, it is best to consider half-life of mercury on the basis of body compartments. Cember (1953) developed a theoretical four-compartment model consisting of: 1) a long-term compartment, the kidney; 2) a short-term compartment, the liver; 3) a tissue compartment, the rest of the body, and 4) an excretion reservoir. By assuming first order kinetics and using parameters established from studies of rats injected with mercuric nitrate, he solved different equations describing the turnover of mercury in each compartment. The model proved reasonably representative of the elimination rates observed in rats injected with mercuric chloride (Cember, 1969).

The elimination rate of mercury from the kidney is dependent upon the dose; excretion rates increased directly with dose. This phenomenon may be the result of an increased turnover of kidney cells resulting from the toxic effect of mercury (Phillips and Cember, 1969).

Hayes and Rothstein (1962) reported that excretion of mercury by rats inhaling mercury vapors was similar to that shown previously following injection of aqueous solutions of mercuric salts. High concentrations of mercury were initially present in the lung, but within 15 days, levels diminished to that found in other tissues. Gage (1961) studied the effect of continuous and intermittent exposure to mercury vapors on excretion rates in rats. Whereas continuous exposure resulted in constant daily excretion, intermittent exposure produced fluctuating daily excretions with peak excretions occurring on days when exposure had not occurred. Excretion rates for phenylmercuric acetate in rats were similar to those observed for mercuric acetate once maximum accumulation of mercury in the kidney had occurred (Ellis and Fang, 1967).

Gage (1964) studied the forms of mercury in the excrement of rats receiving one injection of phenylmercury acetate subcutaneously. Percentages of organic mercury in the urine were about 100, slightly less than 50, and approximately 5% for the first, second and third days, respectively, while feces contained a relatively low percentage of organic mercury even on day one. Cember and Donagi (1953) reported that elimination of phenylmercury with respect to route and rate is dependent upon dose. Larger doses resulted in a higher proportion of mercury being excreted in feces and higher total excretion per day. Most studies indicated that 67% of the dose of phenylmercury excreted by rats is contained in the feces (Ellis and Fang, 1967; Gage, 1964; Prickett et al. 1950; Takeda et al. 1968). Correlation of results of different reports is complicated by variations of dose, length of exposure and time of sample collections.

Excretion of alkoxymercury appears similar to that of arylmercury compounds. Methoxyethylmercury fed to rats was rapidly metabolized to ethylene and inorganic mercury with a half-life of one day. Excretion shortly after administration reflected a high concentration of organic mercury in the urine. In a few days urinary mercury was in the inorganic form. Bile samples collected contained a high proportion of organic mercury. However, the feces contained only inorganic mercury. Because feces content was less than expected on the basis of bilary excretions, resorption and metabolism must occur in the intestine (Daniel et al. 1971). Swensson and Ulfvarson (1968) compared excretion rates of four forms of mercury in Leghorn cocks. Ten days after injection, the percentage of the intial doses remaining in the body were 10% for methoxyethylmercury, 20% for phenylmercury, 40% for mercuric nitrate and 80% for methyl mercury.

Little methyl mercury is excreted from the body through the lungs (Ostlund, 1969), while urine contained 20% of the mercury excreted by rats during the 10-day period following a single injected dose. However, less than 30% of the injected dose was excreted during this time period (Norseth and Clarkson, 1970). In humans, urinary excretion appears even less significant since less than 10% of the mercury excreted within 49 days by this route (Aberg et al. 1969). Miettinen et al. (1971) found less than 0.5% of the excretion during the first 10 days occurred in the urine, but by day 100, the mercury content of urine was approximately 1/4 that of the feces. In both studies, the mercury content of the urine increased during the sampling period (Aberg et al. 1969; Miettinen et al. 1971). The biological half-life of methyl mercury determined by whole body measurements of humans was 76 ± 3 days (Miettinen et al. 1971). The biological half-life of methyl mercury in 48 Iraqi patients contaminated as a result of the 1971–1972 accident was measured from head hair samples. The range was 35 to 189 days, while 90% of the patients in this symmetrical distribution had values of 35 to 100 and a mean of 65 days. The remaining 10% had biological half-life values of 110 to 120 days with one exceptional value of 189 days. There was no correlation of age or diet to these long biological half-lives (Al-Shahristani and Shehab, 1974).

Methyl mercury is excreted principally in the feces (Aberg et al. 1969; Berlin, 1963; Friberg, 1959; Gage, 1964; Miettinen et al. 1971; Norseth and Clarkson, 1970). Approximately 50% of the mercury in feces is in the inorganic form (Norseth, 1971; Norseth and Clarkson, 1970, 1971). Samples collected from rats from the upper part of the small intestine contained significantly lower relative amounts of inorganic mercury and total excretion of mercury per day in the bile greatly exceeded the fecal excretion. Bile, apparently, is the major source of mercury in the small intestine (Norseth and Clarkson, 1971). Approximately 77% of the mercury in bile exists as a methyl mercury-cysteine complex which is rapidly absorbed in the intestine and none is excreted. Mercury which is excreted in feces reportedly arises from exfoliation of intestinal cells, pancreatic excretions, gastric content, plasma catabolism and the non-cysteine bound mercury compounds in bile. Biotransformation of protein-bound methyl mercury to inorganic mercury apparently occurs in the cecum or colon, possibly by microbial action (Norseth and Clarkson, 1971).

Accelerated excretion of methyl mercury has been accomplished by feeding rats reduced hair (Takahasi and Hirayama, 1971) and mice thiol-containing resins (Clarkson et al. 1971). Both materials have a high capacity to bind mercury compounds, and because no absorption of resin or hair occurs from the gastrointestinal tract, absorption of the mercury is markedly decreased.

Takeda *et al.* (1968) indicate that ethyl and n-butyl mercury were excreted in a manner similar to methyl mercury. Dimethyl mercury, however, has a considerably different excretion pattern. Ostlund (1969) reported that 80–90% of a single dose of dimethyl mercury was exhaled by mice within six hours, while the remainder was metabolized to monoethyl mercury.

INDICES OF EXPOSURE

Numerous attempts have been made to identify a body component which could be assayed easily for mercury concentration and thus would serve as a valid index of the extent of exposure. The major criteria for such an index would be that the mercury concentration of that component was proportional to the mercury content of critical organs or tissues of the body. This relationship should hold during both the period of accumulation and the period following exposure. Special attention has been given to blood, urine, feces, and hair as indices of exposure to mercury.

Difficulty has been encountered in identifying an index of exposure for inorganic mercury, or those forms of mercury which are rapidly broken down in the body. As discussed earlier, tissue distribution of these mercurials changes with time. Highest concentrations of mercury accumulate in the kidney with a biological half-time of inorganic mercury in kidney longer than reported for other body organisms and tissues. This may account for observed poor correlations of blood and kidney concentrations of mercury (Permanent Commission and International Association on Occupational Health, 1973). Mercury in urine reportedly correlates with levels in blood, but is not necessarily indicative of the concentration in the kidney (Berlin and Gibson, 1963; Permanent Commission and International Association on Occupational Health, 1973). Goldwater (1964) and Smith *et al.* (1970) reported that a correlation between exposure to mercury vapors and blood and urine mercury levels existed on a group basis, in spite of individual discrepancies. Brown and Kulkarni (1967) reviewed a number of reports showing the concentration of mercury in urine and blood of factory workers exposed to mercury, mercury derivatives and cadmium (Brown and Kulkarni, 1967; Friberg, 1961) showed large daily fluctuations in mercury concentrations in urine. The international committee which met to establish the maximum allowable concentration (MAC Committee) of mercury considered the use of urine and blood as indices. Based on the reports of distribution and excretion of inorganic mercury compounds, it seems reasonable to conclude that blood and urine levels of mercury would serve primarily as an indicator of recent exposure and not necessarily kidney concentrations (Friberg, 1969; Friberg and Nordberg, 1973).

Alkylmercury compounds are more uniformly distributed through the body than other mercury compounds and excretion rates of alkylmercury compounds from the various body tissues are relatively uniform. Thus, greater success has been obtained in locating indices of exposure for alkylmercury compounds. The concentration of mercury in urine is of little value as an index of exposure to alkylmercury compounds. Lundgren and Swensson (1949) studied workers exposed to alkylmercury compounds and reported that while under exposure, mercury was present in urine but shortly after cessation of exposure, mercury could not be detected in urine. Friberg et al. (1971) and Kurland (1973) reported that blood levels of mercury are a good indicator of accumulation of alkylmercury in the brain. Mercury concentration in red blood cells is probably most indicative of alkylmercury exposure. However, whole blood levels can be used, if it is possible to exclude exposure to other forms of mercury (Permanent Commission and International Association on Occupational Health, 1973). A strong correlation between mercury levels of blood and hair during regular, continuous exposure was reported with hair levels approximately 300 times higher than blood levels (Friberg, 1971; Grant, 1971). Research performed in our laboratory showed a high correlation between methyl mercury content of diets and mercury levels in hair of guinea pigs. A correlation was also found between hair levels of mercury and mercury concentration of the optical cortex of the cerebrum. One advantage of hair as an index of exposure is that by clipping sections, it is possible to determine periods of previous exposure. With reference to safe levels for indices of exposure, the MAC Committee (1971) stated that 10 μg Hg per 100 ml whole blood should not be exceeded. The Joint FAO/WHO Expert Committee on food additives (1972) reported that the lowest mercury levels at the onset of neurological symptoms of mercury poisoning were 50 μg/g hair and 0.4 μg/g red blood cells. Suzuki (1976) confirmed the relationship of daily fish consumption to levels found in hair and red blood cells by studying inhabitants of six small Japanese islands. The highest mercury level was hair, 48.8 μg/g and red blood cells, 0.192 μg/g.

TOXICOLOGY

In the study of mercury toxicology, critical concentration and critical organs are extremely important considerations. Critical concentration is reached when undesirable functional changes occur in the cell. A critical organ is that organ in which a critical concentration is first reached and may not necessarily be the organ which accumulates the greatest concentration (Permanent Commission and International Association of Occupational Health, 1973). Also, the clinical toxicity of elemental, inorganic and organic mercury has been reviewed, from the viewpoint of exposure, uptake, distribution, elimination, and clinical effects (Gerstner and Huff, 1977). Koos and Lango (1976) reviewed clinical reports of mercury poisoning in the human fetus in several outbreaks.

Elemental mercury in the liquid state is absorbed slowly and, therefore, is not extremely toxic except when in the vapor state. Readily absorbed by the lungs, elemental mercury vapors penetrate the blood-brain barrier where the critical organ may be either the central and peripheral nervous system or kidney (Permanent Commission and International Association on Occupational Health, 1973). In the ionic form, inorganic mercury does not readily penetrate the blood-brain barrier, and thus the concentration of mercury in the central nervous system is unlikely to reach critical concentration. The kidney is the critical organ for inorganic mercury ions (Permanent Commission and International Association on Occupational Health, 1973). Reaction of mercuric ions with thiol groups appears the major cause of intoxication for mercury vapors or mercuric salts (Friberg, 1969). Changes in membrane permeability, as well as interferences with enzyme activity, can occur as a result of binding of mercuric ions to these ligands.

Almost every toxic action of mercurials is to some extent attributed to an interaction with sulfhydryl groups to form a complex, a mercaptide (Rothstein, 1973). Proteins containing sulfhydryl groups are primary targets for mercurial interaction and the subsequent toxicological effects. Factors which may determine probability of mercurial-sulfhydryl interaction are: presence of neighboring groups on protein molecules which have an attractive or repulsive effect; steric hindrance for mercurials with large organic portions; protein configurations which may shield sulfhydryl groups and properties of the medium such as pH and presence of anions. Effects of mercury on enzymes have been reviewed (Weast, 1974).

Symptoms of acute poisoning by inorganic mercury are gastroenteritis, abdominal pain, nausea, vomiting and bloody diarrhea (Bidstrup, 1964). Within a few days, gingivitis and stomatitis may develop, particularly if the patient maintains a low standard of oral hygiene. Excessive salivation, a metallic taste, swollen salivary glands, ulcerated lips and cheeks, and difficulty in breathing are other symptoms. Severe kidney injury leading to anuria and uremia are noted in acute cases of inorganic mercury poisoning. Tremors and delerium, resulting from damage to the central nervous system can result from exposure to both acute or chronic doses of mercury vapors.

Numerous studies have determined dose-response relationships. The MAC Committee (1971) relied heavily on the studies by Neal *et al.* (1937, 1940) to determine the safe levels of mercury vapor. These studies indicated that exposure to 0.1 mg mercury vapor/m^3 air would not cause poisoning. The MAC Committee acknowledged that Russian studies found typical but nonspecific symptoms of mercury poisoning in workers exposed to 0.1 to 0.3 mg mercury/m^3 air.

The LD_{50} dosage of mercuric mercury received by intraperitoneal injection in mice is 7–12 mg/kg (Swensson, 1952). Davies and Kennedy (1967) reported that a subcutaneous injection of 0.75 mg/ kg body weight of mercuric chloride caused kidney damage in rats. Oral doses of 40 and 160 ppm Hg^{++} given to rats for one year produced changes in kidneys and mercury concentrations of 16 and 49 $\mu g/g$ wet weight, respectively (Fitzhugh *et al.* 1950).

Phenyl- and methoxyethylmercury compounds are readily degraded to compounds similar in toxicity to inorganic mercury. No serious toxic symptoms were reported for humans who consumed approximately 100 mg of phenylmercuric nitrate (Birkhaug, 1933; Webb, 1966). Rats fed diets containing 2.5 ppm phenylmercuric acetate for one year showed only minor kidney lesions. However, pronounced kidney lesions occurred in rats fed a diet containing 10.0 ppm phenylmercury for one year (Fitzhugh *et al.* 1950). Symptoms of poisoning including lesions of the gastrointestinal tract, kidney and liver were seen in pigs fed 2.28–4.56 mg Hg as phenylmercuric acetate/kg/day for 14–63 days (Tryphonas and Nielsen, 1970). Diets containing 5 ppm phenylmercuric acetate fed to rats resulted in reversible senility changes in organs after one year (Goldwater, 1973).

Alkylmercury compounds have presented the greatest toxicological danger to man. A problem encountered in Japan was discussed previously. In Ghana, Iraq and Pakistan, hundreds of people became ill and many died when seed grain treated with fungicides containing methyl and ethyl mercury compounds was consumed (Bakir *et al.* 1973; Derban, 1973; Haq, 1963; Jalili and Abbasi, 1961; Skerfving and Copplestone, 1976; Westöo, 1969). In New Mexico, a family of seven consumed pork from pigs fed methyl mercury dicyandiamide (panogen) treated seed grain. Three children in the family suffered severe brain damage (Daniel and Gage, 1971; Snyder, 1971). In a follow-up study, Snyder and Seelinger (1976) reported that a 6-year old exposed *in utero* was blind, retarded, spastic and subject to myoclonic seizures. Another child is blind and is unable to sit or roll over and has seizures. Two others, now adult, have impaired vision and neuromuscular difficulty. Studies of nerve conduction indicate cortical damage.

Analysis of blood cells of Swedish people consuming fish containing 0.3 to 0.7 mg Hg/kg fish resulted in total mercury levels of 0.008 to 0.39 μg/g of blood cells. Long-term exposure to about 4 μg methyl mercury/kg body weight/day resulted in about 0.3 μg methyl mercury/g blood cells (Skerfving, 1974). Autopsies conducted on humans ranging from 26 weeks of gestation to 88 years showed less than 0.25 g Hg/g wet weight in 70% of organs assayed. As anticipated, the highest levels but also most variable, were found in the kidney. Kevorkian *et al* (1973) found very high mercury concentrations in autopsied tissues collected from 1913–1933. Levels as high as 27.5 ppm were reported. Autopsied tissues collected recently showed up to 1.79 ppm. Statistically, Mottet and Body (1974) showed no increase of mercury burden with age. Moreover, city residents appeared to have higher body burdens than rural counterparts.

In a study of infants born to 15 women exposed during pregnancy to methyl mercury during the Iraq outbreak of 1972, all but one of the mother-infant pairs showed that the blood mercury level of the infant was higher than that of the mother. After birth, transmission of methyl mercury in maternal milk appeared to be a factor in maintaining a mercury level in the infants higher than that in the mothers. Six of the 15 infants had clinical evidence of methyl mercury poisoning and the 5 most severely affected showed gross impairment of motor and mental development. Four were blind and had impaired hearing. The lowest blood mercury level found in an affected infant was 564 ppb. Six infants had blood levels of 200 ppb or higher with no signs of poisoning. Only one affected infant was born to a mother who showed no apparent symptoms of methyl mercury poisoning (Amin-Zaki *et al.* 1974). Other studies (Amin-Zaki *et al.* 1976) involving two Iraqi mother-infant pairs, one infant was mercury contaminated *in utero* while the other was postpartum contaminated by nursing. Both mothers had moderate to severe mercury poisoning symptoms, while no such symptoms were evident in their offspring. Mercury levels in hair of the mothers and its change with time were closely correlated with those in blood, with a hair:blood ratio of about 250:1. At blood mercury levels above 50 ng/ml there was a linear relationship between the concentration in blood and milk, with the level in the milk being about 8.6% of that in blood. Nursing resulted in a blood level of over 600 ng/ml in the infant, while the blood level of mercury in the infant was less than the maximum value attained in the mother's blood. The infant exposed *in utero* had blood mercury concentrations at birth and during the first month post-partum substantially higher than that of the mother (Amin-Zaki *et al.* 1976).

The critical organ for alkylmercury poisoning is the central nervous system (Permanent Commission and International Association on Occupational Health, 1973). The MAC Committee (1971) concluded from a number of animal studies that mercury concentrations of 10 μg/g in brain tissue were indicative of irreversible injury or death. Using a half-life of mercury in man of 70 days and a 15% distribution of alkylmercury in brain, the Committee concluded that intake of 1 mg of alkylmercury per day would lead to toxic concentrations of mercury in the brain. The fetus can be of primary importance in cases of pregnancy. Several reports discussed earlier indicated that higher concentrations of mercury were found in fetal tissue in comparison to the concentrations found in the maternal tissues. Reynolds and Pitkin (1975) showed that the maternal to fetal ratio remained at 10:1 or higher in Rhesus monkeys.

In addition to the usual symptoms and signs of methyl mercury poisoning, Harada (1968) reported that retardation of physical development, abnormal pneumoencephlograms (PEG) and abnormal electroencephlograms (EEG) were detected for patients who developed poisoning symptoms during early infancy. At conception, female rats were placed on control and methyl mercury (2 ppm) contaminated diets. Weaned pups were examined and compared. Those from methyl mercury fed dams exhibited retarded behavioral and learning abilities (Olson and Boush, 1975).

Fetal methyl mercury poisoning can be distinguished from non-fetal poisoning because signs of underdeveloped and malformed central nervous systems are apparent in prenatal poisoning (Takeuchi, 1968). Autopsies revealed a tendency for lesions in brain to be more localized in adults poisoned by methyl mercury, whereas, in the fetal and infantile brains, lesions are more diffuse (Takeuchi, 1968). Atrophy of brain with compensatory increases in fluid are macroscopically noticeable. Monkeys showed neuronal degeneration in the visual cortex and cerebellum with advanced symptoms of methyl mercury intoxication (Truelove *et al.* 1975). Brain mercury levels at autopsy were 22 to 48 μg/g. In other parts of the body, a reduction of red bone marrow, fatty degeneration of liver, and erosion of the duodenum were reported (Takeuchi, 1968).

A latency period of one to several weeks exists between administration of alkylmercury compounds and development of symptoms of poisoning (Norseth and Clarkson, 1970; Peakall and Lovett, 1972). This latency period perhaps represents time required for degradation of alkylmercury compounds followed by toxic action from the mercuric ion (Clarkson, 1972) or that alkylmercury compounds interfere with certain biochemical processes for which the effect is not immediately evident (Shepherd *et al.* 1941). However, in support of the latter theory, Norseth and Clarkson (1970) were unable to detect any significant breakdown of methyl mercury in the brain of rats during a 30-day period following intravenous injections of alkylmercury. The pathology of alkylmercury poisoning indicates that methyl mercury may alter the protein synthesizing processes of the body. Study of the uptake of [U-^{14}C] leucine into brain proteins in brain cortical slices from rats which had been injected with methyl mercury thioacetamide showed a 43% reduction in protein synthesis in animals which were in the latent period of intoxication based on neurological symptoms. Oxygen consumption and aerobic lactic acid formation in brain tissue slices decreased after the animals developed neurological symptoms of methyl mercury poisoning. Succinic dehydrogenase activity decreased in brain homogenates only after neurological symptoms developed (Yamaguchi and Nunotani, 1974).

Results of studies on the subacute toxicity, effects on reproduction and long-term toxicity of methylmercury in rats are reported by Verschuuren *et al.* (1971). In the reproductive and long-term studies, the dietary concentrations of methylmercuric chloride used were 0.1, 0.5 and 2.5 ppm. From these experiments the no-toxic-effect level in rats was found to be between 0.1 and 0.5 ppm. Kidney damage was the most sensitive parameter, with marginal effects seen at 0.5 ppm. Neurological changes were observed only at 25 ppm. In the reproductive study, the viability index was decreased at 2.5 ppm in the F1 and F2 generations.

Rats pretreated with methyl mercury incorporated more of the amino acids into protein than the untreated (Brubaker, 1971). Another report (Lucier *et al.* 1971) indicated that two-day treatment with methyl mercury proliferated the rough endoplasmic reticulum and decreased the amount of smooth endoplasmic reticulum of liver. A depression of the cytochrome P$_{450}$, cytochrome b$_5$ and aminopyrine dimethylase activity was noted in the liver microsomal fractions obtained from rats treated with methyl mercury. In contrast, Pekkanen and Salminen, (1972) reported an increase in cytochrome P_{450} and NADPH-cytochrome C reductase activity in mice livers seven days after injection of methyl mercury nitrate. Rats given methyl mercury showed a decrease in the RNA content in spinal ganglion neurons and an increase in the RNA content of the anterior horn motoneurons of the spinal cord (Chang et al. 1972). Decreased synthesis of RNA by spinal ganglia after methyl mercury intoxication was confirmed in mice which were injected with [^{3}H] uridine (Chang et al. 1972). A significant decrease in the incorporation of the labeled uridine was noted (Chang et al, 1972).

A significant discovery in the reported decrease in the toxic response noted when cystine or selenium was fed in a diet with methyl mercury. Iwata and Okamoto (1970), as reported by Kojima and Fujita (1973), found selenium reduced the toxic effect of methyl mercury. Stillings et al. (1974) reported that 0.4% cystine equaled the protective effect of .6% selenium. Ganther et al. (1972) reported that rats receiving methyl mercury in the drinking water survived longer on basal diets supplemented with 0.5 ppm selenium than rats receiving the same diet without added selenium. Also, Japanese quail fed 20 ppm methyl mercury in diets containing 17% tuna survived longer than quail fed the same level of methyl mercury in a cornsoya diet (El-Begearmi et al 1977; Ganther et al. 1972; Ganther and Sunde, 1974). The selenium content of the diet was 0.49 ppm. Many researchers (Fang, 1977; Fang et al. 1976; Johnson and Pond, 1974; Ohi et al. 1975, 1976; Potter and Matrone, 1964; Prohaska and Ganther, 1977; Sell and Horani, 1977; Stoewsand et al. 1977, 1974; Sumino et al. 1977) have verified the protective effect of selenium against the toxicity of methyl mercury. Pretreatment of rats with selenium followed by feeding ^{203}Hg-methyl mercuric chloride caused a five-fold increase in brain levels of mercury and was located in the crude nuclear fraction (Chen et al. 1975). In fact, selenite increased retention of mercury in tissues regardless of the form of mercury fed. Ganther and Sunde (1974) and Koeman et al. (1973) reported that tuna accumulate selenium concurrently with mercury on about a 1:1 molar ratio. Whereas the mechanism of protection by selenium is unknown, Ganther et al. (1972) stated that natural factors of protection such as with selenium should be considered when conducting toxicity studies (Ganther and Sunde, 1974). Supplemental effect was observed when 500 IV vitamin E were added to a balanced quail diet containing 30 ppm Hg as methylmercury chloride and 0 to 0.1 ppm selenium (Weiss et al. 1971).

Effects of Mercury on Chromosomes

Changes have been reported in chromosomes of onion and were induced by alkyl-, alkoxyalkyl- and arylmercury compounds (Fiskesjo, 1969; Ramel, 1969). Changes reported were C-mitosis, polyploidy, and aneuploidy, with chromosome fragmentation occurring with phenyl and methyl mercury. In addition, nondisjunction of the X̄ chromosome in Drosophila eggs (Ramel and Magnusson, 1969) and with human leukocytes, C-mitotic effects occurred in a medium containing methyl mercury (Fiskesjo, 1970). Skerfving et al. (1970) compared the incidence of chromosome breakage in lympyocyte cells with concentration of mercury in blood of persons eating varying quantities of fish. A statistically significant correlation was found between the number of lymphocyte cells with chromosome breaks or structural rearrangements and the mercury concentration of the red blood cells. However, no significant correlations existed between chromatic breaks, aneuploid or polyploid cells, or endore duplications and the mercury concentration of the red blood cells. Ramel (1972) stated that levels of mercury in food are sufficiently high to cause genetic effects in humans while the medical implications of these effects are unknown.

MERCURY IN FOOD

Environmental exposure of man to mercury is through the air he breathes, the water he drinks, and the food he eats (Magos, 1975). Mercury in foods has been considered the greatest toxicological danger to man but the average level of mercury in food is quite low, approximating 0.02 ppm (Lu, 1974). Considerable variation in levels of mercury occurs, depending on type of food, environmental level of mercury where the food is produced, and possible use of mercury containing compounds for agricultural and industrial production and processing of the food.

Excellent reviews regarding the uptake and movement of mercury in plants were written by Smart (1968) and D'Itri (1972). Translocation of mercury from treated seeds to harvested grain has been questioned. It was questioned whether 20 ppm mercury content of treated seed could contribute significantly to mercury in mature grain. Mercury on seeds would be diluted in soil and transported and deposited throughout the leaves, straw, and grain (Nelson, 1971). No significant differences were noted in mercury content of wheat or barley grown from untreated seed or seed treated with methyl mercury and ethyl mercury compounds (Saha *et al.* 1970). These authors (Saha *et al.* 1970) proved that absorption and translocation of mercury can occur in plants. Smart, (D'Itri, 1972) and the Study Group on Mercury Hazards (1971) concluded that translocation following foliar application of mercury compounds may be more significant than translocation from seeds. Residue levels in crops treated with foliar sprays are four to six times higher than crops not sprayed (D'Itri, 1972; Nelson, 1971).

The low levels of mercury in plant tissue may become significant if the crop is part of a long food chain. Biological species have demonstrated an ability to concentrate mercury which is somewhat dependent upon the form of mercury involved (Peakall and Lovett, 1972). Alkylmercury compounds are biologically concentrated by most species. Mercury content of eggs from chickens fed grain grown from mercury treated seed would represent such a food-related magnification (Westöo, 1969). Biological concentration factors for terrestrial food chains were generally in the order of two to three. However, in aquatic food chains, concentration factors of hundreds and thousands were observed (Peakall and Lovett, 1972). Mercury levels in many species of edible and inedible mushrooms have been studied (Aichberger, 1977; Seeger, 1976, 1977). The highest concentrations were found in the gills with the lowest in the stems. A strong positive correlation was observed between mercury content and protein content of mushrooms. Mercury levels in mushrooms were 30 to 550 times the level found in the soil in which they grew.

Fimreite *et al.* (1971) and Goldwater (1971) observed increasing concentrations of mercury in aquatic food chains and highest concentrations of mercury were found in species at the higher trophic levels. Hannerz (1968) collected specimens of various water plants, algae, moss, and invertebrates and compared ratios of mercury in various tissues to water. Concentration factors at approximately 30 days ranged; 3:560 for emergent portion of plants compared to invertebrates in mercuric chloride treated ponds; 2:995 for emergent portion of plants compared to invertebrates in methoxyethylmercury treated ponds; 3:1400 for emergent portion of plants compared to invertebrates in phenylmercury treated ponds; and 4:4235 for emergent portion of plants compared to invertebrates in methyl mercury treated ponds. Fish concentrate mercury in the various body components at a rapid rate from water or food. After two days, the mean concentration of methoxyethylmercury as mercury in cod was 388 times that of water and 188 times that of food; but 606 times that of water and 1107 that of food, contaminated with methyl mercury. Pike, after 3 days, had a mean concentration of methoxy-ethylmercury as mercury that was 783 times that of water and 70 times that of food; but 750 times that of water and 367 times that of food contaminated with methyl mercury. Johnels and Westermark (1969) reported the concentration factor of pike is at least 3000 when comparing levels of mercury in tissue to water. A linear relationship exists between age or weight of pike and mercury levels in tissues. However, where waters were highly polluted with mercury the extent of exposure was more significant than age or weight in contributing to contamination.

Shellfish, as expected, accumulate mercury at rapid rates. Oysters exposed to 10 or 100 ppb mercuric acetate in seawater accumulated 2,800 times and 1400 times the water level, respectively. On exposure to mercury-free estuarine water, the level fell within 18 days from 18 to 15 ppm and 115 to 65 ppm in the oysters living in 10 or 100 ppb water, respectively (Cunningham and Tripp, 1973). Accumulation was greatest in gill and visceral tissues (Sayler et al., 1975). Moreover, in these latter oysters, a mercury accumulating and metabolizing strain of Pseudomonas was discovered.

The biological magnification of aquatic food chains also can be extended to other higher predator species. Analyses of Great Blue Heron carcasses showed levels of 21.2–23.0 ppm mercury compared to concentrations of 1.8–3.6 ppm mercury in the fish found in their stomachs. In a Common Tern, the carcass level of mercury was 7.5 ppm and fish found in stomach contents contained 3.8 ppm mercury (Dustman et al. 1972).

Concentration of Mercury in Foods

The scientific literature contains enumerable citations to research involving the mercury content of common food items. Some have been discredited after analysis by other laboratories found considerably less mercury (Krenkel, 1973) indicating the need for development of better methods. A listing of the mercury concentration of a number of food items determined over a period of years is shown in Table 11.1. It is interesting to note that the mercury concentration of most of these food items has not changed significantly during this time period. The mercury concentration in foods will vary somewhat, depending on the areas in which it is produced. It is interesting to note the similarity of values between two surveys. Corneliussen (1969) and Kirkpatrick and Coffin (1974) reported the results of a market basket survey from regions of the

United States and Canada. The results of mercury analysis performed on composite samples of market basket items are presented in Table 11.2 and Table 11.3.

It is unfortunate that very few data are available on the chemical form of the mercury which is present in most foods. Methods for differentiating the chemical forms of mercury with reasonable accuracy have been developed only recently and are somewhat more complicated to perform than total mercury analysis. Westöo (1966, 1969) and Stijve and Roschnik (1974) analyzed some food samples for methyl mercury and total mercury. These results are summarized in Table 11.3. Lu (1974) stated that the mercury present in most other food items other than fish is believed to be in the form of inorganic mercury. (see also Table 11.4.)

A major concern has been expressed regarding the concentration of mercury in fish. It is difficult to assess an average level of mercury in fish since the concentration of mercury will vary significantly depending on the following: the species of fish, diet, age, size, and location in its environment (Peterson et al. 1973). The Joint FAO/WHO Committee (1972) stated that 99% of the world's fish catch has a total mercury content not exceeding 0.5 mg/kg and that 95% probably contains less than 0.3 mg/kg. Tuna has been carefully screened; the average tuna apparently contains less than 0.5 ppm mercury. FDA scientists reported large tuna fish, weighing more than 23 kg, average about 0.25 ppm and smaller tuna, weighing less than 12 kg, average 0.13 ppm (Friberg, 1969). Swordfish is the only species of ocean fish which has been banned from U.S. markets, now being sold only after individual fish certification (Anon., 1971). Analysis of longitudinal, depth and transverse sections of three swordfish for total mercury showed distribution to be uniform throughout edible muscle. The report by Miller et al. (1972) indicates that the mercury concentration of these deep sea fish may not have significantly changed in the past one hundred years, since museum specimens contain mercury levels similar to that found in current catches.

Fish caught in the coastal waters and inland waterways of Sweden often contain much higher levels of mercury than deep sea fish (Westöo, 1969). Frequently, these fish were found to contain between 1-5 ppm (Westöo, 1969). Johnels and Westermark (1969) reported that a slight elevation of the mercury content of fish can result from runoff after agricultural uses of mercury. Fungicides used to maintain golf course greens were shown to be accumulated in largemouth bass caught from nearby ponds. Mercury levels of .43 to 7.12 ppm were reported (Koirtyohann et al. 1974). More significant

TABLE 11.2. CONCENTRATION OF MERCURY IN FOODS IN PPM[1].

	Germany[2] 1934	Germany[3] 1938	U.S.[4] 1940	Japan[5] 1964	U.S.[1] 1964	U.S.[6] 1972
Meats	.001-.067	.005-.02	.0008-.044	.31-.36	.001-.15	.001-.007
Fish	.02-.18	.025-.18	.0016-.014	.035-.54	0-.06	
Vegetables, fresh	.002-.044	.005-.035	N.D.	.03-.06	0-.02	
Vegetables, canned			.005-.025		.002-.007	
Milk, fresh	.0006-.004	.0006-.004	.003-.007	.003-.007	.008	0-.009
Butter	.002 (Fats)	.07-.28			.14	
Cheese	.009-.01				.08	
Grains	.02-.036	.025-.035	.002-.006	.012-.048	.002-.025	
Fruits, fresh	.004-.01	.005-.035		.018	.004-.03	
Egg, white				.08-.125	.01	
Egg, yolk				.33-.67	.062	
Egg, whole	.002	.002	N.D.			.002-.005
Beer	.00007-.0014	.001-.015			.004	

[1]Table from Goldwater, 1964.
[2]Data from Stock, 1934.
[3]Data from Stock, 1938.
[4]Data from Gibbs, 1940.
[5]Data from Fujimura, 1964.
[6]Data from Takizawa, 1970.

TABLE 11.3. MERCURY CONCENTRATIONS IN U.S. AND CANADIAN FOODS.

Food class	Boston[1] U.S.	Kansas City[1] U.S.	L.A.,[1] U.S.	Balt-imore[1] U.S.	Minn-eapolis[1] U.S.	Van-Couver[2] Can.	Halifax[3] Can.
	ppm mercury						
Dairy products	0.02	0.004	0.002	0.002	0.01	0.001	0.002
Meat, fish, poultry	0.05	0.04	0.04	0.01	0.04	0.051	0.028
Grain and cereal[2]	0.03	0.02	0.05	0.03	0.04	0.004	0.002
Potatoes	0.01	0.006	0.02	0.006	0.01	0.001	0.001
Legumes	0.01	0.01	0.01	0.005	0.002	0.001	0.001
Beverages	0.002	0.002	0.002	0.006	0.006	N.D.	<0.001
	(1974 samples)[4] ppm mercury						
Dairy products	<.001	<.001	0.002	0.003	0.003		
Meat, fish, poultry	0.027	0.007	0.027	0.030	0.009		
Grain and cereals	0.007	<0.003	<.002	<.008	<.002		
Potatoes	<.005	<.002	<.001	0.005	0.003		
Legumes	0.004	<.002	<.001	0.004	<.001		
Beverages	—	—	—	—	—		

[1]Samples collected June 1967-Apr. 1968. Corneliussen, 1969.
[2]Samples collected in 1970, Kirkpatrick and Coffin, 1974.
[3]Samples collected in 1971. Kirkpatrick and Coffin, 1974.
[4]Samples collected in 1974. Takizawa, 1970.

increases occur in lakes near industrial areas where aerial fallout may occur. Also, the highest levels of mercury occurred in lakes where mercury-containing pollutants were added. Bligh (1972) discussed the mercury concentrations in Canadian freshwater fish, also he reported more than one million pounds of fish were destroyed during the three or four-month period after the Canadians became aware of the mercury problem. Greig and Seagran (1972) reported that fish from Lake St. Clair and the western basin of Lake Erie contain from 0.4–3.0 ppm mercury, depending on the species. In Lake Huron the mercury content of fish reportedly was lower (Greig and Seagran, 1972). Newberne (1974) discussed studies of the mercury concentration of fish in the United States. These reports indicate a substantial number of fish collected from many areas of the United States exceed the 0.5 ppm guideline for mercury established for fish by the Food and Drug Administration. Celeste and Shane (1970) reported that more than 900 samples representing 37 species of fish were collected from 28 bodies of water with known or suspected pollution. About 25% of these samples contained mercury at or above the 0.5 ppm guideline level.

The danger of mercury contained in fish is undoubtedly dependent upon the chemical form in which mercury in fish exists. Westöo (Westermark and Sjostrand, 1960; Westöo, 1966) reported that nearly all the mercury present in fish existed as methyl mercury. Peterson et al. (1973) indicated that several studies have found that a significant portion of the mercury in fish may be in the inorganic form and that the ration of organic mercury to total mercury may not be consistent among or within species. However, the Pacific blue marlin has most of its mercurial burden in the inorganic form (Shultz et al. 1976). Bache et al. (1971) reported the percentage of mercury in the methyl form increased in lake trout in relation to the age of the fish. In fish one-year old, 31–35% of the total mercury was methyl mercury; 67–88% of the mercury in 12-year old fish was methyl mercury. Clarification of the differences is needed. Literature reviews of mercury in fish were published by Newberne (1974) and Peterson et al. (1973).

Established Tolerance Levels for Mercury in Food

The recognition of the toxic nature of mercury compounds and the awareness of their presence in foods have prompted the establishment of regulations by a number of governments to protect their citizens. Because fish seem to present the greatest threat, this food item has received the most attention. Action has ranged from establishing maximum permissible levels of mercury in fish, to recommending restricted consumption of fish. Based on the results of the Swedish studies by Westöo (1966, 1967) which indicated the majority of mercury in fish is in one form, methyl mercury, most regulations are expressed in terms of total mercury. Determination of total mercury is easier and probably more reliable than methods which determine different mercury compounds.

In 1967, the Swedish government established a practical residue limit of 1 mg/kg for mercury in fish. Berglund and Berlin (1969) explained the basis on which this limit was established. Factors considered were the following: the concentration of methyl mercury in different foods, the intake of methyl mercury-contaminated foods by the population, the relationship of intake and body levels of methyl mercury, and the toxic levels of methyl mercury for man. Two allowable daily intakes (ADI) were established: 0.1 mg Hg/day based on the estimated mercury intake via fish, and the mercury concentration of the erythrocytes of a healthy subject to which a 10-fold safety factor was added; 0.06 mg Hg/day based on the critical concentration of mercury in the brain, body distribution, excretion rates, and a 10-fold safety factor. Whereas many have criticized the maximum tolerance permitted in fish by the Swedish government, 1 mg/kg, as being too high, Berglund and Berlin concluded this limit is probably appropriate. They stated that the average Swedish citizen would consume the lower ADI, if all fish contained 1 mg Hg/kg. However, a large percentage of the fish consumed in Sweden are caught from the ocean and contain less than 0.15 mg Hg/kg. The Swedish government has recommended that the population limit consumption of fish captured from potentially dangerous areas, but establishing a limit of less than 1 mg Hg/kg would have necessitated a ban on commercial fishing in more than half of the Swedish fishing area.

This evidence is substantiated by recent evidence (Galster, 1976) that shows the similarity of average levels of mercury in blood, milk placenta and cord blood of Eskimo mothers in Anchorage, Alaska compared to other women who seldom eat fish. Regional high levels exist among those Eskimo women consuming seal oil, seal meat or fish from the Yukon-Kuskokwim coast. However, no blood levels are close to the 200 ng/ml level considered dangerous by Swedish researchers. Skerfving (1974) reported that long-term exposure to an approximate daily intake of 4 μg Hg as methyl mercury per kg body weight resulted in a human blood-cell level of about 0.3 μg Hg/g.

In May 1969, FDA established an action guideline of 0.5 mg

TABLE 11.4.   MERCURY AND METHYL MERCURY CONTENT OF SOME FOODS.

Food	Total mercury, ppm	Methyl mercury % of total mercury		Reference
	Limits	Limits	Average	
Meats				
ox	0.075	92	92	Westöo, 1966
hen	0.023–0.051	73–74	74	Westöo, 1966
pork				
liver	0.096–0.140	68–78	73	Westöo, 1966
liver	0.011–0.049	45–86	67	Westöo, 1968
kidney	0.014–0.086	24–71	50	Westöo, 1968
chop	0.006–0.016	80–100	88	Westöo, 1968
Eggs				
egg yolk	0.010	50–90	70	Westöo, 1968
egg white	0.012–0.025	76–96	89	Westöo, 1966
Fish-muscle tissue				
perch	0.22–3.25	83–93	89	Westöo, 1966
pike	0.56–3.35	88–98	94	Westöo, 1966
haddock	0.033–0.052	76–83	80	Westöo, 1966
cod	0.026–0.036	78–85	82	Westöo, 1966
tuna	0.73		93	Hall, 1974
shrimp	.009	.007	80	Hall, 1974
Plant tissue	0.04	—	—	Miller *et al.* 1960
Mushrooms	4.0–.02	.2–.005	10	Seeger, 1976

Hg/kg in fish and shellfish, a value which was based on the toxicological information and analytical methodology which was available at that time (Celeste and Shane, 1970). Ingestion of 30 μg of methyl mercury daily was considered the highest safe level (Newberne, 1974). Canada has also established the 0.5 mg Hg/kg level for fish. Goldwater (1971) reported that the FAO/WHO recommended a maximum level of 0.05 mg Hg/kg in food items other than fish.

If the ADI of 0.06 mg/day is accurate, a daily consumption of 120 g of fish containing 0.5 mg Hg/kg would be safe. If 0.03 mg/day is the maximum safe level, only 60 g of fish containing 0.5 mg Hg/kg would be safe. The 0.06 mg/day ADI, however, is also higher than the recent FAO/WHO recommended provisional tolerable intake. The Joint FAO/WHO Expert Committee on Food Additives in their sixteenth report (1972) recommended a provisional tolerable intake of 0.3 mg total mercury weekly of which not more than 0.2 mg should be in the methyl mercury form (expressed as mercury). Expressed on a weight basis, tolerable weekly intakes are 0.005 mg/kg body weight for total mercury and 0.0033 mg/kg body weight for methyl mercury expressed as mercury. The Committee based these recommendations on data collected from patients at Niigata, Japan which indicated the lowest mercury levels at the onset of poisoning symptoms were 50 μg/g for hair and 0.4 μg/g for red blood cells. The estimated mercury intake of these patients was

0.3 mg Hg/day, primarily in the methyl mercury form, over a long period of time.

If the major portion of mercury in fish exists as methyl mercury, 57 g of fish containing 0.5 mg/kg could be consumed daily without exceeding the weekly tolerable intake of 0.2 mg methyl mercury. The FAO (1972) reported the per capita fish consumption was 56 g/day in Sweden, 18 g/day in U.S.A. and 88/day in Japan. Mercury poisoning in the U.S.A. does not appear likely unless an individual consumed fish in a large excess to that of the average population.

Finch (1973) published an evaluation of the mercury intakes that a sample population would incur at different hypothetical guidelines. The evaluation was based on the results of a survey of 1586 households, a total of 4,864 persons, who kept accurate records of all fish and shellfish consumptions during a 12-month period. The results showed a total of 52 principal kinds of fish had been consumed in varying amounts. Using data from a number of reliable sources, the average mercury level of each type of fish was calculated on the basis of the following hypothetical guideline: 1) no limit on the mercury content of fish; 2) fish containing more than 1.5 ppm mercury would be removed from the market, 3) fish containing more than 1.0 ppm mercury would be removed from the market; 4) fish containing more than 0.75 ppm mercury would be removed from the market; and 5) fish containing more than 0.5 ppm mercury would be removed from the market. The estimated mercury intakes which would have occurred under each hypothetical guideline were calculated for the members of the survey. Finch also calculated the intakes which would have occurred with no guidelines except that swordfish would be eliminated from the diet. Whereas one family would have exceeded the 30 μg Hg/day ADI limit, if no restrictions were placed on the mercury content of fish, removing swordfish from their diet would have reduced their mercury intake to less than 30 μg Hg/day. Newberne stated that the group surveyed may not have included individuals such as "Weight Watchers" whose fish consumption was far in excess of the average fish consumer (Newberne, 1974).

Recent evidence indicates that some Indians in northern Canada may be suffering from chronic mercury poisoning. Although early symptoms are somewhat nonspecific, such signs as asymmetric constriction of the visual fields and tunnel vision are consistent with methyl mercury poisoning among these fish-eating Indians. In one area of northwest Quebec, daily fish consumption is 0.3 kg and increases to 0.7 kg in the summer. Local fish contain 2 ppm mercury. Indian fishing guides from Northern Ontario have blood levels of 100 to 300 ppb mercury (Shepard, 1976; Spurgeon, 1976).

It is recognized that the safety margin for established tolerance levels of mercury are not as great as those for many other toxic residues. However, the possibility remains that some manifestations of low level consumption of mercury may not be recognized. Strict guidelines have not been established in fear of depriving a number of people of an important source of protein as well as bringing undue hardship on people depending upon the fishing industry (Lu, 1974).

METHODS OF MERCURY ANALYSIS

Voluminous amounts of literature exist on the subject of mercury analysis; among these are many reviews (D'Itri, 1972; Fishbein, 1970, 1971; Friberg, 1971; Krenkel, 1973; Lindstedt and Skerfving, 1972; Nelson, 1971; Smith, 1972). The most common and reliable methods of mercury measurement are colorimetric dithizone procedure, atomic absorption and emission spectrophotometry, neutron activation and gas chromatography.

The major problem of environmental mercury analysis is concentrations in the parts per million or billion range. However, despite low levels, many characteristics of mercury are used in analytical procedures which present problems during other stages of analysis. The strong affinity of mercury for sulfhydryl groups is the basis for colorimetric analysis by the dithizone procedure. Cysteine forms a water soluble cysteine-mercury complex which aids in separating methyl mercury from interfering compounds (Westöo, 1966). Sulfide impregnated filter pads also collect and concentrate mercury compounds for analysis (Ballard et al. 1954; Pappas and Rosenberg, 1966). The affinity of mercury for sulfhydryl compounds necessitates that it is separated from the sample before most methods of detection. This is most often accomplished by a wet oxidation treatment in an aqueous solution or total combustion at elevated temperatures. Because mercury is volatile, extreme care must be taken to avoid losses. Moreover, volatility of mercury is an essential characteristic for detection by cold vapor atomic absorption, emission spectrophotometry and gas chromatography.

The Association of Official Analytical Chemists (AOAC) (1975) and the Working Group of the Bureau International Technique du Chlore (1976) recommended the use of nitric and sulfuric acids for wet oxidation of samples in glass apparatus equipped with a reflux condenser to minimize volatilization losses of mercury.

A flameless atomic absorption method involving digestion with vanadium oxide and nitric and sulfuric acids was recently adopted as official first action and is an alternative to the method now listed as having first action status by the Association of Official Analytical Chemists (AOAC, 1975). Fatty acids and aromatic compounds, which are undigested by this procedure are filtered before analysis. Other wet oxidation mixtures employ sulfuric acid and a 50% solution of hydrogen peroxide (Gorsuch, 1959), sulfuric, nitric, and hydrochloric acids (Teeny, 1975) perchloric acid (Munns and Holland, 1971; Rains and Menis, 1972; Smart and Hill, 1969) bomb digestion with nitric acid (Bureau International Technique du Chlore, 1976), potassium permanganate (Bailey and Lo, 1971; Dusci and Hackett, 1976; Lindstedt 1970; Matsunaga *et al.* 1976; Mushak, 1973), and potassium persulfate-potassium permanganate (Iskander *et al.* 1972). Special care must be taken to prevent losses of mercury. Following digestion the mercury may be concentrated or analyzed directly.

The total combustion techniques utilizes the high volatility of mercury and high heat to separate it quantitatively from the sample matrix. The use of high temperature flame (Hingle *et al.* 1967; Lindstrom, 1959; Longbottom, 1972) or sample combustion in high temperature furnaces has been used (Gage, 1961; Hayes *et al.* 1970; Krenkel, 1973; Muscat *et al.* 1972). Vapors in mercury are either collected by some means, or aspirated directly into an atomic absorption cell for detection.

Mercury collection and concentration has been accomplished by the use of sulfide impregnated filter pads (Ballard *et al.* 1954; Pappas and Rosenberg, 1966), deposition of the mercury on the surface of other metals (Brandenberger and Bader, 1967; Corcoran, 1974; Lidums and Ulfvarson, 1968; Pillay *et al.* 1971), collecting the mercury with ion exchange resins (Becknell *et al.* 1971; Rottschafer *et al.* 1971), and extraction into organic solvents (Westöo, 1966). Should concentration be necessary, the choice of method will determine the detection procedure used.

Dithizone Colorimetric Procedure

Previously, the most commonly used method of mercury analysis was the colorimetric or spectrophotometric measurement of the dithizone complex. Two official methods for mercury analysis, one by the AOAC and the other by the Analytical Methods Committee of

the Society of Analytical Chemistry require addition of hydroxylamine hydrochloride to reduce the remaining oxidizing materials
from the wet oxidation procedure. The AOAC procedure requires
mercury extraction from solution by adding dithizone in chloroform. The latter method recommends dithizone in carbon tetrachloride. Since copper also reacts with dithizone and will interfere,
if present; AOAC precludes interference by requiring the addition
of sodium thiosulfate to the chloroform layer. A water-soluble,
mercury thiosulfate complex is formed, whereas the copper-dithi-
zone complex is stable permitting its removal with chloroform.
After initial separation, the aqueous phase containing mercury-
thiosulfate complex is again oxidized and mercury extracted with
dithizone in chloroform. Mercury-dithizonate is determined immediately in a spectrophotometer at 490 nm against similarly
treated mercury standard solutions. The Analytical Methods Committee recommended nitrite reversion of the mercury rather than
the thiosulfate reversion. After reversion, copper is chelated with
ethylene diaminetetraacetic acid then mercury is reextracted with
dithizone in carbon tetrachloride and the mercury-dithizone complex measured at 485 nm.

The sensitivity of the dithizone procedure is approximately 0.05
ppm using a 10 g sample (Kudsk, 1964). Because of low cost,
simplicity and sensitivity of colorimetric analysis, the dithizone
method has retained some popularity. Time needed for sample
preparation and the possibility of interferences from other metals
principally copper are disadvantages.

Neutron Activation Analysis

The general procedure for neutron activation analysis of mercury
was pioneered by Westermark and Sjostrand (1960) and involves
sealing the sample in a quartz or polyethylene vial, irradiating the
sample with neutrons to produce isotopes, and measuring the radiations from these isotopes by gamma-spectrometry. Either ^{197}Hg or
^{203}Hg may be measured. While the initial radiation flux of ^{197}Hg is
higher, ^{203}Hg has a longer half-life [46.57 days compared to 65 h
(Watts et al. 1976)] and, therefore, may offer advantages for determinations over a period of weeks. Techniques of neutron activation
analysis may be divided into two classes, nondestructive or destructive analysis, depending upon whether the constituents are
separated before measuring radioactivity.

This nondestructive procedure involves irradiating the sample for
two to three days and detecting the 68-K.e.v. x-quanta and 77-K.e.v.

γ-quanta of ^{197}Hg by gamma-spectrometry. Sjostrand (1964) reported the procedure had been applied successfully to the analysis of hundreds of samples of various biological and organic materials, but was limited in its sensitivity to 0.1–0.5 ppm for biological materials. Furthermore, neutron activation appears to be a useful and rapid method for mapping mercury distribution in the central nervous system (Friedman *et al.* 1974). Haller *et al.* (1968) reported a nondestructive neutron activation procedure in which a germanium-lithium detector was utilized to separate the peaks of the many isotopes of irradiated plant tissues. The advantage of this procedure is that mercury and 14 other elements are determined simultaneously without chemical separation. The lowest concentrations of mercury detected were 13 ppb in pears using a 3.317 g sample and 25 ppb in peas using a 0.463 g sample. The authors reported that the statistical deviation was relatively high in mercury analysis, this being caused by Compton and Bremsstrahlung corrections and the necessity to correct for overlap of selenium and mercury peaks.

Sjostrand (1964) described a destructive method for neutron activation analysis which greatly increased sensitivity. Following irradiation, carrier solutions of Hg Cl$_2$ were added to the sample, and the sample was oxidized in hot concentrated nitric acid containing 5–10% concentrated sulfuric acid. After digestion, perchloric acid and aqueous glycine was added, heated, and mercury was collected in the distillate by a gold foil electrode. Mercury in the sample was determined by measuring ^{197}Hg and correcting for chemical yield. Sensitivity limit was estimated as 0.5 ppb mercury for a 0.5 g sample irradiated two to three days in a thermal neutron flux of 10^{12}N/cm^2-second.

Rottschafer *et al.* (1971) dissolved irradiated samples of fish in 1:1 concentrated HNO$_3$-H$_2$SO$_4$, diluted the solution with 1M HCl and added the solution to an ion exchange column where interfering ions like ^{24}Na$^+$ were eluted. The resin-mercury complex was transferred to glass vials and counted. Interfering metal ions were eluted from DEAE cellulose columns with a solution containing 0.01 M HCl abd 0.01 M ammonium thiocyanate. Then, mercury was eluted with 6 M HCl, reduced to elemental mercury and counted (Ishida *et al.* 1970).

Brune (1966) minimized mercury vaporization from aqueous solutions during irradiation by maintaining a temperature of −40 C (Brune, 1969) and separated the radioactive mercury by the isotope exchange technique of Kim and Silverman (1965). Tiny droplets of liquid elemental mercury added to the digest after pH adjustment adsorbed the dissolved mercury isotope.

Neutron activation is probably the most sensitive method available for mercury analysis wherein destructive analysis is more sensitive and probably more reliable than nondestructive analysis (Sjostrand, 1964). The primary disadvantages of both neutron activation procedures are: cost of equipment; need for highly qualified personnel; and analysis time. The Study Group on Mercury Hazards (Nelson, 1971) mentioned that Californium-246 may be utilized as a neutron source in the future and would negate the need for nuclear reactors thus reducing cost.

Spectrophotometry

Atomic absorption spectrophotometry is the most popular method of mercury analysis. Recently developed procedures and equipment for atomic fluorescence analysis also show great promise. Both techniques require mercury atoms to pass between ground and excited energy levels by absorbing or emitting radiation at the characteristic wavelength of 253.7 nm. In atomic absorption analysis the quantity of energy absorbed is measured as a cloud of elemental mercury vapor passes through a beam of ultraviolet light at 253.7 nm while atoms are raised to an excited energy state. In fluorescence analysis the quantity of fluorescent emission from atoms in the excited energy state is measured at 253.7 nm. Both quantities are proportional to the number of mercury atoms present.

Conventional flame atomic absorption spectrophotometry is regaining some popularity since many of the interferences encountered in wet oxidation techniques are avoided (Krenkel, 1973). Sensitivity of this procedure is limited because only a small amount of the sample (estimated at 10^{-5} ml) can be atomized into the optical path at any given moment (Ling, 1968). Apparently, the limit of detection is 1 μg of mercury per liter of solution (Lindstrom, 1959). Hingle *et al.* (1967) studied some variables of the flame atomic absorption techniques, and reported that the sensitivity of detection for mercury could be increased by adding, reducing or complexing agents to convert mercuric ions to mercurous ions before the solution is atomized into the flame.

Mercury compounds are reduced easily to the elemental vapor phase whereby techniques such as "cold vapor" and "flameless" atomic absorption spectrophotometry can be used to quantitatively measure these ions. Although both terms are often used interchangeably, generally "cold vapor analysis" refers to the technique in which mercury ions in solution are reduced chemically to the elemental state, after which the mercury may then be quantitatively removed from the solution and passed through the absorption cell by aerating the solution. The "flameless technique" refers to a method in which heat sufficient to reduce all forms of mercury to the elemental state is applied to a sample causing mercury to vaporize and pass through the absorption cell.

In these two procedures, mercury vapor is passed through a long path absorption cell placed in the path of light between the energy source and the photomultiplier tube. Both also offer greater sensitivity than flame analysis. AOAC has accepted the cold vapor technique as official first action (AOAC, 1975).

Hatch and Ott (1968) published a cold vapor technique for the determination of mercury in ore samples containing nickel and cobalt. Fundamental steps in the procedure involve acid oxidation of the sample, reduction of the mercuric ions by stannous sulfate and aeration of mercury vapor directly through the absorption cell by means of a circulating pump. Dassani et al. (1975) used a cold vapor technique to determine total mercury in samples of seeds, grains, fruits, vegetables, fish and meat. Hoover et al. (1971) reported more accurate results were obtained, if the reduction-aeration-expulsion cycle were repeated until the absorbance value for each solution was near zero. Apparently, an equilibrium existed between mercury in vapor and mercury in solution dependent upon the volume ratios of each and the sample being analyzed. Moreover, addition of a small amount of hydrazine sulfate facilitated the reduction reaction with stannous chloride. The lower limit of detection for mercury in sediment and soils was 0.01 μg Hg per gram of sample (Iskandar et al. 1972). This technique involved digestion and oxidation samples with nitric and sulfuric acids, potassium permanganate and potassium persulfate, reduction of the mercuric ions with stannous chloride and direct measurement of the mercury content of the vapor aerated from the solution.

Clarkson and Greenwood (1964) reported that stannous chloride reduced only inorganic mercury. Magos (1971) reported that a mixture of cadmium chloride, stannous chloride, and sodium hydroxide added to an acidified homogenate of the sample reduced both organic and inorganic mercury ions. Also, Magos and Jawad (1973) expanded upon this procedure and showed a screening method whereby 30 samples per hour could be processed.

The following procedures for "flameless" atomic absorption analysis of mercury were cited by the Study Group of Mercury Hazards. The sample in a large closed flask is burned in the presence of oxygen (Brandenberger and Bader, 1967). The combustion products are absorbed in dilute acid, and the mercury was electro-deposited from the acid solution onto a copper wire which was removed from the solution, placed in a side arm of the absorption cell, and mercury on the wire was vaporized by electrically heating it. Lidums and

Ulfvarson (1968) passed the combustion gases over gold foil which collected mercury. The gold foil was then heated, causing a rapid vaporization of the mercury into the absorption cell. Using sample sizes of less than 1 g, they could detect mercury concentrations below 1 ppb.

Some interferences are encountered when using atomic absorption analysis. Anions of all acids affect the release of mercury by the stannous chloride reduction-aeration technique. Iodide, bromide, sulfate and phosphate cause the greatest reduction in the amount of mercury vaporized (Rains and Menis, 1972). Ballard *et al.* (1954) reported that benzene, pyridine and acetone absorb energy at 253.7 nm. In addition, sulfur dioxide, nitric oxide, nitrogen dioxide, and hydrogen sulfide absorb strongly at 253.7 nm (Welch and Soares, 1976).

The atomic fluorescence determination of mercury appears to have many distinct advantages over atomic absorption methods. Advantages are simplicity since there is no enclosed cell and therefore no fogging of the cell windows, no spurious signal from broad band absorption by organic contaminants, no recirculating system or drying column required, sensitivity is high, and the standard curve is linear over a wide concentration range (Thompson and Reynolds, 1971). The presence of acetone, ethanol, chloroform or benzene in the sample solution resulted in a reduction in the peak height recorded for a given concentration of mercury. The reduction, believed to be caused by quenching of excited mercury atoms by solvent, was larger when argon was used as the carrier gas than when air was employed (Thompson and Reynolds, 1971).

This solvent interference could be removed by first separating the mercury by amalgamation, then revaporizing the mercury by heat (Corneliussen, 1969). In the absence of organic solvents, the peak height with argon was four times greater than that with nitrogen and thirty-five times greater than that with air (Thompson and Reynolds, 1971). Muscat *et al.* (1972) reported good agreement of results obtained by the reduction-aeration procedure using stannous chloride as the reductant and the furnace combustion technique in which the sample was heated to 800 C. In both methods, mercury was collected on a silver amalgamator, revaporized by heat and flushed into the fluorescence cell by carrier gas.

Thomerson and Thompson (1974) reported that instrumentation is available to permit detection of less than 0.5 mg mercury in vapor produced by stannous chloride reduction-aeration technique. He stated that the equipment involved was relatively inexpensive. Corcoran (1974) made five replicate measurements of 0.5 mg mercury which gave a standard deviation of ± 16%.

Chromatography

Use of gas chromatography (GLC) has gained popularity for analysis of organic mercury compounds. However, methods for analysis of inorganic mercury compounds by GLC have not been developed. An excellent review of the literature pertinent to GLC analysis of mercury compounds was recently published (Neal *et al.* 1941).

Most procedures utilize a gas chromatograph equipped with an electron capture detector (Schafer *et al.* 1975; Westermark and Sjostrand, 1960; Westöo, 1966) of either the tritium (^{3}H) or radioactive nickel (^{63}Ni) type with sensitivity approaching 1 ppb with favorable conditions (Westermark and Sjostrand, 1960; Westöo, 1966). For detection to occur with this sytem, it is necessary that the organomercurial form an unsymmetrical halide (bromide, chloride, or iodide). Dialkyl or diaryl mercury compounds will not be detected by electron capture.

Other detectors have been used for GLC analysis of organomercurials. Bache and Lisk (1971) reported detection of mercurials by emission spectrophotometry in a microwave powered inert gas plasma wherein it is possible to detect diorganomercurials. The specificity of the detector for mercury compounds is greater than the electron capture detector. Atomic absorption detectors can be attached to a GLC to detect mercury compounds separated (Longbottom, 1972). Organomercury compounds are reduced to the elemental form by the flame ionization detector or by passing the column gases through a combustion furnace. Then combustion gases are pumped directly through the absorption cell for detection.

A number of different liquid phases have been used for GLC analysis or organic mercury compounds. Good separation was obtained using Carbowax 20 M in a column heated at 130–145 C and purged with nitrogen. Methyl mercury compounds had the shortest retention time, followed by ethyl mercury, methoxyethyl mercury and phenyl mercury compounds (Westermark and Sjostrand, 1960). Bache and Lisk (1971) reported that OV-101 slowed elution of volatile dimethyl mercury to the extent that it was eluted after the monomethylmercury compounds. Tatton and Wagstaffe (1969) reported that the most satisfactory column consisted of 2% polyethylene glycol succinate on Chromasorb G. Elution order was methyl mercury, ethyl mercury, ethosyethyl mercury, methoxyethyl mercury, tolyl mercury and phenyl mercury.

Several authors have noted similar GLC retention times for various salts of a particular organomercurial ion. Johansson *et al.* (1970) used a combined GLC-mass spectrometer to study the phenomena, and concluded that rearrangement of anions occurred on column. Dressman (1972) reported that during GLC analysis of several phenyl mercury compounds, diphenyl mercury and phenyl mercury chloride were formed on column. The relative amount of each compound varied with the concentration of chloride ions present in the injection block. In view of this problem, all aryl mercury compounds must be converted to aryl mercury chloride prior to GLC analysis (Dressman, 1972).

Cappon and Smith (1977) described a general procedure for the gas chromatographic determination of alkyl, aryl, and inorganic mercury compounds. First, methyl-, ethyl-, and phenyl-mercury was extracted as the chlorides, then inorganic mercury was isolated as methylmercury following reaction with tetramethyltin. Total mercury recoveries ranged between 75 and 90% with a minimum detectable level about 1 ppm mercury. A simplified method for the gas chromatographic determination of methylmercury in fish and shellfish is described by Watts *et al.* (1976). Complete extraction of methylmercury and ethylmercury eliminated the need for recovery corrections. The method gave quantitative recoveries for methyl-mercury levels as low as 0.10 ppm in fish and shellfish.

Thin-layer chromatography has received considerable attention for analysis of organic mercury compounds. It has been used to purify organomercury compounds prior to GLC analysis as well as serving as a qualitative verification of the GLC analysis. Ion exchange and column chromatography have also been used for separation of mercury compounds. Chromatography procedures used for mercury analyses has been extensively reviewed (Fishbein, 1971, 1972).

Methods of Separation of Organomercury Compounds

Most procedures for separating organomercury compounds have been based on the methods of Westöo (1966, 1967, 1969). The procedure used is as follows: 1) Hydrolysis of sample homogenate with hydrochloric acid. For some samples, inorganic mercuric chloride is added to the acidified homogenate. Through competition for binding sites this aids in liberating methyl mercury from sulfhydryl groups. With liver samples, proteins are precipitated with molybdic acid rather than adding mercuric chloride, since it is known that methyl mercury is synthesized from mercuric ions by liver (Westöo, 1967). 2) Organic mercury compounds are extracted from the acidified homogenate with benzene. 3) Separate organomercury compounds from interfering compounds co-extracted by benzene. A water soluble cysteine-organomercury complex is formed by adding an aqueous cysteine salt solution to benzene. 4) After separating the aqueous phase, hydrolysis with HCl is repeated and organic mercury is re-extracted with benzene. The final benzene solution is ready for analysis by any of several methods. Westöö (1969) reported that greater than 90% recovery of added methyl mercury in samples of fish, egg white, kidney, blood, meat, bile, and moss. Kamps and McMahon (1972) reported an average recovery of 79.5% in fish samples that FDA analyzed using this procedure. They reported that losses probably result from partition coefficients which are sufficiently reproducible for a given sample to allow correction factors to be used.

Tatton and Wagstaffe (1969) reported alkoxyalkyl mercury compounds are very unstable in even dilute acid solutions, therefore, the above procedure would be unsatisfactory for their analyses. They proposed to use of a slightly alkaline solution of cysteine hydrochloride in isopropanol to extract all organomercurials from the sample homogenate. Extract was washed with diethyl ether or toluene, and finally, the organomercurials were extracted with a diethyl ether solution of dithizone. Recovery of 85–95% of the added organomercury compounds was reported from samples of potatoes, tomatoes, and apples.

Magos (1971) reported a method for determining organic mercury, inorganic mercury and total mercury contents of samples. Treatment of the acidified sample homogenate with a mixture of stannous chloride and cadmium chloride reduced both inorganic and organic mercury compounds to elemental mercury. Then, the total mercury content of the sample was determined by cold vapor atomic absorption spectrophotometry. Treatment of a duplicate sample of the acidified homogenate with stannous chloride only, reduced inorganic mercury to the elemental form but not the organomercurials. Cold vapor atomic absorption analysis of this sample indicated the concentration of inorganic mercury present. Organomercury concentration then is calculated by the difference of total and inorganic mercury concentration. The major advantage of the procedure is that both forms of mercury are determined by relatively the same procedure. However, differentiation of the organic mercury compounds is not possible.

SUMMARY

As a natural element, mercury is distributed throughout the earth and, because of toxicological reasons, it is mandatory that man does not significantly alter the natural distribution. The chemical form as well as the concentration of mercury influence the toxicological response. The alkyl mercury compounds, most dangerous of all in the environment, are produced by chemical reaction. Moreover, some biological organisms can also synthesize methyl mercury from metallic mercury and inorganic forms. The extent of this conversion is unknown. Certainly, aqueous solubility of methyl mercury is markedly greater than some parent compounds.

Extensive information has been compiled on the response of various animal species and various target organs to a wide range of mercury compounds. To assess the extent of exposure of various populations on earth in all sectors is difficult. Further, some reactions caused by chronic levels of some forms of mercury compounds are unknown or unrecognized.

In an attempt to protect humans against the known and unknown effects of mercury, tolerances were imposed for foods. This resulted in both criticism for being too lenient as well as condemnation for being too stringent. Further regulatory control has removed some of the hazards that have contributed to human intoxication. Others persist in unabated form.

ACKNOWLEDGMENT

Contribution from the College of Agricultural and Life Sciences, University of Wisconsin, Madison.

REFERENCES

ABERG, B., L. EKMAN, R. FALK, U. GREITZ, G. PERSSON, and J. O. SNIHS. 1969. Metabolism of methyl mercury (^{203}Hg) compounds in man: excretion and distribution. Arch. Environ. Health 19:478–484.

AICHBERGER, K. 1977. Mercury content of Austrian edible mushrooms and its relation to the protein content of the mushrooms. Z. Lebensm. Unters.-Forsch 163:35–38.

AL-SHAHRISTANI, H. and K. M. SHEHAB. 1974. Variation of biological half-life of methylmercury in man. Arch. Environ. Health 28:342–343.

AMIN-ZAKI, L., S. ELHASSANI, M. A. MAJEED, T. W. CLARKSON, R. A. DOHERTY and M. GREENWOOD. 1974. Intra-uterine methyl mercury poisoning in Iraq. Pediatrics 54:587–595.

AMIN-ZAKI, L., S. ELHASSANI, M. A. MAJEED, T. W. CLARKSON, R. A. DOHERTY, M. R. GREENWOOD and T. GROVANOLI-JAKUB-CZAK. 1976. Perinatal methyl mercury poisoning in Iraq. Amer. J. Dis. Child. 130:1070–1076.

ANON. 1970. Mercury in the environment. U.S. Geol. Surv. Prof. Paper 713:1–5. U.S. Government Printing Office, Washington, D.C.

ANON. 1971. Swordfish banned as U.S. Food-FDA. Food Processing 32(6): 14.

ANON. 1977. Mercury in fish. J. Assoc. Off. Anal. Chem. *60*:470–471.

ASSOCIATION OF OFFICIAL ANALYTICAL CHEMISTS. 1975. Official Methods of Analysis. 12th ed. The Association of Official Analytical Chemists, Washington, D.C.

BACHE, C. A., W. H. GUTENMANN, and D. J. LISK. 1971. Residues of total mercury and methylmercuric salts in lake trout as a function of age. Science 172:951–952.

BACHE, C. A. and D. J. LISK. 1971. Gas chromatographic determination of organic mercury compounds by emission spectrometry in a helium plasma. Anal. Chem. 43:950–952.

BAILEY, B. W. and F. C. LO. 1971. Automated method for determination of mercury. Anal. Chem. 43:1525–1526.

BAILEY, E. H. and R. M. SMITH. 1964. Mercury, its occurrence and economical trends. U.S. Geol. Surv. Circ. 496. U.S. Government Printing Office, Washington, D.C. 11 p.

BAILEY, E. H., P. D. SNAVELY, JR., and D. E. WHITE. 1961. Chemical analysis of brines and crude oil, Cymric Field, Kern County, California, Geol. Surv. Res. 1961 (Article 398), U.S. Geol. Surv. Prof. Paper 424D: D306–309. U.S. Government Printing Office, Washington, D.C.

BAKIR, F., S. F. DAMLUJI, L. AMIN-ZAKI, M. MURTADHA, A. KHA-LIDI, N. Y. AL-RAWI, S. TIKRITI, H. I. DHAHIR, T. W. CLARKSON, J. C. SMITH and R. A. DOHERTY. 1973. Methylmercury poisoning in Iraq. Science 181:230–241.

BALLARD, A. E., D. W. STEWART, W. O. KAMM, and C. W. ZUEHLKE. 1954. Photometric mercury analysis: correction for organic substances. Anal. Chem. 26:921–922.

BECKERT, W. F., A. A. MOGHISSI, F. H. F. AU, E. W. BRETTHAUER, and J. C. McFARLANE. 1974. Formation of methylmercury in a terrestrial environment. Nature *249*, 674–675.

BECKNELL, D. E., R. H. MARSH, and W. ALLIE, JR. 1971. Use of anion exchange resin-loaded paper in the determination of trace mercury in water by neutron activation analysis. Anal. Chem. 43:1230–1233.

BERGLUND, F. and M. BERLIN. 1969. Human risk evaluation for various populations in Sweden due to methylmercury in fish, p. 423–432. *In* M. W. Miller and G. G. Berg (ed.) Chemical Fallout: current research on persistent pesticides. Charles C. Thomas, Springfield, Ill.

BERLIN, M. 1963. Renal uptake, excretion, and retention of mercury. II. A study in the rabbit during infusion of methyl and phenylmercury compounds. Arch. Environ. Health 6:626–633.

BERLIN, M., J. FAZACKERLY and G. NORDBERG. 1969. The uptake of mercury in the brains of mammals exposed to mercury vapor and to mercuric salts. Arch. Environ. Health 18:719–729.

BERLIN, M. and S. GIBSON. 1963. Renal uptake, excretion, and retention of mercury. I. A study in the rabbit during infusion of mercuric chloride. Arch. Environ. Health 6:617–625.

BERLIN, M. and S. ULLBERG. 1963. Accumulation and retention of mercury in the mouse. I. An autoradiographic study after a single intravenous injection of mercuric chloride. Arch. Environ. Health 6:589–601.

BERLIN, M. and S. ULLBERG. 1963. Accumulation and retention of mercury in the mouse. II. An autoradiographic comparison of phenyl-mercuric acetate with inorganic mercury. Arch. Environ. Health 6:602–609.

BERLIN, M. and S. ULLBERG. 1963. Accumulation and retention of mercury in the mouse. III. An autoradiographic comparison of methyl-mercuric dicyandiamide with inorganic mercury. Arch. Environ. Health 6:610–616.

BERTINE, K. K. and E. D. GOLDBERG. 1971. Fossil fuel combustion and the major sedimentary cycle. Science 173:233–235.

BIDSTRUP, P. L. 1964. Toxicity of mercury and its compounds. Elsevier Pub. Co., New York. 112 p.

BIRAM, J. G. S. 1957. Some aspects of handling mercury. Vacuum 5:77–92.

BIRKHAUG, K. E. 1933. Phenyl-mercuric-nitrate. J. Infect. Dis. 53:250–261.

BLIGH, E. G. 1972. Mercury in Canadian fish. Can. Inst. Food Sci. Tech. J. 5(1):A6–A14.

BORG, K., H. WANNTORP, K. ERNE, and E. HANKO. 1966. Mercury poisoning in Swedish wildlife. J. Appl. Ecol. (Suppl.) 3:171–172.

BORNMANN, G., G. HENKE, H. ALFES, and H. MÖLLMANN. 1970. Uber die enterale resorption von metallischem quecksilber. Arch. Toxicol. 26:203–209. (German with Eng. Summary).

BOWEN, H. J. M. 1966. Trace elements in biochemistry. Academic Press, New York. 241 p.

BRADLEY, R. L., JR. 1970. Mercury Hearing before the State of Wisconsin Department of Natural Resources. October 12.

BRANDENBERGER, H. and H. BADER. 1967. The determination of nanogram levels of mercury in solution by a flameless atomic absorption technique. Helv. Chem. Acta. 50:1409–1415.

BROWN, J. R. and M. V. KULKARNI. 1967. A review of the toxicity and metabolism of mercury and its compounds. Med. Serv. J. Can. 23:786–808.

BRUBAKER, P. E., G. W. LUCIER, and R. KLEIN. 1971. The effects of methylmercury in protein synthesis in rat liver. Biochem. Biophys. Res. Commun. 44:1552–1558.

BRUNE, D. 1966. Low temperature irradiation applied to neutron activation analysis of mercury in human whole blood. Acta Chem. Scand. 20:1200–1202.

BRUNE, D. 1969. Aspects of low-temperature irradiation in neutron activation analysis. Anal. Chim. Acta 44:15–20.

BUREAU INTERNATIONAL TECHNIQUE DU CHLORE. 1976. Standardization of methods for the determination of traces of mercury. Part II. Determination of total mercury in materials containing organic matter. Anal. Chim. Acta *84*, 231–254.

BURTON, J. D. and T. M. LEATHERLAND. 1971. Mercury in a coastal marine environment. Nature 231:440–442.

BYFIELD, A. 1970. Mercury hearing before the State of Wisconsin Department of Natural Resources. October 12.

CAMMAROTO, V. A., JR. 1974. Mercury. p. 771–781. *In* U.S. Department of Interior, Bureau of Mines. Minerals Yearbook 1972. Vol. 1. U.S. Government Printing Office, Washington, D.C.

CAPPON, C. J. and J. C. SMITH. 1977. Gas chromatographic determination of inorganic mercury and organomercurials in biological materials. Anal. Chem. *49*:365–369.

CELESTE, A. C. and C. G. SHANE. 1970. Mercury in fish. FDA Papers 4(9):27–29.

CEMBER, H. 1969. A model for the kinetics of mercury elimination. Amer. Ind. Hyg. Ass. J. 30:367–371.

CEMBER, H. and A. DONAGI. 1953. The influence of dose level and chemical form on the dynamics of mercury elimination. XIV International Congress on Occupational Health. Excerpta Med. Int. Cong. Ser. 62:440–442.

CHANG, L. W., P. A. DESNOYERS, and H. A. HARTMANN. 1972. Quantitative cytochemical studies of RNA in experimental mercury poisoning. 1. Changes in RNA content. J. Neuropathol. Exp. Neurol. 31: 489–501.

CHANG, L. W., A. H. MARTIN, and H. A. HARTMANN. 1972. Quantitative autoradiographic study on the RNA synthesis in the neurons after mercury intoxication. Exp. Neurol. 37:62–67.

CHEN, R. W., V. L. LACY and P. D. WHANGER. 1975. Effect of selenium on methylmercury binding to subcellular and soluble proteins in rat tissues. Res. Commun. Chem. Pathol. Pharmacol. *12*:297–308.

CLARKSON, T. W. 1971. Epidemiological and experimental aspects of lead and mercury contamination of food. Food Cosmet. Toxicol. 9:229–243.

CLARKSON, T. W. 1972. Biotransformation of organomercurials in mammals. p. 229–238. *In* R. Hartung and B. D. Dinman (ed.) Environmental mercury contamination. Ann Arbor Science Pub. Inc., Ann Arbor, Mich.

CLARKSON, T. W. 1972. The pharmacology of mercury compounds. Annu. Rev. Pharmacol. 12:375–406.

CLARKSON, T. W. and M. R. GREENWOOD. 1970. Selective determination of inorganic mercury in the presence of organomercurial compounds in biological material. Anal. Biochem. 37:236–243.

CLARKSON, T. W. and A. ROTHSTEIN. 1964. The excretion of volatile mercury by rats injected with mercuric salts. Health Phys. 10:1115–1121.

CLARKSON, T. W., H. SMALL, and T. NORSETH. 1971. The effect of a thiol containing resin on the gastrointestinal absorption and fecal excretion of methylmercury compounds in experimental animals. Fed. Proc. 30:543. (Abstr.)

COOKE, N. E. and A. BEITEL. 1971. Some aspects of other sources of mercury to the environment. p. 53–62. *In* J. E. Watkin (ed.) Mercury in man's environment. Proc. Roy. Soc. Can. Symp., Ottawa, Canada.

CORCORAN, F. L. 1974. Trace mercury determination by atomic fluorescence. Amer. Lab. 6(3):69–73.

CORNELIUSSEN, P. E. 1969. Residues in food and feed. Pesticide residues in total diet samples (IV). Pest. Monit. J. 2:140–152.

CUNNINGHAM, P. A. and M. R. TRIPP. 1973. Accumulation and depuration of mercury in the American oysteer *Crassotrea virginica*. Marine Biol. 20:14–19.

CURLEY, A., V. A. SEDLAK, E. F. GIRLING, R. E. HAWK, W. F. BARTHEL, P. E. PIERCE, and W. H. LIKOSKY. 1971. Organic mercury identified as the cause of poisoning in humans and hogs. Science 172:65–67.

DANIEL, J. W. and J. C. GAGE. 1971. The metabolism by rats of phenylmercury acetate. Biochem. J. 122:24 p.

DANIEL, J. W., J. C. GAGE and P. A. LEFEVRE. 1971. The metabolism of methoxyethyl mercury salts. Biochem. J. 121:411–415.

DASSANI, S. D., B. E. McCLELLAN and M. GORDON. 1975. Submicrogram level determination of mercury in seeds, grains and food products by cold-vapor atomic absorption spectrometry. J. Agr. Food Chem. *23*: 671–674.

DAVIES, D. J. and A. KENNEDY. 1967. The excretion of renal cells following necrosis of the proximal convoluted tubule. Brit. J. Exp. Pathol. 48:45–50.

DAY, F. H. 1964. Mercury, p. 185–187. *In* F. H. Day, The chemical elements in nature. Reinhold Pub. Corp., New York.

DERBAN, L. K. A. 1973. Outbreak of food poisoning due to alkyl-mercury fungicide. Arch. Environ. Health *28*:49–52.

DERRYBURY, O. M. 1972. Investigation of mercury contamination in the Tennessee Valley Region, p. 76–79. *In* R. Hartung and B. D. Dinman (ed.) Environmental mercury contamination. Ann Arbor Science Pub. Inc., Ann Arbor, Mich.

D'ITRI, F. M. 1972. The environmental mercury problem. CRC Press, Cleveland, Ohio. 124 p.

DRESSMAN, R. C. 1972. The conversion of phenylmercuric salts to diphenylmercury and phenyl mercuric chloride upon gas chromatographic injection. J. Chromatogr. Sci. 10:468–472.

DUSCI, L. J. and L. P. HACKETT. 1976. Rapid digestion and flameless atomic absorption spectroscopy of mercury in fish. J. Assoc. Off. Anal. Chem. *59*:1183–1185.

DUSTMAN, E. H., L. F. STICKEL, and J. B. ELDER. 1972. Mercury in wild animals, Lake St. Clair, 1970, p. 46–52. *In* R. Hartung and B. D. Dinman (ed.) Environmental mercury contamination. Ann Arbor Science Pub. Inc., Ann Arbor, Mich.

ECKERT, W. 1970. Statement: Mercury use on Wisconsin golf courses, prepared for mercury hearing before the State of Wisconsin Department of Natural Resources, October 12.

EDWARDS, T. and B. C. McBRIDE. 1975. Biosynthesis and degradation of methylmercury in human feces. Nature *253*:462–464.

EHMANN, W. D. and J. F. LOVERING. 1967. The abundance of mercury in meteorites and rocks by neutron activation analysis. Geochim. Cosmochim. Acta. 31:357–376.

EHRLICH, R. 1970. How mercury pollutes. Chem. Eng. News 48(30):18–19.

EKMAN, L., U. GREITZ, G. PERSSON, and B. ABERG. 1968. Metabolism and retention of methyl-mercurynitrate in man. Nord. Med. 79:450–456. (Swedish with Eng. Summary).

EL-BEGEARMI, M. M., M. L. SUNDE and H. E. GANTHER. 1977. A mutual protective effect of mercury and selenium in Japanese quail. Poultry Sci. *56*:313–322.

ELLIS, R. W. and S. C. FANG. 1967. Elimination, tissue accumulation, and cellular incorporation of mercury in rats receiving an oral dose of ²⁰³Hg-labelled phenylmercuric acetate and mercuric acetate. Toxicol. Appl. Pharmacol. 11:104–113.

ENGEL, G. T. 1966. Mercury, p. 647–656. *In* U.S. Department of Interior, Bureau of Mines. Minerals Yearbook 1965. Vol. I. U.S. Government Printing Office, Washington, D.C.

ERIKSSON, E. 1967. Mercury in nature. Oikos (Suppl.) 9:13.

EVANS, H. L., V. G. LATUS and B. WEISS. 1975. Behavioral effects of mercury and methyl-mercury. Fed. Proc. *34*:1858–1867.

FALK, R., J. O. SNIHS, L. EKMAN, U. GREITZ, and B. ABERG. 1970. Whole body measurements on the distribution of mercury-203 in humans after oral intake of methylradiomercury nitrate. Acta. Radiol. Therapy Phys. Biol. 9:55–72.

FANG, S. C. 1977. Interaction of selenium and mercury in the rat. Chem.-Biol. Interact. *17*:25–40.

FANG, S. C., R. W. CHEN and E. FALKIN. 1976. Influence of dietary selenite on the binding characteristics of rat serum proteins to mercury compounds. Chem.-Biol. Interact. *15*:51–57.

FANG, S. C. and E. FALLIN. 1976. The binding of various mercurial compounds to serum proteins. Bull. Environ. Contam. Toxicol. *15*:110–117.

FAO. 1972. Food Supply, p. 435–462. *In* N. Erus (ed.) Production yearbook 1971. FAO, Rome.

FAO/WHO, Joint Expert Committee on Food Additives. 1972. Evaluation of mercury, lead, cadmium, and the food additives, amaranth, diethylpyrocarbonate, and octylgallate. WHO Food Additive Ser. 1972 (4):1–59.

FEHLING, C., M. ABDULLA, A. BRUN, M. DICTOR, A. SCHUTZ and S. SKERFVING. 1975. Methyl mercury poisoning in the rat: A combined neurological, chemical and histopathological study. Toxicol. Appl. Pharmacol. *33*:27–37.

FIMREITE, N., W. N. HOLSWORTH, J. A. KEITH, P. A. PEARCE, and I. M. GRUCHY. 1971. Mercury in fish and fish-eating birds near sites of industrial contamination in Canada. Can. Field Naturalist 85:211–220.

FINCH, R. 1973. Effects of regulatory guidelines on the intake of mercury from fish. The MECCA project. Fish. Bull. 71:615–626.

FISHBEIN, L. 1970. Chromatographic and biological aspects of organo-mercurials. Chromatogr. Rev. 13:83–162.

FISHBEIN, L. 1971. Chromatographic and biological aspects of inorganic mercury. Chromatogr. Rev. 15:195–238.

FISKESJO, G. 1969. Some results from *Allium* tests with organic mercury halogenides. Hereditas 62:314–322.

FISKESJO, G. 1970. The effect of two organic mercury compounds on human leukocytes in vitro. Hereditas 64:142–146.

FITZHUGH, O. G., A. A. NELSON, E. P. LAUG, and F. M. KUNZE. 1950. Chronic oral toxicities of mercuriphenyl and mercuric salts. Arch. Ind. Hyg. Occup. Med. 2:433–442.

FLEISCHER, M. 1970. Summary of the literature on the inorganic geochemistry of mercury. U.S. Geol. Surv. Prof. Paper 713:6–13. U.S. Government Printing Office, Washington, D.C.

FRIBERG, L. 1959. Studies on the metabolism of mercuric chloride and methyl mercury dicyandiamide. A. M. A. Arch. Ind. Health 20:42–49.

FRIBERG, L. 1961. On the value of measuring mercury and cadmium concentrations in urine. Pure Appl. Chem. 3:289–292.

FRIBERG, L. T. (Chairman). 1969. Maximum allowable concentrations of mercury compounds. Arch. Environ. Health. 19:891–905.

FRIBERG, L. T. (Chairman). 1971. Report from an expert group. Methyl mercury in fish: A toxicologic-epidemiologic evaluation of risks. Nord. Hyg. Tidskrift (Suppl.) 4:1–364.

FRIBERG, L. and G. NORDBERG. 1973. Inorganic mercury—a toxicological and epidemiological appraisal, p. 5–22. *In* M. W. Miller and T. W. Clarkson (ed.) Mercury, mercurials, and mercaptans. Charles C. Thomas, Springfield, Ill.

FRIBERG, L., E. SKOG, and J. E. WAHLBERG. 1961. Resorption of mercuric chloride and methyl mercury dicyandiamide in guinea-pigs through normal skin and skin pre-treated with acetone, alkllarylsulphonate and soap. Acta. Derm. Venereol. 41:40–52.

FRIEDMAN, M. H., E. MILLER, and J. T. TANNER. 1974. Instrumental neutron activation analysis for mercury in days administered methylmercury chloride: use of a low energy photon detector. Anal. Chem. *46*:236–239.

FUJIMURA, Y. 1964. Cited from Goldwater, L. J. 1971. Mercury in the environment. Sci. Amer. 224(5):15–21.

GAGE, J. C. 1961. The distribution and excretion of inhaled mercury vapor. Brit. J. Ind. Med. 18:287–294.

GAGE, J. C. 1964. Distribution and excretion of methyl and phenyl mercury salts. Brit. J. Ind. Med. 21:197–202.

GALSTER, W. A. 1976. Mercury in Alaskan-Eskimo mothers and infants. Environ. Health Perspectives, *15*:135–140.

GANTHER, H. E., C. GOUDIE, M. L. SUNDE, M. J. KOPECKY, P. WAGNERS, S.-H. OH, and W. G. HOEKSTRA. 1972. Selenium: Relation to decreased toxicity of methylmercury added to diets containing tuna. Science 175:1122–1124.

GANTHER, H. E. and M. L. SUNDE. 1974. Effect of tuna fish and selenium on the toxicity of methylmercury: A progress report. J. Food Sci. 39:1–5.

GERSTNER, H. B. and J. E. HUFF. 1977. Clinical Toxicology of mercury. J. Toxicol. Environ. Health 2:491–526.

GIBBS, O. S. 1940. Cited from Goldwater, L. J. 1971. Mercury in the environment. Sci. Amer. 224(5):15–21.

GOLDBERG, A. A. and M. SHAPERO. 1957. Toxicological hazards of mercurial paints. J. Pharm. Pharmacol. 9:469–475.

GOLDWATER, L. J. 1964. Occupational exposure to mercury. The Harben lectures. J. Roy. Inst. Publ. Health Hyg. 27:279–301.

GOLDWATER, L. J. 1964. Cited from Goldwater, L. J. 1971. Mercury in the environment. Sci. Amer. 224(5):15–21.

GOLDWATER, L. J. 1971. Mercury in the environment. Sci. Amer. 224(5): 15–21.

GOLDWATER, L. J. 1973. Aryl- and alkoxyalkylmercurials, p. 56–67. In M. W. Miller and T. W. Clarkson (ed.) Mercury, mercurials, and mercaptans. Charles C. Thomas, Springfield, Ill.

GORSUCH, T. T. 1959. Radiochemical investigations on the recovery for analysis of trace elements in organic and biological materials. Analyst 84:135–173.

GRANT, N. 1971. Mercury in man. Environment 13(4):2–15.

GREENSPOON, G. N. 1970. Mercury, p. 639–652. In U.S. Dept. of Interior, Bureau of Mines, Mineral facts and problems. 1970 Edition. U.S. Government Printing Office, Washington, D.C.

GREIG, R. A. and H. L. SEAGRAN. 1972. Survey of mercury concentrations in fishes of Lakes St. Clair, Erie, and Huron, p. 38–45. In R. Hartung and B. D. Dinman (ed.) Environmental mercury contamination. Ann Arbor Science Pub. Inc., Ann Arbor, Mich.

HALL, E. T. 1974. Mercury in commercial canned seafood. J. Assoc. Offic. Anal. Chem. 57:1068–1073.

HALLER, W. A., L. A. RANCITELLI, and J. A. COOPER. 1968. Instrumental determination of trace elements in plant tissue by neutron activation analysis and Ge(Li) gamma-ray spectrometry. J. Agr. Food Chem. 16:1036–1040.

HAMMOND, A. L. 1971. Mercury in the environment: Natural and human factors. Science 171:788–789.

HANNERZ, L. 1968. Experimental investigations on the accumulation of mercury compounds in water organisms. Rep. Inst. Freshwater Res., Drottningholm 48:120–176.

HANSON, A. 1971. Man-made sources of mercury, p. 22–33. In J. E. Watkin (ed.) Mercury in man's environment. Proc. Roy. Soc. Can. Symp., Ottawa, Canada.

HAQ, I. U. 1963. Agrosan poisoning in man. Brit. Med. J. 5345:1579–1582.

HARADA, Y. 1968. Clinical investigations on Minamata disease. C. Congenital (or fetal) Minamata Disease, p. 93–117. *In* M. Kutsuna (ed.) Minamata Disease. Study group of Minamata disease. Kumamoto University, Japan.

HATCH, W. R. and W. L. OTT. 1968. Determination of submicrogram quantities of mercury by atomic absorption spectrophotometry. Anal. Chem. 40:2085–2087.

HAYES, H., J. MUIR, and L. M. WHITBY. 1970. A rapid method for the determination of mercury in urine. Ann. Occup. Hyg. 13:235–239.

HAYES, A. D. and A. ROTHSTEIN. 1962. The metabolism of inhaled mercury vapor in the rat studied by isotope techniques. J. Pharmacol. Exp. Therapeutics 138:1–10.

HINGLE, D. N., G. F. KIRKBRIGHT, and T. S. WEST. 1967. Some observations on the determination of mercury by atomic absorption spectroscopy in an air-acetylene flame. Analyst 92:759–762.

HIROTA, K. 1969. Distribution of methyl mercury compound in human brain and rat organs, and the effect of D-penicillamine on mercury in rat organs. Clin. Neurol. 9:592–601. (Japanese with Eng. summary).

HOLLINS, J. G., R. F. WILLES, F. R. BRYCE, S. M. CHARBONNEAU and I. C. MUNRO. 1975. The whole body retention and tissue distribution of [^{203}Hg] methylmercury in adult cats. Toxicol. Appl. Pharmacol. *33*: 438–449.

HOOVER, W. L., J. R. MELTON and P. A. HOWARD. 1971. Determination of trace amounts of mercury in foods by flameless atomic absorption. J. Ass. Offic. Anal. Chem. 54:860–865.

HUNTER, D., R. R. BOMFORD, and D. S. RUSSELL. 1940. Poisoning by methylmercury compounds. Quart. J. Med. 33:193–213.

INOUE, K. and S. AOMINE. 1969. Retention of mercury by soil colloids. III. Absorption of mercury in dilute phenylmercuric acetate solutions. Soil Sci. Plant Nutr. *15*(2):86–91. (Chem. Abstr. 71:80273v).

IRUKAYAMA, K. 1968. Minamata disease as a public nuisance, p. 301–324. *In* M. Kutsuna (ed.) Minamata Disease. Study Group of Minamata Disease. Kumamoto University, Japan.

ISHIDA, K., S. KAWAMURA, and M. IZAWA. 1970. Neutron activation analysis for mercury. Anal. Chim. Acta. 50:351–353.

ISKANDAR, I. K., J. K. SYERS, L. W. JACOBS, D. R. KEENEY, and J. T. GILMOUR. 1972. Determination of total mercury in sediments and soils. Analyst 97:388–393.

IVERSON, F. and S. L. HIERLIHY. 1974. Biotransformation of methyl mercury in the guinea pig. Bull. Environ. Contam. Toxicol. *11*:85–91.

IWATA, H. and H. OKAMOTO. 1970. Coefficients of metals on methyl mercury poisoning. Stencil, Dept. of Pharm. Sci. Univ. of Osaka, Japan. (Japanese).

JALILI, M. A. and A. H. ABBASI. 1961. Poisoning by ethyl mercury toluene sulphonanilide. Brit. J. Ind. Med. 18:303–308.

JENSEN, S. and A. JERNELOV. 1969. Biological methylation of mercury in aquatic organisms. Nature 223:753M754.

JERNELÖV, A. 1969. Conversion of mercury compounds, p. 68–74. *In*

M. W. Miller and G. G. Berg. (ed.) Chemical fallout. Current research on persistent pesticides. Charles C. Thomas, Springfield, Ill.

JERNELÖV, A. 1972. Factors in the transformation of mercury to methyl-mercury, p. 167–172. *In* R. Hartung and B. D. Dinman (ed.) Environ-mental mercury contamination. Ann Arbor Science Pub. Inc., Ann Ar-bor, Mich.

JOENSUU, O. I. 1971. Fossil fuels as a source of mercury pollution. Science 172:1027–1028.

JOHANSSON, B., R. RYHAGE and G. WESTÖO. 1970. Identification and determination of methylmercury compounds in fish using combina-tion gas chromatograph-mass spectrometer. Acta Chem. Scand. 24:2349–2354.

JOHNELS, A. G. and T. WESTERMARK. 1969. Mercury contamination of the environment in Sweden, p. 221–241. *In* M. W. Miller and G. G. Berg (ed.) Chemical fallout. Current research on persistent pesticides. Charles C. Thomas, Springfield, Ill.

JOHNSON, S. L. and W. G. POND. 1974. Inorganic vs. organic Hg toxicity in growing rats: Protection by dietary Se but not Zn. Nutr. Rep. Intern. 9:135–147.

JOHNSON, W. C. (Chairman), Analytical Methods Committee, Society for Analytical Chemistry. 1965. The determination of small amounts of mer-cury in organic matter. Analyst 90:515–530.

JONASSON, I. R. and R. W. BOYLE. 1972. Geochemistry of mercury and origins of natural contamination of the environment. Can. Mining Metal-lurgical Bull. 65(1):32–39.

JONES, H. R. 1971. Mercury pollution control. Noyes Data Corp., Park Ridge, N. J. 251 p.

KAMPS, L. R. and B. McMAHON. 1972. Utilization of the Westoo pro-cedure for the determination of methylmercury in fish by gas-liquid chromatography. J. Ass. Offic. Anal. Chem. 55:590–595.

KESSLER, R. H., R. LOZANO, and R. F. PITTS. 1957. Studies on the structure diuretic activity relationships of organic compounds of mer-cury. J. Clin. Invest. 36:656–668.

KEVORKIAN, J., D. P. CENTO, J. F. UTHE, and R. A. HAGSTROM. 1973. Methylmercury content of selected human tissues over the past 60 years. Am. J. Public Health 63:931–934.

KIM, C. K. and J. SILVERMAN. 1965. Determination of mercury in wheat and tobacco leaf by neutron activation analysis using mercury-197 and a simple exchange separation. Anal. Chem. 37:1616–1617.

KIRKPATRICK, D. C. and D. E. COFFIN. 1974. The trace metal content of representative Canadian diets in 1970 and 1971. Can. Inst. Food Sci. Technol. 7:56–58.

KITAMURA, S. 1968. Determination of mercury content in bodies of in-habitants, cats, fishes, and shells in Minamata District and in the mud of Minamata Bay, p. 257–266. *In* M. Kutsuna (ed.) Minamata Disease. Study Group of Minamata Disease. Kumamoto University, Japan.

KLEIN, D. H. 1972. Some estimates of natural levels of mercury in the environment, p. 25–29. *In* R. Hartung and B. D. Dinman (ed.) Environmental mercury contamination. Ann Arbor Science Pub. Inc., Ann Arbor, Mich.

KLEIN, D. H. and E. D. GOLDBERG. 1970. Mercury in the marine environment. Environ. Sci. Technol. 4:765–768.

KOEMAN, J. H., W. H. M. PEETERS, C. H. M. KOUDSTAAL, P. S. TJIOE and J. J. M. DE GOEIJ. 1973. Mercury-selenium correlations in marine mammals. Nature 245:385–386.

KOIRTYOHANN, S. R., R. MEERS and L. K. GRAHAM. 1974. Mercury levels in fishes from some Missouri lakes with and without known mercury pollution. Environ. Res. 8:1–11.

KOJIMA, K. and M. FUJITA. 1973. Summary of recent studies in Japan on methylmercury poisoning. Toxicology 1:43–62.

KONRAD, J. G. 1972. Mercury contents of bottom sediments from Wisconsin rivers and lakes, p. 52–58. *In* R. Hartung and B. D. Dinman (ed.) Environmental mercury contamination. Ann Arbor Science Pub. Inc., Ann Arbor, Mich.

KOOS, B. J. and L. D. LONGO. 1976. Mercury toxicity in the pregnant woman, fetus and newborne infant. Am. J. Obstet. Gynecol. 126:390–409.

KRENKEL, P. A. 1973. Mercury: Environmental considerations, part 1. CRC Critical Rev. in Environ. Control 3:303–373.

KUDSK, F. N. 1964. Determination of mercury in biological materials, a specific and sensitive dithizone method. Scand. J. Clin. Lab. Invest. 16: 575–583.

KUDSK, F. N. 1965. Absorption of mercury vapor from the respiratory tract in man. Acta Pharmacol. Toxicol. 23:250–262.

KUDSK, F. N. 1969. Uptake of mercury vapor in blood in vivo and in vitro from Hg containing air. Acta Pharmacol. Toxicol. 27:149–160.

KUDSK, F. N. 1969. Factors influencing the in vitro uptake of mercury vapor in blood. Acta Pharmacol. Toxicol. 27:161–172.

KURLAND, L. T. 1973. An appraisal of the epidemiology and toxicology of alkylmercury compounds, p. 23–55. *In* M. W. Miller and T. W. Clarkson (ed.) Mercury, mercurials, and mercaptans. Charles C. Thomas, Springfield, Ill.

KUTSUNA, M. 1968. Historical perspective of the study on Minamata Disease, p. 1–4. *In* M. Kutsuna (ed.) Minamata Disease. Study group of Minamata Disease. Kumamoto University, Japan.

LANDNER, L. 1971. Biochemical model for the biological methylation of mercury suggested from methylation studies in vivo with *Neurospora crassa*. Nature 230:452–454.

LEATHERLAND, T. M., J. D. BURTON, M. J. McCARTNEY, and F. CULKIN. 1971. Mercury in North Eastern Atlantic Ocean water. Nature 232:112.

LIDUMS, V. and U. ULFVARSON. 1968. Mercury analysis in biological material by direct combustion in oxygen and photometric determination of the mercury vapour. Acta Chem. Scand. 22:2150–2156.

LINDSTEDT, G. 1970. A rapid method for the determination of mercury in urine. Analyst 95:264–271.

LINDSTEDT, G. and S. SKERFVING. 1972. Methods of analysis, p. 3–13.

In L. Friberg and J. Vostal (ed.) Mercury in the environment; an epidemiological and toxicological appraisal. CRC Press, Cleveland, Ohio.

LINDSTROM, O. 1959. Rapid microdetermination of mercury by spectrophotometric flame combustion. Anal. Chem. 31:461-467.

LING, C. 1968. Portable atomic absorption photometer for determining nanogram quantities of mercury in the presence of interfering substances. Anal. Chem. 40:1876-1878.

LOFROTH, G. 1970. A review of health hazards and side effects associated with the emission of mercury compounds into natural systems. 2nd ed. Swedish Natural Science Research Council, Stockholm, Sweden.

LONGBOTTOM, J. E. 1972. Inexpensive mercury-specific gas chromatographic detector. Anal. Chem. 44:1111-1112.

LU, F. C. 1974. Mercury as a food contaminant. WHO Chron. 28:8-11.

LUCIER, G., O. McDANIEL, P. BRUBAKER, and R. KLEIN. 1971. Effects of methylmercury hydroxide on rat liver microsomal enzymes. Chem. Biol. Interactions 4:265-280.

LUNDGREN, K. and SWENSSON, A. 1949. Occupational poisoning by alkyl mercury compounds. J. Ind. Hyg. Toxicol. 31:190-200.

MAGOS, L. 1967. Mercury blood interactions and mercury uptake by the brain after vapor exposure. Environ. Res. 1:323-337.

MAGOS, L. 1968. Uptake of mercury by the brain. Brit. J. Ind. Med. 25:315-318.

MAGOS, L. 1971. Selective atomic absorption determination of inorganic mercury and methyl mercury in undigested biological samples. Analyst 96:847-853.

MAGOS, L. 1975. Mercury and mercurials. Brit. Med. Bull. *31*:241-245.

MAGOS, L. and A. M. JAWAD. 1973. Method for screening barley and wheat samples for mercury. J. Sci. Food Agric. *24*:1305-1309.

MANSOUR, M. M., D. C. DYER, L. H. HOFFMAN, J. DAVIES, and A. B. BRILL. 1974. Placental transfer of mercuric nitrate and methylmercury in the rat. Am. J. Obstet. Gynecol. *119*:557-562.

MARCH, B. E., R. SOONG, E. BILINSKI and R. E. E. JONAS. 1974. Effects on chickens of chronic exposure to mercury at low levels through dietary fish meal. Poultry Sci. *53*:2175-2181.

MARCH, B. E., R. SOONG, E. BILINSKI and R. E. E. JONAS. 1974. Tissue residues of mercury in broilers fed fish meals containing different concentrations of mercury. Poultry Sci. *53*:2181-2185.

MATSUNAGA, K., T. ISHIDA and T. ODA. 1976. Extraction of mercury from fish for atomic absorption spectrometric detemmination. Anal. Chem. *48*:1421-1423.

MATSUMOTO, A., A. SUZUKI, C. MORITA, K. NAKAMURA, and S. SAEKI. 1967. Preventive effect of penicillamine on the brain of fetal rat poisoned transplacentally with methylmercury. Life Sci. 6:2321-2326.

McALPINE, D. and S. ARAKI. 1958. Minamata disease, an unusual neurological disorder caused by contaminated fish. Lancet 1968 (II):629-631.

McALPINE, D. and S. ARAKI. 1959. Minamata disease, late effects of an unusual neurological disorder caused by contaminated fish. A. M. A. Arch. Neurol. 1:522–530.

McCARTHY, J. H., JR., J. L. MEUSCHKE, W. H. FICKLIN, and R. E. LEARNED. 1970. Mercury in the atmosphere. U.S. Geol. Surv. Prof. Paper 713:37–39. U.S. Government Printing Office, Washington, D.C.

MIETTINEN, J. K. 1973. Absorption and elimination of dietary mercury (Hg^{2+}) and methylmercury in man. p. 233–243. *In* M. W. Miller and T. W. Clarkson (ed.) Mercury, mercurials and mercaptans. Charles C. Thomas, Springfield, Ill.

MIETTINEN, J. K., T. RHOLA, T. HATTULA, K. RISSANEN, and M. TILLANDER. 1971. Elimination of [203]Hg-methylmercury in man. Ann. Clin. Res. 3:116–122.

MILLER, G. E., P. M. GRANT, R. KISHORE, F. J. STEINKRUGER, F. S. ROWLAND, and V. P. GUINN. 1972. Mercury concentrations in museum specimens of tuna and swordfish. Science 175:1121–1122.

MILLER, V. L., P. A. KLAVANO, and E. CSONKA. 1960. Absorption, distribution, and excretion of phenylmercuric acetate. Toxicol. Appl. Pharmacol. 2:344–352.

MOTTET, N. K. and R. L. BODY. 1974. Mercury burden of human autopsy organs and tissues. Arch. Environ. Health 29:18–24.

MUNNS, R. K. and D. C. HOLLAND. 1971. Determination of mercury in fish by flameless atomic absorption: A collaborative study. J. Ass. Offic. Anal. Chem. 54:202–205.

MUSCAT, V. I., T. J. VICKERS, and A. ANDREN. 1972. Simple and versatile atomic fluorescence system for determination of nanogram quantities of mercury. Anal. Chem. 44:218–221.

MUSHAK, P. 1973. Gas-liquid chromatography in the analysis of mercury (II) compounds. Environ. Health Perspectives (Exp. issue) 4:55–60.

NEAL, P. A., R. H. FLINN, T. I. EDWARDS, W. H. REINHART, J. W. HOUGH, J. M. DALLAVALLE, F. H. GOLDMAN, D. W. ARMSTRONG, A. S. GRAY, A. L. COLEMAN, and B. F. POSTMAN. 1941. Mercurialism and its control in the felt-hat industry. U.S. Public Health Serv. Bull. No. 263. U.S. Government Printing Office, Washington, D.C. 132 p.

NEAL, P. A., R. R. JONES, J. J. BLOOMFIELD, J. M. DALLAVALLE, and T. I. EDWARDS. 1937. A study of chronic mercurialism in hatters' fur-cutting industry. U.S. Public Health Serv. Bull. No. 234. U.S. Government Printing Office, Washington, D.C. 70 p.

NELSON, N. (Chairman). 1971. Hazards of mercury; special report to the Secretary's Pesticide Advisory Committee, Department of Health, Education and Welfare, November 1970. Environ. Res. 4:1–69.

NEWBERNE, P. M. 1974. Mercury in fish: A literature review. CRC Critical Rev. Food Technol. 4:311–335.

NISHIMURA, M., N. URAKAWA, and M. IKEDA. 1971. An autoradiographic study on the distribution of mercury and its transfer to the egg in the laying quail. Jap. J. Pharmacol. 21:651–659.

NORDBERG, G. F. and F. SERENIUS. 1969. Distribution of inorganic mercury in the guinea pig brain. Acta. Pharmacol. Toxicol. 27:269–283.

NORDBERG, G. F. and S. SKERFVING, 1972. Metabolism, p. 29–91. *In* L. Friberg and J. Vostal (ed.) Mercury in the environment; an epidemiological and toxicological appraisal. CRC Press. Cleveland, Ohio.

NORSETH, T. 1969. Studies of intracellular distribution of mercury, p. 408–419. *In* M. W. Miller and G. G. Berg (ed.) Chemical fallout. Current research on persistent pesticides. Charles C. Thomas, Springfield, Ill.

NORSETH, T. 1971. Biotransformation of methyl mercuric salts in the mouse studied by specific determination of inorganic mercury. Acta. Pharmacol. Toxicol. 29:375–384.

NORSETH, T. and T. W. CLARKSON. 1970. Studies on the biotransformation of ^{203}Hg-labeled methylmercury chloride in rats. Arch. Environ. Health 21:717–727.

NORSETH, T. and T. W. CLARKSON. 1971. Intestinal transport of ^{203}Hg-labeled methylmercury chloride. Role of biotransformation in rats. Arch. Environ. Health 22:568–577.

NOVICK, S. 1969. Mercury contamination in food, air, and water. Environment 11(4):3–9.

OHI, G., S. NISHIGAKI, H. SEKI, Y. TAMURA, T. MAKI, H. MAEDA, S. OCHIAI, H. YAMADA, Y. SHIMAMURA and Y. YAGYN. 1975. Interaction of dietary methylmercury and selenium on accumulation and retention of these substances in rat organs. Toxicol. Appl. Pharmacol. *32*:527–533.

OHI, G., S. NISHIGAKI, H. SEKI, Y. TAMURA, T. MAKI, H. KONNO, S. OCHIAI, H. YAMADA, Y. SHIMAMURA, I. MIZOGUCHI, and H. YAGYU. 1976. Effect of selenium in tuna and selenite in modified methylmercury intoxication. Environ. Res. *12*:49–58.

OLSON, K. and G. M. BOUSH. 1975. Decreased learning capacity in rats exposed prenatally and postnatally to low doses of mercury. Bull. Environ. Contam. Toxicol. *13*:73–79.

OSTLUND, K. 1969. Studies on the metabolism of methylmercury and dimethyl mercury in mice. Acta. Pharmacol. 27 (suppl 1):1–132.

PAPPAS, E. G. and L. A. ROSENBERG. 1966. Determination of submicrogram quantities of mercury by cold vapor atomic absorption photometry. J. Ass. Offic. Anal. Chem. 49:782–792.

PEAKALL, D. B. and R. J. LOVETT. 1972. Mercury: Its occurrence and effects in the ecosystem. Bioscience 22:20–25.

PEKKANEN, T. J. and K. SALMINEN. 1973. Inductive effects of methylmercury on the hepatic microsomes of mice. Acta. Pharmacol. Toxicol. 32:289–292.

PENNINGTON, J. W. and G. N. GREENSPOON. 1958. Mercury, p. 771–791. *In* U.S. Department of Interior, Bureau of Mines. Minerals Yearbook 1955. Vol. I. U.S. Government Printing Office, Washington, D.C.

PERMANENT COMMISSION AND INTERNATIONAL ASSOCIATION ON OCCUPATIONAL HEALTH, Task Group on Metal Accumulation. 1973. Accumulation of toxic metals with special reference to their absorption, excretion, and biological half-time. Environ. Physiol. Biochem. 3:65–107.

PETERSON, C. L., W. L. KLAWE, and G. D. SHARP. 1973. Mercury in tunas: A review. Fish. Bull. 71:603–613.

PHILLIPS, R. and H. CEMBER. 1969. The influence of body burden of radiomercury on radiation dose. J. Occup. Med. 11:170–174.

PILLAY, K. K. S., C. C. THOMAS, JR., J. A. SONDEL, and C. M. HYCHE. 1971. Determination of mercury in biological and environmental sample by neutron activation analysis. Anal. Chem. 43:1419–1425.

PLATONOW, N. 1968. A study of the metabolic fate of ethylmercuric acetate. Occup. Health Rev. 20(1):1–8.

PLATONOW, N. 1968. A study of the metabolic fate of methylmercuric acetate. Occup. Health Rev. 20(1):9–19.

PLUMMER, F. R. and B. E. BARTLETT. 1975. Mercury distribution in laying hens fed whole meal supplement. Bull. Environ. Contam. Toxicol. 13:324–329.

POTTER, S., and G. MATRONE. 1964. Effect of selenite on the toxicity of dietary methylmercury and mercuric chloride in the rat. J. Nutr. 104:638–647.

PRICKETT, C. S., E. P. LAUG and F. M. KUNZE. 1950. Distribution of mercury in rats following oral and intravenous administration of mercuric acetate and phenylmercuric acetate. Proc. Soc. Exp. Biol. Med. 73:585–588.

PROHASKA, J. R. and H. E. GANTHER. 1977. Interactions between selenium and methylmercury. Chem.-Biol. Interact. 16:155–167.

RAHOLA, T., T. HATTULA, A. KOROLAINEN, and J. K. MIETTINEN. 1971. The biological half-time of inorganic mercury (Hg^{2+}) in man. Scand. J. Clin. Lab. Invest. 27 (Suppl. 116):77. (Abstr.)

RAINS, T. C. and O. MENIS. 1972. Determination of submicrogram amounts of mercury in standard reference materials by flameless atomic absorption spectrometry. J. Ass. Offic. Anal. Chem. 55:1339–1344.

RAMEL, C. 1969. Genetic effects of organic mercury compounds: I. Cytological investigations on Allium roots. Hereditas 61:208–230.

RAMEL, C. 1972. Genetic Effects, p. 169–181. In L. Friberg and J. Vostal (ed.) Mercury in the environment; an epidemiological and toxicological appraisal. CRC Press, Cleveland, Ohio.

RAMEL, C. and J. MAGNUSSON. 1969. Genetic effects of organic mercury compounds. II. Chromasome segregation in the Drosophila melanogaster. Hereditas 61:231–254.

REYNOLDS, W. A. and R. M. PITKIN. 1975. Transplacental passage of methylmercury and its uptake by primate fetal tissue. Proc. Soc. Exp. Biol. Med. 148:523–526.

ROTHSTEIN, A. 1973. Mercaptans, the biological targets for mercurials, p. 68–95. In M. W. Miller and T. W. Clarkson (ed.) Mercury, mercurials, and mercaptans. Charles C. Thomas, Springfield, Ill.

ROTHSTEIN, A. and A. D. HAYES. 1960. The metabolism of mercury in the rat studied by isotope techniques. J. Pharmacol. Exp. Therapeutics 130:166–176.

ROTTSCHAFER, J. M., J. D. JONES, and H. B. MARK, JR. 1971. A simple, rapid method for determining trace mercury in fish via neutron activation analysis. Environ. Sci. Technol. 5:336–338.

ROWLAND, I. R., P. GRASSO, and M. J. DAVIES. 1975. The methylation of mercuric chloride by human intestinal bacteria. Experientia *31*:1064–1065.

SAHA, J. G. 1973. Significance of mercury in the environment. Residue Rev. 42:103–163.

SAHA, J. G., Y. W. LEE, R. D. TINLINE, S. H. F. CHINN and H. M. AUSTENSON. 1970. Mercury residues in cereal grains from seeds or soil treated with organomercury compounds. Can. J. Plant. Sci. 50:597–599.

SAND, P. F., G. B. WIERSMA, H. TAI and L. J. STEVENS. 1971. Preliminary study of mercury residues in soil where mercury seed treatments have been used. Pest. Monit. J. 5(1):32–33.

SAYLER, G. S., J. D. NELSON, JR., and R. R. COLWELL. 1975. Role of bacteria in bioaccumulation of mercury in the oyster *Crassostrea virginica*. Appl. Microbiol. *30*:91–96.

SCHAFER, M. L., V. RHEA, J. T. PEELER, C. H. HAMILTON and J. E. CAMPBELL. 1975. A method for estimation of methyl mercuric compounds in fish. J. Agr. Food Chem. *23*:1079–1083.

SEEGER, R. 1976. Mercury content of mushrooms. Z. Lebensm. Unters.-Forsch. *160*:303–312.

SEEGER, R. 1977. Quicksilver in jungen und atten Pilzen und in Pilzsporen. Deutsch. Lebensm.-Rundsch *73*:160–162.

SELL, J. L. and F. G. HORANI. 1977. Influence of selenium on toxicity and metabolism of methylmercury in chicks and quail. Nutr. Rep. Intern. *14*:439–447.

SHEPARD, D. A. E. 1976. Methylmercury poisoning in Canada. Can. Med. Ass. J. *114*:459–463.

SHEPARD, M., S. SCHUMANN, R. H. FLINN, J. W. HOUGH, P. A. NEAL. 1941. Hazard of mercury vapor in scientific laboratories. J. Res. Nat. Bureau of Standards (U.S.). 26:357–375.

SHULTZ, C. D., D. CREAR, J. E. PEARSON, J. B. RIVERS and J. W. HYLIN. 1976. Total and organic mercury in the Pacific blue marlin. Bull. Environ. Contam. Toxicol. *15*:230–234.

SIBBETT, D. J., R. H. MOYER, and G. H. MILLY. 1972. Emission of mercury from latex paints. Amer. Chem. Soc., Div. Water, Air Waste Chem., General Paper 12(1):20–26. (Chem. Abstr. 80:18957d).

SILBERBERG, I., L. PRUTKIN, and M. LEIDER. 1969. Electron microscopic studies of transepidermal absorption of mercury. Arch. Environ. Health 19:7–14.

SJOSTRAND, B. 1964. Simultaneous determination of mercury and arsenic in biological and organic materials by activation analysis. Anal. Chem. 36:814–819.

SKERFVING, S. 1974. Methylmercury exposure, mercury levels in blood and hair, and health status in Swedes consuming contaminated fish. Toxicology *2*:3–23.

SKERFVING, S. B. and J. F. COPPLESTONE. 1976. Poisoning caused by the consumption of organomercury-dressed seed in Iraq. Bull. W.H.O. 54:101–112.

SKERFVING, S., A. HANSSON, and J. LINDSTEN. 1970. Chromosome breakage in humans exposed to methylmercury through fish consumption. Arch. Environ. Health 21:133–139.

SMART, N. A. 1968. Use and residues of mercury compounds in agriculture. Residue Rev. 23:1–36.

SMART, N. A. and A. R. C. HILL. 1969. Determination of mercury residues in potatoes, grain, and animal tissues using perchloric acid digestion. Analyst 94:143–147.

SMITH, R. G. 1972. Methods of analysis for mercury and its compounds: A review. p. 97–136. In R. Hartung and B. D. Dinman (ed.) Environmental mercury contamination. Ann Arbor Science Pub. Inc., Ann Arbor, Mich.

SMITH, R. G., A. J. VORWALD, L. S. PATIL, and T. F. MOONEY, JR. 1970. Effects of exposure to mercury in the manufacture of chlorine. Amer. Ind. Hyg. Ass. J. 31:687–700.

SNYDER, R. D. 1971. Congenital mercury poisoning. New Eng. J. Med. 284:1014–1016.

SNYDER, R. D. and D. F. SEELINGER. 1976. Methylmercury poisoning: Clinical follow-up and sensory nerve conduction studies. J. Neurol. Neurosurg. Psychiat. 39:701–704.

SOLLMANN, T. and N. E. SCHREIBER. 1936. Chemical studies of acute poisoning from mercury bichloride. Arch. Intern. Med. 57:46–62.

SPURGEON, D. 1976. Mercury poisoning in Ontario. Nature 260:476.

STEWART, D. J. 1970. Mercury hearing before the State of Wisconsin Department of Natural Resources. October 12.

STIJVE, T. and R. ROSCHNIK. 1974. Effects of mercury from fish. Trav. Chim. Aliment. Hyg. 65:209.

STILLINGS, B. R., H. LAGALLY, P. BAUERSFELD and J. SOARES. 1974. Effect of cystine, selenium and fish protein on the toxicity and metabolism of methylmercury in rats. Toxicol. Appl. Pharmacol. 30:243–254.

STOCK, A. E. 1934. Cited from Goldwater, L. J. 1971. Mercury in the environment. Sci. Amer. 224(5):15–21.

STOCK, A. E. 1938. Cited from Goldwater, L. J. 1971. Mercury in the environment. Sci. Amer. 224(5):15–21.

STOEWSAND, G. S., J. L. ANDERSON, W. H. GUTENMANN, and D. J. LISK. 1977. Form of dietary selenium in mercury and selenium tissue retention and egg production in Japanese quail. Nutr. Rep. Intern. 15:81–87.

STOEWSAND, G. S., C. A. BACHE and D. J. LISK. 1974. Dietary selenium protection of methylmercury intoxication of Japanese quail. Bull. Environ. Contam. Toxicol. 11:152–156.

SUMINO, K., R. YAMAMOTO, and S. KITAMURA. 1977. A role of selenium against methylmercury toxicity. Nature 268:73–74.

SUZUKI, T. 1969. Neurological symptoms from concentration of mercury in the brain, p. 245–257. In M. W. Miller and G. G. Berg (ed.) Chemical fallout. Current research in persistent pesticides. Charles C. Thomas, Springfield, Ill.

SUZUKI, T., T. MIYAMA, and H. KATSUNUMA. 1971. Comparison of mercury contents in maternal blood, umbilical cord blood, and placental tissues. Bull. Environ. Contamination Toxicol. 5:502–508.

SUZUKI, T., T. TAKEMOTO, H. KASHIWAZAKI, M. TOGO, H. TOYO-KAWA and T. MIYAMA. 1976. Man, fish and mercury on small islands in Japan. Tohoku J. Exp. Med. 118:181–198.

SWENSSON, A. 1952. Investigations on the toxicity of some organic mercury compounds which are used as seed disinfectants. Acta. Med. Scand. 143:365–384.

SWENSSON, A., K. D. LUNDGREN, and O. LINDSTROM. 1959. Retention of various mercury compounds after subacute administration. Arch. Ind. Health 20:467–472.

SWENSSON, A. and U. ULFVARSON. 1968. Distribution and excretion of various mercury compounds after single injection in poultry. Acta. Pharmacol. Toxicol. 26:259–272.

SWENSSON, A. and U. ULFVARSON. 1968. Distribution and excretion of mercury compounds in rats over a long period after a single injection. Acta. Pharmacol. Toxicol. 26:273–283.

TAKAHASHI, H. and K. HIRAYAMA. 1971. Accelerated elimination of methylmercury from animals. Nature 232:201–202.

TAKEDA, Y., T. KUNUGI, O. HOSHINO, and T. UKITA. 1968. Distribution of inorganic, aryl, and alkyl mercury compounds in rats. Toxicol. Appl. Pharmacol. 13:156–164.

TAKEUCHI, T. 1968. Pathology of Minamata disease, p. 141–228. In M. Kutsuna (ed.) Minamata disease. Study group of Minamata Disease. Kumamoto University, Japan.

TAKIZAWA, Y. 1970. Studies on the Niigata episode of Minamata disease outbreak investigation of causitive agents of organic mercury poisoning in the district along the river Agano. Acta. Med. Biol. 17:293–297. (Chem. Abstr. 74:109862m).

TANNER, J. T. and W. S. FORBES. 1975. Determination of mercury in total diet samples by neutron activation. Anal. Chim. Acta. 74:17–21.

TATTON, J. O. G. and P. J. WAGSTAFFE. 1969. Identification and determination of organomercurial fungicide residues by thin-layer and gas chromatography. J. Chromatogr. 44:284–289.

TEENY, F. M. 1975. Rapid method for the determination of mercury in fish tissue by atomic absorption spectroscopy. J. Agr. Food Chem. 23: 668–671.

THOMERSON, D. R. and K. C. THOMPSON. 1974. Recent developments in atomic absorption spectrometry. Amer. Lab. 6(3):53–61.

THOMPSON, K. C. and G. D. REYNOLDS. 1971. The atomic-fluorescence determination of mercury by the cold vapor technique. Analyst 96:771–775.

THORPE, V. A. 1971. Determination of mercury in food products and biological fluids by aeration and flameless atomic absorption spectrophotometry. J. Ass. Offic. Anal. Chem. 54:206–210.

TOKUOMI, H. 1968. Clinical investigations on Minamata disease. A. Minamata disease in human adult, p. 37–72. In M. Kutsuna (ed.) Minamata disease. Study group of Minamata disease. Kumamoto University, Japan.

TRUELOVE, J. F., R. F. WILLES and I. C. MUNRO. 1975. Toxic effects of methylmercury in infant monkeys (*Macaca irus*). Toxicol. Appl. Toxicol. *33*:125–126.

TRYPHONAS, L. and N. O. NIELSEN. 1970. The pathology of Arylmercurial poisoning in swine. Can. J. Comparative Med. 34:181–190.

TUREKIAN, K. K. and K. H. WEDEPOHL. 1961. Distribution of the elements in some major units of the earth's crust. Geol. Soc. Amer. Bull. 72:175–192.

TURNEY, W. G. 1972. The mercury pollution problem in Michigan, p. 29–31. *In* R. Hartung and B. D. Dinman (ed.) Ann Arbor Science Pub. Inc., Ann Arbor, Mich.

ULFVARSON, U. 1969. The effect of the size of the dose on the distribution and excretion of mercury in rats after single intravenous injection of various mercury compounds. Toxicol. Appl. Pharmacol. 15:1–8.

U.S. ENVIRONMENTAL PROTECTION AGENCY, Office of Air and Water Programs, Office of Air Quality Planning and Standards. 1973. Control techniques for mercury emissions from extraction and chloralkali plants. Pub. AP-118, 60 p.

VERSCHUUREN, H. G., R. KROES, E. M. DEN TONKELAAR, J. M. BERKVENS, P. W. HELLEMAN, A. G. RAUWS, P. L. SCHULLER and G. J. VAN ESCH. 1971. Toxicity of methylmercury chloride in rats. I. Short-term study. II. Reproduction study. III. Long-term toxicity study. Toxicology *6*:85–96; 97–106; 107–123.

VIOLA, P. L. and G. B. CASSANO. 1968. The effect of chlorine on mercury vapor intoxication. Autoradiographic study. Med. Lavoro 59:437–444. (Chem. Abstr. 70:55852e).

WATTS, J. O., K. D. BOYER, A. CORTEZ and E. R. ELKINS, JR. 1976. A simplified method for the gas chromatographic determination of methylmercury in fish and shellfish. J. Assoc. Off. Anal. Chem. 59:1226–1233.

WEAST, R. C. 1974. Handbook of chemistry and physics. 55th ed. CRC Press, Cleveland, Ohio.

WEBB, J. L. 1966. Mercurials, p. 729–985. *In* J. L. Webb, Enzyme and metabolic inhibitors, Vol. II. Academic Press, New York.

WEED, L. A. and E. E. ECKER. 1931. The utility of phenylmercurynitrate as a disinfectant. J. Infectious Dis. 49:440–449.

WEINER, I. M., R. I. LEVY, and G. H. MUDGE. 1972. Studies on mercurial diuresis: Renal excretion, acid stability and structure-activity relationships of organic mercurials. J. Pharmacol. Exp. Therapeutics 138:96–112.

WEISS, H. V., M. KOIDE, and E. D. GOLDBERG. 1971. Mercury in a Greenland ice sheet: Evidence of recent input by man. Science 174:692–694.

WELCH, S. O. and J. H. SOARES, JR. 1976. The protective effect of vitamin E and selenium against methylmercury toxicity in the Japanese quail. Nutr. Rep. Internat. *13*:43–51.

WENNINGER, J. A. 1965. Direct microdetermination of mercury in color additives by the photometric-mercury vapor procedure. J. Ass. Offic. Anal. Chem. 48:826–832.

WERSHAW, R. L. 1970. Sources and behavior of mercury in surface waters. U.S. Geol. Surv. Prof. Paper 713:29–31. U.S. Government Printing Office, Washington, D.C.

WESTERMARK, T. 1965. Kvicksilverfragan i Sverige, p. 25–76. Kvicksilverkonfersensen. Stockholm, Sweden.

WESTERMARK, T. and B. SJOSTRAND. 1960. Activation analysis of mercury. Int. J. Appl. Radiat. Isotopes 9:1–15.

WESTÖO, G. 1966. Determination of methylmercury compounds in foodstuffs: I. Methylmercury compounds in fish, identification and determination. Acta. Chem. Scand. 20:2131–2137.

WESTÖO, G. 1967. Determination of methylmercury compounds in foodstuffs. II. Determination of methylmercury in fish, egg, meat, and liver. Acta. Chem. Scand. 21:1790–1800.

WESTÖO, G. 1968. Determination of methylmercury salts in various kinds of biological material. Acta. Chem. Scand. 22:2277–2280.

WESTÖO, G. 1969. Methylmercury compounds in animal foods, p. 75–93. In M. W. Miller and G. G. Berg (ed.) Chemical fallout. Current research in persistent pesticides. Charles C. Thomas, Springfield, Ill.

W.H.O. Conference on intoxication due to alkylmercury-treated seed. 1976. Bull W.H.O. 53 (Suppl.):1–138.

WILLISTON, S. H. 1968. Mercury in the atmosphere. J. Geophys. Res. 73:7051–7055.

WOOD, J. M., F. S. KENNEDY, and C. G. ROSEN. 1968. Synthesis of methylmercury compounds by extracts of a methanogenic bacterium. Nature 220:173–174.

WORF, G. L. 1970. Mercury hearing before the State of Wisconsin Department of Natural Resources. October 12.

YAMAGUCHI, S. and H. NUNOTANI. 1974. Transplacental transport of mercurials in rats at subclinical dose levels. Environ. Physiol. Biochem. 4:7–15.

YOSHINO, Y., T. MOZAI, and K. NAKAO. 1966. Biochemical changes in the brain in rats poisoned with an alkylmercury compound with special reference to the inhibition of protein synthesis in brain cortex slices. J. Neurochem. 13:1223–1230.

12

Trace Metal Problems with Industrial Waste Materials Applied to Vegetable Producing Soils

G. S. Stoewsand

Developed, as well as developing, countries building sewage and electric plants face the same problem in disposal of the accumulating waste materials. Sludge from sewage treatment plants, and fly ash from coal burning electric power plants account for upwards of 100 million tons (Furr, *et al.* 1976b) and 29 million tons (Brackett, 1970) respectively, annually in the United States. Ocean dumping of these materials, incineration of sludge, lagoons, or simple landfills (Figure 12.1) all present special problems dealing mainly with air and water pollution.

Although disposal of sludge and fly-ash on agricultural soils seems an excellent way of recycling nutrients and trace elements, crop growth may be impaired and/or excessive accumulation may occur creating a subtle health hazard for consumers. A number of studies and in-depth reviews (Lisk, 1971, 1972; Chaney, 1973; Page, 1974; Cottenie, *et al.*, 1976) have been made on the complex interactions of heavy metals in soils and their translocation to plants, but few reports have dealt with the complete food chain, i.e., consumption of these plants by man or animals. Most feeding studies are conducted by adding soluble metal salts to diets at levels greatly exceeding those obtainable in plant tissue. Results obtained for single ion, dose-reponse studies may not be entirely appropriate. A characteristic shared by all heavy metals is the ability to form

423

FIG. 12.1.   FLY ASH PRODUCED AT MILLIKEN STATION, LANSING, NY BEING DUMPED INTO NEARBY LANDFILL

complexes and chelates by reaction with organic ligands in which the ligands donate electrons to form a chemical bond with the heavy metal cation. With the exception of mercury, arsenic, and selenium, very little is known about most of the natural chemical forms of dietary elements and heavy metals such as lead and cadmium (Clarkson, 1977). Chemically, "heavy metals" refer to about 40 metals having a density greater than 5 (Passow et al., 1961). However, by common usage the term includes most toxic metals.

The toxicity of heavy metals is a result of their binding to active sites of important enzyme systems in the cells and to some ligands in the cell membranes of animals and man. Yet, it is difficult to predict the toxic consequences of the presence of dietary heavy metals, since so many factors play a role in determining the binding to active sites (Passow, et al., 1961). Indeed, it has been shown that although young animals appear more resistant than adults to the toxic effects of heavy metals, there is a higher absorption rate, lower excretion, and a special organ distribution in the young (Jugo, 1977). Heavy metal interactions with dietary copper, zinc, iron, calcium, and selenium play an important part in both acute and chronic heavy metal toxicities (Bunn and Matrone, 1966; Bremner, 1974). Dietary factors such as quality and quantity of protein influences lead toxicity (Shackman, 1974; Mylroie, Moore and Erogbogbo, 1977).

The kidneys are a target organ for many metals. Arsenic, cadmium, gold, lead, mercury, and uranium cause tubular necrosis, oliguria and renal failure in cases of acute toxicity. Chronic exposure to lead, cadmium, mercury and uranium has produced significant nephrotoxic effects probably due to malfunctioning sulfhydryl enzymes especially the dehydrogenases (Beliles, 1975). The concern of toxicologists over the past few years has focused on the general environmental pollution and continuous low concentration exposure of heavy metals to entire populations. Vegetables growing on waste materials could contribute to this exposure, if heavy metals move from the soil to the plant and eventually to the consumer.

AVAILABILITY OF METALS

The diversity of metal associations in soils and the number of interrelated parameters which can affect their reactions and mobility result in an exceedingly complex, dynamic and still incompletely understood systems. The availability of metals to plants is altered by the structure, purity and abundance of clay materials; the composition and abundance of organic matter; the presence of essential nutrient elements, moisture, drainage, pH, temperature, the chemistry of the metal of interest, the nature and abundance of other metals, the type of plant, maturity, and nature of the plant growth (Lisk, 1971).

Essential metals as nutrients for plants and animals have been studied in greater detail than non-essential metals. Allaway (1968) has published an extensive account of elements moving from soils through plants and eventually to the consumer. Plants may mature quite normally but contain insufficient cobalt, chromium, copper, manganese, selenium and zinc for adequate animal or human nutrition. Other plants may also mature normally but contain toxic levels of selenium, cadmium, molybdenum or lead. Plants usually exclude arsenic, beryllium, nickel, zinc and mercury due to minimal absorption of the metals from the soil. Lead in soil is taken up more poorly by young than mature plants. Plants may concentrate chromium in their roots rather than tops. Zinc is present in soybeans in rather sufficient amounts but is unavailable to the consumer due to its fixation as a phytate.

Considering metals in soils and plants from either a toxicologic or nutritional standpoint the same questions could be asked (Lisk,

1972). Is the metal in the soil? Is it available to plants? Where in the plant does it accumulate? Is it available to the consumer in an assimilable form? What are the effects of plant maturing or food processing on its digestibility? What other dietary factors, including other metals, can enhance or inhibit absorption and metabolism? Is it stored for long periods of time or excreted rapidly? How is it excreted?

The human body burden and daily intake of selected elements and metals are shown in Table 12.1. In the absence of specific regulations U.S. Public Health Service interim action guidelines for food or beverages of 7 ppm lead and 0.5 ppm cadmium have been established. These guidelines were formulated to prevent excessive

TABLE 12.1. BODY BURDEN AND HUMAN DAILY INTAKE OF SELECTED ELEMENTS

Element	Human Body Burden (mg/70 kg)	Daily Intake (mg)
Aluminum	100	36.4
Antimony	<90	
Arsenic	<100	0.7
Barium	16	16
Boron	<10	0.01–0.02
Cadmium	30	0.018–0.20
Calcium	1,050,000	
Cesium	<0.01	
Chromium	<6	0.06
Cobalt	1	0.3
Copper	100	3.2
Germanium	trace	1.5
Gold	<1	
Iron	4,100	15
Lead	120	0.3
Lithium	trace	2
Magnesium	20,000	500
Manganese	20	5
Mercury	trace	0.02
Molybdenum	9	0.35
Nickel	<10	0.45
Niobium	100	0.60
Potassium	140,000	
Rubidium	1,200	10
Selenium	15	0.06–0.15
Silver	<1	
Sodium	105,000	
Strontium	140	2
Tellurium	600	0.6
Tin	30	17
Titanium	<15	0.3
Uranium	0.02	
Vanadium	30	2.5
Zinc	2,300	12
Zirconium	250	3.5

Data largely from Schroeder, 1965a. Courtesy of Pergamon Press, Ltd.

oral ingestion, which caused poisoning when these metals were extracted from pottery under acidic conditions. A similar guideline of 0.5 ppm has been established for mercury because of accumulation in fish. Tolerances in specific foods related to their intentional use as economic poisons are 1 ppm of lead, 0.1 to 3.5 ppm of arsenic, and 3 ppm of copper (Beliles, 1975).

SEWAGE SLUDGE

The principle objective of municipal wastewater treatment is the removal of suspended and dissolved organic carbon material. Separation by gravity settling is the major method used to remove suspended organic materials. Aerobic biological oxidation is presently used most widely in removing colloidal and soluble organic waste constituents. An alternate process, i.e. a physical-chemical treatment, consists of chemical coagulation followed in series by contact of the wastewater with granular activated carbon. The resulting effluent is comparable in quality with the conventional system effluent.

The practice of spreading sludge on agricultural lands and recreational areas such as parks and golf courses has been used on a limited scale for decades. A number of exhaustive reviews on the agricultural utilization of sewage sludge has been published (Law, 1968; Page, 1974; Council for Agricultural Science and Technology, Report No. 64, 1976). Beneficial effects of sludge added to land include an effective soil conditioner, enhanced plant nutrient composition including nitrogen and phosphorus, and an improvement to the nutritional quality of foods and animal feeds, e.g., elevated levels of zinc, copper, chromium, and selenium. However, besides the contribution of potentially toxic metals in the food supply there are other problems associated with land application of sludge. These include odor, pathogens, parasites, toxic organic chemicals, and contamination of surface or ground waters. Bastian (1976) has estimated the disposition of municipal sewage sludge in the United States to be 15% ocean dumping, 25% in landfills, 35% incineration, and 25% application to land. Environmental regulations will eliminate ocean dumping by 1981 and the greatest relative increase is expected to occur in land application.

A comparison of selected metals and elements occurring in British sludges (Berrow and Webber, 1972) and U.S. sludges (Furr, et al., 1976b) and typical level soils (Swaine, 1955; Bowen, 1966) are shown in Table 12.2.

TABLE 12.2.   RANGE OF SELECTED METALS AND ELEMENTS IN BRITISH AND U.S. SEWAGE SLUDGES (PPM DRIED MATERIAL)

Element	British[1] Sludges	U.S.[2] Sludges	Soils[3] (Typical Levels)
As	—	3–30	—
Cd	60–1500	7–200	0.1
Cr	40–8800	169–14000	100
Cu	200–800	458–2840	20
Hg	—	4–18	0.3
Mo	2–30	1–40	1
Ni	20–5300	36–562	50
Pb	120–3000	136–7627	30
Se	—	2–9	0.2
Sn	40–700	111–492	3
Sr	80–2000	42–360	300
Ti	1000–4500	1080–4580	4000
V	20–400	19–92	100
W	—	1–100	—
Zn	700–49000	560–6890	80
Zr	30–3000	5–92	500

[1]42 sewage sludges from England and Wales. (Berrow and Webber 1972).
[2]16 sewage sludges from U.S. (Furr, *et al*, 1976b)
[3]From Swaine (1955) and Bowen (1966). Courtesy of Academic Press Inc. (London) Ltd.

Metal concentrations in sludges vary over extremely wide limits. It is of interest to note that in the more limited U.S. survey higher maximum levels of chromium and lead were observed in the sludge obtained from Milwaukee and Philadelphia, respectively.

METALS OF PUBLIC HEALTH SIGNIFICANCE

Although excessive intake of any element, essential or not, can be a hazard to the consumer, potential increases of arsenic, antimony, lead, mercury, and cadmium might be thought to be of greatest health concern in vegetables growing on sludge- or fly ash-amended soils.

Table 12.3 shows heavy metal content of several vegetables grown in Alabama on soil amended with sewage sludge (Giordano and Mays, 1976).

In a study where cabbage was grown on straight, dried sewage sludge from Syracuse, NY (Figure 12.2), dried, ground, and fed at 45% of a complete diet to guinea pigs for 100 days, residues of cadmium, lead, and zinc are presented in Table 12.4 (Stoewsand, Babish and Lisk, 1977).

From these recent studies it appears that sludge-grown leafy vegetables concentrate relatively more cadmium and lead than fruit or root-type vegetables. Also, guinea pigs fed sludge-grown cabbage

TABLE 12.3. HEAVY METAL CONCENTRATIONS IN VEGETABLES GROWN ON SANGO SILT LOAM (pH 6.4) IN ALABAMA WITH OR WITHOUT APPLICATION OF SEWAGE SLUDGE

Vegetable	Treatment[1]	Concentration (ppm dry matter)				
		Zn	Cu	Ni	Cd	Pb
Lettuce	C	47.5	5.2	2.4	0.9	2.4
	S	74.2	9.6	1.7	3.6	3.1
Broccoli	C	86.8	7.5	3.3	0.3	2.4
	S	99.4	12.2	2.1	0.5	2.6
Potato	C	15.7	7.8	0.8	0.1	1.3
	S	19.4	8.6	0.9	0.1	1.4
Tomato	C	25.7	5.0	1.3	0.5	1.6
	S	40.4	9.6	1.3	1.2	1.7
Cucumber	C	40.4	7.7	3.4	0.1	2.6
	S	67.5	14.4	2.2	0.4	2.6
Egg Plant	C	14.8	25.1	1.1	0.5	1.2
	S	22.5	26.5	1.0	1.6	1.3
String Beans	C	45.4	8.1	7.6	0.4	2.5
	S	60.5	8.9	2.8	0.4	2.7

From Giordano and Mays, 1976.
[1]C = control; S = Anaerobically digested sludge from Decatur, Alabama applied at a rate equivalent to 224 metric tons of dry matter per hectare. The heavy metal concentrations in the sludge were in ppm: Zn = 1800; Cu = 730; Ni = 20; Cd = 50; Pb = 530.

had higher residues of cadmium in their liver and kidneys; elevated lead levels were found only in their livers. Although these guinea pigs grew normally and appeared healthy, a recent report by Perry, Erlanger and Perry (1977) indicated that rats could develop hypertension following cadmium feeding of only 0.1 ppm within 6 months of age. After 11 months of exposure these rats had only approximately 1 ppm of cadmium residues in their kidneys.

Cadmium content of tissues increase with cadmium content of the diet, and chronic ingestion induces hypertension (Shroeder, 1964b). Injected cadmium salts causes testicular degeneration in male rats, with protection afforded by large doses of zinc (Parizek and Zahar, 1956). This zinc-cadmium interaction mechanism is not completely understood, but may be envolved with mutual binding to an induced protein, metallothionein (Kagi and Vallee, 1961).

Corn grain harvested from soil amended with Chicago sludge had increased levels of zinc, cadmium, nickel, potassium and phosphorus. Feeding this corn, containing 0.59 ppm Cd, to pheasants increased cadmium kidney deposition to 9.4 ppm dry weight (Hinesly, Ziegler, and Tyler, 1976).

Mice were fed diets for 6 weeks containing 45% Romaine lettuce

FIG. 12.2. CABBAGE GROWING ON SYRACUSE NY SEWAGE SLUDGE

grown on various sludge amended soils. Table 12.5 shows cadmium levels in the lettuce, and liver and kidneys of the mice (Chaney, *et al.*, 1977). Milorganite, i.e. dried, commercial sludge from Milwaukee, produced high cadmium containing lettuce, and the highest elevated levels of cadmium in the liver and kidneys of mice consuming this lettuce. In addition, the Milorganite group of mice showed an induction in liver microsomal enzymes. These enzymes, thought to be required for many *in vivo* detoxication processes, are inducible by a number of ingested organic compounds (Conney, 1967). However, most studies have shown that orally administered cadmium alters hepatic microsomal activities by inhibition rather than stimulation (Becking, 1976). Hansen, *et al.* (1976), observed microsomal enzyme induction in the livers of swine fed corn grown on Chicago sludge-amended soils. Stoewsand, Babish and Lisk (1977) showed enzyme induction in the intestine of guinea pigs fed sludge-grown cabbage (Fig. 12.2). Since municipal sludges contain high levels of polychlorinated hydrocarbons, e.g. 23 ppm in dried sludge from Schenectady, N.Y. (Furr, *et al.*, 1976b), enzyme induction could possibly have been caused by these and other organic waste materials translocated into the plants. More research must be done on the relationship of high metal-containing foods and tissue microsomal activity.

TABLE 12.4. CADMIUM, LEAD AND ZINC LEVELS IN SLUDGE-GROWN CABBAGE AND GUINEA PIG TISSUE AFTER CONSUMING THIS VEGETABLE FOR 100 DAYS[1]

	Dietary Levels[2] (ppm)			Liver Residue (ppm)			Kidney Residue (ppm)		
	Cd	Pb	Zn	Cd	Pb	Zn	Cd	Pb	Zn
Soil-grown cabbage	0.04	0.5	8.1	1.2±.3	1.1±1	85.4±2.6	6.3±1.4	0.5±.04	95.2±4.2
Sludge-grown cabbage	2.4	1.8	98.1	6.9±.9[3]	2.5±.4[3]	98.9±2.8[3]	29.9±2.9[3]	0.6±.04	116.1±3.2[3]

[1]From Stoewsand, Babish and Lisk, 1977
[2]Cabbage fed at 45% of diet
[3]Significantly higher levels (p<0.01)

TABLE 12.5. MEAN RESIDUES OF CADMIUM IN SLUDGE-GROWN LETTUCE AND IN MOUSE LIVER AND KIDNEYS (PPM OF DRIED MATERIAL) WHEN FED LETTUCE DIETS

Sludge Treatment	Lettuce	Liver	Kidneys
None	0.6	4.1	10.8
Baltimore	1.7	7.6	17.5
Washington, DC	2.5	7.8	13.8
Washington, DC (compost)	1.1	2.9	6.9
Organogro	11.6	5.5	17.8
Milorganite	26.4	25.0	48.0

Data from Chaney, Stoewsand, Bache, and Lisk, 1977

Several agronomic conditions are presently known to keep the concentration of cadmium low in foods grown on sludge-treated land: Maintain the soil pH at or above 6.5; grow vegetables that accumulate relatively low concentrations; make only small annual sludge applications; use sludges low in cadmium. Composting of digested or raw sludge appears to reduce the crop uptake as well as tissue residue (Table 12.5; CAST, 1976).

Lead, like cadmium, is a non-essential element that shows a low degree of potential toxicity to vegetables and a high degree of potential toxicity to animals and man. As seen in Table 12.2, concentrations of lead can reach in excess of 7,000 ppm in municipal sludge. Plants take up lead in the ionic form in soils, but show decreases with increased available soil phosphorus, increased cation-exchange capacity, and increased soil pH. As seen in Table 12.3, sludge-grown lettuce contains the highest amount of lead, as compared to fruit or root vegetables. However, this specific sludge was relatively low in lead. If sludge is applied as a surface dressing when a crop is actively growing, there could be an increase in the lead content (Chaney, Hornick and Simon, 1976).

Precautions for growing vegetables with relatively low concentrations of lead with sludge-amended soils could generally be on guidelines as previously stated for cadmium.

Lead has been reported to be high in vegetables and other plants grown near heavily trafficed highways (Page et al. 1971). Table 12.6 shows the amount of lead in tissues of guinea pigs fed 45% crown vetch diets containing 89 ppm of lead and grown near busy roadways (Young et al. 1978).

Although the data in Table 12.6 were obtained with natural high-lead containing diets rather than crops from sludge-treated soils, they illustrate one point, i.e., bone lead increases markedly with highest organ lead residues. This, perhaps, is not an immediate health hazard but could increase the already relatively high body burden of lead (Table 12.1). The entire subject of lead as an environ-

TABLE 12.6.   LEAD IN TISSUES OF GUINEA PIGS FED HIGH LEAD-CONTAINING CROWN VETCH[1]

Tissue	Low Lead Vetch[2]	High Lead Vetch [3]
	ppm	ppm
Bone	0.8	15.2
Kidneys	0.3	3.6
Liver	0.3	2.7
Muscle	0.1	0.2
Spleen	0.5	1.0
Adrenals	0.3	0.5

[1]Harvested near a highway with dense traffic
[2]2 ppm
[3]89 ppm
Data from Young *et al.*, (1978).

mental contaminant affecting humans has been reviewed (Hankin, 1972).

Sludge can contain arsenic in concentrations ranging from 10 to 1,000 ppm. The form of combination of arsenic in sludge is unknown (CAST, 1976). Once incorporated into the soil, arsenic reverts to the chemical form of arsenate, which is usually strongly held by the clay fraction of most soils. Studies with inorganic arsenic have indicated that rates in excess of 90 kg of arsenic per hectare must be applied before phytotoxicity is observed (Jacobs, Keeney and Walsh, 1970).

Plants tend to accumulate arsenic in the roots, and the arsenic level of most of the edible portions of plants is well below a concentration of 2.6 ppm considered maximum for safe animal or human consumption. (U.S. Dept. of Agriculture, 1968). However, Schroeder and Balassa (1966), have reported levels of arsenic in plants up to 3.6 ppm.

Arsenic accumulates in shrimp and other shellfish to very high levels. In an experiment published 43 years ago by Coulson, Remington and Lynch (1935), an individual consumed shrimp containing 1,100 micrograms of arsenic. The arsenic was absorbed and completely excreted in the urine within 24 hours. The chemical form of arsenic in shrimp is not known, but dimethylarsenic acid has been identified as a major form of arsenic in the environment and also identified in human urine (Bramen and Foreback, 1973).

Antimony generally occurs in sewage sludge in low concentrations. In the survey of U.S. sludges (Table 12.2) the highest level of antimony, 4 ppm, was found in the Philadelphia sludge (Furr *et al.*, 1976b). Soil investigations have shown antimony moves from soil to plant more readily under neutral conditions. Furr *et al.* (1976c) showed that Swiss chard grown on sludge-fortified soil at pH 6.5 contained 11.5 ppm of antimony as compared to 6.1 ppm of antimony in chard grown at pH 5.5. Guinea pigs fed the "neutral" chard had the highest antimony deposits in their adrenals, i.e., 9.5 ppm of dried tissue.

The preparation of food containers glazed with antimony enamel has given rise to several outbreaks of poisoning. Mortality occurs from cardiac arrhythmia, with gastro-intestinal disturbances and jaundice (Browning, 1969). However, the inorganic form of antimony is generally poorly absorbed with no persistant accumulation of antimony in the body. The trivalent and pentavalent forms of antimony have been used extensively since very early times in medicinal preparations. Tartar emetic (potassium antimony tartrate) has been used intravenously for the treatment of thousands of cases of schistosomiasis (Liu *et al.* 1958). Oral administration of this compound provokes severe vomiting. The minimum lethal dose in rats is 300 mg/kg (Bradley and Frederick, 1941).

Most sludges are relatively low in mercury, and usually there has been little increase in concentration of mercury in plants from sludge application (CAST, 1976). In the U.S. sludge survey, San Francisco had the highest mercury content, 18.0 ppm (Table 12.2). Van Loon (1974) observed mercury concentrations up to 12.2 ppm in tomato fruit after application of a very high mercury sludge to an alkaline soil. Nothing is known about the form in which mercury occurs in sludge. Although mercury undergoes biological methylation in lake sediments (Jernelow, 1970) with uptake in fish, apparently no methylation has been observed in soils, mud or sewage sludge (Rissanen, *et al.* 1970). Additional general information on mercury in foods is presented in Chapter 11.

FLY ASH

Electric power-generating plants burning pulverized coal trap the residual ash in electrostatic precipitators. This material has an estimated total production of 29 million tons and will expectedly increase with the rising use of coal (Brackett, 1970).

Most fly ash is disposed into landfills (Figure 12.1). A small percentage is used in concrete, ceramic, roadbeds, and as an alkaline amendment to coal mine spoils, refuse banks, or forage and pasture crops (Buttermore, Lawrence and Muter, 1972).

Fly ashes contain a broad spectrum of elements, both essential and non-essential (Davison *et al.* 1974; Von Lehmden *et al.* 1974; Furr *et al.* 1977). The elements contained in fly-ash will depend on the parent coal composition, combustion temperature, and trapping efficiency of the electrostatic precipitators.

Table 12.7 presents the range of toxic metals in U.S. fly ashes surveyed from 21 States. In this study 45 elements were analyzed in these fly ashes with information on combustion equipment, mining location, mining methods, and type of coal. The percent ash after the pulverized coal was combusted ranged from 4 to 26%, and the pH ranged from 4.2 to 11.6. In addition, this survey indicated that all of the fly ashes showed an increase in gamma-emitted radioactivity but no attempt was made to identify specific radionuclides (Furr *et al.* 1977).

If we compare the range of metals and elements of fly ash to sewage sludge (Table 12.2) only arsenic, selenium, strontium, and titanium have a higher maximum level in some of the fly ashes.

Fly ash has been added to soils as a nutrient element source of boron, molybdenum, phosphorus, potassium and zinc for crops (Doran and Martens, 1972; Schnappinger, Martens and Plank, 1975). Furr *et al.* (1976a) determined 42 elements in six vegetables grown on potted soils containing 10% fly ash. Table 12.8 lists the concentrations of some of the elements of most toxicologic concern within the edible portion of these vegetables.

Fly ash applications to crops showed relative increases in the concentration of arsenic in beans, cabbage and carrots; copper in potatoes; mercury in cabbage and tomatoes; molybdenum in beans, cabbage, carrots, potatoes and tomatoes; nickel in tomatoes; antimony in cabbage, onions and potatoes; selenium in all vegetables. Cadmium did not increase in any of the vegetables as a result of fly ash amendment. None of these elements were excessively high in the

TABLE 12.7. RANGE OF SELECTED METALS AND ELEMENTS IN U.S. FLY ASH SURVEY FROM 21 STATES (PPM DRIED MATERIAL)

Element	Concentration
As	5–213
Cd	0.1–3.8
Cr	43–259
Cu	45–616
Hg	0.02–0.7
Mo	6–37
Ni	2–115
Pb	3–241
Se	1–17
Sn	27–334
Sr.	59–3245
Ti	2758–8310
V	68–312
W	3–21
Zn	14–406

From: Furr, *et al.* (1977).

TABLE 12.8.   METAL CONCENTRATION OF VEGETABLES GROWN IN POTS ON SOIL AMENDED WITH 10% FLY ASH[1]

Metal	Beans		Cabbage		Carrots		Onions		Potatoes		Tomatoes	
	C[2]	FA[3]	C	FA	C	FA	C	FA	C	FA	C	FA
As	0.01	0.2	0.1	0.2	0.01	0.2	0.1	0.03	0.1	0.1	0.1	0.1
Cd	0.1	0.1	0.2	0.2	1.1	0.6	0.6	0.4	0.4	0.2	0.1	0.1
Cu	3.2	1.3	3.0	1.1	2.0	2.4	3.4	2.0	3.1	5.5	2.2	3.3
Hg	0.2	0.1	0.1	0.3	0.1	0.2	0.3	0.3	0.1	0.2	0.1	0.3
Mo	0.9	3.2	1.0	2.2	0.2	0.4	0.7	0.4	0.2	0.6	0.5	0.8
Ni	3.9	4.3	1.9	1.3	2.7	2.8	2.2	2.9	0.6	0.7	0.5	0.8
Sb	0.4	0.4	1.4	2.0	0.9	0.4	0.8	2.2	0.4	0.7	2.5	1.1
Se	0.02	0.5	0.01	1.0	0.00	0.1	0.00	0.2	0.01	0.5	0.01	0.3

[1]Fly ash from Milliken Station, Lansing, NY; pH = 5.2
[2]C = Control Soil. Ppm dry weight, As = 2.9; Cd = 0.1; Cu = 2.6; Hg = 0.1; Ni = 17.2; Sb = 0.8; Se = 0.3
[3]FA = Fly ash. Ppm dry weight, As = 139; Cd = 0.5; Cu = 327; Hg = 0.3; Mo = 11.5; Ni = 6.6; Se = 16.8
Data from Furr et al. (1976a)

vegetables studied considering that these values are on a dry weight basis.

Selenium is of interest, owing to its somewhat narrow margin between essentiality and toxicity to animals (Browning, 1969). The extent of plant absorption appears approximately proportional to the rate of application of fly ash. White sweet clover voluntarily growing on beds of fly ash has been found to exceed concentrations of 200 ppm (Gutenmann et al. 1976). When guinea pigs were fed diets containing 45% fly ash-grown white sweet clover for 90 days, only rubidium and selenium were elevated in their tissues. These two elements were among the 28 observed to be higher in the fly ash and 19 elements higher in the clover grown on this fly ash as compared to controls, i.e. gravel soil and clover harvested from this soil (Furr et al.. 1975).

It appears that most plants will take up selenium contained in the fly ash-amended soils, and the selenium is readily available to the consumer. Selenium probably exists in fly ash in the elemental form and has been found to be 300 times more concentrated in fly ash than in the corresponding slag produced during coal combustion (Klein et al. 1975). This elemental selenium is possibly oxidized very slowly to a more water-soluble ionic form and thus available to plants (Furr et al. 1976a). It is known that wheat grown on seleniferous soil in South Dakota incorporates much of the selenium as selenomethionine (Olson et al. 1970). Little is known of the form of selenium in vegetables.

Although fly ash could be recommended as a beneficial soil amendment for increasing some soil nutrients for greater crop productivity, certain problems are obvious. Their pH can range from 4 to 11. Alkaline soils could create plant deficiencies in iron, zinc and manganese. Further studies should be made on the increased radioactivity observed in all of the fly ashes surveyed in the U.S. Also, the effect of fly ashes of varying chemical composition applied continuously on different soil types, should be studied in relation to element uptake of vegetables, and element absorption by experimental animals consuming these plants. Organic compounds could also be present in fly ash with subsequent plant uptake similar to some indications with sludge.

ANALYTICAL TECHNIQUES

There are presently 12 instruments used in analyzing biological and environmental trace elements. These are listed in Table 12.9.

Instrument numbers 1, 5, 6, 9 and 12 can analyze samples directly, although 1, 5 and 6 use surface techniques requiring that the biological samples (tissues, soil, sludge, plants) be very thin and homogeneous. Spark source mass spectrometry only requires the samples to be ashed. All of these techniques are not suitable for all metals, and 1, 2, 4, 6, 7 and 10 cannot adequately analyze multi-elements. Numbers 3 and 7 are limited to metals which form thermally stable volatile complexes. Electron spectroscopy for chemical analysis (No. 6) is the only instrument presently available that can measure the oxidation state of the metal. Neutron activation analysis (No. 9) is not sensitive to lead.

Some examples of detection limits of one of the most popular instruments for element analysis, nonflame atomic absorption spectrometry is presented in Table 12.10. It goes without saying that the choice of the analytical technique is based on the requirement of the

TABLE 12.9.   INSTRUMENTAL METHODS OF ANALYSIS OF TRACE ELEMENTS

1. Auger electron spectroscopy
2. Anodic stripping voltammetry
3. Chemical ionization mass spectrometry
4. Differential pulse polarography
5. Electron microprobe
6. Electron spectroscopy for chemical analysis
7. Gas chromatography
8. Inductively coupled plasma sources for atomic emission spectrometry
9. Neutron activation analysis
10. Nonflame atomic absorption spectrometry
11. Spark source mass spectrometry
12. X-ray fluorescence spectrometry

TABLE 12.10.   DETECTION LIMITS FOR SOME ELEMENTS ANALYZED BY NON-FLAME ATOMIC ABSORPTION SPECTROMETRY

Element	Detection Limit
Cd	0.03 ng/ml
Cr	1.2×10^{-12} g
Hg	8×10^{-11} g
Mo	3×10^{-12} g
Ni	4×10^{-12} g
Pb	0.002 ng/ml
Ti	1×10^{-12} g
V	5×10^{-11} g
Zn	2×10^{-14} g

From Dulka and Risby, 1976. Permission to use data granted by the American Chemical Society.

analysis and upon the preference of the analytical chemist (Dulka and Risby, 1976).

CONCLUSIONS

Utilizing the vast amounts of sewage sludge and fly ash by incorporation into vegetable and other crop-producing soils is an attractive means of recycling vast amounts of waste materials. However, to protect the public's health from high levels of cadmium and lead, as well as pathogens and toxic chemicals, Dr. Charles F. Jelinek, Deputy Associate Director for Technology, Bureau of Foods of the FDA, made recommendations regarding the use of sludge on land used to grow human and animal food. In reference to cadmium and lead, these recommendations were: 1) sludges should not contain more than 20 ppm cadmium or 100 ppm lead; 2) maximum total of sludge which may ever be added to an average soil (cation exchange capacity of 5–15) is 9 pounds cadmium per acre and 900 pounds per acre for lead; 3) crops which are customarily eaten raw should not be planted within three years of the last sludge treatment; 4) sludge should not be applied directly to growing or mature crops where sludge particles may remain in or on the food. These recommendations are extremely timely and important in view of the urgency to adequately and safely dispose of sludge from sewage treatment plants. They are re-stated in the Introduction and Overview. Dr. Jelinek warned against waiting until FDA establishes certain levels for contaminants such as cadmium in certain foods. These pollutants should be decreased at its source (Food Chemical News, Dec. 1976). CAST (1976) has recommended that industrial pretreatment of wastewater from highly industrialized areas could decrease substantially the heavy metal content of sludges. Also, grow selected crops, e.g. not leafy vegetables on sludge-amended soils.

In general, sludge can contain more toxic heavy metals, e.g. cadmium, than fly ash. The daily intake of cadmium in the U.S. already approximates the FAO/WHO tolerable daily limit, i.e. 57–71 micrograms per day. In view of the recent report of hypertension in experimental animals following very low dose levels of cadmium (Perry, *et al.*, 1977), caution is advised with any possible increased intake of this metal. Indeed, until more sensitive signals of toxicity on biological systems are devised, caution is advised on increasing the intake of any heavy metal. One of the critical concerns to public health today is that subtle physiological changes caused by trace metals may go completely undetected, or if detected, be attributed to other causes. The late Dr. Henry A. Shroeder of Dartmouth Medical College, who had spent much of his career in trace metal toxicology, noted that it has been only during the past 100 years that man has been polluting his environment with metals from industrialization, smelting, refining and burning of coal and oil for energy. Toxic metals in the environment are a more insidious problem than other forms of pollution. No metal is degradable.

BIBLIOGRAPHY

ALLAWAY, W. H. 1968. Agronomic controls over the environmental cycling of trace elements. Adv. Agron. 20, 235–274.

BASTIAN, R. K. 1976. Municipal sludge management. 8th Annual Cornell Waste Management Conference, Rochester, NY.

BECKING, G. C. 1976. Trace elements and drug metabolism. Med. Clinics of North Amer. 60, 813–831.

BELILES, R. P. 1975. Metals. *In* Toxicology. The Basic Science of Poisons. L. J. Cassarett and J. Doull (Editors) Macmillan Co., New York.

BERROW, M. L. and WEBBER, J. 1972. Trace Elements in sewage sludge. J. Sci. Food Agric. 23, 43–100.

BOWEN, H. J. M. 1966. Trace elements in Biochemistry. Academic Press, New York.

BRACKETT, C. E. 1970. Production and utilization of ash in the United States. Bureau of Mines Circular 8488.

BRADLEY, W. R. and FREDRICK, W. G. 1941. Toxicity of antimony-animal studies. Indust. Med. 2, 15–22.

BRAMAN, R. S. and FOREBACK, C. C. 1973. Methylated forms of arsenic in the environment. Science 182, 1247–1239.

BREMNER, I. 1974. Heavy metal toxicities. Quart. Rev. Biophys. 7, 75–124.

BROWNING, E. 1969. Toxicity of Industrial Metals, 2nd Edition. Butterworth and Co., London.

BUNN, C. R. and MATRONE, G. 1966. In vivo interactions of cadmium, copper, zinc and iron in the mouse and rat. J. Nutr. 90, 395–399.

BUTTERMORE, W. H., LAWRENCE, W. F. and MUTER, R. B. 1972. Characterization, beneficiation, and utilization of municipal incinerator fly ash. Proc. Third Mineral Waste Utilization Symposium. U.S. Bureau of Mines and Illinois Institute of Technology, Chicago, IL.

CHANEY, R. L. 1973. Crop and food chain effects of toxic elements in sludges and effluents. Nat. Assoc. State Universities and Land Grant Colleges Meeting, Washington, DC.

CHANEY, R. L., HORNICK, S. B. and SIMON, P. W. 1976. Heavy metal relationship during land utilization of sewage sludge in the Northeast. 8th Annual Cornell Waste Management Conference, Rochester, NY.

CHANEY, R. L., STOEWSAND, G. S., BACHE, C. A. and LISK, D. J. 1978. Cadmium deposition and hepatic microsomal induction in mice fed lettuce grown on municipal sludge-amended soil. J. Agric. Food Chem. 26, 992–994.

CLARKSON, T. W. 1977. Factors involved in heavy metal poisoning. Fed. Proc. 36, 1834–1839.

CONNEY, A. H. 1967. Pharmacological implications of microsomal enzyme induction. Pharmacol. Rev. 19, 317–366.

COTTENIE, A., DHAESE, A., and CAMERLYNCK, R. 1976. Plant quality response to uptake of polluting elements. Qual. Plant.-Plant Foods - Hum. Nutr. 26, 293–319.

COULSON, E. J., REMINGTON, R. F. and LYNCH, K. M. 1935. Metabolism in the rat of the naturally occurring arsenic of shrimp as compared with arsenic trioxide. J. Nutr. 10, 255–270.

COUNCIL FOR AGRICULTURAL SCIENCE AND TECHNOLOGY (CAST) 1976. Application of sewage sludge to cropland: Appraisal of potential hazards of the heavy metals to plants and animals. Report No. 64, Iowa State University, Ames.

DAVISON, R. L., NATUSCH, D. F. S., WALLACE, J. R. and EVANS, C. A., JR. 1974. Trace elements in fly ash-dependence on particle size. Environ. Sci. Technol. 8, 1107–1113.

DORAN, J. W. and MARTENS, D. C. 1972. Molybdenum availability as influenced by application of fly ash to soil. J. Environ. Qual. 1, 186–188.

DULKA, J. J. and RISBY, T. H. 1976. Ultratrace metals in some environmental and biological systems. Anal. Chem. 48, 640A–652A.

FOOD CHEMICAL NEWS. 1976. December 20, 40–42.

FURR, A. K., KELLY, W. C., BACHE, C. A., GUTENMANN, W. H., and LISK, D. J. 1976a. Multielement uptake by vegetables and millet grown in pots on fly ash-amended soil. J. Agric. Food Chem. 24, 885–888.

FURR, A. K., LAWRENCE, A. W., TONG, S. S. C., GRANDOLFO, M. C., HOFSTADER, R. A., BACHE, C. A., GUTENMANN, W. H., and LISK, D. J. 1976b. Multielement and chlorinated hydrocarbon analysis of municipal sewage sludges of American cities. Environ. Sci. Technol. 10, 683–687.

FURR, A. K., PARKINSON, T. F., HINRICHS, R. A., VAN CAMPEN, D. R., BACHE, C. A., GUTENMANN, W. H., ST. JOHN, L. E., JR., PAKKALA, I. S., and LISK, D. J. 1977. National survey of elements and radioactivity in fly ashes. Environ. Sci. Technol. 11, 1194–1201.

FURR, A. K., STOEWSAND, G. S., BACHE, C. A., GUTENMANN, W. H., and LISK, D. J. 1975. Multielement residues in tissues of guinea pigs fed sweet clover grown on fly ash. Arch. Environ. Health 30, 244–248.

FURR, A. K., STOEWSAND, G. S., BACHE, C. A., and LISK, D. J. 1976c. Study of guinea pigs fed Swiss chard grown on municipal sludge-amended soil. Arch. Environ. Health 31, 87–91.

GIORDANO, P. M. and MAYS, D. A. 1976. Yield and heavy metal content of several vegetable species grown in soil amended with sewage sludge. 15th Annual Hanford Life Science Symposium, Richland, WA.

GUTENMANN, W. H., BACHE, C. A., YOUNGS, W. D. and LISK, D. J. 1976. Selenium in fly ash. Science 191, 966–967.

HANKIN, L. 1972. Lead poisoning - a disease of our time. J. Milk Food Technol. 35, 86–97.

HANSEN, L. G., DORNER, J. L., BYERLY, C. S., TARARA, R. P., and HINESLY, T. D. 1976. Effects of sewage sludge-fertilized corn fed to growing swine. Amer. J. Vet. Res. 37, 711–714.

HINESLY, T. D., ZIEGLER, E. L., and TYLER, J. J. 1976. Selected chemical elements in tissues of pheasants fed corn grain from sewage sludge-amended soil. Agro. Ecosystems 3, 11–26.

JACOBS, L. W., KEENEY, D. R., and WALSH, L. M. 1970. Arsenic residue toxicity of vegetable crops grown on Plainfield sand. Agron. J. 62, 588.

JERNELOV, A. 1972. Factors in the transformation of mercury to methylmercury. In Environmental Mercury Contamination. R. Harting and D. B. Dinman (Editors). Arbor Science Publ. Inc., Harbor, MI.

JUGO, S. 1977. Metabolism of toxic heavy metals in growing organisms: A review. Environ. Res. 13, 36–46.

KAGI, J. H. R., and VALLEE, B. L. 1961. Metallothionein: Cadmium and zinc-containing protein from equine renal cortex. J. Biol. Chem. 236, 2435–2442.

KLEIN, D. H., ANDREN, A. W., CARTER, J. A., EMERY, J. F., FELDMAN, C., FULKERSON, W., LYON, W. S., OGLE, J. C., TALMI, Y., VAN HOOK, R. I., and BOLTON, N. 1975. Pathways of thirty-seven trace elements through coal-fired power plants. Environ. Sci. Technol. 9, 973–981.

LAW, J. P. 1968. Agricultural utilization of sewage effluent and sludge, U.S. Department of Interior, FWPCA, Washington, DC.

LISK, D. J. 1971. Ecological aspects of metals. N.Y. State J. Med. 71, 2541–2555.

LISK, D. J. 1972. Trace metals in soils, plants and animals. Adv. Agron. 24, 267–325.

LIU, J., HSU, C. Y., and LIU, Y. K. 1958. Therapeutic effect of antimony potassium tartrate in the treatment of schistosomiasis. Clin. Med. J. 76, 11–15.

MYLROIE, A. A., MOORE, L., and EROGBOGBO, U. 1977. Influence of dietary factors on blood and tissue lead concentrations and lead toxicity. Toxicol. Appl. Pharmacol. 41, 361–367.

OLSON, E. E., NOVACEK, E. J., WHITEHEAD, E. I., and PALMER, I. S. 1970. Investigations of selenium in wheat. Phytochem. 9, 1181–1188.

PAGE, A. L. 1974. Fate and effects of trace elements in sewage sludge when applied to agricultural lands. EPA-670/2-74-005, Washington, DC.

PAGE, A. L., GANTJE, T. J., and JOSHI, M. S. 1971. Lead quantities in plants, soil, and air near some major highways in Southern California. Hilgardia 41, 1–31.

PARIZEK, J., and ZAHOR, Z. 1956. Effect of cadmium salts on testicular tissue. Nature 177, 1036–1037.

PASSOW, H. A., ROTHSTEIN, A., and CLARKSON, T. W. 1961. The general pharmacology of the heavy metals. Pharmacol. Rev. 13, 185–224.

PERRY, H. M., JR., ERLANGER, M., and PERRY, E. P. 1977. Hypertension following chronic, very low dose cadmium feeding. Proc. Soc. Expt. Biol. Med. 156, 173–176.

RISSANEN, K., ERKAMA, J., and MIETTINEN, J. K. 1970. Experiments on microbiological methylation of mercury (2+) ion by mud and sludge in anaerobic conditions. Conf. Marine Pollution (FAO-FIR WP/70/E-61), Rome.

SHACKMAN, R. A. 1974. Nutritional influences on the toxicity on environmental pollutants. Arch. Environ. Health 28, 105–113.

SCHNAPPINGER, M. G., JR., MARTENS, D. C., and PLANK, C. O. 1975. Zinc availability as influenced by application of fly ash to soil. Environ. Sci. Technol. 9, 258–261.

SCHROEDER, H. A. 1965a. The biological trace elements. J. Chron. Disease 18, 217–228.

SCHROEDER, H. A. 1965b. Cadmium as a factor in hypertension. Ibid 18, 647–656.

SCHROEDER, H. A. and BALASSA, J. J. 1966. Abnormal trace elements in man. Arsenic Ibid 19, 85–106.

STOEWSAND, G. S., BABISH, J. G. and LISK, D. J. 1977. Activity of intestinal aryl hydrocarbon hydroxylase in guinea pigs fed high element containing sludge-grown cabbage. Fed. Proc. 36, 1146.

SWAINE, D. J. 1955. The trace element content of soils. Commonwlth. Bur. Soils, Harpenden Tech. Commun. No. 48, Great Britain.

U.S. DEPARTMENT OF AGRICULTURE, PESTICIDE REGULATION DIVISION. 1968. Summary of registered agricultural pesticide uses.

VAN LOON, J. C. 1974. Mercury contamination of vegetation due to the application of sewage sludge as a fertilizer. Environ. Letters 6, 211–218.

VON LEHMDEN, D. J., JUNGERS, R. H. and LEE, R. E., JR. 1974. Determination of trace elements in coal, fly ash, fuel oil and gasoline - a preliminary comparison of selected analytical techniques. Anal. Chem. 46, 239–245.

YOUNG, R. W., FURR, A. K., STOEWSAND, G. S., BACHE, C. A., and LISK, D. J. 1978. Lead and other elements in tissues of guinea pigs fed crown vetch grown adjacent to a highway. Cornell Veternarian 68, 521–529.

13

Polychlorinated Biphenyls and Polybrominated Biphenyls in Foods

Mary E. Zabik

POLYCHLORINATED BIPHENYLS—USES AND CHEMICAL REACTIVITY

Polychlorinated biphenyls (PCBs) belong to the group of chlorinated aromatic organic compounds which are of public concern because of their wide occurrence and persistence in the environment and their tendency to accumulate in food chains, with possible adverse effects to animals, including man. PCBs are inert, stable chemicals with excellent dielectric properties and adhesivity. Introduction of halogens into the biphenyl ring reduces flammability so that the higher-chlorinated biphenyls are nonflammable, more viscous and have extremely low volatilities. The biphenyl ring has ten positions which can be chlorinated so a total of 210 possible isomers of PCBs exist:

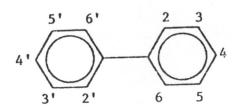

PCBs are manufactured by the addition of gaseous chlorine by weight to the biphenyl ring. Therefore, PCBs are mixtures of compounds with those preparations having a higher amount of chlorine added also having a higher proportion of the more chlorinated isomers. Monsanto was the only American manufacturer and they sold their PCBs under the trade name Aroclor with a four digit number such as 1254. The first two digits indicated the compound was a biphenyl whereas the last two digits represented the percentage of chlorination. Aroclor 1260 had 38% of the biphenyl rings with 6 chlorines and 41% with 7 chlorines while Aroclor 1248 had 40% with 4 chlorines and 36% with 5 chlorines (Thruston, 1971). Foreign manufacturers of PCBs include Germany under the name of Clophen; Italy, Fenchlor; Japan, Kanechlor; France, Phenoclor; United Kingdom, Aroclor; and Russia, Sorol.

PCBs have been used industrially since 1929. Due to their chemical stability, nonflammability, high dielectric constant, and low volatility, PCBs have been used for numerous industrial applications including nonflammable hydraulic and lubricating fluids, heat-exchanger and dielectric fluids, plasticizers for plastics and coolings, and ingredients in caulking compounds, adhesives, paints, printing inks and carbonless carbon paper. Recent EPA estimates indicate that of the roughly 1.25 billion pounds of PCBs introduced into commerce in the United States since 1929, only 55 million pounds have been destroyed by incineration or by environmental degradation, while 290 million pounds are under varying degrees of control in landfills and dumps and 150 million pounds are free in the environment (EPA, 1977).

Environmental contamination by PCBs was first documented in Sweden by Jensen in 1966 but had occurred as unidentified interfering peaks in gas chromatographic traces of organochlorine pesticides in the 1950's and 60's. By 1970, the widespread environmental contamination by PCBs and the possibility of food contamination through leaking heat transfer systems, migration from cardboard manufactured from recycled carbonless carbon paper or paints and silo sealants as well as food chain concentration in aquatic systems was sufficiently documented so that Monsanto voluntarily limited their sales of PCBs to that of only closed-system uses and announced discontinuence of manufacture in 1977. Prior to 1970, 60% of sales had been for closed-systems usage while 25% had been plasticizer applications including carbonless carbon paper. Of the approximately 758 million pounds currently in service, 750 million pounds are in electrical transformers and capacitors (EPA, 1977). The transformers have a 40-year average life while other items have a 15–20 year life so that disposal of PCBs will be a problem for years to come. Destruction of PCBs requires incineration at 1100°C with 3% excess oxygen for 2 seconds or 1500°C with 2% excess oxygen for 1.5 seconds. Improper incineration contributes to widespread distribution by air currents and subsequent fallout, therefore the EPA is proposing strict regulations for disposal and labeling requirements of PCBs.

SOURCES OF PCBs IN THE ENVIRONMENT

Water, Sewage, Sludge and Soil

Water and migration from soil and sewage sludge are possible sources of PCBs in aquatic organisms. Analyses of surface and ground water samples collected in 1971–72 showed only New York, Texas, and Florida had levels above 1 ppb of PCB with a range of 0.1–4.0 ppb (Crump-Wiesner et al. 1974). California, Colorado, Connecticut, Maryland, Massachusetts, Minnesota, New Jersey, Pennsylvania and Virginia had trace levels of PCBs (0.1 to 0.3 ppb) while PCBs were not detected in the waters of the other states. Values for bottom sediments, however, were considerably higher ranging from nondetectable to 3.2 ppm.

Levels of over 400 μg/liter of Aroclor 1254 were reported in June, 1973 in the River Avon, United Kingdom (Kpekata, 1975) but these levels had decreased to <0.3 μg/l by January, 1974. Other industrial rivers in Maine, Massachusetts, and Wisconsin contained levels ranging from 0.2 to 2.8 ppb (Veith, 1972; Veith and Lee, 1971). Green Bay and Milwaukee Harbor had levels of 20 to 120 ppt (Veith, 1972, Veith and Lee, 1971). Nadeau and Davis (1976) reported levels of Aroclor 1016 in the Hudson River of 1.0 to 3.1 ppb with corresponding sediment levels of 6.6 to 2980 ppm. One station, however, had water levels of 2800 ppb and sediment levels of 6700 ppm. Soil migration of Aroclor 1016, induced by percolating water, is affected by both the type of soil and the total volume of effluent (Tucker et al., 1975). Higher effluent volumes resulted in higher soil levels of the PCB while soils high in clay held its PCBs firmly. Iwata and Gunther (1976) reported that between 30 to 50% of the lowest chlorinated PCB component present in a sandy loam soil was translocated to the carrot root, but this was only 3 to 4% for the peak with the greatest chlorination. Municipal refuse can also contribute to PCB contamination (Carnes et al., 1973; Lawrence and Tosine, 1977). Furthermore, Johnson and coworkers (1975b) warn against the potential PCB hazard of feeding aerobically digested municipal garbage to cattle.

The U.S. National Soils Monitoring Program has included analyses for PCBs since 1972 (Carey and Gowen, 1976). Only 0.1% of agricultural soils have had detectable levels of PCBs; nevertheless PCBs have been detected much more frequently in urban and suburban soil. Soil samples from 63% of the 19 metropolitan areas sampled had detectable levels of PCBs. Aroclor 1254 was the most commonly identified PCB, being identified in 40% of the positive samples while Aroclor 1260 was identified in 20% of the positive samples. However, the occurrence of PCBs is less outside the city limits or in the suburban areas of these metropolitan regions.

Industrial Accidents

Nisbet and Sarofim (1972) give several possible modes of transfer of PCBs into the environment, including leaks from sealed transformers and heat exchangers, leaks from hydraulic systems which are only partially sealed, manufacturing spills and vaporization. A leaky heat exchanger resulted in contamination of rice oil in Japan which was responsible for the largest human case of PCB poisoning (Kuratsune et al., 1972); another in contamination of 16,000 tons of chicken feed with levels of 14 to 30 ppm of Aroclor 1242 (Kolbye 1972). Leakage of hydraulic fluid from an air compressor contributed to PCB contamination of Escambia Bay, Florida (Nisbet and Sarofim, 1972).

Silos

Willett and Hess (1975) reviewed the occurrence of polychlorinated biphenyl residues in silos. The coating responsible for most of the PCB-contaminated silos was called Cumar, and was used in the Eastern United States from 1941 to 1970. It was estimated that 5,000 to 6,000 silos have been treated with Cumar although many are no longer in use. Flaking of coating chips containing from 5,000 to 20,000 ppm of Aroclor 1254 resulted in PCBs occurring in silage. Aroclor 1254 is also soluble in silage juices (Skrentny et al., 1971). Once dissolved in the juice, PCBs diffuse into the silage, although most are found within 15 cm of the silo wall or near the base of the silo (Willett, 1974). Fries and coworkers (1974) indicated that it took 40 days for the milk fat of cows fed 200 mg per day of Aroclor 1254 to approach equilibrium levels while the concentration in the milk fat declined 50% within 15 days after discontinuing feeding contaminated feed. Aroclor 1254 elimination rates are similar to DDE (Fries *et al.*, 1972), but the more highly chlorinated PCB isomers are excreted more slowly (Jan *et al.*, 1975).

Paper

A significant portion of PCBs found in paperboard products was contributed by recycled waste paper containing PCB's carbonless "carbon" paper or by printing ink (Stanovick *et al.*, 1973). Thomas and Reynolds (1973) found levels of 0 to 20 ppm PCBs in the 100 samples of paperboard analyzed. Specht (1974) found all but three of the 370 packaging materials analyzed had less than 10 ppm PCB. Storing pork in paper with less than 5 ppm PCBs for 7 days resulted in levels of 0.08 ppm (Tatsuna *et al.*, 1974) but PCB transfer to corn oil, bread, and fried rice crackers was very low. Migration of Aroclor 1242 from food packaging materials to the food has been found to be a vapor-phase phenomenon (Stanovick *et al.*, 1973). Thus, interposing a barrier reduces the migration of 1242 to foods. PVDC-coated paper prevented migration for the 90 day test period while polyethylene film was relatively ineffective as storage time lengthened. Furr and coworkers (1974) reported feeding lactating cows diets of 30% of either newsprint, brown or grey cardboard or computer paper as a cellulose substitute resulted in milk levels of 76 ppb of PCBs and renal fat levels of 15.4 ppm. Nevertheless, a recent study of recycled paper shows that PCB levels in this recycled paper is comparable to that of the first manufacture (D'Arrigo, 1976) so that removal of PCBs from carbonless "carbon" paper has alleviated the problem.

LEVELS OF PCBs IN FOOD

Market Basket Surveys

In 1971–72 a total analyses of 35 market baskets collected in 32 cities with populations ranging from less than 50,000 to one million and more showed trace to 0.012 ppm PCBs in meat, fish and poultry, trace to 0.014 ppm PCBs in cereal grains, and trace to 0.15 ppm in oils, fats, and shortenings. Traces of PCBs were also found in all other food groups, except leafy vegetables and beverages (Manske and Johnson, 1975). From these data mean intakes of PCBs were calculated to be approximately 0.06 mcg/kg/body weight/day or 4.2

mcg/day for a 70 kg man (FDA, 1973). By 1975, the levels of PCBs in the market basket composites had declined to nondetectable levels for all food groups except meat-fish-poultry composites (FDA, 1977). The latter composites contained only trace levels. Nevertheless, individual analyses of various food groups throughout the world has shown measurable levels of PCBs. This is especially true for certain aquatic food sources.

Fish

Table 13.1 summarizes levels of polychlorinated biphenyls found in various locations from 1971 to 1973. Low levels of PCBs ranging from 0.05 to 0.97 ppm have been reported for fish caught in Utah (Smith *et al.*, 1974a). Fish caught in the Great Lakes tended to have much higher levels of PCBs than those in waters connected with the Pacific Ocean or in the Gulf of Mexico. Moreover, there is no evidence of a decline of PCBs in bloaters, Coho salmon, or lake trout from Lake Michigan from 1972 through 1974 (Williford *et al.* 1976). High levels of PCBs, hydrochlorinated hydrocarbon pesticides and the mercury contamination problem in Lake Sinclair led to initiation of the Great Lakes Environmental Contaminants Survey (GLECS) to provide cooperation so that all State and Federal Agencies could better serve the public interest (Bails, 1975). GLECS systematically samples major commercial species, particularly larger specimens which would most likely show highest levels of contamination. The results of the GLECS analyses are compiled in Table 13.2. Not all species are sampled every year and, since it is the objective of this survey to assess the magnitude of the problem, many samples of fish species suspected to have high levels of environmental contaminants are collected from numerous locations and of various sizes while only a few samples of fish species with low levels of contaminants are taken. Food chain magnification of these lipid soluble materials in the aquatic system is evident from the high levels of PCBs found in lake trout. The higher the mean PCB level, the greater the diversity of PCB values reported for fish of varying sizes. For all species, the smaller fish have lower levels of PCBs and other environmental contaminants. The geographic location in which the fish were caught also affects the level of contamination. Fat trout, which are lake trout but with such high levels of body fat and other characteristics that local fishermen recognize them as a distinct entity, caught in the quadrant of Lake Superior North of Grand Marais, Michigan have the highest level of PCB contamination. Since these lipophilic materials concentrate in the adipose tissue,

TABLE 13.1.   PCB LEVELS IN FISH.

Type	Location	Year(s)	Range ppm	Reference
Alewife	Lake Michigan	71	2.5–8.9	Veith, 1975
Bass (white)	Lake Erie	71	1.4–3.5	Carr et al., 1972
Bloaters	Lake Michigan	71	3.4–8.1	Veith, 1975
Carp	Lake Erie	71	0.3–5.3	Carr et al., 1972
Carp	Lake Michigan	71	1.7–11.0	Veith, 1975
Catfish (channel)	Lake Erie	71	1.4–7.8	Carr et al., 1972
Drum (fresh-water)	Lake Erie	71	0.6–1.5	Carr et al., 1972
Mackerel (king)	Gulf of Mexico	71	0.034	Giam et al., 1972
Perch (yellow)	Lake Erie	71	0.2–2.4	Carr et al., 1972
Perch (yellow)	Lake Michigan	71	2.7–10.9	Veith, 1975
Salmon (chinook)	Lake Michigan	71	9.9–24.0	Veith, 1975
Salmon (chinook)	British Columbia	72–73	0.08–0.09	Albright et al., 1975
Salmon (chinook & coho)	Lake Erie	75	2.1–24.6	FDA, 1976
Salmon (coho)	Lake Erie	71	1.0–4.3	Carr et al., 1972
Salmon (coho)	Lake Michigan	71	3.6–17.3	Veith, 1975
Salmon (sockeye)	British Columbia	72–73	N.D.	Albright et al., 1975
Shark	Gulf of Mexico	71	0.008–0.032	Giam et al., 1972
Smelt	Lake Michigan	71	0.7–3.2	Veith, 1975
Squawfish	British Columbia	72–73	N.D.–1.9	Albright et al., 1975
Sucker (large scale)	British Columbia	72–73	N.D.–0.6	Albright et al., 1975
Sucker (red-horse)	Lake Michigan	71	2.8–3.2	Veith, 1975
Sucker (white)	Lake Michigan	71	2.1–10.6	Veith, 1975
Trout (brown)	Lake Michigan	71	6.7–11.9	Veith, 1975
Trout (lake)	Lake Erie	75	10.0–15.0	FDA, 1976
Trout (lake)	Lake Michigan	71	8.1–21.2	Veith, 1975
Trout (lake)	Lake Superior	71	2.7–13.8	Parejko et al., 1975
Trout (lake)	Michigan Hatchery	73	0.1–1.5	Parejko and War, 1977
Trout (rainbow)	British Columbia	72–73	N.D.–0.3	Albright et al., 1975
Trout (rainbow)	Lake Michigan	71	8.8–12.0	Veith, 1975
Trout (steelhead)	Oregon	72	0.10–0.15	Claeys et al., 1975
Tuna	Gulf of Mexico	71	0.036	Giam et al., 1972
Whitefish	Lake Michigan	71	1.5–6.1	Veith, 1975

TABLE 13.2.   SUMMARIES OF MEAN LEVELS OF PCBs FOUND IN FISH ANALYZED BY THE GREAT LAKES ENVIRONMENTAL CONTAMINANT SURVEY.

Species	Lake	1972[a]	1973[a]	1974[b]	1975[c]
Alewife	Michigan	2.4	—[d]	0.8	—
White Bass	Erie	0.8	2.5	2.1	1.5
Brown Trout	Huron	—	3.6	0.9	1.1
Catfish	Erie	4.9	—	3.0	4.0
Catfish	Huron	—	5.0	—	—
Carp	Erie	1.1	9.3	3.8	3.2
Carp	Huron	4.1	—	2.3	1.4
Carp	Michigan	3.9	—	—	—
Chub	Michigan	7.0	3.9	3.7	—
Chub	Superior	—	1.6	0.7	—
Drum	Erie	—	0.9	0.5	0.1
Herring	Superior	—	1.8	0.8	—
Menominee	Erie	—	—	—	—
Menominee	Huron	0.4	—	—	—
Menominee	Michigan	1.1	1.0	—	—
Yellow Perch	Erie	0	0.1	0.03	—
Yellow Perch	Huron	0.02	0.2	0.05	—
Yellow Perch	Michigan	0.4	0.4	—	—
Coho Salmon	Erie	—	—	—	0.4[e]
Coho Salmon	Huron	1.5	—	2.3[e]	—
Coho Salmon	Michigan	11.2	—	1.6[e]	—
Chinook Salmon	Erie	—	—	—	0.3
Chinook Salmon	Huron	—	—	—	1.4
Chinook Salmon	Michigan	12.4	—	—	—
Smelt	Huron	—	0.6	0.6	0.7
Smelt	Michigan	—	1.6	—	—
Suckers	Huron	0.4	—	0.4	0.7
Suckers	Michigan	1.0	0.5	—	—
Lake Trout	Huron	—	—	—	1.6
Lake Trout	Michigan	7.4	5.7	6.1	7.9
Lake Trout	Superior	3.0	2.2	3.6	1.7
Fat Trout	Superior	—	—	—	14.3
Rainbow Trout	Erie	—	—	—	0.7
Walleye	Erie	0.4	0.5	0.2	0.3
Walleye	Huron	—	1.1	0	—
Whitefish	Huron	0.5	0.5	0.2	0.3
Whitefish	Michigan	0.7	0.9	—	—
Whitefish	Superior	—	0.4	0.6	—

[a]Bails 1975.
[b]GLECS 1974.
[c]GLECS 1975.
[d]Species not analyzed that year.
[e]Type of salmon not specified.

fishermen are warned to cut away the fatty tissue before preparing the fish. In addition, the Michigan Department of Natural Resources published a fish eating advisory with sports fishing licenses in 1977 warning fishermen to consume no more than one-half pound per week of fish listed as exceeding the current Food and Drug Administration Temporary Tolerances of PCBs pesticides and mercury and further warned women of child-bearing age against consumption of fish high in PCBs.

PCB levels of 0.02 to 0.1 ppm have been reported for fish from Okinawa while samples from Tokyo Bay had a maximum of 2 ppm (Shimma et al., 1973). Other Japanese fish had levels of 0.07 to 0.11 ppm PCBs (Nakayawa et al., 1975). Fish from Norwegian fjords have PCB levels of 0.3 to 3.9 ppm (Kveseth and Bjerk, 1976) while levels in fish caught in Germany ranged from <0.01 to 0.39 ppm PCBs (Eichner, 1973). Fatty fish from the North Atlantic, the North Sea and the Baltic Sea have been found to contain approximately 1.3 ppm PCBs while non-fatty fish contained 0.2 ppm (Huschenbeth, 1973).

Feeding studies and studies with fish from one source of different ages have established that PCB accumulation increases with age (Bache et al., 1972; Lieb et al., 1974). Zitko and Hutzinger (1976) reported accumulation coefficients of di, tri and tetra chlorobiphenyls as fish uptake from water and food. Accumulation coefficients were much larger from water and decreased with increasing chlorination while those from food increased with increasing chlorination. Unpeeled shrimp have been found to adsorb higher levels of PCBs than peeled shrimp (Khan et al., 1976b).

Dairy Products and Other Foods

Levels of PCBs reported in dairy products are summarized in Table 13.3. Except for two samples, one of Canadian cheese and one of Japanese butter, all levels were less than 0.1 ppm on a total weight basis. A survey of pesticides and PCBs in Canadian eggs showed the average PCBs (calculated as Aroclor 1260) to be 8 ppb with the maximum level to be 27 ppb (Mes et al., 1974). Traces of PCBs were found in 100% of the egg samples. Japanese beef had levels ranging from nondetectable to 0.2 ppm, pork levels from nondetectable to 0.2 ppm, mutton levels from nondetectable to 0.01 ppm while chicken levels ranged from nondetectable to 2.21 ppm (Fujiwara, 1975). Analyses of PCBs in Italian beef indicated ranges of 0.027 to 0.2 ppm (Crisetig et al., 1973, 1975b) with the muscle concentration related to the muscle fat. Norwegian fish oil had PCB levels of 0.64–1.24 ppm (Bjerk, 1973). Few studies have evaluated

TABLE 13.3.   POLYCHLORINATED BIPHENYL LEVELS IN DAIRY PRODUCTS.

Product	Country	Level ppm	Source
Butter	Norway	0.04[a]	Bjerk and Kveseth, 1974
Butter	Japan	N.D.–0.433	Fujiwara, 1975
Butteroil	Canada	0.07–0.17[a]	Frank et al., 1975
Cheese	Italy	0.0–0.169[a]	Crisetig et al., 1975a
Cheese	Italy	0.290–.590[a]	Riva et al., 1973
Cheese	Canada	0.01–0.27	Villeneuve et al., 1973
Cheese	Denmark	0.03–0.05	Villeneuve et al., 1973
Cheese	Holland	0.02–0.08	Villeneuve et al., 1973
Cheese	Greece	0.02	Villeneuve et al., 1973
Cheese	France	0.06	Villeneuve et al., 1973
Cheese	England	0.02–0.03	Villeneuve et al., 1973
Cheese	West Germany	0.05–0.07	Villeneuve et al., 1973
Cheese	Sweden	0.04	Villeneuve et al., 1973
Cheese	Norway	0.05	Villeneuve et al., 1973
Cheese	Switzerland	0.06	Villeneuve et al., 1973
Cheese	Japan	0.02–0.094	Fujiwara, 1975
Milk	Japan	N.D.–0.056	Fujiwara, 1975
Condensed milk	Japan	0.001–0.01	Fujiwara, 1975
Powdered milk	Japan	N.D.–0.02	Fujiwara, 1975

[a]ppm on fat basis.

PCBs in other foods. However, Japanese eggs ranged from non-detectable to 0.143, fruit from nondetectable to 1.0 ppm and un-polished rice from nondetectable to 1.33 ppm (Fujiwara, 1975). Sake had only traces of PCB as the PCBs were absorbed on the steamed rice and were concentrated into the Sake cake (Maekawa and Hiruta, 1974). A study of 25 health foods and comparable "traditional" foods showed that 7 of the health foods contained PCB residues while PCBs were found in only 3 of the traditional foods (Wheeler et al., 1973).

Wildlife

Analyses of wildlife from diverse locations have shown that PCBs are ubiquitous environmental contaminants found in wildlife throughout the globe (Tables 13.4 and 13.5). Fish-eating fowl and mammals tend to have higher levels of these fat soluble contaminants because of food chain magnification. The high levels reported at the upper range in the study by Holden (1973) were found in eggs coming from locations recognized as polluted. City sewage was most often noted as the source for this aquatic contamination. In general, woodcock, pheasant, and chukar had low levels of contamination, although one chukar sample had 7.7 ppm PCBs. PCBs have also

TABLE 13.4.   PCBs IN AVIAN WILDLIFE.

Species	Location	Year(s)	Range ppm	Source
Osprey eggs	Idaho	72–73	N.D.–4.0	Johnson et al., 1975a
Heron eggs	Netherlands	69–71	34–74	Holden, 1973
Heron eggs	United Kingdom	69–71	<.5–7.3	Holden, 1973
Common eider eggs	Denmark	69–71	4.9–9.3	Holden, 1973
Common eider eggs	Finland	69–71	1.8	Holden, 1973
Common eider eggs	Norway	69–71	0.27	Holden, 1973
Common tern eggs	Canada	69–71	2.36–149	Holden, 1973
Diving ducks	Ireland	72	0.13–1.05	Bengtson, 1974
Water fowl	Utah	70	0.004–1.000	Smith et al., 1974b
Chukar	Utah	70	0.030–7.700	Smith et al., 1974b
Pheasant	Utah	71	0.020–0.400	Smith et al., 1974b
Woodcock	United States	70–71	0.0–0.43	Clark & McLane, 1974

been found in arctic fox, seal and polar bears. Although the source of PCB contamination is not documented, migratory animals, as well as ocean, river and air currents are possible sources of PCB contamination. Even though the water solubility of PCBs is extremely low, they do co-distill into the environment and can be transported by air currents to colder regions such as the arctic at which time the water vapor precipitates and the environmental contaminant is there through "fallout". Buhler et al., (1975) studied California seal lions which were found from southern California to the Columbia River in atypical locations, particularly in or near fresh water. Many of these mammals suffered from severe nephritis. PCB levels, however, were found to be similar in the liver, muscle, cerebrum and fat of sick animals to that in healthy animals and, although present, were not thought to have played any role in the 1970–71 leptospirosis epizootic.

LEVELS OF PCBs IN HUMANS

Milk

Human milk from 47 lactating women from industrialized areas of Texas and 32 women from rural areas of New Guinea was reported to be free of PCBs in 1971 (Dyment et al., 1971). Earlier,

TABLE 13.5.   PCBs IN MAMMALIAN WILDLIFE.

Species	Location	Year(s)	Range ppm	Source
Dolphin	California	71–75	147[a]	Taruski et al., 1975
Dolphin	East Coast of U.S.	71–75	37–69[a]	Taruski et al., 1975
Fox	Greenland	72	1.6–3.9	Clausen et al., 1974
Polar bear	Greenland	72	21.0[a]	Clausen et al., 1974
Polar bear	Canada	68–72	0.003–11.8	Bowes & Jonkel, 1975
Porpoise	Greenland	72	1.9–11.4[a]	Clausen et al., 1974
Porpoise	Rhode Island	71–75	74[b]	Taruski et al., 1975
Seal (bearded)	Greenland	72	0.6–3.0[a]	Clausen et al., 1974
Seal (ringed)	Greenland	72	0.6–1.3[a]	Clausen et al., 1974
Seal (hooded)	Greenland	72	0.3–4.9[a]	Clausen et al., 1974
Seal (harp)	Canada	70	Tr–0.42	Frank et al., 1973
Seal (harp)	Gulf of St. Lawrence	71	2–22[b]	Addison et al., 1973
Seal (harbour)	Maine	71	32.0–240.2[b]	Gaskin et al., 1973
Seal (harbour)	Bay of Fundy	71	7.1–63.0[b]	Gaskin et al., 1973
Seals	Canada	68–72	0.05–1.51[b]	Bowes & Jonkel, 1975
Sheep	Greenland	72	1.2	Clausen et al., 1974
Whale	Atlantic Coast	71–75	0.7–114[b]	Taruski et al., 1975
Whale	California	71–75	46[b]	Taruski et al., 1975

[a]ppm of fat basis.
[b]ppm of whole tissue—blubber.

two human adipose tissue samples examined by combined gas chromatography-mass spectrometry had substantial quantities of PCBs ranging from pentachlorobiphenyl to decachlorobiphenyl with at least fourteen isomers and chlorine homologs (Biros et al., 1970). PCBs ranging from 0.04–0.1 ppm were found in eight out of 39 human milk samples from rural Colorado (Savage et al., 1973). In Canada, 96 Ontario women had from 0.1 to 2.5 ppm (fat basis) of PCBs in human milk (Grant et al., 1976) while 43 Japanese women had 0.01–0.31 ppm or 0.5–12.2 ppm (fat basis) of PCBs in their milk and 43 German women had an average of 0.103 ppm or 3.5 ppm (fat basis) (Fujiwara, 1975). Swedish women had low levels of PCBs in human milk with approximately half having 20 ppb and the other half 30 ppb (Westoo, 1974).

Adipose Tissue and Blood

The U.S. National Human Monitoring Program analyzed samples of human adipose tissue collected in the 48 contiguous states on a statistically derived sampling pattern designed to reflect population distributions (Kutz and Strassman, 1976). PCBs exceeded 1 ppm in 35.1 and 40.3% of the tissues collected in 1973 and 1974, respectively, ranging from 0.4 to 3.5 ppm PCBs. People from East North Central, Middle Atlantic, New England, and South Atlantic states generally have average PCB levels exceeding the national average while people from the West North Central, West South Central and Pacific states had average PCB levels below the national average. The gas chromatographic peaks most closely resembled the peak patterns of Aroclor 1254 and 1260 so the PCB isomers with lower degrees of chlorination had apparently been metabolized or excreted. Since the Great Lakes surrounding Michigan represents one of the areas of the United States contaminated with PCBs and significant levels of PCBs have been found in Michigan's rivers and streams, an ancillary study was conducted to correlate fish consumption with PCB levels in the blood of residents of Traverse City, Manistee, Ludington and South Haven, all along Lake Michigan (Humphrey et al., 1976). A highly significant relationship was found between blood PCB levels and fish consumption, with PCB blood levels ranging from 0.007 ppm in persons consuming no fish to a maximum of 0.366 ppm for a person consuming 38.1 kg fish/year. The calculated annual PCB ingestion from fish for the latter person was 84.6 mg/yr. PCBs in these fish were determined from cooked samples provided by the participant. Women living in fishing areas in Japan also had the highest levels of PCBs in their milk, suggesting the major route responsible for PCB accumulation in the human body is the ingestion of fish (Goto, 1975).

Thirty percent of the 172 Canadians studied had 1.0 ppm or more PCBs in their adipose tissue (Grant et al., 1976). Men had 1.020 ± 0.957 ppm PCB, as Aroclor 1260 in their adipose tissue while women had 0.685±0.353 ppm (Mes et al., 1977). Values were lower for people from the central area and were highest for the 26–50 year group. PCBs ranged from 0.1 to 18.04 ppm in the fat of Japanese resident's adipose tissue with levels being higher in fishermen than Japanese with other occupations (Fujiwara, 1975). Japanese blood levels of PCBs ranged from 1.9 to 5.1 ppb (Doguchi and Fukano, 1975) with maternal blood levels being three times that of cord blood (Akiyama et al., 1976).

Yusho Incident

The only major outbreak of human poisoning from ingesting PCB contaminated food occurred in Japan in 1968 and has become known as Yusho. Over 1,000 Japanese were poisoned by rice oil containing 2,000–3,000 ppm of the PCB heat transfer fluid (Kanechlor 400 - 48% chlorine). Yusho victims consumed an average of 2 grams PCB while the minimum dose was 0.5 g (Katasune *et al.*, 1972). Typical clinical findings included chloracne, increased pigmentation, visual impairment due to hypersecretion of the Meibomian glands, and systemic gastrointestinal symptoms, including abdominal pain and disturbances in liver function. Adult patients showed prolonged disease with slow regression of symptoms which suggested PCBs are only slowly metabolized and/or excreted from the human body. Chronic toxicity in man has two distinct effects, that of the skin and that of the liver.

Some babies were born with decreased birth weight and skin discoloration due, presumably, to placental PCB transfer and skin accumulation. The skin discoloration regressed as the children grew but growth rate appeared slower than that of other Japanese children. Average PCB levels in adipose tissue of Yusho victims analyzed from 1969–1972 was 6.3 ppm in the fat with a range of 0.9 to 15.1 ppm (Masuda *et al.*, 1974). This was higher than that in the general population. Much less of the lower chlorinated peaks were seen in Yusho victims at that time. Their adipose tissue, however, showed a very high peak which was thought to be a hexa-or hepta-chloro-biphenyl and had previously been noted in the contaminated rice oil, whereas this peak was low in the normal Japanese population.

TOXIC EFFECTS OF PCBs

Freshly manufactured American PCBs have comparatively low acute toxicity to birds and mammals. In a long term test of various levels of PCBs, numerous effects have been shown. PCBs are moderately toxic to fish and are toxic to some aquatic invertebrates at levels as low as 1 ppb (Nimmo *et al.*, 1975). They are comparatively non-toxic to terrestrial insects, but act synergestically to increase the toxicity of organophosphorus insecticides (Fuhremann and Lichtenstein, 1972).

Effects on Avian Species

Feeding Aroclor 1242 or 1254 at up to 20 ppm does not affect growth or feed efficiency (Hansen *et al.*, 1976a; Britton and Charles 1974). PCBs do not affect egg shell thickness in chickens (Teske *et al.*, 1974, Scott *et al.*, 1975) or doves (Peakall, 1971). Levels of PCBs accumulate in the tissues in proportion to the amount fed. After feeding white Leghorn chickens diets contaminated with Aroclor 1254 from 0.1 to 10.0 mg/kg from 3 days of age to 8 weeks, a linear relationship was found between the level in the extractable fat from adipose tissue, muscle kidney and liver. Depletion studies showed a biological half life of 2.93 weeks (Teske *et al.*, 1974). Clearance rates have been found to be dependent on the particular PCB isomer present and PCBs with 4,4' chlorination have slower clearance rates than ones with chlorination in other postions (Bush *et al.*, 1974).

Hatchability of fertile eggs has been shown to be affected by feeding 20 ppm of 1232, 1242, 1248, 1254 (Briggs and Harris 1973; Lillie *et al.*, 1974). Progeny growth was significantly depressed by feeding all of these PCB mixtures, but only feeding of 1248 to hens caused a significant increase in mortality of the progeny. Feeding 20 ppm of Aroclors 1232, 1242, 1248 and 1254 caused teratogenic effects in chick embryos (Cecil *et al.*, 1974). The most common abnormalities found in the unhatched embryos were edema and unabsorbed yolk. Aroclors 1221 and 1268 did not adversely affect embryonic development, thus adverse effects of PCBs were not directly related to the degree of chlorination. Tumasonis and coworkers (1973) reported leg, toe and neck deformities were present in many of the chicks hatched from egges in which yolk PCB level was 10–15 ppm or more. Hemmorhages present suggested that vascular elements may be targets for PCB action. Feeding SCWL cockerels 50 to 200 ppm affected some cardiovascular and hematological parameters (Iturri *et al.*, 1974). Heart rate was significantly reduced by 1242 and 1254 at ≤100 ppm but was not altered by 1221 or 1260 at dietary levels of 150 ppm. Mean blood pressure was unaffected. Hemoglobin concentrations, HCT, and total erythrocyte count were found to be significantly depressed by 1242 and 1254 at ≤150 ppm but not affected by 1221 or 1260 at any level used. Dietary PCBs potentiate vitamin E-selenium deficiency in the chick by inducing hepatic microsomal benzopyrene hydroxylase and thus decreasing the biological utilization of dietary selenium (Combs, 1975; Combs *et al.*, 1975). Feeding 100 mg of Aroclor reduced serum potassium, calcium and protein values in white pelicans (Greichus *et al.*, 1975) while feeding 100 ppm of Aroclor 1242 to Japanese quail reduced liver vitamin A (Cecil *et al.*, 1973).

Animals

Dietary PCBs, Aroclor 1242 and 1254, fed to swine and sheep at concentrations of 20 ppm from weaning and the time of reaching market weight, reduced feed efficiency and rate of weight gain (Hansen *et al.*, 1976b). Gross and microscopic lesions were few, consisting of increased frequency of pneumonia in swine and sheep and of increased frequency and severity of gastric lesions in swine. In contrast, Platonow and coworkers (1976) found feeding Aroclor 1232, 1242, and 1254 to piglets for a 12-week period appeared to act as a growth stimulant with piglets fed 250 ppm of 1254 gaining 50% more weight after 12 weeks than the controls. No evidence of anomalies were seen with gross and micropathological examinations of treated versus control piglets and the blood coagulation system was unaffected. Although the primary mode of excretion of PCBs in lactating cattle is via the milk (Saschenbrecker *et al.*, 1972), the cow can hydroxylate PCBs and excrete some of the compounds as hydroxylated metabolites (Gardner *et al.* 1976). Most of the metabolites are excreted as conjugates. When a single isomer of PCB was fed, only the 4-hydroxy metabolite was found in the milk. Cows given Aroclor 1242 or 1254 had 10 and 4 monohydroxy, metabolites, respectively, in their milk.

Species differ greatly in their response to PCBs but it appears that nonhuman primates are grossly affected by high levels of PCB intake and show symptoms of PCB toxicity at quite low levels of dietary intake. To study the effect of high levels of PCB intake, adult *Macaca mulatta* monkeys were given diets containing 100 and 300 ppm of Aroclor 1254 for periods ranging from 2 to 3 months. Extreme morbidity was experienced within one month and most died within three months (Allen, 1975). To study the effect of lower levels of PCB intake, six adult female Rhesus monkeys were fed 25 ppm Aroclor 1248 for 2 months (Allen *et al.*, 1974). Within one month, facial edema, alopecia and acne had developed. Feeding this quantity of PCBs was fatal to one monkey which showed signs of anemia, hypoproteinemia, severe hypertrophic, hyperplastic gastritis and bone marrow hypoplasia. Subcutaneous adipose tissue of the five surviving monkeys had an average of 127 μg/g fat of PCBs at the end of the feeding. This level decreased to 34 μg/g fat at the end of eight months but the animals continued to show clinical signs and lesions of PCB toxicity. Metabolism of PCBs with high chlorine content occurs very slowly in Rhesus monkeys; however, di-and trichlorobiphenyls can be totally degraded to hydroxylated metabolites and excreted rapidly (Greb *et al.*, 1975). GLC traces of the accumulated PCBs indicated only the isomers of high chlorine content were stored (Allen, 1975). The one infant born to these females was small in size and had over 25 μg PCB/g of adipose tissue. A third study reports the effect of feeding 2.5 or 5.0 ppm Aroclor 1248 to eighteen female and four male adult Rhesus monkeys (Barsottei *et al.*, 1976). These levels were chosen since they were equal to 50% of the concentration of PCBs allowed in some foods being consumed by humans. By two months, some of the females had developed acne alopecia erythema and swelling of the eyelids. All of the females exhibited some degree of these symptoms at the end of six months. Within four months, menstrual cycles were altered with menostaxis and menorrhagia occuring frequently and sometimes amenorrhoea was also apparent. The animals were allowed to breed after seven months, but their ability to maintain pregnancy was impaired. Resorption and abortions were frequent. Infants which were born were small and skin biopsies confirmed placenta transfer of PCBs. Females also exhibited modifications in serum lipid and protein ratios and enzyme induction. In contrast, all males fed 5.0 ppm Aroclor 1248 developed only slight periorbital edema and erythema and appeared to have no alterations in breeding capacities after fourteen months. Thus, nonhuman primate susceptability to Aroclor 1248 toxicity is much higher in the female.

The degree of chlorination affects accumulation (Grant *et al.*, 1971) and excretion (Matthews and Anderson, 1976) of PCBs in the rat. Hydroxylation via an arene oxide is a prerequisite to the primary mode of excretion of PCBs (Safe *et al.*, 1975; Matthews and Anderson, 1976). However, small levels of up to 0.6% of the dose of a tetrachlorinated biphenyl have been found to be excreted unchanged into the gastrointestinal tract of rats (Yoshimura and Yamamoto, 1975).

Numerous studies have shown that PCBs affect enzyme systems in animals. Feeding Aroclor 1254 at 10 mg/kg induced the drug metabolizing enzymes aniline hydroxylase and amino pyrine n-demethylase in pregnant rabbits and increased protein synthesis (Villeneuve *et al.*, 1971). Increased activity in rats was caused by 10 ppm of Aroclor 1254 (Turner and Green, 1974), 1 mg/kg of Aroclor 1016 or 1242 (Iverson *et al.*, 1975), 20 mg/kg of two isomeric tetrachlorobiphenyls (Chen *et al.*, 1973). PCBs reduce activity of lipid synthesizing enzymes (Holab *et al.*, 1975) and inhibit oxidative phosphorylation in mitochondria (Sivalingan *et al.*, 1973). PCBs have also been found to produce porphyria (Bruckner *et al.*, 1974).

FDA TOLERANCES FOR PCBs IN FOOD

In March of 1972, the Food and Drug Administration published in the Federal Register a notice of a proposed rule limiting the sources by which PCBs may contaminate animal feeds, food and food packaging materials during manufacturing, handling and storage as well as limiting the levels of PCBs that may be present in animal feed, food and food-packaging materials as a result of unavoidable environmental contamination (FDA, 1973). This notice was in response to concern about possible low level, long term toxic effects of humans ingesting PCBs. Temporary tolerances took into account that the minimum dose causing clinical symptoms in the Yusho incident was 500 mg. Applying a safety factor of 10 to 1, this would allow protracted ingestion of 1 μg/kg/day. At that time, FDA market basket survey data indicated that the quantitatively measurable PCBs were equivalent to an intake of approximately 0.06 μg/kg/day but the FDA felt levels could be much higher in persons consuming cerain types of food, even if their diet was moderately well balanced. Concern was expressed for levels of PCBs being found in fish and to a lesser extent in poultry and in grains and cereals being contaminated through transfer of PCBs from packaging materials. In addition to toxological considerations, the FDA took into account the levels of PCBs which could be expected to be found in the food, if the company was following good manufacturing practices and used these lower values for setting the temporary tolerances of PCBs in foods (FDA, 1973). The temporary tolerances for PCBs in foods established in July, 1973 are summerized in Table 13.6.

TABLE 13.6. TEMPORARY FDA TOLERANCES OF PCBs IN FOODS AND RELATED MATERIALS.

Food or Material	1973 Temporary Tolerance[a] ppm	1977 Proposed Reduction[b] ppm
Milk	2.5 (fat basis)	1.5 (fat basis)
Manufactured dairy products	2.5 (fat basis)	1.5 (fat basis)
Poultry	5.0 (fat basis)	3.0 (fat basis)
Eggs	0.5	0.3
Finished animal feed	0.2	—
Animal feed components of animal origin	2.0	—
Fish and shellfish (edible portion)	5.0	2.0
Infant and junior foods	0.2	—
Paper food packaging material	10.0	—

[a]FDA, 1973.
[b]FDA, 1977.

In addition, this FDA action contained special provisions to preclude accidental PCB contamination of food packaging materials (FDA, 1973). The ruling precluded the use of any new equipment or machinery which contained or used PCBs in the manufacture of food packaging materials. In addition, all current equipment was to be assessed for PCBs and any equipment machinery or materials containing PCBs for which there is a reasonable expectation that such articles could cause PCB contamination of food packaging materials should be eliminated as quickly as possible commensurate with good manufacturing practices. Electrical transformers and condensors containing PCBs in sealed containers were exempt from this provision.

More recent toxicological data, particularly the studies showing that low level ingestion of Aroclor 1248 caused toxic effects in monkeys and that infants born to and nursed by these female monkeys were small, had slower growth rates and exhibited behavior abnormalities as well as the reduced levels of PCBs now being found in all foods except fish, led the Food and Drug Administration to announce its intention to reduce the temporary tolerances of PCBs allowed in milk and dairy products, poultry and eggs and fish and shellfish (FDA, 1977). Also of concern is the suspected carcinogenicity of PCBs. However, the FDA report points out that the risk of carcinogenesis following exposure to environmental PCBs is difficult to evaluate since many of the PCB formulations causing carcinogenicity also contained low levels of chlorinated dibenzofurans which may have caused tumors. Thus, the Commissioner has concluded that it is desirable to further reduce human exposure to PCBs and this proposed reduction of temporary tolerances of PCBs in selected foods (Table 13.6) is one of a series of reductions anticipated with the goal being a zero tolerance. The Environmental Protection Agency (1977) has issued strict regulations on labeling and disposal of PCBs to reduce further loss into the environment.

REDUCTION OF PCBs IN FOODS BY PROCESSING AND COOKING

To estimate the potential of reducing PCBs by processing and/or cooking as a method of further reducing human consumption, several studies have followed these lipophilic compounds during food manufacture or home preparation. Products evaluated include milk and other dairy products, poultry and eggs, oils and fermentated grain products, because milk has been known to be contaminated with PCBs from silo coatings, poultry from feed and leaking heat exchangers, fish from environmental contamination magnified through food chain accumulation, and grains from environmental contamination.

Milk and Dairy Products

Since transfer of PCBs from the silo coating Cumar to silage is a source of PCB contamination in milk, which might be a public health hazard, Platonow *et al.*, (1971) studied the feasibility of reducing these PCB levels during normal processing of the milk into various dairy products. Milk from two actively lactating Jersey cows individually given a single oral dose of 10 or 100 mg/kg of Aroclor 1254 was manufactured separately into the following: whole milk, cream, skim milk, heated skim milk, nonfat dry milk, cottage cheese, and whey. PCBs were higher in high fat-containing products but were in no way proportional to the fat content. On a fat basis, PCBs in the extremely low fat whey were over 10 times as high as in the fat of the whole milk. Levels of PCBs in whole milk and cream from cow 2 fed 100 mg PCB/kg were about 10 times higher than the corresponding products from cow fed 10 mg PCB/kg; however, the other products analyzed did not show this relationship. Heating skim milk was the only process which substantially reduced PCBs. Of course, ingestion of low fat products would reduce the likelihood of PCB ingestion. The authors also examined the percentage distribution of the gas chromatographic peaks of the various PCB isomers and found that processing milk into these dairy products did not affect the relative proportion of these gas chromatographic peak areas.

A second study expanded the study of the feasibility of reducing the levels of PCBs during the normal processing of milk into dairy product (*Arnott et al.*, 1977). This study included butter, pasteurized, homogenized cream plus culture, cultured cream, milk plus yogurt culture and yogurt in addition to the previously mentioned dairy products (Table 13.7). PCBs again appeared to decline in skim milk heated to 77°C or higher, supporting the previous results (Platonow *et al.* 1971). PCB levels in yogurt and cultured cream appeared to increase but this apparent anomaly may be due to easier extraction of PCBs from lipoproteins following fermentation. Again processing had little effect on PCB reduction in dairy products.

TABLE 13.7. FAT AND PCB LEVELS IN SEVERAL DAIRY PRODUCTS MANU-
FACTURED FROM MILK OF CONTROL AND TREATED COWS.[1]

	Control		Low Dose		High Dose	
Product	Fat (%)	PCB (μg/g)	Fat (%)	PCB (μg/g)	Fat (%)	PCB (μg/g)
Composite milk	3.83	<0.01	4.38	1.30	3.54	5.90
Skim milk	0.11	<0.01	0.09	0.04	0.36	0.28
Cream	40.22	<0.01	38.90	12.32	30.60	19.11
Butter	80.14	<0.01	86.14	19.78	85.97	34.10
Buttermilk	0.82	<0.01	0.87	0.61	4.11	5.19
Skim milk						
71°C - 1 min	0.11	<0.01	0.09	0.04	0.36	0.45
77°C - 1 min	0.11	<0.01	0.09	0.03	0.36	0.27
82°C - 1 min	0.11	<0.01	0.09	0.03	0.36	0.20
82°C - 10 min	0.11	<0.01	0.09	0.05	0.36	0.23
82°C - 10 min + 60 min to 55°C	0.11	<0.01	0.09	0.05	0.36	0.16
Nonfat dry milk	1.05	<0.01	1.17	0.15	3.75	0.90
Standardized cream	15.40	<0.01	16.30	6.76	17.52	—
Past'd, homo'd cream + culture	14.47	<0.01	15.34	2.07	16.64	7.02
Cultured cream	14.47	<0.01	15.34	3.97	16.64	8.84
Milk + yogurt culture	3.70	<0.01	3.26	1.58	1.19	1.37
Yogurt	3.70	<0.01	3.26	2.72	1.19	3.20

[1]Arnott, *et al.*, 1977.

Poultry and Eggs

Poultry meat and eggs have been seized by the FDA because of PCB contamination (Ramsey, 1973). In some of the instances the source of PCBs was traced to the feed resulting from a leaky heat exchanger while, in other instances, the source of contamination was not located. Two studies are concerned with PCB removal from poultry products, the first relates the effect of cooking on PCB levels in hen tissue and the second the effect of freeze-drying on PCB levels in eggs.

The effect of stewing at 93°C for 2½ hr and pressure cooking at 15 psig for 15 min on PCB levels in chicken pieces was determined by cooking breast pieces, drumsticks, thigh meat and thigh skin from one side of the hen and comparing these levels to the corresponding raw piece. Adipose tissue was divided into half with a portion being stewed or pressure cooked to assess the amount of PCB lost through rendering *per se*. PCB levels in the raw and cooked pieces as well as the broth are summarized in Table 13.8. Cooking brought about a significant decrease in PCBs (p<0.01). PCB levels expressed as ppm of the wet tissue were significantly higher (p<0.01) in stewed than in pressure-cooked pieces but when these values were corrected for differences in fat and were expressed on a fat basis, this was no longer true. Recoveries of PCBs were calculated from the micrograms of PCBs in the cooked meat and in the broth as compared to the level in the raw tissue (Figure 13.1). Except for the adipose tissue, approximately half the recovered PCBs were in the meat itself (Figure 13.2). As expected, most of the PCBs from the adipose tissue was rendered into the broth. Less PCBs were rendered into the broth from meat pieces than had been for the chlorinated hydrocarbon pesticides, lindane, dieldrin and DDT. In the latter case, only 1/4 to 1/3 of the environmental contaminant recovered was found in the meat *per se* (Zabik, 1974).

To study the effect of freeze drying on the level of PCBs in eggs, Khan *et al.* (1976c) spiked two separate samples of homogenized whole eggs (approximately 8 eggs in each sample) with 1 ml of 100 ppm Aroclor 1254 and 1260, respectively. This resulted in a level of 3.5 ppm (solids basis) of Aroclor 1254 in the first lot of eggs and a level of 4.7 ppm (solids basis) of Aroclor 1260 in the second egg homogenate (Table 13.9). Freeze drying resulted in losses of 26.8% of the Aroclor 1254 and 22.6% of the Aroclor 1260 so the less highly chlorinated isomers are apparently lost more easily. Di-, tri- and tetra-chlorobiphenyls were considerably reduced after freeze drying egg samples contaminated with Aroclor 1254, while tri-, tetra-, and pentachlorobiphenyls were successfully reduced from samples containing Aroclor 1260. Hence, the lower chlorinated isomers volatilize more easily. Since these eggs had relatively high levels of PCB contamination, the authors felt that freeze drying might have even greater potential for reducing PCBs.

Fish

Fish, particularly those taken from the Great Lakes and other aquatic areas contaminated with PCBs, have the highest levels of PCBs found in foods. Shellfish taken from contaminated coastal waters can also be sources of PCB ingestion in humans. Two studies have been concerned with the effect of cooking on PCB levels in salmon and trout and two with reduction of PCBs during drying of shrimp.

Smith *et al.* (1973) studied the effect of baking, baking in a nylon bag and poaching in a five percent salt brine on the levels of PCBs and other chlorinated hydrocarbons in salmon steaks. Chinook and Coho salmon were obtained at the Manistee River weir in Michigan as the fish were passing upstream from Lake Michigan to spawn. One-inch salmon steaks were halved before cooking to provide for raw analyses. Results of this study are summarized in Table 13.10.

TABLE 13.8. PCB AND FAT LEVELS IN RAW, STEWED OR PRESSURE COOKED PIECES AS WELL AS IN CHICKEN BROTH.[1]

| Piece | State | Stewed | | | Pressure Cooked | | |
| | | Fat % | ppm 1254 expressed as | | Fat % | ppm 1254 expressed as | |
			Wet Tissue	Fat		Wet Tissue	Fat
Breast	Raw	7.5±2.7	1.56±0.41	28.5±12.2	8.3±3.5	1.45±0.40	19.3±5.8
	Cooked	5.8±1.8	0.72±0.33	15.1±10.7	5.4±1.2	0.75±0.53	13.5±8.2
	Broth	10.9±7.8	1.10±0.50	14.0±5.7	1.7±1.0	0.18±0.11	11.5±3.6
Drumstick	Raw	5.9±2.1	1.57±0.76	28.0±9.4	6.3±1.2	1.32±0.22	20.8±2.5
	Cooked	5.5±0.8	1.02±0.27	19.1±6.2	5.4±1.0	1.05±0.23	20.6±8.6
	Broth	1.5±1.1	0.32±0.12	31.1±11.1	0.5±0.2	0.16±0.03	42.0±29.0
Thigh meat	Raw	7.6±4.2	1.54±0.65	24.9±18.6	5.9±1.2	1.54±0.31	26.6±4.2
	Cooked	6.1±0.5	1.29±0.48	21.0±7.4	5.6±1.0	1.06±0.28	19.4±4.8
	Broth	1.2±0.3	0.21±0.09	16.4±6.6	0.5±0.2	0.11±0.02	27.2±13.9
Thigh skin	Raw	40.9±5.3	6.83±1.78	17.5±6.1	44.2±9.3	7.64±1.38	17.8±4.2
	Cooked	29.6±8.8	5.72±1.47	20.8±7.4	30.2±5.8	5.52±1.62	19.0±6.1
	Broth	2.9±0.7	0.69±0.22	24.8±4.0	1.3±0.3	0.21±0.07	29.9±17.6
Abdominal Adipose Tissue	Raw	61.5±5.9	13.55±3.52	22.9±8.7	66.6±2.0	11.14±4.08	16.8±6.3
	Cooked	42.5±5.0	10.75±3.20	25.5±7.5	38.3±7.3	7.39±2.59	19.4±5.8
	Cooked	10.8±3.3	1.55±0.80	14.8±5.7	3.9±1.2	0.67±0.32	19.7±6.2

[1]From Zabik, 1974.

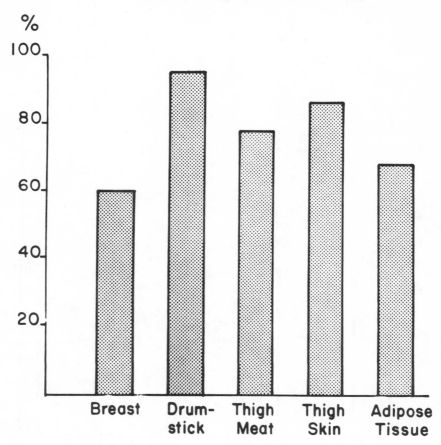

FIG. 13.1.   PERCENTAGE RECOVERY OF PCBs IN COOKED CHICKEN PIECES
AND CHICKEN BROTH (FROM ZABIK, 1974).

Although the major portion of PCB peaks resembled Aroclor 1254, there were some less chlorinated isomers that were quantitated as Aroclor 1248. On a whole tissue basis, the raw chinook salmon had an average of 7.7 ppm of total PCBs and the Coho salmon an average of 6.1 ppm of total PCBs, both of which exceeded the FDA tolerances of 5 ppm. Although some PCBs were found in the cooking drip, very little of the PCBs found in these salmon steaks were lost during cooking. Total micrograms of PCBs in the cooked slices was

DISTRIBUTION OF RECOVERED PCB'S

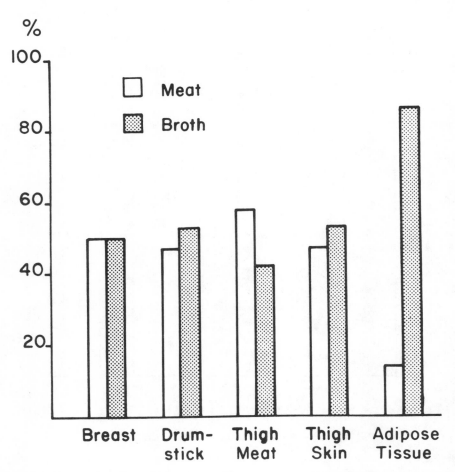

FIG. 13.2. PERCENTAGE DISTRIBUTION OF RECOVERED PCBs BETWEEN COOKED CHICKEN PIECES AND CHICKEN BROTH (FROM ZABIK, 1974).

compared to that in the raw to assess recoveries. Virtually none of the Aroclor 1254 was lost during cooking. In fact, recoveries were sometimes slightly over 100%, perhaps reflecting greater ease of extraction of physiologically incorporated PCBs from cooked flesh. Losses of up to 15% were found for Aroclor 1248 with these losses occurring in salmon steaks baked in nylon bags. None of the Aroclor

TABLE 13.9.   REDUCTION OF PCBs DURING FREEZE-DRYING WHOLE EGGS[1].

Aroclor spiked	Liquid Samples	Freeze-dried Sample	Percent Reduction
	$\mu g/g$ solid	$\mu g/g$ solid	
1254	3.5	2.5	26.8
1260	4.7	3.7	22.6

[1]From Khan, *et al.*, 1976c.

1248 was lost from salmon steaks that were poached. Since moist heat increases cooking losses, this probably facilitated the small reduction of PCBs in the salmon steaks cooked in nylon bags. Steaks were also cooked with and without skin but, as expected, this did not affect the level of PCBs found in the flesh. Even though adipose tissue is associated with the skin, a horizontal transfer of fat and its lipophilic contaminant was not expected during the cooking of these steaks. The results of this study were disappointing in that cooking had been shown to reduce chlorinated hydrocarbon pesticides from fish. It does point out, however, that when the fat content is low and the cooking time short, neither rendering of PCB with the fat nor codistillation of PCB with volatile losses can occurr to a great enough extent to reduce these contaminants in the cooked flesh.

A second study on the effect of cooking on PCB levels in fish used fat trout obtained from Lake Superior (Zabik *et al.*, 1979). The Great Lakes Environmental Contaminants Survey data have shown this type of lake trout to be highly contaminated with PCBs and their high fat levels should be conducive to removal of PCBs via fat rendering. Two experiments were conducted. In the first, the fish were filleted and pieces from the left side of the fish baked at 117°C or broiled 9.8 cm from a broiler source operating at 288°C to an internal temperature of 75°C or cooked by microwave to a similar degree of doneness. The right fillet was analyzed raw. In the second study, center six-inch portions of the whole fish were roasted with and without skin. Pieces adjacent to this larger piece was used for these raw analyses. The results of these studies are summarized in Tables 13.11 and 13.12.

Even though these fish were of similar size ranging from 1.2 to 1.7 kg for the fish which were filleted and 0.5 to 0.8 kg for the trout roasted whole, and were all caught at one time off the Eastern side of the Keweenaw peninsula of Upper Michigan, the PCBs in the raw flesh ranged from 1.0 to 10.7 ppm and resulted in high standard deviations (Tables 13.11 and 13.12). Fat content of these fillets ranged from 25 to 30%. Cooking significantly reduced the levels of PCBs in the fillets. Despite differences noted in cooking parameters,

TABLE 13.10. AVERAGES AND (IN PARENTHESES) RANGES OF PERCENT FAT AND PARTS PER MILLION (FAT BASIS) OF PCBs IN RAW AND COOKED CHINOOK AND COHO SALMON FLESH[1]

Treatment	Chinook			Coho		
	Fat	Aroclor 1248	Aroclor 1254	Fat	Aroclor 1248	Aroclor 1254
Raw	2.65 (0.99–3.9)	18.17 (9.0–53.2)	273.03 (125.6–799.4)	3.59 (2.2–5.2)	14.35 (5.5–21.2)	155.41 (107.5–240.5)
Baked	4.00 (1.6–8.0)	15.40 (6.4–54.6)	274.34 (106.5–1091.4)	4.53 (3.1–6.4)	13.85 (7.5–19.5)	170.69 (87.9–251.2)
Poached	3.43 (1.4–7.1)	18.40 (3.0–51.9)	268.15 (65.2–591.0)	4.08 (1.9–5.2)	15.60 (5.6–24.7)	176.49 (82.5–323.6)
Baked in nylon bag	3.61 (1.0–6.0)	13.77 (5.7–44.1)	223.55 (85.8–542.7)	3.96 (2.9–5.4)	13.48 (12.1–15.2)	165.77 (101.2–191.3)

[1]From Smith et al., 1973.

TABLE 13.11.   PCBs IN FILLETS COOKED BY BROILING, ROASTING OR MICRO-WAVE.[1]

| Cooking | | PCBs (calculated in Aroclor 1254) | | |
Method	State	ppm in flesh	ppm in fat	total μg
Broiled	Raw	3.8 ± 1.7	23.6 ± 11.6	824
	Cooked	3.6 ± 0.6	14.1 ± 7.9	390
Roasted	Raw	4.6 ± 3.2	19.7 ± 12.7	564
	Cooked	3.9 ± 3.6	13.8 ± 10.2	372
Microwave	Raw	4.8 ± 3.5	16.6 ± 8.9	595
	Cooked	4.0 ± 2.8	14.0 ± 7.1	442

[1]From Zabik *et al.*, 1979.

there was not a significant difference in the levels of PCBs in fillets broiled, roasted, or cooked by microwave. However, when the total micrograms of PCBs in the raw and cooked slices (Table 13.11) were used to calculate recoveries of PCBs, broiling was found to reduce the PCBs in the fillets by an average of 53%, roasting by an average of 34% and cooking by microwave by 26%.

The presence or absence of skin was not found to significantly affect the level of PCBs in flesh of roasted trout pieces (Table 13.12). Roasting *per se* brought about a significant reduction in PCBs. Roasting trout pieces with skin resulted in an average loss of 40% of PCBs while roasting without skin resulted in an average loss of 50%. Although not statistically significant, it can certainly be recommended to remove the skin and associated fat before cooking this type of fish.

Khan *et al.* (1976a) studied the effect of sun drying on the levels of PCB in shrimp, half of which were treated by dipping in 1% sodium nitrite solution before being dip-coated in a 100 ppm solution of Aroclor 1254 or 1260. Photolysis at wavelengths similar to sunlight successfully dechlorinates PCBs and thus sun drying might be a feasible method to reduce PCBs in food. Initial levels of PCBs were

TABLE 13.12.   PCBs IN THE FLESH OF WHOLE TROUT PIECES ROASTED WITH AND WITHOUT SKIN[1].

| Roasting | | PCBs (calculated as Aroclor 1254) | | |
Condition	State	ppm in flesh	ppm in fat	Total μg
With skin	Raw	1.82 ± 0.85	10.8 ± 9.6	414
	Cooked	1.17 ± 0.58	7.7 ± 7.2	249
Without skin	Raw	1.87 ± 1.47	10.6 ± 5.1	415
	Cooked	1.34 ± 1.22	7.3 ± 6.3	208

[1]From Zabik *et al.*, 1979.

28.8 to 28.9 μg/g solids for the lot of shrimp dipped in Aroclor 1254 and 31.6 to 33.3 μg/g solids for those dipped in 1260. Levels of PCBs after 12, 24, 36 and 48 hours of exposure to sunlight at ambient temperatures which were fairly constant at about 32°C are in Table 13.13. Sodium nitrite accelerated the photolysis reactions with 26.8 and 32.2% reduction of Aroclor 1254 and 1260, respectively, at the end of the first 12 hours as compared to 14.1 and 18.8%, respectively, in samples without sodium nitrite. Although the percentage reduction was greater for samples dipped in 1260 at the end of the first 12 hours, the reduction was greater for 1254 for all other time periods until all of the PCB was removed. Solar radiation was found to have a two-way impact on PCB degradation: the higher chlorinated biphenyls are dechlorinated, giving rise to increased peak areas of lower chlorinated compounds but these compounds are being simultaneously degraded. The authors stated that preprocessing contamination of shrimp is normally from polluted waters which results in surface contamination of the shrimp. Sun drying of shrimp is practiced in many parts of the world. Since PCBs adhering externally or superficially will be completely degraded during the process, this increases the safety of this food.

Khan *et al.* (1976c) also studied the feasibility of using freeze drying to reduce the levels of PCBs in shrimp homogenates which were contaminated separately with 1 m of 100 ppm Aroclor 1254 or 1260, respectively. Levels in the liquid sample and the sample after freeze drying are summarized in TAble 13.14. Freeze drying reduced the level of 1254 by 40.3% and the level of 1260 by 25.2%. Therefore, freeze drying is another process which offers potential to reduce PCBs in foods.

TABLE 13.13.   EFFECT OF SUN DRYING ON AVERAGE PCB CONTENT OF SHRIMP[1].

PCB Used in Dip	Hr of Exposure	Without Sodium Nitrite		With Sodium Nitrite	
		μg/g solids	% reduction	μg/g solids	% reduction
1254	0	28.9	0	28.8	0
	12	24.8	14.1	21.1	26.8
	24	12.1	58.2	3.5	87.8
	36	1.6	94.4	0	100.0
	48	0	100.0	0	100.0
1260	0	31.6	0	33.3	0
	12	25.6	18.8	22.6	32.2
	24	17.2	45.5	12.0	63.9
	36	10.0	68.2	9.3	72.2
	48	2.7	91.4	0	100.0

[1]From Khan, *et al.*, 1976a.

TABLE 13.14.   EFFECT OF FREEZE-DRYING ON THE PCB CONTENT OF SHRIMP HOMOGENATE[1].

PCB Used	Liquid Sample $\mu g/g$ solids	Freeze-dried Sample $\mu g/g$ solids	Percent Reduction
1254	27.6	16.5	40.3
1260	20.5	15.3	25.2

[1]From Khan et al., 1976c.

Fats and Oils and Other Foods

Oilseed grains and marine oils may be environmentally contaminated with PCBs, thus three studies have been undertaken to assess the effect of processing on PCB levels in margarines and oils. Rice has also been found to contain low levels of PCBs in Japan (Fujiwara, 1975) so a study was also initiated to follow the level of PCB contamination during the production of sake.

Raw marine fish oils containing 3 to 13 ppm PCBs (as Aroclor 1254) were subjected to pilot plant refining, hydrogenation and deodorization for margarine stock production (Addison, 1974). Eight of the samples processed were obtained from the East Coast of Canada and included redfish, flatfish, herring seal and mackerel oil samples while the two West Coast samples were day fish and herring oil. Residues were reduced to below the detectable range of PCBs (0.5 ppm) as a result of processing. Isomers of both low and high levels of chlorination were absent from the gas chromatographic traces of processed oils. The author indicates that the vapor pressures of Aroclor 1254 and 1260 mixtures are less than DDT, it would be expected that processing conditions which remove DDT would also remove PCBs.

Japanese researchers have also studied the effect of processing on PCB levels in oil (Kanematsu et al., 1976a). Oil was deacidified, decolorized and deodorized using laboratory conditions. PCBs were unaffected by deacidification, regardless of the amount of sodium hydroxide used. PCB levels were also unaffected when either activated clay or an absorbant containing activated carbon was used for decolorized and deodorized, using laboratory conditions. PCBs were decrease was proportional to the temperature used for deodorization: 17% at 160°C, 36% at 200°C, and 100% for both 230 and 250°C. The authors concluded commercial deodorization would completely remove PCBs. Hydrogenation with a Ni catalyst had little effect on PCB levels. Use of a copper-chromium catalyst for hydrogenation did reduce PCB levels from 3.590 ppm to 0.014 ppm (Kanematsu et *al.*, 1976b). The authors concluded that this reduction is due to chemical changes such as reduction and dechlorination, rather than to evaporation.

PCBs have been found in the outer rice layers in Japan and thus are largely removed during the milling of rice for sake production (Maekawa and Hirata, 1974). The authors followed the fate of the remaining PCBs during sake fermentation. PCBs were found to be adsorbed on the steamed rice and yeast cells so that sake had little PCBs.

POLYBROMINATED BIPHENYLS

Polybrominated biphenyls (PBBs) are bromine analogs of halogenated biphenyls prepared by adding gaseous bromine to biphenyl so that the isomers have the general structure:

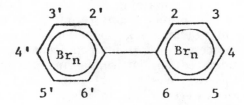

Due to their thermal stability, PBBs are effective and economically feasible plasticizers and fire retardants which can be incorporated into flame resistant polymers. Michigan Chemical Corporation manufactured a PBB product, sold under the trade name Firemaster® BP-6. Firemaster® BP-6 was used primarily for business machines, electrical products and fabricated products (Table 13.15).

Firemaster® BP-6 is a mixture of brominated biphenyls with an average of approximately 63% hexabromobiphenyl, 14% heptabromobiphenyl, 10.5% pentabromobiphenyl, 2% tetrabromobiphenyl and other unidentified isomers (Kerst, 1974). The principle component has been identified as 2,2',4,4',5,5'-hexabromobiphenyl (Sundstrom *et al.*, 1976). BP-6 is a solid at room temperature, softening at 72°C and decomposing at 300 to 400°C. It has a very low water solubility of 1 ppb at 25° and a vapor pressure of 5.2×10^{-8} mmHg at 25°C (Kerst 1974).

PBBs are not a general environmental contaminant as are PCBs. PBBs were not used in food or feed or anything that should come in contact with food or feed. PBBs were also not used in any products

TABLE 13.15.   SPECIFIC INDUSTRIAL USES AND APPROXIMATE ALLOCATIONS OF FIRE MASTER® BP-6 PRODUCED.[1]

Industrial Use	Approx. Allocation of Total Fire Master[R] BP-6 Produced	Examples
	%	
Business machines and industrial equipment	48	Typewriter, calculator and microfilm reader housings; business machine housings
Electrical	35	Radio and TV parts, thermostats, shaver and hand test housings
Fabricated products	12	Projector housings, movie equipment cases
Transportation	1	Miscellaneous small automotive parts, i.e. electrical wire connectors, switch connectors, speaker grills
Miscellaneous	4	Small parts for electrical applications, motor housings, components for industrial equipment

[1]From Kerst, 1974.

that would be expected to come in contact with human skin such as flame retarding fabrics (Kerst, 1974). Thus, PBBs would not be a food toxicology problem, if an incredible error in feed manufacture had not occurred in Michigan in May of 1973.

MICHIGAN INCIDENT

Michigan Chemical Company manufactured the PBB compound, Firemaster® Ff-1 which was Firemaster® BP-6 mixed with the anticaking agent "Flo-Gard" which is manufactured by PPG Industries and which contains 83% silicone dioxide and a maximum of 7% calcium oxide. These Firemaster® Ff-1 compounds were mistakenly used in place of magnesium oxide which was also manufactured by Michigan Chemical Corporation under the trade name Nutrimaster and mixed into high protein dairy pellets (Jackson and Halbert, 1974). The toxic syndrome developed in a 400-cow herd near Battle Creek, Michigan. In late September, 1973 milk production of the 300 lactating cows of this herd decreased dramatically from daily shipments of 13,000 lb/day to 7,600 lb/day within a period of 20 days. At this time, feed consumption also decreased dramtically to one-half the cow's normal dry matter intake.

Starting about one month after the onset of anorexia and decreased milk production, hematomas and later abscesses developed in about 40 cows. Hoof and hair abnormalities developed, pregnancies were prolonged, and weight loss continued even after a change in feed. Feed was tested for and found free of excessive level of heavy metals, mycotoxins and pesticide residues. In April of 1974, the toxic component was identified as a polybrominated biphenyl and the source was traced to the Michigan Chemical Company who supplied magnesium oxide to the feed manufacturer. This initiated a state-wide search for animals with PBBs.

Although the original contamination was a dairy ration, subsequent cross feed contamination of other feeds led to poultry, sheep, swine, and other animals as well as animal products being affected. The Food and Drug Administration set the following guideline tolerances in November, 1974: 0.3 ppm in the fat for milk, meat, and fat of rendered products and 0.05 ppm in whole egg and finished feed and ingredients. Search for contaminated products led to 865 tons of feed, approximately 18,000 lbs of cheese, 2,600 lbs of butter, 34,000 lbs of dry milk and over 400,000 dozen of eggs being destroyed (Mich. Dept. Agr., 1977). Over 23,000 cattle, 4,600 swine, approximately 1,400 sheep, 650 poultry and a small number of goats, rabbits, and horses were destroyed and buried at a common-site at Kalkaska, Michigan. These numbers do not include animals buried on farm sites. In total, 576 premises were quarantined because of this PBB incident.

Food with PBB levels in excess of the FDA guidelines was off the market by the end of 1974 (Wilcox, 1977). Widespread public concern about PBBs intensified in late 1975, 1976 and 1977. The Michigan legislature reduced the allowable level of PBBs in feed to 0.01 ppm in finished feed and ingredients in August, 1976. To increase public confidence in Michigan food and to further safeguard human health, the Michigan legislature reduced the allowable level of PBBs in milk to 5 ppb on a whole milk basis and to 20 ppb in the fat of dairy animals in August 1977. This law provided for state supported analyses of each dairy farm's bulk milk as well as the adipose tissue of all animals born before January 1, 1976 being culled from dairy herds before slaughter. When either milk was found to exceed 5 ppb or an animal was found to have 20 ppb PBBs in the tail fat biopsy, the law provided for state purchase and disposal of the contaminated products. This testing program is in effect until 1982. Product analyses for PBBs conducted at the Michigan Department of Agriculture laboratory and the FDA region laboratory in Detroit had shown that three of several hundreds of samples analyzed from rendering plants in 1976 had trace levels of PBB. Retail milk had non-detectable levels of PBBs but 24 dairy herds were still quarantined for having cows with milk exceeding 0.3 ppm in the fat and the Michigan Department of Agriculture estimated as many as 80,000 animals might have trace levels of PBB.

EFFECT OF PBBs ON ANIMALS

Cattle, Sheep and Pigs

A Holstein cow and a sheep fed 50 ppm PBB for fifteen and thirty days, respectively, had high levels of PBBs stored in their tissue ranging from a low of 0.4 ppm in the spleen and adrenal of the cow to a high of 41.6 ppm in the renal fat of the sheep (Gutenmann and Lisk, 1975). Except for the thyroid, the PBB levels were highest in tissues from the sheep which may be due in part to more efficient intestinal absorption by the sheep which consumed its feed containing PBB as compared to the cow which had to be force-fed with capsules. Levels in the cow's milk peaked at just under 7 ppm at 4½ days. The disappearance rate (t½) for PBB following cessation of dosing was about 10½ days. Fries and Morrow (1975) reported milk concentration of PBBs reached a steady state of 3.07 ppm within 30 days following feeding cows 10 mg/day. After feeding PBBs was discontinued at 60 days, milk concentrations declined 71% within the first 15 days. The reduction thereafter, however, was considerably slower than a half life of 58 days. No adverse effects on health or milk production were noted for 1 year after cessation of PBBs. Gutenmann and Lisk (1975) had reported post-mortem examination showed marked glandular hyperplasia of the major intrahepatic bile ducts of the cow's liver and of the sheep's gall bladder.

High levels of PBBs have been found to be embyrotoxic (Fries *et al.*, 1978; Cook *et al.*, 1978). PBBs were found to concentrate in body and milk fat being 600 and 300 times, respectively, the level in the blood. Levels in perirenal, omental and subcutaneous fat were found to be similar.

Willett and Irving (1975) followed distribution and excretion of PBBs in cows fed PBBs by boluses and calves contaminated with PBBs by placental transfer or via milk. The feces was the major route of excretion with 45% or more being excreted by 168 hr post dosing. The milk was also a significant route of excretion with 23.5% being excreted in 95 days. Placental transfer of PCBs was observed with adipose tissue levels in the calf about half that in the dam. Milk transfer of PBBs resulted in adipose levels in the calf about equal to that in the cow.

Evaluation of the immune system in cattle exposed to PBBs was investigated through in vitro immunoassays of 35 control (unexposed to PBB) and 52 animals exposed to PBB with levels of 0.02 to 30 ppm fat equivalent (Kateley, 1977). PBBs were not found to alter or interfere with lymphocyte surface antigens, complex nuclear and cytoplasmic events required for mitosis and cell division or cellular protein synthesis. PBB exposure at these levels did not predispose the cattle to autoantibody production or leucotoxic serum factors.

Feeding levels of up to 200 ppm PBBs for 16 weeks caused few differences in growing pigs (Ku *et al.*, 1977). The only clinical sign of toxicity was dermatosis on the ventral surface of two pigs receiving 200 ppm PBB. Liver weights were increased and heart, kidney and adrenal of pigs receiving PBBs represented a greater proportion of body weight than in control pigs. Returning the pigs to a control diet resulted in normal growth and reproduction appeared normal.

Poultry

After feeding hens a diet containing 20 ppm Firemaster® BP-6, the steady state level in the eggs was found to be about the same as in the diet but the level in the adipose tissue had been concentrated to approximately four times that in the diet (Fries *et al.*, 1976). This retention and elimination of PBB by hens was qualitatively similar to those of PCBs and the chlorinated hydrocarbon pesticides.

Polin and Ringer (1977) reported the effects of PBBs on laying hens fed dietary levels of 0.2 to 3125 ppm. The accumulation in the yolk is approximately 1.5 times that in the diet and reached a constant concentration as soon as a yolk was fully formed in the ovary or 7–10 days after feeding PBB. The biological half life in the yolk once the hens were removed from PBB contaminated feed was about seventeen days. Nearly 60% of the daily intake went directly into the yolk, 10–11% was excreted and the remaining was distributed throughout the hen. Egg production decreased significantly when the hens were fed 125 ppm PBB. Hatchability also declined and chicks hatched were less viable.

Feeding Japanese quail 0 to 100 ppm PBB for 9 weeks did not affect food intake or growth but egg production decreased and none of the eggs from the group fed 100 ppm hatched (Babish *et al.*, 1975). Males accumulated more PBB in their tissues than females, undoubtedly due to the egg being an additional route of excretion for the females. Feeding male quail 20 ppm PBB brought about the greatest induction of liver microsomal enzymes while the greatest induction in the female occurred when 100 ppm PBB was fed. None of these quails had gross or microscopic lesions.

Fish

Accumulation of bromobiphenyls with up to 4 bromine molecules very closely resembles that of the corresponding chlorobiphenyl (Zitko and Hutzinger, 1976) but at high levels of bromine substitution, less PBBs are accumulated by fish from water than the corresponding PCB (Zitko, 1977). Nevertheless, accumulation from food of these higher brominated isomers is equivalent to or higher than that of the corresponding PCB and highly substituted bromobiphenyls are debrominated in the fish. PBBs also have been found to induce hepatic microsomal monooxygenase reactions in rainbow trout within two days of pretreatment with 150 or 100 mg/kg, 1.p. (Elcombe and Lech, 1977).

Rodents

High levels of PBBs induce hepatic and extrahepatic mixed function oxidases and increase the levels of cytochrome P450 (Dent, 1977) while feeding 1 g/kg to rats resulted in liver enlargement with lipid accumulation and scattered liver cell necrosis (Kimbrough *et al.*, 1977). Using electron microscopy, livers of mice fed 1000 ppm PBBs were found to have marked increase in size of hepatocytes and rough endoplasmic reticulum, a moderate degeneration in mitochondria, a moderate increase in lysozomes and a moderate decrease in glycogen (Corbett *et al.*, 1977).

Feeding levels of 1000 ppm PBBs was weakly teratogenic to mice and rats (Corbett *et al.*, 1975) while levels of 400 and 800 mg/kg were embryo lethal and teratogenic to the young rat (Beaudin, 1977). Young rats nursing on mothers, exposed to 10 mg/kg PBBs did not grow as well as rats nursing on unexposed mothers (Cecil *et al.*, 1977). Pups nursing from mothers fed 100 ppm PBBs showed significant increases in liver weights, microsomal protein, cytochrome P450, and microsomal drug metabolizing enzymes (Moore *et al.*, 1977). These pups appeared more sensitive to the effects of PBBs than did their lactating dams.

Monkeys

During the initial six months of feeding Rhesus monkeys diets with 0.3, 1.5 and 25.0 ppm of a commercial mixture of PBBs, the animals lost weight and sterile abscesses developed on 2 of the 7 animals (Allen *et al.*, 1977). Blood data during this six-month period remained normal but four of the seven animals had flattened and lengthened progestrerone peaks. Two live term infants have been born but appear to have reduced growth rates. Feeding another two monkeys 25 ppm PBB for six weeks resulted in weight loss, abdominal distension and diarrhea. Radiographs suggest an acute hyperplastic gastritis similar to that observed in PCB exposed primates.

HUMAN EFFECTS DUE TO EXPOSURE TO PBBs

Studies are currently underway to determine possible public health effects related to PBB ingestion. The Michigan Department of Public Health began a short term study in the summer of 1974 which involved 165 persons from quarantined farms and 133 persons from non-quarantined farms to serve as a control group (Wilcox, 1977). Most people had detectable levels of PBBs in the blood but persons from quarantined farms had significantly higher levels; 3.7% of the persons from quarantined farms did not have detectable levels of PBB while 28.4% of the control group had nondetectable levels of PBB. No significant differences in any health symptoms of highly exposed persons could be detected so PBBs were not felt to cause an acute disease. Analyses of adipose tissue along with blood samples of over 60 individuals indicated a 300-fold concentration in the fatty tissue. High levels of up to 92 ppm in breast milk fat were found for women living on quarantined farms whereas women representative of the general Michigan population had levels in the breast milk fat from 0 to 1.2 ppm in 1976. Approximately 15% of the women from the Lower Peninsula had non-detectable or trace levels of PBB in the breast milk fat while in the upper Peninsula which had much less contaminated cattle, 82% had non-detectable or trace levels.

To study the possible long term effects of PBB exposure, the health and PBB levels of 4,000 people, approximately 2,000 from quarantined farms and 2,000 from non-quarantined farms in Michigan will be followed and compared to a group of 2,000 persons from Iowa farms. The report on the first years analyses will be published in the spring of 1978.

In addition, Selikoff (1977) and associates from Mount Sinai School of Medicine of the City University of New York examined 1,029 Michigan residents during a clinical field survey in November

1976. This population was made up of invited people from quarantined and non-quarantined farms, consumers from both groups of farms, Michigan Chemical Company workers and "walk-ins" concerned about the PBB episode. A great increase in symptoms of headache, fatigue, dizziness, rash and stiff joints was found to occur late in 1975 and carried on through 1976 and 1977 but these symptoms occurred in all groups and do not seem to correlate with PBB blood levels. Skin rashes and acne, however, were the most common complaint. Liver function abnormalities were found with similar prevalence in dairy farmers and consumers and were strikingly higher than a comparable group of Wisconsin dairy farmers (Lilas et al., 1977). Erthrocyte zinc protoporphyrin levels were all in the acceptable range for all of the 139 Michigan dairy farm residents tested; these individuals all had normal levels of urinary porphyrin excretion (Fishbein et al., 1977).

Possible health effects on children are also being studied. Barr (1977) reported 343 children from 0 to 16 had significantly greater frequency of health symptoms than a control group of 72 from Wisconsin for 1974 to 76 although symptom rates were similar for 1972–73. The quarantined status of the farm appeared to have no bearing on the incidence of symptoms nor did levels of PBB in the blood. Physical, psychological, and neurological examinations of 33 children exposed to PBBs and 20 children who were not exposed, failed to produce any conclusive evidence of impaired health due to PBB exposure (Weil, 1977).

Thus, the effects of PBBs on human health are not known, but the continued health surveillance of persons exposed to PBB contamination which is now underway will determine the effect of the PBB episode on human health.

PBB IN THE ENVIRONMENT AND IN FOOD

Firemaster® BP-6 photolyzes rapidly undergoing reductive dehalogenation and ring methoxylation (Ruzo and Zabik, 1973). The reactivity of the brominated biphenyls is far greater than PCBs so PBBs exposed to sunlight would be expected to photodecompose rapidly. Persistance of $2,2',4,4'5,5'$ hexabromobiphenyl (HHB), two isomers of pentabromobiphenyl, three additional isomers of hexabromobiphenyl and two isomers of heptabromobiphenyl in loam and sandy loam soils was followed during 24 weeks of incubation (Jacobs et al., 1976). Only one penta bromobiphenyl isomer showed a significant decrease. Leeching studies with soils containing 100 ppm of

HBB showed that less than 0.6% loss with leachate quantities equivalent to 20 times the average annual rainfall in Michigan (Filonow *et al.*, 1976). Therefore, PBB contamination of Michigan farm soils due to application of PBB contaminated manure should not leach below the depth of incorporation. Moreover, orchard grass and carrot tops showed no uptake of PBBs (Jacobs *et al.*, 1976). Detectable quantities of PBBs were found in carrot roots from soils with 10 or 100 ppm, but the concentrations in the range of 20 to 40 ppb were too low to quantitate precisely. Moreover, much of this contamination could be removed by rinsing the carrot surface with an organic solvent. Soils on most of the affected Michigan farms are expected to have PBB concentrations much less than 0.1 ppm. Fruits and vegetables have not been found to contain detectable levels of PBBs. This is undoubtedly due to low plant uptake as well as the low levels of PBB in the soil.

Investigation of possible PBB contamination of the Pine River in the vicinity of the Michigan Chemical Company at St. Louis, Michigan in 1974 showed concentrations of up to 3.4 μg/1, 75 yards downstream, to non-detectable levels 12 miles downstream (Hesse and Powers, 1977). Nearshore sediment concentration in the area of the plant outfall were as high as 77,000 μg/kg and declined gradually to 100 μg/kg 24 miles downstream. Elevated PBB levels were found in fish in this vicinity of the Pine River with levels up to a maximum of 1.33 mg/kg in whole fillets of carp. Ducks also had measureable PBBs but most of the contamination was in the high fat-containing skin. Michigan Chemical Corporation discontinued manufacture of PBBs at St. Louis in November, 1974. Follow-up surveys over a 3-year period indicates PBB levels have declined in river water and sediment but fish and duck tissue residues appear unchanged. Since their ingestion could be a source of human contamination of PBB, the Michigan Department of Public Health warning against consumption of Pine River fish remains in effect.

FATE OF PBBs DURING PROCESSING AND COOKING

To study the effect on PBB levels, milk from four dairy herds containing less than .3 ppm (fat basis) of physiologically incorporated PBBs was processed individually into cream, skim milk, butter, and stirred curd cheese (Murata *et al.*, 1977). Pasteurized and freeze-dried whole milk, skim milk, and cream, spray-dried whole milk and skim milk, and condensed whole milk were also made. Examination of PBB levels on a wet weight basis in the total sample shows that the PBB concentration follows the fat in these dairy products (Table 13.16). However, PBB contents (fat basis) of these products were not significantly different.

Spray-drying appeared to promote losses of PBBs from whole milk and skim milk with air outlet temperatures of 85°C or greater. Significant losses of PBB of approximately 30 to 36% from whole milk and 61 to 69% from skim milk of herd 2 were observed. Although the levels in the spraydried products from herd 1 were less than the levels in the pasteurized products, the differences were not significant.

Thus, as had been found with monitoring PCBs in milk, processing had little effect on PBB levels except as the process affected the fat level of the product. Spray drying did appear to have some potential for PBB reduction.

To study the effect of cooking on PBB levels, thigh meat, thigh skin, drumstick and breast (with skin) from half of chickens fed PBBs were analyzed raw, whereas pieces from the other half were analyzed following separate pressure cooking (Smith et al., 1977). After cooking, the level of PBBs in the set tissue decreased slightly (Figure 13.3).

Since rendering of fat is the major mode of removal of these lipophylic compounds, PBBs were also expressed on a fat basis (Figure 13.4). Average values in the fat of cooked pieces were slightly higher than those in the fat of the raw pieces. Thus, while

TABLE 13.16. LIPID CONTENT (%) AND PBBs (PPM) IN MILK AND RELATED DAIRY PRODUCTS.[1]

| Products | Lipid | PBBs | |
		Total Sample	Lipid Basis
Raw whole milk	4.5	.009	.20
Pasteurized milk	3.9	.006	.15
Raw skim milk	.3	.000	.11
Pasteurized skim milk	.2	.000	.16
Raw cream	29.7	.054	.17
Pasteurized cream	32.3	.060	.16
Butter	73.8	.108	.14
Buttermilk	.7	.001	.18
Fresh cheese	28.6	.051	.18
Aged cheese	27.9	.044	.17
Cheese whey	.6	.001	.18
Freeze-dried whole milk	26.1	.054	.20
Freeze-dried skim milk	.8	.001	.08
Freeze-dried cream	80.0	.146	.18

[1]From Murata et al., 1977.

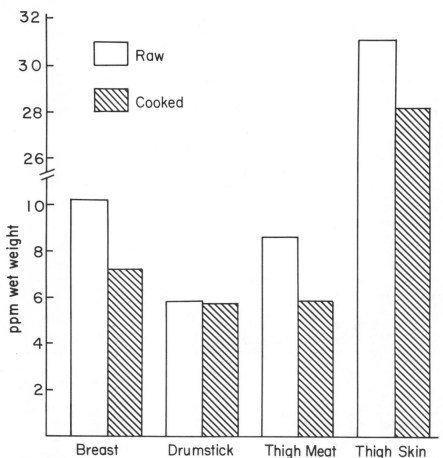

FIG. 13.3. LEVEL OF PBBs EXPRESSED ON A WET WEIGHT BASIS IN RAW AND COOKED CHICKEN PIECES (FROM SMITH *ET AL.*, 1977).

rendering of fat is undoubtedly an important mode of PBB reduction, the amount of PBB reduced is not directly proportional to fat removed.

Total micrograms of PBBs in the cooked chicken and broth were compared to the level in the respective raw chicken piece to calculate the percentage recovery. No significant differences occurred among the percentages of PBB recovered in any of the four pieces; percentage recoveries were 68.1% in thigh skin, 75.8% in the breast piece, 83.9% in the thigh meat, and 84.6% in drumsticks. Recoveries, however, did tend to be higher in the chicken pieces which con-

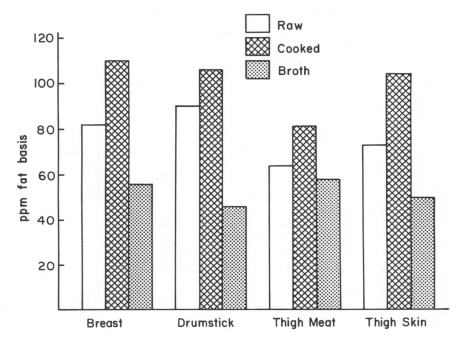

FIG. 13.4. PBBs IN THE FAT OF RAW AND COOKED CHICKEN PIECES AND CHICKEN BROTH (FROM SMITH *ET AL.*, 1977).

tained less fat (i.e. drumstick and thigh meat) even though these pieces had lower percentage meat yields than did the higher fat breast piece or thigh skin. The distribution of the recovered PBBs between the cooked meat and broth is illustrated in Figure 5. The proportion of the recovered PBB in the cooked meat is considerably higher than that found in previous studies with PCBs and chlorinated hydrocarbon pesticides.

Considering ingestion of the cooked meat only, pressure cooking brought about a total loss of PBBs from that which occurred in the raw meat of 44% for the breast piece, 36% for the drumstick, 42% for the thigh meat and 53% for the thigh skin.

Since PCBs and PBBs are lipophilic compounds, they have the potential for food chain magnification of any residues released into the environment. Indeed, they share this capacity for food chain magnification with the halogenated hydrocarbon pesticides. PCBs and PBBs have several similar characteristics many of which are also shared with other halogenated hydrocarbons. PCBs are fairly

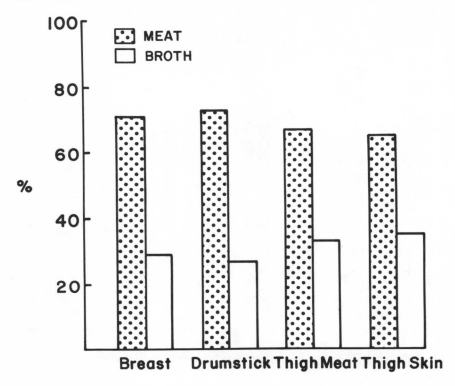

FIG. 13.5. DISTRIBUTION OF RECOVERED PBBs BETWEEN THE COOKED PIECE AND THE CHICKEN BROTH (FROM SMITH *ET AL.*, 1977).

ubiquitous environmental contaminants and are found around the globe. Highest levels are found in food and animal associated with aquatic systems. In contrast, PBB contamination was a localized incident, affecting parts of Michigan. Although it is impossible to legislate against an accident, laws have been enacted to prevent the storage of food chemicals with other chemicals which will, hopefully, minimize the chance of such a castrophe occurring again.

BIBLIOGRAPHY

ADDISON, R. F., KERR, S. R., DALE, J. and SERGEANT, D. E. 1973. Variation of organochlorine residue level with age in Gulf of St. Lawrence harp seals (Pagophilus/groenlandicus). J. Fish. Res. Bd. Can. *30*, 595–600.

AKIYAMA, K., OHI, G., FUJITANI, K., and YAGYU, H. 1976. Polychlorinated biphenyl residues in maternal and cord blood in Tokyo metropolitan area. Bull. Environ. Contam. Toxicol. 14, 588–592.

ALBRIGHT, L. J., NORTHCOTE, T. G., OLOFFS, P. C., and SZETO, S. Y. 1975. Chlorinated hydrocarbon residues in fish, crabs, and shellfish of the lower Fraser River, its estuary, and selected locations in Georgia Strait, British Columbia - 1972–73. Pestic. Monit. J. 9, 134–140.

ALLEN, J. R. 1975. Response of the nonhuman primate to polychlorinated biphenyl exposure. Fed. Proc. 34, 1675–1679.

ALLEN, J. R., CARSTENS, L. A., and BARSOTTI, D. A. 1974. Residual effects of short-term, low-level exposure of nonhuman primates to polychlorinated biphenyls. Toxicol. Appl. Pharmacol. 30, 440–451.

ALLEN, J. R., LAMBRECHT, L., BARISTOW, F., and HSIA, M. T. S. 1977. Pathobiological responses of Rhesus monkeys to polybrominated biphenyl mixture. Workshop on Scientific Aspects of Polybrominated Biphenyls - PBB. Michigan State University, October 25.

BABISH, J. G., GUTENMANN, W. H., and STOEWSAND, G. S. 1975. Polybrominated biphenyls: tissue distribution and effect on hepatic microsomal enzymes in Japanese quail. J. Agric. Fd. Chem. 23, 879–882.

BACHE, C. A., SERUM, J. W., YOUNGS, W. D. and LISK, D. J. 1972 Polychlorinated biphenyl residues: accumulation in Cayagu Lake trout with age. Sci. 177, 1191–1192.

BAILS, J. D. 1975. PCB residues in Lake Michigan fish. Assoc. Fd. Drug Off. Quart. Bull. 39, 181–184.

BARR, M. JR. 1977. Pediatric health aspects of PBBs. Workshop on Scientific Aspects of Polybrominate Biphenyls - PBB. Michigan State University, October 25.

BARSOTTI, D. A., MARLAR, R. J. and ALLEN, J. R. 1976. Reproductive dysfunction in Rhesus monkeys exposed to low levels of polychlorinated biphenyls (Aroclor 1248). Fd. Cosmet. Toxicol. 14, 99–103.

BEAUDOIN, A. R. 1977. Teratogenic studies with polybrominated biphenyls. Workshop on the Scientific Aspects of Polybrominated Biphenyls - PBB. Michigan State University, October 25.

BENGTSON, S-A. 1974. DDT and PCB residues in airborne fallout and animals in Iceland. Ambio. 3, 84–86.

BIROS, F. J., WALKER, A. C. and MEDBERY, A. 1970. Polychlorinated biphenyls in human adipose tissue. Bull. Environ. Contam. Toxicol 5, 317–323.

BJERK, J. E. 1973. DDT and PCB residues in Norwegian fish oil. Norsk VEt. 85, 597–598.

BJERK, J. E. and KVESETH, N. J. 1974. Residues of organochlorine insecticides and PCBs in samples of Norwegian butter in 1972. Nord. Vet. 26, 634–638.

BOWES, G. W. and JONKEL, C. J. 1975. Presence and distribution of polychlorinated biphenyls (PCB) in arctic and subarctic marine food chains. J. Fish Res. Bd. Can. 32, 2111–2123.

BRIGGS, D. M. and HARRIS, J. R. 1973. Polychlorinated biphenyls influence on hatchability. Poultry Sci. *52*, 1291–1294.

BRITTON, W. M. and CHARLES, O. W. 1974. The influence of dietary PCB on the PCB content of carcass lipid in broiler chickens. Poultry Sci. *53*, 1892–1893.

BRUCKNER, J. V., KHANNA, K. L. and CORNISH, H. H. 1974. Effect of prolonged ingestion of polychlorinated biphenyls on the rat. Fd. Cosmet. Toxicol *12*, 323–330.

BUHLER, D. R., CLAEYS, R. R. and MATE, B. R. 1975. Heavy metal and chlorinated hydrocarbon residues in California sea lions (Zalophus californianus californianus). J. Fish. Res. Bd. Can. *32*, 2391–2397.

BUSH, B., TUMASONIS, C. F. and BAKER, F. D. 1974. Toxicity and persistence of PCB homologs and isomers in the Avian system. Arch. Environ. Contam. Toxicol. *2*, 195–212.

CAREY, A. E. and GOWEN, J. A. 1976. PCBs in agricultural and urban soil. National PCB Conference, Chicago Illinois, November 1975. EPA-560/6–75–004, 195–198.

CARNES, R. A., DOERGER, J. U. and SPARKS, H. L. 1973. Polychlorinated biphenyls in solid waste and solid-waste related materials. Arch. Environ. Contam. Toxicol. *1*, 27–35.

CARR, R. L., FINSTERWALDER, C. E. and SCHIBI, M. J. 1972. Chemical residues in Lake Erie fish - 1970–71. Pestic. Monit. J. *6*, 23–26.

CECIL, H. C., HARRIS, S. J., BITMAN, J. and FRIES, G. F. 1973. Polychlorinated biphenyl-induced decrease in liver vitamin A in Japanese quail and rats. Bull. Environ. Contam. Toxicol. *9*, 179–185.

CECIL, H. C., BITMAN, J., LILLIE, R. J. and VERRETT, J. 1974. Embryotoxic and teratogenic effects in\unhatched fertile eggs from hens fed polychlorinated biphenyls (PCBs). Bull. Environ. Contam. Toxicol. *11*, 489–495.

CHEN, P. R., MEHENDALE, H. M. and FISHBEIN, L. 1973. Effect of two isomeric tetra-chlorobiphenyls on rats and their hepatic enzymes. Arch. Eviron. Contam. Toxicol. 1, 36–47.

CLAEYS, R. R., CALDWELL, R. S., CATSHULL, N. H. and HOLTON, R. 1975. Chlorinated pesticides and polychlorinated biphenyls in marine species, Oregon/Washington Coast, 1972. Pestic. Monit. J. *9*, 2–10.

CLARK, D. R. JR. and McLANE, M. A. R. 1974. Chlorinated hydrocarbon and mercury residues in woodcock in the United States, 1970–71. Pestic. Monit. J. *8*, 15–22.

CLAUSEN, J., BRAESTRUP, L. and BERG, O. 1974. The content of polychlorinated hydrocarbons in arctic mammals. Bull. Environ. Contam. Toxicol. *12*, 529–534.

COMBS, G. F. JR., CANTOR, A. H. and SCOTT, M. L. 1975. Effects of dietary polychlorinated biphenyls on vitamin E and selenium nutrition in the chick. Poultry Sci. *54*, 1143–1152.

COMBS, G. F. JR. and SCOTT, M. L. 1975. Polychlorinated biphenyl-stimulated selenium deficiency in the chick. Poultry Sci. *54*, 1152–1158.

COOK, R. M., PREWITT, L. R. and FRIES, G. F. 1978. Effects of activated carbon phenobarbital, and vitamins A, D, and E on polybrominated biphenyl excretion in cows. J. Dairy Sci. *61*, 414-419.

CORBETT, T. H., BEAUDOIN, A. R., CORNELL, R. G., ANVER, M. R., SCHUMACHER, R., ENDRES, J., and SZWABOWSKA, M. 1977. EM changes and other toxic effects of Firemaster BP-6 (polybrominated biphenyls) in the mouse. Workshop on the Scientific Aspects of Polybrominated Biphenyls - PBB. Michigan State University, October 25.

CORBETT, T. H. SIMMONS, J. L., KAWANISHI, H. and ENDRES, J. L. 1977. EM changes and other toxic effects of Firemaster BP-6 (polybrominated biphenyls) in the mouse. Workshop on the Scientific Aspects of Polybrominated Biphenyls - PBB. Michigan State University, October 25.

CRISETIG, G., BORGATTI, A. R. and VIVIANI, R. 1973. Residue of polychlorinated biphenyls in beef. AH: Soc. Ital. Sci. Vet. *27*, 590-595.

CRISETIG, G., BRUSCO, A., MASSA, D. and CARPENE, E. 1975a. Pesticides and polychlorinated biphenyl residues in some Italian cheeses. Industrie Alim. *14*, 112-117.

CRISETIG, G., MORA, A. and VIVIANI, R. 1975b. Residues of persistent organohalogen compounds in tissues of cattle. Folia. Vet. Latin. *5*, 1-26.

CRUMP-WIESNER, H. J., FELTZ, H. R. and YATES, M. L. 1974. A study of the distribution of polychlorinated biphenyls in the aquatic environment. Pestic. Monit. J. *8*, 157-161.

D'ARRIGO, V. 1976. Investigations on the oligoelements (Cr, Co, Cu, As, Se, Cd, Hg and Pb) and polychlorinated biphenyl content in papers from recycled fibres designed for contact with food. Rassenga. Chim. *28*, 12-18.

DENT, J. G. 1977. The characteristics of cytochrome P450 and mixed functions oxidase enzymes following treatment with PBBs - a review. Workshop on the Scientific Aspects of Polybrominated Biphenyls - PBB. Michigan State University, October 24.

DOGUCHI, M. and FUKANO, S. 1975. Residue levels of polychlorinated terphenyls, polychlorinated biphenyls and DDT in human blood. Bull. Environ. Contam. Toxicol. *13*, 57-63.

DYMENT, P. G., HERBERTSON, L. M., GOMES, E. D., WISEMAN, J. S., and HORNABROOK, R. W. 1971. Absence of polychlorinated biphenyls in human milk and serum from Texas and human milk from New Guinea. Bull. Environ. Contam. Toxicol. *6*, 532-534.

EICHNER, M. 1973. Determination of residues of chlorinated insecticides and polychlorinated biphenyls (PCB) in fish and waters of Lake Constance, the Upper Rhine and their tributaries. Zeits. Lebens-Unters-Forsch. *151*, 376-383.

ELCOMBE, C. R. and LECH, J. J. 1977. Induction of monooxygenation in rainbow trout by Polybrominated Biphenyls. Workshop on the Scientific Aspects of Polybrominated Biphenyls - PBB. Michigan State University, October 24.

ENVIRONMENTAL PROTECTON AGENCY. 1977. Polychlorinated biphenyls (PCBs) toxic substances control. Fed. Reg. *42*, 26564-26577.

FILONOW, A. B., JACOBS, L. W. and MORTLAND, M. M. 1976. Fate of polybrominated biphenyls (PBBs) in soils. Retention of hexabromobiphenyl in four Michigan soils. J. Agric. Fd. Chem. *24*, 1201–1204.

FISCHBEIN, A., LAMOLA, A., ANDERSON, K. and LORIMER, W. V. 1977. Investigation of urinary porphyrins and erythrocyte zinc protoporphyrin in Michigan dairy farm residents. Workshop on the Scientific Aspects of Polybrominated Biphenyls - PBB. Michigan State University, October 25.

FOOD AND DRUG ADMINISTRATION, U.S. DEPT. OF HEALTH, EDUCATION AND WELFARE. 1973. Polychlorinated biphenyls (PCBs) contamination of animal feeds, foods and food-packaging materials. Fed. Reg. *38*, 18096–18099.

FOOD AND DRUG ADMINISTRATION, U.S. DEPT. OF HEALTH, EDUCATION AND WELFARE. 1976. Polychlorinated biphenyls (PCBs) in certain freshwater fish. Fed. Reg. *41*, 8409–8410.

FOOD AND DRUG ADMINISTRATION, U.S. DEPT. OF HEALTH, EDUCATION AND WELFARE. 1977. Polychlorinated biphenyls (PCBs). Unavoidable contaminants in food and food packaging materials; reduction of temporary tolerances. Fed. Reg. *42*, 17487–17494.

FRANK, R., RONALD, K. and BRAUN, H. E. 1973. Organochlorine residues in harp seals (Pagophilus groenlandicus) caught in eastern Canadian waters. J. Fish. Res. Bd. Can. *30*, 1053–1063.

FRANK, R., SMITH, E. H., BRAUN, H. E., HOLDRINET, M., and McWADE, J. W. 1975. Organochlorine insecticides and industrial pollutants in the milk supply of the southern region of Ontario, Canada. J. Milk Food Technol. *38*, 65–72.

FRIES, G. F., CECIL, H. C., BITMAN, J. and LILLIE, R. J. 1976. Retention and excretion of polybrominated biphenyls by hens. Bull. Environ. Contam. Toxicol. *15*, 278–282.

FRIES, G. F., MARROW, G. S., JR. and GORDON, C. H. 1972. Similarity of a polychlorinated biphenyl (Aroclor 1254) and DDE in rate of elimination from cows. Bull. Environ. Contam. Toxicol. *7*, 252–256.

FRIES, G. F., MARROW, G. S., JR. and GORDON, C. H. 1973. Longterm studies of residue retention and excretion by cows fed a polychlorinated biphenyl (Aroclor 1254). J. Agric. Fd. Chem. *21*, 117–121.

FRIES, G. F. and MARROW, G. S. 1975. Excretion of polybrominated biphenyls into the milk of cows. J. Dairy Sci. *58*, 947–951.

FRIES, G. F., COOK, R. M. and PREWITT, L. R. 1978. Distribution of polybrominated biphenyl residues in tissues of environmentally contaminated dairy cows. J. Dairy Sci. *61*, 420–425.

FUHREMANN, T. W. and LICHTENSTEIN, E. P. 1972. Increase in the toxicity of organophosphorus insecticides to house flies due to polychlorinated biphenyl compounds. Toxicol. Appl. Pharmacol. *22*, 628–640.

FUJIWARA, K. 1975. Environmental and food contamination with PCBs in Japan. Sci. Total Environ. *4*, 219–247.

FURR, A. K., MERTENS, D. R., GUTENMANN, W. H., BACHE, C. A., and LISK, D. J. 1974. Polychlorinated biphenyl, metals and other elements in papers fed to lactating cows. J. Agric. Fd. Chem. *22*, 954–959.

GARDNER, A. M., RIGHTER, H. F. and ROACH, J. A. 1976. Excretion of hydroxylated polychlorinated biphenyl metabolites in cow's milk. J. Assoc. Off. Anal. Chem. *59*, 273–277.

GASKIN, D. E., FRANK, R., HOLDRINET, M., ISHIDA, K., WALTON, C. J., and SMITH, M. 1973. Mercury, DDT and PCB in harbour seals (Phoca vitulina) from the Bay of Fundy and Gulf of Maine. J. Fish. Res. Bd. Can. *30*, 417–475.

GIAM, C. S., HANKS, A. R., RICHARDSON, R. L., SACKETT, W. M., and WONG, M. K. 1972. DDT, DDE and polychlorinated biphenyls in biota from the Gulf of Mexico and Caribbean Sea - 1971. Pestic. Monit. J. *6*, 139–143.

GOTO, M. 1975. Pollution problems of the Seto Island Sea. Pure Appl. Chem. *42*, 155–166.

GRANT, D. L., PHILLIPS, W. E. J. and VILLENEUVE, D. C. 1971. Metabolism of a polychlorinated biphenyl (Aroclor® 1254) mixture in the rat. Bull. Environ. Contam. Toxicol. *6*, 102–112.

GRANT, D. L., MES, J. and FRANK, R. 1976. PCB residues in human adipose tissue and milk. Presented at the National PCB Conference, Chicago, Illinois. November 1975. EPA-560/6–75–004, 144–146.

GREB, W., KLEIN, W., COULSON, F. GOLDBERG, L. and KORTE, F. 1975. Metabolism of lower polychlorinated biphenyls-^{14}C in the Rhesus monkey. Bull. Environ. Contam. Toxicol. *13*, 471–476.

GREICHUS, Y. A., CALL, D. J., AMMANN, B. M., GREICHUS, A., and SHAVE, A. 1975. Physiological effects of polychlorinated biphenyls or a combination of DDT, DDD and DDE in penned white pelicans. Arch. Environ. Contam. Toxicol. *3*, 330–343.

GUTENMANN, W. H. and LISK, D. J. 1975. Tissue storage and excretion in milk of polybrominated biphenyls in ruminants. J. Agric. Fd. Chem. *23*, 1005–1007.

HANSEN, L. G., BEAMER, P. D., WILSON, D. W. and METCALF, R. L. 1976a. Effects of feeding polychlorinated biphenyls to broiler cockerels in three dietary regimes. Poultry Sci. *55*, 1084–1888.

HANSEN, L. G., WILSON, D. W. and BYERLY, C. S. 1976b. Effects on growing swine and sheep of two polychlorinated biphenyls. J. Am. Vet. Res. *37*, 1021–1024.

HESSE, J. L. and POWERS, R. A. 1977. Polybrominated biphenyl (PBB) contamination of the Pine River, Gratiot and Midland Counties, Michigan. Workshop on the Scientific Aspects of Polybrominated Biphenyls - PBB. Michigan State University, October 25.

HOLDEN, A. V. 1973. International cooperative study of organochlorine and mercury residues in wildlife, 1969–71. Pestic. Monit. J. *7*, 37–52.

HOLUB, B. J., PIEKURSKI, J. and NILSSON, K. 1975. The effect of a PCB (2,4,2′,4′-tetrachlorobiphenyl) on lipid-synthesizing enzymes in rat liver microsomes. Bull. Environ. Contam. Toxicol. *14*, 415–421.

HUMPHREY, H. E. B., PRICE, H. A. and BUDD, M. L. 1976. Evaluation of changes of the level of polychlorinated biphenyls (PCB) in human tissue. Final Report of FDA Contract 223–73–2209.

HUSCHENBETH, E. 1973. Accumulation of chlorinated hydrocarbons in fish. Arch. für Fisch. *24*, 105–116.

ITURRI, S. J., COGGER, E. A. and RINGER, R. K. 1974. Cardiovascular and hematological parameters affected by feeding various polychlorinated biphenyls to the Single Comb White Leghorn Cockerel. Arch. Environ. Contam. Toxicol. *2*, 130–142.

IVERSON, F., VILLENEUVE, D. C., GRANT, D. L. and HATINA, G. V. 1975. Effects of Aroclor 1016 and 1242 on selected enzyme systems in the rat. Bull. Environ. Contam. Toxicol. 13(4), 456–463.

IWATA, Y and GUNTHER, F. A. 1976. Translocation of the polychlorinated biphenyl Aroclor 1254 from soil into carrots under field conditions. Arch. Environ. Contam. Toxicol. *4*, 44–59.

JACKSON, T. F. and HALBERT, F. L. 1974. A toxic syndrome associated with feeding of polybrominated biphenyl-contaminated protein concentrate to dairy cattle. J. Am. Vet. Med. A. *165*, 437–439.

JACOBS, L. W., CHOU, S. F. and TIEDJE, J. M. 1976. Fate of polybrominated biphenyls (PBBs) in soils. Persistence and plant uptake. J. Agric. Fd. Chem. *24*, 1198–1201.

JAN, J., KOMAR, M. and MILOHNOJA, M. 1975. Excretion of some pure PCB isomers in milk of cows. Bull. Environ. Contam. Toxicol. *13*, 313–315.

JENSEN, S. 1966. A new chemical hazard. New Sci. *32*, 612.

JOHNSON, D. R., MELQUIST, W. E. and SCHROEDER, G. J. 1975a DDT and PCB levels in Lake Coeurd' Alene, Idaho, osprey eggs. Bull. Environ. Contam. Toxicol. *13*, 401–405.

JOHNSON, J. C., UTLEY, P. R., JONES, R. L. and McCORMICK, W. C. 1975b. Aerobic digested municipal garbage as a feedstuff for cattle. J. Animal Sci. *41*, 1487–1495.

KANEMATSU, H., MARUYAMA, T., NIIYA, I., IMAMURA, M. MIZUTANI, H., MORITA, Z, and MATSUMOTO, T. 1976a. Behavior of trace components in oils and fats during processing for edible use. I. Removal of organochlorine pesticides and polychlorinated biphenyls (PCB) from oils and fats. Jap. Oil Chem. Soc. J. *25*, 38–41.

KANEMATSU, H., MARUYAMA, T., NIIYA, I., IMAMURA, M., SUZUKI, K., KUTSUWA, Y., MURASE, L., and MATSUMOTO, T. 1976b. Behavior of trace components in oils and fats during processing for edible use. II. Changes in the contents of polychlorinated biphenyls (PCB) and organochlorine pesticides during hydrogenation. Jap. Oil Chem. Soc. J. *25*, 42–46.

KATELEY, J. R. 1977. Immunologic studies in cattle exposed to PBB. Workshop on the Scientific Aspects of Polybrominated Biphenyl - PBB. Michigan State University, October 24.

KERST, A. J. 1974. PBB. Presented to the Michigan Environmental Review Board, September 23.

KHAN, M. A., NORAK, A. F. and RAO, R. M. 1976a. Reduction of polychlorinated biphenyls in shrimp by physical and chemical methods. J. Fd. Sci. *41*, 262–267.

KHAN, M. A., RAO, R. M. and NORAK, A. F. 1976b. Adsorption of polychlorinated biphenyl (Aroclor 1254) on shrimp. Bull. Environ. Contam. Toxicol. *16*, 503–504.

KHAN, M. A., RAO, M. R. and NORAK, A. F. 1976c. Reduction of polychlorinated biphenyls in shrimp and eggs by freeze drying techniques. J. Fd. Sci. *41*, 1137–1141.

KIMBROUGH, R. D., BURSE, V. W. and LIDDLE, J. A. 1977. Persistent liver lesions in rats after a single oral dose of brominated biphenyls (Firemaster FF-1) and concomitant PBB tissue levels. Workshop on the Scientific Aspects of Polybrominated Biphenyls - PBB. Michigan State University, October 25.

KOLBYE, A. C., JR. 1972. Food exposures to polychlorinated biphenyls. Environ. Health Perspec. *1*, 85–88.

KPEKATA, A. E. 1975. Polychlorinated biphenyls (PCBs) in the Rivers Avon and Frome. Bull. Environ. Contam. Toxicol. *14*, 687–691.

KU, P. K., HOGBERG, M. G., TRAPP, A. L., BRADY, P. S., and MILLER, E. R. 1977. Polybrominated biphenyl (PBB) in the growing pig diet. Workshop on the Scientific Aspects of Polybrominated Biphenyls - PBB. Michigan State University, October 25.

KURATSUNE, M., YOSHIMURA, T., MATSUZAKA, J. and YAMAGU-CHI, A. 1972. Epidemological study on Yusho, a poisoning caused by ingestion of rice oil contaminated with a commercial brand of polychlorinated biphenyls. Environ. Health Perspect. *1*, 119–128.

KUTZ, F. W. and STRASSMAN, S. C. 1976. Residues of polychlorinated biphenyls in the general population of the United States. National PCB Conference, Chicago, Illinois, 1975. EPA–560/6–75–004, 139–143.

KVESETH, N. J. and BJERK, J. E. 1976. Organochlorine pesticides and polychlorinated biphenyls in cod from Norwegian fjords. Nordisk Vet. *28*, 170–176.

LAWRENCE, J. and TOSINE, H. M. 1977. Polychlorinated biphenyl concentrations in sewage and sludges of some waste treatment plants in Southern Ontario. Bull. Environ. Contam. Toxicol. *17*, 49–56.

LIEB, A. J., BILLS, D. D. and SINNHUBER, R. O. 1974. Accumulation of dietary polychlorinated biphenyls (Aroclor 1254) by rainbow trout (Salmo gairdneri). J. Agric. Fd. Chem. *22*, 638–642.

LILIS, R., ANDERSON, H. A., VALCIUKAS, J., and FREEDMAN, S. 1977. Comparison of findings among residents on Michigan dairy farms and consumers of produce purchased from these farms. Workshop on the Scientific Aspects of Polybrominated Biphenyls - PBB. Michigan State University, October 25.

LILLIE, R. J., CECIL, H. C., BITMAN, J. and FRIES, G. F. 1974. Differences in response of caged white Leghorn layers to various polychlorinated biphenyls (PCBs) in the diet. Poultry Sci. *53*, 726–732.

MAEKAWA, K. and HIRATA, Y. 1974. Influence of sake quality of rice contaminated with polychlorinated biphenyl. J. Soc. Brew. Jap. *69*, 175–178. Fd. Sci. Technol. Ab. *7*, 2H246 (1975).

MANSKE, D. D. and JOHNSON, R. D. 1975. Pesticide residues in total diet samples VIII. Pestic. Monit. J. *9*, 94–105.

MASUDA, Y., KAGAWA, R. and KURATSUNE, M. 1974. Comparison of polychlorinated biphenyls in Yusho patients and ordinary persons. Bull. Environ. Contam. Toxicol. *11*, 213–216.

MATTHEWS, H. B. and ANDERSON, M. 1976. PCB chlorination versus PCB distribution and excretion. National Conference on PCBs, November 1975, Chicago, Illinois. EPA 560/6–75–004.

MES, J., COFFIN, D. E. and CAMPBELL, D. 1974. Polychlorinated biphenyl and organochlorine pesticide residues in Canadian chicken eggs. Pestic. Monit. J. *8*, 8–11.

MES, J., CAMPBELL, D. S., ROBINSON, R. N. and DAVIES, D. J. A. 1977. Polychlorinated biphenyl and organochlorine pesticide residues in adipose tissue of Canadians. Bull. Environ. Contam. Toxicol. *17*, 196–202.

MICHIGAN DEPARTMENT OF AGRICULTURE. 1977. PBB Contamination stauts Report. September 1, 1977.

MOORE, R. W., DANNAN, G. A. and AUST, S. D. 1977. Effects of fire-master and selected pure brominated biphenyls on microsomal drug metabolizing enzymes. Workshop on the Scientific Aspects of Polybrominated Biphenyls - PBB. Michigan State University, October 25.

MURATA, T., ZABIK, M. E. and ZABIK, M. J. 1977. Polybrominated biphenyls in raw milk and processed dairy products. J. Dairy Sci. *60*, 516–520.

NADEAU, R. J. and DAVIS, R. A. 1976. Polychlorinated biphenyls in the Hudson river (Hudson Falls - Fort Edward, New York State). Bull. Environ. Contam. Toxicol. *16*, 436–444.

NAKAGAWA, H., KAYAMA, M. and ARIJOSHI, E. 1975. Polychlorinated biphenyls in water, sediment and organisms in two rivers in Hiroshima Prefecture. J. Foc. Fish. An. Hub., Hiroshima U. *14*, 253–259.

NIMMO, D. R. *et al.* 1975. Toxicity of Aroclor® 1254 and its physiological activity in several estuarine organisms. Arch. Environ. Contam. Toxicol. *3*, 22–39.

NISBET, I. C. T. and SAROFIM, A. F. 1972. Rates and routes of transport of PCBs in the environment. Environ. Health Perspec. *1*, 21–38.

PAREJKO, R., JOHNSTON, R. and KELLER, R. 1975. Chlorohydrocarbons in Lake Superior Lake Trout (Salvelinus namaycush). Bull. Environ. Contam. Toxicol. *14*, 480–488.

PAREJKO, R. and WU, C. J. 1977. Chlorohydrocarbons in Marquette fish hatchery Lake Trout (Salvelinus namaycush). Bull. Environ. Contam. Toxicol. *17*, 90–97.

PEAKALL, D. B. 1971. Effect of polychlorinated biphenyls (PCBs) on the eggshells of Ring Doves. Bull. Environ. Contam. Toxicol. *6*, 100–101.

PLATONOW, N. S., FUNNELL, H. S., BULLOCK, D. H., ARNOTT, D. R., SASCHENBRECKER, P. W., and GRIEVE, D. G. 1971. Fate of polychlorinated biphenyls in dairy products from milk of exposed cows. J. Dairy Sci. 54, 1305–1308.

PLATONOW, N. S., MEADS, E. B., LIPTRAP, R. M. and LOTZ, F. 1976. Effects of some commercial preparations of polychlorinated biphenyls in growing piglets. Can. J. Comp. Med. *40*, 421–428.

POLIN, D. and RINGER, R. K. 1977. PBB fed to adult female chickens: its effects on egg production, reproduction, viability of offspring and residues in tissues and eggs. Workshop on the Scientific Aspects of Polybrominated Biphenyls - PBB. Michigan State University, October 25.

RAMSEY, L. L. 1973. The Status of PCBs. Assoc. Fd. Drug Quart. Bull. *37*, 43–57.

RIVA, M., CARISANO, A. and DAGHETTA, A. 1973. Rapid GLC determination or organo-chlorine pesticide and polychlorobiphenyl (PCB) residues in foods of animal origin. Riv. Ital. Sost. Grasse *50*, 434–442.

RUZO, L. O. and ZABIK, M. J. 1975. Polyhalogenated biphenyls: photolysis of hexabromo and hexachlorobiphenyls in methanol solution. Bull. Environ. Conta. Toxicol. *13*, 181–182.

SAFE, S., HUTZINGER, O. and JONES, D. 1975. The mechanism of chlorobiphenyl metabolism. J. Agric. Fd. Chem. *23*, 851–853.

SAVAGE, E. P., TESSARI, J. D., MALBERG, J. W., WHEELER, H. W. and BAGBY, J. R. 1973. A search for polychlorinated biphenyls in human milk in rural Colorado, Bull. Environ. Contam. Toxicol. *9*, 222–227.

SCOTT, M. L., ZIMMERMANN, J. R., MARINSKY, S. and MULLENHOFF, P. A. 1975. Effects of PCBs, DDT and mercury compounds upon egg production, hatchability and shell quality in chickens and Japanese quail. Poultry Sci. *54*, 350–368.

SELIKOFF, I. J. 1977. Summary of human health effect. Workshop on the Scientific Aspects of Polybrominated Biphenyls - PBB. Michigan State University, October 24.

SHIMMA, H., ARIMA, S. and NAGAKURA, K. 1973. PCB contents in marine animals in Tokyo Bay. Bull. Jap. Soc. Sci. Fish. *39*, 1151–1162.

SIVALINGAN, P. M., YOSHIDA, T. and INADA, Y. 1973. The modes of inhibitory effects of PCBs on oxidative phosphorylation of mitochondria. Bull. Environ. Cont. Toxicol. *10*, 242–247.

SKRENTNY, R. F., HEMKEN, R. W. and DOROUGH, H. W. 1971. Silo sealants as a source of polychlorobiphenyl (PCB) contamination of animal feed. Bull. Environ. Contam. Toxicol. *6*, 409–413.

SMITH, F. A., SHARMA, R. P., LYNN, R. I. and LOW, J. B. 1974a. Mercury and selected pesticide levels in fish and wildlife of Utah: I. Levels of mercury, DDT, DDE, dieldrin and PCB in fish. Bull. Environ. Contam. Toxicol. *12*, 218–223.

SMITH, F. A., SHARMA, R. P., LYNN, R. I, and LOW, J. B. 1974b. Mercury and selected pesticide levels in fish and wildlife of Utah: II. Levels of mercury, DDT DDE dieldrin and PCBs in chuckors, pheasant and waterfowl. Bull. Environ. Contam. Toxicol. *12*, 153–157.

SMITH, S. K., ZABIK, M. E. and DAWSON, L. E. 1977. Polybrominated biphenyl levels in raw and cooked chicken and chicken broth. Poultry Sci. *56*, 1289–1296.

SMITH, W. E., FUNK, K. and ZABIK, M. E. 1973. Effects of cooking on concentrations of PCB and DDT compounds in Chinook (Oncorhynchus tshawytscha) and Coho (O. kisutch) salmon from Lake Michigan. J. Fish. Res. Bd. Can. *30*, 702–706.

SPECHT, W. 1974. Polychlorinated biphenyls (PCB) in packaging materials. Deut. Lebens.-Rundschau. 70, 136–138.

STANOVICK, R. P., SHAHIED, S. I. and MISSAGHI, E. 1973. Determination of polychlorinated biphenyl (Aroclor 1242) migration into food types. Bull. Environ. Contam. Toxicol. 10, 101–107.

SUNDSTROM, G. O., HUTZINGER, O. and SAFE, S. 1976. Identification of 2,2',4,4'5,5' hexabromobiphenyl as the major component of flame retardent Firemaster® BP-6. Chemosphere 5, 11–14.

TARUSKI, A. G., OLNEY, C. E. and WINN, H. E. 1975. Chlorinated hydrocarbons in cetaceans. J. Fish. Res. Bd. Can. 32, 2205–2209.

TATSUNO, T., SATO, M. and NAKAGAWA, Y. 1974. Transfer of polychlorinated biphenyls to food from packaging materials. Bull. Nat. Inst. Hygen. Sci. 92, 74–78.

TESKE, R. H., ARMBRECHT, B. H., CONDON, R. J. and PAULIN, H. J. 1974. Residues of polychlorinated biphenyl products from poultry fed Aroclor 1254. J. Agric. Fd. Chem. 22, 900–904.

THOMAS, G. H. and REYNOLDS, L. M. 1973. Polychlorinated terphenyls in paperboard samples. Bull. Environ. Contam. Toxicol. 10, 37–41.

THRUSTON, A. 1971. Quantitative analysis of PCBs. PCB Newsletter No. 3, July, 1971.

TUCKER, E. S., LITSCHGI, W. J. and MESS, W. M. 1975. Migration of polychlorinated biphenyls in soil induced by percolating water. Bull. Environ. Contam. Toxicol. 13, 86–91.

TUMASONIS, C. F., BUSH, B. and BAKER, F. D. 1973. PCB levels in egg yolks associated with embryonic mortality and deformity of hatched chicks. Arch. Environ. Contam. Toxicol. 1, 312–324.

TURNER, J. C. and GREEN, R. S. 1974. The effect of a polychlorinated biphenyl (Aroclor® 1254) on liver microsomal enzymes in the male rat. Bull. Environ. Contam. Toxicol. 12, 669–671.

VEITH, G. D. 1972. Chlorobiphenyls (PCBs) in Wisconsin natural waters. Environ. Health Perspec. 1, 51–54.

VEITH, G. D. 1975. Baseline concentrations of polychlorinated biphenyls and DDT in Lake Michigan fish, 1971. Pestic. Monit. J. 9, 21–29.

VEITH, G. D. and LEE, G. F. 1971. Chlorobiphenyls (PCBs) in the Milwaukee River. Water Res. 5, 1107–1115.

VILLENEUVE, D. C., GRANT, D. L., PHILLIPS, W. E. J., CLARK, M. L., and CLEGG, D. J. 1971. Effects of PCB administration on microsomal enzyme activity in pregnant rabbits. Bull. Environ. Contam. Toxicol. 6, 120–128.

VILLENEUVE, D. C., REYNOLDS, L. M. and PHILLIPS, W. E. J. 1973 Residues of PCBs and PCTs in Canadian and imported European cheeses, Canada - 1972. Pestic. Monit. J. 7, 95–96.

WEIL, W. B., SPENCER M. and BENJAMIN, D. 1977. The effects of polybrominated biphenyl (PBB) on infants and young children. Presented to Michigan State Medical Society Annual Scientific Meeting, November 8.

WESTÖÖ, G. 1974. Changes in the levels of environmental pollutants (Hg, DDT, dieldrin, PCB) in some Swedish foods. Ambio 3, 79–83.

WHEELER, W. B., KOBURGER, J. A. and APPLEDORF, H. 1973. PCB content of selected health foods. Presented at American Chemical Society 165th Annual Meeting, Dallas, Texas, April 9–13.

WILCOX, K. R., JR. 1977. The Michigan PBB incident. American Medical Association Congress on Occupational Health, September 20.

WILLETT, L. B. 1974. Coatings as barriers to prevent polychlorinated biphenyl contamination of silage. J. Dairy Sci. *57*, 816–825.

WILLETT, L. B. and HESS, J. F., JR. 1975. Polychlorinated biphenyl residues in silos in the United States. Residue Rev. *55*, 135–147.

WILLETT, L. B. and IRVING, H. A. 1975. Distribution and clearance of polybrominated biphenyls in cows and calves. J. Dairy Sci. *58*, 1429–1439.

WILLFORD, W. A., HESSELBERG, R. J. and NICHOLSON, L. W. 1976. Trends of polychlorinated biphenyls in three Lake Michigan fish. National PCB Conference, Chicago, Illinois, November 1975. EPA 560/6–75–004, 177–181.

YOSHIMURA, H. and YAMAMOTO, H. A. 1975. A novel route of excretion of 2,4,3',4'-tetrachlorobiphenyl in rats. Bull. Environ. Contam. Toxicol. *13*, 681–688.

ZABIK, M. E. 1974. Polychlorinated biphenyl levels in raw and cooked chicken tissue and chicken broth. Poultry Sci. *53*, 1785–1790.

ZABIK, M. E., HOOJJAT, P. and WEAVER, C. M. 1979. Polychlorinated biphenyls, dieldrin and DDT in Lake Trout cooked by broiling, roasting or microwave. Bull. Environ. Contam. Toxicol. *21*, 136–143.

ZITKO, V. 1977. The accumulation of polybrominated biphenyls by fish. Bull. Environ. Contam. Toxicol. *17*, 285–291.

ZITKO, V. and HUTZINGER, O. 1976. Uptake of chloro- and bromobiphenyls hexachloro- and hexabromobenzene by fish. Bull. Environ. Contam. Toxicol. *16*, 665–673.

14

Sources of Pesticide Residues

D. E. Coffin and W. P. McKinley

There are many sources of pesticide residues in foods, and these sources vary from one area to another. They depend upon the method of application, environmental factors, the type of pesticide used, the variety and extent of uses of the pesticide, etc. Many factors have not been studied in sufficient detail to allow us to evaluate adequately the scope of the problem. The food suppy is monitored quite effectively in many countries, but food is not the only source of pesticide residues. To allow a proper appraisal of the situation, the amount of each chemical entering man's body and the combined effect of all of these trace quantites of chemicals on man should be known. Obviously, their effect on wild life, domestic animals, water, soil and air is of concern, but ultimately it is the effect on man that is of prime importance. It is known that persistent pesticides such as the chlorinated hydrocarbon insecticides normally leave residues because of their stability and that they may be concentrated in the food chain. It is also known that they normally present more residue problems than the less-stable chemicals such as the organophosphate insecticides.

Pollution of the environment has been a subject of concern for many years, and in the past decade much information has been obtained to define it. Pesticides are pollutants of the air, water and soil and the degree of pollution by halogenated hydrocarbon pesticides has been relatively well documented during the last few years. However, there is very little information on the degree of pollution by other pesticides.

A partial review of the extensive literature relating to pesticide contamination follows. The role of the various sources of pesticides is

498

discussed and the relative human exposure to pesticides via air, water and food is assessed.

SOURCES

The Atmosphere

Air transport is probably the principal method whereby pesticides are dispersed over wide areas and into bodies of water far removed from the sites of application. The principal means of pesticide entry into the atmosphere are spray drift during application, volatilization from treated surfaces and movement of wind-blown dust particles. Ware *et al.* (1970) reported that less than 50% of aerially applied insecticides in Arizona were deposited on target. Volatilization from soil, plant and water surfaces and vapor phase transport are important processes in the dissipation and movement of pesticides. Caro *et al.* (1971) estimated that losses to the atmosphere of dieldrin and heptachlor incorporated into the soil at the rate of 5.6 kg/hectare just before planting were 157 and 218 g/hectare from the time of planting to harvest of a corn crop. Calculated volatilization rates (Spencer *et al.*, 1973) from surface deposits on soils indicate that DDT would be less volatile than dieldrin but that pesticides such as parathion, lindane and trifluralin would be much more volatile than dieldrin.

It is also evident that wind-blown dust particles may account for an appreciable portion of the pesticides in the atmosphere. Cohen and Pinkerton (1966) reported that a deposit of dust on the Cincinnati, Ohio area in 1965 contained DDT and chlordane. The dust had originated in the High Plains of Texas and New Mexico. Airborne particulates collected at Barbados in the Atlantic Ocean by several workers (Risebrough *et al.*, 1968; Seva and Prospero, 1971; Prospero and Seba, 1972) contained organochlorine insecticides, mainly DDT and its derivatives.

In 1966, Tabor (1966) collected airborne particles near the center of small towns in the United States surrounded by farming activities and in 3 communities with insect control operations. Concentations of DDT in the air ranged up to 23 ng/m^3 for the agricultural towns and to greater than 8000 ng/m^3 for the communities near insect-control applications. Abbott *et al.* (1965) reported residues in rain water collected in the metropolitan area of London. The results for the period February to July 1965 are given in Table 14.1. Abbott (1966) reported concentrations of p,p'-DDE, p,p'-TDE

TABLE 14.1.   ORGANOCHLORINE PESTICIDE RESIDUE LEVELS IN LONDON RAIN-WATER

(Parts per million-million)
Station - Cornwall House, London, S.E.1.

Month 1965	α-BHC	β-BHC	γ-BHC	HEOD (dieldrin)	p,p'DDE	p,p'DDT
Feb.	40	90	90	50	—	400
Mar.	15	—	60	60	—	115
Apr.	30	—	80	50	—	300
May	20	65	70	95	—	190
June	30	—	55	25	15	70
July	25	—	40	10	85	85

Source: Abbott, et al. (1965).

and p,p'-DDT of approximately 4.7, 3.5 and 3.5 ng/m^3, respectively, in the atmosphere in central London. Stanley (1971) reported the levels of 22 pesticide compounds and metabolites in air in nine U.S. localities representative of urban and agricultural areas in 1967 and 1968. p,p'-DDT and o,p'-DDT were found in the atmosphere in all nine localities. p,p'-DDE, o,p'-DDE, 2 isomers of BHC, heptachlor, aldrin, toxaphene, dieldrin, endrin, parathion, methyl parathion, malathion and DEF were also found in some of the rural localities. Heptachlor epoxide, chlordane, TDE and 2,4-D esters were not found in any of the air samples collected. The only pesticides found in the atmosphere in the 4 urban communities involved were p,p'-DDT, o,p'-DDT, p,p'-DDE, 4 isomers of BHC and 2,4-D. Table 14.2 shows the ranges of maximum levels of the pesticides found in the atmosphere at the 4 urban and 5 rural localities. Of particular note

TABLE 14.2.   MAXIMUM PESTICIDE LEVELS FOUND IN AIR SAMPLES (ng/m^3)

	urban (4)		rural (5)	
p,p'DDT	8.6–24.4	(4)	2.7–1560	(5)
o,p'DDT	1.4– 6.2	(4)	2.1– 500	(5)
p,p'DDE	2.4–11.3	(3)	3.7– 131	(4)
o,p'DDE			1.9– 9.6	(3)
α-BHC	4.5– 9.9	(3)	4.4	(1)
β-BHC	1.8– 2.2	(2)		
δ-BHC	9.9	(1)		
γ-BHC	2.6– 7.0	(2)	0.1	(1)
heptachlor			2.3–19.2	(2)
aldrin			8.0	(1)
toxaphene			68.0–2520	(3)
2,4-D	4.0	(1)		
dieldrin			29.7	(1)
endrin			58.5	(1)
parathion			465	(1)
methyl parathion			5.4– 129	(3)
malathion			2.0	(1)
DEF			16.0	(1)

Source: Stanley, et al. (1971).

are the relatively low levels of DDT and DDE found at the urban sites, the greater incidence of the BHC isomers in the urban areas, and the finding of 2,4-D in only one urban location while ten organochlorine and organophosphate pesticides were detected only at the rural locations. The levels of these pesticides in the atmosphere varied with locality and season.

Stanley (1971) estimated that the total human intake from air for the highest levels measured would only approximate the intake from the diet. The human intake from the more typical levels detected in air would be at least two orders of magnitude below the total diet intake. Barney (1969) estimated that the human intakes of DDT and related compounds, dieldrin, endrin, and lindane from air in the United States were 0.227, 0.046, 0.01 and 0.002 ug/kg body weight/day, respectively. The corresponding dietary intakes in the United States were 0.8, 0.08, 0.004 and 0.07 ug/kg body weight/day. These figures indicate that the human intake from air for endrin was much greater than from food, that the intake of DDT and dieldrin from air were approximately 30% and 50% of that from food and that the intake of lindane from air was only about 3% of that from food. In another report (A.R.C., 1970), it was concluded that, in Great Britain, the human intake of chlorinated hydrocarbon pesticides from air was only 2 to 5% of that from food.

There have been isolated reports of other pesticides in the atmosphere but there is not enough information to make any estimate of the human exposure to any pesticides other than the chlorinated hydrocarbons. The chemical and metabolic properties of most other pesticides indicate that they would not persist for a long time in the environment. But the relatively high volatility of some of these pesticides can only lead to the conclusion that they would be present in the atmosphere at relatively high concentrations for short periods. Further information is required to assess the levels of most pesticides in the atmosphere.

Soil

Whether pesticides are applied directly to soils, or as sprays or dust applied to foliage, large amounts of them ultimately reach the soil, which acts as a reservoir from which they move into the atmosphere, water or plants or in which they are metabolized or degraded.

A comprehensive review of the residues of halogenated hydrocarbons in soil reported up to 1972 was compiled by Edwards (1973). With the exception of orchard soils, the reported DDT content of most soils between 1954 and 1972 was less than 3 mg/kg. The mean DDT content of orchard soils reported in 17 separate studies in Canada, Great Britain and U.S.A. between 1950 and 1971 was considerably higher, ranging from 3 to 123 mg/kg. Much lower levels of aldrin,   -BHC, chlordane, dieldrin, endosulfan, endrin, heptachlor and heptachlor epoxide have been reported in a wide variety of soils. Experimental treatments of soil by Nash and Woolson (1967) have shown that DDT and dieldrin persist longest in soils followed by endrin, lindane, chlordane, heptachlor, and aldrin in order of decreasing persistence. Wilkinson et al. (1964) reported toxic residues sufficient to control wireworms in a Canadian silt loam nine years after treatment in 1953 with aldrin or heptachlor at 5.6 kg/hectare. The levels of residues found in this soil nine years after application of the pesticides are shown in Table 14.3.

In contrast to the prolonged persistence of chlorinated hydro-carbon pesticides in soils, organophosphates, with the exception of chlorfenvinphos and fonofos, persist at most for a few weeks in soil and present evidence indicates that the carbamate insecticides are not likely to persist in soils. According to Edwards (1973), the most persistent organic fungicides are quintozene, benomyl, methyl thio-phanate and thiram which may persist in soils for several months to 2 years. Most of the organic fungicides persist in soil for only a few days to a few weeks. Pesticides containing metals such as arsenic, copper, lead, mercury or tin break down to leave residues of these metals which are very persistent in soils. Unlike other pesticides, the presence of significant herbicide residues in soils is usually immediately obvious because of their phytotoxicity to both weed and crop plants. Some of the most persistent herbicides in soils are propazine, picloram, simazine and 2,3,6-trichlorobenzoic acid. Most organic herbicides are rapidly degraded or removed from soils.

TABLE 14.3.   RECOVERIES OF TOXIC RESIDUES FROM SILT LOAM SOIL 9 YR. AFTER TREATMENT WITH ALDRIN OR HEPTACHLOR, APPLIED AT 5.6 KG/HEC-TARE IN 1953

Pesticide	Pesticide Residues Found in 1962 (p.p.m.)		
(2.5 p.p.m.)		GLC	Bioassay
Aldrin dust	Aldrin	0.005	0.230
	Dieldrin	0.098	
Aldrin emulsion	Aldrin	0.006	0.175
	Dieldrin	0.153	
Heptachlor dust	Heptachlor	0.009	0.317
	Heptachlor epoxide	0.169	

Source: Wilkinson et al. (1964).

There is good evidence that appreciable quantities of even the relatively non-volatile chlorinated hydrocarbon pesticides may be lost from soils by volatilization to the atmosphere. Much larger quantities of more volatile pesticides are removed from soils by this process. Wind erosion is another route by which pesticides in soil may reach the atmosphere.

Pesticides are also removed from soils by water and may be transported to bodies of fresh water and eventually to oceans.

Pesticides in a soil may be altered or degraded by chemical and microbiological processes. We have only limited knowledge of the manner of degradation of pesticides in soil, but there is evidence that microorganisms play a major role in their degradation (Edwards, 1973), and some pesticides are readily degraded chemically under the conditions existing in most soils.

Plants grown on soils containing pesticides may take them up from the soils. With the possible exception of some of the metallic elements, herbicides and fungicides are not usually readily translocated from soils into growing crops. There is evidence that some soil fumigants (Wu and Salunkhe, 1971; Newsome et al., 1977) may be taken up by crops grown on soils treated with these chemicals. Data on the absorption of some chlorinated hydrocarbon insecticides from soils into various crops were reported by Lichtenstein (1959 and 1960). Lindane, DDT and aldrin were absorbed into crops. Carrots absorbed more insecticides than any crop and accumulated concentrations of lindane greater than the concentration in soil. The insecticides were most readily absorbed from a sandy loam and least readily from a muck soil. The quantities of residues found in various crops grown on a Carington silt loam soil treated with abnormally high rates of aldrin and heptachlor are shown in Table 14.4. Edwards (1973) presented a tabular review of the relative amounts of chlorinated hydrocarbon pesticides in soil and in various crops grown on these soils. This indicated that chlorinated hydrocarbons are taken up by plants from soil at varying rates depending primarily on the pesticide, the nature of the soil and the particular plant involved.

Water

Pesticides may enter bodies of water by direct application, by drift of pesticides applied to land and foliage, by volatilization from soil and foliage, by wind erosion of soil, by runoff and drainage from pesticide treated land, and by discharge of industrial wastes.

As early as 1960, Middleton and Lichtenberg (1960) demonstrated that small amounts of the chlorinated hydrocarbon insecticides

TABLE 14.4.   RECOVERIES OF ALDRIN (A), DIELDRIN (D), HEPTACHLOR (H), AND HEPTACHLOR EPOXIDE (HO) RESIDUES

| | | Insecticides Applies to Soil, Lb./5-In. Acre | | | |
| | | Aldrin | | Heptachlor | |
		5	25	5	25
		Recovered from Soils, July 21, 1958, p.p.m.			
	A	1.21	9.77	H	2.17	10.78
	D	0.36	1.31	HO	0.21	0.63
		Recovered from Crops of Harvent, p.p.m.			
Radishes	A	0.03	0.28	H	0.00	0.14
	D	0.06	0.24	HO	0.13	0.45
Beets	A	0.00	0.07	H	0.00	0.13
	D	0.07	0.18	HO	0.08	0.29
Potatoes	A	0.05	0.53	H	Traces	0.66
	D	0.09	0.67	HO	0.14	0.56
Onions	A	0.00	Traces	H	Traces	Traces
	D	0.00	0.05	HO	Traces	0.02
Carrots	A	0.15	0.94	H	0.47	3.55
	D	0.09	0.32	HO	0.08	0.43
Cucumbers	A	0.00	0.00	H	0.02	0.04
	D	0.07	0.07	HO	0.08	0.11
Lettuce	A	0.04	0.15	H	0.05	0.27
	D	0.12	0.26	HO	0.05	0.29
Beans	A	0.00	0.00	H	0.00	0.00
(seeds)	D	0.00	0.00	HO	0.00	0.00

Source: Lichtenstein, 1960.

occurred in many waterways. Green, *et al.* (1966) reported that dieldrin occurred in almost all U.S. rivers by 1964, with little consistency in its geographical distribution. Dieldrin, heptachlor and/or its epoxide, DDT and BHC have been found often in rivers and other waterways. There have been only isolated reports of chlordane, endosulfan and toxaphene in water, even in areas where these pesticides have been commonly used.

Edwards (1973) presented a tabulation of the reported chlorinated hydrocarbon pesticide concentrations in water in North America and Europe up to that time. This tabulation illustrates that much higher levels of pesticides have been detected in water, but that the residue levels are usually below 10 ng/ℓ for each of the chlorinated hydrocarbon pesticides found. These pesticides are not completely in solution in the water because they are all of very low solubility; the residues reported are usually primarily of insecticides carried on particulate matter suspended in the water. Keith (1966) and Keith and Hunt (1966) demonstrated that several chlorinated hydrocarbon pesticides rapidly became partitioned between the water, suspended material and bottom sediments in lakes, with much higher concentrations in the particles and bottom sediment than in the filtered water.

There is very little information on the presence of organophosphate and carbamate insecticide residues in water. Harris and Miles (1975) reported on studies by Stevens on the surface erosion of parathion. Parathion was found in runoff water in four of twenty-one samples in 1970 and 1971. Harris and Miles (1975) also examined water samples collected in drainage ditches adjacent to marsh where extensive use of organophosphate insecticides for insect control on vegetable crops had resulted in significant residues of diazinon, ethion and parathion in the organic soils. Residues of parathion were present in the drainage water at levels usually considerably less than 100 ng/ℓ. Striking results were obtained for diazinon which, while low at the beginning of the season, increased to much higher levels. The highest level reported was 2040 ng/ℓ.

A Canadian survey of chlorinated hydrocarbon insecticides in raw water (1977) of over 3000 samples from 333 locations between 1960 and 1975 indicated that approximately 25% of the samples contained lindane, 12% contained DDT, 7% contained endrin while aldrin, dieldrin, endosulfan, heptachlor and methoxychlor were detected in less than 5% of the samples. Chlordane was not detected in any samples. The concentrations detected were variable. Assuming a human intake of 2 liters of drinking water/day, the maximum daily intakes of pesticides would be 6.5 μg DDT, 2.6 μg endrin, 1.3 μg heptachlor plus heptachlor epoxide, 1.0 μg aldrin plus dieldrin, 0.6 μg lindane, and 0.06 μg each of endosulfan and methoxychlor. Average concentrations of these pesticides in water would indicate daily intakes of approximately 20 ng DDT, 3 ng endrin, 1.5 ng each of lindane, aldrin plus dieldrin and heptachlor plus heptachlor epoxide, and 0.1 ng each of endosulfan and methoxychlor. While these calculations of daily human intake may not be particularly accurate they do indicate that, normally, very low levels of chlorinated hydrocarbon residues are ingested in water, but that in some instances the intake from this source may be similar to the normal food intake.

Chlorinated hydrocarbon pesticides in water are concentrated by plants, invertebrates, fish and mammals growing in the water. Levels in water of a few ng/liter of DDT may result in mg/kg levels in organisms in that water. Edwards (1973) has presented an extensive review relating the levels of halogenated hydrocarbon pesticides in water (fresh and marine) to those in invertebrates, fish and mammals. DDT and dieldrin are the most common residues occurring, although the other members of this group of pesticides and their metabolities have been reported in many instances. The levels of DDT in fish do not usually exceed a few mg/kg and dieldrin levels are usually less than 1 mg/kg. Higher levels of these pesticides have been reported in fish-eating mammals. Very large residues of DDT have been found in the blubber of seals from Scotland and Canada (Holden and Marsden, 1967), Sweden and Finland (Jensen *et al.*, 1969) and Alaska (Anas and Wilson, 1970), and in porpoises caught near Scotland (Holden and Marsden, 1967), and Canada (Gaskin *et al.*, 1971). The amounts of chlorinated hydrocarbon pesticides from fish and aquatic mammals may constitute a significant portion of the human intake of these compounds from typical diets. In the case of certain limited populations reliant on fish and marine mammals for a major portion of their diet, the dietary intake of DDT, dieldrin and possibly other chlorinated hydrocarbons will exceed that of populations consuming a greater diversity of food.

Foods

Most individuals are exposed to pesticide chemicals and their metabolites mainly through foods. A large number of papers have appeared in the literature dealing with pesticide residues in a wide variety of agricultural commodities and foods. The majority have dealt with residues of the chlorinated hydrocarbon and organophosphate pesticides and the human exposure to these pesticides can be relatively well estimated. While there is some information available on the levels of many other pesticides in foods, it is not sufficient to permit the estimation of human exposure with any degree of certainty.

Within the past few years, the use patterns of some pesticides have changed dramatically. The use of many of the chlorinated hydrocarbon pesticides such as DDT, aldrin, dieldrin and heptachlor has either been discontinued or severely restricted on a worldwide basis within the past decade. These compounds, which were extensively used for approximately 20 years, have been replaced largely by much less persistent chemicals. However, residues of the chlorinated hydrocarbon pesticides still occur regularly in foods, particularly in those containing animal fats. Recent findings suggest that the residue levels of these very persistent fat-soluble compounds in foods are decreasing, but it is likely that they will continue to be found in foods for several more years even if they are no longer used as pesticides.

Monitoring studies designed to determine the normal levels of pesticide residues in foods and to estimate human dietary exposure to these chemicals have been carried out in several countries in recent years. The results from several Canadian monitoring studies serve to illustrate the situation with respect to the concentration in foods and the dietary exposure to chlorinated hydrocarbon and organophosphate pesticide residues. The average levels of residues found in a survey conducted by Health Protection Branch, Health and Welfare, Canada (unpublished data) of over 2000 randomly selected samples of Canadian food commodities from 1972 to 1975 are shown in Tables 14.5 and 14.6. The major residues found in most products were those of the chlorinated hydrocarbon pesticides with the combination of DDT and its DDE and TDE metabolites being the only residue found in every commodity and usually at levels considerably higher than those of the other chlorinated hydrocarbons. Except for the levels of DDT and dieldrin in carrots and the occurrence of residues of endosulfan and dicofol only in plant products, the chlorinated hydrocarbon residues were concentrated in animal products. No organophosphate residues were detected in any animal products and, generally, they were found only occasionally in plant products. Notable exceptions were the levels of malathion in vegetable oils, ethion and parathion in oranges and phosalone in apples. These two tables illustrate that, although residues of many chlorinated hydrocarbon and organophosphate pesticides may occur in foods, the normal residue levels are low, with none of these pesticides being present at an average level greater than 0.1 mg/kg in any of the 14 food commodities examined. In a companion survey of approximately 500 randomly selected samples of cereal grains,

TABLE 14.5.   MEAN LEVELS OF PESTICIDE RESIDUES IN CANADIAN FOOD COMMODITIES OF ANIMAL ORIGIN, 1972 - 1975 (μg/kg)

	Fat of milk and butter	Fat of beef	Fat of pork	Fat of poultry	Eggs
DDT[1]	34	13	44	60	5
dieldrin	15	6	1	12	<1
BHC	16	17	3	5	
heptachlor epoxide	2	2	<1	<1	<1
toxaphene		38	5		
chlordane		<1			
endrin	<1	<1			

[1]Total of p,p'- and o,p'- isomers of DDT, DDE and TDE.

TABLE 14.6.   MEAN LEVELS OF PESTICIDE RESIDUES IN CANADIAN FOOD COMMODITIES OF PLANT ORIGIN, 1972 - 1975 (µg/kg)

	flour	potatoes	carrots	cabbage	beans	tomatoes	vegetable oils	apples	oranges
DDT[1]	1	4	39	<1	1	1	1	1	1
dieldrin	<1	<1	4	<1			<1		<1
BHC	7		1		<1	1	6		
heptachlor epoxide		<1	<1						
chlordane		2		2					
endrine				3	1		<1		
endosulfan						7			
dicofol								1	<1
azinphosmethyl								5	52
carbophenothion								<1	2
diazinon	<1		3					<1	<1
dimethoate								1	
dioxathion								<1	
disyston		<1	<1						6
ethion								2	26
fenthion							3		
guthion								3	
imidan								10	
malathion	5						39		
methyl parathion			<1						5
parathion			1				1	95	19
phosalone									

[1]Total of p,p'- and o,p'- isomers of DDT, DDE and TDE.

flour, milk, potatoes and vegetables carried out in 1974 and 1975, no chlorophenoxy acid herbicide residues were detected.

While there have been many reports attesting to the presence or absence of many other pesticides in specific instances or under specific conditions, the authors know of no data which could be used as a measure of the levels normally present in the food supply of any country. The manner and time of application and/or the reactivity of most of these compounds indicate that significant residues are unlikely to be present in foods.

However, the levels of pesticides in raw agricultural products are not necessarily a reliable indicator of the dietary exposure of humans to pesticides. Many food preparation and storage processes have been shown to alter pesticide residue levels. Koivistoinen *et al.* (1964) demonstrated major losses of malathion from a variety of fruits and vegetables during canning, cooking, juice production, drying and freezing. Langlois *et al.* (1964, 1965) showed losses of several chlorinated hydrocarbon pesticides from milk during condensing and drying. The complete loss of aldrin, BHC, chlordane, DDT, dieldrin, heptachlor, heptachlor epoxide, kelthane, lindane, methoxychlor, sesone, strobane, DDD and toxaphene during deodorization of vegetable oils was reported by Gooding (1966). Major transfer of chlorinated hydrocarbons to fat drippings may occur by rendering of fat from meats during cooking as has been shown for DDT (Carter *et al.*, 1948). Moffitt and Nelson (1963) demonstrated that malathion, DDT and lindane were present in flour at much lower levels than in the whole grain. Some pesticides may be completely lost while partial losses of others occur during the fermentation process (Painter *et al.*, 1963).

Several countries have conducted total diet or market basket studies to estimate the daily dietary intake of pesticides. While the specific designs of the studies in different countries have been somewhat different, the results obtained in Canada, United States, and Great Britain have been somewhat similar. The results of the Canadian total diet studies for the years 1969 to 1974 are shown in Table 14.7. These studies were designed to estimate the average Canadian dietary intake of pesticides from foods prepared for consumption in a normal manner. This figure demonstrates that DDT (p,p'- and o,p'-isomers of DDT, DDE and TDE) was the major pesticide in the diet over this period and that there was a major decrease in the daily dietary intake of DDT and the total chlorinated hydrocarbon pesticides between 1969 and 1974. In comparison to the 30 μg/person/day intake of chlorinated hydrocarbon pesticides in 1969, the highest intake of organophosphate pesticides was approxi-

TABLE 14.7.   PESTICIDES IN CANADIAN DIET FROM TOTAL DIET STUDIES

Pesticide	Dietary Intake (µg/person/day)					
	1969	1970	1971	1972	1973	1974
Total BHC	2.5	2.2	3.5	3.3	2.3	0.9
heptachlor & heptachlor epoxide		0.04	0.3		0.3	0.1
aldrin & dieldrin	4.0	1.2	2.3	1.5	1.8	0.7
DDT[1]	18.9	7.4	11.6	4.8	4.1	1.7
endrin			0.4		0.2	0.04
dicofol	3.8		1.2	1.4	3.6	
endosulfan	0.4	0.6	0.4	1.1	1.5	0.3
methoxychlor						0.3
diazinon			1.8	2.2	0.2	
ethion		1.4			0.3	1.5
malathion		2.1	3.0	0.7	0.7	0.02
parathion & methyl parathion		0.2	0.2	0.5	0.04	0.02
captan		1.7				
folpet		0.9				
Total chlorinated hydrocarbon	29.6	11.4	19.7	12.1	13.8	4.0
Total organophosphate		3.7	5.0	3.4	1.2	1.5

[1]Total of p,p'- and o,p'- isomers of DDT, DDE and TDE.

mately 5 µg/person/day. By 1974 the daily dietary intake of chlorinated hydrocarbon pesticides had been reduced to approximately 5 µg/person and of organophosphate pesticides to approximately 2 µg/person.

Similar results have been obtained in three surveys of human milk conducted in Canada in 1967, 1970 and 1975 (Ritcey *et al.*, 1972; Mes, 1978). The results of these surveys, designed to determine the levels of chlorinated hydrocarbons in human milk, are shown in Table 14.8. The major residue was DDT and its metabolites. The levels of DDT, dieldrin, γ-BHC and heptachlor epoxide in the human milk decreased dramatically in the 8 year period from 1967 to 1975. β-BHC, oxychlordane and t-nonachlor were not determined in the 1967 and 1970 surveys but were found in the samples collected in 1975. The levels of DDT and dieldrin in human milk in 1975 were still sufficiently high to be a cause of concern in the case of infants whose sole source of food is human milk.

There has been no attempt to appraise all the data in the literature on the topics dealt with in this paper. There is enough information available to make some estimates of the human exposure to the chlorinated hydrocarbon pesticides via air, water and food and of the dietary exposure to organophosphate pesticides. At this time, there is inadequate information available to assess the extent of contamination of the environment and human exposure to other

TABLE 14.8. CHLORINATED HYDROCARBON PESTICIDES IN HUMAN MILK IN CANADA

Pesticide	Mean pesticide level (μg/kg)		
	1967	1970	1975
γ-BHC	3	2	0
β-BHC	N.A.	N.A.	2
heptachlor epoxide	3	4	1
oxychlordane	N.A.	N.A.	1
t-nonachlor	N.A.	N.A.	1
dieldrin	5	3	1
DDT[1]	139	78	44

[1]Total of p,p'- and o,p'- isomers of DDT, DDE and TDE.
N.A. - not analysed.

pesticides. The chemical and biological properties of most of the pesticides in use today suggest that they are unlikely to persist or accumulate in the environment. However, it is probable that relatively large amounts of many of these chemicals might be present in the atmosphere and in water for relatively short times after application.

In conclusion, it can be stated that to the best of our knowledge there is no need for alarm. Pesticides in the environment at this time do not appear to be causing any serious health problems. The very persistent chlorinated hydrocarbon pesticides are decreasing in the environment and, because of their restricted use, this decrease should continue. Except in some localized areas or special dietary situations, the levels of these pesticides in air, water and food do not approach tolerance levels and human exposure to them is only a small fraction of what is considered to be acceptable. However, pesticides must be handled wisely and with caution and much more monitoring information is required to determine the degree of contamination of the environment by the pesticides in use today.

ACKNOWLEDGMENT

The assistance of Josie Butterfield in editing this manuscript is appreciated.

BIBLIOGRAPHY

ABBOTT, D. C., HARRISON, R. B., TATTON, J. O.' G., and THOMSON, J. 1965. Organochlorine pesticides in the atmospheric environment. Nature 208, 1317.

ABBOTT, D. C., HARRISON, R. B., TATTON, J. O.' G., and THOMSON, J. 1966. Organochlorine pesticides in the atmosphere. Nature 211, 259.

ANAS, R. E., and WILSON, A. J., JR. 1970. Organochlorine pesticides in fur seals. Pest. Monit. J. 3, 198.

CARO, J. H., TAYLOR, A. W., and LEMON, E. R. 1971. Measurement of pesticide concentrations in the air overlying a treated field. Proc. Internat. Symp. Identification and Measurement of Environmental Pollutants, Ottawa, Canada.

CARTER, R. H. et al. 1948. Effect of cooking on the DDT content of beef. Science 107, 347.

COHEN, J. M., and PINKERTON, C. 1966. Widespread translocation of pesticides by air transport and rain-out. Adv. Chem. Series 60, 163.

EDWARDS, C. A. 1973. Persistent pesticides in the environment. CRC Press, Cleveland Ohio U.S.A.

ENVIRONMENT CANADA. 1977. Surface Water Quality Data. National Water Quality Data Bank. Inland Waters Directorate, Water Quality Branch.

GASKIN, D. E., HOLDRINET, M., and FRANK, R. 1971. Organochlorine pesticide residues in harbour porpoises from the Bay of Fundy region. Nature 233, 499.

GREEN, R. S., GUNNERSON, G. C., and LICHTENBERG, J. J. 1966. Pesticides in our national waters, In Agriculture and the Quality of Our Environment, American Association for the Advancement of Science, Washington DC, p. 137.

GOODING, C. M. B. 1966. Fate of chlorinated organic pesticide residues in the production of edible vegetable oils. Chem. Ind. 1, 344.

HARRIS, C. R., and MILES, J. R. W. 1975. Pesticide residues in the Great Lakes region in Canada. Residue Reviews 57, 27.

HOLDEN, A. V., and MARSDEN, K. 1967. Organochlorine pesticides in seals and porpoises. Nature 216, 1274.

JENSEN, S., JOHNELLS, A. G., OLSSON, M., and OTTERLUND, G. 1969. DDT and PCB in marine animals from Swedish waters. Nature 224, 247.

KEITH, J. O. 1966. Insecticide contaminations in wetland habitats and their effects on fish eating birds. J. Appl. Ecol. (Suppl.) 3, 71.

KEITH, J. O., and HUNT, E. G. 1966. Levels of insecticide residues in fish and wildlife in California. Trans. 31st. North Amer. Wildlife Res. Conf. 150.

KOIVISTOINEN, P., KANOVEN, M., KORINPAA, A., and ROINE, P. 1964. Stability of malathion residues in food processing and storage. J. Agr. Food Chem. 12, 557.

LANGLOIS, B. E., LISKA, B. J., and HILL, D. L. 1964. The effect of processing and storage of dairy products on chlorinated insecticide residue, I. DDT and lindane. J. Milk Food Technol. 27, 264.

LANGLOIS, B. E., LISKA, B. J., and HILL, D. L. 1965. The effect of processing and storage of dairy products on chlorinated insecticide residue, II. Endrin, dieldrin and heptachlor. J. Milk Food Technol. 28, 9.

LICHTENSTEIN, E. P. 1959. Absorption of some chlorinated hydrocarbon insecticides from soils into various crops. J. Agr. Food Chem. 7, 430.

LICHTENSTEIN, E. P. 1960. Insecticidal residues in various crops grown in soils treated with abnormal rates of aldrin and heptachlor. J. Agr. Food Chem. 8, 448.

MES, J., and DAVIES, D. J. 1979. The presence of polychlorinated biphenyl and organochlorine pesticide residues and the absence of polychlorinated terphenyls in Canadian human milk samples. Bull. Environ. Contam. Toxicol. 21,381.

MIDDLETON, F. M., and LICHTENBERG, J. J. 1960. Measurement of organic contaminants in the nations rivers. Ind. Eng. Chem. Eng. Anal. Ed. 52, 99.

MOFFITT, R. A., and NELSON, J. H. 1963. Chlorinated hydrocarbon residues in cereals. Cereal Sci. Today 8, 72.

NASH, R. G., and WOOLSON, E. A. 1967. Persistence of chlorinated hydrocarbon insecticides in soils. Science 157, 924.

NEWSOME, W. H., IVERSON, F., PANOPIO, L. G. and HIERLIHY, S. L. 1977. Residues of dibromochloropropane in root crops grown in fumigated soil. J. Agr. Food Chem. 25, 684.

PAINTER, R. R., KILGORE, W. W., and OUGH, C. S. 1963. Distribution of pesticides in fermentation products obtained from artificially fortified grape musts. Food Sci. 28, 342.

PROSPERO, J. M., and SEBA, D. B. 1972. Some additional measurements of pesticides in the lower atmosphere of the northern equatorial Atlantic Ocean. Atmospheric Environ. 6, 363.

RISEBROUGH, R. W., HUGGETT, R. J., GRIFFEN, J. J., and GOLDBERG, E. D. 1968. Pesticides: Transatlantic movements in the northeast trades. Science 159, 1233.

RITCEY, W. R., SAVARY, G., and McCULLY, K. A. 1972. Organochlorine insecticide residues in human milk, evaporated milk and some milk substitutes in Canada. Can. J. Publ. Health, March/April.

SEBA, D. B., and PROSPERO, J. M. 1971. Pesticides in the lower atmosphere of the northern equatorial Atlantic Ocean. Atmospheric Environ. 5, 1043.

SPENCER, W. F., FARMER, W. J., and CLAITH, M. M. 1973. Pesticide volatilization. Residue Reviews 49, 1–47.

STANLEY, C. W., BARNEY, J. E., HELTON, M. R., and YOBS, A. R. 1971. Measurements of atmospheric levels of pesticides. Environ. Sci. Technicol. 5, 430.

TABOR, E. C. 1966. Contamination of urban air through the use of insecticides. Trans. N.Y. Acad. Sci. 28, 569.

WARE, G. W., CAHILL, W. P., GERHARDT, P. D., and WITT, J. M. 1970. Pesticide drift. IV. On-target deposits from aerial application of pesticides. J. Econ. Entomol. 63, 1982.

WILKINSON, A. T. D., FINLAYSON, D. G., and MORLEY, H. V. 1964. Toxic residues in soil nine years after treatment with aldrin and heptachlor. Science 143, 681.

WU, M. T., and SALUNKHE, D. K. 1971. Influence of soil fumigation of Telone and Nemagon on the ultrastructure of chromoplasts in carrot roots (Daucus carota). Experientia 15, 712.

15

Antibiotics and Food Safety

Arun Kilara

Chemotherapy or the administration of chemical agents to control and destroy systemic microbial infections is a relatively new field. Ancient folklore medicine advocated the use of moldy cheese to cure cuts and wounds. The ancient Chinese recognized the ability of the juice from moldy soybeans to cure boils. It has also been reported that the South American Indians were cognizant of the medicinal value of cinchona bark in abating malarial fever. It was not until the 19th century, however, that chemotherapy started developing as a science. In this period, the active alkaloids such as quinine and emetine were actually isolated from botanical sources. The next breakthrough did not occur until 1910 when Paul Erlich and his colleagues discovered a compound effective against syphilis and recurring fever, two diseases caused by Spirochetes. The compound called Salvarsan or Compound 606 was arsenobenzene. The success of this treatment prompted a search for other compounds containing elements such as mercury, bismuth, arsenic and antimony. Concurrently, a number of dyestuffs such as a trypan blue and trypan red were being studied for their efficacy against ceratain protozoa. These compounds were not effective in controlling bacterial infections. In 1932 Gerhard Domagk discovered Prontosil, the predecessor of sulfa or sulfonamide drugs, which was effective against bacterial infections causing streptococcal sore throat, bacterial pneumonia, gonorrhea and forms of meningitis. Even as this breakthrough was being made in medicine the antibiotic era was beginning.

[1]Authorized for publication on 2/19/79 as Paper No. 5683 in the journal series of The Pennsylvania Agricultural Experiment Station.

The historic definition of antibiotic was generated nearly 40 years ago as products from living organisms that are not toxic to the organisms producing them but which are capable of inhibiting the growth of, or often killing in very low concentrations, one or more other organisms. The main difference between chemotherapeutic agents and antibiotics was that, while chemotherapeutic agents were synthetic chemicals, antibiotics were secondary metabolites or biosynthetic in origin. These definitions are historic. For example, even though chloramphenicol is now made synthetically it is still referred to as an antibiotic and not a chemotherapeutic agent.

The word antibiosis is derived from Greek (Anti=against, Bios= Life). Pasteur and Jokert in the late 19th century were responsible for the introduction of this word into scientific terminology. They observed that the growth of *Bacillus anthracis* was inhibited by a number of other organisms. At about the same time Bouchard realized that a pure culture of *Pseudomonas aeruginosa* or an extract of its culture medium inhibited *B. anthracis* and, in 1897, Emmerich and Lowe isolated and purified pyocyanase from *P. aeruginosa*. Thus, the first antibiotic was available shortly before the turn of the century but did not gain use because of the toxic side effects. It had also been realized by Gosio in 1896 that antibacterial compounds could be synthesized by organisms other than bacteria. In 1928, a British bacteriologist Alexander Fleming noticed that *Penicillium notatum* growing as a contaminant on an agar culture of *Staphylococcus aureus* was capable of inhibiting the growth of this bacterium. Thus Fleming is credited with the discovery of penicillin. Ten years later, in 1938, Florey, Abraham and Chain successfully isolated and purified the antibacterial compound (penicillin) produced by *P. notatum*. The potency of this compound was remarkable as was its spectrum i.e. efficacy against a wide variety of microorganisms. The Second World War began in Europe and thus further research was shifted from Britain to the United States. In 1944 an American microbiologist, Selman Waksman, isolated streptomycin from the fungus-like bacteria found in soil called actinomycetes. Streptomycin was particularly effective in inhibiting *Mycobacterium tuberculosis*, the organism responsible for human tuberculosis. Penicillin was not effective against this microorganism. The discovery of these two antibiotics catalyzed a systematic search for antibiotics produced by other organisms. In a relatively short span of six years tetracycline and chloramphenicol were added to the list of antibiotics and, after 1950, the list of discoveries of antibiotics grew phenomenally. However, today the list can be narrowed to a few select groups of compounds which are effective and without toxic side effects.

CLASSIFICATION AND MODE OF ACTION

Antibiotics can be classified according to chemical structure and therapeutic properties. One method of classification was provided by Rehm (1967) and is shown in Table 15.1.

Antibiotics may be either bacteriostatic or bactericidal. Bacteriostatic antibiotics inhibit bacterial reproduction but do not kill the organism. Removal of the antibiotic restores normal reproduction of the bacteria. Bactericidal antibiotics, on the other hand, kill the organisms. Antibiotics may also be differentiated on the basis of their breadth or spectrum of activity. Broad spectrum antibiotics, such as tetracyclines and chloramphenicol, affect not only many species within a genus but also unrelated organisms i.e. they are active against bacteria and rickettsiae. Narrow spectrum antibiotics, like penicillin and streptomycin, have limited effects and act only against few bacterial species. It should be clearly understood that no single antibiotic is effective against all microorganisms. Those organisms that are succeptible to an antibiotic are called sensitive and those unaffected by an antibiotic are termed resistant.

The range of activity of an antibiotic depends on many factors among which are the availability of appropriate binding sites on the microorganism and the production of enzymes which inactivate the antibiotic. The mode of action of antibiotics can be classified into four major categories; a) acting on cell wall, b) interfering with protein synthesis, c) affecting cell membranes, and d) affecting nucleic acid metabolism (Carter and McCarthy, 1966). Antibiotics such as penicillin, cephalosporin and vancomycin all differ in chemical structure but have a common mode of action. These antibiotics inhibit the synthesis of mucopeptides in bacteria undergoing reproduction. Mucopeptides are essential components of the bacterial cell wall and, since the synthesis of these compounds are impared or inhibited, the intracellular material of bacteria are lost and cell death ensues. Lack of total envelopment of intracellular constituents by the cell wall renders these organisms succeptible to the environment. Antibiotics that inhibit protein synthesis, such as chloramphenicol, tetracyclines, erythromycin, lincomycin, bind to ribosomes. Tetracycline, for example, prevents the combination of aminoacyl-t-RNA with the initiator site on the 30S component. Streptomycin combines with the 50S component in a way that prevents normal combinations with the 30S components during chain initiation (Weisblum and Davis, 1968).

Actinomycin D is an example of a compound that associates with the surface of DNA in such a way that the formation of messenger RNA (mRNA) cannot proceed. Thus, new mRNA cannot be formed

TABLE 15.1.   CLASSIFICATION OF ANTIBIOTICS

Group	Example
1. Penicillins and Cephalosporins	Penicillin G, Cephalosporin C
2. Oligosaccharide antibiotics	Streptomycin, Kanamycin A
3. Chloramphenicol	Chloromycetin
4. Tetracyclines	Aureomycin
5. Macrolide antibiotics	Erythromycin, tylosin
6. Polyene antibiotics	Nystatin, Pimaricin
7. Polypeptide antibiotics	Tyrothricin, nisin
8. Siderochrome antibiotics	Grisein
9. Griseofulvin	
10. Novobiocin	
11. Sulfonamides	
12. Nitrofurans	

Adapted from Rehm (1967).

but the existing mRNA is unaffected and proceeds with synthesizing new peptides. Puromycin stops protein synthesis but does not interfere with RNA formation. This antibiotic mimics the binding sites of tRNA and contains a free ammonium group to which a growing peptide chain can be transferred. However, puromycin lacks the groups necessary to bind to the other enzymes of the ribosome complex, and protein synthesis ceases.

The antibiotics which interfere with cell membrane functions can be either surface-active agents or steroid binding agents. Polymyxin and colistin are surface active agents which orient themselves between the lipid and protein films. This orientation disrupts bacterial membranes allowing cellular contents to escape or leak out. Polyene antifungal agents such as nystatin and amphotericin B, on the other hand, bind to sterols in cell wall membranes of sensitive yeast and fungi leading to leakage of cellular contents.

The selective toxicity of antibacterial agents has been traced in most cases to specific metabolic or structural differences between microorganisms and mammalian cells. For example, bacteria convert para-amino benzoic acid (PABA) to folic acid and sulfonamides interfere with this conversion. Humans rely on obtaining folic acid through diet and cannot convert PABA. Thus sulfonamides do not interfere with the function of mammalian cell. Similarly, the antibiotics that inhibit DNA or RNA functions are specific for microbial nucleic acids, since there are significant differences between bacterial and mammalian ribosomes.

RESISTANCE TO ANTIBIOTICS

The above discussion indicates that antibiotics exert their effects by various types of interferences in the normal cellular functions of microorganisms. It was also briefly mentioned that organisms not affected by an antibiotic are termed resistant. The term resistant is relative and is applicable only for a specific microorganism and a specific antibiotic. Resistance is not universal, i.e. antibiotic resistance does not occur in all organisms nor with all drugs. After 25 yr of widespread use of penicillin B, for example, no important degree of resistance developed in streptococci that cause throat infections or pneumonia. In contrast, penicillin-resistant staphylococci were isolated shortly after the prolific use of penicillin. Such penicillin-resistant staphylococci are commonly found in hospitalized patients. Some bacterial strains isolated prior to the development of an antibiotic may be resistant to it, other strains apparently develop resistance only after a drug has been used for a period of time. Bacteria develop resistance in one of several ways including mutation, transduction, transformation and conjugation (Benevenisti and Davis, 1973). In the 1950's it was generally hypothesized that the major mechanism for antibiotic resistance was mutation. Some strains of microorganisms which are sensitive to a particular antibiotic may undergo a spontaneous unrelated change or mutation in their genetic material without being exposed to the drug, thereby becoming resistant. Resistance was viewed as a consequence of the presence within a large succeptible population of a few resistant cells that could grow in the presence of an antibiotic. Sometimes a single mutation can make the cell highly resistant, in other cases a number of sequential mutations are required before a sensitive organism develops into a resistant one.

Resistance arising from transduction involves bacteriophages. Bacterial genes containing information necessary to develop resistance to a specific antibiotic may become attached to a bacteriophage as it reproduces in a resistant bacterium. This gene may then be transferred to a sensitive cell by the phage hence converting a sensitive population to a resistant one. During the phenomenon of transformation, a sensitive bacterial cell may become resistant to a specific antibiotic by utilizing a gene that was originally a part of the genetic material of a resistant bacterium. Conjugation is the sexual mating of bacteria. One bacterial cell may transfer genetic material responsible for resistance to another bacterium. In such a process, cellular contact is necessary but resistance to as many as five or six drugs may appear simultaneously and rapidly in a previously sensitive bacterium. Among bacteria known to transfer drug resistance by conjugation are *Salmonella*, *Shigella* and *Escherichia coli*.

During the past decade it has been realized that the emergence of resistant microorganisms is related to the transfer of R-plasmid.

Plasmids are DNA molecules in cellular protoplasm independent of the chromosome, capable of replication but not essential for bacterial survival. There are many types of plasmids but the term "R" is used to denote plasmid responsible for antibiotic resistance. R-plasmids vary in the number of resistance genes they carry by as few as one to as many as ten i.e. the R-plasmid may confer resistance against as few as one antibiotic to as many as ten antibiotics (Watanabe, 1963). The resistance conferred by R-plasmids leads to the synthesis of enzymes capable of attacking antibiotic molecules. The mechanism of transfer of R-plasmids involves conjugation, transformation or transduction process. The R-plasmids are not static entities and are, in fact, constantly evolving within an organism thereby making them resistant to more than one antibiotic. Fortunately, the transfer of R-plasmids by conjugation is not readily favored by environmental factors *in vivo* and hence is not a common occurrence. Plasmid transfer is also dependent on the range of hosts available. Some R-plasmids can be transferred to as many as 36 different species of bacteria and transposons (Tn) are involved in this process. Transposons are segments of DNA within an intact plasmid or bacterial chromosome that have the capacity to detach themselves from one DNA molecule and insert themselves into a recipient DNA molecule. Many plasmid genes which confer antibiotic resistivity reside upon Tn. Since these Tns can be transferred from plasmid to plasmid, the rapid evolution of R-plasmids which possess genes confering resistivity to more than one antibiotic can result. Resistance to Ampicillin, for example, is very common yet plasmids specifying ampicillin resistance can be very different from one another except for a common transposon, TnA, coding for penicillinase (Elwell *et al.*, 1975; Elwell *et al.*, 1977).

From the discussion in the preceeding sections it becomes clear that antibiotics have a very useful role to play in controlling pathogenic microorganisms thereby contributing to the well being of humans and animals. While their uses have increased phenomenally, the potential for the microorganisms to develop resistance to antibiotics is very real. In the next two sections, the use of antibiotics in food preservation and as supplements to animal feeds will be considered.

USE OF ANTIBIOTICS IN FOOD PRESERVATION

The ability of antibiotics to inhibit bacterial growth, even when very minute amounts of these chemicals were used, made them attractive compounds for use in food preservation. On an equal-weight basis, antibiotics were 100 to 1,000 times more effective in controlling bacterial growth than other food preservatives. However, the use of antibiotics as food preservatives is not sanctioned by the U.S. Food and Drug Administration and hence this section will review the literature from a historic perspective. Antibiotics exhibit selective activity i.e. some are effective only against gram-positive bacteria, some against gram-negative bacteria, some against both groups (broad spectrum) and some are strictly antifungal. These compounds are stable over a relatively wide pH range and their effect is mainly bacteriostatic and not bactericidal. Because of this, it is necessary for antibiotics to be present in the food at all times and the stability of the antibiotic becomes an important parameter in determining its use. Hence, Wrenshall (1959) proposed the use of antibiotics in conjunction with other methods of preservation such as refrigeration, freezing and heat processing. The efficacy of antibiotics as preservatives in various foods can be listed as follows:

1) Human Milk:—considerable controversey exists on the merits and demerits of breast feeding. Current thinking of pediatricians seems to favor breast feeding. It was felt that the addition of antibiotics to expressed human milk would enhance the keeping quality of this product. Growth of coliform organisms and lactic acid bacteria may result in reduced shelf-life of human milk. Linneweh (1949a) suggested the addition of 2.0 mg of streptomycin per 100 ml of human milk followed by storage in sterile glass bottles as a means of overcoming the problems of storing human milk. The efficacy of streptomycin in controlling the growth of gram negative pathogens without altering *Lactobacillus bifidus* was also demonstrated (Poetschke, 1949; Poetschke, 1950; Linneweh, 1949b). Several other researchers found essentially similar results (Roos and Kindler, 1959; Bindenwald, 1952; Thomas, 1953). Streptomycin, however, was not an ideal preservative because it did not destroy *Mycobacterium tuberculosis* (Roos, 1951). Additionally, the presence of streptomycin-resistant organisms was also discovered in human milk (Kayser, 1952, Damerow, 1956). Lactating women receiving antibiotic therapy for infections transfer residual levels into the milk they secrete. The duration of the secretion of residual antibiotics into milk varies with the type and dosage of antibiotics (Rozansky and Brzezinsky, 1949; Neuweiler and Kastil, 1953; Teixeira and Scott, 1958).

2) Fruits and vegetables:— Fresh fruits and vegetables have been treated with antibiotics to extend storage times by controlling spotting and sprouting. Radishes (and cole slaw) could be preserved longer with oxytetracycline (Thompson, 1959; Francis, 1960). Cucumber, chicory, escarole and lettuce were treated with equal concentrations (50 ppm) of oxytetracycline, streptomycin and polymyxin B and stored at 30°C, 10°C and 5°C. Oxytetracycline was the most effective in retarding spoilage followed by streptomycin and polymyxin B (Carroll, et al., 1957). Oxytetracycline was also effective in controlling spoilage of fresh leafy vegetables (Kersey et al., 1954). Dihydrostreptomycin and streptomycin could control rots caused in Erwinia species (Bonde, 1953) whereas oxytetracyline could control bacterial soft-rot on potatoes.

Deterioration of fruits in storage is commonly caused by fungal growth. Thus Salunkhe et al. (1962) advocated the use of antifungal antibiotics for the preservation of fruits. A combination of antifungal antibiotics and low level gamma irradiation retarted mold growth on cherries (Cooper and Salunkhe, 1963). Ayres and Denisen (1958) observed that myporzine controlled fungal growth on strawberries and red raspberries whereas candidin, ascosin and nystatin at low levels stimulated the growth of mycotics. Pimaricin has been used to preserve orange juice and orange juice concentrates (Shirk and Clark, 1963; Shirk et al., 1964). Higher levels of pimaricin (20 ppm) were most effective at a storage temperature of 10°C.

The addition of some antibiotics were observed to lower thermal processes with low acid canned foods. For example, addition of nisin at 14 ppm levels to pea puree, cauliflower puree, whole kernel and cream style corn and tomato juice lowered thermal process requirements and enhanced the shelf-life of these products (Campbell et al., 1959). Thermophilic spoilage of canned products could be effectively reduced by the addition of tylosin (Denny et al., 1961). The addition of 20 to 40 ppm of subtilin, to canned foods has been reported to be ineffective in preventing flat-sour or putrefactive deterioriation (Cameron, 1951).

3) Muscle Foods and Eggs:—Muscle foods include meat, fish, and poultry. The antibiotics which have been tested for preserving these foods are predominantly tetracycline and its derivatives. As early as 1954 it was reported that whole, eviscerated or cut-up chickens dipped in a solution of 30 ppm tetracycline extended shelf-life 7–14 days beyond that of untreated controls (Kohler et al., 1954). Upon cooking, the antibiotic absorbed by the muscle tissues was completely destroyed (Ziegler and Stadleman, 1955). It was also shown that antibiotic-treated poultry meat had better shelf-life at elevated storage temperatures of around 20°C (Shannon and Stadelman, 1957).

Turkeys have been treated with 20 ppm chlortetracycline by dipping birds prior to, and after evisceration (Kraft *et al.*, 1958). Most of the antibiotic residue was found in the skin (6.3 ppm) while deeper tissues had some residue (1.6 ppm) (McKee *et al.*, 1959). Beside dipping birds in antibiotic solutions, the use of antibiotics in chill tanks has also been tested (Essary *et al.*, 1958; Broquist *et al.*, 1956; Carey, 1956). It has been shown that chlortetracycline was more beneficial in enhancing shelf-life than is oxytetracycline (Wells *et al.*, 1957; McVicker *et al.*, 1958; Silvestrini *et al.*, 1959). Antibiotics added to ice slush tanks at 10 ppm levels were effective when eviscerated poultry carcasses were dipped and then frozen (Barnes and Shrimpton, 1958; Walker and Taylor, 1960). It has also been suggested that oxygen tension may be another critical factor that may interfere with the effectiveness of antibiotic action (Wells *et al.*, 1958; Jimeniz *et al.*, 1961). Antibiotic treatment does influence the type of flora that survives and generally gram negative organisms are inhibited by antibiotics such as tetracyclines at concentrations greater than 10 ppm and yeast and molds predominate (Eklund *et al.*, 1961; Barnes and Shrimpton, 1968). Sanitation plays an important role in shelf life and antibiotics are not a substitute for good sanitary practices (Janke *et al.*, 1958; Ginsberg *et al.*, 1958; Dawson and Stadelman, 1960; Walker *et al.*, 1960).

Giolotti and Gianelli (1960) used tetracycline to preserve shell eggs held at 18–35°C for eight months post-treatment. Up to 80% of the antibiotic-treated eggs were sterile (microbiologically) while controls were all contaminated. Earlier tests had indicated that chlortetracycline, chloramphenicol, neomycin, polymyxin or streptomycin were not effective in preserving or extending the shelf life of eggs (Cotterill and Hartmann, 1956). Miller and Morrison (1958) had noted that treatment of shell eggs with 20 ppm chlortetracycline hastened spoilage rather that detered it, confirming similar observations made by Schmidt and Stadelman (1957). The loss of shelf-life was attributed to the penetration of antibiotics through the shell where they bind with albumin (Elliott and Romoser, 1957), thereby becoming unavailable.

Most of the work with use of antibiotics to preserve meat pertains to beef. It was observed that penicillin, streptomycin and bacitracin were ineffective in enhancing the shelf life of ground beef, whereas tetracylines at 0.5–2.0 ppm levels nearly doubled the storage life of this product (Goldberg, 1952). Goldberg *et al.*, (1953) also showed that chlortetracycline disappeared from stored beef time within 72 hr. This observation was later confirmed by Weiser *et al.*, (1953, 1954). Chlortetracycline permitted higher aging temperatures of

25°–32°C for 48 hr for beef and mutton and antibiotic residues could be detected up to 3 weeks after treatment (Henneberg and Kissling, 1959). Dipping whole carcasses in tetracycline solutions retarted spoilage one to two days (Oshima and Ujiie, 1959). Pre-mortem injections of oxytetracycline provided protection against microbial deterioration during high temperature aging (Sleth, 1960; Matsui *et al.*, 1959). Similar treatment was accorded to veal with equal success (Gola *et al.*, 1959; Hejlasz *et al.*, 1957). Patents have also been granted for various processes (Williams, 1960; Savich and Jansen; 1957, Deatherage, 1956; Williams 1963; Wrenshall *et al.*, 1960; Williams, 1960; Kohler *et al.*, 1962). Use of antibiotics in sausage manufacture has been investigated also (Deatherage, 1957; Kazakov and Dyklop, 1955; Lorch *et al.*, 1961; 1963).

Information on the addition of antibiotics to pork is limited. Addition of dihydrostreptomycin or penicillin to ground pork retarted spoilage for a mere 24 hr period (Prost, 1955). Many processes have been patented wherein antibiotics are injected intrapertonially or intravenously prior to slaughter (Wrenshall *et al.*, 1960; Williams, 1960). Red and white meats imported into the U.S., mainly from Australia, do not contain internal organs and antibiotic residues in the carcass have not been detected (Dr. Grace Clark, APHIS, personal communications).

Microbial deterioration of fish poses not only quality problems but even public health problems. The overall effect is loss of product which is an economic problem. Thus preservation of fish by using high salt concentration, low pH or by the addition of chemicals has been attempted. These methods, however, have deliterious effects on the delicate flavor of fish. The antibiotics used in the preservation of fish and seafood are chlortetracycline, oxytetracycline, penicillin, streptomycin, oleandomycin, polymyxin, neomycin and subtilin. The various types of seafood and fish treated include whole and shucked oysters, lingcod, crabs, salmon, cod, halibut, haddock, sardine, sole and other fishes. The various processes employed to preserve the above foods have been reviewed extensively by Marth (1966).

4) Milk and Milk Products:—Although the presence of antibiotics in milk and milk products is considered to be an adulteration, experimental attempts have been made to preserve milk with antibiotics. The use of antibiotics in preserving human milk has already been discussed. The addition of streptomycin to milk was observed to inhibit the growth of *E. coli* but does not affect the "pathogenic bacteria" (Nagel, 1950; Poetschke, 1953). Gramicidin and penicillin have also been tested (Lizgunova, 1952; Foley and

Byrne, 1950). In all instances, it was suggested that the antibiotics be added soon after milking in order to prevent the growth of contaminating microorganisms. Tetracyclines and chloramphenicol were found to inhibit the growth of putrefactive bacteria in milk and these two antibiotics, along with oxytetracycline, were also found to inhibit lactic acid bacteria (Inomoto and Hashida, 1952; Hashida, 1953; Hashida and Asai, 1953). The shelf-life of pasteurized milk stored at 30°C, however, was greatly extended. It was pointed out by Dickinson and Poe (1959) that the addition of antibiotic to good quality raw milk, followed by refrigeration, complemented each other and the pasteurization of this milk shortly after antibiotic addition resulted in milk with excellent shelf-life. Evans and Curran (1952) have demonstrated that the addition of subtilin to whole or evaporated milk did not prevent the growth of *C. botulinum*.

5) **Fermentations:**—The growth of undesirable microorganisms in the manufacture of wine and beer has been investigated. Some gram-positive lactic microorganisms such as *Pediococcus cerevisiae* and *Lactobacillus pastorianus* and the gram-negative rod *Flavobacterium proteus* have been observed to grow in wort during fermentation (Strandskov *et al.*, 1953). A small concentration of 0.005 μg polymixin/ml could inhibit the growth of gram-negative rods while stimulating the growth of yeast. Streptomycin, bacitracin, penicillin, and subtilin were needed in excess of 125 μg/ml in order to be effective, while tetracyclines were required at 50 μg/ml levels. It was later observed that polymyxin was suitable for inhibiting but not destroying *P. cerevisiae* (Haas, 1955). Inhibition of *L. pastorianus* by chloramphenicol was also demonstrated by Haas (1955). Polymyxin was found to be stable in finished beer (Kersey *et al.*, 1954). The combined use of 4.0 μg polymyxin/ml was reported to be beneficial in the preservation of beer (Kato *et al.*, 1957). Patents on using antibiotics for the preservation of beer have been granted to Bockelmann and Strandskov (1957) and Strandskov and Bockelmann (1960). Use of antibiotics in wine making has not received much attention. Antimycin A and actidione controlled yeast development in raisin- and sugar-fortified wine (Peynaud and Lafourcade, 1953) while penicillins, tetracyclines and subtilin did not interfere with the development of yeast (Riberau-Gayon and Peynaud, 1952). Biamicina, and Italian antibiotic controlled the fermentation of wine when added at 0.015 – 0.025% concentrations to the must (Dal-Cin, 1948). Control of contaminants during sake fermentation by inhibition of Lactobacilli and Acetobacter species

by penicillin, tetracyclines and cloromycetense has been reported (Sato, 1958; Akiyama *et al*, 1958). Bacterial contaminants in grain mash fermentations for the production of distilled alcoholic beverages has also been successfully attemped by Day *et al*. (1954). The only disadvantage was that the marginal treatment of mash with antibiotics was conducive to acrolein production by some strains of Lactobacilli.

6) **Other Foods:**—The preceding review of the use of antibiotics in foods was based on commodities. The use of antibiotics in prepared foods has also been considered. Food-poisoning strains of *Staphylococcus aureus* may sometimes grow in custards. Godkin and Cathcart (1952) tested the efficacy of several antibiotics in custards containing *S. aureus* when 100 ppm subtilin was added to custard and stored at 37°C (optimum growth temp. for *S. aureus*). The proliferation of microorganisms was retarted from 48–72 hr and spoilage by thermoduric organisms was also absent. Much lower concentrations of oxytetracyclines (1.0 ppm), when combined with 100 ppm subtilin, provided greater protection than when either antibiotic was used singly. Similarly, the growth of staphylococci in ice-cream mix, processed cheddar cheese spread, ham, and sausage could be prevented by the use of tylosin (Greenberg and Silkiker, 1962b; Seitz *et al*., 1963). Rehm (1960) exposed *E. coli* to a combination of several antibiotics and observed that, while tetracycline and chloromphenicol displayed both additive and antigonistic reactions, they did not act synergistically.

The effect of sporeforming bacteria on the spoilage of thermally processed foods is well documented. Heat processing, in conjunction with subtilin treatment, could lead to better food preservation (Andersen and Michener, 1950 a,b). However, 80 ppm subtilin, 100 ppm gramicidin, 40 ppm bacitracin or 40 ppm streptomycin in conjunction with boiling was ineffective in controlling spoilage of canned peas or corn by *C. botulinum* (Burroughs and Wheaton, 1951). Essentially similar results were obtained by Williams and Campbell (1951). Contrary to these reports, LeBlanc *et al*., (1953) reported that spores of putrefactive anaerobe No. 3679 and of *C. botulinum* were appreciably less heat resistant when suspended in foods containing small amounts of subtilin. These results were later verified by Michener (1953 a,b) and O'Brien *et al*. (1956). Denny and Bohrer (1959) reported that antibiotics did not affect the thermal death times (TDT) of PA 3679 or *C. botulinum* but did lower TDT of *B. stearothermophilus* spores. Butyric anaerobes, *B. coagulans*, and *B. stearothermophilus* could be inhibited by tylosin and nisin (Denny *et al.*, 1961). Tylosin has also been observed to be effective against PA 3679 and *C. botulinum* in a variety of foods including meat products (Greenberg and Silliker, 1962, a,b; Poole and Malin, 1964).

Thus, from the foregoing discussion it becomes evident that numerous antibiotics have been tested in foods for their efficacy as a food preservative.

USE OF ANTIBIOTICS IN ANIMAL PRODUCTION

The transmission of antibiotics through the food chain can result in residues in foods. Antibiotics have been used in animal feed to achieve one of many effects. For example, antibiotics have been fed to animals with the aim of improving feed conversions, reducing morbidity and mortality, and increasing growth rate. Livestock feed supplementation is possible through the addition of antibiotics such as chlortetracycline, bacitracin, neomycin, oxytetracycline, oleandomycin, penicillin, tylosin, streptomycin, virginiamycin and flavomycin. It is thought by many that administration of low levels of antibiotics ($<$50 g/ton) does not pose residue problems in edible tissues. However, if tetracyclines are fed in the range of 200 to 1,000 g/ton, residues may be detected in the liver (Bird, 1961). It has also been reported that chlortetracycline added at the rate of 1,000 to 2,000 g/ton resulted in residues in the muscle tissue of poultry while levels of $>$500 g/ton resulted in residues in the egg (Bird, 1961). When 1,000 g/ton chlortetracycline was fed to broilers for 11 days, the following levels of residue were detectable in various fractions: 0.49–0.88 μg/g serum, 1.5–3.0 μg/g liver, 0.68–1.30 μg/g breast and 0.59–0.75 μg/g thigh tissues (Boyd *et al.*, 1960). Generally, the cooking of meats was observed to destroy all antibiotic residues. Many scientists have suggested the use of antibiotics in the diets of swine (Harrington and Taylor, 1955; Kline *et al.*, 1955; Robinson *et al.*, 1956; Ezdakov, 1958; Pivnyak and Obraztsova, 1959; Arpai *et al.*, 1959). However, very little work has been done in this area. In one study, feeding 82 mg chlortetracycline/animal/day for a period of six months resulted in the antibiotic being present in gall bladders, kidney and the liver, with no detectable residues in the spleen or muscle tissue (Kelch, 1960).

Currently, about thirteen different antibiotics are approved by the Food and Drug Administration for use as feed additives or therapeutic agents and approximately 40% of the total antibiotics produced in the U.S. are used in non-medical fields such as feed additives.

The mode of action of antibiotics added to feeds may be due

to a) metabolic effect, b) nutrient sparing effect and/or c) a disease control affect. Some of these effects may be interrelated. The metabolic effects are exerted directly by altering the rate or pattern of metabolic processes, while the nutrient sparing effect of antibiotics is attributed to the alterations in intestinal flora. Some organisms considered to be beneficial to the animal have the capability to synthesize vitamins and amino acids. The growth of these beneficial organisms may be enhanced by an antibiotic-mediated selective destruction of undesirable microflora. The third effect, the disease control effect, is related to the destruction of pathogenic organisms which could otherwise lead to clinical or subclinical manifestations of infection or disease. It has been observed that infections or diseases lead to altered metabolic rates and processes. In a broad sense, then, the action of antibiotics is on the intestinal flora of animals (Hays, 1978).

The amount and the type of antibiotic used in the supplementation of livestock feed depend on factors such as species, age, stage of production, adequacy of diet, sex, environmental conditions, and general health of the animal. In general, the amount of antibiotics used in feed supplementation has not led to problems of residual antibiotics in meat, eggs or milk. Currently, it is thought that 40–50 gm/ton of a broad spectrum antibiotic are needed to obtain "optimal responses" in livestock. When antibiotics were first introduced as additives to feed a level of 10–20 g/ton was recommended—a level perhaps not adequate to elicit "optimal response" (Catron et al., 1951). Growth responses are observed to increase with increasing levels of antibiotics up to a limit of 250 g/ton. However, such high levels may not be economically feasible. Hence a middle-ground between cost and effectiveness has been achieved. Whether the use of antibiotics poses significant public health problems is a subject which is being heatedly debated at this time.

PROBLEMS RESULTING FROM THE USE OF ANTI—BIOTICS

This aspect of the use of antibiotics in food production and processing has to be approached from the view point of public health. The use of antibiotics in medicine has drastically reduced morbidity and mortality, altered epidemiology of diseases by effective control of pathogens, and introduced enhanced prophylaxsis in surgery. In this medical sense, the use of antibiotics involves treatment of patients for short periods of time and can be considered to be a controlled, supervised exposure. Hence, any adverse reactions are readily detectable and avoided. In contrast, exposure of consumers to antibiotics in the food is involuntary and uncontrolled.

The process of resorption, distribution, biotransformation, accumulation and excretion can, individually or in combination, create toxic effects. The intensity of these toxic effects vary between species and is dependent on the dosage, the method of introduction into the body (e.g. oral or dermal) and the period of exposure (single dose or chronic). Depending on the amounts used, the effects can be classified as a) lethal toxicity (LD_{50}), b) tolerated dose (maximum usable amount without inducement of visible toxicity) or c) curative dose (minimum usable amount to eliminate disease-causing factors). The tolerated dose for benzylpenicillin has been reported to be 60 g/day (WHO/FAO, 1969). Because of such high levels required to induce severe toxicity, it is doubtful that any person would ever suffer from acute chronic toxicity from antibiotic residues in foods. However, residues in foods can be capable of inducing immediate, delayed or chronic allergic reactions. Allergic reactions are preceded by a sensitization period. Once an individual is sensitized, immediate allergic reactions may be manifest within half an hour post-exposure while delayed allergic responses may only be evoked 6-8 hr post exposure. Whether exposure of individuals to foods containing antibiotic residues involves allergic reactions or sensitizes them is not clear. It is, however, known that exposure of sensitized individuals to antibiotic residues in food can and does induce eczema or urticaria. Thus, there is really no threat of humans suffering from acute toxicity due to exposure to antibiotic residues in foods. However, allergic responses may be observed in sensitized individuals.

The second aspect of the problem caused by use of antibiotics pertains to microorganisms. The use of antibiotics in the supplementation of animal feeds resulted in antibiotic resistant *Escherichia coli* (Smith and Crabb, 1957) and this resistance was transferable (Watanabe, 1963). The question which arises then is: can these antibiotic-resistant organisms pose health problems to human populations?|In 1969, Britain adopted the practice of avoiding the use of antibiotics effective in veterinary and human disease treatment as feed supplements (Swann, 1969). The U.S. Food and Drug Administration organized a task force in 1970 to evaluate the hazards posed by antibiotics used in animal feeds. The conclusions reached by this task force were similar to those of the British i.e. antibiotics used in human clinical medicine which failed to meet certain guidelines were to be banned from use as growth promoters in animal feeds in July 1973 (Van Houweling, 1972). The sponsors of antibiotics were given until 1975 to prove that the use of antibiotics in animal feeds did not pose hazards to animal or human health (Gardner, 1973). Hazards were established according to the following criteria:

a. If drugs administered to animals caused an increase in animal reservoirs of gram-negative bacilli pathogenic to humans and could be transferred to humans via the food chain, the antibiotics were hazardous to human health.

b. If the antibiotics resulted in increases of antibiotic-resistant gram-negative bacilli, especially if antibiotic resistance pertained to antibiotics used in human medicine, the antibiotics would be prohibited from use.

c. If the antibiotics in question enhanced the pathogenicity of the organisms in animals and these organisms were transmitted to man through the food chain, the antibiotic would not be permitted for use as a feed supplement.

d. If the ingestion of biologically active antibiotic residues in foods lead to an increase in human pathogenic flora resistant to antibiotics, then such antibiotics would not be desirable for use in feeds.

The concerns expressed are legitimate for establishing safety guidelines and have generated many studies to evaluate the seriousness of threats posed by the use of antibiotics in animal feeds. An important genus of the gram-negative bacilli referred to above is *Salmonella* because, systemically, these bacteria are amenable to antibiotic therapy. *Salmonella* originating from animal reservoirs can cause morbidity and mortality in humans and these organisms can contaminate food products. Antibiotics used as feed supplements can conceivably lead to a build-up of antibiotic-resistant *Salmonella* in animals and a large proportion of *Salmonella typhimurium* isolated from humans carry the R factor. Hence, in light of these observations, it becomes important to understand the effects of antibiotics on *Salmonella* (see also chapter 5).

The effects of feeding antibiotics to animals infected with *Salmonella* are varied. The *Salmonella* reservoir decreased, if the infecting organisms were succeptible to antibiotics (Evangelisti *et al.*, 1975; Gutzmann *et al.*, 1976; Jarolmen *et al.*, 1976; Williams *et al.*, 1977), but the reservoir increased, if antibiotic-resistant *Salmonella* were initially present (Dey *et al.*, 1977; Williams *et al.*, 1977). It has been shown that the incidence of antibiotic resistant *Salmonella* isolated from man and animals has been increasing (Neu *et al.*, 1975).

Apart from *Salmonella*, the incidence of antibiotic resistant coliforms is high in animals receiving antibiotics as feed supplements (Mercer *et al.*, 1971; Siegel *et al.*, 1974). It is not, however, clear if antibiotic-resistant coliforms isolated from humans originated from animals receiving antibiotic-supplemented feeds. It has been shown that persons coming in contact with animals or medicated feed do carry higher numbers of antibiotic-resistant coliforms than persons not in contact with such animals (Linton *et al.*, 1972; Fein *et al.*, 1974; Siegel *et al.*, 1975; Levy *et al.*, 1976). In 1976 Levy *et al.* showed that antibiotic resistant coliforms do spread from one animal to another and from animal to man. Such a study poses problems in methodology. Whereas, *Salmonella* have been serotyped, the methods for identifying *E. coli* have not attained such refinements. This organism is present in the intestinal tract of both man and animals. Strains of *E. coli* which are of the "O" serotype and carry the R plasmid are especially common to both man and animals (Hartley *et al.*, 1975; Howe *et al.*, 1976; Howe and Linton, 1976). Hence, the origin of antibiotic-resistant coliforms in humans cannot be proved to be from animals.

The R factors are conferred by plasmids. Compatible plasmids can co-exist in a bacterium whereas incompatible plasmids cannot (Datta, 1975; Falkow; 1975). If two related plasmids enter a bacterium either one of them will be lost or they can recombine to form a single replicon. The R plasmids have been divided into more than 20 compatibility groups (Hartley *et al.*, 1975). The distribution of incompatible R plasmids in animal isolates was the same as those in humans suggesting that the R factors in humans and animals are not separate (Silver and Mercer, 1978). This means that the determination of origin of antibiotic-resistant organisms is not, by any current means, an easy task. Additionally, resistance genes can be transferred from one plasmid to another or from plasmid to chromosome (Berg *et al.*, 1975; Barth *et al.*, 1976; Foster *et al.*, 1975; Gottesman and Rosner, 1975; Hedges and Jacob, 1974; Heffron *et al.*, 1975; Kleckner *et al.*, 1975; Kopecko and Cohen, 1975). The issue of transfer of antibiotic-resistant organisms from animals to man, thus assumes an added dimension because, not only can R plasmids be directly transferred from animal to man, but genes themselves can translocate from plasmid to plasmid. Additionally, it is no longer a requirement that the R-factor be stabilized in a cell prior to transfer to another plasmid or chromosome because even a transient encounter can facilitate such a transfer.

It has been shown that plasmids are responsible for survival of pathogenic organism within a host with the subsequent inducement of disease (Falkow, 1975). The pathogenicity of *E. coli* strains in pigs is governed by a plasmid termed ENT (Smith and Halls, 1968).

Falkow has also reported that ENT and R plasmids are cotransferable i.e. ENT plasmid can acquire resistance genes from R plasmid. Many toxigenic *E. coli* are resistant to more than one antibiotic (Wachsmuth *et al.*, 1976). Thus, a possibility does exist that plasmids contributing to pathogenicity can be more widely dispersed because of availability of a large reservoir of R factors within enteric organisms.

The concerns raised by the Food and Drug administration task force are legitimate. With increased knowledge of bacterial ecology and genetics it should be possible to make a realistic assessment of the risks involved in using antibiotics in livestock production vis-a-vis the benefits accrued by human populations.

The first two problems created by the use of antibiotics pertain directly to public health. The third type of problem created can be considered technoeconomic. These problems pertain to milk and dairy foods, more so than to any other single commodity group. Dairy cattle diseases, including mastitis, are often treated with antibiotics. Such a treatment course often leads to the presence of measurable amounts of the antibiotic in milk. If antibiotic contaminated milk is used for the production of dairy products, the products can also conceivably contain residues. Many of the antibiotics survive pasteurization temperatures, low temperature storage and even freezing. The major problem of antibiotic residues is in manufacture of cultured dairy foods such as sour cream, buttermilk, yogurt, cheese etc. Antibiotics can inhibit the growth of starter cultures. As little as 0.1 International Unit of penicillin per ml of milk can completely inhibit the yogurt culture and in lesser concentrations acidity, flavor and texture of the product can be altered. In cheese production, the residues can prolong the manufacturing time—a critical factor in high-throughput oriented operations. A more detailed discussion of these aspects has been provided by Marth (1966), Mol (1975) and Myers (1964).

Overall then, the technoeconomic problems and allergenic reactions resulting from antibiotic residues in foods are serious and should be controlled. The public health threat from the use of antibiotics in animal feeds requires to be more thoroughly researched and the risk: benefit analysis be performed.

METHODS OF DETECTING ANTIBIOTIC RESIDUES

The problem of analyzing for antibiotic residues in foods is complicated by the fact that the analyst does not know if any antibiotic residues exist and if they exist, the type and quantity are not known.

In clinical and pharmaceutical samples the type of antibiotic and the dosage used are known and specific tests can be devised. The food analyst has to resort to screening techniques which may indicate the presence of inhibitory substances. Whether this inhibitor is an antibiotic or not has to be guestimated after performing some type of presumptive test. The tests can be divided basically into two groups - chemical and microbiological. Chemical methods are very useful for analyzing commercial antibiotic preparations. However, due to the low levels encountered in foods, microbiological assays are more sensitive, thus, more effective. Microbiological tests do suffer from the disadvantage of not identifying the type of antibiotic or inhibitory substance present.

Microbiological techniques rely on the inhibition of the growth of microorganisms by antibiotic residues present in the test sample. Such procedures may involve diffusion processes or rely on turbidity. In the diffusion-dependent tests the material to be assayed diffuses through the "seeded" agar to create a zone of inhibition while in turbidimetric methods retardation of metabolic processes by antibiotics is measured by titration or indicator dyes. The diffusion dependent tests can be of three types: a) cylinder-plate assay methods, b) paper-disc assay methods and c) reverse-phase disc assays. Cylinder-plate assay methods rely on the formation of a zone of inhibition on a surface-inoculated nutrient agar plate. A glass or metal cylinder containing the sample is placed on the surface of the agar. The cone formed is compared with response curves for graded amounts of a known antibiotic. These response curves have to be plotted on semilog paper with antibiotic concentration on the log scale and the diameters of the zone of inhibition on the linear scale or axis.

The principle of the paper disc assay methods is essentially similar to the cylinder-plate assay method except that a paper disc is impregnated with the sample and then placed on the agar surface. Reverse-phase disc assays differ in two major respects from the preceeding two methods. First, the seeded agar used is incapable of supporting the growth of microorganisms. Second, the paper discs are first impregnated with nutrients required for the test organism's growth and dried. For conducting a test the discs are saturated with the sample and placed on the agar. The nutrients required for the growth of the seed or test organisms are leached from the disc and diffused into the agar and the organism can grow. If the sample contains inhibitory substances the seed organism will not grow around the disc. The types of microorganism used for seeding or inoculating the agar were listed by Myers (1964).

The method of sample application can be altered, depending on the physical state of the sample. Solid samples can be applied directly to the surface of the plate and then the plate can be inoculated. For liquids, the holes for application of samples can be cut in a variety of sophisticated ways which offer precise and uniform control of volume of sample used.

There are many factors which affect the size and appearance of the zones of inhibition in diffusion tests.

1. The number of viable organisms used to seed or inoculate the plates is critical because the density of growth and hence visualization of zones is directly related to the initial concentration of bacteria.

2. The temperature of incubation is important in diffusion - dependent assays. Rate of diffusion is directly proportional to temperatures as is the rate of growth of organisms. Both diffusion and rate of growth are responsible for the formation of the border of the zone.

3. Size of the zone of inhibition may also be governed by the sporulation of spore-forming test organisms. Sporulation not only affects the rate of growth but can also influence the size and visibility of zones.

4. Porosity does determine the rate of diffusion and porosity can be altered by the concentration of agar. Lower concentrations of agar lead to larger zones of inhibition.

5. The age of the test plate and innoculation broth are variable factors. It is common practice to pre-pour seeded agar plates and store them in the refrigerator. Temperature shock and storage time may induce organism to sporulate. The inoculation broth should contain organisms in the late log phase and should not be subjected to storage temperatures of below 0°C, otherwise, these conditions could result in sporulation or loss in viability.

6. Other factors influencing the zones are diffusion time, depth of the agar layer, technique of introducing sample onto the plate, inoculation technique and storage time and temperature of the sample.

In these tests diffusion is an important factor and diffusion of the inhibiting substance into the medium is necessary to observe zones. Initial concentration of antibiotics determines when the minimum inhibitory concentration (MIC) reaches a certain point in the test plate. At any point of the solid medium where the concentration of the inhibitor is greater than MIC the organism will not grow. Diffusion is affected, among other factors, by viscosity of the medium, presence of electrolytes, concentration of inhibitor, diffusion coefficient of the inhibitor, time and temperature. Diffusion coefficients can be calculated from Fick's second law in any one direction as

$$\frac{dc}{dt} = \frac{D\ d^2\ C}{d\ x^2}$$

where C is concentration, t is time, D is the diffusion coefficient and x is the distance from the junction of the inhibitor solution and the medium. The diffusion coefficient D varies with the day to day variations in the conditions of the test. The time t can also vary in the routine testing of large samples. Elegant mathematical models and their solutions have been presented by Mol (1975).

In turbidimetric or "test-tube" methods the growth of a test organism in a liquid sample being tested is compared to growth in an antibiotic and inhibitor free liquid. Growth is assessed in a number of ways such as titratable acidity, pH, dye reduction, direct microscopic examination, and by turbidimetry (Myers, 1964).

CONCLUSIONS

As a summary to this discussion, it should be pointed out that the discovery of antibiotics led to a drastic reduction in morbidity and mortality of both humans and animals. The ability of very low concentrations of these substances to inhibit a wide variety of microorganisms made them attractive for use in food preservation. After intensive research the drawbacks of such additives in prolonging shelf-life of foods were evident. The main concerns pertained to cost and effects on public health and these led to the banning of antibiotics in food processing. Parallel to these decisions the use of antibiotics in subtherapeutic doses for animal feed supplementation was increasing. Livestock producers and animal scientists could enumerate the benefits accrued by antibiotic usage in feeds. Regulatory agencies, such as the Food and Drug Administration and USDA's Animal and Plant Health Inspection Service (APHIS), were questioning the management practices of livestock producers. Issues surfacing were the problem of build-up of antibiotic-resistant pools of pathogenic organisms in animal reservoirs and their subsequent transmission to humans. This is an area of legitimate concern. Due to methodological problems in bacterial genetics (e.g. serotyping of E. coli), it is difficult to assess the magnitude of cross contamination of antibiotic-resistant organisms

from animals to humans. Resolutions of such analytical problems are essential to the critical evaluation of present or future threats posed by medicated feeds to the health and well being of mankind. Currently, APHIS checks animal organs, such as livers and kidneys, for the presence of antibiotic residues in domestically grown and slaughtered livestock. Antibiotic residues in food can cause allergic reactions in sensitized people. Apart from public health problems, these residues can cause technoeconomic losses especially in the dairy industry. Antibiotics inhibit the growth of starter cultures and thus prevent or delay the formation of a good quality product. There are numerous microbiological methods available for detecting and quantifying antibiotic residues and methods have been standardized for use in meeting regulatory standards. The use of antibiotics which had proliferated in the 1950's and 1960's has slowed down and their use in human and veterinary medicine is more judicious in the 1970's. On the issue of use of antibiotics in livestock feeds, the words of the 18th Century British poet William Cowper strike a note "And diff'ring judgements serve but to declare
That truth lies somewhere, if we knew but where."

REFERENCES

ANDERSON, A. A., and H. D. MICHENER. 1950a. Effect of subtilin on bacterial spores. Bacteriol. Proc. 1950, p 28.

ANDERSON, A. A., and H. D. MICHENER. 1950b. Preservation of foods with antibiotics. I. The complementary action of subtilin and mild heat. Food Technol., 4:188.

AKIYAMA, H., K. YAMAO, and S. SASAKI. 1958. Antibiotics for controlling contamination in sake brewing. Nippon Jozo Kyokai Zasshi, 53:306.

ARPAI, J., M. BEHUN, Z. LIFKOVA, and D. VRABLICOVA. 1959. The influence of additional feeding of pigs with Aureovit 12 on the quality of meat with reference to its preservation by freezing. Sb. Cesk. Akad. Zimedel. Ved, Zivocisna Vyroba, 5:725.

AYERS, J. C., and E. L. DENISEN. 1958. Maintaining freshness of berries using selected packaging materials and antifungal agents. Food Technol., 12:562.

BARNES, E. M. and D. H. SHRIMPTON. 1958. The effect of tetracycline compounds on the storage life and microbiology of chilled eviscerated poultry. J. Appl. Bacteriol., 23:155.

BARTH, P. T., N. DUTTA, R. W. HEDGES, and N. J. GRINTER. 1976. Transposition of a deoxyribonucleic acid sequence encoding trimethoprim and streptomycin resistances from R 483 to other replicons. J. Bacteriol., 125:800.

BENVENISTI, R., and J. DAVIS. 1973. Mechanisms of antibiotic resistance in bacteria. Ann. Rev. Biochem., 42:471.

BERG, D. E., J. DAVIES, B. ALLET, and J. D. ROCHAIX. 1975. Transposition of R factor genes to bacteriophage λ, Proc. Natl. Acad. Sci. USA, 72:3628.

BINDEWALD, H. 1952. Preservation of mother's milk with streptomycin or citric acid. Deut. Med. Wochschr., 77:1015.

BIRD, H. R. 1961. Additives and residues in foods of animal origin. Amer. J. Clin. Nutr., 9:260.

BOCKELMAN, J. B., and F. B. STRANDSKOV, 1957. Inhibition of microbial growth in beer. U.S. Pat. No. 2,798,811.

BONDE, R. 1953. Preliminary studies on the control of bacterial decay of potato with antibiotics. Amer. Potato J., 30:143.

BOYD, J., H. H. WEISER, and A. R. WINTER. 1960. Influence of high level antibiotic rations, terphthalic acid on antibiotic content and keeping quality of poultry meat. Poultry Sci., 39:1067.

BROQUIST, H. P., A. R. KOHLER, and W. M. MILLER. 1956. Retardation of poultry spoilage by processing with chlortetracycline. J. Agr. Food Chem., 4:1030.

BURROUGHS, J. D. and I. E. WHEATON. 1951. The preservative action of antibiotics in processed foods. Canner, 112:50.

CAMERON, E. J. 1951. Use of antibiotics in preserving foods. Proc. 3d Conf. on Research, Council of Research, Amer. Meat Inst., Chicago.

CAMPBELL, L. L. JR., E. E. SNIFF, and R. T. O'BRIEN. 1959. Nisin sensitivity of Bacillus coagulans. Appl. Microbiol., 7:289.

CAREY, B. W. 1956. The biochemistry of antibiotics on the bacteria associated with food spoilage. Giorn. Microbiol., 2:233.

CARROLL, V. J., R. A. BENEDICT, and C. L. WRENSHALL. 1957. Delaying vegetable spoilage with antibiotics. Food Technol., 11:490.

CATRON, D. V., H. M. MADDOCK, V. C. SPEER, and R. L. VOHS. 1951. Effect of different levels of aureomycin with and without vitamin B_{12} on growing-fattening swine. Antibiotic Chemother, 1:31.

COOPER, G. M. and D. K. SALUNKHE. 1963. Effect of gamma-irradiation chemical packaging treatments on refrigerated life of strawberries and sweet cherries. Food Technol., 17:123.

COTTRILL, J. and P. HARTMAN. 1956. Effect of antibiotics on the incidence and spoilage of shell eggs. Poultry Sci., 35:733.

DAL-CIN, G. 1948. Antibiotics in winemaking. Riv. Viticolt. e enol. (coneglians), 1:335.

DAMEROW, R. 1956. A criticism of human milk preservation with streptomycin. Arch. Kinderheilk., 152:56.

DATTA, N. 1975. Epidemiology and classification of plasmids. In Microbiology - 1974. ed Schlessinger, D., Amer. Soc. Microbiol., Washington, D.C. p 9.

DAWSON, L. E. and W. J. STADELMAN. 1960. Microorganisms and their control in fresh poultry meat. Mich. State Univ., Tech. Bull. No. 278.

DAY, W. H., W. C. SERJAK, J. R. STRATTON, and L. STONE. 1954. Antibiotics as contamination control agents in grain alcohol fermentations. J. Agr. Food Chem., 2:252.

DEATHRAGE, F. E. 1957. Use of antibiotics in the preservation of meats and other food products. Amer. J. Public Health, 47:594.

DEATHRAGE, F. E. 1962. Antibiotics in the preservation of meat. Antibiotics in Agriculture, 9:225.

DENNY, C. B., and C. W. BOHRER. 1959. Effect of antibiotics on the thermal death rate of spores of food spoilage organisms. Food Res., 24:247.

DENNY, C. B., L. E. SHARPE, and C. W. BOHRER. 1961. Effects of tylosin and nisin on canned food spoilage bacteria. Appl. Microbiol., 9:108.

DENNY, C. B., J. M. REED, and C. W. BOHRER. 1961. Effect of tylosin and heat on spoilage bacteria in canned corn and mushrooms. Food Technol., 15:338.

DEY, B. P., D. C. BLENDEN, G. C. BURTON, H. D. MERCER, and R. K. TSUTAKAWA. 1977. Influence of continuous chlortetracycline feeding on experimentally induced salmonellosis in calves. I. Rate and duration of shedding of *Salmonella* (cited by Silver and Mercer).

DICKINSON, L., and C. F. POE. 1959. Preservation of milk with chlortetracycline. Univ. Colo. Stud. Series Chem. Pharm., 2:50.

EKLUND, M. W., J. V. SPENCER, E. A. SAUTER, and M. H. GEORGE. 1961. Effect of different methods of chlortetracycline application on shelf-life of chicken fryers. Poultry Sci., 40:924.

ELLIOT, L. E., and G. L. ROMOSER. 1957. Recovery of antibiotic residue in egg albumin. Poultry Sci., 36:365.

ELWELL, L. P., J. DeGRAAFF, D. SEIBERT, and S. FALKOW. 1975. Plasmid linked ampicillin resistance in *Haemophilus influenzae* Type D. Infect. Immun., 12:404.

ELWELL, L. P., M. ROBERTS, L. W. MAYER, and S. FALKOW. 1977. Plasmid mediated β-lactamase production in *Nisseria gonorrhea*. Antimicrob. Agents Chemother., 11:528.

ESSARY, E. O., E. C. MOORE, and C. Y. KRAMER. 1958. Influence of scald temperatures, chill time, and holding temperatures on the bacterial flora and shelf-life of freshly chilled tray-pack poultry. Food Technol., 12:684.

EVANGELISTI, D. G., A. R. ENGLISH, A. E. GIRARD, J. E. LYNCH, and I. A. SOLOMONS. 1975. Influence of sub-therapeutic levels of oxytetracycline on *Salmonella typhimurium* in swines, calves and chicken Antimicrob. Agents. Chemother., 8:664.

EVANS, F. R., and H. R. CURRAN. 1952. The preserving action of subtilin and mild heat in normal concentrated milk. J. Dairy Sci., 35:1101.

EZDAKOV, N. V. 1959. The fattening of pigs with residues of antibiotic industry. Vses. Akad. Sel'skokhoz. Nauk im V. I. Lenina, Moscow 1958. p 48.

FALKOW, S. 1975. Infection and multiple drug resistance. Pion Ltd., London.

FEIN, D., G. BURTON, R. TSUTAKAWA and D. BLENDEN. 1974. Matching of antibiotic resistance patterns of *Escherichia coli* from farm families and their animals. J. Infect. Dis., 130:274.

FOLEY, E. J., and J. V. BYRNE. 1950. Penicillin as an adjunct to the preservation of quality of raw and pasteurized milk. J. Milk Food Technol., 13:170.

FOSTER, T. J., T. G. B. HOWE, and M. H. RICHMOND. 1975. Translocation of the tetracycline resistance determinant R100–1 to *Escherichia coli* K-12 chromosome. J. Bacteriol., 124:1153.

FRANCIS, F. J. 1960. Discoloration and quality maintenance in cole slaw. J. Amer. Soc. Sci., 75:449.

GARDNER, S. 1973. Statements of policy and interpretation regarding animal drugs and medicated feed. Fed. Regist., 38:9811.

GINSBERG, A., M. REID, J. M. GRIEVE, and K. OGONOWSKI. 1958. Chlortetracycline as a preservative of fresh meat and poultry. Vet. Record, 70:700.

GIOLITTI, G., and F. GIANELLI. 1960. Conservation of eggs by means of antibiotics. Ateneo Parmense 31, (Suppl. 1):73.

GODKIN, W. J., and W. H. CATHCART. 1952. Effect of antibiotics retarding the growth of *Micrococcus pyogenes* var *aureus* in custard filling. Food Technol., 6:224.

GOLA, J., V. SINGER, and M. LUKESOVA. 1959. The use of chlortetracycline (aureomycin) in prolonging the keeping quality of transported veal. Sb. Cek. Akad. Zemedel. Ved. Vet. Med., 5:495.

GOLDBERG, H. S. 1952. Studies on the microflora and deep tissues of beef and their succeptibility to various antibiotics. Bacteriol. Proc. 1952, p 17.

GOLDBERG, H. S., H. H. WEISER, and F. E. DEATHERAGE. 1953. Studies on meat. IV. The use of antibiotics in preservation of fresh beef. Food Technol., 7:165.

GOTTESMAN, M. M. and J. L. ROSNER. 1975. Acquisition of a determinant for chloramphenicol resistance by coliphage lambda. Proc. Natl. Acad. Sci. USA, 72:5041.

GREENBERG, R. A., and J. H. SILLIKER. 1962a. Spoilage patterns in *Clostridium botulinum* inoculated foods treated with tylosin. Bacteriol. Proc., 1962, p 21.

GREENBERG, R. A. and J. H. SILLIKER. 1962b. The action of tylosin on spore-forming bacteria. J. Food Sci., 27:64.

GUTZMAN, F., H. LAYTON, K. SIMKINS, and H. JAROLMAN. 1976. Influence of antibiotic supplemented feed on the occurrence and persistence of *Salmonella typhimurium* in experimentally infected swine Amer. J. Vet. Res., 37:649.

HAAS, G. J. 1955. The influence of antibiotics on some biological contaminants in the brewery. Wallerstein Lab. Commun., 18:253.

HARRINGTON, G. and J. H. TAYLOR. 1955. The effect of antibiotic dietary supplements on the carcass measurements and dressing percentage of bacon pigs. J. Agr. Sci., 46:173.

HARTLEY, C. L., K. HOWE, A. H. LINTON, K. B. LINTON, and M. H. RICHMOND. 1975. Distribution of R plasmids among the O-antigen types of *Escherichia coli* isolated from human and animal sources. Antimicrob. Agents Chemother., 8:122.

HASHIDA, W. 1953. Application of several antibiotics to the food industry. II. Application of several antibiotics to cow's milk. J. Ferment. Technol., 31:15.

HASHIDA, W., and T. ASAI. 1953. Application of several antibiotics to the food industry. III. Application of the preservation of cow's milk. J. Ferment. Technol., 31:122.

HAYS, V. W. 1978. The role of antibiotics in efficient livestock production. *In*. Nutrition and Drug Interrelations, ed. Hathcock, J. N., and J. Coon., Academic Press, New York. p 545.

HEDGES, R. W., and A. E. JACOB. 1974. Transposition of ampicillin resistance from RP4 to other replicons. Mol. Gen. Genet., 132:31.

HEFFRON, F., R. SUBLETT, R. W. HEDGES, A. JACOBS, and S. FALKOW. 1975. Origins of the TEM Beta-lactamase gene found on plasmids. J. Bacteriol., 122:250.

HEJLASZ, Z., M. KOCOT, and Z. ZAWADKI. 1957. Effect of supravital administration of penicillin to meat animals on the prolongation of meat shelf life. Med. Weterynar, 13:657.

HENNEBERG, O. H. and R. KISSLING. 1958. Accelerated ripening of beef and sheep meat at higher temperature when treated with aureomycin. Wein Tierarztl. Monatsschr., 46:697.

HOWE, K. and A. H. LINTON. 1976. The distribution of 0-antigen types of *Escherichia coli* in normal calves compared with man and their R-plasmid carriage. J. Appl. Bacteriol., 40:317.

HOWE, K., A. H. LINTON, and A. D. OSBORNE. 1976. A longitudinal study of *Escherichia coli* in cows and calves with special reference to the distribution of 0-antigen and antibiotic resistance. J. Appl. Bacteriol., 40:331.

INOMOTO, Y., and W. HASHIDA. 1952. Application of several antibiotics to the food industry. I. Preservation of cow's milk. J. Ferment. Technol., 30:287.

JANKE, A., R. G. JANKE, K. BAUER, H. HEILMANN, H. KIRCHMEYER, and W. G. KOSTJAK. 1958. Increase in stability of the characteristics of chickens by treatment with chlortetracycline. Zentr. Bakteriol. Parasitenk. Abt., 2:111.

JAROLMEN, H., R. J. SHIRK, and B. F. LANGWORTH. 1976. Effect of chlortetracycline feeding on Salmonella reservoir in chickens. J. Appl. Bacteriol. 40:153.

JIMENIZ, A. V., R. P. FERNANDEZ, and E. M. GIRAUTA. 1961. Utilization of antibiotic and chelating substances in the treatment and storage of poultry (under refrigeration). Symp. Substances Etranger dans les Alimants 6e Madrid, 1960, p 530.

KATO, S., N. NISHIKAWA, and H. MUNEKATA. 1957. Microbiological studies of beer. I. Characteristics of microorganisms and effects of antibiotics in draught beer. Jozo Kagaku Kenkyu Hokoku, 26:29.

KAYSER, K. 1952. Is it advisable to alter methods hitherto prescribed and adopted in human milk banks? Zbl. Gynak., 74:455.

KAZAKOV, A. M., and V. K. DYKLOP. 1955. Antibiotics for improving the stability of meat and meat products. Tr. Vses. Nauchn. Issled. Inst. Myasn. Prom., (1955):30.

KELCH, F. 1960. Aureomycin added to rations affecting quality of ham. Fleischwritschaft, 12:439.

KERSEY, R. C., F. C. VISOR, and C. L. WRENSHALL. 1954. Residual antibiotic levels in food products during storage and processing. Antibiotic Ann., 1953–1954, p 438.

KLECKNER, N., R. K. CHAN, B. K. TYE, and D. BOTSTEIN. 1975. Mutagenesis by insertion of a drug resistance element carrying an inverted repition. J. Mol. Biol., 97:561.

KLINE, E. A., J. KASTELIC, and G. C. ASHTON. 1955. The effect of vitamin B_{12}, cobalt and antibiotic feeding on the composition of pork tissue of 100-pound pigs. J. Nutr., 56:321.

KOHLER, A. R., H. P. BOROQUIST and W. H. MILLER. 1954. Chlortetracycline and the control of poultry spoilage. Food Technol., 8:19.

KOHLER, A. R., W. H. MILLER and H. M. WINDLAN. 1962. Preservation and improved color retention of meats. U.S. Pat. No. 3,050,401.

KOPECKO, D. J., and S. N. COHEN. 1975. Site-specific Rec A independent recombination between bacterial plasmids:involvement of palinodromes of the recombinant loci. Proc. Natl. Acad. Sci. USA, 72:1373.

KRAFT, A. A., L. E. ELLIOT, and A. W. BRYANT. 1958. Effect of antibiotic treatment on storage life of turkeys. Food Technol., 12:660.

LeBLANC, F. R., K. A. DEVLIN, and C. R. STUMBO. 1953. Antibiotics in food preservation. I. The influence of subtilin on the thermal resistance of spores of *Clostridium botulinum* and the putrefactive anaerobe 3679. Food Technol., 7:181.

LEVY, S. B., G. B. FITZGERALD and A. B. MACONE. 1976. Changes in the intestinal flora of farm personnel after introduction of tetracycline - supplemented feed on a farm. N. Engl. J. Med., 295:583.

LINNEWEH, F. 1949a. On the preservation of human milk. Med. Klin., 44:166.

LINNEWEH, F. 1949b. Streptomycin for the preservation of human milk: clinical experiences. Med. Klin., 44:666.

LINTON, K. B., P. A. LEE, M. H. RICHMOND, W. A. GILLESPIE, A. J. ROWLAND, and V. BAKER. 1972. Antibiotic resistance and transmissable R factors in intestinal coliform flora of healthy adults and children in an urban and rural community. J. Hyg., 70:99.

LIZGUNOVA, A. V. 1952. Antibacterial properties of gramicidin in milk. Gigiena i Sanit., 1952, p 27.

LORCH, L. V., R. NEGRI and G. PENSO. 1961. Occurrence of fermentations in the presence of microorganisms with induced antibiotic resistance and under antibiotic action. Sci. Rept. 1st Super Sanita., 1:397.

LORCH, L. V., R. NEGRI, G. PENSO, and P. SAVI. 1963. Process to treat food products by use in presence of antibiotics and either antibiotic resistance or antibiotic dependent bacteria strains. U.S. Pat. No. 3,098,744.

MARTH, E. M. 1966. Antibiotics in foods - naturally occurring, developed and added. Residue Rev., 12:65.

MATSUI, T., G. TOKUTOMI, A. TAKASE, Y. AKAO, R. SANO, T. BITO, M. NAKAJIMA, H. WATANABE, and T. FUKAZAWA. 1959. Putrefaction prevention experiment on meat by using an antibiotic (oxytetracycline). Bull. Inst. Public Health (Tokyo), 8:135.

McKEE, R. C., J. CONKEY, and J. A. CARLSON. 1959. A study on the comparative shelf-life of wet and dry packed poultry. Poultry Sci., 38:260.

McVICKER, R. J., L. E. DAWSON, W. L. MALLMANN, S. WALTER, and E. JONES. 1958. The effect of certain bacterial inhibitors on shelf-life of fresh fryers. Food Technol., 12:147.

MERCER, H. D., D. POCURULL, S. GAINES, S. WILSON, and J. V. BENNETT. 1971. Characteristics of antimicrobial resistance of *Escherichia coli* from animals: Relationship to veterinary and management uses of antimicrobial agents. Appl. Microbiol., 22:700.

MEYERS, R. P. 1964. Antibiotic residues in milk. Residue Rev., 7:9.

MICHENER, H. D. 1953a. Effect of subtilin or ungerminated bacterial spores. Bacteriol. Proc. 1953, p 29.

MICHENER, H. D. 1953b. The bacterial action of subtilin on *Bacillus stearothermophilus*. Appl. Microbiol., 1:215.

MILLER, W. A. and R. W. MORRISON. 1958. The microbiology of formerly clean and dirty, non-segregated, washed, early spring and early summer eggs. Poultry Sci., 37:1022.

MOL, H. 1975. Antibiotics and Milk. A. A. Balkema, Rotterdam.

NAGEL, V. 1950. Preservation of raw milk with streptomycin. Arztl. Wochenschr., 5:393.

NEU, H. C., C. E. CHERUBIN, E. D. LONG, B. FLOUTON, and J. WINTER. 1975. Antimicrobial resistance and R factor transfer among isolates of Salmonella in the northeastern United States: A comparison of human and animal isolates. J. Infect. Dis., 132:617.

NEALWEILER, W. and P. KASTLI. 1953. Aureomycin treatment of puerperal mastitis. Gynecologia, 135:1.

O'BRIEN, R. T., D. S. TITUS, K. A. DEVLIN, C. R. STUMBO, and J. C. LEWIS. 1956. Antibiotics in food preservation. II. The influence of subtilin and nisin on the thermal resistance of food spoilage bacteria. Food Technol., 10:352.

OSHIMA, T., and F. UJIIE. 1959. The use of antibiotics in the preservation of beef. Japanese J. Vet. Sci., 21:61.

PEYNAUD, E., and S. LAFOURCADE. 1953. The study of a new antibiotic and antiseptic agent against Saccharomyces. Compt. Rend., 236:1924.

PIVNYAK, L. G., and A. S. OBRAZTSOVA. 1959. Fungal mycelia in the raising and fattening of agricultural animals. Vses. Akad. Sel'skokhoz. Nauk. Moscow. 1958, p 60.

POETSCHKE, G. 1949. Disinfection and preservation of human milk with streptomycin. Klin. Wochschr., 27:476.

POETSCHKE, G. 1950. Disinfection and preservation of raw milk with streptomycin. Mschr. Kinderheilk, 98:177.

POETSCHKE, G. 1953. Milk Product containing an antibiotic. Brit. Pat. No. 688,968.

POOLE, G. and B. MALIN. 1964. Some aspects of action of tylosin on *Clostridium* species PA 3697. J. Food Sci., 29:475.

PROST, E. 1955. Evaluation of the usefulness of antibiotics in the Production of canned meat, esp. chopped pork. Ann. Univ. Mariae Curie-Skledowska Sect. DD - Vet. Sci., 8:49.

REHM, H. J. 1960. Action of preservatives with some antibiotics on *Escherichia coli*. Z. Lebensm. Untersuch. u Forsch., 113:144.

REHM, H. J. 1967. Industrielle Mikrobiologie. Springer-Verlag, Berlin.

RIBERAU-GAYON, J., and E. PEYNAUD. 1952. Inhibition of wine yeasts by vitamin K_5 and other antibiotics. Compt. Rend. Acad. Agr. France, 39:479.

ROBINSON, W. L., L. E. KUNKLE, and V. R. CAHILL. 1956. The use of an antibiotic in rations for hogs. Ohio Agr. Expt. Sta., Wooster, Res. Bull. No. 769.

ROOS, H. 1951. The preservation of raw human milk. Mschr. Kinderheilk, 99:66.

ROOS, H., and M. KINDLER. 1950. Preservation of human milk with citric acid and streptomycin. Mschr. Kinderheilk, 97:494.

ROZANSKY, R., and A. BREZEZINSKY. 1949. The excretion of penicillin in human milk. J. Lab. Clin. Med., 34:497.

SALUNKHE, D. K., G. M. COOPER, A. S. DAHLIWAL, A. A. BOE, and R. L. RIVERS. 1962. On storage of fruits; effects of pre- and post-harvest treatments. Food Technol., 16:119.

SATO, M. 1958. Effects of antibiotics on contamination of sake -moromi mash. Nippon Nogei Kagaku Kaishi, 32:119.

SAVICH, A. L., and C. E. JANSEN. 1957. Improving the color of meat. U.S. Pat. No., 2,798,812.

SCHMIDT, F. J., and W. J. STADELMAN. 1957. Effects of antibiotics and heat-treatment of shell eggs on quality after storage. Poultry Sci., 36:1023.

SEITZ, E. W., J. A. ELLIOTT, and C. K. JOHNS. 1963. Inhibition of staphylococci by the antibiotic tylosin in cheddar cheesemaking. Bacteriol. Proc. 1963, p 16.

SHANNON, W. G., and W. J. STADELMAN. 1957. The efficacy of chlortetracyline at several temperatures in controlling spoilage of poultry meat. Poultry Sci., 36:121.

SHIRK, R. J., and W. L. CLARK. 1963. The effect of pimaricin in retarding the spoilage of fresh orange juice. Food Technol., 17:108.

SHIRK, R. J., E. F. KLINE, and W. L. CLARK. 1964. The effect of pimaricin in control of yeast growth in orange juice concentrates. Proc. 24th. Ann. Meeting Inst. Food Technol., p 104.

SIEGEL, D., W. G. HUBER, and F. ENLOE. 1974. Continuous therapeutic use of antibacterial drugs in feed and drug resistance of the gram-negative enteric flora of food producing animals. Antimicrob. Agents Chemother., 6:697.

SIEGEL, D., W. G. HUBER, and S. DRYSDALE. 1975. Human therapeutic and agricultural uses of antibacterial drugs and resistance of the enteric flora of humans. Antimicrob. Agents Chemother., 8:538.

SILVER, R. P., and H. D. MERCER. 1978. Antibiotics in animal feeds: An assessment of the animal and public health aspects. *In*. Nutrition and Drug Interrelations. *ed.* Hathcock, J.N. and J. Coon, Academic Press, New York, p 649.

SILVESTRINI, D. A., G. W. ANDERSON, and E. S. SNYDER. 1959. Effect of tetracycline on the preservation of poultry processed by two commercial methods. Poultry Sci., 38:132.

SLEETH, R. B. 1960. Effect of environmental conditions during aging on shrinkage and organoleptic characteristics of beef and the efficacy of oxytetracycline for aging beef. Dissertation abstr., 30:3686.

SMITH, H. W., and W. E. CRABB. 1957. The effect of continuous administration of diets containing low levels of tetracyclines on the incidence of drug-resistant *Bacterium coli* in the feces of pigs and chickens: The sensitivity of *Bact. coli* to other chemotherapeutic agents. Vet. Rec., 69:24.

SMITH, H. W., and S. HALLS. 1968. The transmissible nature of the genetic factor in *Escherichia coli* that controls enterotoxin production. J. Gen. Microbiol., 52:319.

STRANDSKOV, F. B., and J. B. BOCKELMANN. 1953. Antibiotics as inhibitors of microbiological contamination in beer. J. Agr. Food Chem., 1:1219.

STRANDSKOV, F. B., and J. B. BOCKELMANN. 1960. Inhibiting microbial growth in brewery yeast fermentations. U.S. Pat. No. 2,964, 406.

STRANDSKOV, F. B., J. A. BRESICA, and J. B. BOCKELMANN. 1953. The application of selective bacteriostatis of brewery bacteriology. Wallerstein Lab. Commun., 16:31.

SWANN, M. M. 1969. Report on the joint committee on the use of antibiotics in animal husbandry and veterinary medicine. H. M. Stationery Office, London.

TEIXEIRA, G. C., and R. B. SCOTT. 1958. Further clinical and laboratory studies with novobiocin. I. Treatment of staphylococcal infections in infancy and childhood. II. Novobiocin concentrations in blood of new-born infants and in the breast milk of lactating mothers. Antibiotic Med., 5:577.

THOMAS, H. 1953. Tests on human milk preserved with streptomycin and citrates in new-born children. Mschr. Kinderheilk, 100:457.

THOMPSON, B. D. 1959. Post harvest treatment of strawberries and raddishes with cobalt-60 irradiation. Proc. Fla. State Hort. Soc., 72:114.

VAN HOUWELING, C. D., 1972. Report to the Commissioner of the Food and Drug Administration by the FDA Task on the use of antibiotics in Animal Feeds. Food and Drug Admin., Rockville, Maryland.

WACHSMUTH, I. K., S. FALKOW, and R. W. RYDER. 1976. Plasmid-mediated properties of enterotoxigenic *Escherichia coli* associated with infantile diarrhea. Infect. Immun., 14:403.

WALKER, E. A. and J. H. TAYLOR. 1960. The effect of chlortetracycline on viscerated frozen chickens commercially processed in Great Britain. J. Appl. Bacteriol., 23:155.

WALKER, E. A., and J. H. TAYLOR. 1960. The effect of chlortetracycline

chlortetracycline on eviscerated, chilled chickens commercially processed in Great Britain. J. Appl. Bacteriol., 23:145.

WATANABE, R. 1963. Infective heridity of multiple drug resistance in bacteria. Bacteriol. Rev., 27:87.

WEISBLUM, B., and J. DAVIES. 1968. Antibiotic inhibitors of bacterial ribosomes. Bacteriol. Rev., 32:493.

WEISER, H. H., L. E. KUNKLE and F. E. DEATHERAGE. 1954. The use of antibiotics in meat processing. Appl. Microbiol., 2:88.

WEISER, H. H., H. S. GOLDBERG, V. R. CAHILL, L. E. KUNKLE, and F. E. DEATHERAGE. 1953. Observations on fresh meat processed by infusion of antibiotics. Food Technol. 7:495.

WELLS, F. E., J. V. SPENCER, and W. J. STADELMAN. 1958. Effect of packaging material and techniques on shelf life of fresh poultry meat. Food Technol., 12:425.

WELLS, F. E., J. L. FRY, W. W. MARION, and W. J. STADELMAN. 1957. Relative efficacy of three tetracyclines with poultry meats. Food Technol., 11:656.

WHO/FAO Expert Committee (12th report) on Food Additives. 1969. Specifications for the identity and purity of food additives and their toxicological evaluation: Some Antibiotics. WHO Tech. Rept. Series, Geneva, p 430.

WILLIAMS, B. E. 1960. Tenderizing and preserving meat. U.S. Pat. No. 2,942,986.

WILLIAMS, B. E. 1963. Tenderizing and processing meat. U.S. Pat. No. 3,076,712.

WILLIAMS, O. B. and L. L. CAMPBELL, JR. 1951. The effect of subtilin on thermophilic flat sour bacteria. Food Res., 16:347.

WILLIAMS, R. D., L. D. ROLLINS, M. SELWYN, D. W. POCURULL, and H. D. MERCER. 1977. The effect of feeding chlortetracycline on fecal shedding of Salmonella typhimurium by experimentally infected swine (Cited by Silver and Mercer, 1978).

WRENSHALL, C. L. 1959. In. Antibiotics - Their Chemistry and Non-Medical Uses, ed. Goldberg H.S., Van Nostrand Co., Princeton, p 449.

WRENSHALL, C. L., J. R. McMAHAN and R. C. OTTKE. 1960. Preserving fresh meat. U.S. Pat. No. 2,942,982.

ZIEGLER, F., and W. J. STADELMAN. 1955. The effect of aureomycin treatment on the shelf-life of fresh poultry meat. Food Technol., 9:107.

16

Safety and Wholesomeness of Irradiated Foods

Horace D. Graham

For years, man has preserved his food by such established methods as drying (heat), freezing and fermentation. Recently, the use of ionizing radiations has been proposed and experimented with as a new physical method to preserve food. This relatively new concept in the technology of food preservation started soon after World War II. At this time feasibility studies were done in both U.S.A. and Great Britain on the peaceful use of the atom. A vigorous and broad program in food irradiation was initiated by the U.S. Army in 1953. This involved the treatment of various foodstuffs of plant, animal and marine sources and covered biochemical, microbiological, organoleptic and wholesomeness assessments of such irradiated foods. In the U.S.A. itself, the program was, in 1960, divided into two phases each of which was carried out by a different agency. One phase, termed the Radiation Sterilization of Foods Program, was carried out by the U.S. Army. The other phase, termed The Low-Dose Pasteurization of Foods Program, was carried out by the then U.S. Atomic Energy Commission. Since then, countries in all parts of the world have engaged in or taken a keen interest in the preservation of food by irradiation. Even some of the less opulent nations have acquired the equipment and techniques to carry out experimentation.

According to a fairly recent summary report (Wierbicki *et al.*, 1975) over fifty countries have some form of food irradiation research and application.

Early work indicated that this process had a very great potential value for the conservation of agricultural products, including fish and meats. However, despite years of extensive experimentation, much publicity and heavy expenditures, the process has not yet gained universal acceptance.

USEFUL TYPES OF RADIATIONS

Ionizing radiations used for food processing are of three basic types—gamma rays produced from Cobalt-60 and Cesium-137, electrons with a maximum energy of 10 million electron volts (Mev) in the U.S.A. and of 5 Mev in the United Kingdom and X-Rays of 5 Mev maximum energy produced by electrons in an x-ray target. When a naturally radioactive element or an isotope induced artifically decays, alpha particles, beta particles, gamma rays and neutrons are produced, along with other radiations and energy particles.

Alpha particles are helium atoms devoid of two outer electrons while beta particles or rays are high energy electrons and are also called cathode rays. Gamma rays or photons are a type of X-rays.

Alpha particles have little penetrating power, being unable to even penetrate a sheet of paper while beta particles or electrons are more penetrating and will be stopped by a sheet of aluminum. Gamma rays have great penetrating power, being able to go through a block of reasonably thick lead. Neutrons have a great penetrating power, being of such great energy that they can alter atomic structure and so elements which they strike become radioactive. These elements will in turn, then, emit their own high energy radiations.

For purposes of food irradiation, emmissions with a good penetrating power are needed so that the deep, inner portion of the food material will be subject to the irradiation treatment and, at the same time, microorganisms on the surface of the food will be destroyed to as great an extent as possible. Gamma rays possess these properties and so have been used principally for food irradiation. Beta particles have been used to some extent. Neutrons, on the other hand, because of their extreme penetrating power, would rupture atomic structures within the food and make them radioactive. Their use, therefore, would be unsuitable for the preservation of food.

SOURCES OF GAMMA AND BETA RAYS:

Spent uranium fuels after their use in a nuclear reactor can provide the gamma and beta rays, used in food irradiation. The spent fuel elements, with a still high enough radioactivity for food preservation purposes, are put into a proper enclosure which is shielded to prevent stray emissions. The food material for preservation is placed in the radiation path and exposed for a given period so as to allow it to absorb the necessary dose of radiation.

Artificially induced radioactive elements such as Co^{60} has been used more frequently over the past years as the radiating fuel. Here, gamma rays are mainly operative.

Beta particles can be produced quite efficiently by electronic machines such as the Van deGraaf electrostatic generator or by a linear accelerator. Details on the mechanices of the procedures for irradiating a food material and for the generation of Beta particles can be found in articles or texts on the general topic of Food Irradiation (Desrosier and Desrosier, 1977; Potter, 1977; Godblith, 1967; *Urbain, 1978*).

HOW IONIZING RADIATIONS ACT:

In order to carry out their action of preserving the food material or any other biological material, the ionizing radiations must penetrate the material. The degree or extent of penetration depends upon the nature of the radiation and on the nature of the material itself. As stated previously, each type of particle has a different penetrating power. Beta particles have less penetrating power than gamma particles, while neutrons have great penetrating power and are so high in energy that they can bring about alterations in atomic structure and render elements with which they come in contact (strike) radioactive.

Two principal factors determine the efficacy of radiations in manifesting their effects: one is the capability of the radiation to alter molecules. The other is their ability to knock out (remove) electrons from the atoms of the materials through which they pass. Each type of radiation differs in one or both of these respects. Beta particles, for example, generally are better able to produce ionizations in materials through which they pass than are gamma rays. The higher the energy of electron beams the deeper will be the penetration of material traversed by them and the greater will be the alteration in molecules and ionization. Neutrons being of extremely high energy will alter atomic nuclei making them radioactive. Ionizating radiations are thought to elicit their actions through direct effects or indirect effects or by a combination of both, on the food product or biological material.

Direct effects are thought to occur through direct contacts or interactions between the high energy rays and particles with vital centers of molecules of the cells or product. A bullet-target situation has been envisaged. Although such collision reactions do occur, it is thought that, at a given radiation dose, their frequency is not high enough to account fully for the effects of radiations observed in a substrate; and so indirect effects are thought to contribute substantially, if not, principally to these effects.

Indirect effects occur mainly in the aqueous phase of the food material resulting in the radiolysis (decomposition) of water to produce, hydrogen and hydroxyl radicals.

The possible reactions involved are outlined below:

$$HOH \xrightarrow{\text{Irradiation}} \cdot OH + H \cdot$$

The $\cdot OH$ and $H \cdot$ radicals can react with themselves or with O_2 to form hydrogen peroxide, hydrogen or the peroxide radical.

Two hydrogen radicals can react to form H_2

$$\cdot H + \cdot H \longrightarrow H_2$$

The hydrogen radical can react with O_2 to form the peroxide radical.

$$\cdot H + O_2 \longrightarrow \cdot HO_2$$

Hydrogen peroxide may be formed by two reactions:

$$\cdot OH + \cdot OH \longrightarrow H_2O_2$$

$$\cdot HO_2 + \cdot HO_2 \longrightarrow H_2O_2 + O_2$$

The products formed through the indirect effects of irradiation are of primary importance with respect to the ultimate reactions which may occur in the product and with respect to the wholesomeness of the product.

Hydrogen peroxide is a strong oxidizing agent and a poison to the biological systems, while hydroxyl radicals are strong oxidizing agents and hydrogen radicals are strong reducing agents. These two radicals can react with and bring about severe changes in molecular structure or organic matter. The formation of free radicals in radiation reactions is a very important phenomenon because, having an unpaired electron and being rather unstable and highly reactive, they tend to react with one another and with other molecules to pair their odd electrons and thus stabilize themselves.

Changes in irradiated foods are brought about through the abovementioned reactions namely, the formation of ion pairs and free radicals, the recombination of free radicals as well as by other accompanying and related chemical phenomena.

Over the years several foods of plant, animal and marine sources have been treated with ionizing radiations of varying doses to bring about various effects. Several of these applications and their objectives are summarized in Table 16.1. Irradiation sources used were Cobalt-60, Cesium 137 in most cases and, in a few cases, electrons.

TERMS AND UNITS USED IN FOOD IRRADIATION STUDIES.

Several terms such as roentgen, electron volts, roentgen equivalent, rad, G ray and kG ray, are used to express, quantitatively, the intensity of radiation and radiation doses. A breif description of each term is given below.

An *erg* is the unit of energy expended when a force of one dyne acts through a distance of one centimeter. One erg = 10^{-7} Joules (j).

An *electron volt* (ev) is the energy acquired by any charged particle carrying unit electronic charge, when it falls through a potential difference of one volt. One electron volt = 1.60207×10^{-12} erg. Customarily, calculations are made in terms of a million electron volts, (Mev) since the electron volt is such as small amount of energy (1 Mev = 10^6 ev.)

A *roentgen* (r) is defined as that quantity of X-or gamma radiations which will produce one electrostatic unit (e.s.u.) of charge of either sign (positive or negative) per cubic centimeter (cc;cm^3) of air, under standard conditions of temperature and pressure.

Other difinitions given by Glasstone (1950) and summarized by Desrosier and Desrosier (1977) are:

(A) The quantity of radiation which will produce 2.083×10^9 ion pairs per cm^3 of dry air or

(B) The radiation received in 1 hour from a 1 gram source of radium at a distance of one yard.

When referring to the energy absorbed by materials, the terms roentgen equivalent. physical (rep) and rad are used most frequently.

The roentgen equivalent physical is defined as the quantity of ionizing radiation absorbed by soft tissues in gaining 93 ergs of energy

TABLE 16.1.   SUMMARY OF SOME IRRADIATION-TREATED FOODS AND DOSES USED

Product	Purpose of Irradiation Treatment	Dose Limit (K rad)
Potato	Sprout inhibition during storage and marketing.	5–15
Onion	Sprout inhibition during storage and marketing.	5–15
Garlic	Sprout inhibition during storage and marketing.	7–15
Wheat and Ground Wheat Product	Control insect infestation in stored product.	100 max.
Rice	Control insect infestation in stored product.	100 max.
Cocobeans	Control insect infestation in stored product.	70 max.
Spices and Condiments	Reduce number of non-spore-forming pathogens (radicidation).	800–1000
Papaya	Control insect infestation, delay ripening, improve keeping quality.	100
Mangoes	Control ripening, improve keeping quality.	25–35
Oranges	Reduce spoilage, extend shelf life.	75–200
Strawberry	Prolong storage life of fresh fruit, partial elimination of microbes.	250 max.
Mushroom	Retard cap opening, increase shelf life.	250 max.
Bananas	Delay of ripening	35
Asparagus	Reduce number of spoilage organisms (radurization), improve keeping quality.	200 max.
Chicken (eviscerated)	Prolong storage life and/or retard pathogens like salmonella.	700
Cod and red fish	Reduce microbial spoilage, reduce pathogens in packaged and unpackaged fish.	100–220
Shrimps	Reduce number of spoilage organisms (radurization). Improve keeping quality.	50–200
Ham	Sterilization	3,500–5,600
Bacon	Sterilization	4,500–5,600

per gram of soft tissue. Originally, a value of 83 ergs of energy per gram was assigned to the rep but this has been upgraded to 93 since energy absorption is greater in tissues and in bone than air. The term rep is not used frequently nowadays.   Instead, the term rad is used.

The *rad* is a measure of energy absorbed. It is the quantity of irradiation which results in the absorption of 100 ergs (10^{-5} joules) per gram of irradiated material. Because it is based upon energy absorbed, the rad is a very useful term and is the one most commonly employed in food irradiation studies.

A newly introduced, so-called standard irradiation (SI) unit is known as the Gy (g ray), $1Gy = 10^2$ rad. The term k Gy is being used now, also. One k Gy = 1,000 Gy.

TERMINOLOGIES IN FOOD IRRADIATION

In the field of radiation processing of foods, the terms radurization, radicidation and radappertization are used. These terms are explained below and the FAO/IAEA/WHO definitions are given in quotation marks.

Radurization.—The term radurization is derived from the words "radiare" which means to radiate and the word "durare" which means to prolong or extend. It is now used instead of the terms radiation pasteurization and irradiation by non-sterilizing doses. Definition: "Treatment of food with a dose of ionizing radiation sufficient to enhance its keeping quality by causing a substantial reduction in the numbers of viable, specific spoilage microorganisms."

Radicidation.—The term radicidation is derived from the words "radiare" to radiate and -cida, caedere, to kill. Definition: "Treatment of food with a dose of ionizing radiation sufficient to reduce the number of viable, specific non-spore-forming, pathogenic bacteria to such a level that none is detectable in the treated food when it is examined by any recognized bacteriological testing method."

Radappertization.—The term "radappertization or radiation appertization is derived from the word "Appert", the name of the French confectioner who is credited with suggesting thermal processing as a method for the preservation of canned foods. Definition: "Treatment of food with a dose of ionizing radiation sufficient to reduce the number and/or activity of viable microorganisms to such a level that very few, if any, are detectable by any recognized bacteriological or mycological testing method applied to the treated food. The treatment must be such that no spoilage or toxicity of microbial origin is detectable, no matter how long or under what conditions the food is stored after treatment, provided it is not recontaminated."

Inasmuch as it results in commercial sterilization of foods by irradiation, it resembles commercial heat sterilization closely.

Applications

Each of these methods of irradiation processing of foods has been credited with advantages in specific or general areas. Radurization of meats of various animals, meat dishes, eviscerated poultry, ground meat, food ingredients and fish etc. at appropriate doses and depending on the food material, radiation and post-irradiating conditions, has increased the shelf life or storage life of these products significantly (Wierbicki et al., 1975).

Equally important advantages have been claimed for radicidation. Low-dose treatment of foods can eliminate parasites such as Trichinae spiralis, tapeworms, and salmonella. A special application is in the disinfestation of spices where heat would remove desirable volatiles and the usual treatment of ethylene oxide might leave undesirable residues.

Radappertization has been used widely in the processing of ham, bacon, beef, etc. Heat inactivation of enzymes at 65°–75°C and proper packaging over partial vacuum in a sealed, air-tight, moisture proof, light-proof container, which is impermeable to microorganisms, precedes irradiation at the appropriate dose.

Table 16.2 summarizes possible applications of ionizing radiation in the meat industry under radurization, radicidation and radappertization conditions. Table 16.3 notes the minimum doses required for radappertization of several meats and meat products.

TABLE 16.2. POSSIBLE APPLICATIONS OF IONIZING RADIATION IN MEAT INDUSTRY

Application	Dose Range (Mrad)	Irradiation Temp. (°C)
Radurization for extension of refrigerated storage (0°C to 5°C), e.g., meat, poultry, and fish.	0.05 to 0.5	5°±5°
Radicidation - destruction of specific pathogens and parasites, e.g., salmonellae from meat, poultry, and animal feeds; trichinae, tapeworms, and liver flukes in meats.	0.1 to 1.0	5°±5°
Sterilization of food ingredients, e.g., spices.	1 to 2	Ambient
Radappertization (sterilization) to allow long-term unrefrigerated storage, e.g., for meats, meat products, poultry and fish.	2 to 6	−30°±10°C
Reduction of nitrite in cured radurized and radappertized meats.	0.5 to 4.0	5° to −40°C

From Wierbicki et al., (1975)

TABLE 16.3. MINIMUM REQUIRED DOSES (MRD) FOR RADAPPERTIZATION (PRELIMINARY DATA)

Food	Irradiation temperature (°C)	MRD (Mrad)
Bacon	5 to 25	2.3
Beef	−30 ± 10	4.7
Beef	−80 ± 10	5.7
Beef[b]	−30 ± 10	3.7
Chicken	−30 ± 10	4.5
Ham	5 to 25	2.9
Ham	−30 ± 10	3.7
Ham[c]	−30 ± 10	3.3
Pork	5 to 25	4.6
Pork	−30 ± 10	5.1
Shrimp	−30 ± 10	3.7
Codfish cakes	−30 ± 10	3.2
Corned beef	−30 ± 10	2.4
Pork sausage	−30 ± 10	2.7

[a]Based on 10^{12} reduction in numbers of spores of *Cl. botulinum* (12-D)
[b]With the additives: 0.75% NaCl and 0.375% sodium tripolyphosphate
[c]Low $NaNO_2/NaNO_3$ (25/100 mg/kg)
Source: Mr. Abe Anellis, US Army Natick Laboratories, from: Josephson *et al.*, (1975a)

QUALITY OF FOOD TO BE IRRADIATED:

As in any other method of food preservation, the quality of the final product is governed largely by the quality of the initial raw material. Irradiation should be done only on foods which meet the appropriate pre-irradiation standards (FAO/WHO, 1977). This is particularly important with respect to the destruction of micro-organisms of public health significance, which is the main objective of irradiation. If, for example, a food is contaminated with fecal matter, although the common fecal contaminants and enteric pathogens would be destroyed, any pathogenic viruses present may survive. This would not only pose a public health hazard but would ill-represent the capability of the method to accomplish its claimed potential under normal circumstances. The hygienic quality of foods to be irradiated should, therefore, be fully established.

Chemical and physiological conditions of materials at the time of irradiation are important also. In fruits, for example, maturity at the time of irradiation can be critical. Bananas and mangoes irradiated at the improper state will not exhibit the typical delay in ripening or will exhibit abnormal ripening patterns as well as biochemical characteristics. Varying patterns of response will be exhibited by fruits irradiated at different stages of maturity, and moisture content of grains at the time of irradiation will affect final product quality. "To obtain a product of high quality, start with raw material of high quality."

Enzymes in foods must be inactivated prior to irradiation because these, being much more radiation-resistant than most of the ordinary microorganisms thereon, will not be destroyed by the irradiation. Usually, enzyme inactivation is accomplished thermally. Enzyme inactivation is particularly important in the irradiation preservation of meats and flesh foods.

MICROBIOLOGICAL ASPECTS OF IRRADIATED FOODS AND THEIR PUBLIC HEALTH SIGNIFICANCE

The safety of any processed food is generally measured, by among other things, the absence of certain microorganisms and/or their metabolic toxic products which are known to cause illness of any kind to the consumer.

Among such organisms, *Clostridium botulinum*, Salmonella species, *Staphylococcus aureus* and other pathogenic types have always demanded alertness and attention. Fungi which produce mycotoxins are also important as well as pathogenic yeasts and foodborne viruses. Of equally great concern is the problem of radiation resistance in microorganisms, since these may survive treatments sufficient to eradicate even the most potentially dangerous toxin producers such as *Clostridium botulinum*. Additionally, the possibility of mutagenesis must be considered since this can alter the virulence, toxigenesis, adaptation and antibiotic resistance of microorganisms in food. Resistance to irradiation itself may be altered, also. (See Chapters 1 and 3) *Clostridium botulinum* produces a lethal toxin and is the causative agent of the deadly disease known as botulism. The spores and toxin of this organism are very resistant to irradiation. Since it is the most radiationresistant pathogenic organism encountered in foods, and, in view of the public health significance of toxin, the dosage of irradiation necessary to destroy its spores has been established. This has been termed the D_M value and refers to the radiation dose which gives or produces a 90% reduction in the microbial load of the particular substrate. In beef substrate (above pH 4.5) for example, the D_M value for this organism is 0.4M rad. Figure 16.1 shows the experimental curve for determining the D_M value for *Clostridium botulinum* in beef substrate. *Clostridium botulinum* will not be a concern in foods with a pH value below 4.5, and other spoilage organisms which may be present are much less resistant, being destroyed by about 0.2M rad. In general, it is considered safe to use a sterilization process of $12D_M$ values or an equivalent of 2.4M rad. This arbitrary 12-D concept is

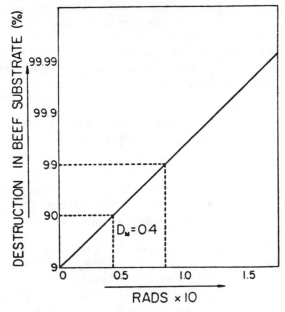

From: Desrosier & Desrosier (1977)

FIG. 16.1.   THE D_M VALUE FOR CL. BOTULINUM
Unit of destruction of microorganisms

the accepted standard of safety for radiation sterilization, stipulating that the radiation process reduces a hypothetical population of spores of *Clostridium botulinum* by 12 decimal logarithmic cycles. *Clostridium botulinum* type A is believed to be the most resistant to irradiation. Both types A and B can germinate and produce toxin at ambient temperatures and so are of primary significance in the radiation sterilization of foods. *Clostridium botulinum* type E is of importance in the preservation of marine products by low dose irradiation (pasteurization). This is particularly critical when the irradiated products are held for extended periods at temperatures in the refrigeration range. Type E *Clostridium botulinum* has been found to be quite common, having been detected in almost all the coastal areas of U.S.A. Type F *Clostridium botulinum* has been detected, also. Desrosier and Desrosier (1977) making salient comments on the experimental determination of D values, point out that almost all the work on this topic has been done using gamma rays and that similar experiments with other types of radiations, particularly electrons and especially at high levels of inactivation, should be done. They noted, also, that the quality of the product may be unfavourably altered even at such low doses of irradiation (below 5M rad.)

Approximate doses of irradiation necessary to destroy various organisms of public health significance in foods are summarized in Table 16.4. A cursory survey of these data indicate that several Salmonellae are more resistant than the more common coliform

TABLE 16.4. APPROXIMATE MINIMAL DOSES OF GAMMA IRRADIATION FOR THE DESTRUCTION OF SPECIFIC MICROORGANISMS AND TOXINS OF PUBLIC HEALTH SIGNIFICANCE

Microorganisms or Toxin	Medium	Inactivation Factor	Dose (Mrad)
C. botulinum, type A	canned meat	10^{12}	4.5
C. botulinum, type E (toxic strain)	broth, minced lean beef	10^6	1.5
C. botulinum, type E (non-toxic strain)	broth, minced lean beef	10^6	1.8
Toxin (*C. bot.*, type A)	cheese	10^3	>7.0
Toxin (*C. bot.*, type A and B)	broth	10^6 (based on mouse-units)	>3.0
Staphylococci (6 phage patterns	broth, minced lean beef	10^6	0.35
Toxin (Staph. emetic-factor)	pork sausage	?	<1.0,<2.0
Toxin (Staph. emetic-factor)	water	?	0.72
Toxin (Staph. alpha lysin)	pork sausage	32	2.1
Salmonella (*pullorum anatum, bareilly, manhattan, oranienburg, tennessee, thompson, typhimurium*)	broth	10^6	0.32–0.35
Aerobacter (Enterobacter)	broth	10^6	0.16
E. coli	broth, minced lean beef	10^6	0.18
E. coli, adapted strain	broth, minced lean beef	10^6	0.35->1.2
M. tuberculosis	broth	10^6	0.14
Streptococcus faecalis	broth, minced lean beef	10^6	0.38
Streptococcus faecalis (adapted strain)	broth	10^6	0.6–1.3
Streptococcus faecalis		10^2	0.08–0.24
Streptococcus faecalis		10^2	0.17–0.65
Viruses: Herpes, mumps, influenza A and B, polio, type 2, vacinnia	tissue extracts	10^9 [1]	1.0
Polio, type 2	tissue-culture medium	10^9 [1]	2.0
Mumps	0.5% albumen	10^9 [1]	1.5
Influenza A	saline	50% reduction from 10^9 [1]	1.0
Influenza A	saline + 1% tryptophan	0	1.0
Polio, encephalitis	brain tissue	10^6 (L.D. 50)	3.5–4.0
Vaccinia	buffer	10^6 (L.D. 50)	1.5–3.0

[1]Virus particles in suspension at a concentration of 10^9 per ml.
From: Desrosier and Desrosier (1977)

organisms and that the faecal Streptococci are known to have the highest resistance reported for enteric bacteria. These facts have been interpreted (Desrosier and Desrosier, 1977) to be important since, under such conditions, an ordinary coliform diagnostic test would be inapplicable to irradiated foods and a check for faecal Streptococci would be unacceptable because their overall resistance may be unusually high.

RADIATION RESISTANT ORGANISMS.

The existence of radiation resistant microorganisms might constitute a potential health hazard and pose diagnostic difficulties. This should be thoroughly explored. Pathogenic yeasts, in particular which can grow at 5°C or lower are of particular concern since they survive radiation pasteurization treatment.

Extermely and unusually resistant to radiation one organism, *Micrococcus radiodurans*, will survive radiation doses which will destroy *Clostridium botulinum*. Radiation resistance is exhibited also by certain other micrococci and faecal streptococci, which will survive radiation pasteurization. Some yeasts and molds will also survive this treatment.

Recent studies by Snyder and Maxcy (1979) has shown that highly radiation-resistant Moraxella-Acinetobacter (gram-negative coccobacilli) do not grow on meat. Therefore, these microorganisms, though present do not pose any public health hazard, even if more sensitive organisms were destroyed by non-sterilizing doses of irradiation, thus eliminating competition. Up to now, the source of these organisms and their true significance in meat supplies are not known.

Radiation induced mutation or selection has been proved and, recently, Maxcy (1977) using gamma radiation injured cells of *Escherichia coli, Salmonella typhimurium* and of *Moraxella sp.* found that they formed smaller colonies on surface plates and had longer lag phases than unirradiated cells. This possibility of the developement of radiation resistant forms of microorganisms is one strong reason why good manufacturing practices should be adhered to.

Grains and cereals have been treated with low doses of irradiation to eliminate fungi. Since some of these organisms can produce mycotoxins, the fate of mycotoxin-producing molds, their resistance to irradiation and their ability as potential survivors to produce toxins, are of public health significance. This is particularly important since, in the treatment of these products, low doses of irradiation are employed. Aspergillus and Penicillium species are important here. There have been conflicting reports on the production of aflatoxin by *Aspergillus flavus* after irradiation doses of less than one Krad and, apart from work on the aflatoxins, there is little information of the production of other mycotoxins. *Aspergillus ochraceus* after irradiation at 150 and 200 K rad, reportedly produced more ochratoxin and patulin production by strains of *Penicillium patulum*, originally isolated from bread, were investigated for the effects of low level gamma radiation on their growth. More information on the effects of low level irradiation or mycotoxin-producing molds is needed, particularly if this process will be used for the widespread preservation of food products in general and grains and cereals in particular.

Viruses are highly resistant to irradiation and their elimination from food by irradiation would be difficult. In view of this, if any irradiated food of animal origin is to be consumed directly i.e. without post-irradiation cooking, such food, prior to irradiation should be given sufficient heat treatment to eliminate viruses. Much more work is need to elucidate the susceptibility of food-borne viruses to irradiation.

Cliver (1965) suggested the possibility of mutations occurring in food-borne viruses which had survived gamma irradiation of products. In order to determine if viruses which contaminate food can produce mutants of public health significance, four enteroviruses were subjected to levels of gamma radiation which would most likely be used in food processing. Gamma radiation from ^{60}Co in doses of 2–5K Gy (200–500 K rad was used).

Using conditions which would favor any variants over the parent type, selection tests were done on each irradiated virus type for each of the following properties: serotype, neurotropism, replication at increased temperature and acidity and host species specificity. From the results obtained it was concluded that significant mutations among viral survivors of gamma irradiation at recommended food preservation doses would be unlikely.

RESISTANCE OF ENZYME TO IRRADIATION:

Enzymes are extremely resistant to irradiation, much more so than even the spores of *Clostricium botulinum* . Such relatively high radioresistance is attributed to the presence in the biological systems of free radical scavengers which act as natural protectors.

Generally, it may be said that complete inactivation of enzymes requires about five to ten times the dose required for the destruction of microorganisms. As seen in Figure 16.2 the D_E value for enzymes is about 5. Though almost complete enzymatic destruction would be achievable by four D_E values or about 20M rads., this high dose would destroy many vital food components and also probably render the product unwholesome and unsafe. Preliminary blanching of the food product at about 160°F for a few minutes, would obviate this.

Enzyme inactivation in irradiated foods is of great importance for several reasons. First, since irradiation itself will not destroy them, irradiated foods will be unstable because during storage, enzymatic deterioration will cause spoilage. Second, irradiated foods are more susceptible to enzymatic attack than unirradiated foods. Apparently, irradiation brings about changes which expose more reaction sites. Irradiation is thought to bring about unfolding of the protein molecule, leading to the availability of more reaction sites. A parallel situation is found in the cooking of foods, where heat denaturation of proteins any/or gelatinization of starch render these substrates more readily degradable by digestive enzymes.

High resistance of enzymes to irradiation has been demonstrated with milk phosphatase, which was not destroyed by irradiation doses sufficient to sterilize milk. (Proctor and Goldblith, 1951). In the destruction of bacteria, there is a complementary effect between heat and ionizing radiation but this does not occur in the inacti-

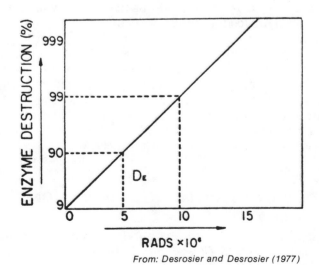

From: Desrosier and Desrosier (1977)

FIG. 16.2.   THE D_E VALUE FOR ENZYMES
Unit of destruction of enzymes.

vation of enzymes. This has been demonstrated with pea peroxidase (Baker and Goldblith, 1961). Thus, two separate operations must be carried out in the irradiation preservation of meats and other foods in which enzymatic action is detrimental in the post-irradiation handling: (1) Enzyme inactivation by heat and (2) Irradiation processing.

As succinctly stated by Goldblith (1967):

"Irradiation doses in the sterilizing region do not markedly reduce the thermal energy requirements for enzyme inactivation, nor does the relatively small amounts of heat needed to inhibit enzymes markedly affect the doses needed for sterilization."

Stability of irradiated products can be achieved only by pre-irradiation thermal inactivation of enzymes. Proteolytic enzymes are of particular importance since their post-irradiation activity will lead to the development of strong off flavors and odors.

It is thought that, although both direct and indirect effects of irradiation may be involved, enzymes are affected principally by the indirect effects i.e. by the free radicals formed in the solvent phase which, in foods, is water. Therefore, dilute solutions of enzymes are relatively more sensitive to irradiation than are concentrated solutions. Enzymes in their natural environments, as in foods, are relatively very resistant.

The necessity to thermally inactivate enzymes prior to irradiation does not enhance food irradiation as a rapid, convenient method of food preservation. Whereas, in canning, for example, moist heat inactivates enzymes with concomitant destruction of at least the vegetative phase of microorganisms, this dual simultaneous effect is not achieved with irradiation.

POST-IRRADIATION HANDLING OF FOODS:

Irradiated foods, if not properly handled after treatment, are subject to deterioration, spoilage and loss of nutritive value. Storage conditions and packaging are, therefore, important post-irradiation considerations. A general specification of international agencies and organizations is that "Foods to be irradiated and their packages should be of suitable quality, acceptable hygienic condition and appropriate for this purpose and shall be handled, *before* and *after* irradiation according to good manufacturing practices—".

It has been suggested that perishable irradiated foods should be stored at 0°C, but since foods and their microbial flora vary so much, a single specific temperature cannot be recommended. A somewhat general procedure would be to store irradiated foods at a temperature sufficiently low to prevent the growth of any microbe, particularly those of public health significance, and without, at the same time, causing physiological damage to the product. For example, by storing irradiated fish at 3°C, the growth of *Clostridium botulinum* Type E would be prevented but irradiated bananas stored at such low temperatures would suffer cold damage. This would also be true for many other fruits. The relative humidity of the storage atmosphere is very important. Important, too, is the water content of the food product. There is a relationship between the temperature of storage permissible, the relative humidity and the water content (Behere *et al.*, 1978).

Irradiated grains can be stored at below 10°C since aflatoxin forming strains of Aspergillus will not grow well at this temperature. If the water content is low, a temperature of below 20°C may suffice.

Post-irradiation handling of any food, including grains should be such as to preclude the outgrowth of toxin-forming or any other type of spoilage organism. The regrowth of fungi on irradiated wheat was reported by Ingram (1975) to be slower than on unirradiated wheat. Behere *et al.* (1978) reported that irradiated wheat deliberately infected with conidia of *Aspergillus flavus* had less aflatoxin than the unirradiated control. These studies indicate that not only are properly stored irradiated grains, particularly wheat, stable but that they are also quite resistant to re-infection.

VARIOUS TYPES OF PACKAGING MAY BE USED FOR IRRADIATED FOOD.

To be suitable as a packaging material, irradiation should not affect its functional properties nor should it cause the production and/or release of deleterious or non-desirable substances which may migrate into the food. Desrosier and Desrosier (1977) give a useful summary of packaging of radiation processed foods. For sterile products, hermetically sealed containers must be used and the food is irradiated in the container. Therefore, not only must the material of which the container is made be assessed but also the shape and flexibility. For pasteurized products, special packaging for each type of product is necessary. Irradiation may affect different packaging materials in different ways. In the case of rigid, metal containers, tin-coated containers and aluminum containers, sterilizing doses of irradiation do not affect steel metal base of the former and aluminum containers are not affected either. At doses of 6 million rads (6M rad) and higher, some damage to both may occur.

The enamels used in the inferior of tin plated cans must be of the proper type also. For foods with a high fat content, oleoresin enamels are unsuitable, but seem suitable for enzyme inactivated foods. The shape of the container is considered to be very important, and a cubical form is thought to be the most satisfactory for optimum dose distribution and use of the source of radiation. However in actuality, cylindrindrical cans are encountered most frequently and for such specialized items as meat loafs and sardines, rectangular cans are used.

Compounds used for the end sealings of containers react in different ways. Irradiation apparently depolymerizes butyl-rubber sealing compounds, but, on a whole, compounds used for end sealing are somewhat improved by irradiation.

Can coating is not destroyed by irradiation since it does not promote the so-called tin rot or disease—i.e. the transition of rhombic to crystalline structure of tin. This condition is prevented by traces of bismuth. Therefore, tin coatings over base steel is suitable for irradiation.

High moisture, perishable foods may be stored, after irradiation, in flexible containers at room temperature. Extremely high doses of irradiation of over 3M rads will cause brittleness in cellophanes, saran and plioform packaging material while 2M rads or more will cause inconsequential physical changes in mylar, polyethane, vinyl and polyethylene plastic films. At doses less than 2M rads, changes in the physical characteristics of flexible containers are negligible.

An important problem in the irradiation of foods in plastic containers is the production of off-odors in the foods. Most foods treated in this way suffer from this problem.

At sterilizing doses, nylon gives rise to little off-odor production, but with polyethylene short fragmentations of the polymer are produced and enter the food. Also, strong malodorous compounds are given off.

Migration of microorganisms through thin plastic films can be an additional problem because these can develop fine holes during salting and rough handling. Such thin films should not be used.

The compatability of various containers with radiation processes and processing equipment has been tabulated (Desrosier and Desrosier, 1977 p.432). Reportedly, Scotchpack and similar containers can tolerate as rough a handling as the tin can.

EFFECTS OF IRRADIATION ON NUTRITIONAL QUALITIES OF FOODS.

Both macro- and micro-nutrients are affected by the treatment of foods with ionizing radiations. The type and extent of such changes depend on the type or nature of the food itself and on the dose of radiation. Proteins, lipids, carbohydrates and vitamins are affected to varying extents.

Since foods differ so widely in their chemical composition, the significance of changes due to any treatment, including irradiation must be weighed carefully. The possible adverse effects of such changes and the bio-availability of the nutrients need close considerations. For foods consumed in small quantities, small changes in nutrient composition or bio-availability may be less significant than for foods eaten in large amounts daily. These considerations become serious for large populations in developing areas where single items like wheat, rice, millet, etc. are staples. Overall, it appears that irradiation, at the recommended doses, does not cause any significant impairment in the nutritional qualities of foods so treated.

Effect on Proteins

Proteins can be severely affected by ionizing radiations. Overall, the primary changes are those affecting odor, taste and color. Generally, it is thought that irradiation at required dose levels has little influence on biological response other than alteration in the antigenicity of specific proteins, especially at high dosages. Low dosages may cause molecular uncoiling, coagulation, unfolding, even molecular cleavage and splitting of amino acids. Molecular rearrangement has been demonstrated by subjecting aqueous solutions of casein and egg albumin to cathode rays. Apparently, peptide linkages were not attacked since no marked increase in amino nitrogen was detected. Radiation effects were concentrated around the sulfur linkages and hydrogen bonds were broken, also. Desrosier and Desrosier (1977) list the sequence of protein bonds attacked by ionizing radiations as follows: -S—CH_3, -SH, imidazol, indol, alpha amino, peptide and proline. Odorous compounds and ammonia may be produced by irradiated proteins or proteinaceous substances. The development of off flavors in foods due to irradiation at high doses has been attributed to the presence of benzene, phenols and sulfur compounds formed from phenylalanine, tyrosine and methionine, respectively. These amino acids are sensitive to irradiation and are cleaved to produce the undesirable compounds.

One very sensitive protein is egg white, which is a mixture of proteins. An egg irradiated with 0.6M rad will show a thin, watery condition, even though this dose is insufficient to destroy bacterial spores and effect sterilization. This reduction in thickness is thought to be due probably to the destruction or alteration of ovomucin which is mainly responsible for the thickness of egg albumin. Other subtle changes, detectable electrophoretically, occur also. These total gross and minor changes potentially affect egg quality which is determined by the thickness of egg albumin.

In the case of milk, the casein may be affected with a resulting increase in rennet coagulation time and reduced heat stability. Irradiation of milk results also in flavor changes and off-flavors resembling the burnt flavor of overheated milk.

Irradiation studies on wheat revealed no appreciable effect with low doses (Rao *et al.*, 1978a). However, at 1M rad, significant change occurred in the level of free amino acids. There was an overall increase of about 8.5% in total free amino acids. Most of this increase was attributed to the rise in the levels of glycine, alanine, valine, methionine, lysine, isoleucine, leucine, tyrosine and phenylalanine. The results are summarized in Table 16.5. At the higher dose level, such increases were attributed to depolymerization of the proteins. Data for beef and seafoods are shown in Tables 16.5–16.7, respectively.

Effects on Carbohydrates

Irradiation of complex carbohydrates leads to depolymerization. In general, simple carbohydrates undergo no dramatic changes during the irradiation of foods, especially with respect to their metabolizable energy. Irradiation of pure carbohydrates has produced degradation products which allegedly have mutagenic and cytotoxic effects. However, here again, the undesirable effects were produced using very high dose levels. Moreover, reactions in pure solutions do not necessarily reflect realistic situations in complex foods.

Studies on wheat have shown that irradiation at 0.02–1M rad, increased the initial levels of water-soluble reducing sugars by 5–92% compared to control samples. (Table 16.8). Such overall increase in initial total reducing sugars resulted from the step-wise and random degradation of starch. Diastatic activity, expressed as

TABLE 16.5.   EFFECT OF IRRADIATION ON FREE AMINO ACID CONTENT OF WHEAT[a]

	Wheat irradiated (Mrad) at						
	0	0.02		0.2		1.0	
Amino acid	(mg/g N)	(mg/g N)	% inc[b] over control	(mg/g N)	% inc[b] over control	(mg/g N)	% inc[b] over control
Aspartic acid	4.03	3.97	—	3.97	—	4.01	NS
Threonine	0.37	0.35	—	0.39	5.4	0.39	NS
Serine	3.47	3.51	1.1	3.35	—	3.28	—
Glutamic acid	3.04	2.88	—	2.77	—	2.90	—
Proline	0.44	0.45	2.2	0.47	6.8	0.46	4.5
Glycine	0.84	0.83	—	0.86	2.3	0.87	3.5
Alanine	1.92	2.11	9.8	2.36	22.9	2.6	35.4
Valine	0.76	0.78	2.6	0.82	7.7	0.84	15.2
Methionine	0.33	0.34	3.0	0.34	3.0	0.37	12.1
Isoleucine	0.52	0.56	7.6	0.53	1.9	0.04	23.0
Leucine	0.62	0.64	3.2	0.66	6.4	0.65	4.8
Tyrosine	0.48	0.49	2.0	0.54	12.5	0.60	25.0
Phenylalanine	0.47	0.49	4.2	0.50	6.3	0.51	9.0
Lysine	0.46	0.45	—	0.45	—	0.47	2.1
Histidine	0.22	0.21	—	0.23	4.5	0.21	—
Arginine	1.73	1.79	3.4	1.77	2.3	1.80	4.0

[a]Results are averages of three determinations.
[b]Inc = Increase
From: Rao et al. (1978a)

"maltose value" was significantly increased after an incubation (fermentation) period of 1 hour, indicating the increased availability of the reducing sugars to amylase action. In view of the importance of reducing sugar-amino acid reactions in the generation of bread flavor and aroma, the changes observed here are highly advantageous.

Effects of Lipids

With respect to lipids, much of the studies on the effects of irradiation have been done using unrealistically high doses. However, the findings do indicate trends and/or susceptibility. Generally, peroxidation may occur and this, in turn, can affect certain labile vitamins such as vitamins E and K. Such effects parallel those observed in thermal processing. The formation of peroxides and volatile compounds, and the development of rancidity and off-flavors have been reported by, among others, Merrit (1972) and Nawar (1972).

The effects of irradiation on wheat lipids have been assessed by several investigators, (Chung et al. 1967; Rao et al., 1978b). In general, lipids in cereals seem to be degraded only at high doses of irradiation and no significant effects on iodine value, acidity or color

TABLE 16.6. EFFECTS OF DIFFERENT PROCESSING METHODS UPON THE AMINO ACID AND VITAMIN CONTENT OF ENZYME-INACTIVATED BEEF

	Treatment and Length of Storage							
	Frozen Control		Thermally Sterilized ($F_0 = 5.8$)		^{60}Co (4.7–7.1 Mrad)[1]		Electron (10 MeV) (4.7–7.1 Mrad)[1]	
Nutrient	0	15	0	15	0	15	0	15
	(Months)		(Months)		(Months)		(Months)	
Amino acids (wt %)								
Aspartic acid	2.12	1.99	2.22	2.11	2.27	2.07	2.15	2.24
Threonine	0.91	1.02	0.95	0.99	1.02	1.00	0.91	0.98
Serine	0.78	0.92	0.81	0.91	0.84	0.88	0.80	0.87
Glutamic acid	3.84	3.63	3.94	3.60	4.10	3.70	3.89	3.91
Proline	1.06	1.17	1.08	1.11	1.12	1.06	1.07	1.05
Glycine	1.39	1.44	1.41	1.40	1.45	1.45	1.28	1.39
Alanine	1.43	1.44	1.47	1.42	1.50	1.46	1.40	1.48
Valine	1.00	1.06	1.07	1.06	1.09	1.09	1.04	1.08
Isoleucine	0.95	0.97	1.02	1.00	1.03	1.04	0.98	1.03
Leucine	1.78	1.84	1.87	1.84	1.95	1.91	1.76	1.94
Tyrosine	0.77	0.84	0.82	0.85	0.85	0.86	0.80	0.88
Phenylalanine	0.89	0.96	0.93	0.96	0.94	0.95	0.89	1.00
Lysine	2.02	2.05	2.09	2.09	2.12	2.01	2.09	2.06
Histidine	0.76	0.80	0.83	0.78	0.81	0.78	0.75	0.83
Arginine	1.43	1.57	1.33	1.52	1.61	1.39	1.55	1.62
Cystine[2]	—[3]	0.16	0.36	0.16	0.24	0.24	0.24	0.32
Methionine	0.53	0.54	0.57	0.54	0.62	0.54	0.56	0.55
Tryptophan	0.33	0.26	0.28	0.26	0.30	0.25	0.27	0.26
Total amino acids	21.99	22.66	23.05	22.60	23.86	22.68	22.43	23.49
Vitamins (%)								
Thiamin (mg)	0.05	0.056	0.02	0.017	0.02	0.015	0.03	0.019
Riboflavin (mg)	0.51	0.099	0.44	0.120	0.49	0.120	0.49	0.085
Niacin (mg)	4.74	4.75	4.92	4.75	4.97	4.75	5.00	6.19
Pyridoxine (mg)	0.49	0.099	0.41	0.065	0.44	0.030	0.41	0.060

Source: Office of the Surgeon General, Department of the Army, Contract No. DADA 17-71-C-1030. Industrial Bio-Test, Inc., Contractor, from: Josephson et al. (1975b)

[1]Packaged beef air-evacuated to internal pressure (IP) of approximately 100 mm Hg. IP at start and after irradiation and thawing was approximately 250 and 350 mm Hg, respectively. Temperature of product was −40 to −5°C during irradiation.

[2]Not including cysteic acid.

[3]Chromatographic peak not resolved from methionine.

TABLE 16.7. EFFECT OF IRRADIATION ON THE TOTAL AMINO ACID (% OF PROTEIN) CONTENT OF SEA FOODS

Amino Acid	Clams[1] 0[3]	Krad 450 AP[4]	Krad 350 VP[5]	Haddock[2] 0	Krad 250 AP	Krad 150 VP
Tryptophan	1.10	1.24	1.15	1.27	1.18	1.18
Lysine	6.89	6.69	7.35	10.78	9.40	9.82
Histidine	1.31	1.74	1.35	2.26	1.75	2.22
Threonine	3.49	4.05	4.15	4.02	3.72	4.54
Valine	3.89	4.12	3.99	4.50	4.70	4.89
Methionine	2.18	2.30	2.12	3.00	3.11	3.31
Isoleucine	3.75	4.00	3.68	4.64	4.76	5.35
Leucine	6.27	6.50	5.89	5.32	7.54	8.47
Phenylalanine	2.88	3.43	2.68	3.32	3.40	4.15
$1/2$ Cystine	1.09	1.02	1.05	0.99	0.83	1.13
Ammonia	1.42	1.78	2.01	1.51	1.43	1.31
Arginine	6.24	6.79	6.93	6.66	6.07	5.13
Aspartic acid	7.46	7.60	7.75	9.30	9.78	11.05
Serine	3.47	4.08	3.81	3.91	3.79	4.97
Glutamic acid	11.35	12.11	12.41	13.33	11.12	15.75
Proline	2.85	3.24	3.14	2.97	3.14	3.57
Glycine	6.45	6.85	7.01	3.91	4.22	4.55
Alanine	7.62	7.76	7.97	5.41	5.73	6.08
Tyrosine	2.88	3.11	2.51	2.83	3.17	3.76
Protein (%)	10.75	8.97	10.05	19.00	17.49	17.63
Moisture (%)	85.50	88.28	86.30	79.56	81.14	79.10

Source: Derived from Brooke et al. (1964, 1966), from: Josephson et al. (1975b)
[1]Stored 30 days at 0°C postirradiation.
[2]Stored 30 days in ice.
[3]Fresh.
[4]Air packed.
[5]Vacuum packed.

intensity of wheat flour lipids have been observed. Rao et al. (1978b), showed that total lipid content of wheat was not changed due to irradiation treatment. However, at 1M rad, there was a significant increase (20%) in total free lipids and a decrease (46%) in bound lipids. The results summarized in Tables 16.9 and 16.10 do attest to the general claim that the recommended doses of irradiation do not cause any significant, undesirable changes, if any. In fact, loaf volume was increased at doses of up to 0.2M rad. Lipid-protein complexes which are of critical importance in baking were not noticeably affected at low doses as shown in Table 16.11. The purothionines are low molecular weight wheat proteins which are associated with wheat lipids. As is well-known, the gas-retaining complex in dough is thought to be composed of protein units bound to polar lipids.

In another study, Rao et al. (1978b) evaluated bread prepared from irradiated wheat. Here again satisfactory performance of the flour and acceptability of the product were very good when wheat irradiated at low doses was used. (Table 16.12).

TABLE 16.8.   REDUCING SUGARS AND DIASTATIC ACTIVITIES IN IRRADIATED WHEAT[a]

Dose level (Mrad)	Initial reducing sugars		Maltose value	
	mg maltose/ 10g wheat flour	% inc over control	mg maltose liberated/10g wheat flour at 30°C for 1 hr	% inc over control
0	90	—	150	—
0.02	95	5.5	172	40.6
0.2	125	38.2	211	78.0
1.0	172	91.1	259	78.0

[a]Aqueous extracts of wheat flour were analyzed for total reducing sugars. 'Maltose value' was measured as described in text. Results are averages of three independent experiments.

From: Rao *et al.* (1978a)

Effects on Vitamins

Some destruction of vitamins does occur during the irradiation of foods. Vitamins C, E and K are destroyed, to varying extents, depending on the dosage used. Thiamine is very very labile to irradiation.

As would be expected, losses are lower with low-dose (pasteurizing) doses than with sterilizing doses. Such losses are thought to parallel those experienced in the heat processing of foods.

Ascorbic acid in solution is quite labile to irradiation but in fruits and vegetables seems quite stable at low doses of treatment. Retention in several fruits is shown in Table 16.13. Studies by Cuevas Ruiz *et al.* (1972) and Jirvatana *et al.* (1970) showed similar good retention of ascorbic acid in irradiated mangoes and papayas. Retention of ascorbic acid in several vegetables is summarized in

TABLE 16.9.   EFFECT OF GAMMA-IRRADIATION ON LIPID COMPOSITION OF WHEAT[a]

Wheat lipid component	Radiation dose (Mrad)			
	0	0.02	0.2	1
Total lipids (g%)	1.78	1.84	1.72	1.74
Total free lipids (TFL) (g%)	1.20	1.25	1.42	1.44
Bound lipids (g%)	0.50	0.52	0.28	0.27
Non-polar lipids (% of TFL)	60.00	60.00	57.00	56.00
Polar lipids (% of TFL)	48.00	47.00	45.00	45.50
Nonpolar/polar lipids	1.25	1.19	1.26	1.22

[a]Total, free and bound lipids were extracted from wheat flour and weighed to constant weights as described in the text. Polar and nonpolar lipids were separated from TFL by silicic acid column chromatography and eluted with methanol and chloroform, respectively. Results are averages of three determinations.

From: Rao *et al.* (1978b)

TABLE 16.10.   LIPID CONSTITUENTS IN IRRADIATED WHEAT[a]

Lipid constituent	Radiation dose (Mrad)			
	0	0.02	0.2	1
Iodine number (mg I_2 absorbed/g fat)	114	114	105	—
Saturated fatty acids (g%)	0.32	0.31	0.32	0.31
Unsaturated fatty acids (g%)	1.21	1.17	1.10	1.03
Phospholipids (% of total lipids)	12.0	11.9	12.0	11.9
Triglycerides (% of total lipids)	31	31	30	30

[a]Lipid composition was determined by methods described in the text. Results are averages of three independent estimations, carried out in triplicate.
From: Rao et al. (1978b)

Table 16.14. Tables 16.15–16.18 show the effect of irradiation on the vitamin content of wheat and wheat products, ham, porkloin and seafoods, respectively. Effects on the vitamin content of enzyme-inactivated beef are shown in Table 16.5.

Effects on Meat Pigments

Radiation sterilization of meats results in changes of the pigments. These changes, it is generally agreed, are accompanied by a reduction of heme pigments to the ferro form. Recently, Kamarei and Wierbicki (1979) showed that ionizing radiation reduces the heme iron of the brown pigment of cooked meat (globin myohemichromogen) to an unstable red pigmet globin myohemochromogen. On exposure to air, the globin myohemochromogen reverts to the original ferric (brown) pigment.

Detailed discussions on the nutritional as well as other aspects of

TABLE 16.11.   DISTRIBUTION OF PUROTHIONINES IN IRRADIATED WHEAT[a]

Radiation dose (Mrad)	Total lipids (g/100g wheat flour)	Purothionine I (mg protein/g lipid)	Purothionine II
0	1.2	6.0	3.0
0.02	1.2	6.0	3.0
0.2	1.2	5.0	3.0
1.0	1.3	4.5	4.0

[a]Proteins were isolated from total lipids, extracted with C:M mixture and dissolved in WSB. Insoluble fraction was designated as purothionine I and soluble as purothionine II. Values are averages of three experiments.
From: Rao et al. (1978b)

TABLE 16.12.   EFFECT OF IRRADIATION ON BAKING PROPERTIES OF WHEAT [a]

Radiation dose Mrad	Specific loaf vol cc/g	Crust color % Transmittance	General appearance
0	3.0	98	Satisfactory
0.02	3.5	95	Satisfactory
0.20	3.8	90	Satisfactory
1.00	2.7	76	Not satisfactory

[a]Specific loaf volume was measured by mustard seed displacement method. Bread crust was extracted with 70% ethanol and color intensity measured at 550 nm in a Bausch and Lomb Spectronic-20. General appearance of bread was assessed by subjective evaluation by panel members. Results are averages of three experiments. From: Rao *et al.* (1978a)

food irradiation are provided by Josephson *et al.* (1975b) and by a recent ACS Symposium (1978).

TOXICOLOGICAL STUDIES.

Toxicological studies, properly executed, constitute a critical exercise leading to the possible government certification of an irradiated food as being safe and wholesome. Much attention has been given to this by various study groups and international bodies.

Toxicological studies are done by animal feeding tests. Traditionally, they extend over three generations in one species or, in many requirements, for one or more generations in at least three different species. Thus, studies may be of a short term (8–13 weeks) or long term (two years) nature.

Experimental diets used in toxicological testing should be previously tested to assure that they are suitable. The maximum amount of food which can be incorporated into the diet without the

TABLE 16.13.   EFFECT OF RADIOPASTEURIZATION ON ASCORBIC ACID RETENTION (%) IN FRUIT

Product	Dose Krad	Retention Percent
Oranges, Temple	100	97
	200	72
Tangerines	40	104
	80	94
	160	94
Tomatoes	100	86
	200	86
	300	91
Papayas	125	110

Source: Calculated from Dennison and Ahmed and Wenkam and Moy by Thomas, Nutrition Div., Natick Laboratories, from: Josephenson *et al.* (1975a)

TABLE 16.14. ASCORBIC ACID AND CAROTENE RETENTION IN IRRADIATED VEGETABLES

Product	Dose	Percentage Retention Ascorbic Acid	Carotene
	Krad		
Carrots	10		102
	20		115
	30		113
	40		109
	80		99
	Mrad		
Beans, green	4.8	73	169
Carrots	4.8	78	87
Corn	4.8	71	56

Source: Calculated from Thomas and Calloway (1961) and Kuzin and Abdurakhmanov (1970), from: Josephenson et al. (1975b)

impairment of optimum growth survival and reproduction of the test animals should be determined, also. Such a maximum, will be conditioned by the physical nature of the food, the nutritional quality of the food and also by the presence of natural components of the foods which themselves may cause notable toxicological effects when present in high levels in the test diet. The cases of onions and mushrooms have been cited. These points are of relevance, since, if natural components of the diet itself cause toxicological problems, then any minor effects of irradiation on a component of the diet, could possibly accentuate this stress, leading to undeserved blame on the irradiation process per se.

The increasing practice of irradiating prepared feed for laboratory animals also drew comment of the FAO/WHO group, which expressed concern over the possible effect of irradiated feed on control groups of animals used in toxicological studies. This, it was rightly pointed out, could obscure differences which would have been noted had a non-irradiated feed been used. Another possible

TABLE 16.15. EFFECT OF IRRADIATION ON VITAMIN RETENTION (%) OF WHEAT AND WHEAT PRODUCTS

Krad	Thiamin	Riboflavin	Niacin	Pyridoxine
20[1]	88	91	88	—
200[1]	88	87	91	—
30–50[2]	100	100	89	100
30–50[3]	100	100	117	100

Source: Calculated from Vakil et al. (1973) and Heiligman et al. (1973), from: Josephenson et al. (1975a)
[1]Wheat.
[2]Flour.
[3]Bread from irradiated flour.

TABLE 16.16.   EFFECT OF PROCESSING ON THE VITAMIN CONTENT OF SHELF-STABLE CANNED HAM

Vitamin	Treatment	mg/100 g[a]	% retention
Thiamine	Control	3.82 ± 0.38[b]	—
	4.5 Mrad at $-80° \pm 5°$C	3.25 ± 0.79	85
	Thermally processed	1.27 ± 0.36	32
Riboflavin	Control	1.01 ± 0.18[b]	—
	4.5 Mrad at $-80° \pm 5°$C	1.25 ± 0.09	123
	Thermally processed	1.10 ± 0.24	109
Niacin	Control	31.5 ± 0.81[b]	—
	4.5 Mrad at $-80° \pm 5°$C	23.8 ± 2.92	76
	Thermally processed	14.6 ± 4.49	46
Pyridoxine	Control	1.11 ± 0.15[b]	—
	4.5 Mrad at $-80° \pm 5°$C	1.02 ± 0.12	92
	Thermally processed	0.64 ± 0.03	57

[a]Moisture, fat salt-free basis.
[b]Average $\pm$ S.D. Three samples per treatment.
Data furnished by Thomas, Nutrition Div., Natick Labs., from: Josephenson *et al.* (1975a)

important contribution to the safety assessment of irradiated foods would be the availability of data on growth rate, reproductive capacity, haematology, tumor incidence mortality and morbidity of the animal colonies, before and after being irradiated foods. Such comparative basic biological data were, in the opinion of the committee, grossly lacking and would fortify claims for wholesomeness and safety.

TABLE 16.17.   EFFECT OF PROCESSING ON THE VITAMIN CONTENT OF SHELF-STABLE CANNED PORK LOIN

Vitamin	Treatment	Mg/100 Gm[1]	Retention (%)
Thiamin	Control	3.69 ± 0.22[2]	
	4.5 Mrad @ $-80°$C $\pm 5°$	3.14 ± 0.25	85
	Thermally processed	0.76 ± 0.08	20
Riboflavin	Control	1.02 ± 0.28	
	4.5 Mrad @ $-80°$C $\pm 5°$	0.79 ± 0.06	78
	Thermally processed	0.82 ± 0.02	81
Niacin	Control	20.3 ± 5.1	
	4.5 Mrad| @ $-80°$C $\pm 5°$	15.9 ± 2.6	78
	Thermally processed	13.2 ± 1.8	65
Pyridoxine	Control	0.76 ± 0.05	
	4.5 Mrad @ $-80°$C $\pm 5°$	0.75 ± 0.07	98
	Thermally processed	0.63 ± 0.07	84

Source: Thomas and Josephson (1970), from: Josephenson *et al.* (1975b)

[1]Moisture, fat, salt-free basis.
[2]Mean $\pm$ S.D., three samples per treatment.

TABLE 16.18.   EFFECT OF IRRADIATION ON VITAMIN RETENTION (%) OF SEA FOODS

Vitamin	Clams[1] Krad 450 AP[3]	350 VP[4]	Haddock[2] Krad 250 AP	150 VP
Thiamin	80	67	37	78
Riboflavin	99	111	105	100
Niacin	84	97	106	100
Pyridoxine	63	93	125	115
Pantothenic acid	115	115	164	178
Vitamin B-12	92	91	110	90

Source: Calculated from Brooke *et al.* (1964, 1966), from: Josephson *et al.* (1975b)
[1]Stored 30 days at 0°C postirradiation.
[2]Stored 30 days in ice.
[3]Air packed.
[4]Vacuum packed.

CARCINOGENICITY AND TOXICOLOGICAL STUDIES

With respect to carcinogenicity, the general opinion is that, when foods are irradiated under the recommended or approved conditions no hazardous levels of carcinogens are produced. Studies on the production of carcinogens from sterols normally found in foods have shown that no carcinogens are produced.

Several agencies and groups have conducted short term (8–13 weeks) as wll as long term (two-year) animal feeding studies to determine the carcinogenicity of irradiated foods. Results of short-term (8–12 weeks) subacute toxicity animal feeding studies conducted by the U.S. Army Medical Service, using hundreds of rats, dogs, mice and monkeys, led investigators to the general, but equivocal conclusion that no carcinogenicity resulted from the consumption of the irradiated foods. However, there were some doubtful cases, particularly when long-term feeding studies were done. Despite the large number of food items and animals of several species used, and the continued experimentation along these lines, it appears that more work needs to be done before convincing conclusions can be arrived at.

In conjunction with carcinogenicity studies, evidence for the presence of other toxic or harmful substances were sought in the short term and long term animal studies and are still being investigated in other areas. In the Army Medical Services studies, using foods irradiated with [60]Cobalt or electrons up to 10 million electron volts at doses up to 5–6 M rad., it was concluded that such foods were wholesome. Though such tested foods may be safe and nutritionally adequate, Whitehair and Hilmas (1968) have made the cogent observation that it is difficult to predict what chemical compounds will be formed in a given complex food after irradiation because the test diets themselves may be unrealistic and the irradiation doses used are far in excess of the ones which would be actually applied. To complicate this further, any extrapolation of animal data to man generally raises questions of reliability and realism: The joint FAO/ IAEA/WHO Expert Committee, in discussing the relevance of test materials to food as consumed by man, states "In testing for potential toxicity (and in other biological testing, such as for nutritional quality) it is necessary to establish that the material being tested corresponds to the food product that will be consumed in the diet." Sometimes, the relatively high amount of the experimental food item included in the diet could, even in the control, lead to problems which may very well be accentuated by small and insignificant consequences of irradiation. Examples of the diets used in some studies are given by Whitehair and Hilmas (1968) and by Wierbicki et al. (1975).

FORMATION OF MUTAGENIC SUBSTANCES IN IRRADIATED FOODS.

Irradiation of foods could possibly result in the formation of small amounts of decomposition products with mutagenic and/or cytotoxic properties. These could possibly induce genetic changes including chromosome abberrations in biological systems. In view of such possibilities, much in vitro work has been done with various biological systems grown on irradiated media. Some of these have been summarized by Whitehair and Hilmas (1968). Among mutations noted are those of Staphylococcus aureus grown on ultra-violet irradiated nutrient broth, Drosophila melangaster reared on irradiated media. Barley embryos grown on irradiated potato mash have been shown to have an increased number of micronuclei. Perhaps the experiment most directly related to food irradiation studies was that of Holsten et al. (1965). They found that irradiation breakdown products of sucrose caused growth inhibition of carrot tissue culture system. Later, (Shaw and Hayes, 1966) noted that the mitotic rate was severely depressed and chromosome fragmentation in human leukocyte cultures increased as a result of adding irradiated sucrose at a final concentration of greater than 0.2% to the cultures. Since sucrose is a natural component of many foods and may be added to others e.g. curing brines for meats, this elicited much public concern about irradiated foods. As pointed out by Whitehair and Hilmas, such in vitro experiments demonstrate interesting possibilities but definitive claims can be substantiated only by the execution of properly designed in vivo experiments using animals fed the irradiated food itself.

Although there is still no general approval of irradiated foods, experienced workers in this field, based on the data available, have arrived at the general conclusions regarding their wholesomeness and safety: That

(1) Irradiated foods are wholesome and safe. No incidence of chronic toxicity nor carcinogenicity have been reported, attributable to the irradiated food itself.
(2) Some destruction of nutrients in foods does occur but generally these are thought to be within the same range and of the same type incurred by heat processing of foods.

INDUCED RADIOACTIVITY IN FOODS.

This topic has received much attention and, rightly so, because as emphasized by Desrosier and Desrosier (1977) all food is radioactive. The background radioactivity of foods varies with the agricultural origin. Since there is a natural background of radioactivity on earth, all foodstuffs consumed by man are radioactive. However, the radioactivity level varies widely, sometimes by a factor of 10 or more.

Induction of radioactivity is a function of the type of radiation, the energy level of the incident radiation, the dose applied to the food, the percent isotopic abundance of the elements and the half-life of the radioactive nucleotide produced, (Goldblith, 1967) and the amount of radioactivity induced is a direct function of the total dose of the incident electron beam. The maximum radioactivity induced in food or other material which has been irradiated can be estimated from the Herschman equation which is as follows:

$$Q \cong \frac{30 \ D \ E_n}{T}$$

Where Q = The activity in micro-microcuries
            D = The dose is megarad
            E = Energy of radiation in Mev
            n = The decay constant of the isotope formed
            T = Half-life in years of the isotope formed.

Radioactivity can be induced in foods by neutrons since these are deep penetrating particles and so foods should be treated with a radiation source which has a low neutron flux. Radiation of foods in air rather than under water results in a lowered possibility of inducing radiation. This difference is allegedly due to the content of natural heavy water and tritium of water.

Radioactivity can be induced in foods with electron energies of over 2.3 Mev but such activity is short-lived. Higher energy levels of 25 Mev or more will, of course, cause induction but even then the induced activity is much less than that normally present in the food-i.e. much lower than the natural background irradiation itself.

On a whole, it might be generalized that the preservation of foods with gamma photons and high energy electrons will not lead to any induced radiation in the material which is detectable over the background levels, as long as there is a low neutron flux in the gamma radiation source. Therefore, it has been claimed that no health hazards from induced radiation are posed from the consumption of foods preserved by gamma radiation.

DOSIMETRY AND DOSE SPECIFICATIONS IN IRRADIATION OF FOODS.

Wholesomeness and safety of irradiated foods, irrespective of the proposed use of such material, depends heavily on the availability of means and equipment for reliable dosimetry.

Some means of accurately measuring radiation dose is necessary to know the exact amount of radiation being absorbed by the sample under treatment, to assure reproducibility of experimental data between laboratories, to assure or to know dose distribution during irradiation within large packages and package-to-package interface etc. Dose distribution within the product is especially important when the product density varies, for example in meats, the lean to fat ratio:

Dosimeters have been categorized into three groups:

(1) Primary standards or operating standards
(2) Secondary or dosimeters (actinometers)
(3) Production control devices.

Ionizing chambers and calorimeters are primary dosimeters or standards. Ionizing chambers are based on the total dose equivalent of the radiation beam in terms of ionizing energy (Ionization). Calorimeters are based on the total dose equivalent of the radiation beam in terms of thermal energy. (Calorimetry) Although both of these types of dosimetery are well suited for exact measurements and serve well as primary standards, they are unsuitable for routine work because they are complicated methods and require skilled personnel. Moreover, they are expensive and time consuming. Modified ionization chambers for continuous monitoring of the radiation source itself have been developed.

Secondary dosimeters or actinometers have proven very useful. These must always be calibrated against a primary standard. Such dosimeters may be categorized as solid state, liquid state, chemical type, biological type, etc.

Examples of solid state dosimeters are cobalt glass, photographic films, polyvinyl chloride films, sivler phosphate glass and luminescence degradation. Water solutions of dyes which change color due to irradiation have been suggested for use as dosimeters. The three secondary standards recommended for use in dosimetry are the cobalt-glass, Fricke Ferrous-Ferric dosimeter and the Ceric-Cerous dosimeter. If cobalt glass is attached to a food container, the color will change in proportion to the dose received and the change in optical density can be measured spectrophotometrically against a standard. The Ferrous sulfate or Fricke dosimeter is one of the most important and most widely used in irradiation studies. It is named after H. Fricke who, along with S. Morse first studied the reaction in 1929. In this mode of dosimetry, Ferric ions which are formed are determined spectrophotometrically by measuring the absorption at 305 nm. Beer's Law is obeyed and there is a direct proportionality between absorbance at 305 nm and the concentration of Fe^{+++}. An acid solution containing a ferrous salt and some chloride is used. The chloride inhibits side reactions and the effects of impurity. Water is ionized by the irradiation to produce H_2O^+, which dissociates to give a hydroxyl radical ($\cdot OH$) and a proton (H^+). The hydroxyl radical oxidizes the ferrous ion to produce the hydroxyl ion and the ferric ion as follows:

$$H_2O \xrightarrow{\text{Irradiation}} H_2O^+$$

$$H_2O^+ \longrightarrow OH + H^+$$

$$OH + Fe^{++} \longrightarrow OH^- + Fe^{+++} \quad \text{(detected by measuring absorbance at 305 nm)}$$

Fricke dosimeters are accurate and precise. Each 100 electron volts of absorbed energy results in the production of 15.45 ± 0.11 ferric ions. A direct measure of the energy absorbed is obtained, therefore, by measuring the ferric ion concentration. The yield of ions or molecules is termed the G value. In the example here G = 15.45. The G value is used widely in irradiation studies.

Less precise, inexpensive so-called go-no-go or "yes or no" dosimeters can be used also for indicating that a sample was given an irradiation treatment. They consist of dyed plastic tapes or colored markings or containers which change color on exposure to irradiation. Color changes are related to the dose received. They may also consist of liquid plastics which gel on irradiation. With such simple devices and more accurate monitoring of the radiation source with reliable and precise dosimeters, required doses for individual situations can be readily assured.

Technical specifications for the irradiation of foods are stated in "ranges", indicating that no part of the irradiated food shall receive less than the minimum or more than the maximum dose indicated. The use of ranges rather than fixed values is done because attainment of completely uniform dose distribution in an irradiator is impractical. Moreover, an optimum radiation dose cannot be fixed because, even for a specified food product, dose requirements may vary with variety, climatic conditions, or infestation intensity in a particular locale.

In principle, the applied dose should be no higher than that necessary to achieve the desired effect, such as sterilization, disinfestation, sprout inhibition etc. For this reason, maximum doses are set. The setting of a minimal dose is necessary and critical where elimination of pathogenic organisms or plant pests in case of quarantine regulations must be met.

Reportedly, (FAO/WHO, 1977) reliable chemical methods for the detection of irradiated foods are few. However, since irradiation has to be done in government licensed installations, clandestine operations are hardly likely.

INNOVATIONS AND VARIATIONS IN IRRADIATION PROCEDURES

Several innovative variations have been introduced in irradiation processing in attempts to lower the doses required, to stabilize food and to eliminate or minimize undesirable side effects such as the formation of free radicals. Accomplishment of any or all of these would result in a more wholesome and safe product. Thermoirradiation, the use of free radical scavengers, irradiation in the frozen state, irradiation in a vacuum or in an inert atmosphere, lowering the moisture content of the product to be irradiated, lowering the microbial load of the product to be irradiated and the use of chemicals to sensitize microorganisms on/or in the food products are among the main approaches which have been tried or advocated.

Thermoirradiation is based on the demonstrated complementary action of thermal energy and ionizing energy. Several workers have shown that substerilizing doses of ionizing energy can lower the amount of heat necessary to sterilize foods and vice versa. Using spores of *Bacillus subtilis*, it has been shown that a combination of heat and irradiation allowed the use of low doses for almost complete destruction, whereas much higher doses of radiation or much higher temperatures, when applied singly, accomplish less destruction. Figure 16.3 depicts the results. Based on microbiological work by certain research groups, it has been suggested that such synergistic effect could be advantageously applied in the sterilization of foods. To demonstrate such a possibility, a theoretical lethality curve showing the effects of thermoirradiation on the spores of *Clostridum botulinum* has been developed. Examination of such a curve, as shown in Figure 16.4, reveals that between Zero°C-lower temperatures and between Zero°C to about 80°C the resistance of the spores to irradiation increases. However, beyond 80°C resistance decreases drastically. These phenomena have led to the suggestion of the possibly of sterilizing foods in hermetically sealed containers using lower temperatures which are less damaging to the foods. Low acid foods could be hot filled and then irradiated at low doses. Foods and containers which could not stand sterilization temperatures could be processed by thermoirradiation. Meat products packed in very large containers would be quite amenable to this treatment. Due to the slow heat penetration to the center of such large cans, the outside layers of the product would be overprocessed, if sterilization is to be achieved.

Secondary effects in the irradiation process result in the formation of free radicals. Since the presence of free radicals is not desirable, attempts have been made to keep such a possibility to the minimum by carrying out irradiation in the frozen state, irradiating in a vacuum or by the addition of free radical scavengers to the food system. Non-toxic compounds such as sodium ascorbate are used. These compounds compete with food components, such as flavor molecules, for the free radicals. There is, therefore, a sparing effect. A perfect scavenger should be non-toxic, should not itself contribute

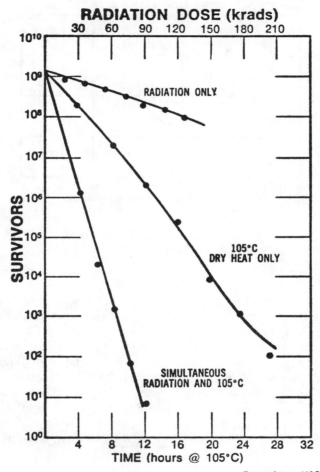

From: Anon. (1973)

FIG. 16.3 A COMPARISON OF INACTIVATION OF *BACIL-LUS SUBTILIS* VAR. *NIGER* USING GAMMA RADIATION AT ROOM TEMPERATURE, DRY HEAT ALONE AT 105°C, AND THEN SIMULTANEOUS DRY HEAT AND GAMMA RADIATION.

off-flavors, should not cause any significant increase in the required dose and should allow for its thorough mixing with the food material. The last quality is not readily achievable in solid foods and so such an ideal scavenger would be limited to liquid or highly disintergrated materials. Irradiation in the frozen state, immobilizes

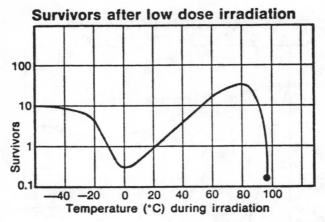

FIG. 16.4. LETHALITY CURVE SHOWING THE EFFECT OF TEMPERATURE DURING IRRADIATION ON SPORES OF *CL. BOTULINUM*.

free radicals which are formed even under these conditions, thus preventing their migration to food components beyond the locale of the free radical formation. Thus, undesirable reactions are grossly limited. Though highly efficacious, the approach does not be itself alone completely eliminate the problem in radiosterilized foods. Irradiation in a vacuum or inert atomosphere is effective in avoiding or minimizing the action of free radicals, because less oxidative free radicals are formed. The medium will be devoid of oxygen with which hydrogen radicals can react to form an oxidizing radical such as $HO_2\cdot$ or hydrogen can react with O_2 to form H_2O_2 as follows:

$$H\cdot + O_2 \longrightarrow HO_2$$

$$H_2 + O_2 \longrightarrow H_2O_2$$

The $H\cdot$ is formed through the decomposition of water and subsequent reactions, as previously outlined. This technique is more applicable to liquid foods than to solid foods.

Inherent drawbacks in this approach are that liquid foods can be more easily handled this way than solid foods and microorganisms are much more sensitive in the presence of oxygen than in its absence. Therefore. although a better product will result, this approach does not totally resolve the problems when sterilization doses are used.

Off-flavor development in products irradiated in a dry state is less than if the moist product was irradiated, because the reduction of moisture content minimizes the formation of free radicals. A consequence of this possible flavor improvement is that higher sterilization doses may be required, because solvent reduction reduces the formation of free radicals through the "indirect action" of irradiation through which bacteria would be destroyed.

Small, but meaningful reduction in dose requirements may be achieved by reducing the initial microbial load of the product to be processed. This effect parallels the established principle of thermal processing, especially in canning where commercial sterilization is more effective, if the microbial load is lowered.

Sensitizing of microorganisms to irradiation by non-toxic compounds which do not themselves contribute off-flavors or off-odors has been proposed to reduce the dose of radiation necessary. Vitamin K_5 and antibiotics have been suggested. The latter was considered promising, but the possibility of antibiotic residues in the food makes their use impractical (see Chapter 15). If ever such a practical compound is available, it has to be specific for important food spoilage and pathogenic organisms, especially *Clostridium botulinum*.

Other aspects of the radiation processing of foods which may reflect on wholesomeness and safety are irradiation conditions and repeated irradiation.

Foods for wholesomeness studies should be irradiated under the same conditions as would be used if they were to be processed for human consumption. This will allow for greater validity interpretation of results.

Irradiation conditions vary for individual products and depend, too, on the proposed use or fate of the irradiated commodity. The case of potato has been cited. Here, the time after harvesting, packaging (packaged or loose), post-irradiation storage temperature, post-irradiation storage time before chipping or household use, all dictate different irradiation conditions.

Repeated irradiation is not recommended because degradation of the food product may ensue with lowered acceptability and nutritional quality, even though it may not be toxicologically hazardous. Toxicological, nutritional and microbiological assessments are promulgated for foods treated with specific dose ranges and the product should be presented as such. There are no readily available tests for detecting repeated irradiation (FAO/WHO, 1977). Proper labelling, record keeping and surveillance are recommended deterrents. The irradiation of secondary products from primary irradiated products, e.g. flour from irradiated grains, should not be done and should be detectable from well-kept, comprehensive records.

IS FOOD IRRADIATION A PROCESS OR AN ADDITIVE?

Differing opinions exist as to whether the irradiation of foods should be considered a process or whether it is a food additive. The Joint FAO/IAEO/WHO Expert Committee (1977) pointed out that irradiation, being a physical process for treating foods, is comparable to the preservation of foods by heat or by freezing and differs from these only with respect to the type of energy used. The committee thinks that the assessment of wholesomeness, as gauged by microbiological, nutritional and toxicological considerations, should be based on the concept of food irradiation being a process.

Another view holds that food irradiation is an additive because it adds something to the food. The Food Additives Amendment to the Food and Drug Law passed in 1958, states clearly that irradiation is an additive. Based on the food additive concept, the wholesomeness of irradiated foods would then have to be assessed in the same way that it is done for other food additives and pesticides, that is, on the basis of acceptable daily intake and on safety factors. Furthermore, the Delaney Amendment, passed also in 1958, places further strains on the food irradiation picture with respect to carcinogenicity. Breakdown products with carcinogenic potentials in the most minute quantities would render the product non-permissible for human consumption. Furthermore, classifying food irradiation as a food additive dampens the interest of commercial concern in being invovlved since the required labelling would awaken suspicion in customers who are now superconscious and critical of additives in foods. Irrespective of the arguments re "process" vs "additive" categorization, the wholesomeness and safety of the processed product are of supreme importance to the public. This continues to be the prime concern of food processors and public health officials as well, and should convincing evidence be presented on behalf of irradiated foods, it is hoped that logical and sustaining decisions will be made.

FUTURE AND POTENTIALS OF FOOD IRRADIATION

Ten to fifteen years ago, there was great optimism that food irradiation, particularly at pasteurizing doses, would represent man's greatest peaceful use of the atom. In spite of much earnest experimentation in U.S.A., Europe, Asia and other parts of the world, millions of dollars in expenditure and vigorous promotion, such august views have perhaps been severely diluted by recent unfortunate developments in the irradiation of beef in U.S.A. Gross laxity in long-term feeding experiments carried out in studies contracted out by the U.S. Army has dashed the hopes of obtaining US FDA clearance for such products. This severe setback (Anon., 1978) will not only choke off further heavy governmental funding but will certainly engender disappointment among other groups which normally follow very closely developments in the U.S.A.

Another serious impediment to the governmental approval of irradiated foods in the U.S.A. is the classification by the FDA of food irradiation as an additive. As forementioned, this makes the field unattractive to commercial food processors who would have to state this on the labels. Such labelling would naturally deter purchasing by consumers who, at this time, are highly sensitive to the food additive question.

Most investigators and international bodies seem convinced that irradiated foods are free of carcinogenic and mutagenic substances. It has been pointed out that, even in the cooking of foods, minute quantities of carcinogens are formed. There is no requirement to prove their absence in cooked or canned foods but it must be done in irradiated foods. This is so because, as a food additive, The Delaney Amendment to the Food and Drugs Act applies. Proof of a zero level of carcinogens in any processed food would be almost an impossibility with present-day sophisticated analytical techniques. Again, in toxicological studies, animal (rats, mice, monkeys, dogs etc.) must be used and not humans. Long-term feeding studies, though revealing, are expensive and time consuming. Edible materials preserved by heating have been consumed by man for thousand of years and have gained general acceptance for wholesomeness and safety. Thus, by accident or divine timing, heat-preserved foods gained "automatic" exemption for the Food and Drugs Act and the Delaney Clause. It has been said (Anon., 1978) that "if canning and cooking had to go through the FDA procedures, they would probably never manage to qualify."

Irradiation does have potential to rid foods of microorganisms of public health significance. Poultry and poultry products are highly contaminated with salmonella, a rather radiation-sensitive organism. Low doses of irradiation can destroy salmonella without affecting the fresh character of the meat. Thus, the elimination of salmonellosis and its reduction in loss of work time by the patients by such mild processing treatment would be advantageous. Also, the U.S. Army's program on the irradiation of poultry still holds promise. Long-term feeding studies seem to be progressing well and there are hopes for FDA approval by 1983.

Codfish fillets treated with pasteurizing doses of irradiation have been shipped to Holland and, reportedly, well accepted by the populace there. This bodes well for the irradiation of fish in general in so far as acceptability is concerned. Extension of the shelf life of fresh fish would allow for long distance transhipments and, once clearance for processing and exportation are obtained, this would be a giant step forward (Brynjolfsson *et al.*, 1977).

Claims have been made that irradiation as a means of food preservation could mean much for developing countries which lack refrigeration and energy and where there are excessive losses of agricultural products. Certainly, the potential exists, but such prospects should be entertained with guarded enthusiasm. Until and unless irradiated foods are approved by governmental agencies and accepted by the public in the developed countries, particularly the U.S.A., the possibility of its use in the developing countries is not very great. Big Macs (Hamburgers) and Kentucky Fried Chicken have gained worldwide popularity because of their acceptance in U.S.A. "As the U.S.A. goes, so goes the world," it has been said. To expect acceptance and extensive use by developing countries of a food preservation process which still lacks full certification and support in the developed areas, would probably appear to be using less fortunate humans as experimental animals in an ambitious experiment. Investigators and leaders in developing countries are not only fairly abreast of developments but are quite sensitive and critical. This opinion is fortified by the piercing questions directed at the author by food scientists in a certain developing country while he was visiting there, concerning the results of irradiating sucrose solutions.

In addition to the above-mentioned preoccupations would be the cost of facilities for extensive food irradiation. Unless such facilities were obtained by way of foreign aid, the equipment would have to be imported, using scarce foreign exchange.

One of the greatest potentials of food irradiation is its ability to reduce the quantity of nitrite necessary in cured meat products.

Great success has been achieved in this aspect by workers at U.S. Army Natick R and D Command, laboratories (Wierbicki et al., 1976; Shults et al., 1977).

As is well-known nitrate and nitrite are components of the curing brine in the preparation of ham, bacon and corned beef. These chemicals are responsible for the characteristic organoleptic properties of cured meat (color, flavor) and also inhibit outgrowth of the dangerous toxin-producing Clostridium botulinum (See chapt. 10).

It has been demonstrated that, in radappertized meats, the amount of nitrite added to cured smoked ham and bacon could be reduced from the conventionally used level of 156mg/Kg to 25mg/Kg and still maintain the characteristic color, odor and flavor. High overall acceptance of the products was attained while still assuring control of the dangerous toxing-forming Clostridium botulinum. In the case of the radappertized ham, 25mg/Kg nitrite, 50–100 mg/Kg nitrate and ascorbate/erythrorbate were needed to obtain the characteristic color, flavor, odor and taste in the cured meat. Inasmuch as residual nitrite in cured meat products has been incriminated in the formation of carcinogenic nitrosamines, through reaction with secondary and tertiary amines in the product or in the body itself, this finding could lead to significant and highly beneficial ends. If such findings are further enforced and irradiated products gain official clearance, the process would be hailed as a major "protective" step for the consumer against botulism and the probable intake of a carcinogenic precursor-nitrite.

Another very promising use of irradiation is in the preparation of diets for hospital patients who have had organ transplants or who are suffering from leukemia: These patients, in order to minimize rejects, have had their immune responses to foreign material suppressed and so are extremely susceptible to bacterial infections. They live in a sterile environment and are fed sterile foods. Here, heat sterilization is unsuitable for some foods and restricts the variety of foods which may be eaten by the patient. A much wider selection can be presented if the food is radappertized (Josephson et al., 1975).

Stimulation of appetite, morale upliftment and improved nutritional condition are the resultant advantages claimed here for irradiated foods.

Irradiated foods, it is thought, hold great potential for the armed forces feeding program and for outer space feeding. Under such emergency or stress conditions, approval, even if temporary and limited, would be much more easily obtained. The options of choice would also be more limited, if not restricted. In the complete absence of any other method to preserve or hold foods in such situations, stable irradiated foods would be the most appropriate to carry along. In terms of taste, they would be superior to dehydrated foods, less bulky than canned foods, provide no waste-disposal problems, and certainly be as nutritious, if not more so. Since space travel must, of necessity be short-termed and military service itself need not be for life, wholesomeness and safety questions need not be overemphasized. Scientific evidence attests to the wholesomeness and safety of irradiated foods, including meats despite the lack of official certification.

The successful performance of irradiated meats for feeding astronauts in space flights lends credence to its potential here and is a good omen for its uses under such and similar conditions.

Small indentations have been made in some areas for the use of irradiated foods, particularly in the case of poultry. Limited clearances have been granted by the Netherlands, the Soviet Union and the Canadian governments for low-dose irradiation of fresh, eviscerated chicken in order to extend shelf life and /or for the control of salmonella. In the Netherlands and the Soviet Union, experimental batches were prepared with doses of 300 K rad and 600 K rad, respectively. The Canadian government approved test marketing of fresh and frozen eviscerated poultry irradiated with a maximum of 0.75M rad. Radicidation of poultry feed using 1.5M rad was approved by health authorities of the Israeli government.

Treatment of plant products to extend storage and shelf life or to arrest undesirable physiological and/or biochemical changes presents a wide avenue for the irradiation preservation of foods. Low-dose irradiation of onions and potato to inhibit sprouting of papaya, mangoes and bananas, and of grains seems to accomplish the tasks well (see table 1) without any significant damage to the products or derivatives and without significant loss in nutritive substances. Most of the irradiated materials in this category have been considered safe, wholesome and acceptable by the international agencies (FAO/IAEA/WHO, 1977) and by several countries (Anon. 1976).

Wierbicki et al. (1975) list the main obstacles toward commercialization of irradiated foods as:

(1) The legal definition of irradiation as a food additive rather than as a food processing method.
(2) The stigma attached to the word irradiation and
(3) The lack of factual economical data upon which to base the planning of the industrial operations.

Consumer education by government agencies has been suggested to overcome the fear of irradiation. In the U.S.A., at least, the effectiveness of any such program will hinge heavily on the resolution of obstacle Number 1.

From a close examination of all the varying facets, arguments pro and con, data and evidences available up to date, it may be concluded that irradiation as a tool for processing foods does have vast potentials. Realization of such potentials depend heavily upon the planning and execution of experiments in such a way to convince governments and experts that the process does produce wholesome and safe products. If the guidelines for the process, presented by the IAEA and quoted below are adhered to, such convincing evidence can be amassed. In addition, government agencies may need to reassess their stands on the legal aspects pertaining to the process. Commercial enterprise, should, also, keep a close watch on developments and do their part in developing those aspects which will benefit science and mankind.

GENERAL REQUIREMENTS FOR THE PROCESS

Gamma rays from the isotope ^{60}Co or ^{137}Cs or electrons generated from machine sources operated at or below an energy level of 10 Mev shall be used.

In order to meet the requirements of safety and efficacy of food processing, the dose absorbed by the food shall be within the range specified for each individual food irradiation treatment.

Radiation treatment of foods shall be carried out in facilities licensed and registered for this purpose by the competent national authority. In this respect, the following are relevant:

—Such facilities shall be designed to meet the requirements of safety and efficacy of food processing.
—The facilities shall be staffed by adequately trained and competent personnel.
—Control of the process within the facility shall include the keeping of adequate records including quantitative dosimetry.
—Premises and records shall be open to inspection by appropriate authorities.
—Control shall be carried out in accordance with the recommended Code of Practice for the Operation of Radiation Facilities used for the Treatment of Foods.

Any future development of food irradiation should give serious consideration to the economics of the process. Few reliable data are available on the unit cost of products processed by this method and compared to commercial processing of the same products by heat, etc. However, it is certain, that since irradiation will not replace established food processing methods, it will have to compete with them, if it is approved. Certainly, to be profitable, an irradiation processing unit will need large, adequate supplies of raw materials on a regular basis. High volume food irradiation plants will involve input of capital and such capital outlay, if not met by government agencies, will have to be provided by companies or organizations interested in the process.

REFERENCES

AMERICAN CHEMICAL SOCIETY, 1978: Symposium on Current Studies on the Chemistry of Food Irradiation: J. Agr. Food Chem. *26*:1–35 (Contains 6 articles on Various Aspects of Food Irradiation).

ANON. 1973. New Advances in irradiated foods. Foods of tomorrow. Food Processing, Spring 1973.

ANON. 1976. Meeting on Food Irradiation. IAEA PR 76/18. Vienna, 8th September 1976.

ANON. 1978. Army still plugging for FDA approval of irradiated meat. Science Vol. 202, 3rd Nov. 1978 p. 500.

BAKER, R. W. and GOLDBLITH, S. A. 1961. The complementary effects of thermal energy and ionizing energy on peroxidase activity in green beans. Food Science *26*:91–94.

BEHERE, A. G., SHARMA, A., PADWALDESAI, S. R. and NADKARNI, G. B. 1978. Production of aflatoxins during storage of gamma-irradiated wheat. J. Fd. Science. *43*:1102–1103.

BRYNJOLFSSON, A., WIERBICKI, E. and SANDERS, C. 1977. Irradiation update. Food Processing May 1977.

CLIVER, O. 1977. Unlikelihood of mutagenic effects of radiation on viruses. Annex 2 in - Wholesomeness of Irradiated Food. FAO/IAEA/WHO Expert Committee Report - WHO Geneva 1977.

CUEVAS-RUIZ, J., GRAHAM, H. D. and LUSE, R. A. 1972. Gamma radiation effects on the biochemical components of Puerto Rican mangoes. J. Agr. Univ. of Puerto Rico. *56*:26–32.

CHUNG, O., FINNEY, K. F. and POMERANZ, Y. 1967. Lipid in flour from gamma irradiated wheat. J. Food Sci. *30*:577.

DESROSIER, N. W. and DESROSIER, J. N. 1977. Technology of Food Preservation. 4th Edition Avi Publishing Co.

FAO/WHO 1977. Wholesomeness of Irradiated Food. Report of a Joint FAO/IAEA/WHO Expert Committee. World Health Organization Geneva Technical Report 604. Series 6. 44pp.

GLASSTONE, S. 1950. Sourcebook on Atomic Energy. D. Van Nostrand Co. New York, N.Y.

GOLDBLITH, S. A. 1967. Radiation processing of foods and drugs. Chapter 12 in Fundamentals of Food Processing Operations. Heid, J. L. and Joslyn, M.A. Eds. Avi Publishing Co. Westport, Conn. 730 pp.

HOLSTEN, R. D. and SUGII, M. and STEWARD, F. C. 1965. Direct and indirect effects of radiation on plant cells: Their relation to growth and growth induction. Nature 208: 850–856.

INGRAM, M. 1975. Technical report series, International project in the field of food irradiation. IFIP R 33:14.

JIRVATANA, V., CUEVAS-RUIZ, J. and GRAHAM, H. D. 1970. Extention of storage life of papayas grown in Puerto Rico by gamma irradiation treatments J. Agric. Univ. of Puerto Rico. 44:314–319.

JOSEPHSON, E. S., BRYNJOLFSSON, and WIERBICKI, E. 1975a. The use of ionizing radiation for preservation of foods and feed products in radiation research. Biomedical, Chemical and Physical Perspectives Academic Press Inc. pp. 96–117.

JOSEPHSON, E. S., THOMAS, M. H. and CALHOUN, W. K. 1975b. Effects of treatment of foods with ionizing radiation. Chapter 14 in Nutritional Evaluation of Food Processing. Eds. Harris, R. S. and Karmas, E. Avi Publishing Co. 690 pp.

KAMAREI, A. R., KAREL, M. and WIERBICKI, E. 1979. Spectral studies on the role of ionizing radiation in color of radappertized beef. J. Food Sci. 44: 25–32.

MAXCY, R. B. 1977. Comparative viability of unirradiated and gamma irradiated bacterial cell. J. Fd. Science 44: 1056–1059.

MERRITT, C. JR., ANGELINI, P., WIERBICKI, E. and SHULTS, G. W. 1975. Chemical changes associated with flavor in irradiated meat. J. Agr. Food Chem. 23: 1037–1041.

NAWAR, W. W. 1972. Radiolytic changes in fats. J. Rad. Res. Rev. 3. 327.

POTTER, N. N. 1977. Food Science 3rd Edition. Avi Publishing Co. Westport, Conn. 780 pp.

PROCTER, B. E. and GOLDBLITH, S. A. 1951. Food processing with ionizing radiations. Food Technol. 5: 376–380.

RAO, V. S., VAKIL, U. K., BANDYOPADHYAY, C. and SREENIVA-SAN, A. 1978a. Effect of gamma-irradiation of wheat on volatile flavor components of bread. J. Fd. Sci. 43: 68–71.

RAO, V. S., VAKIL, U. K. and SREENIVASAN. 1978b. Effects of gamma-irradiation of composition of wheat lipids and purothionines. J. Fd. Science 43: 64–67.

SHAW, M. W. and HAYES, E. 1966. Effects of irradiated sucrose on the chromosomes of human lymphocytes in vitro. Nature 211: 1254–56.

SHULTS, G. W., COHEN, J. S., HOWKER, J. J. and WIERBICKI, E. 1977. Effects of sodium nitrate and sodium nitrite additions and irradiation processing variables on the color and acceptability of corned beef brisket. J. Food Sci. 42: 1506–1509.

SNYDER, L. D. and MAXCY, R. B. 1979. Effect of a_w of meat products on growth of radiation resistant *Moraxella-Acinetobacter.* J. Food Sci. *44:* 33–36.

URBAIN, W. M. 1978. Food Irradiation. Advances in Food Research *24:* 155–227.

WHITEHAIR, L. A. and HILMAS, D. E. 1968. Radiological problems in the national food irradiation program. Chapter 12 in Safety of Foods. H. D. Graham. (Ed.) Avi Publishing Co.

WIERBICKI, E., BRYNJOLFSSON, A., JOHNSON, H. C. and ROWLEY, D. B. 1975. Preservation of meats by ionization radiation - An update: Rapporteurs' Papers - Paper 14 21st European Meeting of Meat Research Workers, Berne, Switzerland, 31 August- 5 September, 1975. 21pp.

WIERBICKI, E., HEILIGMAN, F. and WASSERMAN, A. E. 1976. Irradiation as a conceivable way of reducing nitrites and nitrates in cured meats. Proc. 2nd. Symp. Meat Prod., Zeist, 1976. Pudoc Wageningen. pp. 75–81, Recommendations at conclusion of this meeting re nitrite to be found on page 92.

17

Toxins in Plants

Richard F. Keeler

The most impressive aspect of toxic plants, their toxins, and their deleterious effects on man and animals is the sheer magnitude of the problem. If plant toxins are considered in the broad sense as any toxic compound elaborated by plants or toxic elements incorporated by them and we cursorily survey some of the recent general references on the subject (Kingsbury, 1962; Watt and Breyer-Brandwijk, 1962; Everist, 1974; Hardin and Arena, 1974; and Keeler *et al.*, 1978), we begin to appreciate the magnitude of the problem. Literally thousands of plants in the world are toxic, and thousands of separate toxins are present in these plants. However, most plants in the world are not hazardous, and from among them, in the main, plants for food or feed are selected. There are exceptions. Some common plant foods are toxic, but the hazard is reduced through selection of edible parts or time of harvest, by genetic selection, or by processing. Most animal feeds are also relatively nonhazardous. Where, then, does the problem with plant toxins lie?

Some poisonings in man and animals result from lack of care in the selection or pretreatment of these potentially hazardous foods or feeds. Some poisonings in man result from accidental ingestion of misidentified plants or of plants of unknown toxicity. Poisonings in livestock result primarily when they graze native forage in range areas of the world. Livestock eat certain palatable toxic plants with relish and eat others when forced to do so by shortage of other forage or by poor management.

With some exceptions, a plant toxin hazardous to one animal species will be hazardous to another, so a knowledge of toxic effects in livestock is useful information for understanding toxicosis in man. However, a compelling reason for worrying about the effects of plant toxins in livestock is that cases of serious toxicoses in livestock exceed those in man by orders of magnitude. Although one can find exceptions in the older literature (Kingsbury, 1964 and Watt, and Breyer-Brandwijk, 1962), in this present enlightened age and in developed regions of the world, epidemic outbreaks of toxicoses in man from plant toxins are rare. Among livestock grazing native forage, such outbreaks are still common. Epidemics of plant toxicoses in livestock involving hundreds or thousands of animals continue to occur every year. The hazardous plants are known, and the information is often understood by owners of these animals. So why do these poisonings continue? The answer lies in the logistics of the animal grazing industry and in the value of the native forage.

An estimate has been made of the forage value of the native range of the contiguous 48 states of the United States and the losses due to plant toxins from poisonous plants on these ranges (Keeler, 1978). The production value of the grazing land harvested by foraging livestock in 1970 was about $2.5 billion, a value that compares very favorably with the production value of two of the principal agricultural crops of the U.S.—corn at $5.5 billion and soybeans at $3.2 billion. Clearly, the value of native forage is high and cannot be left unused, despite the fact that possibly 5% of all grazing livestock have serious encounters with poisonous plants each year. The absolute number of animals involved is astounding when one considers that the range carrying capacity is 213 million animal unit months. Hundreds of thousands of animals are affected by plant toxins each year in the United States alone. One can consider the staggering numbers worldwide and then begin to grasp the magnitude of the problem with plant toxins.

What plant families and genera contain toxins? General reference works on the subject suggest extensive botanical variation among members carrying toxins (Kingsbury, 1962; Watt and Breyer-Brandwijk, 1962; Everist, 1974; Hardin and Arena, 1974; and Keeler et al., 1978) and suggest phylogenetic interrelationships or affinities where they exist. Everist (1978) recently considered the botanical affinities of toxic Australian plants, and his conclusions show the complexity of the problem. He reports that 800 species of poisonous plants are known in Australia alone; of these 469 are native, 244 are naturalized, and the other 87 are not naturalized but are nonetheless cultivated. The native species are distributed throughout 72 families, most of which contain fewer than 10 toxic species each. Often, toxic species have few toxic family relatives so the correlation between toxicity and phylogenicity is not great. However, in some families, many species have similar toxins. Clearly, a taxonomy based on presence of similar toxic compounds differs from one based on morphology. Plant toxins are found worldwide in thousands of species, and correlation between toxicity and phylogenicity is often unclear.

What are the plant toxins from a chemical point of view? They comprise an extensive list that is hardly surprising when one considers the array of available metabolic processes in which effects could center in affected animals. The chemical classes include alkaloids, proteins, polypeptides, aminoacids, amines, cyanogenic glycosides, various miscellaneous nitrogen-containing compounds, terpenes, phenolics, alcohols, steroidal compounds, coumarins, and inorganic compounds and elements to name a few (Kingsbury, 1962; Watt and Breyer-Brandwijk, 1962; Everist, 1974; Hardin and Arena, 1974; and Keeler *et al.*, 1978).

Because of the extent of the subject matter, there has been arbitrarily selected for this discussion of plant toxins (A) a group of plant toxins from food and feed plants which cover a range of toxin types and which include an example or two of plants not considered valuable feed, but nonetheless grazed and (B) a group of plant toxins with the special propensity of producing teratogenic effects or birth defects in offspring from animals grazing these plants.

TOXINS FROM FOOD AND FEED PLANTS

A. Steroidal alkaloids

Certain members of the *Solanum* genus are perhaps the most widely consumed toxin-containing food plants in the world. Two *Solanum* spp. of particularly widespread use are *Solanum tuberosum* (potato) and *Solanum melongena* (eggplant). Figures cited by Schery (1952) for world production in the late 1940's suggest that potato production exceeded that of any of the cereals and was greater than 200 million metric tons. More recent estimates place production at more than 300 million metric tons (Jadhav and Salunkhe, 1975). Eggplant is used less, although it is common fare in many parts of the world. Both potatoes and eggplant as well as certain other exotic *Solanums* and *Lycopersicon* spp. (tomato) contain toxic steroidal alkaloid glycosides and occasionally give rise to toxicoses (Kingsbury, 1964 and Jadhav and Salunkhe, 1975) in man.

Some of these same *Solanum* spp. and certain *Liliaceae* family plants that also contain steroidal alkaloids are included as serious hazards for livestock. *Zygadenus* and *Veratrum* spp. both common range lilies, are often grazed by livestock with serious toxic consequences (Kingsbury, 1964). Crop residues, as well as the fruit, from potatoes and tomatoes may cause toxicosis in livestock when they eat these materials (Kingsbury, 1964, Keeler, 1978 and Jadhav and Salunkhe, 1975).

Among the toxic steroidal alkaloids in these various plants are a range of compounds with considerable structural variation, including glycosides, esters, and parent alkamines or aglycones (Kupchan and By, 1968 and Schreiber, 1968). The aglycone ring systems (Fig. 17.1) in most of the steroidal alkaloids in the *Veratrum* and *Zygadenus* genera, referred to as alkaloids of the veratrum group, are C-nor-D-homo steroids with an additional ring or two. The terminal piperidine ring is one methylene removed from the steroid. The methylene at C_{17} of the steroid is attached α to the nitrogen of the piperidine. Many compounds have, in addition, a methylene bridge attached between the nitrogen of the piperidine and carbon 13 of the steroid. These are six-ring alkaloids with tertiary nitrogen atoms (ring system A). Protoverine, germine, veracevine, zygadenine, and a variety of esters are examples. Compounds with an ether bridge between the steroid and the piperidine attached at C_{17} of the steroid and β to the nitrogen of the piperidine are furanopiperidines (ring system B). Examples are jervine and its glycoside pseudojervine and cyclopamine (11-deoxojervine) and its glycoside cycloposine. Com-

RING SYSTEMS OF VERATRUM ALKALOIDS

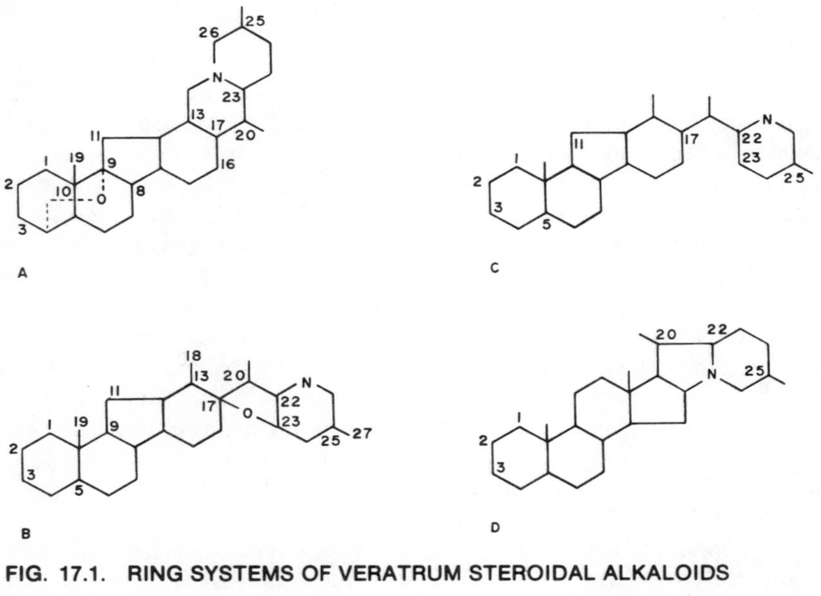

FIG. 17.1.   RING SYSTEMS OF VERATRUM STEROIDAL ALKALOIDS

pounds with but a methylene bridge (C_{17-22}) between the piperidine and the steroid (as in ring system C) include veratramine, its glycoside veratrosine, and muldamine. Alkaloids of *Veratrum* spp. that have the conventional steroid ring system (ring system D) with a resemblance to the solanidines from *Solanum* spp. are rubijervine, isorubijervine, and the glycoside of the latter, isorubijervosine. Alkaloids from members of the *Solanum* and *Lycopersicon* genera, commonly referred to as the solanum alkaloids (Figure 17.2), invariably have the common C_{27}–carbon steroid skelton of cholestane. Most have either the spirosolane or solanidane skeletons. Examples include solasodine, tomatidine, solanidine, demissidine, and their glycosides.

Toxicologic effects of steroidal alkaloids vary considerably (Jadhav and Salunkhe, 1975; Krayer, 1958 and Nishie *et al.*, 1971) as a function in part of the wide variation in structure of these compounds. Tertiary amine veratrum esters produce a transient hypotensive effect, a decrease in arterial resistance, a decrease and irregularity in heart rate, a slowing of the rate of respiration, and an emetic action. Salivation and vomition are toxic signs one most frequently observes from overdose of *Veratrum californicum* in sheep and arise principally from the tertiary esters. Secondary amines are much less active on a weight basis, and the effects are somewhat different. In sheep, the most characteristic effect from the secondary amines, veratramine and jervine, is convulsions. Solanum glycosides such as solanine from potatoes produce gastrointestinal disturbances such as nausea, vomition, and diarrhea and hemolytic and hemorrhagic lesions in the gastrointestinal tract.

Conditions that precipitate toxic effects from steroidal alkaloids vary. For example, *Zygadenus* spp. are among the first plants to appear on spring ranges and, therefore, frequently are grazed by livestock with disastrous results (Kingsbury, 1964). Hundreds of animals die yearly from consuming *Zygadenus* spp. In some areas, it kills more livestock than does any other plant. Concentration of steroidal ester alkaloids is highest in the overwintered bulbs that

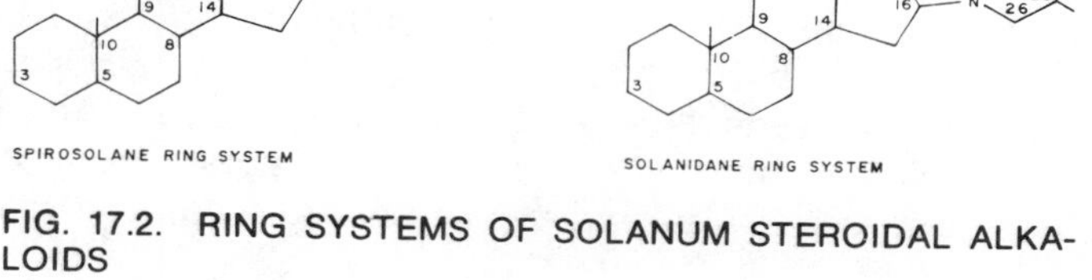

SPIROSOLANE RING SYSTEM   SOLANIDANE RING SYSTEM

FIG. 17.2. RING SYSTEMS OF SOLANUM STEROIDAL ALKALOIDS

are frequently pulled from soft, moist soil and then eaten as the animal grazes the plant. Unlike *Veratrum* spp., which sheep seem to prefer over many nontoxic forage plants, *Zygadenus* is not usually selected by well-fed livestock. Nonetheless, it is often consumed because other green forage may be sparse in early spring.

Cases of poisoning from potato ingestion in man center on use of potatoes with an exceptionally high alkaloid content. Ordinarily, potato ingestion is not hazardous. Jadhav and Salunkhe (1975) summarized results of some of the reports of severe illness and death due to potatoes with a high alkaloid content. These potatoes, recognized by an extremely bitter taste, may produce gastrointestinal disturbances, inhibit cholinesterase activity, and induce death in severe cases. Alkaloid levels in incriminated potatoes may be over 10 times the levels in "safe" tubers, which seldom exceed 5–10 mg/100 g of fresh weight of tuber (Jadhav and Salunkhe, 1975). Careless storage or handling of tubers, particularly exposure to light, and use of cultivars of naturally high alkaloid levels are conditions that give rise to poisoning in man. Livestock are usually poisoned by eating potato waste from commercial processing plants that was exposed to light, by eating peel or sprout material with a high alkaloid content, by eating tubers left in the fields, or by eating potato vines (Kingsbury, 1964 and Jadhav and Salunkhe, 1975).

B. Glucosinolates

A number of plant genera in the Cruciferae family, including some that are common foods or feeds, contain a group of compounds known as glucosinolates. These glucosinolates are the source of certain toxins arising from chemical modification during processing and subsequent metabolism of the glucosinolates. The deleterious products are nitriles, isothiocyanates, thiocyanates, or related compounds derived from hydrolysis of the glucosinolates catalyzed by thioglucosidases released when the plant material is crushed (Fig. 17.3). Factors that determine which toxic product will form include identity of the original glucosinolate, hydrolysis conditions, and treatment prior to hydrolysis (Van Etten and Tookey, 1978).

Plants included among those containing glucosinolates (Van Etten and Tookey, 1978) are food plants from *Brassica oleracea* (cabbage, broccoli, brussels sprouts, cauliflower, collards, kale, and kohlrabi); various rapes from both *Brassica campestris* and *Brassica napus* used as foods and feeds as well as a source of semidrying oils; and *Crambe abyssinica*, another oil source. Seed meals from which the oil has been expressed also find use as animal feed. The semidrying

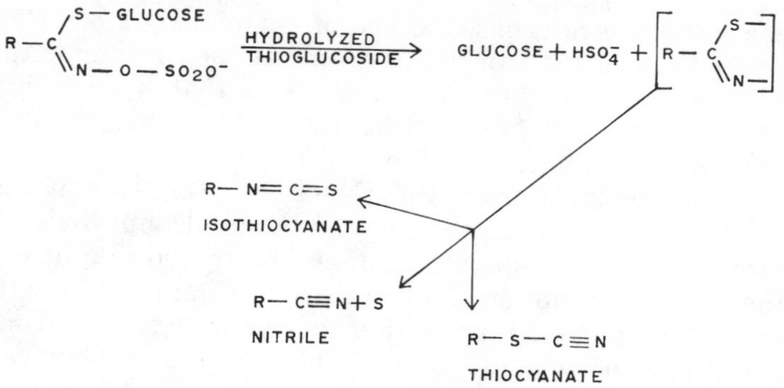

$$R-C \overset{S-GLUCOSE}{\underset{N-O-SO_2O^-}{\diagup \diagdown}} \xrightarrow[\text{THIOGLUCOSIDE}]{\text{HYDROLYZED}} GLUCOSE + HSO_4^- + \left[R-C \overset{S-}{\underset{N-}{\diagup \diagdown}}\right]$$

$$R-N\!\!=\!\!C\!\!=\!\!S$$
ISOTHIOCYANATE

$$R-C\!\!\equiv\!\!N + S$$
NITRILE

$$R-S-C\equiv N$$
THIOCYANATE

FIG. 17.3.   CONVERSION OF GLUCOSINOLATES TO TOXIC PRODUCTS

oils from *Brassica* spp. and *Crambe* spp. are less important than corn or cottonseed oils but are nonetheless important items of commerce (Schery, 1952).

Toxicity problems stem from the metabolic products of the glucosinolates present in the plants or oil-free residues (Rutkowski, 1971). The biologic effects differ for each of the toxins. When (R)-2-hydroxy-3-butenyl-glucosinolate is hydrolyzed, the compound 5-vinyl-oxazoladine-2-thione (commonly called goitrin) is formed. This compound initially suppresses iodine uptake (Greer, 1962), and a type of goiter results (Carroll, 1949). This antithyroid effect of goitrin is not reversed by administration of iodine (Van Etten and Tookey, 1978).

When indolylmethyl-glucosinolates from cabbage are hydrolyzed, the thiocyanate ion is formed (Van Etten and Tookey, 1978). Thiocyanate ions can also produce an antithyroid effect, but the mechanism is different from the goiter produced by goitrin. Thiocyanate evidently exerts its effect by preventing thyroid accumulation of inorganic iodide; and in this case, the thyroid enlargement can be prevented by increasing the iodine content of the ration.

When conditions of hydrolysis are favorable, more particularly with certain of the Cruciferae such as *Brassica nigra*, the wild mustard of livestock ranges, isothiocyanates are the hydrolysis products. These compounds are highly vessicant and produce severe gastroenteritis, pain, salivation, diarrhea, and irritation of the mouth (Kingsbury, 1964 and Van Etten and Tookey, 1978). The condition is common in livestock grazing mustard (Kingsbury, 1964). In addition, iodine responsive goiter is produced when isothiocyanates are converted to thiocyanate ion (Langer and Stole, 1965).

Under certain conditions, the use of crambe seed meal has produced pathologic effects in the liver and kidney rather than in the thyroid. These effects are due to the fact that hydrolysis of the glucosinolate yields organic nitriles rather than thiocyanate or isothiocyanate (Van Etten and Tookey, 1978).

Thus, glucosinolates from the Cruciferae family can induce a variety of deleterious effects in man and animals, but little evidence exists to suggest that goiter in man can be traced to glucosinolates with the possible exception of a few cases in Finland (Van Etten and Tookey, 1978). In fact, man seldom develops toxicoses from these plants. But livestock do, and efforts to counter the problem center on keeping dose levels low, using heat or other pretreatment to destroy the hazardous compounds, or plant breeding to reduce the glucosinolate content of the plants (Van Etten and Tookey, 1978).

C. Isoflavones and coumestans

Two interesting classes of plant toxin from plants cultivated as animal feed are the isoflavones from clover and the coumestans from alfalfa and barrel medic. These compounds have estrogenic activity and produce a serious decrease in fertility in sheep grazing the plants (Cox, 1978 and Gardner and Bennetts, 1956). The situation with clover will serve to illustrate.

In an effort to improve pastures in western Australia, subterranean clover (*Trifolium subterraneum*) has been sown in these grazing areas for about 50 years. However, workers discovered that the plant was responsible for reduced fertility in ewes. This estrogenic effect proved to be due to certain isoflavones present in the plant. The problem is now less severe because of widespread use of cultivars of subterranean clover low in estrogenic compounds obtained through selection and breeding. Even now, some one million ewes in Australia alone fail to produce lambs each year because of estrogenic compounds in clover (Cox, 1978).

The direct cause of the infertility is failure of fertilization because of poor sperm penetration to the oviduct (Turnbull *et al.*, 1966 and Fels and Neil, 1968). This poor penetration results from cystic glandular hyperplasia of the cervix and uterus of affected ewes. Cervical mucus in affected ewes is increased in amount and fluidity, and fewer sperm enter the cervix (Smith, 1971). In addition, the level of early abortions is higher than normal.

The estrogenic compounds responsible for the condition are isofavones, of which biochanin-A, genistein, and formononetin (Fig. 17.4) are examples (Cox, 1978). The relative estrogenicity of these

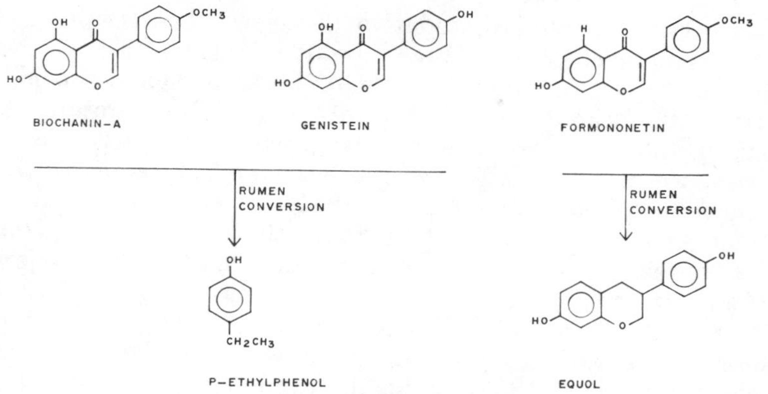

FIG. 17.4. RUMEN CONVERSION OF NATURAL ISOFLAVONES

compounds illustrates a very important aspect of plant toxins—that metabolic alteration in the ingesting animal can either increase or decrease the toxicity of the parent toxin from the plant. Metabolic alteration is particularly likely in ruminant animals whose rumen microflora produce many unexpected conversions (Allison, 1978). Other metabolic conversions occur as well in the liver of all animals.

Both of the 5-hydroxy-isoflavones, genistein and biochanin A, are very estrogenic when administered parenterally. However, when sheep graze plants with high levels of these compounds, rumen microorganisms adapt rapidly and within a few days almost all is converted to p-ethylphenol and other inactive compounds (Cox, 1978 and Braden and McDonald, 1970).

By contrast, the third isoflavone mentioned above, formononetin, has almost no estrogenic activity when administered parenterally. Yet, it is largely converted by rumen microorganisms to the compound equol, which is very estrogenic (Cox, 1978 and Braden and McDonald, 1970). Indeed, in sheep, the permanent infertility resulting from subterranean clover ingestion is apparently produced by equol derived from rumen conversion (Cox, 1978).

D. Cyanogenic glycosides

A group of plant toxins nearly ubiquitous in distribution are the cyanogenic glycosides. These compounds which release hydrogen cyanide (HCN) upon hydrolysis, are found in about 1,000 plant species from some 250 genera in 80 families (Conn, 1978). The released cyanide ion binds strongly to cytochrome oxidase and thus

prevents normal cellular oxidative processes. The resulting cytotoxic anoxia accounts for the lethality of cyanogenic glycosides
(Conn, 1978 and Montgomery, 1969). Signs of cyanide poisoning
include early stimulation of respiration, rapidly changing to dyspnea, excitement, gasping, staggering, paralysis, prostration, convulsions, coma, and death. The mucous membranes of mouth and
eye may be cyanotic (Kingsbury, 1964).

Cyanogenic glycosides are found in both food and feed plants as
well as in many native plants on livestock ranges. Toxicoses in both
man and animals have been reported (Kingsbury, 1964, Everist,
1974, Fels and Neil, 1968 and Montgomery, 1969). The principal
sources of cyanogenic glycosides in human food plants that cause
toxicoses (Braden and McDonald, 1970 and Conn, 1978) are cassava
(*Manihot esculenta*), and certain varieties of lima beans (*Phaseolus
lunatus*), and bamboo shoots (*Bambusa* spp). Ingestion of seed kernels of certain fruits such as apple, apricot, bitter almond, lemon,
lime, peach, pear, plum, and prune likewise causes occasional toxicoses. Among plants cultivated as feed plants in the United States,
and to some extent as food plants in the U.S. and elsewhere, and
rich in cyanogenic glycosides are the various millets, *Sorghum
vulgare* (sorghum), *Sorghum sudanense* (Sudan grass), and *Zea maize*
(maize). Native plants on livestock ranges that are very high in
cyanogenic glycosides and frequently cause livestock deaths include
Triglochin maritima (arrowgrass), and *Prunus virginiana*, or *Prunus melanocarpa* (chokecherries). Toxicity of each of these species
and the hundreds of others containing cyanogenic glycosides is
highly varied, and selection and breeding methods often have been
applied to obtain varieties low in toxicity.

The toxins in these cyanophoric plants are glycosides of α-hydrox-
ynitriles or cyanohydrins. More than 20 cyanogenic glycosides are
known (Conn, 1978), but others doubtlesly exist because few of the
known cyanophoric plants have been examined carefully. Some of
the common glycosides (Conn, 1978) are: amygdalin, a glycoside of
D-gentiobiose and (R)-mandelonitrile, from almonds, apricots, and
chokecherry; dhurrin (D-glucose and p-hydroxy-(S)-mandelonitrile)
from the sorghums; toxiphyllin (D-glucose and p-hydroxy-(R)-man-
delonitrile) from bamboo and the *Taxus* spp. and linamarin (D-
glucose and α-hydroxyisobutyronitrile) from cassava and lima beans.

These relatively innocuous compounds become toxic when free
HCN is released through enzymatic conversion (Conn, 1978 and
Montgomery, 1969). The glycoside (illustrated by amygdalin in Fig.
17.5) is hydrolyzed by a β-glucosidase released from the plant tissue
by disruption of cells as the plant is chewed or macerated. Among

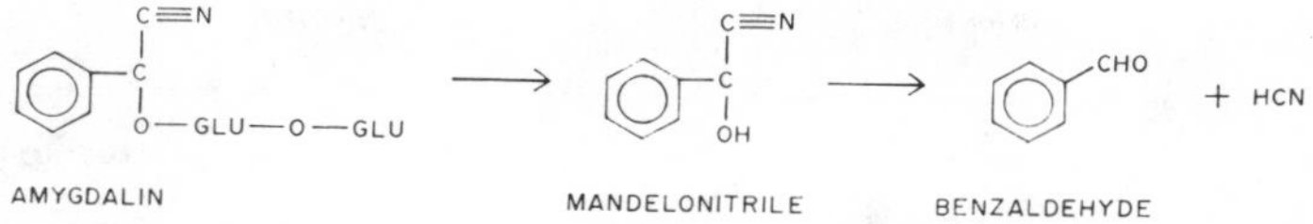

FIG. 17.5. FREE, TOXIC HYDROGEN CYANIDE RELEASED FROM
THE GLYCOSIDE AMYGDALIN AFTER CELLULAR DISRUPTION

ruminant animals β-glucosidases from rumen bacteria can also catalyze this reaction. The cyanohydrins that are products of this reaction are either converted nonenzymatically or by the action of hydroxynitrilelyases to free HCN and the appropriate aldehyde or ketone.

Although the cyanogenic glycosides can be hydrolyzed by acid, the pH of human stomach acid (pH 2.0) is not low enough to produce hydrolysis of amygdalin (Conn, 1978). The risk of cyanide toxicosis in man is reduced if the β-glycosidases of the plant are inactivated by heat or if the cyanogen and the released HCN are leached out with water. Both procedures are used for making cassava root material safe. Either the cassava material is cooked thoroughly and the cooking water discarded before use, or the material is scraped, grated, and extensively soaked to produce leaching (Everist, 1978, Conn, 1978 and Montgomery, 1969). Fermentation is another detoxifying method in use (Everist, 1978 and Montgomery, 1969).

E. Methylazoxymethanol

Another toxic food plant that presents minimal hazard in man today because of removal of toxin by processing is *Cycas circinalis* and related plants (cycads) (Yang and Mickelsen, 1969). The principal food value of cycads stems from the high concentration of starch they contain. However, without processing to remove toxins, the plant is not a suitable food source. The glycoside cycasin from cycads is hydrolyzed by β-glucosidases in the gut to yield methylazoxymethanol (Fig. 17.6). Methylazoxymethanol is both a neurotoxin and a carcinogen (Yang and Mickelsen, 1969).

Two or three procedures now in use lower the hazard. When starch is prepared from cycads, the starch granules are extracted from the crushed cycad. Then either prolonged soaking, water washing, heating, or fermentation is used to detoxify the starch. Prolonged soaking is the usual method and causes hydrolysis of the cycasin. The toxin methylazoxymethanol is released but readily decomposes in the soaking solution, or on heating, to nontoxic

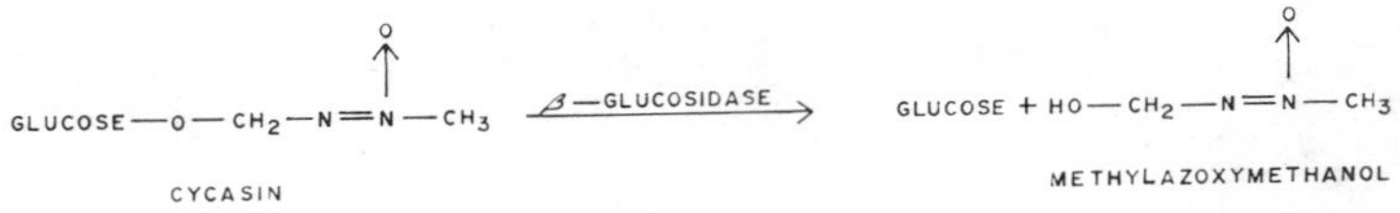

GLUCOSE—O—CH₂—N≡N—CH₃ →β—GLUCOSIDASE→ GLUCOSE + HO—CH₂—N≡N—CH₃

CYCASIN   METHYLAZOXYMETHANOL

FIG. 17.6 HYDROLYSIS OF CYCASIN TO THE TOXIN METHYLA-
ZOXYMETHANOL

compounds (Yang and Mickelsen, 1969). No increased prevalence of cancer in humans has been reported from areas where cycad is consumed (Hirono, Kachi and Kato, 1970). However, a higher than expected incidence of certain neural disorders common in those areas (Kurland, 1972) may reflect incomplete removal of toxin.

However, the significant problem with this plant toxin is related to the extensive grazing of toxin-containing palms of the *Cycas*, *Macrozamia* and *Bowenia* spp., by domestic livestock, particularly in Australia (Hooper, 1978) where outbreaks of toxicosis sometimes are massive (Hooper, 1978 and Everist, 1974).

In sheep, two forms of liver disease are common with cycad toxin—an acute liver necrosis and a chronic megalocytosis and veno-occlusive disease (Kurland, 1972). Cattle are more frequently affected with neural symptoms than are sheep, but both develop central nervous system (CNS) lesions and show an ataxia resulting from those necrotic CNS lesions (Hooper, 1978). Some question remains whether methylazoxymethanol is responsible for the CNS lesions, but it is known to cause neural disease in fetuses and newborn mice (Hooper, 1978). A third syndrome produced by methylazoxymethanol is an experimental carcinogenic effect in laboratory animals. It causes kidney and intestinal tumors in rats and liver tumors in mice, hamsters, and rabbits (Hooper, 1978 and Everist, 1974).

The biochemical mechanism by which methylazoxymethanol produces liver and CNS lesions is not known, but the carcinogenic effects may result from methylation of deoxyribonucleic acid (Matsumoto and Higa, 1966). Some protection against the carcinogenic effect is afforded by the radioprotective compound cysteamine (Miwa, 1975).

F. Sesquiterpene lactones

A number of Compositae plants found on ranges where animals graze that are readily consumed by them owe their toxicity to one or more of nearly 700 known sesquiterpene lactones (Herz, 1978 and Yoshioka, *et al.*, 1973). The 700 compounds vary significantly in structure, but many are very toxic (Herz, 1978), and some account for the known toxicoses produced by a number of Compositae. Among plants in this category are members of the *Hymenoxys* and *Helenium* or *Dugaldia* genera, which have caused significant poisoning in livestock for as long as grazing has been practiced.

Bitterweed (*Hymenoxys odorata*) is common on semi-arid ranges from Kansas through Texas and into Mexico. A few pounds of bitterweed is a lethal dose in sheep, and thousands of sheep are poisoned yearly by the plant (Kingsbury, 1964). Poisoned sheep usually show salivation and vomition, and there are generally gastrointestinal lesions (Kingsbury, 1964). In phytochemical studies summarized by Herz (1978) on members of the *Hymenoxys* genus, he and his coworkers isolated a number of sesquiterpene lactones and speculated they might be responsible for the toxicity. Subsequently, a sesquiterpene lactone, hymenovin (Fig. 17.7), was isolated from bitterweed by Ivie and coworkers (Ivie, *et al.*, 1975) and was shown to account for the toxicity. Sheep poisoned with hymenovin had the same toxicosis signs and lesions as those found in plant-poisoned sheep. Another group of researchers reported similar results and designated the same compound hymenoxon (Kim, *et al.*, 1975).

Orange sneezeweed (*Helenium hoopesi* or *Dugaldia hoopesi*) is abundant in nearly all the western United States at higher elevations in areas that comprise prime summer sheep range. When sheep consume a few kg per day of sneezeweed for 2 or 3 weeks they become violently ill. They develop what sheepmen call "spewing sickness." Symptoms include frequent vomition, excess salivation, belching, frothing, and irregular pulse and respiration. Lesions include gastrointestinal irritation, liver and kidney congestion, and necrotic areas and adhesions in the lung—the latter resulting from ingesta entering the lungs via the trachea from frequent vomition. Thousands of sheep die yearly from ingestion of this plant (Kingsbury, 1964). Results from recent research suggest that sneezeweed contains the same sesquiterpene lactone as that found in bitterweed, hymenovin (Ivie, *et al.*, 1976) [hymenoxon Hill, *et al.*, 1977]. Because

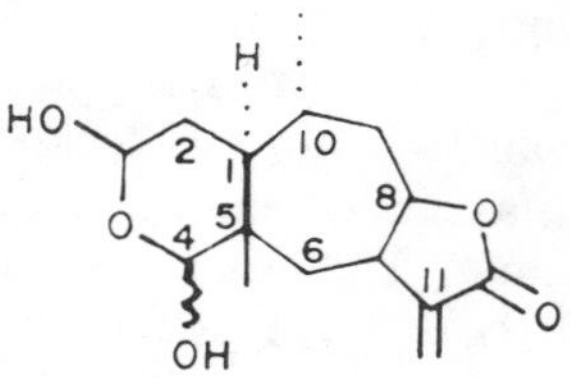

FIG. 17.7.   THE SESQUITERPENE LACTONE TOXIN, HYMENOVIN

toxicologic symptoms and lesions in animals poisoned by both bitter-weed and sneezeweed are the same, the researchers concluded that sneezeweed also owes its toxicity to the presence of this sesquiter-pene lactone.

The biochemical mechanism by which this sesquiterpene lactone exerts its effect is not known but may relate to alkylation of sulfhy-dryl groups or to an effect on rumen flora (Herz, 1978 and Ivie *et al.*, 1975).

G. Cardenolides

Toxicities vary considerably among the group of milkweeds (*Asclepias* spp.), but a few of them are among the most toxic of all poisonous plants and represent a significant hazard to livestock (Kingsbury, 1964). In some range areas, livestock practices have been drastically altered to attempt to solve the problem. Ranchers have changed from sheep to cattle operations or resorted to nonuse of certain areas by sheep. Three of the most toxic of the milkweeds are *Asclepias labriformis*, *Asclepias subverticillata*, and *Asclepias eriocarpa* which can kill livestock with as little as 0.05, 0.2, and 0.25%, respectively, of an animal's body weight of green plant (Kingsbury, 1964). That amount is about one or two mouthfuls for a grazing sheep.

In the early days of sheep grazing operations in the United States, sheep were "trailed" twice each year for long distances between summer mountain ranges and winter desert ranges. Toxic plant problems occurred frequently along these trails because of heavy grazing pressure. *Asclepias labriformis* grew along certain of these trails, and hungry sheep often ate a few mouthfuls of the plant and then died. In those areas, ranchers had to change to cattle operations to avoid bankruptcy.

Animals poisoned by milkweeds, according to Marsh and Clawson (1924), became weak and depressed and had labored respiration and died without convulsions or coma. The lungs, kidney, and gastroin-testinal tract were usually congested and had microscopic lesions. In our experience (Benson, *et al.*, 1978), affected animals had, in addi-tion, a marked slowing of respiration rate. Expiration was forced and accompanied by "raleing" sounds. Mucous membranes were cyanotic. Animals were depressed, weakened, unsteady, were in considerable distress, walked with difficulty, and could not stand unassisted before death. There was no vomition, convulsions, or coma. The lungs and kidneys had marked lesions, and often other organs were affected—rumen, abomasum, and intestines. Lungs were congested with petechial or massive hemorrhages and showed atelectasis and emphysema.

In 1929, Couch (1929) isolated a "resinoid" from *A. eriocarpa* that he reported was lethal to guinea pigs. Roeske (1971) did not find that type preparation lethal to mice. However, she did find a positive correlation between the cardenolide content of four milkweeds and their toxicity to mice (Roeske, 1971), and in purification of extracts, toxicity paralleled cardenolide content. Three cardenolides were isolated from *A. eriocarpa* and two from *A. labriformis* extracts (Roeske, 1971 and Seiber *et al.*, 1975). Preliminary structures were worked out (Seiber *et al.*, 1975); Figure 17.8 shows the structure for one of the compounds, labriformin, found in both plants.

Sheep dosed orally with crude cardenolide preparations from these plants or with pure labriformin (Benson, *et al.*, 1978) had toxicoses signs and lesions indistinguishable from those produced by plant material. These findings showed that cardenolides do account for the toxicity of milkweed. Effects were observed in sheep with as little as 1.1 mg of labriformin/kg of body weight. The cardenolide labriformin is a very potent toxin, indeed.

H. Photosensitizing toxins

Hundreds of toxic plants cause a photosensitivity reaction in animals when the plants are consumed. Grazing livestock encounter and readily eat many of these plants regularly with severe consequences. Affected animals become restless, exposed areas of the skin become red, and affected areas may become edematous from capillary leakage of plasma. These parts may then have plasma leakage through the skin with subsequent necrosis. Secondary infection than frequently develops, with sloughing of tissue (Kingsbury, 1964 and Everist, 1974).

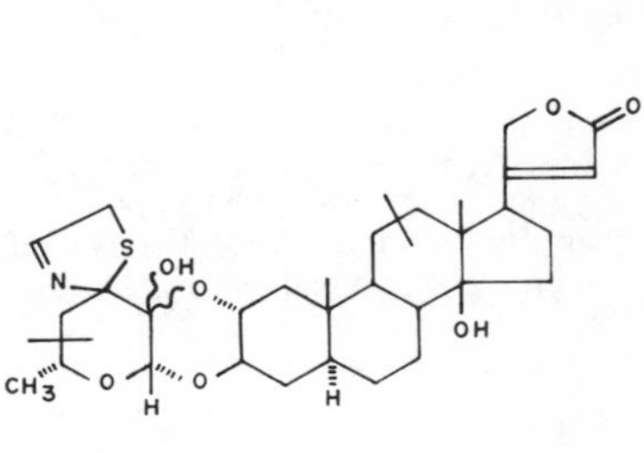

FIG. 17.8. THE TOXIC CARDENOLIDE, LABRIFORMIN

LABRIFORMIN

+20—4H

The hypersensitivity to light is caused by presence of a photo-reactive compound in the blood. A variety of compounds produce the effect. Some are photodynamic compounds synthesized by the plant, and others are metabolites that have not undergone biliary elimination because of liver damage produced by a plant toxin. Because of these two differing modes of photosensitization, the terms primary and secondary photosensitization have been coined to designate, respectively, the former and the latter compounds (Kingsbury, 1964). In primary photosensitization, the initial lesion results from the effect of sunlight directly on the photodynamic plant compound circulating in the vascular system. In secondary, or hepatogenic, photosensitization, the initial lesion is the liver lesion produced by the toxin, and the photosensitivity results from the effect of sunlight on an uneliminated metabolite in the vascular circulation, such as phylloerythrin, a chlorophyll metabolite (Kingsbury, 1964).

An example of primary photosensitization by a toxin from *Cymopterus watsonii* and an example of secondary photosensitization by a toxin from *Tetradymia glabrata*, will be considered here.

Cymopterus watsonii is a common plant in sagebrush and juniper foothill belts in Colorado, Nevada, Oregon, and Utah. Grazing sheep are in these regions in early spring as the plant begins to get good growth. It is quite palatable and readily eaten. Few sheep that ingest the plant die, but the photosensitized skin areas cause problems, particularly to nursing ewes because of swelling and necrosis of teats and udder. Loss of lambs from starvation and dehydration can be frequent (Binns, *et al.*, 1964). Williams reported that xanthotoxin and bergapten (Fig. 17.9) were the responsible primary photo-

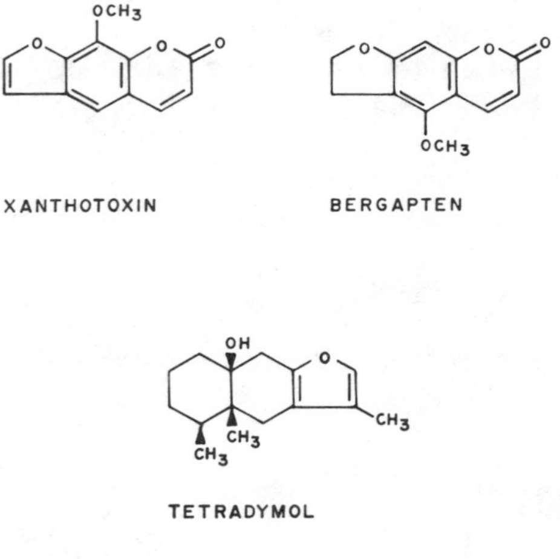

XANTHOTOXIN

BERGAPTEN

TETRADYMOL

FIG. 17.9. THE PRIMARY PHOTO-SENSITIZING TOXINS, XANTHO-TOXIN AND BERGAPTEN; AND THE COMPOUND TETRADYMOL, AN HEPATOTOXIN CAUSING SEC-ONDARY PHOTOSENSITIZATION

sensitizing compounds (Williams, 1970). These furocoumarins directly absorb long wavelength light with resultant damage to cellular constituents, perhaps because they are able to form photoadducts with pyrimidine bases (Ivie, 1978).

Among plants with toxins that cause secondary photosentization are members of the *Tetradymia* genus. *Tetradymia glabrata* (little horsebrush), The most important of the lot, grows quickly in the spring and is available to sheep being trailed from winter to summer ranges. The plant has evidently caused toxicosis in sheep for as long as grazing has been practiced in western United States. Sometimes hundreds of sheep are affected and develop in mild cases what ranchers term "bighead" resulting from edema of exposed skin areas of the face and ears. Other signs, particularly in severely affected animals, indicate they are suffering from liver damage as the primary insult. They have depression, anorexia, tremors, incoordination, prostration, dyspnea, and coma and may subsequently die. Photosensitizing signs develop in animals, when livers are not so severely damaged that the animal dies. The livers in acute cases appear enlarged with marked necrosis (Kingsbury, 1964, Jennings, *et al.*, 1978 and Johnson, 1978). For an as yet unexplained reason (Johnson, 1974), toxicosis develops even more readily in animals preconditioned with other range plants including *Artemisia nova* than in animals ingesting *Tetradymia* alone.

A hepatotoxin from *T. glabrata* has been isolated and characterized by Jennings and coworkers. The assay work was done in nonruminant laboratory animals, so applicability to ruminants is yet to be demonstrated. Jennings and coworkers summarized the chemistry and toxicologic findings (Jennings, *et al.*, 1978). The principal toxin is a furanoeremophilane they named tetradymol. Its ester was also present and nearly as toxic. The structure of tetradymol, the most abundant of the toxins, is shown in Figure 9. His studies have suggested that tetradymol is metabolized by the mixed function oxidases of the liver to a metabolite that is even more toxic and that the furanoeremophilanes express their toxicity by uncoupling oxidative phosphorylation from electron transport (Jennings, *et al.*, 1978).

I. Other Toxins

Hundreds of other interesting and important toxins exist in plants that comprise our food and feed sources. They range from inorganic elements—copper, lead, calcium, flourine, selenium, etc., to compounds of very exotic organic structures. Their mechanisms of action in producing toxicosis are equally diverse. The reader may consult treateses on the subject of poisonous plants to see their range and diversity (Kingsbury, 1962; Watt and Breyer-Brandwijk, 1962; Everist, 1974; Hardin and Arena, 1974; and Keeler *et al.* 1978).

Among inorganic elements, selenium ranks high in importance as a toxicant because of its widespread presence in soils and because it is readily taken up by plants such as certain members of the *Astragalus, Xylombia, Stanleya,* and other genera (Van Kampen and James, 1978). A variety of pathologic conditions of livestock have been attributed to selenium intoxication including conditions called "blind staggers," "alkali disease," acute selenosis, etc. All were said to result from either chronic or acute toxicoses from excessive selenium intake. There is little doubt that selenium can cause a variety of pathologic conditions, although some workers have questioned whether "blind staggers" can be attributable to selenium since the neurovisceral lesions (cytoplasmic vaculation) resemble those from true locoism caused by locoweeds, devoid of selenium.

A toxin of unusual structure is hypoglycin A from *Blighia sapida* (the Ackee tree) native to Africa but cultivated in this hemisphere (Hegarty, 1978). It is cyclopropane amino acid [β - (methylene-cyclopropyl) - alanine]. The fruit is commonly eaten after parboiling and frying which destroys the toxin. Improperly prepared fruit, however, still retains the toxin which produces severe hypoglycemia in humans. The effect evidently results from a reduction in the rate of fatty acid β-oxidation.

Not only are important food and feed toxins found in higher plants, but many are found in lower forms such as fungi. Two of the most important from an economic as well as toxicoses incidence point of view, are toxins from mushrooms and from ergot.

There are no poisonous plants more famous than mushrooms. And as the common verse reminds us, among mushroom gatherers there are no ancients who are careless:

> There are old mushroom hunters,
> And there are bold mushroom hunters,
> But there are no old, bold mushroom hunters.

Very few mushroom species are poisonous, but difficulties in distinguishing edible from poisonous forms is the cause of much trouble. Hardin and Arena (1974) have reviewed the difficulties and also described the 6 basic groups of poisonous compounds and syndromes produced. For example, problems arise from ingestion of *Amanita phalloides* or *Amanita verna* (the destroying angel). These two and others contain complex polypeptides such as amanitine and phalloidine which cause extreme abdominal pain, and sweating, distorted vision, degenerative changes in kidney, liver and cardiac musculature, oliguria and anuria. Mortality is 50–90%. *Amanita muscaria* and others containing muscarine cause watering of eyes and mouth, sweating, slowing of heart beat, pupilary contraction, breathing difficulty, abdominal cramps, and vomition. Fatalities are rare.

Ergot poisoning is due to parasitization of grains of certain cultivated and wild grasses including rye, wheat, barley and some native and cultivated pasture grasses by various species of the genus *Claviceps*. Kingsbury (1964) has provided an excellent summary of ergot poisoning including a description of the fungal types, life cycles, production of toxin, host species, toxin types, syndromes and pathologic effects. Humans are usually poisoned when they ingest uncleaned cereal grains, particularly rye, and products made from them. Livestock are poisoned from ergotized pasture grasses, hay, grain, or screenings from grain. The toxins are principally indole alkaloids. All are interrelated and yield lysergic acid upon alkaline hydrolysis.

Symptoms of ergotism are either gangrenous or convulsive. In gangrenous ergotism, vasoconstriction occludes blood circulation in the extremities and results in necrosis and subsequent sloughing of tissue. Convulsive ergotism, on the other hand, is characterized by muscular spasms of varying severity. Outbreaks in humas are rare now that grains are well cleaned, but outbreaks in livestock continue to occur because of ergot infestations on grasses grazed by the animals (see also chapter 7).

PLANT TOXINS THAT ARE TERATOGENIC

Until recently, studies on toxic plants considered only the direct effects of the toxin on humans or animals ingesting the plants. No thought was given to the possibility that plant toxins could produce congenital deformities (i.e., were teratogenic). The common congenital deformities were all believed by researchers to be related to genetics. But now it is known that many common congenital deformities in livestock are produced by plant toxins, and some toxins have been identified and characterized. One wonders whether some of the congenital deformities in humans can be attributed to plant toxins.

In this section, 3 classes of plant toxins that produce congenital defects in livestock are considered as examples of plant toxins now known to be teratogenic.

A. Quinolizidine alkaloids

A congenital deformity in calves   crooked calf disease (Fig. 17. 10) is prevalent in Alaska, Canada, and western United States. The disease is recognized by deformities that include twisted or bowed limbs, spinal or neck curvature, or cleft palate. Sometimes as many as 30% of calves born in a given herd will be affected. The deformities are permanent and even increase in severity because of the stress of added weight (Shupe et al., 1967a,b).

Because of observations in areas where the disease was prevalent in California, Wagnon (1960) reported ranchers believed ingestion of Lupinus laxiflorus by pregnant cows was responsible for the condition. He tried unsuccessfully to produce the condition by feeding the plants to pregnant cows (Wagmon, 1960). Our research group also tried to produce it by feeding lupines. Feeding Lupinus sericeus or Lupinus caudatus from the 40th to 70th days of gestation produced deformities (Shupe et al., 1967b).

Quinolizidine alkaloids present in the lupines are very toxic, and so it seemed possible that the teratogen might be a quinolizidine alkaloid.

We compared, by gas chromatography, lupine alkaloid profiles in lupine plants that were on hand from collections during the course of about 10 years. We looked for correlation between profiles

FIG. 17.10 LUPIN-INDUCED CROOKED CALF DISEASE SHOWING TYPICAL LIMB DEFORMITIES (ARTHROGRYPOSIS)

and incidence of crooked calf disease produced by those plants. Profiles were similar in all teratogenic plants and unlike those in nonteratogenic plants (Keeler, 1973). Four major alkaloid peaks were generally present in teratogenic plants, but only the fourth peak was invariably present in teratogenic plants and invariably absent or of low concentration in nonteratogenic plants (Keeler, 1973). The identity of the compound producing that peak was anagyrine based on mass, nuclear magnetic resonance, and infrared spectroscopy (Keeler, 1973). So, epidemiologic evidence suggested that the teratogen was likely anagyrine if an alkloid were responsible (Keeler, 1973).

Results from feeding trial experiments established (Keeler, 1976a) that alkaloid extracts from teratogenic lupines produced the disease. Preparations rich in anagyrine were teratogenic, but other alkaloids present in the semipurified anagyrine preparations did not produce the disease when fed to other cows. Both feeding trials and epidemiologic evidence, therefore, ascribed the teratogenicity to anagyrine. Severity of malformations was directly related to the level of anagyrine present in the preparations fed.

The structures of the alkaloids of the four peaks (Keeler, 1973) are shown in Figure 17.11. They are all similar. All have a carbonyl at carbon atom 2 in ring A, but only anagyrine has two double bonds in ring A. Anagyrine is different in configuration at carbon atom 6 from lupanine and epimethoxylupanine. Perhaps the unsaturation of ring A or configuration are important in conferring teratogenicity.

If plant concentration of anagyrine varied as a function of plant

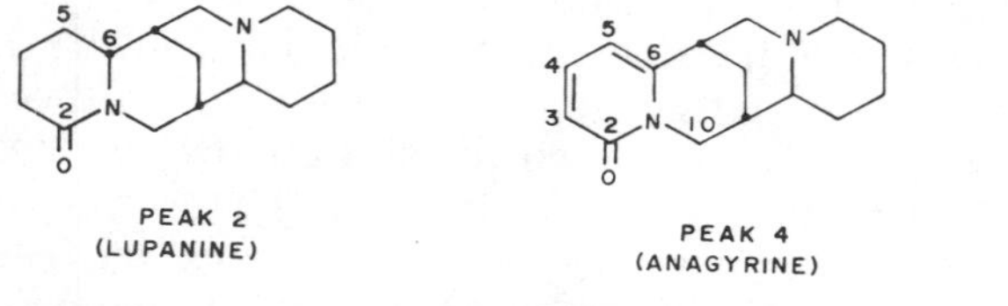

FIG. 17.11. STRUCTURES OF THE PRINCIPAL ALKALOIDS FROM *LUPINUS CAUDATUS*, A PLANT INDUCING CROOKED CALF DISEASE

maturity, the author believed one could design a breeding and grazing strategy to reduce disease incidence. Since cows are susceptible principally from the 40th through the 70th days of gestation, incidence could be reduced if one would prevent cows from grazing lupine when anagyrine content was high and cows were in the susceptible gestational period.

In *L. sericeus* and *L. caudatus* plant samples collected through the grazing season, concentration in above-ground parts of all alkaloids, including anagyrine (Keeler, 1976), was high early in growth and decreased markedly as plants matured. Concentration in mature, intact seeds was also very high. Consequently, a pregnant cow in the 40th- to 70th-day gestation range is at greatest hazard when she is grazing teratogenic lupine early in growth or during seeding. Hazard is low when flowering and postseeding stage plants are grazed.

B. Piperidine alkaloids

Most crooked calf disease in western United States that our laboratory has investigated was attributable, to maternal ingestion of lupine plants during gestation. It was speculated (Keeler, 1971a) in 1971 that *Conium maculatum* might be the cause of some of the cases not due to lupine ingestion. That speculation was based on observations of an outbreak of limb deformities in calves in Utah that year. The deformities included limb rotation, permanent carpal or elbow joint flexure, spinal curvature, or cleft palate and lip, and thus were similar to the deformities induced by lupine. No lupine was found in pastures where deformed calves were observed, nor did cows have access to lupine at any time during gestation. Considerable *C. maculatum* was found, however, and cows had had access to that during the probably susceptible first trimester. *Conium* was the only plant that was present in quantity that could be classified as a known poisonous plant. Scattered reports in the literature on the teratogenicity in laboratory animals of compounds moderately related structurally to the piperidine alkaloids of coniine were also found. It was speculated that, if *Conium* were responsible, then, perhaps, one of its piperidine alkaloids might be the teratogen.

Pregnant cows were gavaged the *Conium* alkaloid coniine during the gestation period determined for similar deformities induced by lupine. Coniine produced deformities like those of the field outbreak (Keeler, 1974). Pregnant cows were then gavaged fresh, young *C. maculatum*. Deformities resulted when cows were gavaged the plant from the 50th through the 75th days of gestation (Keeler and Balls). Both arthrogryposis and spinal curvature, like the defects produced by the piperidine alkaloid coniine, were produced.

More than 98% of the total alkaloid of fresh *Conium maculatum* of young age, as gavaged, was γ-coniceine. Evidently, both coniine, the principal alkaloid of mature plant, and γ-coniceine, the principal alkaloid of young plant, are teratogenic (Fig. 17.12). The double bond in the ring between the nitrogen and the α carbon atom evidently did not reduce activity.

C. Steroidal alkaloids

Until the early 1960's, epidemics of cyclopia and related con-genital deformities in sheep were common in Idaho. Hundreds of lambs were afflicted with these deformities each year (Binns, *et al.*, 1963). Deviations from normal included a single or double globe cyclopia (Fig.17.13), and a shortened upper jaw and protruding and curved lower jaw, sometimes with a peculiar skin-covered proboscis above the single eye. Mildly affected cases had normal eyes and a shortening of the upper jaw or cebocephaly. *Veratrum californicum* caused the condition when pregnant ewes ingested it on the 14th day of gestation (Binns, *et al.*, 1965). Results from studies suggested that the teratogen was an alkaloid. A number of veratrum alkaloids were tested in pregnant sheep to check their teratogenicity. The com-pounds jervine and two others we named cyclopamine and cyclo-posine produced deformities like natural cases (Binns, *et al.*, 1965). Other compounds tested did not produce cyclopia.

Results from infrared, nuclear magnetic resonance and mass spectrometry analysis showed that cyclopamine was 11-deoxojervine (Keeler, 1969a) and that cycloposine was glucosylcyclopamine (Kee-ler, 1969b), the glucoside of cyclopamine. These two compounds and jervine are closely related, but cyclopamine is the teratogen of natural importance because of plant concentration.   Figure 17.14 shows three compounds and two closely related compounds that did not produce cyclopia in sheep, veratramine and muldamine (Keeler, 1971b). The data suggested that an intact furan ring was required for activity (Keeler, 1970), perhaps conferring some essential config-uration on the molecule.

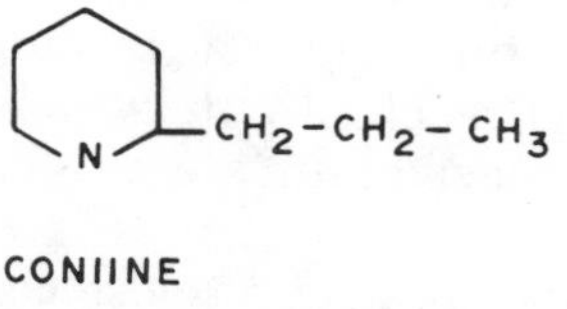

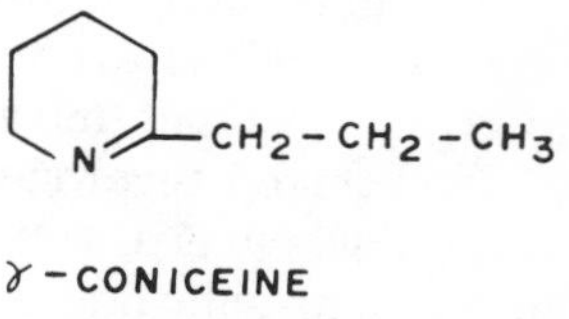

CONIINE   γ-CONICEINE

FIG. 17.12. THE PIPERIDINE ALKALOIDS, CONIINE AND Y CONICEINE

FIG. 17.13.   THE CYCLOPIC LAMB DEFORMITY IN-
DUCED BY MATERNAL INGESTION OF CYCLOPAMINE,
THE VERATRUM STEROIDAL ALKALOID TERATOGEN

These steroidal alkaloid teratogens from *Veratrum* spp. are structurally related to a number of solanum alkaloids. We have had an
interest since the late 1960's in the possible teratogenicity of solanum alkaloids (Keeler, 1970). It was, therefore, with considerable
interest that we learned of an hypothesis by J. H. Renwick (1972) in
1972 that maternal consumption of blighted potatoes by pregnant
humans produced congenital anencephaly and spina bifida (ASB).
Although Renwick believed as one possibility that the teratogen was
elaborated by potatoes as a consequence of blight, we wondered
whether the alleged teratogen could be a solanum steroidal alkaloid
if the hypothesis were valid.

We found that solasodine, a solanum alkaloid of close configuration to cyclopamine, was teratogenic in hamsters. It produced various deformities, including spina bifida and exencephaly (Keeler, *et*

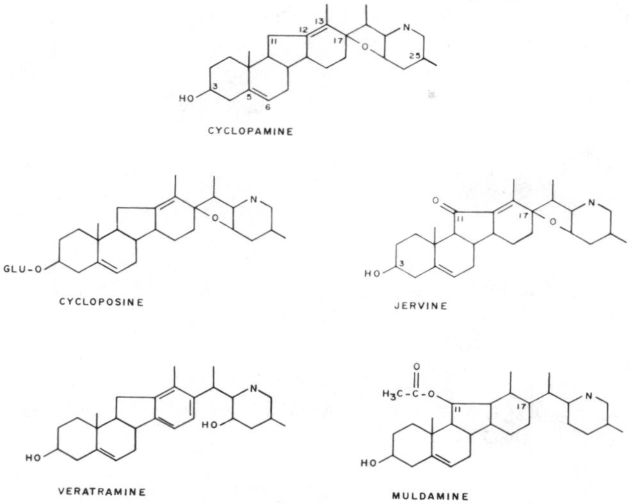

FIG. 17.14. STRUCTURES OF FIVE STEROIDAL ALKA-
LOIDS FROM *VERATRUM* SPP.

al., 1976). However, solasodine has not been reported in potatoes. Tomatidine, with a piperidine ring opposite that of solasodine, and diosgenin, the non-nitrogen containing analog of solasodine, were not active. Apparently, the terminal ring had to be a piperidine with the nitrogen projecting α with respect to the steroidal plane (Fig. 17.15).

A steroidal solanum alkaloid common to most potatoes is the glycoside of solanidine (Fig. 17.15), solanine. My colleague, Dr. D. Brown, wondered whether solanidine and related epimers would be teratogenic, if their tertiary nitrogen atoms had a free electron pair constrained toward the α side of the molecule. He synthesized such epimers (Brown and Keeler, 1978), and feeding trials showed those with the free electron pair α, but not those constrained β, were teratogenic. Interestingly, in the known solanidans of potatoes, the free electron pair is constrained β.

Many researchers (Elwood, *et al.*, 1973 and 1974) were not convinced that the epidemiology upon which Renwick based his hypothesis was best interpreted as Renwick suggested. In addition, the birth of ASB children to mothers on potato-avoidance trials during pregnancy (Nevin and Merrett, 1975) was particularly strong evidence against the hypothesis.

Many research workers have examined whether laboratory animals would give birth to deformed offspring if mothers were fed

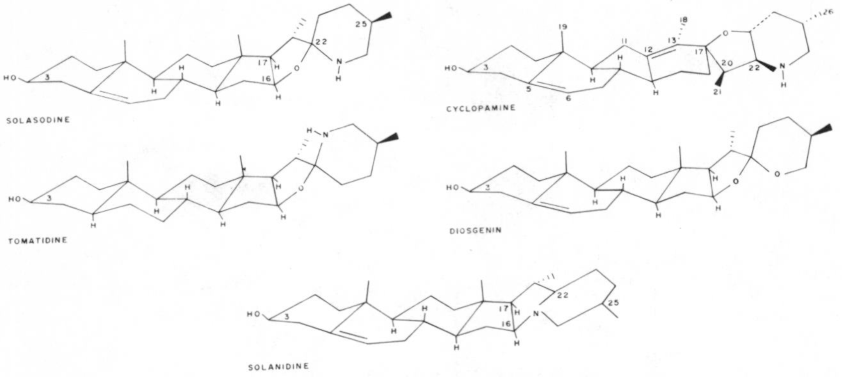

FIG. 17.15.   RELATED VERATRUM AND SOLANUM STEROIDAL ALKALOIDS AND THE COMPOUND, DIOSGENIN

during pregnancy with various potato preparations or compounds, including alkaloids, derived therefrom. Poswillo and coworker (Poswillo, *et al.*, 1972) found cranial osseus defects in neonatal marmosets from dams dosed with certain potato preparations, but they were later unable to confirm the observation (Poswillo, *et al.*, 1973). Swinyard and Chaube (Swinyard and Chaube, 1973) injected extracted potato alkaloids or pure solanine in pregnant dams and produced no neural tube defects in rats and rabbits. Ruddick and coworkers (Ruddick, *et al.*, 1974) reported that freeze dried, blighted 'Kennebec' potatoes produced no terata when fed to rats. We showed (Keeler, *et al.*, 1974 and 1975) that neither early nor late blighted 'Russet Burbank' potatoes, control healthy tubers, nor extracts therefrom were teratogenic in rats, rabbits, hamsters, or mice. Nishie and coworkers reported (Nishie, *et al.*, 1975) that tomatime, α-chaconine, and α-solanine did not produce significant teratogenic effects in chicks. Intraperitoneal injection of solanine to pregnant mice failed to produce terata, according to Bell and associates (Bell, *et al.*, 1976). Chaube and Swinyard (1976) reported no neural tube defects in offspring from rats subject to acute or chronic intraperitoneal injection of α-chaconine or α-solanine.

Mun and associates (Mun, *et al.*, 1975), however, showed that both pure solanine and glycoalkaloids extracted from *Phytophthora* blighted potatoes produced "rumplessness or trunklessness" in chick embryos when eggs were treated during early development. Jelinek and coworkers (Jelinek, *et al.*, 1976) verified and extended those observations.

We (Keeler, *et al.*, 1978) tested the teratogenicity of dried, ground potato sprout preparations of high alkaloid content (35–55 mg/g) from seven potato varieties. All produced congenital deformities in one strain of hamsters. Incidence of affected litters ranged from 8 to 25% among the seven varieties. Deformities included primarily exencephaly, cranial bleb (encephalocele), microphthalmia, and spina bifida. Neither peel nor tuber material from sprouted or control tubers was teratogenic. Alkaloid content in peel and tuber preparations tested were 2 and 0.03 mg/g, respectively.

Oral doses of sprout material gavaged the hamsters ranged from 2500 to 3500 mg/kg. By extrapolation, the average 50 to 70 kg pregnant woman would have to eat from 125 to 250 g of dry sprout material (900–1750 g of fresh weight) in a single dose to consume comparable amounts. Sprouts, of course, are not generally consumed in any amount. Peel and tuber material were not teratogenic at 3–5 times that dose level.

BIBLIOGRAPHY

ALLISON, M. J. 1978. The Role of Ruminal Microbes in the Metabolism of Toxic Constituents from Plants. *In* "Effects of Poisonous Plants on Livestock," R. F. Keeler, K. R. Van Kampen, and L. F. James, Eds. Academic Press, New York. pp. 101–118.

BELL, D. P., GIBSON, J. G., McCARROL, A. M., McCLEAN, G. A., and GERALDINE, A. 1976. Embryotoxicity of Solanine and Aspirin in Mice. J. Reprod. Fertil. 46:257–259.

BENSON, J. M., SEIBER, J. N., KEELER, R. F., and JOHNSON, A. E. 1978. Studies on the Toxic Principle of *Asclepias Eriocarpa* and *Asclepias labriformis*. *In* "Effects of Poisonous Plants on Livestock," R. F. Keeler, K. R. Van Kampen, and L. F. James, Eds. Academic Press, New York. pp. 273–284.

BINNS, W., JAMES, L. F., SHUPE, J. L., and EVERETT, G. 1963. A congenital cyclopian-type malformation in lambs induced by maternal ingestion of a range plant, *Veratrum californicum*. Amer. J. Vet. Res. 24:1164–1175.

BINNS, W., JAMES, L. F., and BROOKSBY, W. 1964. *Cymopterus watsonii*: A photosensitizing plant for sheep. Vet. Med. Small Anim. Clin. 59:375–379.

BINNS, W., SHUPE, J. L., KEELER, R. F., and JAMES, L. F. 1965. Chronologic evaluation of teratogenicity in sheep fed *Veratrum californicum*. J. Am. Vet. Med. Assoc. 147:839–842.

BRADEN, A. W. H., and McDONALD, I. W. 1970. *In* "Australian Grasslands," R. M. Moore, Ed. Aust. Nat. Univ. Press, Canberra. p. 381.

BROWN, D., and KEELER, R. F. 1978. Structure-activity relation of steroidal teratogens. III. Solanidan epimers. J. Agric. Food Chem. 26:566–569.

CARROLL, K. K. 1949. Isolation of an Antithyroid Compound from Rape Seed (*Brassica napus*). Proc. Soc. exp. Biol. Med. 71:622–624.

CHAUBE, S., and SWINYARD, C. A. 1976. Teratological and Toxicological Studies of Alkaloids and Phenolic Compounds from *Solanum tuberosum*. J. Toxicol. App. Pharmacol. 36:227–237.

CONN, E. E. 1978. Cyanogenesis, the Production of Hydrogen Cyanide by Plants. *In* "Effects of Poisonous Plants on Livestock," R. F. Keeler, K. R. Van Kampen, and L. F. James, Eds. Academic Press, New York. pp. 301–310.

COUCH, J. F. 1929. Experiments with Extracts from the Wooly-Pod Milkweed (*Asclepias eriocarpa*). Am. J. Pharmacol. 101:815–821.

COX, R. I. 1978. Plant Estrogens Affecting Livestock in Australia. *In* "Effects of Poisonous Plants on Livestock, R. F. Keeler, K. R. Van Kampen, and L. F. James, Eds. Academic Press, New York. pp. 451–464.

ELWOOD, J. H., and MACKENZIE, G. 1973. Associations between the incidence of neurological malformations and potato blight outbreaks over 50 Years in Ireland. Nature, 243:476–477.

EMANUEL, I. S., and SEVER, L. E. 1973. Questions concerning the possible association of potatoes and neural tube defects, and an alternative hypothesis relating to maternal growth and development. Teratology, 8:325–331.

EVERIST, S. L. 1974. "Poisonous Plants of Australia." Angus and Robertson, Cremorn Junction, N.S.W.

EVERIST, S. L. 1978. Botanical affinities of Australian poisonous plants. *In* "Effects of Poisonous Plants on Livestock," R. F. Keeler, K. R. Van Kampen, and L. F. James, Eds. Academic Press, New York. pp. 93–100.

FELS, H. E., and NEIL, H. C. 1968. Effects on Reproduction of Prolonged grazing on Oestrogenic Pastures by Ewes. Aust. J. Agric. Res. 19:1059-1068.

FIELD, B., and KERR, C. 1973. Potato blight and neural tube defects. Lancet, 2:507–508.

GARDNER, C. A., and BENNETTS, H. W. 1956. The Toxic Plants of Western Australia. West Australian Newspapers Ltd., Perth.

GREER, M. A. 1962. Natural Occurrence of Goitrogenic Agents. Recent Prog. Hormone Res. 18:187–219.

HARDIN, J. W. and ARENA, J. M. 1974. Human Poisoning from Native and Cultivated Plants. Duke University Press, Durham, North Carolina.

HEGARTY, M. P. 1978. Toxic Amino Acids of Plant Origin. *In* "Effects of Poisonous Plants on Livestock." R. F. Keeler, K. R. Van Kampen, and L. F. James, Eds. Academic Press, New York. pp. 575–585.

HERZ, W. 1978. Sesquiterpene Lactones from Livestock Poisons. *In* "Effects of Poisonous Plants on Livestock," R. F. Keeler, K. R. Van Kampen, and L. F. James, Eds. Academic Press. New York. pp. 487–497.

HILL, D. W., KIM, H. L., MARTIN, C. L. and CAMP, B. J. 1977. Identification of Hymenoxon in *Baileya multiradiata* and *Helenium hoopesii*. J. Agric. Food Chem. 25:1304–1307.

HIRONO, I., KACHI, H., and KATO, I. 1970. A Survey of Acute Toxicity of

Cycads and Mortality Rate from Cancer in the Miyako Islands, Okinawa. Acta Pathol. Jpn. 20:327–337.

HOOPER, P. T. 1978. Cycad Poisoning in Australia—Ethiology and Pathology. In "Effects of Poisonous Plants on Livestock," R. F. Keeler, K. R. Van Kampen, and L. F. James, Eds. Academic Press, New York. pp. 337–347.

IVIE, G. W. 1978. Toxicological Significance of Plant Furocoumarins. In "Effects of Poisonous Plants on Livestock," R. F. Keeler, K. R. Van Kampen, and L. F. James, Eds. Academic Press, New York. pp. 475–485.

IVIE, G. W., WITZEL, D. A., HERTZ, W., KANNAN, R., NORMAN, J. O., RUSHING, D. D., JOHNSON, J. H., ROWE, L. D., and VEECH, J. A. 1975. Hymenovin: Major Toxic Constituent of Western Bitterweed (Hymenoxys odorata D. C.) J. Agric. Food Chem. 23:841–844.

IVIE, G. W., WITZEL, D. A., HERZ, W., SHARMA, R. P., and JOHNSON, A. E. 1976. Isolation of Hymenovin from Hymenoxys richardsonii (Pingue) and Dugaldia hoopesii (Orange sneezeweed). J. Agric. Food chem. 24:681–682.

JADHAV, S. J., and SALUNKHE, D. K. 1975. Formation and control of chlorophyll and glycoalkaloids in tubers of Solanum tuberosum L. and evaluation of glycoalkaloid toxicity. In "Advances in Food Research," C. O. Chidchester, Ed. Vol. 21, pp. 307–354.

JELINEK, R., KYZLINK, V., and BLATTNY, C. JR. 1976. An Evaluation of the Embro-toxic Effects of Blighted Potatoes on Chicken Embryos. Teratology, 14:335–342.

JENNINGS, P. W., REEDER, S. K., HURLEY, J. C., ROBBINS, J. E., HOLIAN S. K., HOLIAN, A., LEE, P., PRIBANIC, J. A. S., and HULL, M. 1978. Toxic Constituents and Hepatotoxicity of the Plant Tetradymia glabrata (Asteroceae). In "Effects of Poisonous Plants on Livestock," R. F. Keeler, K. R. Van Kampen, and L. F. James, Eds. Academic Press, New York. pp. 217–228.

JOHNSON, A. E. 1974. Predisposing influence of range plants on Tetradymia—related photosensitization in sheep: Work of Drs. A. B. Clowson and W. T. Huffman. Am. J. Vet. Res. 35:1583–1585.

JOHNSON, A. E. 1978. Tetradymia Toxicity—A New Look at an Old Problem. In "Effects of Poisonous Plants on Livestock," R. F. Keeler, K. R. Van Kampen, and L. F. James, Eds. Academic Press, New York. pp. 209–216.

KEELER, R. F. 1969a. Teratogenic compounds of Veratrum californicum (Durand). VI. The structure of cyclopamine. Phytochemistry 8:223–225.

KEELER, R. F. 1969b. Teratogenic compounds of Veratrum californicum (Durand). VII. The structure of the glycosidic alkaloid cycloposine. Steroids, 13:579–588.

KEELER, R. F. 1970. Teratogenic compounds of Veratrum californicum (Durand). IX. Structure activity relationships. Teratology, 3:169–174.

KEELER, R. F. 1971a. Known and suspected teratogenic hazards in range plants. Coll. Vet. Toxicol. Ann. Meeting. Detroit. (Abstracts).

KEELER, R. F. 1971b. Teratogenic compounds of Veratrum californicum

(Durand). XIII. Structure of muldamine. Steroids, 18:741–752.

KEELER, R. F. 1973. Lupin alkaloids from teratogenic and nonteratogenic lupins. I. Correlation of crooked calf disease incidence with alkaloid distribution determined by gas chromatography. Teratology, 7:23–30.

KEELER, R. F. 1973. Lupin alkaloids from teratogenic and nonteratogenic lupins. II. Identification of the major alkaloids by tandem Gas chromatography—mass spectrometry in plants producing crooked calf disease. Teratology, 7:31–35.

KEELER, R. F. 1974. Coniine, a teratogenic principle from *Conium maculatum* producing congenital malformations in calves. Clin. Toxicol. 7:195–206.

KEELER, R. F. 1976a. Lupin Alkaloids from Teratogenic and Nonteratogenic Lupins. III. Identification of Anagyrine as the Possible Teratogen by Feeding Trials. J. Toxicol. Environ. Health, 1:887–898.

KEELER, R. F. 1976b. Lupin alkaloids from teratogenic and nonteratogenic lupins. IV. Concentration of total alkaloids, individual major alkaloids, and the teratogen anagyrine as a function of plant part and stage of growth and their relationship to crooked calf disease. J. Toxicol. Environ. Health 1:899–908.

KEELER, R. F. 1978. Toxins and teratogens of the lily and nightshade families. *In* Society of Economic Botany Symposium on Poisonous Plants, Miami, 1977, (In Press, Columbia University Press).

KEELER, R. F., and BALLS, L. D. Teratogenic Effects in cattle of *Conium maculatum* and conium alkaloids and analogs. Clin. Toxicol. 12: 49–64.

KEELER, R. F., and BINNS, W. 1968. Teratogenic compounds of *Veratrum californicum* (Durand). V. Comparison of cyclopian effects of steroidal alkaloids from the plant and structurally related compounds from other sources. Teratology, 1:5–10.

KEELER, R. F., DOUGLAS, D. R., and STALLKNECHT, G. F. 1974. Failure of Blighted Russet Burbank Potatoes to Produce Congenital Deformities in Rats. Proc. Soc. Exp. Biol. Med. 146:284–286.

KEELER, R. F., DOUGLAS, D. R., and STALLKNECHT, G. F. 1975. The Testing of Blighted, Aged, and Control Russet Burbank Potato Tuber Preparations for Ability to Produce Spina Bifida and Anencephaly in Rats, Rabbits, Hamsters, and Mice. Am. Potato J. 52:125–132.

KEELER, R. F., YOUNG, S., and BROWN, D. 1976. Spina Bifida, exencephaly, and cranial bleb produced in hamsters by the solanum alkaloid solasodine. Res. commun. Chem. Pathol. Pharmacol. 13:724–730.

KEELER, R. F., VAN KAMPEN, K. R. and L. F. JAMES, EDS. 1978. Effects of Poisonous plants on Livestock. Academic Press, New York. 600p.

KEELER, R. F., YOUNG, S., BROWN, D., STALLKNECHT, G. F., and DOUGLAS, D. R. 1978. Congenital Deformities Produced in Hamsters by Potato Sprouts. Teratology (In Press).

KIM, H. L., ROWE, L. D., and CAMP, B. J. 1975. Hymenoxon, A Poisonous sesquiterpene Lactone from *Hymenoxys odorata* (Bitterweed). Res. Commun. Chem. Pathol. Pharmacol. 11:647–650.

KINGSBURY, J. M. 1964. Poisonous Plants of the United States and Canada. Prentice-Hall, Inc., Englewood Cliffs, New Jersey.

KINLEN, L., and HEWITT, A. 1973. Potato blight and anencephaly in Scotland. Brit. J. Prev. Soc. Med. 27:208–213.

KRAYER, O. 1958. Veratrum Alkaloids. In "Pharmacology in Medicine." V. Drill. Ed. 3rd ed. McGraw-Hill, New York. pp. 33/1–33/10.

KUPCHAN, S. M., and A. W. BY. 1968. Steroidal Alkaloids: The Veratrum Group. In "The Alkaloids," R. H. F. Manske, Ed., Vol. x, pp. 193–285.

KURLAND, L. T. 1972. An Appraisal of the Neurotoxicity of Cycad and the Etiology of Amylotrophic Lateral Sclerosis on Guam. Fed. Proc. 31:1540–1542.

LANGER, P., and STOLE, V. 1965. Goitrogenic Activity of Allylisothiocyanate—A Widespread Natural Mustard Oil. Endocrinology, 76:151–155.

MARSH, C. D., and CLAWSON, A. B. 1924. The Wooly-Pod Milkweed (Asclepias eriocarpa) as a Poisonous Plant. U.S. Dept. Agric. Bull. 1212.

MATSUMOTO, H., and HIGA, H. H. 1966. Studies on Methylazoxymethanol, the Aglycone of Cycasin: Methylation of Nucleic Acids in vitro. Biochem. J. 98:20c.

McMAHON, B. S., YEN, S., and ROTHAM, K. J. 1973. Potato Blight and Neural Tube Defects. Lance, 1:598–599.

MIWA, T. 1975. Protective Effect of Cysteamine and Glutathione against the Toxicity and Carcinogenicity of Methylazoxymethanol. Gifu Daigaku Igakubu Kiyo, 23:495–500.

MONTGOMERY, R. D. 1969. Cyanogens. In "Toxic Constituents of Plant Foodstuffs," I. E. Liener, Ed. Academic Press, New York. pp. 143–157.

MUN, A. M., BARDEN, E. S., WILSON, J. M., and HOGAN, J. H. 1975. Teratogenic Effects in Early Chick Embryos of Solanine and Glycoalkaloids from Potatoes Infected with Late Blight Phytophthora infestans. Teratology, 11:73–78.

NEVIN, N. C., and MERRETT, J. D. 1975. Potato Avoidance During Pregnancy in Women with a Previous Infant with either Anencephaly and/or Spina Bifida. Brit. J. Prev. Soc. Med. 29:111–115.

NISHIE, K., NORRED, W. P., and SWAIN, A. P. 1975. Pharmacology and Toxicology of Chaconine and Tomatidine. Res. commun. Chem. Pathol Pharmacol. 12:657–668.

NISHIE, K., GUMBMANN, M. R., and KEYL, A. C. 1971. Pharmacology of Solanine. Toxicol. Appl. Pharmacol. 19:81–92.

POSWILLO, D. E., SOPHER, D., and MITCHELL, S. J. 1972. Experimental Induction of Foetal Malformations with "Blighted" Potato: A Preliminary Report. Nature. 239:462–464.

POSWILLO, D. E., SOPHER, D., MITCHELL, S. J., COXON, D. T., CURTIS, R. F., and PRICE, K. R. 1973. Investigations into the Teratogenic Potential of Imperfect Potatoes. Teratology, 8:339–348.

RENWICK, J. H. 1972. Hypothesis: Anencephaly and Spina Bifida are usually preventable by avoidance of a specific but unidentified substance present in certain potato Tubers. Brit. J. Prev. Soc. Med. 26:67–68.

ROESKE, C. 1971. Correlation of Cardenolide Content and Mammalian Toxicity in Four Species of Milkweed (*Asclepias* spp.). Masters Thesis. University of California, Davis.

RUDDICK, J. A., HARWIG, J., and SCOTT, P. M. 1974. Nonteratogenicity in Rats of Blighted Potatoes and Compounds Contained in them. Teratology, 9:165–168.

RUTKOWSKI, A. 1971. Feed value of Rapeseed Meal. J. Am. Oil Chem. Soc. 48:863–868.

SCHERY, R. W. 1952. Plants for Man. Prentice-Hall, Inc., Englewood Cliffs, New Jersey. p. 425.

SCHREIBER, K. 1968. Steroidal Alkaloids: The Solanum Group. *In* "The Alkaloids," R. H. F. Manske, Ed., Vol. x, pp. 1–192.

SEIBER, J., BENSON, J., ROESKE, C., and BROWER, L. 1975. Qualitative and Quantitative Aspects of Milkweed Cardenolide Sequestering by Monarch butterflies. Am. Chem. Soc. Abstracts, Pest. 103.

SHUPE, J. L., BINNS, W., JAMES, L. F., and KEELER, R. F. 1967a. Crooked calf syndrome. a plant-induced congenital deformity. Zuchthygiene, 2:145–152.

SHUPE, J. L., BINNIS, W., JAMES, L. F., and KEELER, R. F. 1967b. Lupine. A cause of crooked calf disease. J. Am. Vet. Med. Assoc. 151: 198–203.

SMITH, J. F. 1971. Ewe Fertility—Production of Cervical Mucus. J. Reprod. Fertil. 24:132.

SPIERS, P. S., PIETRZYK, J. J., PIPER, J. M., and GLEBATIS, D. M. 1974. Human Potato Consumption and Neural Tube Malformation. Teratology, 10:125–128.

SWINYARD, C. A., and CHAUBE, S. 1973. Are Potatoes Teratogenic for Experimental Animals? Teratology, 8:349–358.

TURNBULL, K. E., BRADEN, A. W. H., and GEORGE, J. M. 1966. Fertilization and Early Embryonic Losses in Ewes that had Grazed Oestrogenic Pastures for 6 Years. Aust. J. Agric. Res. 17:907–917.

VAN ETTEN, C. H., and TOOKEY, H. L. 1978. Glucosinolates in Cruciferous Plants. *In* "Effects of Poisonous Plants on Livestock," R. F. Keeler, K. R. Van Kampen, and L. F. James, Eds. Academic Press, New York. pp. 507–520.

WAGNON, K. A. 1960. Lupine poisoning as a possible factor in congenital deformities in cattle. J. Range Manage. 13:89–91.

WATT, J. M., and BREYER-BRANDWIJK, M. B. 1962. The Medicinal and Poisonous Plants of Southern and Eastern Africa. E&S Livingston Ltd, Edinburgh & London.

WILLIAMS, M. C. 1970. Xanthotoxin and bergapten in spring parsley. Weed Sci. 18:479–480.

YANG, M. G., and MICKELSEN, O. 1969. Cycads. *In* "Toxic Constituents of Plant Foodstuffs," I. E. Liener, Ed. Academic Press, New York. pp. 159–167.

YOSHIOKA, H., MABRY, T. J., and TIMMERMAN, B. N. 1973. "Sesquiterpene Lactones," University of Tokyo Press, Tokyo. 544 pages.

18

Poisonous Marine Animals

H. Dymsza, Y. Shimizu, F. E. Russell and H. D. Graham

INTRODUCTION

Early man, who lived by the sea and fished for food, had to be no less skilled in toxicology than his inland counterpart. While teeming with potential sources of food, clothing, medicine and other materials, the seas, then as now, contained thousands of dangerous organisms, which could bite, sting or be poisonous when eaten. However, early inhabitants learned by experience to distinguish between edible and poisonous seafoods. Furthermore, they communicated their findings among themselves and to future generations. For example, in order to prohibit the eating of poisonous fish by the ancient Israelites, Mosaic law contained passages of advice, which are still sound today: "Whatsoever hath not fins and scales ye may not eat; it is unclean to you" (Deuteronomy 14:9–10; ca 1451 B.C.).

Hieroglyphics on some of the tombs of ancient Egypt depict the puffer *Tetraodon stallatus*, a fish recognized as being deadly poisonous (Gaillard, 1923). Ancient literature on therapeutic agents shows that as early as 618 A.D. the Chinese were aware of the fact that many fish contained toxic components, which could be used as potent drugs (Read, 1939).

Fascinating accounts of historical happenings involving toxins of marine origin have been compiled by Halstead (1965). Perhaps best known is the story of puffer fish intoxication experienced by Captain James Cook in 1774 in New Hebrides. Shortly thereafter in the 1757–98 period, Captain George Vancouver described symptoms of an illness resulting from eating shellfish from the west coast of

North America which apparently were contaminated with paralytic shell fish poison (PSP). During World War II, Japanese troops were warned of the dangers of eating poisonous fish, but about 400 deaths occurred. As a result of U.S. military activity in the South Pacific, Fish and Cobb (1954), of the University of Rhode Island, under contract with the U.S. Office of Naval Research, published the first extensive U.S. report on toxic marine animals. However, the classic work in the field is the comprehensive three-volume series authored by Halstead (1965, 1967, 1970).

The overall impact of ichthyosarcotoxism, the specific term used to designate intoxication resulting from consumption of the flesh of poisonous fish, is difficult to estimate. From a public health aspect, while some morbidity and mortality figures are available and will be discussed, they are believed to be grossly incomplete. Since many outbreaks occur in scattered insular areas or are rather mild, they are not reported to local public health agencies nor to the World Health Organization. Russell (1968), estimated that fewer than 20,000 cases of poisoning occur each year, of which less than 200 are fatal.

The extreme potency of fish and mollusk toxins have given man good reason to beware, and to study them. They are among the most toxic biological materials known to man. Various natural poisons have been compared by Mosher and co-workers (1964). They showed that whereas *Clostridium botulinum* toxin, Type A was estimated to be lethal to man at an I.P. dose level of 0.0003 µg/kg, saxitoxin from contaminated shellfish or tetrodotoxin isolated from the liver or ovaries of the puffer-like *Spaeroides rubruoes* was estimated to be lethal at doses of 9 and 8–20 ug/kg, respectively. In contrast, the fatal dose for sodium cyanide was given as 10,000 µg/kg.

Currently, it is not possible to assess the total economic losses attributed to ichthyosarcotoxism. Severe financial losses can occur from outbreaks of paralytic shellfish poisoning (PSP) causing shellfish to be kept off the market and possibly influencing consumers to stay away from purchasing all types of seafoods. It is also known that a major deterrent to the development of fisheries, based on shallow-water food fish, in some areas of the world, is the fear of ciguatera poisoning. Where this toxin is a problem, improvement of the food supply through an inshore fishery would be facilitated by development of methods for testing large numbers of fish, or an ability to accurately predict which fish are safe to eat. Such knowledge would be particularly important for the increased utilization of ordinarily edible species which can become toxic at certain times.

Should there be renewed interest in whole fish protein concentrate (FPC) made from mixed species, especially from tropical seas, the presence of poisonous fish such as puffers could present a problem. Some factors affecting toxicity would be the nature of the toxin, amount of toxin, the solvent used in extraction of the protein and processing temperature.

With the continual depletion of land-based resources and dependence on intensive high-energy consuming forms of agriculture, increased exploitation of the oceans becomes imperative. No doubt, many unused marine plants and organisms are potential sources of food, drugs and chemicals. Yet, not a great deal is known about them or the toxins associated with the 1,000 or so marine species of organisms known to be poisonous or venomous. Overall lack of knowledge has, to some extent, limited the full utilization of the potential of food from the sea. Of some 25,000 recognized species of fish, only about 25 currently constitute the major portion of the world's food fishery catch.

The focus of this review is on certain species of fish and mollusks and their ichthyotoxins which could pose a threat to human health, a food source or economic stability. Thus, the reader should be aware of the fact that, by narrowing the scope of this review to fish poisons, many important aspects of the toxicology of marine food sources such as microbiological and environmental contaminations and their possible deleterious effects are not discussed. These are dealt with in Chapters 1–15, to some extent, and in much detail by the excellent compendium by Speck (1978).

FISH POISONS AND SPECIES

Illness can result from ingestion of the flesh, certain organs or eggs of approximately 500 species of salt water fish. Since only a few of these species are major sources of human food, the problem of ichthyosarcotoxism is generally confined to a few species. A list of the major fish poisons implicated in ichthyosarcotoxism and representative causative species are given in Table 18.1.

CIGUATERA TOXIN

Of all the marine fish toxins, ciguatera is the most injurious from both public health and economic aspects. It is often a serious problem in that intoxication can cause morbidity and mortality, but

TABLE 18.1. REPRESENTATIVE FISHES AND MAMMALS RESPONSIBLE FOR OUTBREAKS OF MARINE FOOD POISONING

Poison	Common Name	Species	Habitat
Ciguatoxin	Barracuda	*Sphyraena barracuda*	Caribbean, Indo-Pacific
	Grouper,		
	Yellowfin	*Myeteroperca venenosa*	Caribbean
	Jack	*Caranx hippos*	Caribbean
	Moray Eel	*Cymnothorax undulatus*	Hawaii to East Africa
	Parrotfish	*Scarus caeruleus*	Caribbean
	Pompano	*Alectis crinitus*	Caribbean
	Snapper, Red	*Lutjanus bohar*	Tropical Pacific
	Surgeonfish	*Acanthurus glauco-pareicus*	Tropical Pacific
		Acanthurus hepatus	Atlantic Coast
	Triggerfish	*Balistes capriscus*	Atlantic
		Canthidermis sobaco	Florida, Cuba
	Wrasses	*Lachnolaimus maximus*	Florida, Caribbean
Tetrodotoxin (Puffer poison)			
	Puffer	*Arothron hispidus*	Tropical Seas
		Fugu rubripes nubripes	China, Japan
		Fugu vermicularis vermicularis	China, Japan
		Sphaeroids annulatus	California, Mexico
		Sphaeroides maculatus	U.S. Atlantic Coast
		Sphaenoides tesudineus	U.S. Atlantic Coast
Scombrotoxin			
	Albacore	*Thunnus alalunga*	Atlantic Ocean
	Bonito	*Sarda sarda*	Warm Atlantic Ocean
	Mackerel, Atlantic	*Scomber scombrus*	Warm Atlantic Ocean
	Mackerel, Pacific	*Scomber japanicus*	Warm Seas
	Skipjack	*Euthynnus pelamis*	Tropical Seas
	Tuna	*Thunnus thynnus*	All Warm Seas
Clupleotoxin			
	Anchovy	*Engraulis ringens*	Peru, Chile
	Herring	*Opisthonema olginum*	Caribbean to Cape Cod
	Sardine	*Harengula ovalis*	Indo-Pacific
	Sardine	*Sardinella perforata*	Malaya
Ichthyoallyeinotoxin (*Hallucinory Mullet*)			
	Mullet	*Mugil cephalus*	Cosmopolitan
		Neomyxus chaptalli	Indo-Pacific
Turtle, Marine			
	Green Sea Turtle	*Chelonia mydas*	All Tropical Seas
	Hawksbill Turtle	*Eretmochelys imbricata*	All Tropical Seas
	Leatherback Turtle	*Dermochelys coriacea*	Tropical Seas

(Continued)

TABLE 18.1.   (Continued)

Poison	Common Name	Species	Habitat
Vitamin A Intoxication			
	Bearded Seal (Liver)	*Erignathus barbatus*	Arctic Alaska and Arctic Eurasia
	Polar Bear (Liver)	*Thalarctos maritimus*	Arctic Ocean ice pack

equally important is the impediment to the development or expansion of commercial fisheries in many tropical areas of the world where fish can and should make a significant contribution to the food supply. Analyses of the ciguatera problem in the Caribbean and South Pacific areas have been prepared by Sylvester *et al.* (1977) and Banner (1965), respectively.

Public Health Significance

Since the early 1600's when Spanish explorers first sailed in Caribbean waters and coined the term "ciga" for a marine mollusk intoxication, it has come to mean a specific type of fish toxicity. Over the years, ciguatera poisoning has affected many people living or traveling between latitudes 35°N and 34°S. From rather inadequate records for the period 1601 to 1963, Halstead (1967) compiled data on more than 4,497 individuals who suffered from ciguatera poisoning. Of this number, there were 542 deaths, making the fatality rate 12%. Russell (1975) has reported on 35 more recent cases referred to him for consultation from a wide distribution of states ranging from Massachusetts to Hawaii. Estimates are that about 2,000 outbreaks of ciguatoxications are reported to health authorities in the South Pacific, but with few fatalities, many cases are unrecorded.

Species and Distribution

While as many as 300 species of fish have been implicated in or assumed to have caused ciguatera poisoning, relatively few species are usually involved. As shown in Table 18.1, the species considered to be most frequently associated with outbreaks include barracudas (Sphyyraenidae), jacks (Carangidae), morays (Muraenidae), parrotfishes (Saridae), sea basses (Serranidae), snappers (Lutjanidae), surgeonfishes (Acanthuridae), triggerfishes (Balistidae) and wrasses (Labridae). In an epideminological study of outbreaks in French

Polynesia, Bagnis (1968) found that over 50% of the cases could be attributed to surgeonfishes (Acanthuridae). The leading offenders in the Caribbean were listed by Sylvester *et al.* (1977) as jacks or carangids followed by snappers and groupers. These and other fish known to cause ciguatera usually inhabit shallow waters, with depths of less than 200 feet. As adults, the fish appear to be carnivorous rather than plankton feeders. Other ecological aspects are poorly understood; thus, ciguatera is described as "unpredictable" and "sporadic." Edibility of fish can change suddenly from safe to toxic or vice versa. Edibility can also vary from area to area within an archipelago.

Toxin—Sources, Isolation and Chemistry

The etiology of ciguatea is becoming unraveled. Earlier, certain algae as *Schizothnix calciola* present in areas infested with toxic fish, were implicated. More recently, Yasumoto *et al.* (1977) reported that the primary agent in ciguatera is most likely a dinoflagellate known as *Diplopsalis*. It is believed that the *Diplopsalis* attaches to algae and detritus on coral reefs. After being consumed by herbivorous organisms which are, in turn, food of predatory fish, the toxins become concentrated, and end up in edible species. In most instances, the viscera is more toxic than the flesh, the liver being the most toxic. The livers of the larger species such as barracudas and sharks should be avoided, as a high percentage may carry potent amounts of ciguatera poison. Apparently, ciguatoxin can accumulate and be stored in the visceral organs of large carnivores for many years (Yasumoto and Scheuer, 1969).

Scheuer (1970), studied and compared ciguatera toxin from a number of sources, including red snapper (*Lutjanus bohar*) and moray eel (*Gymnothorax javanicus*). All sources produced similar symptoms in mice, but each gave evidence of being a mixture of toxic factors. A light yellow oil isolated by Scheuer *et al.* (1967) from the muscle and liver of the moray eel (*Gymnothorax javanicus*) was named ciguatoxin. The eel muscle was extracted with acetone and diethyl ether and purified by multiple column chromatography on silicic acid and alumina. Following elution with chloroform-methanol, final separation was achieved by preparative thin-layer chromatography on silica gel. The resulting toxin, having an empirical formula of $C_{35}H_{65}NO_8$ had characteristics of a lipid with a nitrogenous moiety. Further study showed ciguatoxin to be a hydroxylated lipid molecule with an approximate molecular weight of 1500, which existed in the form of a viscous oil (Scheuer, 1977). The toxin is stable to heat used in cooking.

Assay

There are no distinguishing characteristics which can be used to detect ciguateric fish by appearance. Furthermore, there is no simple and rapid chemical or biological test suitable for screening very large numbers of fish. It is known that the cat and the mongoose fed raw toxic fish for assay purposes, will show characteristic ciguatorid symptoms. A mouse bioassay technique has been described by Banner *et al.* (1961) which involves I.P. injection of fish tissue extracts. Purified preparations were found to be lethal at levels of 0.4–25 μg. More recently, success has been achieved in the development of immunological assays. As the ciguatoxin molecule does not produce immunochemical reactions, it had to be coupled with protein molecules. In their radioimmunoassay, Hokama *et al.* (1977) conjugated ciguatoxin to human serum albumin. The resulting assay procedure has been reported as practical and relatively specific for detection of ciguatoxin. Further studies are being conducted to increase specificity. McFarren (1967) has described a bioassay procedure to differentiate between ciguatera poison, saxitoxin (paralytic shellfish poison), and tetrodotoxin (puffer poison). The procedure starts with the grinding or blending of the flesh of the fish to be assayed. A representative 100-gm sample is placed in a beaker with 100 ml of 0.15N HCl and boiled for 5 min. After cooling, the pH is adjusted to 2–4, the mixture made up to a volume of 200 ml and clarified by centrifuging or settling. Appropriate dilutions are determined by I.P. injection of 1 ml of the original or diluted supernatant to give death times in mice weighing approximately 20 gm each, of 5–7 min for saxitoxin, 20–25 min for tetrodotoxin and 110–360 min for ciguatera-like toxin. Three 20-gm mice are then injected with the determined dilution. Concentration of the toxin present, in terms of mouse units, is calculated from the mean death time of the three mice, using a dose response curve or table for the particular toxin. Mouse units are converted to μg poison per ml by multiplying by the proper correction factor (CF). Detailed assay procedures have been published (McFarren *et al.*, 1965; McFarren, 1967). Where careful assay is impossible, a crude estimate of safety can be made by observing the reactions of the cat, dog or mongoose after being fed suspected fish.

SYMPTOMS AND TREATMENT

Fortunately, the mortality rate, even in severe ciguatera poisoning, is rather low. It is rarely fatal. However, severely afflicted patients may take many months to recover fully. After consumption of raw, cooked or dried ciguatera-producing fish, symptoms may develop immediately or as late as 30 hours thereafter. The first symptom is usually a tingling about the lips, tongue and throat, followed by numbness. Other accompanying symptoms may include nausea, abdominal cramps, diarrhea, visual disturbances, headache and muscle pains. Usually, there is progressive weakness, which in severe cases can lead to muscle paralysis, convulsions and death.

As a neurotoxin, ciguatera disrupts the ionic balance of excitable cell membranes of the nervous and muscular systems. Aspects of the pharmacological action of ciguatoxin have been discussed by Rayner *et al.* (1972). There does not appear to be any specific treatment for human victims. The stomach is emptied, if possible. In some cases, a 10% calcium gluconate solution, given I.V., has relieved nervous symptoms. Often, tracheotomy, oxygen inhalation and I.V. administration of fluids may be necessary. In testing antidotes on mice it was found that tetrodotoxin, a more potent fish poison, had a curative action.

TETRODOTOXIN

Since the time of ancient China and Egypt, it has been known that the liver, ovaries and roe of pufferfish are poisonous. Of approximately 100 species belonging to the order Tetraodontiformes, about 80 can be toxic. The toxic fish are commonly known as puffers, blowfish, globefish and porcupine fish. In Japan, where the fish is considered a delicacy, licensed chefs prepare *fugu* dishes from the flesh and testes. The licensed handlers are required to have a knowledge of toxic species, seasonal variations in toxicity, and be skilled in proper evisceration. Nevertheless, several hundred cases of intoxications are apparently recorded in Japan each year, of which about one-half are fatal.

Species

Puffers can be recognized by their characteristic shape, large teeth and the distinctive odor of freshly dressed tissue. Their distribution in tropical seas is worldwide, and some species such as Porcupine fish. *Diodon hystrix* Linnaeus, enter temperate waters. Tetrodotoxin is, however, not limited to the puffers. It has been isolated from an unrelated fish, *Gobius criniger*, which caused sporadic poisonings in Taiwan and the Philippines (Noguchi and Haskimoto, 1973). Tetrodotoxin was found also in the eggs of the American salamander *Tapicha torosa* and the skin of Costa Rican frogs of the genus *Atelopus*, specifically *Atelopus chiriquiensis* (Wakely *et al.*, 1966; Mosher *et al.*, 1964; Kim *et al.* 1975). A list of fish known to be sources of tetrodotoxin has been given by Halstead (1967).

Tetrodotoxin concentrates in the eggs, ovaries and liver of puffer fish and newts. Skin, intestine and muscle contain lower but variable concentrations, depending upon the species and the time of the year. During fall and winter when the ovaries increase in size, the amount of toxin in the eggs, ovaries and liver increases.

Chemistry and Toxicity

The story of the characterization of the chemical structure of tetrodotoxin in four different laboratories by 1964 and its synthesis in 1972 has been told by Scheuer (1977). Tetrodotoxin has a cyclic hemilactal structure with a molecular formula of $C_{11}H_{17}N_3O_8$ (Fig. 18.1). The crystalline toxin is practically insoluble in water and most organic solvents. It is quite stable to heat or autoclaving. Partial or complete inactivation occurs in acid solution at pH 4 or below, or in alkaline solution at pH 9 or above. Thus, ordinary cooking or commercial canning procedures cannot be expected to sufficiently inactivate the toxin to produce a safe product.

Along with man, all vertebrates, except those which produce tetrodotoxin, are affected by the toxin. It is one of the most potent of all natural poisons. The LD_{50} by I.P. injection in mice has been given as 11 μg/kg, with the oral LD_{50} approximately 20 times as much. Intravenous administration of doses as low as 1–5 μg/kg of crystalline tetrodotoxin to experimental animals quickly produced weakening of muscle contractions, depression of respiration and hypotension.

In man, the fatal dose is estimated to be equivalent to approximately 1 to 2 mg of crystalline tetrodotoxin. Usually, consumption of 10 gm or more of puffer roe would be required to be fatal. However, with some highly toxic species, during the winter season, as little as a gram of ovary may produce fatal poisoning.

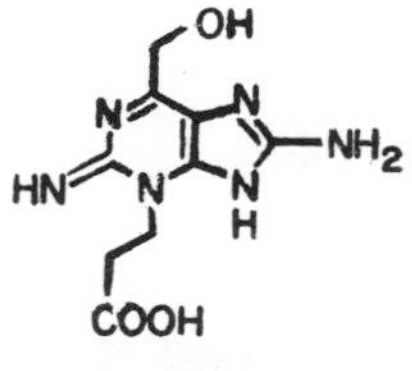

FIG. 18.1.   TETRODOTOXIN

Assay

Due to the lack of specific chemical methods, tetrodotoxin is detected by bioassay. In the procedure described by Wakely *et al.* (1966), the test is based on the relationship between the time required for death after I.P. injection into mice and the dose. Specificity was increased by dialysis of weakly acidic crude extracts against water, and concentration of the aqueous extract. The dose is adjusted to give a death time between 1 to 15 minutes. Sensitivity of the method is sufficient to detect a level of about 0.5 μg of tetrodotoxin.

Symptoms and Treatment

Within one hour after consumption of tetrodotoxin, toxic symptoms appear in recognizable stages. The first symptom is numbness of the lips, tongue, and often of fingers. These symptoms are accompanied by nausea and vomiting. In the second stage, the numbness becomes more marked, and there is muscular paralysis of the extremities. If prognosis is unfavorable, severe ataxia and paralysis develops with loss of consciousness. Finally, death results from respiratory paralysis.

The physiological mechanism of tetrodotoxin action is to inhibit the passage of an impulse, a blocking of propagated action potentials in nerves. It does this by selectively blocking the early transient increase in cell membrane permeability to potassium ions. The blocking of nerve excitation by tetrodotoxin is similar to that obtained with saxitoxin (paralytic shellfish poison) which has a different chemical structure. Tetrodotoxin also has a blocking effect on skeletal muscle fibers, causing the weakening of voluntary muscles that precedes paralysis in severe intoxication. The pharmacology and mechanism of action of tetrodotoxin and saxitoxin have been reviewed by Kao (1972) and Narahashi (1972).

The treatment of tetrodotoxin poisoning consists of trying to remove the toxin from the gastrointestinal tract by such means as gastric lavage and catharsis. Measures are taken to support respiration and prevent circulatory failure.

SCOMBROID POISONING

Scombroid poisoning, also referred to as histamine toxicity, can result from ingestion of scombroid species of fish which are usually edible. The toxicity is, most often, neither fatal nor severe. The etiology, although not fully understood, is related to the large levels of free histidine, an amino acid, normally contained in the muscle tissue of scombroid fish. Under certain conditions, such as storage at too high temperatures, certain bacteria may decarboxylate the free histidine to produce toxic levels of histamine. However, there is uncertainty as to whether histamine is the only factor involved in scombroid toxicity.

Public Health Significance

Outbreaks of scombroid poisoning have occurred following consumption of "fresh," canned, salted and dried fish. Since most cases produce symptoms which are mild and of rather short duration, many outbreaks are unreported. The comprehensive review of histamine toxicity from fish products by Arnold and Brown (1978) documents a number of outbreaks. Since the U.S. Center for Disease Control began its foodborne surveillance program, a total of 30 outbreaks of scombroid-type fish poisonings have been listed for the 1966–1975 period. However, a 1973 outbreak attributed to canned tuna afflicted 254 people in 8 states. As may be expected because of a larger per capita consumption of fish, outbreaks have been more common in Japan. For example, during 1951–1954, there were 14 reported scombroid toxicity outbreaks involving 1215 individual cases (Arnold and Brown, 1978).

Species

Scombroid fish belong to the families Scomberesocidae and Scombridae which normally contain free histidine in their flesh. Typical species are the tuna (*Thunnus thynnus*), skipjack tuna (*Katsuwonus pelamis*), mackerel pike (*Cololabis saira*), mackerel (*Scomber japonicus*), sardines (*Sardinops sagax*) albacore (*Thunnus alalunga*), and bonito (*Sarda chiliensis*). Toxicity has also been reported from eating mahi-mahi (*Coryphaena hippurus*). Dolphins belong to the family Coryphaenidae, but like the Scombridae, contain large amounts of histidine in their tissues.

Chemistry and Biology

Along with others, Geiger *et al.* (1944) and Geiger (1944a, 1944b, 1955) have made considerable contributions to our understanding of the mechanism of histamine formation in scombroid fish. It is now known that histamine formation depends on the free histidine content of fish flesh and the histidine decarboxylase activity of normal and contaminant bacteria. The microorganisms implicated as histamine formers are *Proteus* species including *Proteus morganii*, *Proteus vulgaris* and *Proteus mirabilis*. Activity of bacterial histidine decarboxylases was found to be greatest at temperatures below 30°C, with an optimum temperature of 20°C–25°C. Little histamine was formed at 40°C, and histamine-forming bacteria were killed at 60°C.

At low temperatures, there are conflicting reports on the ability of microorganisms to produce histamine. It is generally accepted, however, as shown by Edmunds and Eitenmiller (1975) in Spanish mackerel and other fish that little bacterial histamine is formed at near-freezing or freezing temperatures. Another factor influencing histidine decarboxylase activity is an optimum pH range of 2.5–6.5. Since the pH of fresh scombroid fish muscle ranges from pH 5.5–6.5, the slightly acidic environment is inducive for histamine production by bacteria capable of decarboxylating histidine.

Assay

The classical bioassay based on contraction of guinea pig ileum has largely been replaced by fluorometric procedures. A review of fluorometric methods for histamine has been made by Shore (1971). Histamine can be determined also by colorimetry, gas-liquid chromatography and thin-layer chromatography, but there is a search for simple and rapid procedures (Lerke and Bell, 1976).

From a practical viewpoint, the production of histamine is not considered an index of quality deterioration, until levels in excess of 10 mg of histamine per 100 gm of fish are reached. Toxic samples implicated in scombroid poisoning have contained from 100 mg% (100 mg free base of histamine per 100 gm fish flesh) to 4000 mg%. While the 100 mg% level of histamine or above has been generally associated with illness, a limit of 10 mg of histamine-like substances per 100 gm tissue has been suggested as a quality standard for regulatory purposes. Current quality control procedures employed by processors of scombroid fish include physical inspection of raw and precooked fish and histamine analysis of individual lots of canned products.

Symptoms and Treatment

Poisonous scombroid fish flesh often has a sharp or pepper-like taste. After consumption, symptoms characteristic of a severe allergy may develop from a few minutes to three hours. Typical symptoms are intense headache, dizziness, throbbing of the large vessels of the neck, accompanied by a "hot" flushing of the face and neck. The flushing results from the dilating action of histamine on blood vessels. There may be dryness of the mouth, thirst, difficulty in swallowing, palpitation of the heart and gastrointestinal distress. The victim may develop a rash with intense itching. In severe cases, there is a possible danger from shock. Generally, however, the acute phase lasts only 8–12 hours, followed by complete recovery (Halstead, 1967).

Treatment consists of evacuation of the stomach and catharsis. Anti-histamine drugs are also administered. As for prevention, scombroid fish should be consumed soon after capture or promptly canned or frozen. It is dangerous to consume scombroids which have been allowed to stand at tropical temperatures or show evidence of not being absolutely fresh.

Histamine as a Cause of Scombroid Poisoning

Reports that large amounts of histamine can be administered orally to man without effect has cast doubt on the theory that histamine alone, causes scrombroid toxicity (Douglas, 1970). At the same time, I.V. injection of as little as 0.007 mg of histamine base elicited changes in facial blood vessels and cardiac rate, thus indicating the presence of an efficient mechanism to block intestinal absorption and entry into the circulating blood (Weiss et al. 1932). Thus, attention has focused on the presence in tuna and other fish, and on the role of such compounds as anserine and carnosine. A recent report by Bjeldanes et al. (1978) concludes that it is likely that cadaverine, along with histamine, is of importance in the aetiology of scrombroid poisoning.

OTHER FISH TOXINS

Clupeoid Poisoning

Clupeotoxin takes its name from tropical sardines or herrings belonging to the genus *Clupea*, but there is reason to believe that it may be the same as ciguatoxin. The evidence for considering clupeotoxin as a distinct toxin is based on the observation that clupeotoxic fish appear to exist only during certain months of the year. Halstead (1967) has compiled a list of over 181 cases resulting in more than 77 fatalities for the period from 1741–1962. The species involved include *Clupea thrissa, Clupea harengus, Sardinella perforata* and *Sardinella longiceps*, all from tropical waters. Sporadically and unpredictably, it can affect the herring *Opisththonema olginum* in the Caribbean and the Peruvian anchovy, *Engraulis ringens*. Ingestion of toxin-laden fish brings about rapid and severe symptoms. A sharp metallic taste and dryness of the mouth are followed by nausea, vomiting, diarrhea, abdominal pain and weakness. In severe cases, death may occur from cardiovascular collapse as soon as 20 minutes after ingestion. Little is known about the clupeotoxin chemistry except that the toxin is not destroyed by cooking temperatures.

Hallucinogenic Mullet Poisoning (Ichthyoallyeinotoxin)

Over the years, a number of reports have indicated that eating the flesh of members of the Mugilidae and Mullidae families, from certain Pacific locations, can result in hallucinations. These reports do not appear to have been verified by carefully controlled scientific testing (Helfrich and Banner, 1960).

Marine Turtle Poisons

At times, for unknown reasons, certain species of marine turtles may become dangerous to consume. Affected species reported as poisonous are Green Sea Turtle (*Chelonia mydas*); Hawksbill Turtle (*Eretmochelys imbricata*) and Leatherback Turtle (*Dermochelys coriacea*). After ingestion, symptoms develop within a few hours to several days. The victim usually experiences nausea, vomiting, diarrhea and abdominal pain. A dry burning sensation of the lips and tongue may be followed by a white coating and ulceration of the tongue. If severely poisoned, the subject may lapse into sleep and death. Death rate is about 44% of those intoxicated. Thus, turtles native to the Indo-Pacific region, and especially turtle liver, should be consumed only with caution (Halstead, 1970).

Vitamin A Intoxication

Numerous stories have been told about poisonings from eating the liver and kidneys of polar bears (*Thalarctos maritimus*) and/or the bearded seal (*Erignathus barbatus*). Toxicity is believed to be due to the high concentration of vitamin A in the livers. Rodahl and Moore (1943) and Rodahl (1949), have reported finding levels of over 20,000 I.U. per gram of liver. After a single ingestion of 1 to 3 million or more I.U. of vitamin A, the symptoms appearing 2 to 12 hours after ingestion, are throbbing headaches, nausea, vomiting, diarrhea, abdominal pain, drowsiness and convulsions. Physiologically, there is a rise in cerebrospinal fluid pressure. The affliction is rarely fatal, and recovery usually occurs within a week (Halstead, 1970).

PARALYTIC SHELLFISH POISONING

Paralytic shellfish poisoning is caused by certain mollusks which have ingested toxic dinoflagellates and which are subsequently eaten by man. The causative organisms are members of the order Dinoflagellata. Dinoflagellates are widely distributed throughout neritic waters and in the high seas, from the polar oceans to the tropics. "Blooms" of toxic dinoflagellates sometimes occur during weather disturbances, or under certain other conditions, and result in the phenomenon frequently referred to as "red water," "red tide," or "brown water." However, blooms may appear yellowish, greenish, bluish, or even milky in color, depending on the protistan involved and a number of ecological factors. When excessive numbers of these animals collect, there may be a mass mortality of fishes in the area (Brongersma-Sanders, 1957; Hughes and Merson, 1976).

The cause of the mass mortality of marine life during and following plankton blooms has been the subject of considerable discussion. Among the factors that have been implicated are: oxygen depletion in the water, due either to the number of plankton present or to the release of decay products by these organisms and the dying fish; asphyxiation through a blanketing of the fish with a mass of plankton; or the production of a toxin by the protistan. It is known that the early larvae of certain mollusks and crustaceans that respire through their body walls are not affected by blooms, while those forms with specialized respiratory gills die when the organs become covered with the disintegrated bodies of the plankton (Motoda, 1944). The loss of many tons of food fish each year can be attributed to these blooms of dinoflagellates.

However, the most serious food problem relating to the protistan is paralytic shellfish poisoning. The relationship between blooms of plankton and shellfish poisoning is well established (Sommer et al. 1937; Needler, 1949), although most early works attributed the poisoning to other causes: copper salts, putrefaction processes, diseases of the shellfish, a "virus," other toxic marine animals, contaminated water, industrial wastes, and bacterial pathogens.

Sommer and Meyer (1937), published the results of their intensive investigation of paralytic shellfish poisoning. They demonstrated a direct relationship between the number of *Gonyaulax catenella* in the sea water and the degree of toxicity in the mussel *Mytilus californianus*. These workers also established methods for extracting and assaying the poison and suggested an experimental and clinical approach to the problem that has served as a guide for subsequent workers in the field. Much of this work and that of subsequent investigators has been done with *Gonyaulax catenella*, which is widely distributed along the Pacific coast of North America, *Gonyaulax tamarensis* (or *G. excavuta*) found along the Atlantic coast of North America and *Pyrolidinum bahamense* in the South Pacific waters.

It is not known how the protistan poison accumulates or concentrates in the mollusk. It can be stored in certain organs of the animal without deleterious effects. The site of concentration of the toxin may vary with the different species of shellfish, with the different seasons of the year, and with certain other factors. In most instances, the poison is concentrated in the digestive glands of the mollusk. In several claims, the toxin may accumulate in the gills, while in at least one clam, *Saxidomus*, it is found in the siphons. Table 18.2 lists most of the molluscan species of the Western Hemisphere that have been reported to transvect dinoflagellate poison.

TABLE 18.2. MOLLUSCA OF WESTERN HEMISPHERE IMPLICATED IN PARALYTIC SHELLFISH POISONING

Species	Distribution
Mopalia muscosa	Pacific Coast of North America
Acmaea pelta	Pacific Coast of North America
Donax denticulatus	West Indies
Mactra solidissima	Atlantic Coast of North America
Schizothaerus nutalli	Pacific Coast of North America
Myaarenaria	Atlantic Coast of North America, Alaska, British Columbia, Oregon, and Northern California
Mytilus californianus	Aleutian Islands East and South to Socorro Islands
Mytilus edulis	World-wide
Modiolus demissus	Virginia to Florida
Crassostrea gigas	Pacific Northwest
Placopecten magellanicus	Labrador to North Carolina
Penitella penita	Pacific Coast of North America
Ensis directus	Canada to Florida
Siliqua patula	Alaska to Central California
Spondylus americanus	North Carolina to West Indies and Gulf of Mexico
Macoma nasuta	Kodiak Island to Baja California
Protothaca staminea	Aleutian Islands to Baja California
Saxidomus giganteus	Alaska to Central California
Saxidomus nuttalli	Northern California to Baja California

A nonparalytic type of shellfish poisoning following ingestion of certain clams, oysters, and gastropods has been reported in Japan and is thought to be due to a toxic plankton, but similar episodes have not been reported in the Western Hemisphere.

Chemistry and Toxicology

A mass poisoning during 1885 was the impetus for studies on the chemistry of shellfish toxin (Salkowski, 1885). Since then, a number of investigations on the chemistry of Gonyaulax, preparations of the mussel *Mytilus californianus*, and the Alaskan butter clam *Saxidomas giganteus* have been carried out. These studies indicate that most of the physical, chemical, and gross physiopharmacological properties of the three toxins are identical (Table 18.3).

TABLE 18.3. SOME CHEMICAL PROPERTIES OF MUSSEL, CLAM, AND *GONYAULAX CATENELLA* POISONS

Property	Mussel Poison	Clam Poison	*Gonyaulax catenella* Poison
Molecular formula	$C_{10}H_{17}N_7O_4 \cdot 2HCl$	$C_{10}H_{17}N_7O_4 \cdot 2HCl$	$C_{10}H_{17}N_7O_4 \cdot 2HCl$
Molecular weight	372	372	372
N content	25.9	26.8	26.1
Diffusion coefficient	4.9×10^{-6}	4.9×10^{-6}	4.8×10^{-6}
Specific optical rotation	$130° \pm 5$	$130° \pm 5$	128
pKa	8.3, 11.5	8.3, 11.5	8.2, 11.5
Molecular extinction of oxidation product	6,000–7,000 At 235 and 333 rm		
Solubility	Soluble in water, less so in alcohol; insoluble in lipid		
Form	Two tautomers		
Acid	Stable		
Base	Stable		
Sakaguchi test	Negative	Negative	Negative
Benedict-Behre test	Positive	Positive	Positive
Jaffe test	Positive	Positive	Positive
Ring structure	Present	Present	
Aromatic structures	None		
Carbonyl groups	None		
Reduction with H_2	Nontoxic	Nontoxic	Nontoxic

The Structure of Saxitoxin is:

Saxitoxin is one of the most lethal biological toxins known. The human appears to be twice as susceptible as the mouse. The intraperitoneal minimal lethal dose of the toxin for the mouse is approximately 9.0 µg./kg. body weight. The minimal lethal oral dose for man is thought to be between 1.0 and 4.0 mg.

The toxin is absorbed slowly from the gastrointestinal tract and is excreted rapidly by the kidneys. It depresses respiration, the cardio-inhibitory center, and conduction in the myocardium (Prinzmetal, 1932). Work by Evans (1967) indicates that the toxin has a dual paralyzing action. It has a direct effect on skeletal muscle, blocking the muscle action potentials without depolarizing the cells. It also abolishes conduction in peripheral nerves, but has no curare-like action at the neuromuscular junction. Evans' work indicates that the poison exerts its deleterious effects by blocking peripheral nerve and muscle, rather than activity in the medullary respiratory center, as previously believed. The basic mechanism of this change involves alterations in sodium conductance.

Saxitoxin is the neurotoxin which has been isolated from toxic butter clams, toxic mussels, axenic cultures of *Gonyaulax catenella* and in aged extracts of scallops collected during a bloom of *G. tamarensis*. However, recent works, have shown that several toxins are actually responsible for most paralytic shellfish poisonings (Shimizu, 1978). For example, the Atlantic causative organism, *Gonyaulax tamarensis*, was found to contain gonyautoxin - II, - III and newsaxitoxin as major toxins (Shimizu, *et al* 1975, and Oshima, *et al*, 1977). Similarly, toxic mussels from Alaskan waters afforded gonyautoxin - I, - II, - III, - IV, and a small amount of saxitoxin or neosaxitoxin (Shimizu, *et al*, 1978). Up to date, eight new toxins (gonyautoxin - I - VII and neosaxitoxin) have been isolated from various organisms (Shimizu, 1979).

All the newly isolated toxins seem to be closely related to saxitoxin. The structures of gonyautoxin - II and neosaxitoxin have been proposed as 11 - hydroxysaxitoxin sulfate and 1 - hydroxysaxitoxin, respectively. (Shimizu, *et al*, 1976; Boyer, *et al*, 1978 and Shimizu, *et al*, 1978). The pharmacological properties are also similar to that of saxitoxin.

Clinical Problem

Three types of shellfish poisoning are recognized: *gastrointestinal*, *allergic*, and *paralytic*.

Gastrointestinal shellfish poisoning is characterized by nausea, vomiting, abdominal pain, weakness, and diarrhea. The onset of symptoms generally occurs 8 to 12 hr. following ingestion of the offending mollusc. This type of poisoning is caused by bacterial pathogens and is usually limited to gastrointestinal symptoms and signs. It rarely persists for more than 48 hr.

Allergic or erythematous shellfish poisoning may vary from one individual to another. The onset of symptoms and signs occurs 30 min. to 6 hr. after ingestion of the mollusc to which the individual is sensitive. The usual presenting manifestations are diffuse erythema, swelling, urticaria, and pruritus, involving the head and neck and then spreading to the body. Headache, flushing, epigastric distress, and nausea are occasional complaints. In the more severe cases, generalized edema, severe pruritus, swelling of the tongue and throat, respiratory distress, and vomiting sometimes occur. Death is rare, but persons with a known sensitivity to shellfish should avoid eating them.

Paralytic shellfish poisoning is known variously as gonyaulax poisoning, paresthetic shellfish poisoning, or mussel poisoning. Symptoms usually develop within 30 min. following ingestion of the offending mollusc or crab. Paresthesia, described as tingling, numbness, or burning, is noted first about the mouth, lips, and tongue; it then spreads over the face, scalp, and neck, and to the finger tips and toes. Sensory perception and proprioception are affected to the point that the individual moves incoordinately and in a manner similar to that seen in another, more common form of intoxication. Ataxia, incoherent speech, or aphonia are prominent signs in severe poisonings. The case fatality rate varies from 1 to 10%.

In fatal cases, death usually occurs within 12 hours as a result of respiratory paralysis and cardiovascular collapse.

Prevention and treatment of paralytic shellfish poisoning have received much attention. Attempts to find an antidote have not been successful and treatment is mainly symptomatic. Apomorphine is thought to be more effective than lavage in removing pieces of shellfish from the stomach. Lloyd's reagent (adsorbent charcoal) may be administered to adsorb the poison. Emptying of the bowels and removal of the toxin from the intestinal tract are important and so a catharic or an edema are effective treatments. Vomiting, if not voluntary, should be induced. Artificial respiration should be administered promptly.

Avoidance of paralytic shellfish poisoning can be effected to a great extent if the gatherer and/or consumer have (has) knowledge of the areas where toxic shellfish occur, what species are involved, what portions or parts of the fish are hazardous and the effects of cooking and other methods of food preservation on toxin stability or degradation. Public education will do much to disseminate this type of information.

Government agencies like the Food and Drug Administration and the National Shellfish Sanitation Program units have strong and effective means to stop the sale and distribution of shellfish from contaminated areas. Contaminated areas are closed to fish and strict surveillance is done along with periodic assay of shellfish from the affected areas.

A comprehensive discussion of symptomatology, treatment, prevention and other pertinent aspects, is provided by Prakash *et al.* (1971).

Assay Methods for Paralytic Shellfish Toxins

In order to detect and determine, quantitatively, the toxins involved in paralytic shellfish poisons (PSP), assay methods have been developed and/or proposed. By far the best known, most used and universally preferred, method is the mouse bioassay test.

Several chemical and serological tests have been known for a long time but, because of their complex or unreliable nature, they have not gained wide acceptance. Within the last few years, several methods based on chemical or chemico-physical approaches have been advanced and at least one chemical method has been promoted as being of promise for the assay of PSP toxin(s).

The Mouse Bioassay Method

In this method, mice are injected intraperitoneally with an acidic-aqueous extract of the shellfish and the time from the moment of injection to death is noted.

Somner and Meyer in 1937, identified *Gonyaulax catenella* as the source of the paralytic shellfish poison (PSP) found in California mussels and developed the first practical bioassay for the toxin. These workers defined the *mouse unit* as "the minimum amount of poison which is required to kill a 20-gram mouse in 15 minutes when 1 ml of an acid-aqueous shellfish extract is injected intraperitoneally into the animal." The time to death is defined as "the time from injection to the last gasping breath of the mouse" (Prakash *et al.*, 1971). The time is measured to the nearest 5 seconds. These workers reported toxicity in terms of the number of mouse units in 100 grams of shellfish flesh.

Several improvements and/or modifications of the original method were made by Canadian workers, as noted by Prakash *et al.* (1971) and, finally, Schantz *et al.* (1958) purified the shellfish poison and used it as a reference standard. The use of purified toxin as a standard permitted the results to be expressed in terms of the amount (weight in micrograms) of poison instead of in mouse units. Investigations had shown that poisons isolated from shellfish from different geographical areas such as California, Alaska, British Columbia and the Canadian Atlantic Coast, as well as from dino-flagellates of these regions exhibited chemical and physical simi-larities.

Various laboratories have carried out collaborative studies on the bioassay method and good agreement was obtained (McFarren, 1959). As a consequence, the bioassay method was published in 1960 as an official method of the Association of Official Agricultural Chemists (A.O.A.C., 1970).

In the actual test, each mouse is injected intraperitoneally with 1 ml of an acid extract prepared from clams, oysters or mussels by extracting 100 grams of the properly prepared material with 0.1N hydrochloric acid. The time of inoculation is noted carefully and the time of death, as indicated by the last gasping breath, is determined by observing the mice carefully. A good stop watch is recommended for noting times. It is recommended to use 2 or 3 mice per deter-mination and a dilution of the extract which will result in death times of 5–7 minutes. Animals weighing approximately 20 grams each are used.

To calculate the toxicity of the sample, the median death times of the mice are determined and with the aid of a table, the corres-ponding mouse units are determined. If the test animals weigh less than 19 grams or more than 21 grams, a correction for each mouse must be made. This is done by multiplying the mouse units corres-ponding to the death time for that mouse by the weight correction factor for the said mouse. The correction factor is obtained from a table; then, the median mouse unit for the group is determined. The death time of survivors is taken as >60 min or equivalent to <0.875 mouse units in calculating the median. Finally, mouse units are converted to micrograms of poison.

micrograms of poison per 100 grams of meat
= (micrograms per ml × dilution factor) × 200

If the value obtained is greater than 80 micrograms per 100 grams, the material should be considered as hazardous and unsafe for human consumption. The mouse bioassay test is the most widely used one because it is simple to perform and does not require very expensive equipment. Moreover, the cost of mice is relatively low.

Several factors affect the accuracy of the test. These are discussed by Prakash *et al.* (1971). Female mice were found to be slightly more susceptible to the poison than male mice and the intraperitoneal toxicity of the poison was reduced if the salt concentration was greater than 0.1M. It was estimated by Schantz (1960) that, if the assay solution injected into the mice had a 1% concentration of sodium chloride, the "death time" could be increased enough to lower the assay values of the poison by as much as 50%. The strain of mice used is important, also, and proper standardization of the mouse strain is a critical preliminary step. Other possible causes of variations include, individual differences even in a standardized colony, varying sensitivity depending on the time of the day the injection is done, variations due to age, weight and sex, as mentioned previously, and errors in the technique of intraperitoneal injection. Schantz *et al.* (1958) and McFarren (1959), have estimated the accuracy of the standard test as ± 20%. However, it is thought that in bioassays of shellfish of low toxicity (80 μg/100 gram), the actual toxicity may be underestimated by as much as 60%. Such lowered estimates probably result because an undiluted extract has to be injected into the mice and such an extract would have a high salt concentration. Much more accurate results are obtained when the toxicity is high and so allows for greater dilution of the extract, thus lowering the salt concentration.

Median times of 5–7 minutes are preferred for the most accurate results. Since, occasionally, some animals might survive, median death times instead of average death times are used. Details of the procedure are given in the Recommended Procedures for the Examination of Sea Water and Shellfish, American Public Health Association (1970) and by the A.O.A.C. (1975). These references should be consulted by anyone requiring more information since only a cursory indication of methodology is intended in this presentation.

Chemical Methods

Of the chemical methods used to measure paralytic shellfish poison, the one proposed by Bates and Rapoport (1975) and later modified by Bates *et al.* (1978) seems promising. It involves the oxidation of saxitoxin, the paralytic shellfish poison, by H_2O_2 to produce a fluorescent purine (Fig. 18.2). The concentration of the fluorescent purine may be determined by ultraviolet absorbance or fluorescence.

Essentially, the method involves the following: the shellfish is extracted with aqueous trichloroacetic acid, neutralized and centrifuged. The supernatant is placed on a properly prepared Bio-Rex 70 (50–100 mesh) ion-exchange resin. The column is eluted (washed) with 30 ml of water and 1 ml of 0.24M H_2SO_4 and the eluents (washings) are discarded. The column is then eluted (washed) with 3.9 ml of 0.25M H_2SO_4 and the eluent collected in a centrifuge tube. The material is mixed and divided into two equal portions in centrifuge tubes. To one portion, 2 ml of 1.3M NaOH and 0.05 ml of 10% H_2O_2 are added and the ingredients mixed well. To the second portion, water, instead of H_2O_2 is added. After centrifugation at 1000G for 1 minute, the supernatants are transferred to cuvettes. Forty minutes after the addition of the H_2O_2, the mixture is adjusted to pH 5.0 by the addition of 0.16 ml of glacial acetic acid. Finally, the fluorescence of the two portions is measured at 380 nm. The fluorescence of the control (portion without H_2O_2) is subtracted from that of the oxidized portion (portion containing H_2O_2). Toxin concentration is determined from a suitable standard. According to the authors, this method constitutes a sensitive chemical assay for saxitoxin and is superior to the mouse bioassay in several respects. They noted further that the method is being considered for routine analysis of West Coast shellfish samples.

A recent article by Shimizu and Ragelis (1978) discusses alternates to the mouse bioassay.

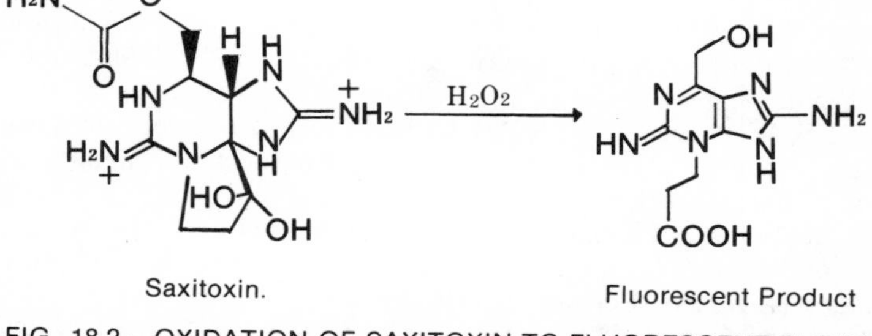

Saxitoxin.   Fluorescent Product

FIG. 18.2. OXIDATION OF SAXITOXIN TO FLUORESCENT PURINE COMPOUND.

REFERENCES

AMERICAN PUBLIC HEALTH ASSOCIATION, INC. 1970. Recommended Procedures for the Examination of Sea Water and Shellfish. 4th Ed. American Public Health Assn., Washington, D.C. 20036.

ARNOLD, S. H. and BROWN, W. D. 1978. Histamine toxicity for fish products. In. "Advances in Food Research." C. O. Chichester, E. Mrak and G. Stewart, Editors. Vol. 24, Academic Press, New York.

ASSOCIATION OF OFFICIAL AGRICULTURAL CHEMISTS. Paralytic Shellfish Poison, Biological Method (20). *Official Methods of Analysis of the Association of Official Agricultural Chemists* (10th ed.) Washington, D.C., 1970, pp. 305–307.

BAGNIS, R. 1968. Clinical aspects of ciguatera fish poisoning in French Polynesia. Hawaii Nurs. Bull. 28:25.

BANNER, A. H. 1965. Ciguatera in the Pacific. 1965. Hawaii Med. J. 24:353.

BANNER, A. H., SASAKI, S., HELFRICH, P., ALENDER, C., and SCHEUER, P. 1961. Bioassay of ciguatoxin. Nature 189:229.

BATES, H. A., and RAPOPORT, H. 1975. A chemical assay for saxitoxin, the paralytic shellfish poison. J. Agr. Food Chem. 23:237–239.

BATES, H. A., KOSTRIKEN, R., and RAPOPORT, H. 1978. A chemical assay for saxitoxin. Improvements and modifications. J. Agr. Food Chem. 26:252–254.

BJELDANES, L. F., SCHUTZ, D. E. and MORRIS, M. M. 1978. On the aetiology of scombroid poisoning. Cadaverine potentiation of histamine toxicity in the guinea pig. Food Cosmetic Toxicol. 16:157.

BOYER, G. L., SCHANTZ, E. J., and SCHNOES, H. K., 1978. Characterization of 11-Hydroxysaxitoxin sulfate; a major toxin in scallops exposed to blooms of the poisonous dinoflagellate *Gonyaulax tamarensis*. Chem. Commun. 889–890.

BRONGERSMA-SANDERS, M. 1957. Mass mortality in the sea. *In Treatise on Marine Ecology and Paleoecology*, J. W. Hedgpath (Editor). Waverly Press, Baltimore.

DOUGLAS, W. W. 1970. Histamine and antihistamines:5-hydroxytryptamine and antagonists. In "The Pharmacological Basis of Therapeutics." L. S. Goodman and A. G. Gilman, eds., 4th Ed., pp. 621–662. Macmillan, New York.

EDMUNDS, W. J. and EITENMILLER, R. R. 1975. Effect of storage time and temperature on histamine content and histidine decarboxylase activity of aquatic species. J. Food Sci. 40:516.

EVANS, M. H. 1967. Block of sensory nerve conduction in the cat by mussel poison and tetrodotoxin. *In* Animal Toxins, F. E. Russell and P. R. Saunders (Editors). Pergamon, London.

FISH, C. J. and COBB, M. C. 1954. Noxious Marine Animals of the Central and Western Pacific Ocean. Research Report No. 36, U.S. Fish and Wildlife Service.

GAILLARD, C. 1923. Recherches sur les poissons representes dans quelques tombeaux Egyptiens de l'ancien empire. Mem. Inst. Francais Arch. Orient 51:97.

GEIGER, E. 1944a. Histamine content of unprocessed and canned fish. A tentative method for quantitative determination of spoilage. Food Res. 9:293.

GEIGER, E. 1944b. On the specificity of bacterium decarboxylase. Proc. Soc. Exp. Biol. Med. 55:11.

GEIGER, E. 1955. Role of histamine in poisoning with spoiled fish. Science 121:865.

GEIGER, E., COURTNEY, G. and SCHMAKENBERG, G. 1944. The content and formation of histamine in fish muscle. Arch. Biochem. 3:311.

HALSTEAD, B. W. 1965 (Vol. I), 1967 (Vol II), 1970 (Vol III). Poisonous and Venomous Marine Animals of the World. U.S. Govt. Printing Office, Washington, D.C.

HELFRICH, P. and BANNER, A. H. 1960. Hallucinatory mullet poisoning. J. Trop. Med. Hyg. 1960:1.

HOKAMA, Y., BANNER, A. H. and BOYLAN, D. B. 1977. A radio-immunoassay for the detection of ciguatoxin. Toxicon 15:317.

HUGHES, J. M., and MERSON, M. H. 1976. Current concepts-fish and shellfish poisoning. The New England Jour. of Medicine, 295, 1117.

KAO, C. Y. 1972 (Suppl.). Pharmacology of tetrodotoxin and saxitoxin. Federation Proc. 31:1117.

KIM, Y. H., BROWN, G. B., MOSHER, H. S. and FUHRMAN, F. A. 1975. Tetrodotoxin: Occurrence in Atelopid Frogs of Costa Rica. Science 189: 151.

LERKE, P. A. and BELL, L. D. 1976. A rapid fluormetric method for the determination of histamine in canned tuna fish. J. Food Sci. 41:1282.

McFARREN, E. F. 1959. Report on collaborative studies of the bioassay for paralytic shellfish poison. J. A. Offic. Agr. Chem. 42:263.

McFARREN, E. F. 1967. Differentiation of the poisons of fish, shellfish and plankton. In Russell, F. E. and Saunders, P. E., Ed. Animal Toxins. Pergamon Press, Oxford, England.

McFARREN, E. F., TANABE, H., SILVA, F. J., WILSON, W. B., CAMP-BELL, J. E. and LEWIS, K. H. 1965. The occurrence of a ciguatera-like poison in oysters, clams and Gymnodinium breve cultures. Toxicon 3:111.

MOSHER, H. S., FUHRMAN, F. A., BUCHWALD, H. D. and FISCHER, H. G. 1964. Tarichatoxin-Tetrodotoxin: A Potent Neurotoxin. Science 144:1100.

MOTODA, S. 1944. Sea and Plankton. Kawade Shobo Publ., Tokyo.

NARAHASHI, T. 1972 (Suppl.). Mechanism of action of tetrodotoxin and saxitoxin in excitable membranes. Federation Proc. 31:1124.

NEEDLER, A. B. 1949. Paralytic shellfish poisoning and Gonyaulax tamarensis. J. Fish. Res. Bd. Can. 7, 490–504.

NOGUCHI, T. and HASHIMOTO, F. 1973. Isolation of tetrodotoxin from a goby Gobius criniger. Toxicon 11:305.

OSHIMA, Y., BUCKLEY, L. J., ALAM, M. and SHIMIZU, Y., 1977.

Heterogeneity of paralytic shellfish poisons. Three new toxins from cultured *Gonyaulax tamarensis* cells, *Mya arenaria* and *saxidomus giganteus*. Comp. Biochem. Physiol. 56:31–34.

PRAKASH, A., MEDCOF, J. C. and TENNANT, A. D. 1971. Paralytic shellfish poisoning in eastern Canada. Bulletin 177. Fisheries Research Board of Canada. Ottawa, Canada. 87 pages.

PRINZMETAL, M., SOMMER, H., and LEAKE, C. D. 1932. The pharmacological action of "mussel poison." J. Pharmacol. Exptl. Therap. *46*, 63–74.

RAYNER, M. D. 1972. Suppl. Mode of action of ciguatoxin. Federation Proc. 31:1139.

READ, B. E. 1939. Chinese materia medica. Fish drugs. Peiking Nat. Hist. Bull. 136:190.

RODAHL, K. 1949. Toxicity of polar bear liver. Nature 164:530.

RODAHL, K. and MOORE, T. 1943. The vitamin A content and toxicity of bear and seal liver. Biochem. J. 37:166.

RUSSELL, F. E. 1968. Poisonous Marine Animals. In "Safety of Foods." H. Graham, Ed., Avi Publ. Co., Westport, Conn.

RUSSELL, F. E. 1975. Ciguatera poisoning. A report of 35 cases. Toxicon 13:383.

SALKOWSKI, E. 1885. Zur Kenntniss des Giftes der Miesmuschel (*Mytilus edulis*). Arch. Pathol. Anat. Physiol. *102*, 578–592.

SCHANTZ, E. J. 1960. Biochemical studies on paralytic shellfish poisons. Ann. N.Y. Acad. Sci. 90:843–855.

SCHANTZ, E. J., E. F. McFARREN, M. L. SCHAFER, and K. H. LEWIS. 1958. Purified shellfish poison for bioassay standardization. *J. A. Offic. Agr. Chem.* 41:160.

SCHEUER, P. J. 1970. Toxins from fish and other marine organisms. In Advances in Food Research. Chichester, C. O., Marak, E. M. and Stewart, G. Ed. Vol. 18. Academic Press, N.Y.

SCHEUER, P. J. 1977. Marine toxins. Accounts of Chem. Research 10:33.

SCHEUER, P. J., TAKAHASHI, W., TSUTSUMI, J. and YOSHIDA, T. 1967. Ciguatoxin: Isolation and chemical nature. Science 155:1267.

SHIMIZU, Y. 1978. Dinoflagellate toxins, in Marine Natural Products, Chemical and Biological Perspectives, Vol. I, P. J. Scheuer, ed., Academic Press pp. 1–42.

SHIMIZU, Y. 1979. Developments in the Study of Paralytic Shellfish Toxins, in 'Toxic Dinoflagellate Blooms' D. Taylor and H. Seliger, ed. Elsevier North Holland, Inc. pp. 321–326.

SHIMIZU, Y., ALAM, M., OSHIMA, Y. and FALLON, W. E. 1975. Presence of four toxins in red tide infested clams and cultured *Gonyaulax tamarensis* cells. Biochem. Biophys. Res. Commun. 66:731–737.

SHIMIZU, Y., BUCKLEY, L. J., ALAM, M., OSHIMA, Y., FALLON, W. E., KASAI, H., MIURA, I., GULLO, V. P. and NAKANISHI, K. 1976. Structures of gonyautoxin-II and -III from the east coast toxic dinoflagellate *Gonyaulax tamarensis*. J. Am. Chem. Soc. 98:5414–5416.

SHIMIZU, Y., FALLON, W. E., WEKELL, J. C., GERBER, D., JR. and

GAULITZ, E., JR., 1978. Analysis of Toxic Mussels (*Mytilus* sp.) from the Alaskan Inside Passage, J. Agric. Food Chem. 26:878–881.

SHIMIZU, Y., HSU, C. P., FALLON, W. E., OSHIMA, Y., MIURA, I. and NAKANISHI, K. 1978. The structure of neosaxitoxin, J. Am. Chem. Soc. 100:6791.

SHIMIZU, Y., and RAGELIS, E. 1978. Alternatives to the mouse bioassay. *In* Toxic Dinoflagellate Blooms (Eds.) D. L. Taylor and H. H. Seliger, Proc. of the Second International Conference on Toxic Dinoflagellate Blooms, Key Biscayne, Florida, October 31-November 5.

SHORE, P. A. 1971. The chemical determination of histamine. Methods Biochem. Anal. Suppl. Vol., pp. 89–97.

SOMMER, H., and MEYER, K. F. 1937. Paralytic shellfish poisoning. Arch. Pathol. *24*, 560–598.

SOMMER, H., WHELDON, W. F., KOFOID, C. A., and STOHLER, R. 1937. Relation of paralytic shell-fish poison to certain plankton organisms of the genus Gonyaulax. Arch. Pathol. *24*, 537–559.

SPECK, M. L. (Ed.) 1978. Compendium of methods for the microbiological examination of foods. American Public Health Association, 1015 Eighteenth Street, N.W., Washington, DC.

SYLVESTER, J. R., DAMMANN, A. E. and DEWEY, R. A. 1977. Ciguatera in the U.S. Virgin Islands. Marine Fisheries Review 39:14.

TAYLOR, D. L. and SELIGER, H. H. (Eds.) 1978. Proc. of the Second International Conference on Toxic Dinoflagellate Blooms, Key Biscayne, Florida, October 31-November 5, 1978.

WAKELY, J. F., FUHRMAN, G. J., FUHRMAN, F. A., FISCHER, H. G. and MOSHER, H. S. 1966. The occurrence of tetrodotoxin (tarichatoxin) in amphibia and the distribution of the toxin in the organs of newts (taricha). Toxicon 3:195.

WEISS, S., ROBB, G. P. and ELLIS, L. B. 1932. The systemic effects of histamine in man. Arch. Inter. Med. 49:360.

YASUMOTO, T. and SCHEUER, P. J. 1969. Marine toxins of the Pacific-VIII Ciguatoxin from Moray eel livers. Toxicon 7:273.

YASUMOTO, T., NAKAJIMA, J. and BAGNIS, R. 1977. Finding a dinoflogellate as a likely culprit of ciguatera. Nikon Suisan-Gakkai Shi. 43:1021.

19

The Need of Additives in Industry

D. J. Jorgensen

INTRODUCTION

It is generally the needs and wants of the consumer which dictate the use of additives by industry.

What people eat is governed as much or more by cultural, psychological and social values as by reasons of nutrition and health. One of the major functions and responsibilities of the food processor and formulator, although this is sometimes questioned, is to supply the healthful foods we need while also meeting the ethnic requirements and making food visually and organoleptically appealing. Although we eat and drink to live, perhaps the greatest pleasure of living is to eat and drink. Perhaps most people will agree that it would be a horrible existence to gulp down a formless, colorless, flavorless concentrate of the necessary nutrients in order to merely support life.

Food additives have been developed to provide many functions. Consumers, especially those in the more developed areas of the world can select from a variety of attractive, tasty, safe, healthful and convenient foods due largely to the availability of food additives, which make it possible to prepare, formulate, store or ship the foods themselves or their components.

The legal meaning of the term "food additive" differs significantly from country to country. In the United States "food additives" are defined by the "Federal Food, Drug and Cosmetic Act" approved June 25, 1936, and as currently amended.

For the purpose of this chapter, the generally accepted definition of the Food Protection Committee of the National Academy of Sciences—National Research Council, Washington, D.C., will be used. It states: "A food additive is a substance or a mixture of substances, other than a basic foodstuff, which is present in a food as a result of any aspect of production, processing, storage or packaging. The term does not include chance contaminants."

Since the Food Protection Committee definition includes only substances intentionally added, the myriad of chemical components inherent in the composition of the raw foodstuffs or substances which have entered the food unintentionally such as pesticides, microbial contaminants, and other foreign matter resulting from handling, storing and transporting are eliminated.

Each additive used in food processing serves one or more purpose such as improving nutritional value, inhibiting spoilage, providing unique texture or physical characteristics, facilitating preparation, enhancing quality or consumer acceptability, and/or for reasons of economy.

Additives are needed to extend the range and to supplement methods of handling, packaging and processing of foods. These methods would include wrapping, coating, hermetically sealing in plastic film, bottling, packing in an array of boxes and jars, canning, dehydrating and freezing. While preservation from biological spoilage first comes to mind, and indeed is one of the more important functions in food handling and processing, food additives provide many other added benefits to the processes for getting food to the consumer in the best possible condition.

Examples of Functions of Food Additives

1. Antioxidants to protect nutrients, color and flavor—
   —Butylated hydroxytoluene (BHT) and butylated hydroxyanisole (BHA) are added to potato chips to prevent rancidity.
   —Ascorbic acid is added to canned fruit to improve nutritional value and prevent browning.
2. Preservatives to retard spoilage—
   —Sorbic acid or potassium sorbate is added to packaged cheese to retard mold growth calcium propionate to bread for the same reason.
   —Lactic acid is added when packing olives to provide clarity by inhibiting fermentation and spoilage.
3. Flow agents to prevent lumping and caking—

—Tricalcium phosphate is added to salt so it will flow freely from the package on the shelf and from the shaker on the table.
4. Acidulants, flavors, colors and sweeteners to improve the aesthetics—
—Citric acid, orange oil and orange colorings are formulated into gelatin dessert to provide a pleasing flavor and appearance.

These are only a few examples of how food additives act as processing and packaging aids. More will be found in the summary of functions of food additives at the end of this chapter.

Additives make it possible for food manufacturers to provide a large variety of foods and beverages from which the consumer may choose. The word "fabricated" was created to describe foods which are formulated or compounded from selected ingredients and additives to give a finished product of desired properties, characteristics and nutritional qualities. These foods may be fashioned to resemble a natural product such as an orange drink, to simulate orange juice, or maybe a completely unique creation such as a snack bar with a balanced concentrate of nutrients to substitute as a quick meal. Most products lie somewhere between these extremes and are designed to give one or more special or extra properties when they are compared to a natural counterpart. Generally, one of the major contributions of these products is convenience. Such convenience may be in storing, preparing and consuming. However, the fabricated foods may have other very special properties such as low calories for those who want to maintain their weight at a certain level, high vitamin content for those who feel the need to supplement their diet and low carbohydrates for the diabetics. They may be a low cost substitute for an expensive food, may replace scarce commodities and certainly, not least of all, they may be a novelty to perk up a monotonous diet.

The foregoing discussion of fabricated foods has set the stage for showing the needs of an area which food additives have made possible. By their very nature, fabricated foods are a blend of food ingredients with selected additives to give the tailored functional characteristics desired. The best way to show this is by examining the ingredient lists on the labels of three of these products, as outlined below.

Breakfast Drink

"INGREDIENTS: Sugar, citric acid (for tartness), calcium phosphates (regulate tartness and prevent caking), modified starches (provide body), potassium citrate (regulates tartness), cellulose gum (vegetable gum-provides body). Natural orange flavor, vitamin C, hydrogenated coconut oil, artificial flavor, artificial color, vitamin A palmitate, alpha tocopherol and BHA (preservatives)."

Snack Crackers

"INGREDIENTS: Enriched flour (wheat flour, niacin, iron, thiamine mononitrate and riboflavin), vegetable and animal shortening (coconut oil and lard), salt, sugar, corn syrup, enzyme modified cheddar cheese solids, modified butterfat, yeast extract, whey, monosodium glutamate, monocalcium phosphate, hydrolyzed vegetable protein, lactic acid, malto dextrin, sodium phosphates, spices, artificial flavoring and artificial coloring."

Pimento Spread with Jalapeno Peppers

"INGREDIENTS: Pimento pasteurized process cheese food (American cheese, water, skim milk cheese for manufacturing, sodium citrate, whey, pimentos, salt, cream, citric acid, and artificial color). Salad dressing (containing calcium disodium EDTA), jalapeno peppers, pimentos, sweet peppers, sugar, citric acid, spices. 1/10 of 1% sodium benzoate and sorbic acid as preservatives."

Mention has been made of the relationship between nutrition and health and the use of additives. However, this is so important that it deserves a more in-depth examination. As discussed in chapters 1–6,8 and 24, food spoilage organisms can be very dangerous. While heat processing, dehydrating and freezing are helpful in inhibiting dangerous levels of biocontamination, they are not the complete answer. Residual *Clostridium botulinum* spores can produce deadly toxin levels if a low pH is not maintained by acidulants or if a preservative (e.g., sodium nitrite in cured meats) is not used. Sorbic acid is showing promise as a preservative in this area as well as its recognition as a mold inhibitor. Baking temperatures for making bread are incapable of killing the spores of the bacterium *Bacillus subtilis*. These bacteria can become active under summer storage conditions and produce a growth called "rope" which makes bread inedible. For this reason, propionate is added to bread as a preservative. Molds can produce toxins such as *aflatoxins* in peanuts, soybeans and cereal grains if they are not kept dry or protected by a preservative.

All foodstuffs begin losing nutritional value soon after harvest. Processing, shipping and storage can result in further losses. Added vitamins and minerals restore and supplement the nutritive value of such foods.

Synthetic sweeteners can be substituted for sugar in the special diets for the diabetic and the weight conscious.

Calcium phosphates and calcium carbonate supplements in food can prevent deficiencies of calcium for those who cannot tolerate dairy products.

Polyunsaturated vegetable oils are used in foods for those who are concerned with elevated cholesterol levels in their blood. These oils, being unsaturated, are highly susceptible to oxidation. Therefore, sequestrants and antioxidants are used to retard rancidity.

There is increasing concern about dental health particularly in the developing teeth of children. For this reason, so-called sugarless chewing gum were produced with sugar substitutes such as saccharin and sorbitol. In a number of countries, including the U.S., fluorides are added to the water supply to reduce caries. The Pan American Health Organization has been conducting tests in Columbia, S.A wherein small amounts of fluorides admixed with calcium phosphates are added to table salt. These tests are being conducted with a sample of rural children to determine if this is a good approach for inhibiting tooth decay in areas where fluoridated central water systems are not available. Under the sponsorship of the American Dental Association, studies of food-induced caries and methods for reducing the causes have been initiated. A number of studies on phosphates added to foods, particularly dicalcium phosphate, conducted by academic and industry researchers, have shown some impressive results. Such products appear to have a promising future in this area.

Food is a very precious commodity. One reads almost daily of shortages, malnutrition and starvation in many parts of the world. It is a serious problem to maintain food production to meet the ever-increasing needs of an expanding population, but it is discouraging to realize that over a third and perhaps as much as a half of the food produced may never reach the stomachs of the consumer. Even in the United States, it has been estimated that about 10% of the crops are lost in the field, 10% in shipping, handling and processing, and 15% in the home, where it is prepared and eaten. In some countries, field and storage losses run much higher, although little may be lost to the hungry stomach in the place where it is prepared and eaten. It is easy for one to recognize the losses in the field, in shipping and in storage due to weather, insects, birds and animals, biological spoilage, etc., and to understand how preservatives can and do save food in the delivery chain to the table. However, food additives reduce waste in other ways. One very interesting development is the use of fine tricalcium phosphate in cereal products to inhibit insect infestation. The mechanism of how a non-toxic food additive such as this can destroy insects is not too well understood, but it is believed that it interferes with the respiratory or dehydration protective systems in some manner. Tricalcium phosphate is formulated into the nutritive concentrates shipped by the United States Government to the developing regions of the world and reduces losses from a variety of flour beetles as well as serving as a mineral nutrient and a flow agent. The protection and the restoring of nutrients in food by additives has been discussed already. However, this deserves mentioning again since loss of nutrients, in a real sense, is a loss of food. Foods can generally be prepared more efficiently and with less waste in the processing plant than in the household. Thus, more subtly perhaps, the ready-to-eat and convenience foods of today are less wasteful. Potatoes are a good example of this. The potato processor throws away much less of the potato in his peelers than does the home preparer with the peeling knife. Portions from the package are more readily gauged by the meal maker than when trying to select the correct size and number of potatoes from a bag and, of course, convenience packaged items are protected against spoilage loss and cannot sprout. The packaged frozen varieties contain disodium phosphate and dextrose to prevent discoloration and for retention of moisture and flavor. The ingredient label on a package of dehydrated potatoes of the "hash brown" variety reads as follows:

"Dried potatoes (with color and freshness preserved by sodium phosphate, sodium sulfite and BHT), corn starch, dried onion, dextrose, mono and diglycerides, sodium stearyl fumarate, calcium stearoyl-2-lactylate."

In addition to the preservatives, emulsifiers are used to assist in reconstitution of the potato structure and for tenderness when the water is added back.

Finally, but no less important, food products must appear, smell and taste good to the diner when put on the table or they will not be eaten and thus be wasted. This is where additives, used to maintain and supplement the esthetic values of foodstuffs, can play a very important part—preservatives and antioxidants to protect original flavor and color, color and flavor additives for improving these characteristics, hydrocolloids and emulsifiers to give structural properties, etc.

The foregoing discussion on reducing waste by use of additives also serves to describe the economic savings resulting from their use. Loss through waste of food is also a loss in dollars represented by the value of seed, fertilizer, labor, etc., and the cost of energy.

There are those additives which are absolutely fundamental to the existence of certain food types just as vinegar is to the pickle. Examples of these are: (1) Carbon dioxide to the carbonated beverages, (2) leavening agents such as the acid phosphates with sodium bicarbonate to the bakery products, cakes, biscuits, donuts, cookies, etc., (3) pectin to jellies and jams, (4) gelatin to gelatin desserts, (5) starches to puddings, (6) artificial sweeteners such as saccharin to nonsugar or low sugar sweet goods, (7) emulsifiers to salad dressings, (8) butter flavor and yellow coloring to margarine, (9) acidulants such as citric acid to powdered drink mixes, and (10) enzymes to cheese. No claims are made that these are the top ten and the reader can certainly lengthen the list.

A much longer list could be made of additives and the food with which they are associated. These have become standards of the industry because they are so important to the characteristics and properties of the end product. A few of these will be discussed briefly but no attempt will be made to provide a major listing because there are so many. Examples are: (1) Vitamins A&D to supplement milk, (2) monosodium glutamate to enhance flavor of soups, (3) sorbates to protect and phosphates to emulsify process cheese, (4) propionates to preserve bread, (5) nitrites in ham and bacon for color, flavor and preservation, (6) sodium benzoate to preserve catsup, (7) ascorbic acid in canned and powdered fruit drinks as a nutrient, (8) BHT and BHA in cooking oil to protect against rancidity, (9) mono- and diglycerides in cakes and cake mixes to improve texture, (10) red coloring in cocktail cherries.

BENEFITS VS RISKS IN USE OF FOOD ADDITIVES

There is much discussion these days about "risk/benefit" decisions on food additives and on the question of "how safe is safe." Chapters 20–23 covering regulations and safety of food constituents have addressed this subject, but since such judgements are an integral part of industry's decisions to use or not to use additives, some comment on the subject, even at the risk of being repetitious, seems warranted here.

Everyone makes choices based on risk/benefit judgments daily. Your child wants to climb on boulders at the park. Should you allow it? If the rocks are low and surrounded by soft sand, perhaps you would. But if the rocks are near a cliff, definitely not.

That decision uses firsthand information to balance the risks of falling against the enjoyment of climbing. People make decisions that way every day—whether or not to drive when it is icy, use a shaky ladder, or go outside without an umbrella. Answers depend on the situations.

Other decisions balancing risks and benefits are not so easy. The problem of making a rational judgment of what is reasonably safe for the benefits derived from a food additive is a difficult one. Because the risk of health and life is such an emotional issue, the tendency of regulatory agencies is to be ultra-conservative, if there is any doubt of safety and, unfortunately, there are no tests which will prove absolute safety. To paraphrase how one regulator views the situation: We are making twentieth century decisions on nineteenth century test procedures and they do not tell us precisely whether something is "safe or unsafe".

In the case of cancer, there is no benefit great enough that it will overcome the slightest indication of risk. The much debated Delaney clause rigidly states that nothing that is judged to be cancer-causing can be added to food. The Delaney Clause of the 1958 Food Additive Amendment to the Federal Food, Drug and Cosmetic Act, Section 409 (C) (3) (A) provides: "that no additive shall be deemed to be safe if it is found to induce cancer when ingested by man or animal, or if it is found after tests which are appropriate for the evaluation of the safety of food additives, to induce cancer in man or animal. . . . ."

There is, therefore, no room for measuring the benefits of saccharin, cyclamates, nitrites, or any other substance to be added to food, if it is judged to cause cancer in a mouse, rat, etc., or man, at any concentration, whether representative of levels as consumed or many time these levels.

Prospects for development of new food additives at the present time are uncertain. Only two new chemicals as direct food additives have been approved in the U.S. since 1969. These are the anti-oxidant TBHQ, closely related to BHA, and BHT in 1972 (CFR, Title 21, 172.185) and the sweetener aspartame in 1974 (CFR, Title 21, 172.804). At this writing, aspartame is not yet allowed in food because of a stay order (40 FR 56907, 12/5/75) pending validation of toxicity test data.

It is the feeling of some that the present regulation of food additives is unnecessarily restrictive and deprives the public of benefits which they could safely enjoy. It is argued, also, that this is unreasonable and unfair since the natural chemical constituents of food are not judged in the same manner.

As for the Delaney Clause, per se, in a democratic form of government, the public should make the final judgement of risks and benefits—either through personal choice or through the decisions of its elected representatives. But as its critics note, the Delaney Clause as now worded eliminates the ability to make judgments. It tries to reach the unattainable goal of absolute safety.

For that reason, many scientists and industrial companies support efforts by the Congress to modernize the Clause. There must always be control over the food we eat and the additives used in food. When human or animal health is in danger, chemicals and processes should be carefully regulated, and in some cases, prohibited.

However, regulations should not exclude competent, scientific evidence. There must be realistic standards and the realization that, in the complex equation that establishes the public good, there must be room for benefits as well as risks.

These decisions will never be as easy as determining if a child should play on rocks at the park. However, they should be made in an atmosphere as free from fear and emotionalism as possible, and with careful consideration of all the evidence.

It is hoped that a more rational judgment of acceptable risks will not only be possible through advancement of sciences, but, that it will also prevail in the regulation of food additives.

A perspective of the usage of the various food additives is given by the following list, which lists the functions and quantities used in descending order of usage. This is not to suggest that the larger functions are more important. Since accurate data for use are hard to obtain, the quantities in pounds should be considered as ranges of magnitude only.

This is not intended as a comprehensive compendium of food additives. There are a number of publications which serve this function well, but the best way to visualize the "needs for additives" is to examine what they are and what they do. The following summary pages have been prepared for this purpose. Many of the additives are multifunctional, filling several needs and so will appear in more than one listing. The phosphates are an excellent example of multifunctional products as illustrated in Table 19.1. A word of caution— the listing of additives on these pages should not be construed as a recommendation for use of any product in any food. Before any additive is used in food, it first should be determined that such use meets all applicable government regulations.

U.S. Food Additive Consumption

Function	Quantity (Millions of Lbs./Yr.)
Thickeners/stabilizers (Hydrocolloids)	430 – 470
Flavors & enhancers	290 – 320
Emulsifiers (Surfactants)	270 – 300
Acidulants	180 – 200[1]
Chemical leavening agents	175 – 200
Colors	80 – 85[2]
Humectants	60 – 70
Nutritional supplements (Vitamins)	55 – 65
Preservatives	50 – 60
Enzymes	>25 (sold basis of activity)
Dietary sweeteners	> 5
Antioxidants	5 – 8
Others	>200
Total	2,000

[1] 80% Citric acid
[2] 90% Caramel color

THICKENERS/STABILIZERS

Needs Filled:

These hydrocolloids modify and form the physical character of foods by: thickening, binding, suspending, gelling, coating, bulking, forming films, or stabilizing foams.

They can be used to produce a cloud in a liquid, function as a whipping agent or just produce a specific mouth sensation.

Additives:

Tree Exudates:

Gum arabic, gum tragacanth, gum karaya, larch gum, ghatti gum,

Seed or Root Gums:

Locust bean gum, guar gum, psyllium seed, quince seed, corn starch, arrowroot starch, tapioca starch wheat starch, potato starch.

Starch derivatives:

Hydroxyethyl starch, hydroxypropyl starch

Cellulose derivatives:

Carboxymethyl cellulose (CMC), methyl cellulose, hydroxypropyl-methyl cellulose, hydroxyethyl cellulose, hydroxypropyl cellulose, ethylhydroxyethyl cellulose.

Seaweed extracts and their derivatives:

Agar, alginates, carrageenan, furcellaran, propyleneglycol alginate, triethanolamine alginate.

Others

Gelatin
Dextran
pectin

Examples

Carrageenan is used to stabilize the suspension of chocolate in chocolate flavored milk.

Several of the gums; CMC, alginates, guar, locust bean and others, are used as stabilizers in ice cream.

Gelatin forms the base for desserts and salads.

Gum arabic is used as a foam stabilizer in beer.

FLAVORS/ENHANCERS

Needs Filled:

Flavors and enhancers (or potentiators) are added to foods to supplement, imitate, modify or strengthen flavors found in nature and/or cultivated in ethnic environments. Thus, their use is to develop and improve the sensory appeal of a food so that it will be

TABLE 19.1. FOOD PHOSPHATES—NATURE AND FUNCTIONS.

Produce	Chem. Formula	pH 1% Sol.	Neutralizing Value as Applicable	Needs Filled
Sodium acid pyrophosphate (SAPP)	$Na_2H_2P_2O_7$	4.2	73–74	Leavening acid, protein interaction, sequestrant (Iron), preservative (Prevents darkening of potatoes)
Typical Dough Rate of Reaction of Various Grades				
24 (very slow) 26 **28 (baking powder)** 37 **40 (dough-nuts)** **43 (fast)**				
Sodium aluminum phosphate, basic	$Na_8Al_2(OH)_2 (PO_4)_4$	9.3		Used as a cheese emulsifier with DSP.
Sodium aluminum phosphate, acidic	$Na_3Al_2H_{15}(PO_4)_8$ (Anhydrous)	2.4	100	Slow reacting leavening, particularly useful in white cake and pancake mixes.
Blend of sodium aluminum acid and monocalcium phosphate, anhydrous	$Na_3Al_2H_{15}(PO_4)_8$ & $Ca(H_2PO_4)_2$		80–83	Leavening acid for self-rising flour, biscuit, cake and pancake mixes.
Blend of sodium aluminum phosphate and monocalcium phosphate monohydrate	$Na_3Al_2H_{15}(PO_4)_8$ & $Ca(H_2PO_4)_2 \cdot H_2O$		94–96 94–96	Slow reacting leavening acid for cake and pancake mixes.
Anhydrous monocalcium phosphate stabilized	$Ca(H_2PO_4)_2$	4.6	84	Delayed leavening acid for self-rising flour, corn meal and all mixes.
Monocalcium phosphate monohydrate (MCP) (Grades available varying in particle size)	$Ca(H_2PO_4)_2 \cdot H_2O$	4.6	81	Very fast leavening acid for double acting baking powder and for mixes requiring early gas release, e.g., pancake mixes. Also used in dough conditioners for pH adjustment and as a buffer in powdered beverages.

(Continued)

TABLE 19.1.   (Continued)

Product	Chem. Formula	pH 1% Sol.	Neutralizing Value as Applicable	Needs Filled
Disodium phosphate (DSP) Anhydrous	Na_2HPO_4 $Na_2HPO_4 \cdot H_2O$	9.0		Mildly alkaline. Used for pH buffering, protein interaction and calcium precipitatio, i.e., farina, meat, instant pudding, evaporated milk and process (cheese emulsifier).
Monosodium phosphate (MSP)	NaH_2PO_4	4.6		Very soluble, fast reacting acid salt for specialized uses.
Phosphoric acid Available in conc. of 75, 80, 85, and 105%.	H_3PO_4	1		Used as acidulant in beverages and as a scavenger of metal ions in shortenings.
Trisodium phosphate (TSP) Anhydrous and crystalline	Na_3PO_4 $Na_3PO_4 \cdot 12H_2O$	12.0		Very alkaline. Used with DSP as an emulsifier in cheese and in detergent formulations for cleaning equipment.
Dicalcium phosphate (DCP) Anhydrous Dihydrate	CaH_2PO_4 $CaH_2PO_4 \cdot 2H_2O$			Relatively insoluble. Used as a dentifrice polishing agent, calcium supplement and special leavening (very slow). Has potential as additive to reduce tooth decay.
Tricalcium phosphate (TCP)	$3\ Ca_3(PO_4)_2 \cdot$			Very insoluble. Used at 1–2% level. Prevents lumping and improves flow of powder and granular products. Also used as a nutritive supplement and to provide turbidity to synthetic citrus beverages. Used to inhibit insect infestation in cereal products.
Tetracalcium pyrophosphate (TCPP)	$Ca_2P_2O_7$			Very insoluble. Used as a dentifrice polishing agent and as a nutritive supplement.

Sodium tripolyphosphate (STP)	$Na_5(P_3O_{10})$	9.9	Good sequestrant for calcium and magnesium. Because of its protein interaction it is widely used in meat curing formulations. Very effective in binding water to protein.
Tetrasodium pyrophosphate (TSPP)	$Na_4P_2O_7$	10.3	Major use as a sequestrant. Widely used in meat and sausage products and with DSP in instant puddings due to its protein interaction, sequestrant properties and pH control.
Monoammonium phosphate (MAP)	$NH_4H_2PO_4$	4.7	Acid Salt for acidifying.
Diammonium phosphate (DAP)	$(NH_4)_2HPCA$	7.9	Basic salt. Used as a yeast food and in dough improvers.
Insoluble sodium metaphosphate (IMP)	$(NaPO_3)n$ $n=100-500$	5.5 5.5	A mixture of IMP, DSP and TSP is used as a cheese emulsifier.
Assay sodium polyphosphate (S.Q.) (Hexameta)	$Na_8P_6O_{19}$ (average chain length) $Na_{16}P_{14}O_{43}$ (average chain length)	7.9 6.9	Primary use as a sequestrant in water softening. Some use in meats and cheeses.

eaten. Flavor is probably the most important factor influencing preference for one food item over another.

There are probably over 2000 natural extracts and synthetic chemicals used to formulate flavors. Only the families and types will be listed here.

Additives:

Essential oils (botanical extracts from fruits, seeds and plant
                      parts)
Civet and musks (from animal sources)

Synthetic Chemicals:

Cyclic Compounds (aromatics)

"Acyclic Compounds:

(Aliphatic alcohols, aldehydes, esters and acids)"
Amino acids

Miscellaneous: Nitrogen compounds and substances of undefined structures

Examples:

Soft drinks are made in a wide variety of flavors from cola to lemon-lime. Candies, confectionery products and chewing gum are major users of flavors. Monosodium glutamate enhances meaty soup flavors.

EMULSIFIERS (SURFACTANTS)

Needs Filled:

The surface active agents disperse or emulsify immiscible liquids into more stable, continuous mixtures. Usually, the two phases are oil and water which would exist as separate layers in the food product, were it not for the addition of emulsifiers. They provide wetting of surfaces to improve solution or dispersion of solids in liquid and help to stabilize suspended solids in a liquid. They modify viscosity and lubricate. In combination with the thickeners, fats, solids and liquids of foods, emulsifiers function to give pleasing, smooth, mouth sensations.

Additives:

Esters of fatty acids—capric, caprylic, lauric, myristic, oleic, palmitic and stearic: Ascorbyl, Glyceryl citrate, Glyceryl lactylate, Polyoxyethylene, Polyoxyethylene sorbitan, Sorbitan, Propyleneglycol, Mono & diglycerides.
Others:
Sodium phosphate glycerides, Sodium sulfoacetate glycerides, Acetylated monoglycerides, diacetyl tartaric glycerides, lecithin, stearyl-2-lactylic acid, calcium and sodium stearyl-2-lactylate.

Examples:

Mono and diglycerides are used in cake mixes to blend fats and maintain a tender moist texture. Salad dressing emulsions are stabilized with emulsifiers such as polyoxyethylene sorbitan mono-stearate. Mono and diglyceride emulsifiers are incorporated into peanut butter to maintain oil stability and improve spreadability.

ACIDULANTS

Needs Filled:

The food acids and acid salts are probably the most useful multi-purpose additives of all classes of products. A major use is to provide the sour taste and a stimulating sensation to foods and beverages. In nature, they supply a major flavor characteristic to most fruits and many vegetables. They act as buffers to control the pH of foods and preserve foods by inhibiting growth of microorganisms. Their function as *Leavening Agents* in baked goods is covered separately as is their chelating function as *Antioxidants* and their use as *Preservatives*.

Additives:

Organic acids, citric, adipic, fumaric, lactic, malic, tartaric, succinic, acetic (vinegar), propionic, sorbic, benzoic.
Ascorbic, erythorbic, stearic, oleic, other fatty acids, glucono D-lactone.

Inorganic Acids: Phosphoric, hydrochloric.

Acid Salts: Potassium bitartrate, Monocalcium phosphate, sodium aluminum phosphate, sodium acid pyrophosphate, monosodium phosphate, sodium aluminum sulfate.

Examples:

Phosphoric acid is used in cola drinks because it complements the cola flavor. Citric and fumaric acid may be blended to provide the acidulant for powdered beverage mixes. Tartaric acid is favored for grape-flavored drinks. Adipic acid is used in gelatin desserts because of its favorable action on protein.

CHEMICAL LEAVENING AGENTS

Needs Filled:

Leavening agents are the backbone of the baking industry. The word "leavening" is derived from the Latin word *levo* meaning to raise or make light. The characteristic sponge or cellular structure of breads, cakes, biscuits, donuts, cookies, crackers, etc., are the result, during baking, of expanded gases—air incorporated by mechanical mixing and/or carbon dioxide generated by fermentation or chemical reaction. It is the latter system, chemical leavening, which is concerned with food additives—a bicarbonate, usually sodium bicarbonate and an acid, usually an acid phosphate salt. The cellular structure of baked goods contributes not only to their attractive appearance and pleasing mouth sensations but is also an aid to digestion in that a greatly extended surface is presented to the action of digestive juices.

Additives:

Carbon dioxide donors; Sodium bicarbonate, potassium bicarbonate, calcium carbonate.

Leavening acids: Sodium acid pyro phosphate (SAPP), sodium aluminum phosphate (SALP), Monocalcium phosphate (MCP). Dicalcium phosphate dihydrate (DCPD), sodium aluminum sulfate, potassium acid tartrate (Cream of tartar), glucono-delta-lactone, Fumaric acid.

Examples: (See Food Phosphates Table 19.1, page 663)

Sodium bicarbonate, monocalcium phosphate and sodium aluminum sulfate mixtures are used as double-acting household baking powders.

Sodium bicarbonate and sodium acid pyrophosphate are used to leaven cake donuts.

Sodium bicarbonate and sodium aluminum acid phosphate are used to leaven white cake and pancake mixes.

COLORS

Needs Filled:

Colors provide an appealing appearance to foods and identify them. Thus, orange drinks are orange, cherries are red and cola beverages are brown.

Additives:

Caramel, annatto extract, carotene, cochineal, saffron, xanthophyll, turmeric, paprika, tagetes, carrot oil.

FD&C Colors: Blue No. 1, Orange B, Citrus Red No. 2, Red No. 3, Red No. 40, Yellow No. 5.

Examples:

Caramel is used to provide an amber color to soft drinks. Red No. 40 is used to color cocktail cherries.

Hard candies are colored with a wide range of colors which identify them with their flavors, e.g., yellow - lemon, green - lime, violet - grape, etc. Moist pet foods are colored red to imitate raw meat.

HUMECTANTS

Needs Filled:

These additives, as their functional name implies, maintain the moisture balance of foods and thus affect the textural characteristics of foods providing a moist mouthfeel or a smooth creamy sensation. They may modify crystal formation and/or jointly with the emulsifiers and stabilizers, act as plasticizers or softeners.

Additives:

Glycerin, propylene glycol, sorbitol, mannitol, triacetin, sodium tripolyphosphate, tetrasodium pyrophosphate, dextrose, corn syrup.

Examples:

Sorbitol is incorporated into marshmallows to maintain softness. Propylene glycol or glycerine is used in moist shredded coconut. One or more humectants may be used in chewing gum to stabilize moisture.

NUTRITIONAL SUPPLEMENTS

Needs Filled:

These additives provide a balance of essential components in foods needed for body development and maintenance of health. They are added to foods to supplement the nutrients in foodstuffs which nature has failed to supply in adequate amounts or which have been lost during processing and/or storage. They can be added in varying amounts to meet special requirements such as the tailoring of infant, geriatric and convalescent foods.

Additives:

Vitamins: Ascorbic acid (C), thiamine (B), alpha tocopherol (E), cyanocobalamin (B_{12}), riboflavin (B_2), calciferol (D_2), cholecalciferol (D_3), pyridoxine (B_6), niacin, pantothenic acid, biotin (H), folic acid, retinol (A), vitamin K.

Amino Acids: lysine, tryptophan, methionine, threonine, leucine, phenylalanine, tyrosine, valine, glutamic acid, proline.

Minerals: iodine and its salts, calcium carbonate, calcium phosphates, calcium citrate, calcium sulfate.

Salts of: iron, magnesium, cobalt, boron, copper, molybdenum, nickel, potassium, molybdenum, nickel, zinc, potassium.

Examples:

Vitamins A and D are added to milk, vitamin C is added to fruit drinks. Breakfast cereals may be fortified with calcium by adding calcium phosphate as well as supplemented with various vitamins.

PRESERVATIVES

Needs Filled:

Preservatives maintain freshness, or original appearance, flavor and consistency. They protect foods from biological spoilage during transport and they lengthen shelf life. Some inhibit the growth of dangerous microorganisms and the formation of toxins, thus enhancing the safety of the food.

Others protect food nutrients from degradation and loss, reduce waste due to spoilage and further decrease cost by supplanting or improving other more costly food protective methods.

Additives:

Sorbic acid, potassium sorbate, benzoic acid, sodium benzoate, propionic acid, calcium propionate, sodium propionate, acetic acid, sodium acetate, calcium acetate, sodium diacetate. Methyl p-hydroxybenzoate, propyl p-hydroxybenzoate n-Heptyl p-hydroxybenzoate, sulfur dioxide, sodium sulfite, potassium sulfite, sodium bisulfite, potassium bisulfite, sodium metabisulfite, potassium metabisulfite, sodium nitrite, potassium nitrite, sodium chloride, stannous chloride, sugars, various acids for pH control.

Examples:

Potassium sorbate solution is sprayed on English muffins to improve shelf life. Sorbic acid is used in cheese to prevent mold spoilage. Sodium benzoate is used in catsup to inhibit spoilage and calcium propionate is used in bread to control mold growth and rope.

ENZYMES

Needs Filled:

These biocatalysts, produced by living cells, promote reactions in foodstuffs to produce the predetermined characteristics and properties of the particular food. They are, therefore, fundamental to the preparation of many foods familiar to us today.

Only a few of the most important enzymes and types of enzymes will be listed.

Additives:

Pepsin, rennet, amylases, lipases, papain, invertase, lactase, maltase, glucose oxidase, catalases, ficin.

Examples:

Amylase is used to produce syrups and dextrose from corn starch, rennet to coagulate milk to produce cheese and papain for tenderizing meat.

DIETARY/NON-CARBOHYDRATE SWEETENERS

Needs Filled:

The synthetic or non-carbohydrate sweeteners serve to provide sweetened foods to meet three needs:
1. For diabetics who must limit sugar intake.
2. For the obese and others who wish to reduce carbohydrate calorie intake.
3. For those who desire to reduce food-induced dental caries.

Additives:

Sodium saccharin, calcium saccharin, sorbitol, mannitol, xylitol, aspartame,[1] calcium cyclamate.[2]

Examples:

Sugarless chewing gum used to contain sorbitol and/or saccharin and diet soft drinks are sweetened with saccharin.

ANTIOXIDANTS

Needs Filled:

As the name implies, antioxidants are needed in foods to stabilize and protect against degradation from oxidation. Their major func-

---

[1]FDA stayed approval of aspartame at 40 FR56907 December 5, 1975, pending validation of toxicity test data.
[2]Use of cyclamates is not now permitted in the US but they are still used in some countries.

tion is to inhibit rancidity (off-flavor), discoloration and polymerization of fats. Fats which become rancid by oxidation not only result in food waste because they are organoleptically unacceptable but toxic by-products of oxidation can also be hazardous to health. Reduction of the nutritional value of foods due to loss of vitamin activity and of essential fatty acids is another penalty of oxidation.

Additives:

Butylated hydroxytoluene (BHT), butylated hydroxyanisol (BHA), 4-hydroxymethyl-2,6-di-*t*-butylphenol, 2-(1,1-dimethylethyl)-1,4-benzenediol, (TBHQ), 2,4,5-trihydroxybutyrophenone (THBP), ethoxyquin (primarily in animal feed), tocopherols (Naturally occurring but also added as vitamin E), ascorbic acid and salts (Vitamin C), erythorbic acid and salts, sulfites and bisulfites (Anti-browning), thiodipropionic acid, stannous chloride.

Chelating Agents: These serve as scavengers for metal ions which catalyze oxidation. Citric acid, phosphoric acid, disodium - ethylenediamine tetra acetic acid (EDTA), calcium disodium - ethylenediamine tetra acetic acid (EDTA).

Examples:

BHT with citric acid is used in lard to prevent rancidity, BHA in packaged cereals to protect flavor and sodium erythorbate in processed meats to assist in stabilizing color.

OTHERS

Solvents to dissolve food ingredients for processing or formulating liquid food products.

Additives:

Ethanol (ethyl alcohol), propylene glycol, glycerin, triacetin, acetylated monoglycerides, oils and fatty acids.

Example:

Ethanol and propylene glycol are used to produce liquid flavorings such as vanilla extract.

Dough Conditioners are used to provide a number of special properties for processing and handling dough products. Table 19.2 lists some of these and their functions.

Defoaming Agents are used to eliminate foaming during commercial processing of foods, to prevent foaming of packaged foods and to prevent foaming in preparation of foods.

Additives:

Dimethylpolysiloxane, silicon dioxide, oxystearin, octanoic Acid, various fatty acids and salts.

TABLE 19.2.   DOUGH CONDITIONER FUNCTIONS

(Frequently used)

Component	Function	Relative Amount	Compounds Allowed in Bread
Calcium	Improves dough handling Strengthens dough >H_2O Enrichment	Large	Monocalcium phosphate Dicalcium phosphate Calcium sulfate Calcium carbonate Calcium lactate
pH control (Buffering)	Optimize pH for yeast fermentation. Strengthens (matures) dough.	Large to Small	Lactic acid Monocalcium phosphate Dicalcium phosphate Calcium carbonate
NH$_4$	Yeast nutrition	Small	Ammonium sulfate Ammonium chloride
Oxidant	Strengthen dough (KIO$_3$). Larger volume bread (KBrO$_3$).	100ppm	Potassium iodate Potassium bromate Calcium iodate Calcium bromate Calcium peroxide
Salt	Flavor	Large	Sodium chloride
Filler	Separate components	Large	Starch, flour

(Used less frequently)

Reducing agents	Easier and faster mixing	Small	L-cystine Ascorbic Acid
Proteases	Easier and faster mixing	Small	Fungal or bromelain
Amylases (low temp)	Reduce amount of sugar Increase loaf volume and quality.		Fungal and cereal
Amylases (high temp)	Keep bread soft		Bacterial

Example:

Aluminum stearate is used as a defoamer in processing beet sugar and yeast.

Alkaline buffers/neutralizing agents are used to maintain the pH of a product.

Additives:

Calcium carbonate, calcium oxide, magnesium oxide, disodium phosphate, trisodium phosphate, diammonium phosphate, other alkaline sodium, potassium, calcium and, magnesium salts.

Example:

Calcium carbonate is used to reduce excessive acidity in wines. It it also used as a neutralizer for ice cream.

Firming Agents are used to maintain the firm texture of canned fruit and vegetables.

Additives:

Calcium carbonate, sodium aluminum sulfate, calcium citrate, calcium chloride, monocalcium phosphate, magnesium chloride, other calcium salts.

Example:

Calcium citrate is used in canned tomatoes to maintain firmness.

Carbonating agents are used to provide gaseous carbonation of soft drinks, malt liquors and alcoholic beverages.

Additives:

Carbon dioxide, sodium bicarbonate/acid, calcium carbonate/acid.

Example:

Bottled and canned carbonated soft drinks.

Anticaking/flow agents are used to prevent lumping and/or to provide a free flowing product.

Additives:

Tricalcium phosphate, calcium silicate, calcium stearate, microcrystalline cellulose, calcium carbonate, magnesium carbonate, silicon dioxide.

Example:

Tricalcium phosphate is used in packaged table salt to keep it free flowing.

Clouding agents are used to produce a haze or cloudiness in a liquid food product.

Additives:

Tricalcium phosphate, microcrystalline cellulose, carboxymethyl cellulose, various gums (hydrocolloids), starch.

Example:

Gum arabic is used in an orange-flavored powdered beverage mix to produce a cloud in the drink.

Clarifying agents are used to remove haze or cloudiness from a liquid food.

Additives:

Polyvinylpyrrolidone, tannic acid, gelatin,

Example:

Tannins may be added to wines to help remove insolubles.

Summary

Additives are used in foods by industry to provide functional appeal, to protect them from spoilage, to improve safety, to supplement nutrition and to reduce cost. Their usage is generally required to meet the varied needs and wants of consumers. A large variety of tasty, healthful foods are available due largely to the food additives which make them possible.

Additives may be classified according to the functions which they serve. For example—emulsifiers and thickeners provide physical form and texture; preservatives and antioxidants protect the commodity and lengthen its shelflife; flavors and colors contribute esthetic values; vitamins and minerals add nutrients for health; enzymes and leavening agents serve as processing aids. Many of the additives are multifunctional and serve several needs; thus ascorbic acid will provide nutrition and act as an antioxidant, citric acid will give a sour taste and also act as preservative, and tricalcium phosphate will function as a flow agent and as a mineral supplement.

Additives have made the so-called "fabricated foods" possible by allowing a food technologist to compound selected ingredients in order to obtain a finished product of desired properties, characteristics and nutritional qualities. They permit formulation of foods convenient to prepare or to meet special dietary requirements.

Approximately 2 billion pounds of food additives are used annually in the United States. This amounts to about 10 pounds per person or less than 1% of the 1,500 pound per capita consumption of foods each year in this country.

BIBLIOGRAPHY

COMMITTEE ON SPECIFICATIONS OF THE COMMITTEE ON FOOD PROTECTION 1972. Food Chemicals Codex 2nd Ed., National Academy of Sciences - National Research Council, Washington, D.C.

DESROSIER, N. W., 1977. The Technology of Food Preservation 4th Ed., The AVI Publishing Company Inc., Westport, Connecticut.

ELLINGER, R. H., 1972. Phosphates as Food Ingredients, The Chemical Rubber Company, Cleveland, Ohio.

FOOD PROTECTION COMMITTEE/FOOD and NUTRITION BOARD, 1965. Chemicals Used in Food Processing, National Academy of Sciences National Research Council, Washington, D.C.

FURIA, THOMAS E., 1968. Handbook of Food Additives, The Chemical Rubber Company, Cleveland, Ohio.

JOSLYN, M. A. and HEID, J. L., 1963–64. Food Processing Operations, The AVI Publishing Company, Inc., Westport, Connecticut.

PANEL ON CHEMICALS and HEALTH OF THE PRESIDENTS' SCIENCE ADVISORY COMMITTEE, 1973. Chemicals and Health, Science and Technology Policy Office - National Science Foundation, Washington, D.C.

UNITED STATES GOVERNMENT, 1977. Code of Federal Regulations, Title 21, Food and Drugs.

NOTE: The contents of this chapter represents the views of the author and not necessarily those of Monsanto Company.

20

The Proper Use of Food Additives

Horace D. Graham

Food as well as food additives are chemicals. All chemicals if properly used, can lead to highly desirable and favorable results. However, if not properly used, they can produce undesirable and even quite deleterious effects. There are, unquestionably, improper ways of using food additives, just as there are improper ways of using common foods, drugs, tools, vehicles, or indeed anything that is normally expected to serve a beneficial purpose. The ways in which food additives can be properly used to the advantage of growers, manufacturers, and consumers are numerous.

Proper use can lead to the adequate nutrition of individuals, a resultant healthy, productive population, a safe, sufficient and maybe even a substantial reserve of food. Many outstanding examples may be cited here. The fortification of milk with vitamins, of flour with vitamins and minerals, and salt with iodine, are but some of the many ways in which the proper use of additives has led to improved nutrition and better health of populations in many parts of the world. Food, no matter how nutritious, if it is not eaten serves no nutritive purpose. The proper use of additives to stabilize chocolate and other drinks, and to impart smooth texture to ice cream and other products are very good examples here. What, then, constitutes the proper use of a food additive?

Proper use of a food additive stipulates that:
1) It must be safe and should pose no hazard to the consumer.
2) It must perform the useful function claimed for it.
3) It must not conceal or cover up any fault or deficiency with respect to ingredients or poor manufacturing practices—in short, there must be no deception of the consumer and good manufac-

uring practices should not be sacrificed by the use of an additive. 4) There should be no significant reduction in the nutritive value of the food due to the presence of the additive. All of the above-mentioned points relate directly to the safety of foods, and government agencies, manufacturers and civic (activist) groups are all vigilantly seeing to it that the consumer is well protected.

FUNCTIONALITY

The functions of food additives are manifold. They are expected to confer benefits not only to producers and manufacturers of food, but ultimately to consumers, both individually and collectively. The purposes are directed toward facilitation or improvement of production, toward protection against spoilage or deterioration caused by microorganisms or pests, toward better and more sanitary handling and transportation, toward enhancement of nutritional value and organoleptic and esthetic acceptability, and toward greater convenience of preparation both institutionally and in the home.

None of these objectives could be justified if the end product turned out to be unsafe or hazardous. Hence, in contrast to days gone by when food chemicals could be used more or less indiscriminately and with relatively little regulatory control, users are nowadays required by law to establish with reasonable certainty that no harm will result from their intended use The chemical, physical, and technological procedures that serve as a background for the toxicological evaluation of the safety of food additives are beyond the scope of this manuscript. However, the general principles upon which approval for use rests, either in the form of a permissive regulation or as an exemption from such a regulation, will be discussed briefly.

GRAS

One of the basic conditions upon which such approval is predicated in the United States is the general recognition by properly qualified scientific experts that any substance in question is safe under the conditions of its intended use. This opinion may be based either on toxicological or other scientific evidence, or on experience arising from the common use of the substance in food. Judgments based on human experience, like all epidemiological assessments, however, are frequently limited by the absence of adequate control data and of sufficiently objective criteria. Foods themselves are composed of an infinite number of chemical substances of varying degrees of complexity, and their safety, like that of added chemical substances, should be evaluated on the same basis in so far as experience is concerned. This may not be too difficult when dealing with acutely toxic products, but it is extremely difficult when only moderate, subtle, or chronic degrees of hazard are being evaluated. Hence, long-term animal feeding studies must be relied upon for predicting potential hazard under conditions of use, rather than evidence based on human usage.

Feeding studies may extend over periods as long as two years or more for small animals, such as rodents, and even longer in the case of larger mammals, such as dogs or monkeys. Despite the many parameters of physiological and pharmacological effect and the prolonged periods during which tests and observations are made in the course of these studies, adequate allowance must be made in the transition from "no-effect" dose levels in animals to acceptable intake levels in man. By experiments of this type, reasonable certainty of safety has been assured in the permissive use of food additives. It cannot be said, however, that safety is ever absolutely guaranteed, even when animal studies are followed by investigations in human subjects. Tests in man are necessarily limited not only as to the number of subjects and the short duration, but by ethical and legal considerations that restrict the frequency and type of observations and, of course, prohibit sacrificing for postmortem examination. Individuals vary in their physiological states, in their nutritional requirements and habits, and in their idiosyncracies and hypersensitivities. Hence, studies with relatively small numbers of normal human subjects cannot be regarded as definitive or as having universal application.

DIRECT AND INDIRECT ADDITIVES

Some substances are ingested as nonintentional components of foods, present by virtue of the treatment of raw agricultural commodities with pesticides, or the feeding or medicating of food-producing animals with growth promoters or drugs. Incidental, so-called indirect additives may be present in foods by transfer during contact with equipment, containers, or packaging materials. Hundreds of constituents of packaging materials, such as polymeric resins, paper, can lacquers, adhesives, etc. are regarded as potential migrants into foods with which they might come in contact. The extent to which they occur in foods depends upon such factors as solubility, surface area, temperature, and time of contact. The presence of industrial, animal, and food wastes in streams and the chemical treatments to which water supplies are subjected are also responsible for the intake by man of various "foreign" substances, albeit in very low concentration. To round out the picture, attention must also be directed to the accidental ingestion of chemical substances, often due to ignorance or carelessness in the use of the great variety of chemical preparations employed in the home for cleansing, sanitary, decorative, or hobby purposes.

Proper use of additives, though apparently more directly pointed at the intentional ones, can be extended to indirect additives too, because the careless use of insecticides, pesticides, etc., could result in their entry to foods causing some maleffects and concern. At times such improper use may arise not deliberately but through carelessness and mistakes. The case of the PCB episode is a classical but unfortunate example. Pollution of streams and waters from which fish are taken for human consumption has been well documented and the consequences of pesticides and mercury and other heavy metals in food sources, attest to improper disposal and/or handling of these substances.

The proper use of chemicals (indirect additives) on crops is important, too. An example is the application of nematicides to certain crops like plantains and bananas. This should not be done after the plant is six months old so as to avoid uptake and residual chemicals in the fruit. Similarly, when pesticides are applied to certain vegetables, these crops should not be used as food for at least seven days or more, as specified on the instructions.

The allegation, either expressed or implied, that the use of food additives is contrary to public interest, in that they permit deception of the consumer, is unwarranted. Under our law, a food is deemed to be adulterated "if damage or inferiority has been concealed in any manner; or if any substance has been added thereto or mixed or packed therewith so as to increase its bulk or weight, or reduce its quality of strength, or make it appear better or of greater value than it is." These rigidly enforced requirements are taken into account when permission is sought for the use of a food or color additive.

GOOD MANUFACTURING PRACTICES

Certain of the basic requirements for the proper use of food additives are encompassed in what has been designated "good manufacturing practice".

While it has never been possible to define it in sufficient detail to apply to all aspects of food prcduction or processing, it is useful to point out that in relation to food additives, good manufacturing practice has been defined as follows: (Sec. 121.101(b) of the Food Additive Regulation)

(b) For the purposes of this section, good manufacturing practice shall be defined to include the following restrictions: (1) The quantity of a substance added to food does not exceed the amount reasonably required to accomplish its intended physical, nutritional, or other technical effect in food; and (2) The quantity of a substance that becomes a component of food as a result of its use in the manufacturing, processing, or packaging of food, and which is not intended to accomplish any physical or other technical effect in the food itself, shall be reduced to the extent reasonably possible. (3) The substance is of appropriate food grade and is prepared and handled as a food ingredient . . .

Until recently, there were no standards adopted, especially for foodgrade chemicals. However, through the efforts and sponsorship of the Food Protection Committee of the National Academy of Sciences-National Research Council, a Food Chemicals Codex was developed and published in 1966. It consists of monographs describing the specifications and standards of identity, quality, and purity of about 500 chemical substances known to be employed as direct additives in food production and processing. This compendium covers most of the principal food chemicals in common use. Though a great many of the less frequently used substances (including flavoring agents and indirect additives) are not included, it should not be assumed that these substances are either unsafe or unlawful.

Regardless of how small the risk may be in the use of chemicals in foods, it must be balanced against the potential benefit to be derived. In countries where food production lags behind the needs of the population, the chief reasons for using chemicals in agricultural and industrial production of foods are to improve the yield and quality of crops and livestock; to prevent or control diseases, parasites, and insects that prey upon plants and animals, and to minimize post-harvest spoilage and deterioration. In the light of present-day knowledge and techniques of pest control, it seems inexcusable that man should yield so much of nature's bounty in insects, fungi, rodents, weeds, and other pests. In many developing countries food crops are exposed to great damage and loss due to the ravages of such pests. The yields of the principal economic crops of the countries could be greatly enhanced by improvement of cultural practices, including the proper use of pesticides.

Fungal diseases, insect infestation of crops in developing countries, especially those in the humid tropics, can effectively be controlled by the appropriate application of pesticidal sprays and dusts, which could result in substantial economic and nutritional benefit to the population. Increased production of poultry and livestock and higher yields of milk, eggs, and meat through the use of nutritional concentrates, growth stimulants, and veterinary drugs (where applicable) can effect substantial improvement in the protein content and quality of the diet in countries that now depend so heavily on low protein foods, such as rice, plantain, and cassava. Any uses of chemicals or drugs to promote better nutrition must certainly be considered advantageous, provided residues or metabolic intermediates present no hazard to the consumer.

Among the most serious obstacles to the distribution of foods from producing to consuming areas are the problems of transportation and storage. The lack of adequate highways and refrigerated vehicles and the practice of shipping in bulk in truckloads, instead of in easily handled and stored containers, are well-recognized obstacles to the utilization of existing food resources in many developing communities. In such areas, the use of protective agents against insects and mold, and preservatives, would be beneficial.

It is apparent from these remarks that, whereas the criteria for acceptable (i.e., safe) dietary intake of food chemicals may have international applicability, the justification for use of these substances may vary in different parts of the world. Such factors as the extent to which indigenous resources meet the nutritional needs of the population, the local technological and distribution facilities, and cultural practices and religious preferences are among the considerations that determine the propriety of using a given food additive, even though its use would be safe. A country that enjoys the benefits of a temperate climate, excellent transportation and storage facilities can hardly justify the need for chemical preservatives to the degree that a less favored, tropical country might.

Yet, such more advanced and industrialized countries may be more liberal in permitting the use of food additives for less vital purposes, such as contributing greater eye or taste appeal and for promoting convenience in home preparation. The number of permitted food additives in the United States, for example, number hundreds, comprising vitamins, minerals and other substances of both a natural and synthetic nature. It seems unlikely, therefore, that food laws and regulations will ever be adopted on a uniform, international basis.

THE NEED

In general, it may be stated that in a rapidly expanding world, the unbridled advance of industrialization with the consequent dependence of a proportionately larger segment of the world's inhabitants on fewer agricultural workers, it seems possible or even inevitable that safe chemicals will be necessary to protect crops in the field, during storage and transportation and, during preparation, to enhance the final quality, flavor, aroma and acceptability of foods.

Current efforts to supplement the world's food protein supply by the use of unconventional resources such as single cell protein, and leaf protein concentrate will undoubtedly necessitate the proper use of safe additives for functional properties and to promote flavor, aroma and consumer acceptance.

Accelerating activity of weight concious groups will, most likely, lead to increased use of such additives as supplements and fortifiers (vitamins, minerals, amino acids, etc.). At the same time, the upsurge of natural food addicts, could produce more and more criticism about some of the even allegedly perfectly safe and nutritive substances added to foods.

As more women, especially mothers, join the working force, home cooking will become a drudgery rather than the hithertofore family pleasure, and the so-called "liberated women" will turn more heavily to convenience foods and instant mixes. The formulation and/or preservation of these will depend very much on the availability of safe food additives. Fastfood outlets in metropolitan areas, college campuses and around large industrial sites will likewise be using such foods, including drinks and desserts.

Over the past few years, there has arisen national as well as international sensitivity to the use of additives in foods. Aggressive consumer groups have questioned and challenged the use of certain food additives, and have even, on certain occasions, attributed malfunctions and maleffects to their presence in specific foods. At the same time, government agencies, fully aware of their mandate to protect the public, has conducted extensive re-evaluations of the permitted food additives and have even withdrawn many from the GRAS list. Industry, likewise, keeping closely abreast of consumer awareness and government watchfulness, has shared in the responsibility to supply safe and nutritious food by doing, promoting and financing fundamental as well as applied research aimed at producing the safest, most useful and functional additives.

Flavorings, spices and preservatives such as salt, have been used in foods since ancient times. However, the true functions of these were not recognized until fairly recently. More and more additives were discovered and used. Then, as climate, political and economic situations affected the supply of natural spices, flavorings, colorants, etc., ingenious chemists and Food Scientists pioneered in the synthesis of identical compounds for use on a commercial scale. Not long after, the great argument of "Natural" vs. "Artificial" arose and has not yet abated.

Often it is assumed that a substance, as it exists in the "product" itself is natural. For example, vitamin C as it occurs in an orange, grapefruit or in 'rose hip' is natural. This identical product, with the same chemical properties and physiological functions, if it is synthesised in the laboratory, may be considered artificial. A wave of criticism has at times arisen over the use of artificial additives in foods. Such artificiality has often been associated with and even equated to toxicity. Such accusations have led scientists to point out that "natural" does not necessarily mean safe and synthetic or artificial does not necessarily mean toxic. In fact, many foods which are eaten daily contain potent toxins (Strong, 1973). Examples include cyanogenetic glycosides (cassava, almonds, lima beans, etc.), nitrates (vegetables, beets, broccoli, cabbage), mushroom poisons, antivitamins, enzyme inhibitors (trypsin inhibitors, etc.). The history of mankind has demonstrated most unequivocally that the proper use (selection and/or preparation) of foods containing "natural toxins" results in their daily, unrestricted and safe use for the alimentation of mankind. The cassava, for example, contains the potent cyanogenetic glycoside Linamarin. However, by a short, natural fermentation and the subsequent heat in cooking, the potent cyanide therein is totally eliminated. Few foods are eaten in such quantities universally as the white potato. This tuber contains solanine alkaloids but by reaping them at the proper stage of maturity, storing them properly and allowing for the necessary curing, the dangers of toxicity are completely eliminated. Rhubarb, sorrel, spinach and certain other leafy vegetables contain oxalate which interferes with calcium metabolism. Likewise, phytates in cereals will reduce the assimilation of calcium from the diet.

Excessive use of Vitamins A and D, two essential vitamins, will lead to toxicity and excessive use of sodium in diets will be deleterious for humans, especially for those who suffer from hypersensitivity. Excesses of the most essential minerals may be bad. However, the proper use, i.e. amounts, of these will assure good health. All chemicals, therefore, including foods and food additives, if properly used, can satisfactorily benefit the consumer. Proper usage rests with the manufacturer, formulator and the consumer. It is to the credit of manufacturers, government agencies and consumer groups that concerted and wholehearted efforts have been and are being made to assure that additives, both direct and indirect are being properly used so that man, his food and his environment will be healthy and safe.

Only recently, the HEW established a National Toxicology Program within the Public Health Service to strengthen HEW's activities in the testing of chemicals of health concern, as well as in the development and validation of new and better test methods. (Federal Register, Nov. 15, 1978 through Food Technology, January, 1979).

On all fronts, efforts are being made to assess possible hazards of food additives (Cramer *et al.*, 1978, Oser and Hall, 1977), and it is certain that the proper use of additives, practiced by all concerned, will undoubtedly, result in benefits to mankind.

REFERENCES

CRAMER, G. M., FORD, R. A. and HALL, R. L. 1978. Estimation of Toxic Hazard - A decision Type Approach. Fd. Cosmet. Toxicol. 16: 255.

OSER, B. L. and HALL, R. H. 1977. Criteria employed by the expert panel of FEMA for the GRAS evaluation of flavouring substances. Fd. Cosmet. Toxicol. *15*. 457.

STRONG, F. M. ED. 1973. Toxicants Occurring naturally in Foods. 2nd Edition. National Academy of Sciences. Washington, D.C.

21

Regulating Additives From Food Contact Materials

G. W. Ingle

THE PRESENT REGULATORY STRUCTURE

Manufacturers of food-contact materials have now had two decades of experience with the 1958 Food Additive Amendment to the U.S. Federal Food, Drug, and Cosmetic Act. This period had produced 150 FDA regulations comprising 6,000 individual chemical compounds prior to the April 1977 realignment of these regulations, in 21 CFR 171–186 to provide a simpler classification by broad category of composition or use. These compounds are identified as "indirect" food additives for specified food-contact uses. Some compounds are legally exempt from the requirement for regulation in certain uses because of "prior sanction",[1] or general recognition of their safety for those uses.

The 1977, realignment improved the structure and organization of these regulations, which are largely specifications. Most of these describe specific uses for which lists of certain chemical compounds are permitted. For purposes of enforcement, control is exerted by a limiting analytical test, such as maximal total extractables determined under given conditions. Relatively few are concerned with the safe use of individual chemical compounds—to provide some guidance to the formulation of other new materials. Almost no regulation defines or refers to actual or safe levels of such com-

---

[1]Before 1958, FDA "sanction" was indicated informally by letters of considered "no objection" to specific chemical compounds or polymeric or other compositions in designated uses.

pounds entering the food supply through the prescribed uses. This lack of reference to quantitative levels in the diet, for these and for prior sanctioned, or generally recognized as safe ("GRAS") compounds, has been frequently cited as the justification for requiring new petitions for further regulations of new or different uses of the same compounds.

By contrast, the levels of the "direct" or intentional additives in food are usually far greater and generally, but not always, explicitly defined (National Academy of Sciences—National Research council 1965). Again, in contrast, the description of the uses of these direct additives is nearly always subordinated to identification of each compound. In round numbers, 200 regulations for the same number of *direct* additives use two-thirds the verbiage used in the regulations (since April 1977) for 6,000 *indirect* additives. This overt emphasis on use of indirect additives frustrates systematic structuring. This greater emphasis on relatively few uses—rather than identities—of many indirect additives thus requires regulating new or other uses of chemical compounds frequently already listed. Aside from requiring repetition of the entire petition-regulation procedure, this emphasis diverts attention from the significant problem of safe levels of specific chemical compounds in the diet.

There is widespread opinion that indirect additives present no hazard which justifies the type and extent of control applied to intentional or direct additives. The Canadian Law embodies this concept.

NEW OBJECTIVES

A systematic approach is needed that will provide an explicit basis for minimizing concern for food-additives where this is not needed. At the same time, attention must be directed to those relatively few cases where assurance of safety requires effective controls. Basic to this should be the rational concept so well expressed by Holland's Dr. De Wilde (1966) that toxicity is a continuum on which each chemical compound is characterized by a quantitative level. Below this "no effect" level, long-term use by humans is safe. With regard to carcinogens, the "Delaney Clause" in the 1958 Food Additives Amendment effectively denies the existence of such threshold levels. More recently, however, some pharmacokinetics studies suggest that carcinogens at low levels—higher than those characteristic of relevant uses as indirect additives, especially residual monomers,—may be safely metabolized by test animals, and thus, by extrapolation, by humans. Confirmation of these threshold values, by experimental means more explicit than epidemiological evaluation, remains to be demonstrated.

On September 14, 1966, in New York City, a Symposium on indirect food additives was presented by the American Chemical Society's Division of Organic and Plastics Coatings, and of Agricultural and Food Chemistry. Several appraisals of the effectiveness of the U.S. system for controlling indirect food additives were made by representatives of affected industries, especially the plastics industry. If this industry continues to be most critical of this system, it is probably because of the uniquely rapid development of new plastics for food-contact purposes. A common theme at that meeting and since, is that this regulatory framework is not well-constructed to meet effectively, and at reasonable cost, its original objectives. The question was then raised as to how closely the U.S. legislators really coordinated with toxicologists, packaging technologists, and analytical chemists to define food additives. The critical phrase: ". . . intended use which results or may reasonably be expected to result in (a substance) becoming a component or otherwise affecting the characteristics of any food . . ." was originally intended to provide perspective in defining and distinguishing between direct and indirect additives. Later, however, FDA's legal staff took the position that the manufacturer's act of performing extraction tests was proof in itself that migration (of an indirect additive) was reasonably expected! The need remains for a more quantitative basis.

As a result, excessive costs and delays are incurred by manufacturers in "proving safety" of new food-contact materials. Even so, present procedures provide no evident reliable basis for determining the overall input of chemical compounds into our food supply, or of their safety to consumers, whether these compounds be direct, indirect, helpful, or objectionable additives. It follows that there can be no reasonable basis for comparing the levels of these food-additives with those of the chemical compounds occurring naturally in our foods. Some of these compounds are in both categories.

A PROPOSAL

A proposal is made here to re-orient this regulatory scheme. Primary emphasis is on defining safe and actual levels of individual chemical compounds acting as direct or indirect additives from all sources.

Safe levels have not been and need not be determined for many presently regulated chemical compounds. Many compounds, as in the now obsolete "Adhesives" regulations, formerly 21 CFR 121.2520, are explicitly not migratory in such uses. Thus, safe levels need not be established for these or other compounds which are actually not food additives because they remain in the food-packaging material, because of experimentally demonstrated "no-migration" or barrier films, or other construction or design. Further, energetic exploitation of the concept of "general recognition of safety (of specific chemical compounds) for the intended use", coupled with periodic review of new toxicological information, would eliminate the need for determining safe levels for many other chemical compounds already so characterized for specific uses, such as certain flavoring substances.

The proposed scheme begins with an alphabetical order of relevant chemical compounds available commercially. Provision would be made for accumulating the estimated or determined orders of magnitude of concentration in food of each presently sanctioned, regulated, or petitioned chemical compound. For indirect additives, this is to be expressed in terms of probable levels developed by migration into the food supply. The estimated or determined orders of magnitude of safe levels are compiled on the same scale.

This list—excluding levels—might be treated as a subset of the annual inventory of chemical substances created under section 8 (b) of the 1976 Toxic Substances Control Act (P.L. 94–469).

PROBLEMS TO BE SOLVED

Determining such levels is a major and continuing problem. Of the two major unknowns, the first comprises the actual levels of migration into real foods. This is generally not the same as laboratory measures of extraction by food-simulating solvents. The second is the frequency of occurrence of these specifically packaged (or otherwise contacted) foods in the entire diet. For proposed new food-contact materials, order-of-magnitude estimates are the best type of data that can be expected. For established uses, data from packaging material manufacturers and food packers could be made more accurate by periodic analyses of markets and diets. Allowance must be made for various categories of diets, such as those for infants, children, adults, those on diets controlled for medical reasons, etc.

Until the early 1970's, the usual evaluation of the safety of a chemical substance, for example, intended for use in a plastics formulated for food-contact, involved in essence, the development and comparison of two classes of information. First, the amounts of the substance extracting in food-simulating solutions at equilibrium conditions, were determined, following the general analytical schemes published by FDA in the then regulations CFR 121.2514, and in its 1976 "Guidelines for Chemistry and Technology Requirements of Indirect Food Additive Petitions". In most cases, all other migrating ingredients in the formulation were deliberately selected from among those earlier sanctioned or regulated. Also, such tests generally evaluated the food-contact material for specified types of food.

Second, in the absence of reliable animal-feeding data showing a "no-effect" level, such information would be obtained from studies designed to include a range of concentrations (in the test animal (s) diet) related to those found in the extraction-tests. While this implied that the entire test-diet consistently contained these concentrations of the substance being tested, ultimate conversion to "no-effect" levels for humans would be provided by dividing no-effect levels found in 90-day animal feeding studies by 1000, and by 100 for two-year studies.

Thus, if the least non-interpolated dietary concentration showing "no-effect" in the animal studies exceeded the equilibrium extraction level, the substance would be a candidate for likely regulation, under the test-conditions, for the intended use. Conversely, if the reverse were true, and could not be changed by reformulating and repeated testing, the substance could not be considered for regulation.

In the early 1970's the finding that vinyl chloride monomer was carcinogenic created the possibility that other vinyl monomers, and, in general, other classes of substances, would prove to be carcinogenic, if tested adequately. Thus, the established procedure, described above in oversimplified terms, has been increasingly recognized as valid only to the extent that the substance could be proved to be non-carcinogenic. Due to the "Delaney Clause" in the 1958 Food Additives Amendment to the Pure Food and Drug Act, FDA may not regulate, for use directly or indirectly in food, at any concentration, no matter how low, any substance shown to be carcinogenic at any concentration, no matter how great. It follows that there is great impetus for recent and current programs to develop relatively simple, rapid and inexpensive tests for carcinogenicity, so far via bacterial and other mammalian tissue tests for mutagenicity. These are part of the widening range of activities to understand and control chemical-carcinogenicity, in food-packaging materials and in all other impacts of chemical substances on health and the environment. One critical, and so far unresolved, question is whether experimental proof can be found for the existence of a "threshold" level of a chemical carcinogen, below which human and other animal detoxification and repair mechanisms will tolerate exposure to such substances.

A Format

A format for organizing such information, and in quantitative terms, where appropriate, would comprise these headings:

(1) Name of chemical compound.

(2) Safe level in Human Diet (ppm), with significant literature references.

(3) Incremental levels assignable to each use, with designation for the corresponding direct or indirect food additive regulation.

(4) The sum of the individual levels listed in item 3 (above), and subject to periodic revision.

The major implications of this format are these:

(1) The safe levels are experimentally determined for individual chemical compounds that are, in fact, additives by virtue of demonstrated migration from contact-materials to foods.

(2) Once these are determined by acceptable procedures, these safe levels are officially publicized for use by all concerned investigative, manufacturing and consuming sectors of the public.

(3) The total numbers of compounds of interest overwhelms the total available toxicological resources—thus, bases for judgment, analogy, comparison, and interpolation must be identified and exploited, in appropriate balance.

(4) The cumulative levels are merely estimated here, to an order of magnitude; in actual practice, their determination by periodic statistical inspection of markets.and diets becomes a matter of first importance.

(5) Comparison of safe with cumulative levels will indicate an estimated degree of hazard and the corresponding degree of effort appropriate to their control, for specific compounds and combinations. If justified by this comparison, regulation could require nothing more than listing the new use.

To define these factors, several analytical schemes need to be developed and standardized. (A schematic plan with such calculation has been presented [Ingle, 1967]). The direct measure of migration of radioactive-tagged ingredients into real foods under actual conditions of packing and storage has been described (Booher and Stringer, 1967). A less costly alternative proposed by several investigators is to develop correlations between such data and established extraction tests using food-simulating solutions. Such correlations would probably be more valid, if they were developed for specific categories of food. These groups could be those established in important existing regulations such as 21 CFR 177 Indirect Food Additives: Polymers Sub part B, 177.1010—.1980 Substances Used as Basic Compounds of Single and Repeated Use Food Contact Surfaces. Perhaps those classifications used in the USDA (Household Food Consumption Survey 1975) or FAO (Rept. Food Composition Tables) studies of diet-components would be better. In his analysis of the extraction process, De Wilde (1966) points to the work of Gharlanda and others in determining the effects of diffusion coefficients of polymers on their extraction by various solvents. What may be equally important is defining migration into foods in which liquids are minor components. This would determine the diffusion, distribution, and limiting concentrations—in such foods—of chemical compounds from contiguous plastics and other materials.

Validation of Extractant

In such attempts to correlate extraction and migration, explicit attention should be given to a long-recognized, but rarely quantified factor mentioned by De Wilde, viz., validation of extractant by showing that it does not attack the packaging material. In packaging practice, of course, long-term trial and error, if not explicit tests, will identify and eliminate specific uses that involve mechanical failure of the package in contact with food. To avoid the waste, if not hazard, in such uses, the proponent should convince himself of the utility of the intended use by environmental tests for the onset of brittleness, changes in taste and odor, etc. Tests for the latter with intended foods are well established. Tests for onset of brittleness of plastics have been devised and published (Cohen and Haslett, 1964), but are apparently not yet used for these purposes. Showing that a given extractant—for a specific time-temperature exposure—produces early embrittlement of a given plastic is evidence that this extractant and, conceivably, the foods it simulates should have only explicitly limited, if any, contacts with that plastic. These limitations should be used to avoid use of overly aggressive extractants or extracting conditions or even packaging of certain reactive foods.

Determination of Migration Into Total Diet

Because of the central nature of this problem, the development of standard procedures and of factors determining actual migration into foods and the distribution of such packaged foods in the diet might be funded by Federal, if not by broader industry groups. Supplemental research by individual manufacturers or industry groups could provide further detail for more specific materials and intended end uses.

A striking proximate analysis of this type, involving some assumptions, was presented by Frawley (1967) at the American Chemical Society Symposium mentioned previously. He aimed to compute the *"probable* consumption (of an indirect additive) by man"—the focus of the 1958 Food Additives Amendment—as a function of the concentration of a specific chemical compound in paper used for food packaging. The formulations tested were 1, 2 and 4% of rosin size in paper. The rosin size was radioactively tagged to facilitate measuring its migration under realistic conditions into real foods representing the recognized diet- or commodity-groups as an average U.S. diet. He found that each 1% of rosin size in the paper contributed, on the average, 2 ppm of rosin into a diet so packaged. Assuming that no more than 25% of a real diet would be so packaged, each 0.2% of rosin size would contribute no more than 0.1 ppm in a total real diet. Thus, with one critical assumption that could be verified by methods above, methodology is given for relating concentrations of an ingredient in a food packaging material to that in an average or specific U.S. diet.

With Specific Exceptions, Ingredients of Packaging Materials Are Safe

Frawley combined this result with that of an analysis of the probability of toxicity of such ingredients. Diligent search found 143 two-year feeding studies. With the help of fellow toxicologists, he later (Frawley, 1967) increased this number to 220, perhaps 90% of all such studies performed up to that date, representing perhaps $15,000,000 in toxicologic research costs.[2] Excluding carcinogens,

---

[2]No more complete analysis is currently available, but in the latter years preceding the enactment of the 1976 Toxic Substances Control Act, it is estimated that the number of such chronic feeding studies increased substantially, and will continue to increase, in response to regulations to be promulgated, especially under section 4 of this Act.

heavy metals and pesticides, which average 100 times as toxic as the average of other compounds so tested, all *packaging ingredients* below 0.2% in the packaging material—or below 0.1 ppm in the diet—could be considered safe, based on these studies.

This sweeping conclusion would be a highly desirable simplification. Translated to our proposed scheme, it would mean that all uses of ingredients of food-contact materials providing no more than 0.1 ppm in the diet, could be quickly recorded and permitted without further regulatory requirements, such as proof of safety. Excluded, of course, are heavy metals, pesticides, and carcinogens, if any, for which threshold levels have not been adopted.

While this simplification is indeed desirable, broaderscale verification is needed for other food-contact materials, especially plastics, in different types of food-contact. These should include, for example, processing of whole packaged meals at cooking temperatures and liquid foods. Nor does this particular analysis provide for the occasional, in some cases frequent, possibility of one ingredient being used in more than one of the five classes of packaging materials listed by Frawley: glass (closure), metal (coated), paper, cellophane, and plastics. Nor does it account for the food faddist, or other individual with unusual food preferences or requirements, who chooses a diet so unbalanced as to greatly disproportionate the role of any one food-packaging material and its migrating compounds, if any, in his diet.

POSSIBILITIES FOR IMPROVEMENTS IN REGULATORY PROCEDURE

The proposed system provides a framework for accumulating and organizing for effective analysis, statistics of use, as well as the required toxicological and analytical data, to show incremental and total levels of each additive, direct and indirect, in the diet.

Once established and publicized, it would serve as a basis for faster administrative decisions by the proponent of the additive and by the regulating agency. With attention directed to the estimated incremental increases in the level in the diet, the petition for another use of a compound already in use, and the response to the petition, could be a highly abbreviated procedure. The new use would be listed and its incremental contribution to the total level identified. Far greater consideration could then, as it should be, be given to petitioned uses that increase the total level in the diet close to the established safe level and to uses of entirely new compounds without prior reference.

Inherent in this simplified procedure is eliminating the present time-consuming and otherwise costly analytical work that is frequently less concerned with proving safety than it is with the design of a regulation broadly applicable, for enforcement reasons to a wide range of products of different manufacturers within the scope of the described use. As Heckman (1966) has written so vividly, such writing means "providing FDA with assurance that no one (meaning another prospective proponent or manufacturer) with less integrity will be able to produce something called by the same generic name in a way that might lead to strange consequences (toxic hazard)."[3] The proposed procedure would clearly separate "proof of safety" from analytical procedures for enforcement and direct the latter to only one objective, i.e., determining incremental "probable consumption by man." There should be far less occasion for apparently conflicting, competitive concepts, unless these are explicitly rationalized. For example, there should be no controversy over specifying a food-contacting plastic by tests defining its content of a residual extractable ingredient. This can be related to migration to foods. If migration varies so sharply between aqueous and fatty foods that this relationship is not valid, or if migration is such as to approach the safe level, then more critical requirements, perhaps specific to individual foods or types of foods, must be established.

In the case of new uses of new or established compounds, provisional levels in the diet would be based on estimates made on the basis suggested. Subsequently, these provisional levels would be subject to adjustment on the bases of actual practice. An equally important result of continuous monitoring of our food supply could be the *reduction* of levels of chemical compounds obsoleted or restricted or banned for whatever reason.

Further Implications

This proposal is considered as an objective that could be fully functional in two years or less after adequate preparation and publicity. The first step towards its achievement could be taken more rapidly in terms of analyzing, refining, and establishing the concept and framework. Present regulations are largely compatible with systematic sequential changes.

---

[3] Words in parentheses supplied

Among the many implications of this proposed system are these:

(1) Systematic re-evaluation of the real and quantitative importance of direct and indirect additives, for both new and traditional food-contacting materials, as these affect the safety of our diet. This should provide a broadly acceptable basis for dismissing from concern broad classes of materials and uses, to confirm Frawley's well-documented thesis that the very nature of packaging materials and their manufacture reduces hazard to a very low order of probability. This should identify compounds and uses of serious or growing concern.

(2) Identification of more critical problems in toxicology to which the limited resources in manpower and funds can be dedicated. Aside from centering attention on those compounds where usage has greatly increased—perhaps beyond the no-effect levels validated by earlier toxicological testing—the sheer number of new compounds to be tested requires improving the effectiveness of tests shorter than two years in duration. For example, De Wilde cites a systematic basis, proposed for use in Holland, using 90-day, no-effect levels with a safety factor of 1,000, as an alternative to two-year, no-effect levels with a safety factor of 100. Study needs to be emphasized of even shorter-term alternatives for indication of longer term-effects. These are the several hierarchal testing schemes announced by Dr. Bruce Ames, Upjohn Co., Battelle Institute, and the U.S.A. Environmental Protection Agency, beginning with bacterial and mammalian tissue for tests of mutagenicity and thence proceeding sequentially to more lengthy and detailed tests, including those for animal carcinogenicity, if necessary, to reach decision.

(3) With particular respect to plastics, shorter-term lower-cost tests need to be devised to prove the safety of extractable lower molecular weight fractions. As little as 0.01% extractable oligomer of a "new" copolymer, for example, justifies, by present thinking, demand for 90-day or two-year feeding studies. These require pounds of oligomer. The cost of isolating this amount of material equals that of the feeding study itself. As a result, many newer polymers useful in protecting or improving our food supply are not commercialized.

(4) A truly unified international approach to speed the effective management of indirect food-additives is a necessity. An excellent example of such effective, international cooperation has been that of the European Economic Community Countries, whose International Technical Bureau for Plastics Materials has been unifying the several concerned national tests prior to E.E.C. standardization. With the advent of several new national laws for controlling "toxic

substances" of all kinds and in all important uses, the beginnings of such international concern and control are occurring. Lest one becomes complacent about the effectiveness of these laws, the conflicts and ambiguities in their developments [of such laws] on the one hand, and in related areas of science, on the other, must be recognized and considered in making judgments about safety, in the absence of perfection in laws and in science.[4]

[4]"Public Participation in Toxicology Decisions," by Peter Barton Hutt, Esq., Food, Drug, Cosmetic Law Journal, pp 275–285 (June, 1977).

Food Regulations

22

Food Regulations in the Americas

Albert H. Nagel

THE RATIONALE FOR FOOD STANDARDIZATION

In any consideration of food regulations in the Americas, it should be recognized that most Latin American Countries of the Western Hemisphere base their food standards on Roman Law while the English-speaking countries have their basis in Anglo-Saxon Law. The Roman Law differs from the Anglo-Saxon approach as follows:

1. Foods and drugs are treated separately.
2. Food inspection and food analysis are separate but both are under the Department of Health.
3. Registration of manufactured food products is required in most Latin American Countries. Registration generally provides funds to cover the cost of control.

If safety regulations in the Western Hemisphere are to be harmonized, thought must be given to differences in regional philosophy. Because the principles of food regulation in North, Central and South America all had their origins in Europe in Anglo-Saxon or Roman Law, the variations which exist at present are the result, first of all, of the fundamental difference in approach of these two systems. In addition, local and regional adaptation of regulations to meet particular needs have resulted in further obstacles to international trade, particularly trade between nations in the Western Hemisphere. Total standardization of food regulations is probably an impossibility in the forseeable future. However, harmonization (the removal of contradictions, duplications, and ambiguities) is an achievable goal.

To administer food regulations, two approaches used are the development of positive lists (i.e. only those substances mentioned in the regulations are permitted) or the negative list concept which considers that all foodstuffs not expressly prohibited can be used. In addition to these two philosophies, a third approach is a combination of some of the elements of the positive and negative lists (Bigwood and Gerard, 1967). In the world-wide harmonization of food regulations, the reconciliation of these different systems is a major consideration.

The basic intent of food regulations is to prevent fraud and to protect the public health.

In considering food standards, the stage of development of the countries concerned must be taken into account, since needs are different. The degree of industrialization; an urban or rural population; the existence of nutritional deficiencies; problems in food production, transportation, storage, and distribution must all be identified. Furthermore, it is possible for a country to develop or to adopt food standards and, yet, be physically unable to enforce such regulations for reasons of geography, economics or because of the difficulty of controlling a large number of small producers. Throughout the world, there is an increasing awareness of standardization in regulations and the role of standards is an ever-changing one.

Nations such as the United States which have long had food regulations are now refining and developing stricter interpretations of their regulations. Developing nations which cannot afford to establish their own standards are obtaining them readymade from organizations such as Codex Alimentarius. The Joint FAO/WHO* Food Standards Program being undertaken by the Codex Alimentarius Commission under the aegis of the United Nations is a most appropriate and effective approach to the development of international standards.

Since the world's governments are in a state of regulatory ferment, the food industries of the world find it difficult to forecast the future. Multi-national companies must, at present, develop different versions of the same product to meet differing national regulations. There are obvious economies to be realized, if a single formulation for a food product would have acceptance, if not throughout the world, then at least on a regional basis.

Mention has already been made that regulations should be developed only if a mechanism exists for their enforcement. In this regard, it is important to make regulations as general as possible. If necessary, supplementary, more detailed laws can be provided.

A common problem throughout the world is the difficulty of

---

*FAO - Food and Agriculture Organization
WHO - World Health Organization

maintaining food regulations up to date, as new technology emerges and consumer habits change.

An effort should be made to simplify regulations and reflect current consumer needs and current technological practices. While every effort should be made toward elimination of barriers, there may be certain regional and national differences which must be maintained.

If the basic food laws are kept general, then the goal of maintaining regulations current is more achievable. Foods which are perishable or which must be carefully handled in order to avoid health hazards require more detailed specifications. Foods which are not sensitive or which are processed to render them stable do not require as much detail. The main objective is to provide a wholesome food product which is safe and which is informatively labeled. A harmonized system must make provision for such cases.

Work toward uniformity of food standards is a long-term activity involving much cooperation between nations and regulatory bodies (Barrera-Benitez, 1975). Unless the task is undertaken objectively, and approached from an international basis, there is a risk of creating problems where none now exist, resulting in retrogression rather than progress.

MICROBIOLOGICAL STANDARDS

In the development of international food standards and in the harmonization of those standards, one of the most difficult areas of reconciliation is that of microbiological criteria for foods. The International Commission on Microbiological Specifications for Foods (ICMSF) of the International Association of Microbiological Societies (IAMS) has been engaged in a program to develop microbiological specifications for foods (ICMSF, 1976). In this effort, they have appraised the public health aspects of the microbiological content in foods, particularly in those products of international interest. They have considered recommendations for international analytical methods and guides to the interpretation of the significance of microbiological data. While the Commission has dealt principally with the needs of countries with more advanced food technology, it has not overlooked the needs of the developing countries.

The ICMSF which is chaired by Dr. F. S. Thatcher* of Canada also has representation from the U.S., Argentina, and Paraguay,

---

*Formerly of the Health Protection Branch (now retired)

and many other countries. In addition, there is a Latin American Subcommission with representation from Paraguay, Ecuador, Uruguay, Peru, Venezeula, Chile, Argentina, and Brazil.

Besides dealing with microbiological methods, a comprehensive sampling plan has been developed which takes into account the more centralized production of foodstuffs by large food companies as well as the international commerce in foods.

The Commission has so designed its recommendations as to be within the range of competence of its members and advisors and, therefore, excluded the fields of parasitology, virology, and mycology. National food control and public health authorities, food manufacturers and international organizations dealing with microbiological food questions will benefit from the recommendations of the commission. The work of the commission is the result of five years of study by ICMSF members and consultants.

A workshop on microbiological standards was held in Mexico City in 1970 and in Ottawa, Canada in 1973 with three other workshops being held in other parts of the world in the intervening years.

The proposed criteria provide a rational basis for microbiological standards which can be made more or less stringent depending on the circumstances. The thoroughness of the examination of a food would be related to the severity of the potential hazard anticipated. The proposed plans represent the best judgment that can be made with available experience and data.

An important innovation is the adoption of a plan involving three classes of quality (acceptable, marginal, defective). Most plans are necessarily limited by the number of samples which it is practical to examine and by the ability of producers to meet more rigorous criteria. The Committee feels that, while control of processing and handling at the source offers the best consumer protection, the long-range prospects for sampling at ports of entry are good.

At present, the primary need in the attainment of microbiological food safety is improved plant quality control. In the meantime, microbiological analysis of the product as received remains essential and the tendency to increase such sampling will continue. Faster analytical methods are needed which will permit the examination of very large numbers of samples rapidly and at low cost. Commission members are already working on such developments.

A principal objective is the acceptance or rejection of food shipments on a statistical basis, eliminating the need for making such decisions on a casual sampling basis. Because of lack of adequate data, not all foods in international trade have identified sampling plans. The Commission has initiated long-term computer projects to collect such data on the microbiological content of foods and to estimate their relationship to such factors as disease involvement, quality of processing and production, and the effects of statistical quality control on food quality and health. Regulatory agencies, according to the Commission publication, will not protect consumers if they fail to accept laboratory findings or fail to act on them.

NORTH AMERICAN FOOD STANDARDS

The United States of America and Canada are the northern members of the Western Hemisphere. In the U.S., regulation of food-stuffs which cross state borders is a federal government responsibility. Each state possesses its own laws regarding foodstuffs but the principle of federal preemption exists in the event that state and federal regulations conflict on food products moving in interstate commerce.

The basic food regulation in the U.S. is the Federal Food, Drug, and Cosmetic Act of 1938 and its important amendments, the Food Additives Amendment of 1958 and the Color Additive Amendment of 1962 (U.S. Code, 1962). The use of food additives and the regulation thereof has been the subject of considerable interest worldwide in recent years and justification for their use has been viewed from different perspectives depending on national requirements.

The U.S. in many respects has a unique body of regulations found nowhere else in the world. Examples are such approaches as the GRAS List and the Delaney Clause. GRAS is an acronym for Generally Recognized As Safe. This is a device which the U.S. Food and Drug Administration has adopted to give endorsement to those substances which have had many years of use and for which there is no evidence of any harmful properties. This GRAS approach has not received acceptance outside of the U.S. It is currently undergoing toxicological and technological review to support its continued use.

Another U.S. idiosyncrasy is the Delaney Clause of the U.S. Food, Drug and Cosmetic Act which states, "No additive shall be deemed to be safe if it is found to induce cancer when ingested by man or animal, or if it is found, after tests which are appropriate for the evaluation of the safety of food additives, to induce cancer in man or animal".

The United States has recently experienced another metamorphosis in the enforcement of its food regulations.

A proposed ban on saccharin by the Food and Drug Administration became the subject of Congressional hearings and review. The result was an 18-month moratorium on the ban while additional scientific data is obtained. At the same time the risk/benefit issue will be considered with respect to food additives in general.

The Congressional action points to a relatively recent problem in the development of regulations. As methods of analysis (chemical and toxicological) become more and more sophisticated, the detection of hazardous substances becomes more likely. When the point is reached that vital foods are rendered suspect, and this mere suspicion, by law, causes its removal from the food supply, the question arises whether the regulatory process is obsolete or at least static.

All regulations, as written, are subject to interpretation. Most laws can be adapted to fit changing regulatory situations. In the case of the Delaney Clause, its terms are nearly absolute and allow for very little flexibility.

The Commissioner of the FDA has pointed out on many occasions that he could have banned saccharin on the basis of the provisions of the general federal code of regulations without resorting to the Delaney Clause. However, that alternate route would have been far more circuitous and open to more challenge.

For many years U.S. food regulations had a group of products which had standards of identity. The elimination of this category of foods is under consideration. A principal feature of the class is a much more liberal declaration of ingredients than is required for most products.

The United States has for many years certified artificial food colors as to purity. Abandonment of this burdensome approach is under consideration. A classification of provisionally listed food colors is now undergoing study and by 1980 the provisional listing will be abandoned.

The U.S. is also unique in the field of nutritional labelling in which the percentage of certain key ingredients is prominently listed on the label of foods. The problem now appears to be that only a small part of the U.S. public understands this declaration. An educational program is in progress.

In Canada, the first food regulation was enacted by the Canadian Parliament in 1874 and prohibited the sale of food stuffs harmful to health. The Food and Drug Act of May 14, 1953 is now the basic statute for Canadian Food Regulations. Food regulations in Canada are discussed in details in Chapter 23.

Canada, as a former British possession, has considerable regulatory similarity with the United Kingdom. Because of its geographical proximity to the United States and some common interests, it not only is influenced in food regulation by its neighbor to the

South but often influences U.S. regulatory decisions. However, in general, Canada has followed an independent course from the U.S. in regulating its food products. It has chosen to ban saccharin from food products. It allows FD&C Red #2 (banned in the U.S.) but has not accepted Allura Red (FD&C Red #40) which is allowed in the U.S.

The regulatory bodies of the U.K., Canada and the U.S. have a tripartite arrangement for consultation when serious toxicological questions arise. The objective is to share information not necessarily to reach a common decision.

Canada has a special regulatory problem shared with certain European countries such as Belgium in that it is bilingual. Label declarations need more space. Regulations must take into account that literal translations do not always convey the same meaning.

Canada has many U.S. subsidiaries operating within its borders. Differences between U.S. and Canadian regulations can affect standardization by multinational food companies.

Other particular problems in Canada relate to food advertising U.S. television stations cover most of southern Canada. Food products sold both in the U.S. and Canada may comply with U.S. law but might not be in accord with Canadian policy. This problem, to be sure, is also seen in Europe and in Latin America. International Food Advertising standards have been discussed by various groups including Codex Alimentarius. However, except for the International Chamber of Commerce Voluntary Code of Advertising, no such standards have been developed nor are any likely on a worldwide basis in the foreseeable future.

This brief section devoted to the two North American countries is not intended to minimize the importance of their regulations. It rather reflects the fact that their procedures were developed long ago and are undergoing refinement. The Latin American countries, on the other hand, are undergoing tremendous growth and are changing rapidly from an agrarian to an industrial economy. The change calls for different regulations and a dramatic change in the regulatory process.

The problems of yesterday in the U.S. and Canada are the problems of today in Latin America.

LATIN AMERICAN FOOD REGULATIONS

Latin America (Central and South America) contains a diversity of people, foods, and environment. These differences are reflected in the national regulations of the southern sector of the Western Hemisphere. Yet, though there is some variation there is a basic similarity which makes harmonization a distinct possibility.

The Codex Alimentarius Commission conducted a survey of food standards in Latin America in 1969 (CAC, 1970). Codex Alimentarius divided Latin America into four major sub-regions based on their food standards, food processing procedures, and pattern of consumption of food products. These regions are:

A. Mexico, Central America and the Caribbean Islands
B. Northern and Western Countries of South America
C. Brazil
D. The River Plate Countries

Utilizing the Codex classifications, a review of Latin American regulations follows:

Mexico, Central America and The Carribbean Islands

Mexico.—Issued standards for some food stuffs and for food additives. These were published in the Diario Oficial, Toma CCXXVI, No. 38, 15 February, 1958. A Sanitary Code was issued by the Mexican Congress and was made effective in March, 1975. The Congress legislates matters affecting the public health in accordance with the Mexican Constitution. It is left to the Executive Branch of Government and the Council of Sanitation to rule on sanitary matters. The Secretariat of Sanitation and Social Welfare handles the federal enforcement and interpretation of laws. Due for revision in the Mexican food regulations are those dealing with food hygiene and with labelling. In Mexico, there are additional food regulations under the jurisdiction of each state.

Cuba.—Established a series of food standards called "Ordenanzas Sanitarias". These standards were published in the Gaceta Official No. 233, 29 November, 1956. The Ordenanzas Sanitarias cover not only definitions of acceptable and unacceptable foods but the penalties imposed for substances detrimental to the public health. Cuban Law covers not only substances which have been proven to be harmful but also substances which are the subject of scientific uncertainty.

Caribbean Countries and Guyana.—At present, Barbados, Guyana, Jamaica and Trinidad and Tobago have a considerable number of food standards relating principally to milk, butter, margarine, coffee, and sugar. The member countries of the Caribbean Free Trade

Area are considering the harmonization of food standards and labelling within the region. Such legislation will probably be based on the Canadian approach to regulations.

Northern and Western Countries of South America

Bolivia.—Issued a number of regulations concerning food but no Pure Food Laws have been fully developed.

Chile.—Issued food legislation, "Reglamento Sanitario de Alimentos" published in the Diario Oficial, No. 24789 in November, 1960. The Chilean regulations cover the prevention of the sale of adulterated meat, fish, fruit, milk, and fermented beverages which may be harmful to the public health. It gives authority to perform sanitary inspections of manufacturing plants as well as retail establishments. It covers both foods destined for national trade and for international trade.

Colombia.—The Ministry of Public Health has issued a general regulation for foodstuffs published in the Diario Oficial on April 16, 1964, No. 31343.

Ecuador.—Adopted the Latin American Food Code.

Peru.—Has its own food code (Codigo Peruano de Alimentos) published in 1963, and is undertaking the drafting of a number of other standards through the Instituto Nacional de Normas Tecnicas Industriales y Certificacion (INANTIC). The Peruvian Pure Food Code contains definitions not only for acceptable foods but for adulterated, contaminated, and spoiled foods and for mislabeled foods. A spoiled food is defined as one having undergone deterioration as a result of physical, chemical or biological change. A contaminated food is considered one which contains pathogenic organisms, toxins, or parasites, chemical or radioactive substances which could prove harmful to man or animals. An adulterated food is one which contains impurities dangerous to health, or contains any organic substance which is decomposed and, therefore, unfit for food. In determining penalties, Peruvian health authorities take into account the degree of danger to the public health.

Food imported into Peru must meet more severe requirements than food products produced within Peru. Imported foods must meet all the requirements for consumption in the country of origin and must be registered in advance with the Peruvian health authorities, as well as having a guarantee from the country of origin.

Venezuela.—The Minister of Health and Welfare has issued standards to control the quality of processed foods of animal origin. Such regulations have been published in the Gaceta Oficial de la Republica de Venezuela, April 3, 1968, #28597.

Brazil

Brazil.—Thus far, Brazil has no national food legislation. The regulations are developed by each state in Brazil depending upon individual requirements. The city of Rio de Janeiro has in force a food regulation, Municipal Decree #9688 issued in April, 1949. A similar regulation exists in the State of Sao Paulo. Normally, other states will recognize permits granted by individual states covering food products. Brazil established the basic standards for foodstuffs with Decree Law 986. The law covers the registration of food stuffs, labelling, additives, and standards of identity.

River Plate Countries

Argentina.—Law #14439 of the Organization of Ministers was promulgated in June, 1958 and contains Sanitary and Food Codes for provincial and municipal authorities for protection of the health of the consumer. The code covers all matters related to manufacture and distribution of dietetic foods, biological medicines, drugs, and mineral waters as well as foods themselves. This code constitutes a national food legislation. Argentina has issued an official Codex Alimentarius Argentino which deals with the production and transportation of foodstuffs for human consumption. The code was published in the Boletin Oficial de la Republica Argentina, #21732 in July, 1969.

Uruguay.—Food regulations consist of a series of By-Laws known as municipal regulations. The most important is that of Montevideo which is called the General By-Law for the supervision and control of foods and was issued as Executive Order #5504 (26–VII) in 1947. Executive Order 27–III was issued for dietetic products in 1953. Executive Order 16–II governing the use of food additives such as colors and preservatives was issued in 1959.

Regional Latin American Standards

In addition to the individual national standards efforts, Latin American countries have been quite active in efforts to standardize and harmonize regulations to permit for freer interchange of goods between their nations. The Latin American Free Trade Association was established by the Treaty of Montevideo in 1960 with the following members: Argentina, Brazil, Chile, Mexico, Paraguay, Peru and Uruguay. Since then Colombia, Ecuador, Bolivia and Venezuela have joined. The Central American Common Market has also been established with the following members: Costa Rica, El Salvador, Guatemala, Honduras, and Nicaragua.

An agreement was signed in 1967 in Uruguay to develop a Latin American Market by 1985 (Pacto Andino) which would incorporate the Latin American Free Trade Association and the Central American Common Market. In 1954, the first edition of the Latin American Food Code was published. A second edition was released in 1964. The Latin American Food Council is charged with keeping the Code up to date. This Code is non-mandatory but has been adopted, thus far, only by Ecuador, Paraguay, and Peru.

The Latin American Food Code contains 19 sections:

1. General Provisions
2. General Stipulations on Food Preparation and Marketing
3. Food Storage, Preservation and Processing
4. Utensils, Containers, etc.
5. Labeling
6. Meat
7. Fats and Oils
8. Dairy Products
9. Cereals
10. Sugar and Sweets
11. Vegetable Products
12. Non-alcoholic Beverages
13. Fermented Beverages
14. Liquors, Distilled
15. Stimulating Drinks (Coffee, Tea, Cocoa)
16. Additives
17. Dietetic Products
18. Animal Feeds
19. Household Goods

The Pan American Health Organization, the Latin American Regional Office of the World Health Organization, in 1965, issued a series of food health standards (Normas Sanitarias de Alimentos) for the countries of the Central American Common Market and for Panama. Some 380 standards have been submitted covering essen-

tially the same products as listed under the Latin American Food Code but with the following additional categories:
- Fish, Crustaceans, Molluscs, and Shellfish Products
- Spices and Condiments
- Salt
- Pectins
- Organic and Inorganic Chemicals

In developing these standards, a number of resolutions have been adopted including a recommendation for a common market department of legislation and control of food products.

CODEX COORDINATING COMMITTEE FOR LATIN AMERICA

In addition to the foregoing, Latin American countries have become very active in the development of world-wide standards through the Codex Alimentarius Commission.

Any effort to provide for harmonization of food regulations in the Americas will undoubtedly involve the Codex Alimentarius. Currently, a Codex regional committee for Latin America has been formed and a conference of the Codex members in Latin America was held in Mexico City in September, 1978. A most comprehensive presentation of Latin American food standards can be found in Codex document cx/Latin America 78/12.

Dr. Eduardo Mendez of Mexico has served as a Vice-Chairman of the Codex Commission giving recognition to the importance of the Latin American countries in the development of world-wide food standards.

At the first session of the Codex Coordinating Committee for Latin America, held in Rome in March, 1976, eight countries were represented: Argentina, Brazil, Chile, Cuba, France, Mexico, Uruguay and Venezuela. Dr. Mendez of the Delegation of Mexico was elected Chairman and Dr. V. Gonzalez Marval of the Venezuelan Delegation was elected Vice-Chairman. The Committee noted that a questionnaire had been sent to all the countries in the Latin America area in 1974 which attempted to define those regulatory topics which might be discussed by the Coordinating Committee. Only four countries responded to that questionnaire. The Committee also prepared to make recommendations regarding priorities for a Latin American Regional Conference.

The Delegation of Brazil suggested that particular attention should be paid to the problem of contaminants in food such as pesticides, inorganic substances and mycotoxins. In addition, suitable codes of hygienic practice and the development of microbiological specifications for foods, taking full account of the specific interests of Latin American region, should be considered.

Argentina stressed the need to develop worldwide rather than regional standards. Cuba suggested that because of the advanced state of many Codex Commodity Standards, the Latin American Regional Coordinating Committee should devote its attention to standardization of work procedures and food control services which would allow products to meet the quality levels required by the Codex Standards. Cuba recognized the need in the regional standards for methods of analysis and sampling.

The Codex Secretariat emphasized that FAO surveys already carried out, identified a number of problems to which the Latin American Coordinating Committee might address itself. The Secretariat noted that the versatility of food laws and regulations existing on a regional basis would split responsibility for jurisdiction between different ministries in various countries. The Secretariat pointed out that the Coordinating Committee for Africa and the Food Standards Council for Asia had endorsed a model food law prepared by the Codex Secretariat which is intended for member countries to compare with their existing food laws. The model law would improve the harmonization of food laws on a regional or worldwide level. A similar model food law has been suggested for the Latin American Food Standards Conference when it is held. An action plan was outlined which might be implemented following the Latin American Regional Conference:

1. Harmonization of Food Legislation
   a. Examination of Codex worldwide standards as they would apply to particular conditions prevailing in the Latin American Region.
   b. Consideration of standards for foods of particular interest to the Latin American Region including, where appropriate, regional standards.
2. Promoting Uniformity in Food Inspection Services
3. Promoting Uniformity of Food Legislation, including a Model Food Law
4. Specific Consideration of Pesticide Residues, Mycotoxins, and other chemical residues in food as well as the use of food additives, codes of hygienic practice and microbiological specifications.

It was pointed out that any regional food standards should take into account existing technical work at the regional level.

The question of membership in the Latin American Coordinating

Committee was also reviewed. At the 1978 FAO/WHO Regional Food Standards Conference for Latin America, it was the general view that countries with territories partly in the region should be observors and not members of the committee.

The provisional agenda for a regional Food Standards Conference for Latin America has, as a proposed agenda item, cooperation with Latin American economic groups, the Economic System of Latin America (SALE) and with the Latin American Free Trade Association (LAFTA/ALALC).

Many national legislations are designed to deal with local needs but, at the same time, they must be compatible with the standards of the countries with which they trade.

Latin America has one of the highest rates of population growth in the world. Consequently, food standards to assure good quality and good nutrition are a very pressing need.

In addition to the requirement for food standards to facilitate trade and to protect the public health is the necessity to adequately label foods. The labels should properly represent the contents and should not be misleading.

At the present time, national standards in the Latin American area still contain many technical and non-technical trade barriers. Nevertheless, at least some of the Latin American nations are combining their efforts to obtain more uniform standards (Zimmerman, 1972).

The organizations dealing with inter-regional regulations in Latin America are the Latin American Food Code Council, the Latin American Free Trade Association, the Pan American Standards Bureau, the Pan American Commission for Technical Standards, and the Central American Institute for Technical and Industrial Research (ICAITI).

The need for some uniformity in Latin America is great (CAC, 1976). While some progress has been made toward harmonization, the lack of unanimous agreement continues to pose problems. For example, food colors among Latin American nations have been standardized between Brazil, Chile, Colombia, Venezuela, Peru, Costa Rica. Uruguay, Bolivia and Guatemala but have not been accepted in Argentina and Mexico.

A number of Latin American Countries are revising their laws and are paying more attention to requirements for foods imported into their country. They also recognize that products produced in their nation for export must comply with foreign regulations. It, therefore, becomes important for regulatory officials in each country to advise their local food industries regarding requirements in other countries in order to avoid rejection.

CODEX WORLD WIDE STANDARDS - THEIR ACCEPTANCES IN THE AMERICAS

The Codex Alimentarius Commission is a joint activity of the Food and Agriculture Organization (FAO) and the World Health Organization (WHO). The Commission was established in Geneva in October, 1962. The membership of the Commission is open to all member states of the FAO and the WHO who are interested in international food standards. The Codex Alimentarius has a ten-step procedure for the elaboration of world-wide food standards and an eleven-step procedure for the elaboration of regional standards.

The Codex Standards which have been developed are minimum standards and thus have universal application. The acceptance procedure allows for several types of acceptance as well as for rejection. At this time, about 114 nations are members of the Codex Commission.

Representatives from the member countries include every nation in the Western Hemisphere. As worldwide standards are discussed in Codex Commission and Codex Committee Meetings, the delegates reach agreements which satisfy the international requirements while protecting the national interests.

A significant added benefit is the creation of standards which can be adopted outright or with slight modification by the smaller countries which do not have the resources for standards development.

The Codex Alimentarius Commission in March, 1977, issued a summary of acceptances of recommended worldwide and regional Codex Standards as well as recommended Codex maximum limits for certain pesticide residues (CAC, 1977). This list has been reviewed as it applies to the Western Hemisphere member nations and a brief summary of the food standards of more general interest is provided. No attempt at completeness has been made in the following summary. Standards have been selected for comment to demonstrate the trend of acceptances and some of the significant reasons for acceptance, specific deviations or for non-acceptance.

RECOMMENDED INTERNATIONAL GENERAL STANDARDS FOR LABELING OF PREPACKAGED FOODS CAC/RS/1-1969

Argentina, Bolivia, Canada and the USA have accepted this standard. Bolivia gave full acceptance and the other three accepted with specified deviations. Argentina requires declaration of the country of origin for all foods, both domestic and imported.

Class names for ingredients are not acceptable in Canada except for milk solids, specified vegetable oils, flavors and spices. Class titles for food additives are not acceptable in Canada with the exception of colors and sodium phosphate. Products labeled and packaged outside of Canada require declaration of the country of origin. All mandatory information must appear on the label in both English and French. There are other deviations, the specific details of which are contained in Codex Document CAC/Acceptance.

Commodity - White Sugar CAC/RS/4-1969

Argentina and Canada have accepted this standard with specified deviations. Argentine Law #11275 on the identification of merchandise requires the country of origin to be shown.

Canada requires white sugar to contain 99.8% sucrose and expressed a reservation regarding lead content and color requirements.

Commodity - Anhydrous Dextrose - CAC/RS/7-1969

Argentina, Canada and the USA have accepted this standard with specified deviations. Argentina does not authorize use of sulphur dioxide and requires declaration of the country of origin.

Canada expressed a reservation regarding lead as a contaminant.

The USA accepts only AOAC Methods as official in U.S. Standards.

Commodity - Tomato Concentrate - CAC/RS/57-1972

Bolivia has given this recommended Codex Standard target* acceptance. Costa Rica has given full acceptance.

Commodity - Canned Grean Peas - CAC/RS/58-1972

Honduras has given this standard full acceptance. Bolivia has given target acceptance. Costa Rica has indicated that the use of Wool Green BS color is not permitted in their country.

---

*Target acceptance means that the country concerned indicates its intention to accept the general standard after a stated number of years.

Commodity - Canned Plums - CAC/RS/59-1972

Bolivia has given this standard target acceptance. Costa Rica has indicated that the use of the color Ponceau 4R is not permitted in their country. The USA has accepted with deviations not yet specified.

Commodity - Table Olives - CAC/RS/66-1974

Ecuador and El Salvador have given full acceptance to this recommended Codex Standard.

Commodity - Raisins - CAC/RS/67-1974

Eduador and El Salvador have given full acceptance.

General Standard for Fats and Oils - Not Covered by Individual Standards - CAC/RS/19-1969

Argentina has accepted this standard with specified deviations. The use of coloring agents is not permitted. Certain emulsifiers such as Polysorbate 60 and 80 are not permitted. The country of origin must be declared.

Trinidad and Tobago gave target acceptance. Their Edible Oil Standards which were amended in March, 1972 do not reflect the criteria specified in the proposed Codex Standard.

Commodity - Edible Soy Bean Oil - CAC/RS/20-1969

Trinidad and Tobago has given full acceptance.

Canada has accepted with specified deviations reserving the right to use more sophisticated methods to determine authenticity of the oil. Certain coloring agents and flavors are not permitted in Canada.

The USA indicated non-acceptance of the Codex Standards.

Commodity - Edible Cottonseed Oil - CAC/RS/22-1969

Trinidad and Tobago gave full acceptance.

Argentina gave acceptance with specified deviations relating to the saponification value. Use of color and flavors is not permitted.

Canada accepted with specified deviations reserving the right to use more sophisticated methods to determine authenticity, among other reservations.

Commodity - Lard - CAC/RS/28-1969

Trinidad and Tobago gave target acceptance.
Argentina registered specified deviation with respect to essential composition and quality factors.

Commodity - Edible Tallow - CAC/RS/31-1969

Trinidad and Tobago gave target acceptance.
Argentina did not accept pointing out that the description of the product is inappropriate since tallows are not edible having lost their suitability for human consumption. Argentina felt that the product should be described as an edible fat with declaration of the name of the animal from which the product was derived.

Commodity - Margarine - CAC/RS/32-1969

Trinidad and Tobago gave full acceptance.
Argentina had the following specified deviations: Vitamin A and its esters are considered colors by the Argentina food code. Also the use of Vitamins D and E are not permitted. Argentina permits the use of flavors such as diacetyl and any other synthetic flavors permitted by the competent authorities. Country of origin must be identified. All margarine sold in Argentina must contain a "tracer" substance to help in the detection of prohibited mixtures.

The USA noted the following specified deviations: In the USA no provision is made for the use of marine oils. The USA deviates in essential component and quality factors. The USA Standard provides for the use of calcium disodium EDTA as a preservative. The USA Standard requires label declaration of characterizing flavors, colors and/or spices. A mandatory addition of Vitamin A makes the product subject to the U.S. nutritional labelling regulations. The USA Standard requires pasteurization of milk ingredients which is not required by the Codex Standard. The Codex Standard does not provide for harmless bacteria starters after pasteurization which is permitted by U.S. regulation for flavor enhancement.

Commodity - Olive Oil - CAC/RS/33-1970

Trinidad and Tobago has given target acceptance as has Argentina. Colombia has specified deviations as follows: The free acidity values are considered too high. One percent acidity for virgin oil and two percent for refined oil is recommended.

Commodity - Special Dietary Foods with Low Sodium Content

Canada had indicated non-acceptance because in Canada a low sodium content is defined as a food containing no more than 50% of the sodium that would be present normally in the food, if the sodium were not reduced.

The Codex Document, CAC/Acceptance, March, 1977 also lists acceptances for Codex maximum limits for pesticide residues. Among the countries in the Western Hemisphere providing responses are Canada, Ecuador, and El Salvador who have responded on Aldrin, Dieldrin, Chlordane, DDT, Dichlorovos, Inorganic Bromide Pesticides; pesticides on which Argentina and Bolivia have also responded are Lindane, Malathion, Parathion, Piperonyl Butoxide and Pyrethrins.

SUMMARY

The criteria for the evaluation of the safety of foods has changed markedly since 1968 when the International Symposium on the Safety and Importance of Foods in the Western Hemisphere was held at the University of Puerto Rico (Graham, 1968).

In particular, chemical, microbiological and toxicological methodology has become far more sophisticated.

Population growth has also been dramatic, focusing attention on the food supply, adequate nutrition and the use of food additives to extend the shelf-life and durability of foods.

Technological advances in product development have increased significantly the number of processed food products available.

Concerns of obesity on the one hand and malnutrition on the other hand have directed attention to special purpose foods.

Health concerns relating to the safety of food have historically been connected with prevention of fraud through adulteration and with the elimination of contaminants hazardous to health. In recent years, there has also been a preoccupation with cancer and its causes that has impacted on food regulations.

In such an atmosphere, it is obvious that regulations must keep pace, always adhering, however, to the tenets of protecting the public health and facilitating trade.

Outdated regulations, inadequate or unenforced food laws should be reviewed in the national interest.

Any revisions should be made in consultation with the nation's food processors to assure that the changes are realistic.

Another dramatic change which is evident in the last decade is the international approach to regulations. In the past, regulations have been tailored to the domestic needs of a nation which were seen as separate and distinct from those of other countries. As trade between nations increased, conflicting food laws became an obstacle, very often with serious economic consequences.

The competition between nations for a share of the world food market emphasized that national food legislation, in many cases, needed revamping to achieve a harmonization between trading nations. Such trade very often is greatest between contiguous countries. Therefore, the regional approach to harmonization became a first step. It was soon obvious, however, that differing regional standards are a continuation of the problem on a larger scale rather than a solution.

Fortunately, the Joint FAO/WHO Food Standards Program is actively developing worldwide food standards and, within that framework, is giving special attention to the needs of the developing nations in several regions of the world.

The Codex Alimentarius Commission is carrying out the program. The Codex Standards developed are minimum standards and thus have universal application. The acceptance procedure allows for several types of acceptance as well as for rejection. At this time, about 114 nations are members of the Codex Commission.

Representatives from the member countries include every nation in the Western Hemisphere. As worldwide standards are discussed in Codex Commission and Codex Committee Meetings, the delegates reach agreements which satisfy the international requirements while protecting the national interests.

A significant added benefit is the creation of standards which can be adopted outright or with slight modification by the smaller countries which do not have the resources for standards development.

Uniformity in standards recognizes that people of all races and nationalities deserve a safe and nutritious food supply.

Food standardization also recognizes that better and less expensive food can be made available to the peoples of the world because of economies in production resulting from formula simplification, larger volumes and broader markets.

A safe food supply is a worldwide objective which is assured by scientific research, good manufacturing practices and uniform food regulations which set minimum standards of quality.

BIBLIOGRAPHY

BIGWOOD, E. J. and A. GERARD, 1967. Fundamental Principles and Objectives of a Comparative Food Law, S. Karger, AG, Basel, Switzerland, 1, 36–38.

BARRERA-BENITEZ, H. 1975. Legislation and Regulations Affecting Food Safety and Quality in Latin America and Their Need for Harmonization in this Region. American Chemical Symposium on Food Legislation, Chicago, Illinois.

CODEX ALIMENTARIUS COMMISSION, 1970. Information on Food Standards in Latin America. Alinorm 70/31, Document CX 2/7.3 Rome, Italy.

CODEX ALIMENTARIUS COMMISSION. 1978. Report of the Joint FAO/WHO Food Standards Regional Conference for Latin America, cx/Latin America 78/12.

CODEX ALIMENTARIUS COMMISSION, 1976. Report of the First Session of the Coordinating Committee for Latin America, Rome, Italy, Alinorm 76/17.

CODEX ALIMENTARIUS COMMISSION, 1977. Summary of Acceptances of Recommended Worldwide and Regional Codex Standards and Recommended Codex Maximum Limits for Pesticide Residues. CAC/Acceptances, Rome, Italy.

CODEX ALIMENTARIUS COMMISSION PROCEDURAL MANUAL-FOURTH EDITION, 1975. Rome Italy.

CODIGO LATINO-AMERICANO DE ALIMENTOS, 1964. Latin American Food Council. La Plata, Argentina.

FOOD AND DRUGS ACT OF CANADA, 1953. Ottawa, Canada.

GRAHAM, H. D. 1968. Editor. The Safety of Foods. Avi Publishing Company, Westport, Connecticut.

INTERNATIONAL COMMISSION IN MICROBIOLOGICAL SPECIFI—CATIONS FOR FOODS (ICMSF) 1976. Microorganisms in Foods, Ottawa, Canada.

U.S. CODE OF FEDERAL REGULATIONS, TITLE 21 FOOD AND DRUGS, 1938 and Amendments 1958 and 1962.

ZIMMERMAN, J. G. 1972. Harmonization of food laws and food standards in Latin America, Food and Drug, and Cosmetic Law Journal 27, 645–650.

23

Food Control Under the Canadian Food and Drugs Act

A. B. Morrison

Prior to 1867, food control in Canada consisted of a few requirements governing the quality, grading, packing and inspection of certain basic foods. The first great impetus to improve food control in Canada, however, occurred in 1874, as a result of widespread public indignation over the purported evils of strong drink. A committee of parliament was appointed to study the problem, and after rejecting the possibility of prohibition of the sale of all liquor, decided it was not liquor itself but only *bad* liquor which ought to be banned. After parliamentary deliberation, an act entitled, "An Act to Impose License Duties on Compounders of Spirits and to Amend the Act Respecting Inland Revenue and to Prevent the Adulteration of Food, Drink and Drugs" became operative on January 1, 1875. The Act provided for the bonding and licensing of liquor manufacturers, and for the appointment of analysts, who were to be persons possessing "competent medical, chemical or microscopical knowledge". More significantly, the Act, which became known as "The Inland Revenue Act of 1875", provided the first national control in Canada and indeed, in North America, over adulteration of liquor, foods and drugs. The adulteration aspects of the Canadian Act were patterned closely after an English Act passed in 1872. Adulterated foods were defined as those containing deleterious ingredients or any material which resulted in the food being of less value than was understood by the name. Penalties for adulteration of food were severe. For willful adulteration the first offence involved the penalty of $100 and the second, prison with hard labor for up to 6 months.

722

Throughout the years, food legislation in Canada has evolved in scope and complexity, as the food industries have developed, consumers have become better informed, and scientific advances have provided a sound background for regulation. A notable advancement in food control in Canada was provided by the Food and Drugs Act of 1953. In addition to provisions for consumer protection embodied in previous legislation, this Act provides authority in respect of inspection of food manufacturing plants, and makes it an offence to manufacture, package or store for sale a food under unsanitary conditions. The Food and Drugs Act is based upon the authority of the Federal Government to legislate on criminal matters. As criminal law, it applies to food manufactured, processed, distributed or sold anywhere in the country and is not subject to any requirement of interprovincial trade. Domestic and imported foods are treated in an identical manner under the Act.

Food is defined as any article manufactured, sold or represented for use as food or drink for man, chewing gum, and any ingredient that may be mixed with food for any purpose whatever. The most important sections of the Act which deal specifically with foods include the following:

Section 4. No person shall sell an article of food that
    (a) has in or upon it any poisonous or harmful substance;
    (b) is unfit for human consumption;
    (c) consists in whole or in part of any filthy, putrid, disgusting, rotten, decomposed or diseased animal or vegetable substance;
    (d) is adulterated; or
    (e) was manufactured, prepared, preserved, packaged, or stored under unsanitary conditions.

Section 5. (1) No person shall label, package, treat, process, sell, or advertise any food in a manner that is false, misleading, or deceptive, or is likely to create an erroneous impression regarding its character, value, quantity, composition, merit, or safety.

Section 7. No person shall manufacture, prepare, preserve, package, or store for sale any food under unsanitary conditions.

It is obvious that Sections 4 and 7 of the Act relate directly to food safety. Section 5 (1), which is aimed at preventing fraud or deception, may also bear on the safety of foods under certain circumstances. For example, sale of a food product represented as being an excellent source of one or more nutrients, but which in fact did not contain the nutrient(s) in question in the purported amounts, would not only be fraudulent, but the product itself would also potentially be unsafe for use under certain circumstances.

The Act provides broad powers for inspectors. Under authority of Section 21, an inspector may at any reasonable time enter any place where on reasonable grounds he believes any food to which the Act or Regulations apply is manufactured, prepared, preserved, packaged or stored, examine any such article and take samples thereof, and examine anything that he reasonably believes is used or capable of being used for such manufacture, preparation, preservation, packaging or storing. He may also open and examine any receptacle or package that on reasonable grounds he believes contains any food to which the Act or Regulations applies, examine books, documents or other records, and seize and detain for such time as may be necessary any article by means of or in relation to which he reasonably believes any provision of the Act or Regulations has been violated.

The Act also provides authority for the Governor-in-Council (in effect the Federal Cabinet) to make regulations for carrying the purposes and provisions of the Act into effect. The most important sections of the Act relating to the regulation making power respecting foods are the following:

Section 24 (1) The Governor-in-Council may make regulations for carrying the purposes and provisions of this Act into effect, and in particular, but not so as to restrict the generality of the foregoing, may make regulations

> (a) declaring that any food or class of food is adulterated if any prescribed substance or class of substances is present therein or has been added thereto or extracted or omitted therefrom;
> (b) respecting
> > (i) the labelling and packaging and the offering, exposing and advertising for sale of food,
> > (ii) the size, dimensions, fill and other specifications of packages of any food,
> > (iii) the sale or condition of sale of any food,
> > (iv) the use of any substance as an ingredient in food to prevent the consumer or purchaser thereof from being deceived or misled as to its quantity, character, value, composition, merit or safety or to prevent injury to the health of the consumer or purchaser;

(c) prescribing the standards of composition, strength, potency, purity, quality or other property of any article of food.

Soon after enactment of the Inland Revenue Act of 1875, it became apparent that the absence of legal standards of composition for foods seriously hampered efforts to provide adequate control over food safety and quality. Early attempts to devise food standards were limited by the absence of satisfactory analytical methodology. As scientific advancements have permitted, steady progress has been made in defining standards for a large number of staple foods. At present, the Food and Drug Regulations contain specifications for the composition of nearly 300 foods. The standard for ice cream, for example, reads as follows:

B.08.062 Ice Cream

(a) shall be the frozen food made from ice cream mixed by freezing

(b) may contain cocoa or chocolate syrup, fruit, nuts or confections

(c) shall contain not less than
   (i) 36% solids;
   (ii) 10% milk fat, or where cocoa or chocolate syrup, fruit, nuts or confections have been added, 8% milk fat, and
   (iii) 1.8 pounds of solids per gallon of which amount not less than 0.50 pounds shall be milk fat, or, where cocoa or chocolate syrup, fruit, nuts or confections have been added, 1.8 pounds of solids per gallon of which amount not less than 0.40 pounds shall be milk fat; and

(d) shall contain not more than
   (i) 100.000 bacteria per gram,
   (ii) 10 coliform organisms per gram as determined by the official method.

In all instances, where a standard for a food is prescribed, the food in question shall contain only the ingredients included in the legal standard for the food; each ingredient shall be incorporated in the food in the quantity within any limits prescribed for that ingredient; and, if the standard includes an ingredient to be used as a food additive for a specified purpose, that ingredient shall be a food additive set out in the regulations for use as an additive to that food for that purpose.

Food additives are defined in the regulations as follows:

B.01.001 Food additive means any substance, including any source of radiation, the use of which results, or may reasonably be expected to result in it or its byproducts becoming a part of, or affecting the characteristics of a food, but not including

(a) any nutritive material that is used, recognized, or commonly sold as an article or ingredient of food,

(b) vitamins, mineral nutrients and amino acids,

(c) spices, seasonings, flavouring preparations, essential oils, oleoresins and natural extractives,

(d) agricultural chemicals,

(e) food packaging materials and components thereof, and

(f) drugs recommended for administration to animals that may be consumed as food.

A request that a food additive be included in the tables of acceptable compounds must be accompanied by a submission from the manufacturer including:

(a) a description of the food additive;

(b) a statement of the amount of the food additive proposed for use and the purpose for which it is proposed;

(c) data establishing that the food additive will have the intended physical or other technical effect;

(d) detailed reports of tests made to establish the safety of the food additive under the conditions of use recommended;

(e) data to indicate the residues that may remain in or upon the finished food, and

(f) a proposed maximum limit for residues of the food additive in or upon the finished food. If the submission is considered satisfactory, the additive is included in the list of acceptable compounds.

There are now several hundred additives which may be used for various purposes - as anti-caking agents, bleaching, maturing and dough-conditioning agents, colouring agents, emulsifying, gelling, stabilizing and thickening agents and so on. For example, calcium aluminum silicate is permitted in or upon salt at a maximum level of use of 1%; in flour salt, garlic salt, and onion salt at a maximum level of use of 2%; and in unstandardized dry mixes at levels required for good manufacturing practice. Similarly, lactylated mono- and di-glycerides may be added to shortenings at a maximum level of use of 8% (except that the total combined mono-and diglycerides and lactylated mono- and diglycerides must not exceed 20% of the shortening).

During the past few years there has been increasing interest in the sale of simulated meat products, based upon soybean protein. These products are the result of advanced technology which permits the production of soybean products which look and taste not dissimilar to meat products. Standards have been defined in the Food and Drug Regulations to require that such products be of acceptable nutritional quality, comparable to that of the meat product they are intended to replace. For example, simulated meat products that resemble fresh sausage must have a total protein content of not less than 9%, acceptable protein quality as determined by an official method, a fat content of not more than 40%, and specified vitamin and mineral levels. If isolated essential amino acids have been added to such a product, they must be present in amounts not exceeding that which improves the nutritional quality of the protein. These regulations have been put in place because of the need to make certain that simulated products are not of inferior nutritional quality, as compared to the traditional product which they replace.

As will be noted from the above, increased attention now is being given to nutritional aspects of food standards. Addition of nutrients to foods in Canada may be carried out under the following guidelines:

1. Vitamins, minerals and protein added to traditional foods should be present in amounts related to the purpose of the addition:
    (a) to replace those nutrients lost in the course of good manufacturing practice , if the amount originally present provided at least 10% of the daily requirement of those nutrients in a reasonable daily intake of the food. The amount added should compensate for that loss in processing.
    (b) to correct a demonstrated deficiency of one or more of these nutrients in some segment of the population, if such addition is the most effective means of correcting the deficiency. The amount should only be sufficient to correct the deficiency.
2. A food sold or used as a substitute for a traditional food that normally provides, in a reasonable daily intake of the food, at least 10% of the daily requirement of a nutrient (including calories) should contain an equivalent amount of that nutrient.
3. Foods sold or represented as meal replacements should contain essential nutrients, including calories, in amounts related to the purpose of the meal, e.g. instant breakfasts and infant formulas.
4. Snack foods for which nutritional claims are made and which

supply at least 200 calories in a reasonable daily intake, should contain essential nutrients in proportion to their caloric content. No nutritional claims should be made for unfortified snack foods or "fun-foods".

The results of a national survey of the nutritional status of Canadians, conducted recently, showed that significant numbers of Canadians are consuming less than adequate quantities of Vitamin D, Vitamin A, Vitamin C, iron, and in some instances, thiamin. As a result, the Food and Drug Regulations now require Vitamin D be added to milk, including skim milk, whole milk, condensed milk and evaporated milk, and to margarine. Vitamin C must be added to fruit nectars, fruit drinks, evaporated milks, fruit juices, and tomato juice. Vitamin A must be added to skim milk and partially skimmed milks and to margarine. Thiamin, riboflavin, niacin and iron must be added to flour and alimentary pastes. In addition, potassium iodide must be added to salt for table or general household use, in order to prevent iodine deficiency.

A wide variety of chemicals are used in the production, storage or transport of food. Pesticides, plant growth regulators, and fertilizers are important members of this large group of chemicals. In Canada, pesticides used on food crops must be registered for use under the Pest Control Products Act, administered by the Canada Department of Agriculture. Tolerances for pesticides in foods are established under regulations to the Food and Drugs Act. In establishing the tolerances, evidence on the safety of the pesticide involved is carefully considered, but the tolerance is not set any higher than the level which may be found in the food under the conditions of good agricultural practice, even when the "safe" level may be higher. Tolerances for nearly 100 pesticide chemicals on a wide variety of food products have been established under the Food and Drug Regulations. Over the past few years a concerted effort has been made to eliminate tolerances or reduce them, as increasing evidence has accumulated on the safety and conditions of use for the products involved. Particular attention has been paid to the organochlorine family of pesticides, including DDT, because of their persistance in the environment. Permitted food uses for DDT have now essentially been eliminated in Canada. As a result of marked restrictions on the use of DDT on food crops, there has been a steady decrease in the amounts of DDT and related compounds on food crops, and in human breast milk samples collected during the last decade.

The control of harmful microorganisms and their toxins is provided for primarily under the authority of Sections 4 and 7 of the

Act. Increasing emphasis is being placed on the development of microbiological standards for processed foods. Because of the large number of such foods on the market, it has been necessary to develop priorities, based upon an assessment of the likelihood the food may become a hazard from a microbiological point of view.

Sections 4 and 7 of the Act also provide the legislative base for control of heavy metals such as lead and mercury, and other environmental contaminants such as polychlorinated biphenyls (PCB's). Extensive research on the toxicology of heavy metals is conducted as the basis for detailed regulations to control these contaminants in foods. Environmental contamination with PCB's has markedly been curtailed, and it can be expected that PCB levels in certain freshwater fish species and in human milk will begin to drop during the next few years. Because of the environmental persistence of PCB's, however, it will be necessary to continue monitoring foods to ensure excessive contamination with PCB's does not occur.

It should be emphasized that the Food and Drugs Act and Regulations do not in themselves ensure the safety of the nation's food supply. Foods offered for sale in Canada have not been "approved" by the Federal Government. Regulations have been promulgated under the authority of the Food and Drugs Act which should achieve this objective. But the responsibility for placing on the Canadian market foods which do not contain any poisonous or harmful substance and which are sold in a manner that is not deceptive rests squarely on the shoulders of the food processor, distributor and retailer.

Responsibility for administering the Food and Drugs Act is shared by two Federal Agencies. The Department of Consumer and Corporate Affairs administers food labelling, advertising and fraud requirements of the Act and Regulations, with medical and health advice from the Health Protection Branch of the Department of National Health and Welfare. The Health Protection Branch is responsible for ensuring that the safety aspects of the Food and Drugs Act and Regulations are met. The Health Protection Branch has a number of organizational entities, but there are two which are concerned with food control. The Food Directorate has approximately 270 staff, and conducts extensive research investigations into various aspects of food safety, nutrition, food microbiology and food composition, as well as detailed review of manufacturers' submissions and establishment of standards and regulations. The Field Operations Directorate carries out food inspection activities from its five regional offices across the country. At the end of 1977, there were approximately 100 food inspectors on the staff of the Branch.

The Food and Drugs Act and Regulations have proven able, over the years, to provide the legislative and regulatory base required for proper food control in Canada. With the passage of time and inevitable increases in complexity of the food system in Canada, modifications will no doubt be necessary, but the basic aims of the Act, to protect consumers against health hazards and fraud related to foods, can be expected to be of enduring worth.

Food Supply Systems

24

Safety of Food Service Delivery Systems in Schools

John S. Avens

A number of school lunch food preparation and service systems have developed from the more conventional on-site preparation and service system. Preparation of food in a centrally located kitchen facility with any of a variety of forms of packaging for transportation to and service at satellite schools are systems currently being used in the United States. Food can be transported in bulk to be portioned at the satellite schools into individual servings. Food can be portioned into insulated containers at the central preparation site and transported hot or cold to the satellite school. Meal items can be obtained commercially either canned and ready to heat and serve in the can, or frozen preportioned to be thawed and heated. Van Egmond (1974) described these more common school lunch food systems. Any of these systems have characteristic differences which present unique problems in terms of proper management to assure safe food served to children. The control of bacterial contamination of food through sanitation, and the control of bacterial growth and multiplication and possible toxin production in the food must be accomplished in any food preparation and service system. How food safety is assured in any system depends on a thorough understanding of all potential food safety hazards inherent in each system. However, many potential hazards are common to more than one system.

Microorganisms of food safety significance and foodborne infections and intoxications as well as food spoilage have been adequately discussed in recent scientific and popular literature. The basics of food service sanitation and technology related to food safety has been adequately presented in various texts on general food service related to restaurants and institutions and in recent scientific literature. However, a few specific in-field studies to determine the relative food safety of different school lunch preparation and delivery systems have only recently been conducted and have yet to be reviewed in a single text. The purpose of this chapter is to review the state of knowledge relative to specific aspects of food safety affecting food service delivery systems in schools. Food service sanitation related to personal hygiene, equipment and facilities sanitation and food handling sanitation as well as temperature control of food will be covered as the topics relate to problems unique to school food service systems, even though much of this information is applicable to food service in general.

FOOD SAFETY IN SCHOOLS

The National School Lunch Program in the United States underwent considerable expansion during the mid-1970s. As a result of this Federal Government effort, many school districts were faced with the option of establishing new facilities. Decisions were made, often based on limited knowledge, relative to the most suitable type of food service system for the school's particular situation.

One important parameter to consider when selecting a food service system for a school lunch program or evaluating a system already in use, is the potential food safety hazards associated with the system. These potential food safety hazards relate not only to equipment and facilities but also to food service managers and every food service employee using the equipment and facilities.

Bryan *et al.* (1971) reported that on numerous occasions foodborne disease outbreaks have occurred which may have been caused by turkey meat prepared and served by school food service operations. They cited one outbreak involving students and teachers in an elementary school. Those affected experienced mild gastroenteritis with abdominal cramps and diarrhea 8 to 24 hours after consuming a lunch served at the school. *Clostridium perfringens* was isolated from victims cultured. Attack rates determined from food histories of those consuming the suspected meal identified either turkey or dressing as the contaminated food.

Bryan and McKinley (1974) reviewed factors contributing to foodborne outbreaks relative to use of turkey meat in school lunch kitchens. They outlined the following contributing factors:

1. Raw turkey meat can be contaminated with salmonella, *Staphylococcus aureus* and *Clostridium perfringens* before it arrives in the school kitchen.
2. Turkey meat requires relatively frequent handling (deboning, slicing, etc.) during preparation.
3. Inadequate cooking may allow pathogenic bacteria to survive.
4. Cooked turkey meat is often exposed to potential recontamination by facilities, equipment and personnel which became contaminated by raw turkey and were not thoroughly washed and disinfected. Facilities, equipment and personnel may also be sources of other bacterial contamination not associated with the raw turkey.
5. Turkey meat is usually prepared at least one day before serving. This allows plenty of time for any pathogenic bacteria which survived cooking to multiply in meat or stock not rapidly cooled to and maintained at 7°C (45°F) or less. Cooling cooked turkey meat is seldom done adequately in school lunch kitchens, due to limited refrigerator space, necessitating storage of warm meat in large volume containers. Turkey meat is sometimes even left overnight in warm ovens.
6. Before serving, often many hours after cooking, turkey meat, gravy, and dressing may not be adequately reheated to at least 73°C (165°F), or may even be served cold. Thus, surviving or contaminating bacteria which may have multiplied to dangerous quantities will not be killed before being consumed by school children.

Rollin and Matthews (1977) determined cooling effectiveness of walk-in refrigerators as used in cook/chill food service systems. Entree is cooked in a central kitchen 1 to 3 days prior to service and cooled in bulk until portioned and transported to satellite school on day of service to be reheated to serving temperature. Time-temperature determinations during chilling of nine entrees under actual operating conditions and during chilling of barbecue ground beef in a controlled laboratory experiment, indicated it was not possible to cool the cooked food in standard $12 \times 20 \times 4$ inch pans to 7°C (45°F) within 4 hours in walk-in refrigerators. Also, it was not possible to cool the cooked barbecue ground beef through a 49–16°C (120–60°F) temperature range within 2 hours in a typical walk-in refrigerator.

A number of studies have been conducted related to food safety hazards of school lunch preparation and delivery systems.

Bryan and McKinley (1972, 1974) and Bryan *et al.* (1971) addressed the problem of foodborne illness prevention by time-temperature control of thawing, cooking, chilling, and reheating turkeys in school lunch kitchens. They conducted a bacteriological survey and a time-temperature evaluation of the thawing, cooking, chilling and reheating of turkey, stock and dressing in a school lunch kitchen. Preparation practices that could contribute to food poisoning outbreaks were studied in three schools. They found thawing frozen turkey carcasses at room temperature in only their plastic bag resulted in spoilage bacteria multiplying on the skin surface. Skin surface contamination was destroyed by all cooking methods used for thawed turkeys, which included gas ovens, electric ovens, convection ovens, pressure-type steamers, steam kettles and pots on a range. Whole cooked turkeys cooled in the refrigerator took 15 to 18 hours to adequately cool to below the dangerous bacterial growth temperature zone. Stock stored in one-gallon glass jars in a refrigerator took over 12 hours to cool to 10°C (50°F); in 14-gallon pots refrigerated overnight, temperature was still high enough to support bacterial growth. Three-inch layers of cut turkey meat took 7 hours to cool adequately under refrigeration; halved turkey rolls took 10 hours. These cooling time periods were long enough to allow bacteria to multiply to dangerous levels in the food, potentially resulting in decomposition of the food, toxin production and/or infectious levels of pathogenic bacteria. Varying reheating methods employing steamers, kettles and pans on a range and in ovens, were observed among various schools. These findings led to specific recommendations for heat destruction of foodborne pathogenic bacteria in turkey meat, stock and dressing, and inhibition of bacterial growth and multiplication during thawing, chilling and holding, and for cleaning and sanitizing equipment. These recommendations have been incorporated into general procedures discussed later in this chapter.

Tuomi, et al. (1974) conducted a laboratory study simulating practices of handling precooked chilled ground beef gravy in school kitchens to determine potential food safety hazards. They found reheating for a specific period of time did not insure that the minimum temperature of 74°C (165°F) was reached. They, therefore, recommended measuring the temperature of food to indicate adequacy of heating. They warned that holding precooked chilled food at room temperature may contribute to foodborne disease outbreaks, if food has been mishandled before delivery to the school, and recommended refrigerated holding. Their findings indicated the temperature of chilled food subjected to holding before preparation and serving should not exceed 15.5°C (60°F) at the end of a maximum holding period of 5 hours. They emphasized the importance of rapid cooling of precooked chilled food, since the greatest increase in number of aerobic bacteria in the gravy occurred during cooling rather than holding.

Cremer and Chipley (1977a) studied a satellite food service system relative to time-temperature conditions during product flow and microbiological and sensory quality of spaghetti and chili. They identified critical phases of processing indicated by wide and unacceptable time-temperature ranges, particularly after food was heated for service. Other critical phases were holding, assembly and beginning storage at the satellite school. After heating for service there was high variability in internal temperature of food among all containers in every replicate determination for both spaghetti and chili. This they attributed to the efficiency of the convection oven used for reheating food at the school. They attributed variable temperatures during holding and assembly at the central kitchen to inadequate manual adjustment of the temperature control on the jacketed chill tank. Variability in reheating oven temperature was attributed to oven heating efficiency and human variability in setting controls. The microbial quality of both food items was generally good but variable with various pathogens identified in specific samples. Cooking and reheating caused reductions in the total microbial population, and storage resulted in a slight increase. A wide variety of microorganisms were isolated from all samples of both raw and precooked meat, including yeasts and fungi (cooked meat), *Clostridium sporogenes*, *Clostridium perfringens*, micrococci including *Staphylococcus aureus* and *Staphylococcus epidermidis*, *Escherichia coli*, *Enterobacter aerogenes*, *Salmonella typhimurium* (1 raw meat sample only), *Proteus* sp., *Pseudomonas* sp. and *Achromobacter* sp. Their presence in the precooked meat indicated underprocessing by the meat supplier and/or post-cooking contamination. Coliforms, staphylococci and clostridia were found in both chili and spaghetti after cooking and reheating indicating inadequate time exposure to adequate end-point heating temperatures. They recommended a more thorough cooking of the food products prior to assembly, a more thorough reheating of meal items for service and a more precise control of time and temperature during all phases of product preparation. Their study indicated a potential public health hazard, if food is mishandled in the satellite food service system, but that this system can be effective in providing safe food to large numbers of students, if good quality control is exercised.

Cremer and Chipley (1977b) also studied precooked frozen hamburger patties in a satellite food service system relative to time-temperature conditions and microbiological and sensory quality during product flow. They identified critical phases in preparing this precooked frozen entree for consumption. They found that time-temperature conditions and microbiological quality of the hamburger patties were variable but generally acceptable. Internal temperatures after heating the patties for service ranged from 67°C (152°F) to 89°C (192°F). Internal temperatures less than 80°C (176°F) were considered low enough to warrant careful monitoring of product when purchased and strict control of mishandling of product during preparation and heating for service. Pathogenic bacteria including *Clostridium* and *Staphylococcus* as well as non-pathogenic indicator bacteria were detected in the precooked hamburger patties, indicating either inadequate cooking of the patties by the supplier or post-processing contamination. Therefore, a potential for food safety hazards exists in a satellite food service system, if precooked food is mishandled and receives time-temperature abuse during preparation for serving. They warn that preventing microbial multiplication and further contamination during product assembly and transport in the satellite system, and careful monitoring of internal product temperature during reheating to eliminate variability, are important factors determining the safety of precooked frozen hamburger patties.

Many cases of salmonellosis have been traced to salmonella contamination of precooked roast beef from 1975 through 1977 (Gregg, 1977a). Precooked roasts of beef are often cooked to less than 54°C (130°F) which is not hot enough to destroy salmonellae. Although these outbreaks did not involve school food service systems, they point out the potential danger of inadequately cooking large meat roasts or inadequately reheating commercially precooked meat roasts for any institutional food service. Effective September 2, 1977, the U.S. Department of Agriculture issued a regulation requiring precooked roast beef be cooked to an internal temperature of 63°C (145°F) to destroy salmonellae (Gregg, 1977b).

Bunch, *et al.* (1977) determined the fate of *Staphylococcus aureus* in beef-soy loaves subjected to procedures used in hospital chill food service systems. This system of storing partially cooked food refrigerated for greater than 24 hours is also used in some school food service systems. Heating the loaves in a gravity convection oven to an internal loaf temperature of 60°C (140°F) was not lethal to *Staphylococcus aureus*. Refrigerated storage of the partially cooked beef-soy loaves for 24, 48, or 72 hours followed by heating portions to 80°C (176°F) in a microwave oven resulted in no detectable *Staphy-*

lococcus aureus in any samples. However, since the loaves were in a temperature range between 6.5 and 45.5°C (44 and 114°) that permits growth of *Staphylococcus aureus* for over 7 hours of cooking and cooling, any enterotoxin produced in the loaves during this time would not be destroyed by the short heat exposure in the microwave oven. They concluded that food prepared in a chill food service system must be handled carefully during all phases of preparation to minimize microbial contamination and time-temperature abuse. Perhaps, cooking smaller loaves and cooling in smaller portions would be advisable for school food service operations.

A study was conducted by Avens, *et al.* (1977, 1978) to determine potential food safety hazards associated with four different school lunch preparation and delivery systems and to determine if any system(s) was (were) more or less susceptible to these hazards than others. Potential food safety hazards were determined by microbiological indicators of unsanitary food handling and time-temperature abuse.

Test schools included small and large elementary schools from various locations in the United States. The specific objective was to compare four food preparation and delivery systems regarding the microbiological content of foods prepared, processed or handled by the systems relative to amount of contamination or mishandling. The four school lunch delivery systems studied were:

1. Conventional on-site school food preparation and service.
2. Food preparation in a school system's central kitchen followed by hot bulk transport to satellite schools.
3. Food preparation in a school system's central kitchen followed by chilled transport of preportioned food to satellite schools. Some of the menu items produced by this system required heating prior to service.
4. Purchase of frozen preportioned meals which were heated to serving temperature in individual schools.

Four schools representing each of the four delivery systems were selected for a total of 16 schools from which food samples were collected.

All schools served ten specified menu items during the week of the in-school test: (1) ground beef and spaghetti; (2) oven fried chicken; (3) fish sticks; (4) meatloaf; (5) green peas; (6) canned carrots; (7) instant mashed potatoes; (8) chocolate pudding; (9) canned peaches and (10) canned baked beans. Line samples were collected at sites which bracketed major steps of the food preparation and service operation:

A. On-site preparation and service.

    1. After ingredients of each formula were combined or after individual items such as vegetables were opened preparatory to cooking.

    2. After formulas or individual items were heated (cooked).

    3. After specified time in hold cabinet (hot or cold).

    4. After specified time on serving counter.

B. Central preparation and hot bulk delivery.

    1. After ingredients of each formula were combined or after individual items such as vegetables were prepared for cooking.

    2. After formulas or individual items were heated (cooked).

    3. Immediately after food arrived at the satellite school.

    4. After specified time in hold cabinet (hot or cold).

    5. After specified time on serving counter.

C. Central preparation and chilled preportioned delivery.

    1. After ingredients for each formula were combined or after individual items such as vegetables were prepared for cooking.

    2. After formulas or individual items were heated (cooked).

    3. After the preportioned menu items were chilled, held and transported (upon arrival at satellite school).

    4. After food was reheated to serving temperature at satellite school.

    5. After specified time in hold cabinet (hot or cold).

    6. After specified time on serving counter.

D. Frozen preportioned delivery.

    1. Immediately after the frozen food arrived at the school (or from the school's freezer storage).

    2. After foods had been reheated to serving temperature.

    3. After specified time on serving counter.

Microbiological count data were obtained on each of triplicate samples for every sampling point, food item, school and delivery system. Microbiological count data included total aerobic mesophilic plate count (APC) per gram of food, coliforms most probable number (MPN) per gram of food, *Escherichia coli* MPN per gram of food, and coagulase-positive staphylococci MPN per gram of food. Temperature of food at time of sampling was measured for every sample taken. Elapsed time from the first sampling point was determined for each subsequent sampling point. The microorganism and time-temperature data were used to indicate time-temperature abuse of food or contamination of food due to unsanitary conditions which could involve facilities, equipment, utensils and/or personnel.

Analysis of variance of aerobic mesophilic plate count (APC) data indicated differences among the four food preparation and delivery systems were not statistically significant (P<.05), but differences among schools and sampling points over all systems were statistically significant (P<.01). Differences in APC among sampling points within schools would be expected, since sampling points were based on treatment or handling of food. It was, therefore, meaningful to examine these differences within each school for every food item to see if these differences were explained by time-temperature effects or handling. Here, it was also meaningful to look at indicator microorganism data to indicate possible abuse in handling the food item prior to each sampling point. Since analysis of variance of APC data indicated a highly significant difference (P<.01) among schools, it was meaningful to look at indicator microorganism data and time-temperature data within each school to determine if food preparation and handling abuse may have accounted for school differences in APC.

Schools and systems were compared on the basis of relative frequency of improper food temperature and objectionably high bacterial counts. These frequencies were indicators of potential microbiological problems in each system and should not be confused with the finding that no significant differences existed in the microbiological quality of the food delivered by the four preparation and delivery systems tested. The International Commission on Microbiological Specifications for Foods (ICMSF, 1974) has published maximum levels (m) of various microorganisms in various foods which are acceptable and attainable, and minimum levels (M) which are unacceptable and indicate defective quality. Microbiological data from this school lunch system study were compared to the smaller m value for foods and microorganisms, where published m values were available. Maximum microbiological limits for school lunch menu items with no available m values were derived from other published microbiological specifications for foods. Recommended microbiological limits were used for comparison with data derived only from the first sampling point, after ingredients of each formula were combined or individual items were prepared for cooking. Bryan (1974) indicated time-temperature limits related to cooling, cooking, reheating and hot holding of food; and sanitation requirements. These limits and recommendations were used as guidelines in evaluating microbiological and time-temperature data for this school lunch system study.

Schools within the frozen preportioned delivery system more consistently met or exceeded food temperature recommendations and had low frequency of objectionably high bacterial counts. However, this system did not have as many opportunities at the school for objectionably high bacterial counts to occur, since the food was not handled as much as in the other three systems. The microbiological quality of the incoming food would depend on the supplier of the frozen meals.

One school in the central preparation and chilled preportioned delivery system had the second highest freqency of improper food temperature occurences and the highest frequency of objectionably high bacterial counts of the 16 total schools.

The ground beef and spaghetti (meal item 1) from this school was adequately cooked (to 93°C) before the first sampling point (Table 24.1). It lacked 1°C of being properly chilled ($\leqslant$7°C) after transport to the satellite school (25.7 hours from sampling point 1). It was only reheated, however, to 50°C (at least 71°C required), cooled on the serving line and never did reach the recommended minimum hot holding temperature of 55°C. The adequate cooking temperature resulted in only 49 APC/g and no detectable indicator bacteria. Thus, subsequent time-temperature abuse did not result in an actual food safety hazard.

The raw chicken (meal item 2) breaded with flour from this school had higher than suggested maximum levels of APC, coliforms, E. coli and coagulase-positive staphylococci before cooking (Table 24.1). Adequate cooking at 93°C internal muscle temperature reduced APC from 120,000/g to 6/g and eliminated all viable indicator organisms. The chicken lacked 4°C of being properly cooled over 25.8 hours prior to arrival at the satellite school, but was adequately reheated and held well above 55°C during serving. A potential food safety hazard introduced by the raw chicken was thus averted.

The meatloaf (meal item 4) from this school was within suggested maximum limits for bacterial counts in frozen ground beef before cooking, but did contain coliforms and E. Coli (Table 24.1). It received less than adequate cooking which left it with relatively high APC/g, but no indicator organisms. It lacked 4°C of being properly cooled after 26.4 hours and had apparently been recontaminated with E. coli and other coliforms prior to arrival at the satellite school. The meatloaf was not adequately reheated nor was it held at 55°C while serving. E. coli was not detected after inadequate reheating, but APC increased slightly during serving and coliforms were possibly introduced during preportioning and survived the inadequate reheating. The source of E. coli and other coliforms could have been recontamination with raw ground beef, perhaps from utensils or during preportioning, holding or transport, or could have come from human or rodent feces through lack of sanitation prior to arrival at the school. Coliforms probably were not introduced after reheating or on the serving line since the food item was sealed in a foil-covered tray. Cross-contamination before arrival at the school and time-temperature abuse with the meatloaf presented a potential food safety hazard in this system.

The frozen peas (meal item 5) from this school had higher than suggested maximum levels of APC which increased significantly prior to cooking (Table 24.1). However, adequate cooking greatly reduced APC from > 300,000/g to 45/g and eliminated coliforms, and adequate holding temperature maintained APC at a low level.

Canned food such as carrots and beans should contain no viable mesophiles before the can is opened. Slight contamination would be unavoidable during handling and preparation since equipment is not expected to be sterile. Large increases in APC and coliforms or the introduction of $E.$ $coli$ or coagulase-positive staphylococci in a canned food during preparation, particularly accompanied with time-temperature abuse, would indicate a potential food safety hazard. There seemed to be no problem with canned carrots (meal item 6) from this school (Table 24.1). However, the canned beans (meal item 10) increased significantly in APC from 3,000/g to 13,000/g over 26.1 hours of inadequate cooling, prior to inadequate cooking which reduced the APC to 2,800/g, which then did not increase significantly at adequate holding temperature.

The dry mashed potato flakes (meal item 7) from this school contained 56,000 APC/g, in excess of the suggested maximum limit of 20,000 APC/g and contained no indicator microorganisms (Table 24.1). After mixing with dry milk and warm water, but not cooking, the APC of the mixture was reduced to 11,000/g, probably due to dilution. However, 34 coliforms (MPN)/g and 4.9 $E.$ $coli$ (MPN)/g were detected in the mixture which may have come from the dry milk or water which may have been recently indirectly contaminated with human or rodent feces through unsanitary procedures. During possibly 26 hours of inadequate cooling including transport to satellite school, APC increased slightly to 28,000/g, coliforms increased slightly to 46 (MPN)/g and $E.$ $coli$ decreased to less than 3.2 (MPN)/g. The mashed potatoes then received inadequate cooking to only 64°C at the satellite school which did reduce the APC to 2,900/g, coliforms to < 3.2 (MPN)/g and $E.$ $coli$ to less than detectable levels. Warm mixing and holding of mashed potatoes at greater than 7°C for 26 hours prior to inadequate cooking, and cooling on the serving line indicated a potential food safety hazard, particularly if any poor sanitation practice allowed $E.$ $coli$, indicative of recent fecal contamination, to be introduced into the food product.

One school in the central preparation and hot bulk delivery

TABLE 24.1. COMPARISON OF MEASURED LEVELS OF INDICATOR MICROORGANISMS AND FOOD TEMPERATURE WITH RECOMMENDED LIMITS—CENTRAL PREPARATION AND CHILLED PREPORTIONED DELIVERY. ONE SCHOOL

Meal item	Sampl. point	APC/gram Geom. Mean ($\times10^{-1}$)	APC/gram Sug. Max. ($\times10^{-1}$)	Coliforms (MPN)/gram Geom. Mean	Coliforms (MPN)/gram Sug. Max.	E. coli (MPN)/gram Geom. Mean	E. coli (MPN)/gram Sug. Max.	Staph (MPN)/gram Geom. Mean	Staph (MPN)/gram Sug. Max.	Time fr. Sampl. pt. 1	Temp. (°C) Measured	Temp. (°C) Recom.
1	1	0.049	1000	<3		<3	100	<3	100	0	93	⩾74
	2	not cooked at this point—cooked at point 1 and chilled									8	⩽7
	3	0.35		<3		<3		<3		25.7 h	50	⩾71
	4	0.17		<3		<3		<3		26.6 h	42	⩾55
	5	0.15		<3		<3		<3		26.8 h	52	⩾55
	6	0.096		<3		<3		<3		27.2 h	14	none
2	1	120	100	170		69		130		0	93	⩾74
	2	0.0062		<3		<3		<3		0.7 h	11	⩽7
	3	2.3		<3		<3		<3		25.8 h	80	⩾71
	4	0.0047		<3		<3		<3		26.7 h	73	⩾55
	5	0.013		<3		<3		<3		27.0 h	70	⩾55
	6	0.0068		<3	12/cm²	<3	6/cm²	<3	3/cm²	27.5 h	10	none
4	1	390	1000	220		8.9	100	<3	100	0	70	⩾74
	2	4.5		<6.3		<3.2		<3		1.5 h	11	⩽7
	3	4.4		<3		<3		<3		26.4 h	44	⩾71
	4	1.9		<3		<3		<3		27.3 h	52	⩾55
	5	1.4		<3		<3		<3		27.8 h	40	⩾55
	6	3.4		<6.3		<3		<3		28.0 h		

	No.	10	<7.4	10	<3	<3	0	–17	
5	1	56		10			0	10	none
	2	not cooked at this point							<7
	3	>300	4.9	<3	<3	<3	25.4 h	90	>74
	4	0.045	<3	<3	<3	<3	26.3 h	76	≥55
	5	0.055	<3	<3	<3	<3	26.7 h	68	≥55
	6	0.068	<6.2	<3	<3	<3	27.2 h	18	none
6	1	0.013		0	0	0	0	10	none
	2	not cooked at this point							<7
	3	0.083	<3	<3	<3	<3	26.3 h	64	>74
	4	0.0050	<3	<3	<3	<3	27.2 h	66	≥55
	5	0.0050	<3	<3	<3	<3	27.4 h	60	≥55
	6	0.0087	<3	<3	<3	<3	27.9 h	20	none
7	1	56	34	20	100	0	0	16	none
	2	11	46	4.9	<3	<3	0.1 h	11	none
	3	28	<3.2	<3.2	<3	<3	26.0 h	64	<7
	4	2.9	<3	<3	<3	<3	26.9 h	66	>74
	5	3.9	<3	<3	<3	<3	27.4 h	48	≥55
	6	5.4	<3	<3	<3	<3	27.6 h	20	≥55
10	1	3.0		0	0	0	0	16	none
	2	not cooked at this point							<7
	3	13	<3	<3	<3	<3	26.1 h	57	>74
	4	2.8	<3	<3	<3	<3	26.7 h	66	≥55
	5	2.1	<3	<3	<3	<3	27.1 h	57	≥55
	6	3.5	<3	<3	<3	<3	27.4 h		

system had the highest frequency of improper food temperature occurences and the 5th highest frequency of objectionably high bacterial counts of the 16 total schools.

The ground beef and spaghetti (meal item 1) from this school was only heated to 45°C (at least 74°C is recommended) and was never measured at higher than 46°C during service (Table 24.2). APC increased from 230/g to 3,100/g during transport. The mixture before cooking contained grated cheese. This cheese ingredient probably contributed coagulase-positive staphylococci resulting in 6.6 (MPN)/g in the mixture prior to cooking. Cheese could be expected to have 1,000 coagulase-positive staphylococci (MPN)/g even under good manufacturing practice (ICMSF, 1974). Therefore, 6.6 (MPN)/g in the mixture was not alarming. However, they should have been killed by cooking but were not, probably because of the low cooking temperature, which never approached 74°C. After transport, coagulase-positive staphylococci numbered 9.1 (MPN)/g and increased to 11 (MPN)/g toward the end of serving. At this point, the spaghetti was at room temperature (26°C), 4.2 hours after sampling point 1. This was a case of severe time-temperature abuse which with the added cheese ingredient, could have allowed staphylococcus enterotoxin production, resulting in food poisoning. Perhaps, the acidity of the sauce and short period of time at optimum temperature restricted staphylococcus enterotoxin production. Once produced, however, staphylococcus enterotoxin is not destroyed by cooking. Thus, a potential food safety hazard was encountered here which could have been made worse by longer holding of food at improper temperature.

The raw chicken (meal item 2) from this school had excessive coliforms, *E. coli* and coagulase-positive staphylococci (Table 24.2). Adequate cooking temperature destroyed them and significantly lowered the APC from 55,000/g to 68/g, thus allowing no problem to develop during transport and holding on the serving line at close to incubation temperatures. This is an example of severe time-temperature abuse (30–34°C for 2.5 hours) for meat which could have caused food poisoning from the staphylococci or enteric pathogens, had they not been destroyed by more than adequate cooking.

This school showed frequent time-temperature abuse with other foods after close to adequate cooking resulting in very slight microbiological buildup in the foods over less than 3 hours time. This school did not have adequate hot holding equipment on the serving line, resulting in a continuous potential food safety hazard for children eating food prepared at that school.

One school in the on-site preparation and service system had the

TABLE 24.2. COMPARISON OF MEASURED LEVELS OF INDICATOR MICROORGANISMS AND FOOD TEMPERATURE WITH RECOMMENDED LIMITS—CENTRAL PREPARATION AND HOT BULK DELIVERY, ONE SCHOOL

Meal item	Sampl. point	APC/gram Geom. Mean ($\times 10^3$)	APC/gram Sug. Max. ($\times 10^3$)	Coliforms (MPN)/gram Geom. Mean	Coliforms (MPN)/gram Sug. Max.	E. coli (MPN)/gram Geom. Mean	E. coli (MPN)/gram Sug. Max.	Staph (MPN)/gram Geom. Mean	Staph (MPN)/gram Sug. Max.	Time fr. Sampl. pt. 1	Temp. (°C) Measured	Temp. (°C) Recom.
1	1	1.0	1000	<3.2	100	<3	100	6.6	100	0	29	none
	2	0.23		<3		<3		<3.4		1.5 h	45	≥74
	3	3.1		<3.2		<3		9.1		2.4 h	38	≥55
	4	0.62		<3		<3		<8.5		3.3 h	46	≥55
	5	1.3		<3		<3		11		4.2 h	26	≥55
2	1	55	100	98	12/cm²	45	6/cm²	<4.6	3/cm²	0	0	none
	2	0.068		<3		<3		<3		1.3 h	82	≥74
	3	0.32		<3		<3		<3		2.2 h	30	≥55
	4	0.034		<3		<3		<3		3.0 h	34	≥55
	5	0.12		<3		<3		<3		3.8 h	30	≥55

second highest frequency of objectionably high bacterial counts, but the 6th lowest frequency of improper food temperature occurrences of the 16 total schools. This was due partly to four of the meal items (ground beef and spaghetti, green peas, instant mashed potatoes and canned baked beans) having excessive bacterial counts before cooking, which were reduced by adequate cooking temperatures (Table 24.3). However, the mashed potatoes (meal item 7) appeared to be the most objectionable food item at this school. Potato granules had excessive APC and coliforms prior to mixing with hot milk. Cooking temperature was measured in the potatoes at 52°C (74°C is recommended), as APC decreased slightly from 27,000/g to 18,000/g and coliforms increased slightly, from <4.4/g to <8.5/g. Serving temperature of one sample taken from the hot hold cabinet prior to serving was 36°C, an optimum temperature for the incubation of food poisoning pathogens.

The frozen preportioned delivery system seemed to significantly lower levels of bacteria in most food items by adequate cooking (Table 24.4). The major problem in this system appeared to be chocolate pudding (meal item 8), which was not heated at the school, but allowed to thaw before serving. In one school the APC and coliform counts in the chocolate pudding were objectionably high as received from the supplier and remained high during serving (Table 24.4). There was less handling and manipulation of food in this system with less chance for contamination and time-temperature abuse at the school and apparently more attention given to adequate cooking resulting in perhaps a potentially safer meal item. However, this system was dependent on the food supplier to produce a safe food product. Perhaps, the food would have more opportunity for contamination and time-temperature abuse during preparation, although this would depend on the individual supplier. As indicated by the data for chocolate pudding (Table 24.4), the greatest problem appeared to exist for foods which were not intended to be cooked or were not adequately cooked at the school. Perhaps the greatest food poisoning hazard would be staphylococcus enterotoxin produced in the food during manufacture which would not be destroyed by cooking at the school.

Data from every sampling point, menu item, and school were studied and could be discussed in a similar manner. The above examples are representative. There were cases where time-temperature abuse correlated with objectionably high bacterial counts and cases where it did not. Most of the data, however, were explainable, as in these examples discussed. Indicator microorganism data and time-temperature data were variable among schools with-

TABLE 24.3. COMPARISON OF MEASURED LEVELS OF INDICATOR MICROORGANISMS AND FOOD TEMPERATURE WITH RECOMMENDED LIMITS—ON-SITE PREPARATION AND SERVICE, ONE SCHOOL

Meal item	Sampl. point	APC/gram Geom. Mean (×10³)	APC/gram Sug. Max. (×10³)	Coliforms (MPN)/gram Geom. Mean	Coliforms (MPN)/gram Sug. Max.	E. coli (MPN)/gram Geom. Mean	E. coli (MPN)/gram Sug. Max.	Staph (MPN)/gram Geom. Mean	Staph (MPN)/gram Sug. Max.	Time fr. Sampl. pt. 1	Temp. (°C) Measured	Temp. (°C) Recom.
2	1	470	100	160	$12/cm^2$	79	$6/cm^2$	<4.6	$3/cm^2$	0	10	none
	2	3.8		<3		<3		<3		1.8 h	90	>74
	3	0.45		<3		<3		<3		3.5 h	66	≥55
	4	0.41		<3		<3		<3		3.8 h	65	≥55
5	1	<20	10	28	10	<3		<3		0	−20	none
	2	0.490		<3		<3		<3		2.7 h	90	>74
	3	2.6		<3		<3		<3		3.6 h	55	≥55
	4	0.630		<3		<3		<3		4.0 h	64	≥55
7	1	<27	20	<4.4	<3	<3	0	<3	100	0	19	none
	2	18		<8.5		<3		<3		1.3 h	52	>74
	3	19		<3.2		<3		<3		2.4 h	36	≥55
	4	14		<3		<3		<3		3.0 h	55	≥55
10	1	0.26	0	<3	0	<3	0	<3	0	0	19	none
	2	0.13		<3		<3		<3		2.1 h	84	>74
	3	0.033		<3		<3		<3		3.5 h	50	≥55
	4	0.026		<3		<3		<3		3.8 h	65	≥55

TABLE 24.4. COMPARISON OF MEASURED LEVELS OF INDICATOR MICROORGANISMS AND FOOD TEMPERATURE WITH RECOMMENDED LIMITS—FROZEN PREPORTIONED DELIVERY, ONE SCHOOL

Meal item	Sampl. point	APC/gram Geom. Mean (×10³)	APC/gram Sug. Max. (×10³)	Coliforms (MPN)/gram Geom. Mean	Coliforms (MPN)/gram Sug. Max.	E. coli (MPN)/gram Geom. Mean	E. coli (MPN)/gram Sug. Max.	Staph (MPN)/gram Geom. Mean	Staph (MPN)/gram Sug. Max.	Time fr. Sampl. pt. 1	Temp. (°C) Measured	Temp. (°C) Recom.
3	1	13	100	<3	100	<3	<3	<3		0	-23	<0
	2	0.0027		<3		<3		<3		1.4 h	90	≥71
	3	0.0017		<3		<3		<3		2.6 h	70	≥55
5	1	1.1	100	9.1	100	<3		<3		0	-23	<0
	2	0.0096		<3		<3	<3	<3		1.3 h	86	≥71
	3	0.014		<3		<3		<3		2.9 h	62	≥55
6	1	4.2	100	<3.2	100	<3		<3		0	-23	<0
	2	0.0021		<3		<3	<3	<3		1.3 h	64	≥71
	3	0.0030		<3		<3		<3		2.9 h	58	≥55
7	1	1.6	100	<3	100	<3	<3	<3		0	-23	<0
	2	0.0093		<3		<3		<3		1.3 h	90	≥71
	3	0.019		<3		<3		<3		2.9 h	67	≥55
8	1	>46	10	200	100	<3	<3	<3	<10	0	-23	<0
	2	>100		140		<3		<3		1.0 h	0	≤7
	3	7.8		220		<3		<3		2.6 h	2	≤7

in systems and supported the highly significant difference ($P<.01$) among schools indicated by analysis of variance of APC data. Food preparation and handling abuse was a characteristic of individual schools and was not uniquely characteristic of any particular food preparation and delivery system or systems. This was shown by indicator microorganism data and time-temperature data as well as by analysis of variance of APC data. The data pointed out potential food safety hazards exemplified by some of the schools in each food preparation and delivery system. These potential problems could be reduced through application of better food handling practices and indicate areas where all systems could be improved. The data also demonstrated that all four food preparation and delivery systems are capable of producing microbiologically safe meals for children to consume in the school lunch programs. Since no one food preparation and delivery system was found inferior to the others relative to microbiological food safety, decisions may be made on the relative desirability of the four systems based on other factors.

CONTROL OF FOOD SAFETY IN SCHOOLS

The following specifications apply specifically to school food service systems and are aimed at assuring safe meals for children in mass feeding situations.[1]

Equipment and Facilities Sanitation

The building premises should be kept clean and organized. Waste and garbage cans should be outside and away from building. They should be covered and never allowed to overflow. A hose and drain should be available outside of the building to allow periodic cleaning and sanitizing of trash and garbage containers.

There should be no cracks nor other openings in the building where rodents may enter. All pipe openings entering the building should be sealed. Doors and windows should not be left open unless they have effective screens. Rodents and insects can be controlled by eliminating all accessible food through properly covered storage. Traps and insect catching devices may be used, if necessary. Fly swatters should be avoided since they could spread pathogenic

---

[1]Adapted to school food service from: Westhusin, B. and J. S. Avens. 1978. Food Safety in the kitchen. Cooperative Extension Service. Colorado State University, Fort Collins, Colorado (Westhusin and Avens, 1978).

bacteria to food. All containers of food should be stored well above the floor and out away from the walls to allow cleaning and prevent rodent harborage. Pesticides should be used to control rodents and insects only as a last resort, since they are harmful chemicals which must not contact human food. Professional exterminators should be employed, if infestation cannot be otherwise controlled.

Proper equipment and materials for cleaning and sanitizing facilities and equipment should be obtained. All cleaning and sanitizing chemicals and materials should be stored in containers in which they were purchased, and in a separate location away from food. Instructions for their use should be carefully followed to prevent accidental contamination of food, either directly or indirectly, such as by residues left on food contact equipment.

General cleaning steps for food contact equipment and utensils are: (1) prerinse, (2) soap and scrub, (3) rinse, and (4) sanitize. In automatic dishwashers, prerinse water should be about 49°C to 54°C (120°F to 130°F). Soap or detergent in wash water should be at a higher temperature, 60°C to 71°C (140°F to 160°F), as should the rinse water. The final sanitizing step can be done with approved commercial sanitizers according to instructions or with extremely hot water, 77°C to 88°C (170°F to 190°F).

All appliance-type equipment or other multipart equipment should be unplugged from the electrical outlet and/or disassembled before cleaning. Each part, as well as the base of the appliance, should be cleaned. Food slicers, food mixers, grinders, can openers, scales, chopping blocks and cutting boards, carts, toasters, beverage dispensing machines, steam kettles, grills, broilers and fryers are a few examples of equipment that should be cleaned daily. Larger, more stationary equipment or part of the facilities, such as tables, shelves, cabinets, refrigerators, ovens, dishwashers, sinks, floors, walls and doors, may intentionally or unintentionally contact food. These items or areas must constantly be maintained in a clean and sanitary condition. Of particular concern in cleaning these items is that food lodged in cracks, corners or joined parts must be completely removed, so that bacteria will not multiply and/or produce toxins and thus contaminate other food and equipment. The dining area should be clean, as well as the kitchen and serving area. Tables and chairs should be thoroughly cleaned as soon as the meal is finished and the area is vacated. The same cleaning cloth should not be used to wipe table tops as was used to wipe chairs. This common undesirable practice observable in many types of food service establishments, particularly restaurants, indirectly facilitates the "fecal-oral route" of transmitting pathogenic microorganisms to humans.

Food Sanitation

The sanitation of food starts with purchasing from reliable purveyors. However, even at best purveyors occasionally supply bad food, so time should be taken to inspect each shipment of food received at the school kitchens. All cases of food should be opened and inspected immediately upon receipt. Perishable and semiperishable foods should be checked for signs of spoilage. Spoiled meat products may be discolored, slimy or have a putrid odor. Spoiled fruits and vegetables may be mushy or soft, discolored, moldy or have a decomposed odor. In canned products, signs to watch for are rust or corrosion on the can, dented seams and swollen lids, and abnormal odor or appearance when opened just prior to preparation.

After inspection, food should be stored immediately in covered containers in clean facilities at proper temperature and relative humidity. The following food storage recommendations will help prevent food poisoning and lengthen shelf life:

	Temperature	Relative Humidity
Shelf-stable foods	10°C to 21°C (50°F to 70°F)	50% to 60%
Dairy products and eggs	3°C to 4°C (38°F to 40°F)	75% to 85%
Red meats and poultry	–1°C to 2°C (30°F to 36°F)	75% to 85%
Fresh fish	–2°C to –1°C (28°F to 30°F)	75% to 85%
Fresh vegetables and fruit	2°C to 4°C (35°F to 40°F)	85% to 95%
Frozen foods	–29°C to –18°C (–20°F to 0°F)	

Note: Fresh fish can be stored with meats, if packed in ice. If only two refrigerators are available, red meat, poultry and fish should be stored in one; fresh fruits, vegetables, dairy products and eggs should be stored in the other. Cooked foods would require a third refrigerator.

Fresh cut red meat can be stored, refrigerated 3 to 5 days, fresh ground red meat and fresh poultry 1 to 2 days. Processed meat such as sausage, frankfurters, ham and bacon can be safely stored refrigerated up to 7 days. Cooked red meats and cooked entrees containing red meat can be stored refrigerated 3 to 4 days, gravy and meat broth 1 to 2 days. Cooked poultry and cooked entrees containing poultry can be stored refrigerated 1 to 2 days.

Quantity food preparation is an important step where sanitation problems can occur. When cutting and proportioning raw red meat, poultry or fish; hands, food containers, utensils, table, and cutting board that contact the raw meat should be thoroughly washed and disinfected before contacting other food. Failure to do this will cause cross-contamination of other foods that come in contact with hands, food containers, utensils and working surfaces. After handling any raw food materials, food handlers *must* wash their hands and all contaminated food contact surfaces before working with other foods, particularly cooked foods. Food deposits left in corners and cracks of equipment and tables will harbor bacteria which can multiply rapidly at 4°C to 60°C (40°F to 140°F). When a food product is dropped on the floor, it should be thrown away or at least trimmed if possible. It should never be put back with other foods which could cause contamination of much larger quantities of food. The floor, even when properly cleaned, is probably the dirtiest place in the kitchen and no food should be kept that has had contact with it.

Temperature Control of Food

Since, even under sanitary conditions, most food normally contains some bacteria, it must be handled in such a manner to prevent multiplication of bacteria present. It is important in food preparation to keep foods out of the dangerous temperature zone, 4°C to 60°C (40°F to 140°F), to prevent bacterial multiplication. There are two processes which take place in school food service kitchens that unavoidably allow portions of food to be in this temperature zone for a certain period of time. These processes are *thawing* and *cooling*. The time that foods are in this dangerous temperature zone during these processes must be minimized by implementing specific procedures.

Thawing is safest when done in the refrigerator. This procedure does not allow the food temperature to rise any higher than the refrigerator temperature which should be 7°C (45°F). If this method is used, there must be some planning in advance. A frozen 20 lb turkey carcass will take three days or more to thaw in a refrigerator.

Although it is generally recommended to thaw frozen meat, even large turkey carcasses, in the refrigerator, it may take a few days for a large turkey carcass to thaw completely. Under proper refrigeration the skin surface would be at a low enough temperature to inhibit bacterial multiplication and prevent multiplication of pathogenic bacteria. However, in many school lunch programs, it would not be practical to utilize that much refrigeration space for such a long period of time. Therefore, alternate thawing procedures may be safely used for frozen vacuum packaged meat, if certain precautions are followed.

Frozen vacuum packaged meat may be thawed much faster than in a refrigerator by immersing it in cold running water for a few hours depending on the size of the meat or carcass. This is conveniently done in school kitchens by placing the vacuum packaged meat in one section of a clean double sink and allowing tap water to fill the sink and slowly overflow into the other section. This thawing procedure can be started late in the day prior to day of cooking the meat and continued overnight in the running tap water. The sink then will be available for normal use during the day. Turkey carcasses will be thawed by morning and should be prepared immediately for cooking. Smaller pieces of frozen meat would take much less time to thaw and ideally should not be left overnight. It is important that the tap water be cold and flowing. Even cold water is warm relative to the frozen meat and it is adding heat to thaw the meat while maintaining the meat surface at a low enough temperature to inhibit bacterial multiplication. Water must be kept flowing slowly to continuously provide additional heat necessary to thaw the entire piece of meat and to maintain cold enough water to inhibit bacterial growth on the thawed meat. If this thawing method can be employed during the workday, water in the sink can be frequently drained and replenished rather than continuously flowing, particularly if water conservation is a factor. In this method of thawing the plastic bag should be left on the carcass or piece of meat. If any food is placed in a sink, it is essential that the sink be provided with a suitable drain trap to prevent any accidental backup of sewer waste which could contaminate the food.

If the sink has no drain trap and, if water conservation is a major factor to consider, another alternative thawing method may be used for large frozen turkey carcasses or large frozen packages of meat. The frozen meat in its plastic package may be enclosed in a double-walled paper bag and left at room temperature to thaw. Placing the meat in a paper shopping bag and then placing a second paper bag over the opening of the first or using a single paper bag provides enough insulation to allow the meat to thaw more uniformly, while maintaining air temperature in the immediate environment of the meat surface low enough to inhibit bacterial multiplication. Klose *et al.* (1968) determined the effect of this thawing method on the interior and surface temperatures of thawing turkey carcasses. A single double-walled paper bag increased the minimum thawing time for 12–14 lb. frozen turkey carcasses at 21°C (70°F) by about 4 hours, while the time between thawing and four doublings of surface bacteria was more than twice that of thawing at 21°C (70°F) without the paper bag. The insulation maintained a low surface temperature without greatly prolonging thawing time. At 28.6°C (84°F) air temperature, the skin surface temperature of the carcasses thawed in paper bags remained below 12.6°C (55°F) for several hours after carcass was thawed. They suggested for thawing frozen turkey carcasses in an insulating paper bag, total thawing time in room air at 21–26°C (70–80°F) should not exceed 15 hours for 4 to 6 lb. carcasses and 20 hours for 12 to 14 lb. carcasses. Bryan and McKinley (1974) recommended when thawing frozen turkey carcasses at room temperature in paper bags, not to allow them to remain at room temperature more than one hour per pound, if one double layer bag is used, or 1¼ hour per pound if two double layer bags are used. Carcasses thawed by this method should be cooked immediately after thawing. This thawing procedure should never be used for frozen prestuffed turkey carcasses.

An alternate method of thawing is by microwave energy. Some of the newer units have a defrost cycle which provides a satisfactory method of thawing smaller pieces of frozen meat. Preferably, the meat should be cooked immediately after thawing by this method.

Foods requiring cooking, including all entrees containing raw meat should be cooked to an internal temperature of 73°C (165°F). It is often recommended to cook poultry to an internal muscle temperature as high as 84°C (185°F) at an oven temperature of at least 161°C (325°F). It may take considerable additional time to adequately cook unthawed food. Large frozen turkey carcasses used in school food service operations require too much time to adequately cook without prior thawing (Bryan and McKinley, 1974). They should be completely thawed before cooking. Poultry dressing cooked inside the carcass or separately must reach at least 73°C (165°F). Precooked food should be reheated to an internal temperature of at least 70.5°C (160°F), meat to at least 73°C (165°F).

Hot foods for serving should be held above 54°C (130°F) after preparation and cooking (Bryan, 1974). This can be done in school food service systems by placing the pans full of hot food into a steam bath or steam table, or hot holding cabinet for transportation, holding and serving. Insulated and/or heated cabinets should be used to transport hot food from a central kitchen to satellite schools to maintain food temperature above 54°C (130°F) during transit. Whatever the equipment available, it must provide or conserve enough heat to maintain all parts of the food above 54°C (130°F). Hot food should not drop below this temperature at any time between cooking and serving. Steam holders or hot cabinets are *not* capable of cooking the food, but are designed to maintain hot food at its proper temperature for a limited period of time after it has been prepared and cooked. Cooked food should never be left in an unheated oven for any period of time.

Slow cooling is a process which enables bacteria to multiply in food. If large containers of stock, soup or any liquid or semi-solid food, or large roasts or large containers of cut meats are put into the refrigerator hot, warm, or even at room temperature, it may take a day or more before the center is cooled to a safe temperature. All portions of hot food should cool to 21°C (70°F) within one hour and 7°C (45°F) within 3 hours (Bryan, 1974). There are various methods of properly cooling large quantities of food. A refrigerator with forced circulation of air speeds cooling. Whether or not the refrigerator has a fan to circulate air, hot food will cool faster if it is not packed tightly into the refrigerator. Adequate space should be left between hot food containers to allow for cold air circulation. Hot food should be transferred from large containers into many small shallow pans, and large pieces of hot solid food should be cut into smaller pieces stacked no more than 3 inches thick in shallow pans. This increases the surface area exposed to the cold air, thus decreasing the time needed to adequately cool the food. Another method is to set the hot liquid food container in a sink and either cold water is run on the outside of the container or it is surrounded by ice. The food product must be stirred periodically to allow it to cool faster. Once the food is cooled by this method it should be stored in the refrigerator. One other method that is effective is to place a shallow pan full of hot or warm solid food on ice in a larger tray in the refrigerator. Hot foods should never be allowed to "temper" at room temperature before storing in the refrigerator. This undesirable but still common practice is a carryover from the old days of ice boxes and old refrigerators which required periodic defrosting due to frost buildup. Modern "frost-free" electrical refrigeration units recover low temperature rapidly when hot food is placed in them and do not require shut-down for "defrosting". Hot foods will not cool fast enough at room temperature and should always be immediately placed properly in an adequate refrigerator. Refrigerators should be maintained between 0°C and 7°C (32°F and 45°F). Insulated and/or refrigerated cabinets or trucks should be used to transport cold food from a central kitchen refrigerator to satellite schools to maintain food temperature below 7°C (45°F) during transit.

Personal Hygiene

Good personal hygiene is a must for all food handlers in any school food service system. Food handlers are an important reservoir for pathogenic bacteria. Personal hygiene starts with a bath or shower before going to work. All clothes should be clean and free of soil. A freshly laundered uniform is recommended rather than working in "street clothes". Clothing that is not freshly cleaned before work can harbor large amounts of bacteria. *Clean* head covers should be worn because hair is a source of staphylococcus. Also, hair is aesthetically objectionable in food. However, the head cover also keeps long hair back out of the face when leaning over the work area, so the hands are not contaminated each time hair is pulled back. Long hair should be fastened back so it does not hang over the face. Hairnets are the most effective and otherwise suitable head covers and should be worn by all food handlers with long or even medium length hair. For people with *very short* hair, a paper cap positioned on the forehead below the hairline is satisfactory. Cloth or plastic caps or other hats which cannot be laundered or disposed of daily are not acceptable.

Hand washing is a very necessary practice and should be done frequently by all food handlers. Hands and lower arms should be washed thoroughly immediately before handling food after returning to the kitchen from any other area, and particularly after using the toilet, smoking, coughing, sneezing, blowing the nose, touching sores or bandages, or touching raw foods. A hand washing sink should be conveniently located and supplied with disinfectant hand soap and individual disposable paper towels. Individual single-use paper towels are important because it does no good to wash hands and then dry them on a dirty towel. It may take at least 20 to 30 seconds to wash hands adequately. If a food handler were to wash the hands 10 times a day, which should be a minimum, it would require only 5 minutes of time. This is a small price to pay for the proper sanitation necessary for disease prevention.

Workers who have diseases that may be transmitted by foods and show symptoms of diseases that would allow the contamination of food or food contact equipment or who have been diagnosed as carriers of pathogenic microorganisms, should not be allowed in the school food service kitchen or serving and dining area; they should stay home until completely recovered. A few examples of many such symptoms and diseases to be aware of are diarrhea, fever, nausea, runny nose, sinus congestion, sore throat, salmonellosis, typhoid fever, flu and colds. Pathogenic viruses and bacteria are unavoidably transferred to food by food handlers.

Workers who handle food should not have infected sores, such as boils or any pus-containing lesions, on their hands or anywhere on their body. Such sores should be treated and covered to promote rapid healing. Infected sores should be healed before the person is allowed to work in the food handling operation. Staphylococci are found in very large quantities in sores and pimples and cause food poisoning when they contaminate food and are allowed time, at the necessary temperature, to produce toxin in the food. Staphylococci are also naturally present on the hair and skin and in the mouth and nose of healthy people. Playing with hair, touching mouth and nose and scratching the body should be avoided during food handling. Touching cooked or prepared food with the hands should be avoided. Plastic disposable gloves, or preferably clean utensils, should be used to transfer, dispense or serve prepared food, rather than touching food with bare hands. Eating food, smoking or chewing tobacco or gum in food preparation areas should be prohibited due to possible contamination of food being prepared, by hands, ashes, smoke, chewed food, gum or tobacco particles or saliva from the mouth. Handling dishware and stainless steel utensils in a sanitary manner is very important. Drinking glasses should be picked up by the base and plates and dishes handled by the bottom or rim only. Eating or serving utensils should be picked up by the handle, never touching the eating or serving food-contact surface.

This section on control of food safety in schools has discussed some of the major factors contributing to food safety at four critical control points: (1) equipment and facilities sanitation, (2) food spoilage, storage and sanitation, (3) temperature control of food, and (4) personal hygiene. These critical control points correspond rather closely to the four *critical control points* identified by Bobeng and David (1977) for the Hazard Analysis Critical Control Point (HACCP) models they developed for entree production in conventional, cook/chill, and cook/freeze food service systems. In these three food service systems they have identified *"control points"* which are points in the process (process stages of entree production) where a potential food hazard may exist. Although individual schools within a school food service system may differ slightly, in general, the *control points* for entree production and distribution are: procurement, preparation, cooking or heating, storage in bulk either hot, chilled or frozen, portioning, assembly, distribution, and service. In cook/chill and cook/freeze systems, entrees are heated just before service; in cook/freeze systems bulk food product is thawed before portioning. They defined *"critical control points"* as those points in a process that eliminate or reduce a microbiological hazard. The four *critical control points* they identified for their HACCP models were: (1) ingredient control and storage, (2) equipment sanitation, (3) personnel sanitation, and (4) time-temperature. In their discussion of factors contributing to microbiological control at each of these four *critical control points*, they covered many of the factors contributing to food safety discussed in this section on control of food safety in schools, as well as some additional factors not covered here. They have provided an excellent review of controllable factors determining food hazard or safety in food service operations. This review by Bobeng and David (1977) refers to a number of excellent sources of specific information necessary for the school food service worker to use in implementing proper food handling procedures. The effectiveness of control measures at *critical control points* is determined by sensitive monitors. Continuous or periodic surveillance of time-temperature and comparison with established or recommended standards is a suitable monitor of control effectiveness. Effective control of equipment and personnel sanitation should be monitored by continuous or periodic inspection of adherence to established sanitation standards. The person doing the monitoring should be a food service manager or supervisor, adequately trained in all aspects of food service safety.

The HACCP system, as developed for application to quality control in food service systems by Bobeng and David (1977), should be seriously considered for implementation by all school food service systems. The HACCP system is a preventive approach to quality control related to food safety.

In this section, no attempt has been made to provide a complete list of rules and regulations, including all specifications and recommendations, related to operating a safe food service operation. Such information has been and will continue to be published in numerous forms available to food service managers and employees. The following are recommended as examples of many such references presently available:

Focus on Kitchen Sanitation and Food Hygiene (Axler, 1974)
Identifying Foodborne Disease Hazards in Food Service Establishments (Bryan, 1974)

Food Service Sanitation. Proposed Uniform Requirements
(Food and Drug Administration, 1974)
Food Sanitation (Guthrie, 1972)
Food Service Sanitation (Litsky, 1973)
Quantity Food Sanitation, 2nd Ed. (Longree, 1972)
Sanitary Techniques in Food Service (Longree and Blaker, 1971)

Also, referring to appropriate state, county and/or city health
department published rules and regulations, governing the safety
and sanitation of food service establishments under their regulatory
jurisdiction is recommended. Some states and school systems
publish handbooks on food service sanitation and safety specifically
applicable to their own local food service operations.

EDUCATIONAL PROGRAMS IN FOOD SERVICE SANITATION AND SAFETY

In two particular school food service operations there were a few
objectionable practices and/or conditions noted.[2] Workers used bare
hands to handle food when utensils could have been used. Food was
inadequately covered and stored near the delivery door, allowing
contamination from the outside environment. Food containers were
stored directly on the floor. Food temperatures were not determined
during preparation or serving. Sandwiches, cream pies and other
à la carte items were not refrigerated on the serving line. However,
no serious problem would develop if only a few items were placed
out at one time and consumed rapidly. This was the case and the
rest were kept refrigerated until set out for immediate serving.
Cleaning compounds were stored in the same room and some even
on the same shelf next to food items. Flies were a problem since
screens on windows were left open and the screen door did not close
adequately. These were the only objectionable occurrences observed
in an otherwise very sanitary, well-managed school food service
operation. This public school system has a training program for food
service managers and other employees. Employees attend state
health department training workshops in food safety every two
years; managers more frequently. Thus, there is an awareness of
food safety among food service workers and managers in this public
school system.

In another particular public school food service operation inspected, there were even fewer objectionable practices and/or con-

---

[2]Adapted from an inspection report written by Ms. Jane Macy, student, Colorado
State University, Fort Collins Co, 1975.

ditions noticed.[3] Almost all food safety recommendations related to food service establishments were strictly adhered to. They had managers and employees who were obviously well trained in food service safety and followed the recommendations of the Department of Health precisely. The overall operation was run very efficiently and the employees seemed to take pride in their work. There was high worker morale and good manager-employee relationship. Although the equipment and facilities were considerably older than in more recently constructed school food service operations, the well-trained, knowledgeable and concerned employees resulted in safe food for hundreds of school children.

Educational programs in food service sanitation and food safety should be developed and made available to all school food service programs. Short courses, workshops and training materials should be developed for continuous on-the-job education of food service employees as well as food service supervisors and managers. There have been enough indications of unsanitary food handling and time-temperature abuse from reported studies and inspections of school food service operations, to conclude that some schools have food service workers that need this type of training. They lack complete knowledge of food service sanitation and technology specifications and often do not understand the reasons behind them. There are many educational materials available for training food processing and food service employees in the area of sanitation including: personal hygiene, microbiology, foodborne illness, cleaning and sanitizing, food safety, and food plant construction (Schuler, 1975). These materials include bulletins, pamphlets, posters, magazines and books, speeches, mimeographed materials, courses, films, film-strips and slide sets. However, such available materials must be used in effective training programs. Many school food service operations are involved in effective food safety training programs, but many are not.

A national food service sanitation training program has been developed in Canada aimed at education of all personnel in the Canadian food service industry (Davies, 1976). In the United States the Food and Drug Administration has recently promoted the concept of training and certifying food service managers, and in 1976 announced recommendations for a uniform training course (Davis, 1977a). Several training programs have been conducted in various parts of the United States. We are now entering an era of interest in food service sanitation of national scope (Davis, 1977b).

---

[3]Adapted from an inspection report written by Mr. Gregory Pappas, student, Colorado State University, Fort Collins CO, 1976.

Government food regulatory agencies will exert more control of all types of food service establishments, including schools. More uniform regulations related to food service sanitation and safety will be developed for use on a national basis. In 1977 the National Institute for the Foodservice Industry (NIFI) developed a plan for a uniform national program of sanitation education and certification for food service owners, operators, managers and supervisors (Hall, 1977). Hall predicts a continuing and growing network of mandatory and voluntary training and certification programs with uniform course content, administration and examinations, for the food service industry. Such programs will help professionalize the food service industry and result in safer food for consumers dining away from home.

School food service will undoubtedly be affected by this national trend toward uniform training and certification of food service managers. The trained and certified manager will be responsible for training all food service employees and conducting continuous inspection of the school food service operation. This should ultimately increase the knowledge and awareness of all school food service personnel, develop in them a more professional attitude, and result in even greater assurance of safe food for all school children to consume. Every food service employee in the school is ultimately responsible for the safety of food prepared and served to every child or adult in that school, consuming the food. A child can get very sick or even die because of inadequate food safety knowledge and improper procedure of a school food service employee. Every school food service manager and employee should be knowledgeable of all aspects of food safety and should understand the reasons for required proper procedures for preparing and handling food.

BIBLIOGRAPHY

AVENS, J. S., PODUSKA, P. J., SCHMIDT, F. P., JANSEN, G. R., and HARPER, J. M. 1977. Food Safety Hazards Associated With School Food Service Delivery Systems. Research Report. Colorado State University, Fort Collins, Colorado.

AVENS, J. S., PODUSKA, P. J., SCHMIDT, F. P., JANSEN, G. R., and HARPER, J. M. 1978. Food safety hazards associated with school food service delivery systems. J. Food Sci. 43, 453–456.

AXLER, B. H. 1974. Focus on Kitchen Sanitation and Food Hygiene.

Howard W. Sams & Co., Inc., Indianapolis, IN and ITT Educational Publishing, Indianapolis, IN/New York, NY.

BOBENG, B. J., and DAVID, B. D. 1977. HACCP models for quality control of entree production in foodservice systems. J. Food Protect. 40, 632–638.

BRYAN, F. L. 1974. Identifying foodborne disease hazards in food service establishments. J. Environ. Health 36, 537–540.

BRYAN, F. L., and McKINLEY, T. W. 1972. Prevention of foodborne illness from turkeys served in schools. 7F0142972. Center for Disease Control, Public Health Service, U. S. Department of Health, Education and Welfare, Atlanta, Georgia.

BRYAN, F. L., and McKINLEY, T. W. 1974. Prevention of foodborne illness by time-temperature control of thawing, cooking, chilling, and reheating turkeys in school lunch kitchens. J. Milk Food Technol. 37, 420–429.

BRYAN, F. L., McKINLEY, T. W., and MIXON, B. 1971. Use of time-temperature evaluations in detecting the responsible vehicle and contributing factors of foodborne disease outbreaks. J. Milk Food Technol. 34, 576–582.

BUNCH, W. L., MATTHEWS, M. E., and MARTH, E. H. 1977. Fate of *Staphylococcus aureus* in beef-soy loaves subjected to procedures used in hospital chill foodservice systems. J. Food Sci. 42, 565–566.

CREMER, M. L., and CHIPLEY, J. R. 1977a. Satellite food-service system assessment in terms of time and temperature conditions and microbiological and sensory quality of spaghetti and chili. J. Food Sci. 42, 225–229.

CREMER, M. L., and CHIPLEY, J. R. 1977b. Satellite food service system: time and temperature and microbiological and sensory quality of precooked frozen hamburger patties. J. Food Protect. 40, 603–607.

DAVIES, R. J. 1976. The national sanitation training program: Canadian Restaurant Association's answer to safe and sanitary foodservice. J. Milk Food Technol. 39, 367–369.

DAVIS, A. S. 1977a. Sanitation training for food service managers - a must! J. Food Protect. 40, 198–199.

DAVIS, A. S. 1977b. A new era in food service sanitation. Food Technol. 31(8), 69.

FOOD and DRUG ADMINISTRATION. 1974. Food service sanitation. Proposed uniform requirements. Department of Health, Education and Welfare. Federal Register 39, 35438–35449.

GREGG, M. B., ED. 1977a. Multi-state outbreak of *Salmonella newport* transmitted by precooked roasts of beef. MMWR 26, 277–278.

GREGG, M. B., ED. 1977b. Follow-up on salmonellae in precooked roasts of beef. MMWR 26, 394 & 399.

GUTHRIE, R. K. 1972. Food Sanitation. The Avi Publishing Co., Inc., Westport, CT.

HALL, C. G. 1977. National program for food service sanitation. Food Technol. 31(8), 72.

ICMSF. 1974. Microorganisms in Food. 2. Sampling for Microbiological

Analysis: Principles and Specific Applications. International Commission on Microbiological Specifications for Foods. University of Toronto Press, Toronto, Ontario.

KLOSE, A. A., LINEWEAVER, H., and PALMER, H. H. 1968. Thawing turkeys at ambient air temperatures. Food Technol. 22, 1310–1314.

LITSKY, B. Y. 1973. Food Service Sanitation. Modern Hospital Press, Chicago, IL.

LONGREE, K. 1972. Quantity Food Sanitation. Second Ed. Wiley-Interscience, New York, NY.

LONGREE, K., and BLAKER, G. G. 1971. Sanitary Techniques in Food Service. John Wiley and Sons, Inc., New York, NY.

ROLLIN, J. L., and MATTHEWS. M. E. 1977. Cook/chill foodservice systems: temperature histories of a cooked ground beef product during the chilling process. J. Food Protect. 40, 782–784.

SCHULER, G. A. 1975. A survey of educational materials available for training food processing and food service employees in the area of sanitation including: personal hygiene, microbiology, foodborne illness, cleaning and sanitizing, food safety, and food plant construction. Poultry Sci. 54, 927–944.

TUOMI, S., MATTHEWS, M. E., and MARTH, E. H. 1974. Temperature and microbial flora of refrigerated ground beef gravy subjected to holding and heating as might occur in a school foodservice operation. J. Milk Food Technol. 37, 457–462.

VAN EGMOND, D. 1974. School Foodservice. The Avi Publishing Co., Inc., Westport, CT.

WESTHUSIN, B., and AVENS, J. S. 1978. Food Safety in the kitchen. Bulletin 499A. Cooperative Extension Service, Colorado State University, Fort Collins, CO.

Index

Aw, 28
Acids, organic, 21-22
Acidic and acidified foods, 79-80
Acidulants, 79, 87, 668
Aerobic mesophilic plate count (APC),
        742, 743, 746, 747, 749, 751, 752
Aflatoxins, 202-207
Air, 56
    sources of contamination, 56
    numbers of bacteria in, 56
Amines in food, 327
    in nitrosamine formation, 320
    in spices, 333
Amanita spp, 610
Amoebic dysentery, 308
Anagyrine, 613
Animal products, controlled atmospheric
        storage of, 64
    microbial contamination in abattoir,
        63
Anthrax, 300
Antibiotics, detection methods for, 533-
        535
    in animal production, 527-528
    in eggs, 522-524
    in fermentations, 525-526
    in fruits and vegetables, 521-522
    in human milk, 524-525
    in milk and milk products, 524-525
    in muscle foods, 522-524
    in other foods, 526
    mode of action, 517-518
    Public health aspects, 528-532
    resistance of microorganisms to, 518-
        520
    transfer resistance, 519-520
Antimony, 433-434
Antioxidants, 673
Arboviruses, group B, 308

Arsenic, 433
Asclepias spp, 606
Astragulus spp, 610

Bacillus, heat resistance of, 23
    in food spoilage, 17
    in raw milk, 17
    proteases, 26
B. anthracis, 301
B. cereus, role in
        food infections, 39-40
B. stearothermophilus, 23
B. subtilis, 40
Bergapten, 608
Bighead, 609
Biochanin-A, 600
Blighia spp, 610
Botrytis cinerea, 201
Botulinal spores, in meats, 77-78, 88
        in the environment, 71-73
Botulism, 68-70
Brassica spp, 599
Brevianamide A, 238
Brewers yeast, 303
Brucella, B. abortus, 38-39, 300, 301
        B. melitensis, 38, 300-301
        B. suis, 38, 300-301, 304
Brucellosis, 300-301, 304
Byssochlamys, 213
B. nivea, 213
Buffers, 674

Cadmium, 429, 431, 432

Canadian Food and Drug Act, 706, 723-725
Canned crab meat, 80
Carbonating agents, 676
Cardenolides, 606
Carmine dye, 303
Carrion, extent of microbial contamination, 57
    parameters influencing spoilage, 58
    sources of contamination, 57
Caviar, 93
Central American Common Market, 711
Citrinin
    occurrence, 199, 215
    structure, 216
    toxicity, 215
Claviceps spp, 611
Claviceps purpurea, *199, 234*
Clarifying agents, 677
Clonorchis sinensis, cycle in fish, 43-44
    Clonoretis, 44
Clostridium, 18, 24
Clostridium botulinum
    12-D concept, 22, 27
    F value, 77
Clostridium perfringens, 736
Clouding agents, 676
Codex alimentarius, 702
Codex secretariat, 713
Coliforms, 742, 746-747, 749, 751-752
Colors, additive amendment, 705
    certified, 706
    provisional, 706
γ- coniceine, 615
    coniine, 614-615
    conium spp, 614
Cooling of cooked food, 738-739, 756, 759-760
Coordinating Committee for Africa, 713
Coxiella burnetti, 305
Crambe spp, 597
Critical phases of food processing, 739-742, 761-762
Crooked calf disease, 612-614
Cryptosporiopsis curvispora, 201
Cyanide, 601
Cyanogenic (Cyanogenetic) glycosides, 8, 601
Cycad, 603
Cycas spp, 603
Cycasin, 604
Cyclochlorotine, 221-222
Cyclopamine, 615
Cyclopia, 615
Cyclopiazonic acid
    structure, 234
    toxicity, 234

Cycloposine, 615
Cymopterus spp, 608
Cysticercosis, 309, 311-312
Cysticercus bovis, 311
Cysticercus cellulosae, 311, 312
Cytotoxic anoxia, 602

Decumbin
    occurrence, 232
    structure, 232
    toxicity, 232-233
Defoaming agents, 674
Delaney Clause, 10, 584, 660, 705
11-deoxyjervine, 615
Diarrhea, infantile, 306
Dibothriocephalus, 46-47
D. latrium, 46
Dietary foods, 719
Diethynitrosamine in foods, 329
Dimethylnitrosamine in foods, 329
Diosgenin, 617
Diploidia, 202
Dough conditioners, 674, 675
Dugaldia spp, 605

Echinococcus spp, 310
Economic system for Latin America, 714
Educational programs in food service sanitation and safety, 763-765
Egg plant, 595
Emulsifiers, 667
Enterobacteriaceae, 33, 34, 37
Enzymes, 672
    resistance to irradiation, 559-561
E. coli, 32-33, 306, 739, 746-747, 751-752
    outbreaks in cheese, 33
    role in food infections, 32
Equipment cleaning, 754, 756
Ergot, 199, 611
    occurrence, 234, 235
    structure, 236
    toxicity, 234, 235
Erysipeloid, 301-302
Erysipelothrix, 42
E. rhusiopathiae, 42-43

Fabricated foods, 652, 654, 655
Favism, 8

Firming agents, 675
Fish, 302, 306, 625-651
Flavors/enhancers, 666
Flow agents, 676
Fly ash, 7, 434-437
Food additives
    amendment, 705
    benefits vs. risks, 659-661, 706
    consumption, 661
    definitions, 653
    direct, 680
    extraction from contact materials,
        693
    functions, 654-661, 662-667, 679
    good manufacturing practices and,
        681
    GRAS, 679, 705
    indirect, 680
    migration from contact materials,
        694
    need of an industry, 652-677
    per capita consumption, 652
    proper use of, 678
Food advertising, 707
Food and Agriculture Organization
    (FAO), 715
Foodborne diseases, control of, 265-296
    education and training in relation
        to, 293-294
    factors contributing to outbreaks of,
        265-268
    measures to minimize outbreaks, 268-
        273
    microbiological surveillance of foods
        in relation to, 288-291
    public health or quality control acti-
        vities in relation to, 285-288
Food and Drug Administration (FDA),
    763-764
Food losses and waste, 657, 658
Food phosphates, 663-665
Food preservation
    by heat, 77-79
    by other means, 80-94
Food Safety Council, 2
Food Safety evaluation, 11
Food safety in schools, 736-765
Food standards, Canada, 726
Food standards council for Asia, 713
Food standards program, 702
Food spoilage, 17, 755
    intrinsic, extrinsic factors, 20
    principal organisms, 17

Foods, definition of, 4
    environmental contamination of, 5
    faddism, 5

    hazards, 3-5
    Low-acid canned, 77
    natural toxicants in, 8-9
    nutritional value of, 728-729
    organic, 5
Formoronetin, 600
Fumagillin, structure, 229
    toxicity, 229
Fumigatin, structure, 222
    toxicity, 222
Fumitremorgen A, 227, 228
Fumitremorgen B, 227, 228, 230
Fungal spoilage of various foods, 200-202
Fusarium spp, 199-224

Galerina, 235
Genistein, 600
Geotrichum, 201
Gibberella zea, 199, 215, 223
Glucosinolates, 8, 598
Goiter, 599
Gossypol, 8
Griseofulvin, occurrence of, 219
    structure, 220
    toxicity, 219-220
Gymnoascus, 213

Hazard Analysis Critical Control Point
    (HACCP), 761-762
Heavy metal toxicity, (see also metals),
    424, 730
Helvolic acid
    structure, 226
    toxicity, 226
Helenium spp, 605
Histamine, 26
Honey, 8-9
    natural toxicants in, 9
Humectants, 670
Hydatid Cyst, 310-311
Hydatidosis, 309, 310-311
Hydrogen cyanide, 611
Hyenanchin, 9
Hygiene, 760-763
Hymenovin, 605
Hymenoxy spp, 605
Hypoglycin, 8, 610

Inhibitors, 27

chemical, 27-28
natural in foods, 27
Inorganic elements, 609
International Association of Microbiological Societies (IAMS), 703
International Chamber of Commerce voluntary code of advertising, 707
International Commission on Microbiological Specification for Foods (ICMSF), 703
International Symposium on Safety and Importance of Foods in the Western Hemisphere, 720
Irradiated foods, carcinogens in, 574-575
FDA's view of, 585
future and potentials of, 585-590
handling after irradiation, 561-562
microbiological aspects of, 555-558
nutritional quality of, 564-571
packaging materials for, 562-564
radiation-resistant organisms in, 558-559
toxicological and mutagenic aspects of, 571-575
Islanditoxin, structure of, 222
toxicity of, 221-222
Isoflavones, 600
Isothiocyanate, 599

Jervine, 615

Kidneys
target organ for metals, 425
Klebsiella, 33-34
role in food infection, 34
K. pneumoniae, 34
Kojic acid, structure, 225
toxicity, 225

Labriformin, 607
Lactic acid bacteria, 22
vitamins required for growth, 26
Lactobacillus, 17-18
Latin American Food Code, 711
Latin American Food Council, 711
Latin American Free Trade Association, 711

Latin American Regional Conference, 712
Lead, 432
Leavening agents, 663-665, 668
Leptospira, 306
Leptospirosis, 306
Limb deformities, 612
Listeria, 29, 42
in raw milk, 42
L. monocytogenes, 42, 304-305
Listerosis, 304-305
Locoism, 610
Lungfluke disease, 44
Lupinus spp, 612
Luteoskyrin, structure of, 221
toxicity of, 221

Macrozamia spp, 604
Manihot spp, 602
Margarine, 719
Meat animals, parasites of, 309-312
Meat, inspection, 312-315
Mercury, allowable Daily Intake (ADI), 390-392
biotransformation of, 360-375
absorption, 362-365
biochemistry, 362
excretion, 372-375
tissue distribution, 367-371
transport and transformation, 365-367
Chlor-alkali plants, mercury losses, 355-357
fungicides, 357-358
in Eskimos, 390, 392
in foods, 385-393
indices of exposure, 375-377
interconversion of, 362
Iraq accident, 374
man's uses of, 354-355
methods of analysis, 393-402
Minimata Bay accident, 350, 360-361
Natural levels of, 351-354
selenium, effect on toxicity, 382
toxicity of, 377-383
uses of, 355-360
Metals, 423-433 (see also specific ones)
analysis, 437-438
body burden of, 426
guideline for in foods, 426-427
in sewage sludge, 428
in vegetables, 429, 431, 436
tissue residues, 431, 433-434
Methylazoxymethanol, 603
Microbacterterium, 26

M. thermosphactum, 24, 25
Microbiological standards, 703
Milkweed, 606
Model food law, 713
Monilinia fructicola, 201
Monoacetoxyscirpenol
    occurrence, 238-239
    toxicity, 239
Mushrooms, 610
Mycobacterium spp, 304
Mycophenolic acid
    structure, 226
    toxicity, 226
Mycotoxins, control of, 239-242
    decontamination of, 240-245
    detection of, 243-245
    detoxification of, 245-246
    inactivation of, 239
Myiasis, intestinal, 309

National Institute for the Food Service
    Industry (NIFI), 765
National school lunch program, 736
Neutralizing agents, 674
Newcastle disease, 306-307
Nitrate and nitrite, 326
    as food additives, 326
    average daily intake, 327
    effect on nitrosamine formation in
        bacon, 333
Nitriles, 559
Nitrite, 82-93, 326
Nitrosamines, 28, 319
    bacterial synthesis of, 325
    carcinogenicity due to, 325
    formation *in vivo* and *in vitro,* 324
    in spice-nitrite premixes, 333
    kinetics of formation of, 320
    occurrence in foods, 328
    precursors of, 326
    toxicity to sheep and mink, 320
Nitrosohydroxyproline in cured meats,
    329
Nitrosohydroxypyrrolidine in cured
    meats, 329
Nitrosoproline in cured meats, 329
Nitrosopyrrolidine in foods, 329
Nitrososarcosine in foods, 329
Nutrition labelling, 706

Obesity, 4

Ochratoxin(s), occurrence, 199
    structure, 212
    toxicity, 209, 211
Ornithosis, 306
Oxidation-reduction potential, 24-25
    role in food spoilage, 24

Pacto Andino, 711
Pan American Commission for Technical
    Standards, 714
Pan American Standards Bureau, 714
Paracolon infection, 305, 308
Paragonimiasis, 44
Paralytic shellfish poisoning
    assay methods for, 644-647
    cause of, 639-641
    chemistry and toxicology of, 641-642
    clinical aspects of, 642-644
Paratyphoid infections, 308
Pasteurella multocida, 307
Pasteurella pseudotuberculosis, 307
Pasteurella tularensis, 307
Pasteurellosis, 307
Patulin, occurrence, 199, 213-215
    structure, 218
    toxicity, 217
Penicillic acid
    occurrence, 199, 216-218
    structure, 218
    toxicity, 217
Penicillium spp, 219-238
Penitrems A, B and C, 227, 228
Perigo type factor, 85-86
Pesticide residues, 9, 498-514
    degradation by food processing, 509
    human exposure, 506
    in atmosphere, 499-501
    in foods, 507-511
    in human milk, 9, 510-511
    in soil, 501-503
    in water, 503-506
Phallotoxins, 238
    structure, 238
    toxicity, 235-237
Phosphate salts, 663-665
Photosensitization, 607
Phylyctaenea vagabunda, 201
Piperidine alkaloids, 614
Plant toxins, 593-595
Poisonous marine animals, 625-651
    ciguatera toxin in, 627-632
    clupeoid poisoning from, 637-638
    hallucinogenic mullet poisoning
        from, 638

histamine toxicity from, 634-637
ichthyoallyeinotin (*see* hallucinogenic
    mullet poisoning)
marine turtle poisoning from, 638
scombroid poisoning from (*see* hista-
    mine toxicity)
tetrodotoxin in, 632-634
toxocity due to consumption of, 626-
    627
vitamin A intoxication from, 538-539
Poisonous plants, 594-624
Polybrominated Biphenyls, 474-486
    Cooking of poultry, 483-485
    Effects on cattle, 477
        fish, 479
        humans, 480-481
        monkeys, 479-480
        pigs, 477
        poultry, 478-479
        rodents, 479
        sheep, 477
    Levels in environment and food, 481-
        482
    Processing of dairy products, 482-483
    Uses, 474-475
Polychlorinated Biphenyls, 444-474
    Cooking of fish, 465-471
        poultry, 464-465
    FDA tolerances, 460-461
    Levels in dairy products, 452-453
        fish, 449-450
        food, 448-449
        human adipose tissue, 455-456
        human blood, 455-456
        human milk, 454-455
        wildlife, 453-455
    Paper, 448
    Processing of diary products, 463-464
        eggs, 465
        fats and oils, 473-474
        sake, 474
        shrimp, 471-473
    Reactivity, 444-446
    Sewage, 446-447
    Silos, 447
    Sludge, 446-447
    Toxic effects on animals, 458-460
        avian species, 457-458
        humans, 456-457
    Uses, 444-446
    Water, 446-447
    Yusho Incident, 456-457
Postmortem inspection, 313
Potato sprouts, 619
Prepackaged food standards, 715
Preservatives, 671
Proteus vulgaris, 37

Proteus morganic, 37
Pseudomonas, 17
    in food infections, 36
    in food spoilage, 17
    in meats, 24
    in raw milk, 19
    temperature requirements of, 22
P. aeruginosa, 36
P. cocovenans, 36
Psittacosis, 306
Psoralens, structure, 232
    toxicity, 229, 231
Ptomaine poisoning, 15, 26

Q fever, 305
Quinolizidine alkaloids, 612

Raoult's law, 18
Raw, fermented dry sausages, 92-93
Regional Latin American Standards, 710
Relative humidity, 19
Rhizopus, 201
Rhizopus oryzae, 36
Roman law, 701
Rubratoxins, occurrence, 218
    structure, 219
    toxicity, 218-219

Salted fish, 80-82
Salmonella, 22, 28-31, 120-175, 737, 739
    antibiotic resistant plasmids, 130, 131
    biochemical characteristics, 121, 122
    control, 146, 147
    effect of chemical treatment, 136, 137
    effect of drying, 135
    effect of freezing, 24, 135, 136
    effect of heating, 132-135
    effect of ionizing radiation, 136, 137
    effect of pH, 131
    effect of temperature, 132
    effect of ultraviolet radiation, 136
    genetic characteristics, 130, 131
    growth characteristics, 131-132
    heat resistance, 132-134
    in animal by products, 138, 139
    in bakery products, 146
    in confectionery, 145
    in dairy products, 30, 144, 145

in eggs, 141-143
in fish, 145
in fish meal, 139, 140
in food service products, 146
in grain, 145, 146
in meat, 143, 144
in milk, 144, 145
in poultry and poultry products, 140, 141
in sandwiches, 146
methods for enumeration, 148
methods for identification, 147, 148
methods for isolation, 147
phage typing, 147, 148
rapid methods, 148, 149
selective enrichment, 147
selective plating media, 147
serological typing, 122, 149
serotypes in food-borne disease, 126-128
sources of infection, 128-130
transferrable drug resistance, 130, 131
Salmonella food poisoning, 120, 122-127, 302-303, 315, 740
causative organisms, 29, 30, 120-122, 302, 308
clinical course, 122, 123
food borne outbreaks, 126-128
incidence in U.S., 124, 125
incidence world-wide, 125, 126
symptoms, 30, 122
Salmonella surveillance unit, 303
Sanitation, equipment and facilities, 753-755

Sanitation, educational programs, 763-765
equipment and facilities, 753-755
food, 755-756
Sarcocystis spp, 309
Sarcosporidiosis, 309
School food preparation and service systems, 735, 737, 739, 741, 743, 744, 745, 748, 750, 753, 761
Sclerotinia sclerotiorum, 229
Selenium, 436, 610
Sesquiterpene lactones, 604
Sewage sludge, 6-7, 427-428
Shelf-stable canned cured meats, 82
Shigella, 34-35, 308
role in food infections, 34
S. flexneri, 34
S. sonnei, 34, 308
Simulated meat products, 728
Slaughterhouse standards, 313
Smoked fish, 80-82

Sodium nitrate, 28
Soil, 54
metal content of, 428
types of microoganisms in, 55
Solanidine, 618
Solanine, 618
Solanum spp, 595
Solasodine, 616
Sorghum spp, 602
Stachybotryotoxicosis, 199, 223
Standards of identity, 706
Staphylococcus aureus, 16, 22-23, 27, 28, 37, 737, 740, 746-747, 749, 751-752
Staphylococcal enterotoxins, amount to cause illness, 111
pharmacological action, 112
properties, 111
relation to coagulase, 109-110
relation to DNase, 110
specific antibodies, 110
stability, 112-113
types, 109-111
Staphylococcal food poisoning, 108-119, 305
cause, 108-113
foods involved, 108, 16, 21, 25
incrimination of specific foods, 115-116
phage typing, 109
prevention, 113-115
symptoms, 108
Starter cultures, 91
Steroidal alkaloids, 595, 615, 616
Sterigmatocystins, biosynthesis, 207-208
structure, 210
toxicity, 206
Streptococcus, 40-42
role in food infections, 41, 305
S. faecalis, 40
S. faecium, 40
S. lactis, 21
Surfactants, 667
Sweeteners, synthetic, 672

Taeniasis, 311, 312
Taenia saginata, 46, 311
Taenia solium, 46, 311, 312
Tapeworms, 46-47
Teratogens, 611
Tetradymol, 608
Tetradymol spp, 608
Terreic acid, structure, 225
toxicity, 225

Thickners/stabilizers, 662
Thiocyanates, 599
Tomatidine, 617
Toxicity, definition of, 3
Toxic plant affinites, 594
Toxic plant genera, 594
Toxins in plants, 594-624
Toxoplasmosis, 309, 310
Toxoplasma gondii, 309
Treaty of Montevideo, 711
Tremorgenic mycotoxins, occurrence, 227-228
    structure, 230-231
    toxicity, 227-229
Trichinella spiralis, 44-46, 309, 310
Trichinosis, 44, 309, 310, 315
Trichothecenes, occurrence, 223-224
    structure, 223
    toxicity, 223-224
Trifolium spp, 600
Triglochin, spp, 602
Tryptoquivaline, 227, 229
Tryptoquivaline-related mycotoxins, 238
Tryptoquivalone, 227, 229
Tuberculosis, 304
Turkey X disease, 202-203
Tutin, 9

U.S. Department of Agriculture (USDA), 740
U.S. Federal Food, Drug and Cosmetic Act, 1938, 705
U.S. Food and Drug Administration, 740

Vegetables, metal uptake by, 429, 431, 432, 433
Veratrum spp, 595, 615
Veridicatin, 233-234
Verruculogen TR-1, 227, 228
Verruculogen TR-2, 227, 228
Vesicular stomatitis, 307
Vibrio foetus, 308
Vibrio parahaemolyticus, 35, 36, 37
Viral diseases (foods), 181-186, 192
    encephalitis (tick borne) 178, 192
    encephalitis (symptoms of), 192
    gastroenteritis, 186
    infectious hepatitis, 183-186, 192
    infectious hepatitis (symptoms of), 183
    poliomyelitis, 181-183, 192

Viruses, assaying (testing) for, 188-192
    concentration techniques, 189, 192
    dilution techniques, 189, 192
    disadvantages of, 191, 192
    elution techniques, 189, 192
Viruses, control of (foods), 178-181
    miscellaneous foods, 181
    shellfish, 179, 181
    vegetables, 178, 179
Viruses, definition of, 176, 191
Viruses, inactivation (destruction) of, 186-188
    drying, 187, 188, 192
    freezing, 187, 192
    ionizing radiation, 188, 192
    miscellaenous methods, 188
    thermal, 186, 187, 192
Viruses, prevention of (foods), 191, 192
Viruses, transmission of, 176, 177, 192
    host to host, 176
    to foods, 179
    (*see also* Viral diseases (foods), 181-186)
Viruses, types in foods, 177, 178
    infectious to animals, 177
    infectious to humans, 178
    zoonoses, 177, 178

Water, 2, 55
    microorganisms in, 56
    role in food quality, 2-3
    sources of, 56
Water activity (*see also* Aw), 18, 90
    minimum levels for microorganisms, 20-21
White scours, 306
World Health Organization (WHO), 715

Xanthotoxin, 608

Yersinia, 37-38
    sources of in foods, 38
Yersina spp, 37

Z value, 77
Zea spp, 602
Zygadenus spp, 595

Other AVI Books

BASIC FOOD MICROBIOLOGY
  Banwart
FOOD AND BEVERAGE MYCOLOGY
  Beuchat
FOOD COLLOIDS
  Graham
FOOD ENGINEERING SYSTEMS, VOL. 1
  Farrall
FOOD MICROBIOLOGY: PUBLIC HEALTH AND SPOILAGE
ASPECTS
  deFigueiredo and Splittstoesser
FOOD PROCESSING WASTE MANAGEMENT
  Green and Kramer
FOOD QUALITY ASSURANCE
  Gould
FOOD SANITATION
  Guthrie
FUNDAMENTALS OF FOOD MICROBIOLOGY
  Fields
HANDBOOK OF PACKAGE MATERIALS
  Sacharow
MICROBIOLOGY OF FOOD FERMENTATIONS
  2nd Edition *Pederson*
PACKAGING REGULATIONS
  Sacharow
PRACTICAL FOOD MICROBIOLOGY AND TECHNOLOGY
  2nd Edition *Weiser, Mountney and Gould*
QUALITY CONTROL FOR THE FOOD INDUSTRY
  3rd Edition VOLS. 1 and 2 *Kramer and Twigg*